Mechanics and Thermodynamics of Continua

BERNARD D. COLEMAN

Mechanics and Thermodynamics of Continua

A Collection of Papers Dedicated to B. D. Coleman on His Sixtieth Birthday

Invited by
H. Markovitz, V. J. Mizel, and D. R. Owen

Reprinted from
Archive for Rational Mechanics and Analysis
edited by C. Truesdell

With 55 Figures

Springer-Verlag

Berlin Heidelberg New York
London Paris Tokyo
Hong Kong Barcelona

Professor Dr. Hershel Markovitz
21 Disraeli St., 92222 Jerusalem, Israel

Professor Dr. Victor J. Mizel
Professor Dr. David R. Owen

Department of Mathematics, Carnegie Mellon University
Pittsburgh, PA 15213, USA

ISBN-13: 978-3-540-52999-6 e-ISBN-13: 978-3-642-75975-8
DOI: 10.1007/978-3-642-75975-8

Library of Congress Cataloging-in-Publication Data. Mechanics and thermodynamics of continua: a collection of papers dedicated to B. D. Coleman on his sixtieth birthday / invited by H. Markovitz, V. Mizel, and D. Owen. p. cm. "Reprinted from Archive for rational mechanics and analysis, edited by C. Truesdell." Includes bibliographical references and index. "Published works of Bernard D. Coleman": p. ISBN 3-540-52999-3 (Springer-Verlag New York Berlin Heidelberg: alk. paper). - ISBN 0-387-52999-3 (Springer-Verlag New York Berlin Heidelberg: alk. paper). 1. Mechanics, Analytic. 2. Mathematical physics. I. Coleman, Bernard D. (Bernard David), 1930-. II. Mizel, Victor J. III. Owen, David R. IV. Markovitz, Hershel, 1921-.
QA807.5.M43 1991 531-dc20 90-25323

55/3140-543210 – Printed on acid-free paper

Preface

This volume collects papers dedicated to *Bernard D. Coleman* on his sixtieth birthday, July 5, 1990.

In 1954 *Coleman* received his doctorate in chemistry at Yale with two theses, one under the aegis of *Kirkwood,* the other, of *Fuoss.* The former concerned theory and experiment on dilute polymer solutions. It was at Yale that *Coleman* read selections from the writings of *Gibbs* as well as standard texts on thermodynamics. He recalls being awed by the generality of the subject but uneasy about its claimed completeness and discouraged by its foggy enunciations.

In 1954–1957 *Coleman* was a research chemist at the du Pont Company. His first task was to explain why nylon appeared weaker than a competitor's product. He studied the dependence of the strength on the speed of loading, and he predicted that nylon would be the stronger in most uses. He received a bonus and came to be regarded as an expert on the strength of artificial fibres; his desk was covered with spools of yarn sent in by men in the field who wished him to test them. He preferred to construct a theory of the dependence of lifetimes of bundles on the strength of the individual fibres (papers (3), (4), (6)–(15), (17)). This theory was ignored at the time but was taken up in industry many years later.

Coleman and his friends at the du Pont Company were regarded playfully as "the trouble makers" because they kept late rather than early hours and did not eat lunch in the company cafeteria. *Flory,* originally a research chemist at du Pont and later a consultant for it, in 1957 had been named first executive director of research at the Mellon Institute. On a visit to the du Pont Company he met *Coleman,* and shortly thereafter he appointed him senior research fellow. *Coleman* was then twenty-seven. In those days ambitious young men welcomed pay but did not know there was such a thing as tenure. A typist seemed more valuable, and *Coleman* had that, too. Perhaps he did not recognize his good fortune in not needing to struggle up the academic rungs, to endure the flatulent dolts on faculty committees, or to spoon out pablum to roomfuls of hostile catatonics and (for the sake of social democracy) give passing grades to the failed majority. For ten years *Coleman* could put all his energy into study, thought, and writing. That good fortune was not only his: For what he did with this great chance, the community of rational mechanics will always be grateful.

At Yale *Coleman* received little training in mathematics; at the du Pont Company, while watching testing machines, he could read *Leaderman's* book on visco-elastic filamentous materials, in which he encountered quotations from *Maxwell* and *Boltzmann* on the rate-dependence of material response. At the Mellon Institute *Markovitz,* too, was a senior fellow. In September, 1957, the two began to follow courses on elasticity, tensor analysis, and continuum mechanics by *Noll,* then scantly known, at the Carnegie Institute of Technology. An invitational meeting on visco-elasticity in the following April at Lancaster, Pennsylvania, brought *Coleman* and *Noll* together. In those days a person went to a meeting so as to learn from a few competent lectures without having to be himself one more "invited speaker" or to listen to many multiples of ten minutes of trivial trash. *Ericksen* lectured on "laminar shear flows" of incompressible, Rivlin-Ericksen fluids. That class of flows contains all those for which *Rivlin* and others had obtained exact solutions. *Ericksen's* paper, with *Criminale & Filbey* as co-authors, was to appear soon in Volume 1 of the *Archive.* At the meeting, *Coleman* and *Noll* found that they had similar views on thermodynamics. The rheologists there, like those we had encountered elsewhere, told us that classical thermodynamics was a complete, closed, perfect science, all in *Gibbs's* paper, and they laughed at us. We laughed at them, but silently, for we had read fundamental parts of *Gibbs's* work, especially that on the isothermal and isentropic theories of three-dimensional elasticity, which, surely, the rheologists could not understand. We knew also the basic inequality for increase of entropy asserted by *Duhem* (1901) and in "The Mechanical Foundations" (1952) called "the Clausius-Duhem inequality" (Eq. (28.5)), from which *Eckart* (1940) had drawn consequences by guessing the signs of two terms ("Mechanical Foundations", text following Eq. (31.1)). *Noll* pointed out the issue: You cannot vary every term independently, and so how can we get information from the inequality as it stands? From that time onward *Coleman* worked toward ideas and methods of a theory of constitutive structure.

On July 28, 1958, *Noll* wrote to me that he was collaborating with *Coleman,* "a bright young physical chemist... . [Y]ou may remember him from the Lancaster meeting." They were working on what I had called the "Hauptproblem of finite elasticity" (1955). The famous series of papers by *Coleman & Noll* began to appear in 1959, some in the *Archive,* some elsewhere. They will be familiar to many readers because much of their contents has subsequently gone into textbooks and surveys. One of them was paper (18), in which, by a direct, functional approach to continuum mechanics on the basis of *Noll's* paper in Volume 2 of the *Archive,* they analysed directly, almost without calculation, the laminar shear flows (later called "viscometric flows") and emphasized the material functions later called "viscometric". In paper (20), on helical flows, appears the first application of an anholonomic basis in continuum mechanics. Even while writing these papers, their principal joint efforts were directed toward understanding and clarifying thermodynamics.

Paper (32), on the thermodynamics of elastic materials with heat-conduction and viscosity (1963), written while *Coleman* and *Noll* held visiting appointments at the Johns Hopkins University, showed how to use the "postulate of irreversibility" proposed in *The Classical Field Theories* (and unfortunately, then and now, confused with the Clausius-Duhem inequality). The heat supply appearing in it, extending the Clausius-Duhem inequality so as to incorporate *Stokes's* treatment of the decay of sound through radiant heat, provided the additional term to be varied. While *Coleman & Noll's* paper of 1963 set up the machinery but applied it only to two classical instances, in the next issue of the *Archive* appeared paper (33), by *Coleman & Mizel,* "Thermodynamics and departures from Fourier's law of heat conduction", in which for the first time the rule of equipresence was applied effectively to eliminate superfluous variables.

Use of thermodynamic principles to derive restrictions upon constitutive functions seemed new to everyone in 1963, but in fact *Sadi Carnot* had done precisely that in his theory of specific heats. His outcome was faulty, even inconsistent, but that was because the theory of heat accepted in 1824 would not allow an ideal gas to have constant specific heats. Most historians of thermodynamics have failed to recognize *Carnot's* conceptual achievement because they do not know (except for the term "caloric theory") the common store that *Carnot* drew upon, while in their comparisons with "modern" thermodynamics they interpret "modern" as the ossified rituals of the nineteenth century, still droned in elementary physics texts.

In 1964 *Coleman* published two major studies (papers (42) and (43)) on the thermodynamics of materials with memory. These show him as the principal architect of rational thermodynamics. The sequence on waves in materials with memory (1965/6) (papers (50)–(53), (58), with various co-authors) are now classics of acoustics. Major works on stability, birefringence, and electromagnetism followed. In 1974 *Coleman* & *Owen* in "A mathematical foundation for thermodynamics" (paper (95)) set up abstract "basic concepts and assumptions which appear to be present in all branches of thermodynamics" and showed that smooth energy and entropy functions, which were laid down axiomatically in *Coleman's* earlier work on materials with memory, can be deduced as theorems in the new and inclusive framework. Such generality calls for difficult mathematics. What may be the culmination of this line of thought is "The second law of thermodynamics for systems with approximate cycles" (paper (129), co-authors *Owen* & *Serrin,* 1982). As the title suggests, the treatment includes materials incapable of undergoing cyclic processes.

During all this time and after it *Coleman* completed works, some of them with co-authors, on thermodynamics of particular materials, on physical chemistry, on polymer physics, on neural networks (with a view to measured response of the compound eye of the horseshoe crab), on the convergence of learning programs for computers, on population dynamics, on the cold drawing of polymeric fibres, on the dynamics of elastic rods, and on several other subjects.

In 1968 the Mellon Institute was absorbed by the Carnegie Institute. *Coleman,* stripped of his privileges and reduced to being a mere professor, joined his friends in the mathematics department. While in other countries there are in all intellectual fields institutes for research alone, in the United States, while physicians, biologists, economists and political scientists may enjoy them, the electorate seems to regard mathematicians as school teachers and nothing more, unless they can help develop weapons. At Carnegie Mellon *Coleman* acquired the titles of professor of biology and professor of chemistry, but it would be impolite for me to ask if emoluments came with them. In 1988 he moved to Rutgers with the splendid and long-deserved title, not empty, of J. Willard Gibbs Professor of Thermomechanics as well as Professor of Mathematics.

C. Truesdell

Contents

The Homogeneous Field Approximation of Classical Thermodynamics

W. A. DAY

For BERNARD COLEMAN *on his sixtieth birthday*

Preamble

Classical thermodynamics draws heavily upon arguments about the work done by a deformable body operating in a cycle; the purpose of the present paper is to re-examine the conclusion of an argument of this type and to point out that, in at least one important respect, classical theory fails to represent the behaviour of real bodies.

The argument I have in mind runs as follows. Suppose that $\theta(t)$, $p(t)$, and $v(t)$ are, respectively, the absolute temperature, the pressure, and the volume, of a body at the instant t, and that these quantities are connected by the gas law

$$pv = r\theta,$$

r being a positive constant. Let T and a be any positive numbers, let E be any number, and let the temperature and the pressure be

$$\theta(t) = \theta_0 + a\cos\left(\frac{2\pi t}{T}\right),$$

$$p(t) = p_0 \exp\left(-b\sin\left(\frac{2\pi t}{T}\right)\right),$$

where b is to be chosen, θ_0 and p_0 are positive constants, and, in order to ensure that $\theta(t)$ is positive, θ_0 exceeds a. Each of $\theta(t)$ and $p(t)$ is periodic, with a period which coincides with T, and the maximum temperature difference $\operatorname{Max}(\theta - \theta_0)$ coincides with a. Moreover, the work done by the body in a period is

$$\begin{aligned}\int_0^T p\dot{v}\,dt &= -\int_0^T v\dot{p}\,dt = -\int_0^T \frac{r\theta}{p}\dot{p}\,dt\\ &= -r\int_0^T \left(\theta_0 + a\cos\left(\frac{2\pi t}{T}\right)\right)\left(-\frac{2\pi b}{T}\cos\left(\frac{2\pi t}{T}\right)\right)dt\\ &= \pi abr,\end{aligned}$$

and, hence, by choosing

$$b = E/\pi a r,$$

we can arrange that the work done by the body in a period has the value E. In brief: classical thermodynamics predicts that, for a body which obeys the gas law and operates with a given period and within a given range of temperatures, it is possible to adjust the pressure in such a way that the work done in a period takes any assigned value.

In the course of this argument, as in all of classical thermodynamics, the homogeneous field approximation is in force; that is to say it is assumed that the fields of temperature and stress are spatially homogeneous at each instant.

The homogeneous field approximation excludes two factors which influence the behaviour of real bodies: heat conduction within the body is ruled out if the temperature is spatially homogeneous, and inertial effects are ruled out if the stress is spatially homogeneous. Indeed, so drastic is the homogeneous field approximation that classical thermodynamics cannot justly be described as a field theory.[1]

I propose to ask if the conclusion of the argument can be sustained in the context of a genuine field theory – linear thermoelasticity – which incorporates both heat conduction and inertia.

What the field theory implies turns out to be strikingly different from what the homogeneous field approximation implies: the field theory predicts that, for a given period which is sufficiently large and for a given range of temperatures, heat conduction and thermomechanical coupling combine to place a finite upper bound on the work that the body can do in a period, no matter how the pressure is adjusted. There is, though, no lower bound on the work that the body can do in a period, or, to put matters another way, the pressure can be adjusted so that an arbitrarily large amount of work is done *on* the body in a period; such a disparity between the conclusions concerning the work done by the body and the work done on the body arises from the irreversible nature of heat conduction.

The linear theory of thermoelasticity which lies behind the arguments of this paper is approximate at best and it would be of considerable interest if the results could be extended to an exact nonlinear theory.

A field theory

I shall consider an isotropic slab which occupies the region

$$\{(x, y, z): A \leqq x \leqq B, \quad -\infty < y < \infty, -\infty < z < \infty\}$$

and undergoes a motion in which particles are displaced in a direction orthogonal to the faces $x = A$ and $x = B$. All the fields encountered depend, at most, upon the single Cartesian coordinate x and the time t.

---

[1] It is noteworthy that the homogeneous field approximation retains a foothold within certain modern theories of continuum thermodynamics which employ a thermodynamic universe containing an ideal gas as a distinguished member. See, for example, SERRIN [1, 2] and COLEMAN, OWEN, & SERRIN [3].

If there are no body forces nor external sources of heat, the temperature $\theta(x, t)$ and the displacement $u(x, t)$ satisfy the equation of energy[2]

$$\frac{\partial}{\partial x}\left(k(x)\frac{\partial\theta}{\partial x}\right) = c(x)\frac{\partial\theta}{\partial t} + \theta_0\mu(x)\frac{\partial^2 u}{\partial x\,\partial t},$$

and the equation of motion

$$\frac{\partial\sigma}{\partial x} = \varrho(x)\frac{\partial^2 u}{\partial t^2},$$

the stress $\sigma(x, t)$ being

$$\sigma = -\mu(x)(\theta - \theta_0) + \beta(x)\frac{\partial u}{\partial x}.$$

In these equations the positive constant θ_0 is the temperature of a stress-free reference state of the body.

The coefficients are: the thermal conductivity $k(x)$, the specific heat at constant strain $c(x)$, the stress-temperature modulus $\mu(x)$, the mass density $\varrho(x)$, and an elastic modulus $\beta(x)$. Each of the coefficients may vary with x and, hence, the body is permitted to be materially inhomogeneous; I shall take it that $k(x)$, $c(x)$, $\mu(x)$, $\varrho(x)$, and $\beta(x)$ are positive and continuously differentiable in the interval $[A, B]$.

The constitutive relation for the stress enables us to substitute

$$(\mu(\theta - \theta_0) + \sigma)/\beta$$

for $\partial u/\partial x$ in the equation of energy, and when we do so we deduce the equation

$$\frac{\partial}{\partial x}\left(k\frac{\partial\theta}{\partial x}\right) = \left(c + \theta_0\frac{\mu^2}{\beta}\right)\frac{\partial\theta}{\partial t} + \theta_0\frac{\mu}{\beta}\frac{\partial\sigma}{\partial t}.$$

Likewise, if we divide the equation of motion through by ϱ, differentiate with respect to x, and substitute once again for $\partial u/\partial x$, we arrive at the equation

$$\frac{\partial}{\partial x}\left(\frac{1}{\varrho}\frac{\partial\sigma}{\partial x}\right) = \frac{\mu}{\beta}\frac{\partial^2\theta}{\partial t^2} + \frac{1}{\beta}\frac{\partial^2\sigma}{\partial t^2}.$$

In accordance with the considerations of the preamble, the faces of the slab are required to be subjected to a variable surface temperature $\tau(t)$ and a variable surface pressure $p(t)$; thus the boundary conditions

$$\theta(A, t) = \theta(B, t) = \tau(t),$$

$$\sigma(A, t) = \sigma(B, t) = -p(t)$$

are in force.

Our interest is in what happens when $\tau(t)$, $p(t)$, $\theta(x, t)$, and $\sigma(x, t)$ are periodic in their dependence upon t, with a common period T. In that case the work done

---

[2] CARLSON's article [4] provides a full account of linear thermoelasticity. My forthcoming tract [5] approaches the subject from a somewhat different starting point.

by the body in a period is[3]

$$\int_0^T p(t)\left(\frac{\partial u}{\partial t}(B,t)-\frac{\partial u}{\partial t}(A,t)\right)dt=\int_0^T p\left(\int_A^B \frac{\partial^2 u}{\partial x\,\partial t}\,dx\right)dt$$

and, hence, is

$$\int_0^T\int_A^B p\left(\frac{\mu}{\beta}\frac{\partial\theta}{\partial t}+\frac{1}{\beta}\frac{\partial\sigma}{\partial t}\right)dx\,dt.$$

I shall study only the simplest case in which the surface temperature, the surface pressure, and the temperature and stress fields within the body, are harmonic in their dependence upon t, with frequency $\omega/2\pi$ and period $T=2\pi/\omega$. Thus I suppose τ and p to have the forms

$$\tau(t)=\theta_0+\operatorname{Re}[a\exp(i\omega t)],$$

$$p(t)=p_0+\operatorname{Re}[b\exp(i\omega t)],$$

where a and b are any complex numbers and p_0 is any real number, and I suppose that the temperature and stress fields are

$$\theta(x,t)=\theta_0+\operatorname{Re}[\Theta(x)\exp(i\omega t)],$$

$$\sigma(x,t)=-p_0+\operatorname{Re}[\Sigma(x)\exp(i\omega t)],$$

where Θ and Σ satisfy the boundary value problem

$$(k\Theta')'=i\omega\left(\left(c+\theta_0\frac{\mu^2}{\beta}\right)\Theta+\theta_0\frac{\mu}{\beta}\Sigma\right),$$

$$\left(\frac{1}{\varrho}\Sigma'\right)'=-\omega^2\left(\frac{\mu}{\beta}\Theta+\frac{1}{\beta}\Sigma\right),$$

$$\Theta(A)=\Theta(B)=a,\quad \Sigma(A)=\Sigma(B)=-b.$$

In this problem the primes denote derivatives with respect to x, and the complex-valued Θ and Σ depend upon ω, a, and b as well as upon x. The existence and uniqueness of Θ and Σ for sufficiently small values of ω—which is all that is required here—can be established by standard means. Existence and uniqueness for general values of ω are established in [6].

Since

$$p=p_0+\operatorname{Re}[b]\cos\omega t-\operatorname{Im}[b]\sin\omega t,$$

$$\theta=\theta_0+\operatorname{Re}[\Theta]\cos\omega t-\operatorname{Im}[\Theta]\sin\omega t,$$

$$\frac{\partial\theta}{\partial t}=-\omega\operatorname{Re}[\Theta]\sin\omega t-\omega\operatorname{Im}[\Theta]\cos\omega t,$$

$$\sigma=-p_0+\operatorname{Re}[\Sigma]\cos\omega t-\operatorname{Im}[\Sigma]\sin\omega t,$$

$$\frac{\partial\sigma}{\partial t}=-\omega\operatorname{Re}[\Sigma]\sin\omega t-\omega\operatorname{Im}[\Sigma]\cos\omega t,$$

[3] In strict terms, what is defined here is the work done in a period by a cylindrical part of the slab, the cross-sectional area of the cylinder being equal to unity and its generators being parallel to the x-axis.

it must be that

$$\int_0^{2\pi/\omega} \int_A^B p \left(\frac{\mu}{\beta} \frac{\partial \theta}{\partial t} + \frac{1}{\beta} \frac{\partial \sigma}{\partial t} \right) dx \, dt$$

$$= \pi \int_A^B \left(- \operatorname{Re}[b] \operatorname{Im} \left[\frac{\mu}{\beta} \Theta + \frac{1}{\beta} \Sigma \right] + \operatorname{Im}[b] \operatorname{Re} \left[\frac{\mu}{\beta} \Theta + \frac{1}{\beta} \Sigma \right] \right) dx$$

and, hence, that the work done by the body in a period is

$$-\pi \operatorname{Im} \left[\bar{b} \int_A^B \left(\frac{\mu}{\beta} \Theta + \frac{1}{\beta} \Sigma \right) dx \right],$$

where $\bar{b}$ is the complex conjugate of b.

The question to be considered is how the work done by the body in a period depends upon b when ω and a are held fixed.

What the homogeneous field approximation predicts

The homogeneous field approximation ignores the fact that the fields of temperature and stress within the body must satisfy a pair of coupled partial differential equations and assumes that it will do to take

$$\theta(x, t) = \tau(t), \qquad \sigma(x, t) = -p(t).$$

In the present context this is tantamount to the approximation

$$\Theta(x) = a, \qquad \Sigma(x) = -b,$$

which, in turn, leads to the approximation

$$-\pi \operatorname{Im} \left[\bar{b} \left(a \int_A^B \frac{\mu}{\beta} dx + b \int_A^B \frac{1}{\beta} dx \right) \right]$$

$$= -\pi \int_A^B \frac{\mu}{\beta} dx \operatorname{Im}[a\bar{b}]$$

for the work done by the body in a period. Thus, if E is any number, the choice

$$b = \frac{iEa}{\pi |a|^2 \int_A^B \frac{\mu}{\beta} dx}$$

ensures that the work done by the body in a period coincides with E. Hence: *the homogeneous field approximation predicts that, for any given positive ω and any given nonzero complex number a, it is possible to choose b so that the work done by the body in a period takes any assigned value.*

Such a prediction is in line with that of the preamble.

What the field theory predicts

By contrast with what has just been established, the field theory predicts that: *there is a positive ω_0 such that, for any given ω which lies in the interval $0 < \omega < \omega_0$ and for any given complex number a, the work done by the body in a period is bounded above as a function of b.*

In order to prove this assertion we start by replacing the boundary value problem which determines Θ and Σ by the pair of integral equations

$$\Theta(x) = a - i\omega \int_A^B K(x, y) \left[\left(c(y) + \theta_0 \frac{\mu(y)^2}{\beta(y)} \right) \Theta(y) + \theta_0 \frac{\mu(y)}{\beta(y)} \Sigma(y) \right] dy,$$

$$\Sigma(x) = -b + \omega^2 \int_A^B R(x, y) \left[\frac{\mu(y)}{\beta(y)} \Theta(y) + \frac{1}{\beta(y)} \Sigma(y) \right] dy,$$

in which the kernels are

$$K(x, y) = \begin{cases} \displaystyle\int_A^x \frac{ds}{k(s)} \int_y^B \frac{ds}{k(s)} \Big/ \int_A^B \frac{ds}{k(s)}, & A \leqq x \leqq y \leqq B \\ \displaystyle\int_x^B \frac{ds}{k(x)} \int_A^y \frac{ds}{k(s)} \Big/ \int_A^B \frac{ds}{k(s)}, & A \leqq y \leqq x \leqq B, \end{cases}$$

$$R(x, y) = \begin{cases} \displaystyle\int_A^x \varrho(s)\, ds \int_y^B \varrho(s)\, ds \Big/ \int_A^B \varrho(s)\, ds, & A \leqq x \leqq y \leqq B, \\ \displaystyle\int_x^B \varrho(s)\, ds \int_A^y \varrho(s)\, ds \Big/ \int_A^B \varrho(s)\, ds, & A \leqq y \leqq x \leqq B. \end{cases}$$

The fact that the number

$$\theta_0 \int_A^B \int_A^B K(x, y) \frac{\mu(x)}{\beta(x)} \frac{\mu(y)}{\beta(y)}\, dx\, dy$$

is strictly positive turns out to be the key to our argument. The argument would fail if heat conduction were absent from the theory, for then $k(x) = 0$ and $K(x, y)$ is not defined, and it would fail if thermomechanical coupling were absent, for then $\mu(x) = 0$.

The next step is to derive crude estimates which confirm the expectation that

$$\Theta(x) = a + O(\omega), \quad \Sigma(x) = -b + O(\omega^2),$$

as $\omega \to 0$, and to ascertain how the constants associated with these order relations depend upon b.

In what follows $M_1, \ldots, M_{14}$ are positive constants which are independent of ω, a, and b. Let

$$M_1 = \operatorname{Max} \int_A^B K(x, y)\left(c(y) + \theta_0 \frac{\mu(y)^2}{\beta(y)}\right) dy,$$

$$M_2 = \theta_0 \operatorname{Max} \int_A^B K(x, y)\frac{\mu(y)}{\beta(y)}\, dy,$$

$$M_3 = \operatorname{Max} \int_A^B R(x, y)\frac{\mu(y)}{\beta(y)}\, dy,$$

$$M_4 = \operatorname{Max} \int_A^B \frac{R(x, y)}{\beta(y)}\, dy,$$

the maxima being taken with respect to points x of the interval $[A, B]$. Then the integral equations imply the inequalities

$$\|\Theta - a\| \leqq \omega(M_1 \|\Theta\| + M_2\|\Sigma\|),$$

$$\|\Sigma + b\| \leqq \omega^2(M_3 \|\Theta\| + M_4 \|\Sigma\|),$$

in which the norms are supremum norms, that is to say

$$\|\Theta\| = \operatorname{Max} |\Theta(x)|$$

and so forth. Hence

$$\|\Theta - a\| \leqq \omega M_1(\|\Theta\| + M_5 \|\Sigma\|),$$

$$\|\Sigma + b\| \leqq \omega^2 M_3(\|\Theta\| + M_5 \|\Sigma\|),$$

where

$$M_5 = \operatorname{Max}\left(\frac{M_2}{M_1}, \frac{M_4}{M_3}\right)$$

and it follows that

$$\|\Theta\| + M_5 \|\Sigma\| \leqq |a| + M_5 |b| + (\omega M_1 + \omega^2 M_3)(\|\Theta\| + M_5 \|\Sigma\|).$$

Thus

$$\|\Theta\| + M_5 \|\Sigma\| \leqq \frac{|a| + M_5 |b|}{1 - \omega M_1 - \omega^2 M_3}$$

and

$$\|\Theta - a\| \leqq \frac{\omega M_1(|a| + M_5 |b|)}{1 - \omega M_1 - \omega^2 M_3},$$

$$\|\Sigma + b\| \leqq \frac{\omega^2 M_3(|a| + M_5 |b|)}{1 - \omega M_1 - \omega^2 M_3}$$

so long as $\omega M_1 + \omega^2 M_3 < 1$.

If we return to the integral equations that Θ and Σ satisfy we see that

$$\int_A^B \left(\frac{\mu}{\beta}\Theta + \frac{1}{\beta}\Sigma\right) dx = a\int_A^B \frac{\mu(x)}{\beta(x)} dx$$
$$- i\omega \int_A^B\int_A^B K(x, y)\frac{\mu(x)}{\beta(x)}\left[\left(c(y) + \theta_0\frac{\mu(y)^2}{\beta(y)}\right)\Theta(y) + \theta_0\frac{\mu(y)}{\beta(y)}\Sigma(y)\right] dx\, dy$$
$$- b\int_A^B \frac{dx}{\beta(x)} + \omega^2 \int_A^B\int_A^B \frac{R(x, y)}{\beta(x)}\left[\frac{\mu(y)}{\beta(y)}\Theta(y) + \frac{1}{\beta(y)}\Sigma(y)\right] dx\, dy,$$

or, on replacing Θ by $a + (\Theta - a)$ and Σ by $-b + (\Sigma + b)$ on the right-hand side, that

$$\int_A^B \left(\frac{\mu}{\beta}\Theta + \frac{1}{\beta}\Sigma\right) dx = a\left[\int_A^B \frac{\mu(x)}{\beta(x)} dx\right.$$
$$\left.- i\omega \int_A^B\int_A^B K(x, y)\frac{\mu(x)}{\beta(x)}\left(c(y) + \theta_0\frac{\mu(y)^2}{\beta(y)}\right) dx\, dy + \omega^2 \int_A^B\int_A^B \frac{R(x, y)}{\beta(x)}\frac{\mu(y)}{\beta(y)} dx\, dy\right]$$
$$+ b\left[-\int_A^B \frac{dx}{\beta(x)} + i\omega\theta_0 \int_A^B\int_A^B K(x, y)\frac{\mu(x)}{\beta(x)}\frac{\mu(y)}{\beta(y)} dx\, dy - \omega^2 \int_A^B\int_A^B \frac{R(x, y)}{\beta(x)\beta(y)} dx\, dy\right]$$
$$+ Z = a(M_6 - i\omega M_7 + \omega^2 M_8) + b(-M_9 + i\omega M_{10} - \omega^2 M_{11}) + Z,$$

where, in view of the estimates already derived, the remainder Z satisfies

$$|Z| = \left| -i\omega \int_A^B\int_A^B K(x, y)\frac{\mu(x)}{\beta(x)}\left[\left(c(y) + \theta_0\frac{\mu(y)^2}{\beta(y)}\right)(\Theta(y) - a)\right.\right.$$
$$\left. + \theta_0\frac{\mu(y)}{\beta(y)}(\Sigma(y) + b)\right] dx\, dy$$
$$\left. + \omega^2 \int_A^B\int_A^B \frac{R(x, y)}{\beta(x)}\left[\frac{\mu(y)}{\beta(y)}(\Theta(y) - a) + \frac{1}{\beta(y)}(\Sigma(y) + b)\right] dx\, dy \right|$$
$$\leq \frac{\omega^2 M_1(|a| + M_5|b|)}{1 - \omega M_1 - \omega^2 M_3}\int_A^B\int_A^B K(x, y)\frac{\mu(x)}{\beta(x)}\left(c(y) + \theta_0\frac{\mu(y)^2}{\beta(y)}\right) dx\, dy$$
$$+ \frac{\omega^3\theta_0 M_3(|a| + M_5|b|)}{1 - \omega M_1 - \omega^2 M_3}\int_A^B\int_A^B K(x, y)\frac{\mu(x)}{\beta(x)}\frac{\mu(y)}{\beta(y)} dx\, dy$$
$$+ \frac{\omega^3 M_1(|a| + M_5|b|)}{1 - \omega M_1 - \omega^2 M_3}\int_A^B\int_A^B \frac{R(x, y)}{\beta(x)}\frac{\mu(y)}{\beta(y)} dx\, dy$$
$$+ \frac{\omega^4 M_3(|a| + M_5|b|)}{1 - \omega M_1 - \omega^2 M_3}\int_A^B\int_A^B \frac{R(x, y)}{\beta(x)\beta(y)} dx\, dy$$
$$= \frac{(\omega^2 M_{12} + \omega^3 M_{13} + \omega^4 M_{14})}{1 - \omega M_1 - \omega^2 M_2}(|a| + M_5|b|).$$

Hence, the work done by the body in a period is

$$-\pi \operatorname{Im}\left[\bar{b} \int_A^B \left(\frac{\mu}{\beta}\Theta + \frac{1}{\beta}\Sigma\right) dx\right]$$

$$= -\pi \operatorname{Im}\,[(M_6 - i\omega M_7 + \omega^2 M_8)\, a\bar{b}]$$

$$- \pi\omega M_{10}\,|b|^2 - \pi \operatorname{Im}\,[\bar{b}Z]$$

$$\leqq \pi\,|M_6 - i\omega M_7 + \omega^2 M_8|\,|a|\,|b| - \pi\omega M_{10}\,|b|^2$$

$$+ \pi \frac{(\omega^2 M_{12} + \omega^3 M_{13} + \omega^4 M_{14})}{1 - \omega M_1 - \omega^2 M_3}(|a|\,|b| + M_5\,|b|^2).$$

The coefficient of $|b|^2$ in this last expression is *negative* provided that ω satisfies the inequalities

$$\omega M_1 + \omega^2 M_3 < 1,$$

$$\frac{(\omega M_{12} + \omega^2 M_{13} + \omega^3 M_{14})}{1 - \omega M_1 - \omega^2 M_3} M_5 < M_{10}.$$

Thus, for any fixed ω chosen in this way and for any fixed complex a, the work done by the body in a period is bounded above as function of b, and the proof of our assertion is complete.

As has been remarked already, the argument hinges upon the fact that the constant

$$M_{10} = \theta_0 \int_A^B \int_A^B K(x, y) \frac{\mu(x)\,\mu(y)}{\beta(x)\,\beta(y)}\, dx\, dy$$

is positive.

The argument establishes more than has been claimed, for it is clear that the work done by the body in a period is not bounded below as a function of b. Since the work done on the body in a period is the negative of the work done by the body in a period it must be that: *for any given ω which lies in the interval $0 < \omega < \omega_0$ and for any given complex number a, it is possible to choose b in such a way that the work done on the body in a period exceeds any assigned number.*

In this last assertion the number a is permitted to vanish and, once again, the field theory is in conflict with the homogeneous field approximation, for the homogeneous field approximation implies that the work done by the body in a period must vanish whenever a vanishes.

The implications of an inertialess field theory

There is a third theory—the inertialess field theory[4]—which is intermediate in difficulty between our field theory and the homogeneous field approximation, and which nicely avoids the crudity of the latter and the complexity of the former.

---

[4] CARLSON [4] calls this "the quasi-static theory".

The inertialess field theory is obtained by neglecting the term

$$\varrho \frac{\partial^2 u}{\partial t^2}$$

in the equation of motion. Within the inertialess field theory, therefore, the stress field, which is

$$\sigma(x, t) = -p(t),$$

is spatially homogeneous, but the temperature field, which now satisfies the partial differential equation

$$\frac{\partial}{\partial x}\left(k \frac{\partial \theta}{\partial x}\right) = \left(c + \theta_0 \frac{\mu^2}{\beta}\right) \frac{\partial \theta}{\partial t} - \theta_0 \frac{\mu}{\beta} \dot{p}$$

and the boundary conditions

$$\theta(A, t) = \theta(B, t) = \tau(t),$$

is not so.

According to the inertialess field theory, the work done by the body in a period is

$$\int_0^T \int_A^B p \left(\frac{\mu}{\beta} \frac{\partial \theta}{\partial t} - \frac{1}{\beta} \dot{p}\right) dx\, dt$$

and, by virtue of the periodicity of p, this reduces to

$$\int_0^T \int_A^B p \frac{\mu}{\beta} \frac{\partial \theta}{\partial t} dx\, dt.$$

Once again, I consider only the simplest case in which the surface temperature and the surface pressure have the forms

$$\tau(t) = \theta_0 + \mathrm{Re}\,[a \exp(i\omega t)],$$

$$p(t) = p_0 + \mathrm{Re}\,[b \exp(i\omega t)],$$

and the temperature field is

$$\theta(x, t) = \theta_0 + \mathrm{Re}\,[\Theta(x) \exp(i\omega t)],$$

where Θ is the solution of the boundary value problem

$$(k\Theta')' = i\omega\left(\left(c + \theta_0 \frac{\mu^2}{\beta}\right) \Theta - \theta_0 \frac{\mu}{\beta} b\right),$$

$$\Theta(A) = \Theta(B) = a.$$

The work done by the body in a period is

$$-\pi\, \mathrm{Im}\left[\bar{b} \int_A^B \frac{\mu}{\beta} \Theta\, dx\right].$$

The interesting feature of the inertialess field theory is that the work done by the body in a period is bounded above, for a given ω and a given a, whatever the size of ω: *if ω is any positive number and a is any complex number, there is a finite number $\Omega(\omega)$ such that the work done by the body in a period cannot exceed*

$$\Omega(\omega)\,|a|^2$$

no matter how b is chosen.

Our argument will provide an explicit formula for $\Omega(\omega)$.

We begin the proof by remarking that, for given values of ω, a, and b, the existence and uniqueness of Θ is guaranteed. For, on multiplying the differential equation that Θ satisfies by the complex conjugate $\bar{\Theta}$, integrating with respect to x, and appealing to the boundary conditions, we obtain the identity

$$\bar{a}(k(B)\,\Theta'(B) - k(A)\,\Theta'(A))$$

$$= \int_A^B k\,|\Theta'|^2\,dx + i\omega \int_A^B \left(c + \theta_0 \frac{\mu^2}{\beta}\right) |\Theta|^2\,dx - i\omega\theta_0 b \int_A^B \frac{\mu}{\beta}\bar{\Theta}\,dx.$$

In order to show that Θ is unique, if it exists, it is enough to show that $\Theta(x)$ must vanish identically in the interval $[A, B]$ when $a = b = 0$. On setting $a = b = 0$ in the identity just obtained we deduce that

$$\int_A^B k\,|\Theta'|^2\,dx + i\omega \int_A^B \left(c + \theta_0 \frac{\mu^2}{\beta}\right) |\Theta|^2\,dx = 0$$

and it follows immediately, because ω and $c + \theta_0\mu^2/\beta$ are positive, that Θ does indeed vanish identically when $a = b = 0$. Thus, the uniqueness of Θ, for general a and b, is established.

In order to show that Θ must exist it is enough to observe that the boundary value problem which determines Θ is equivalent to the integral equation

$$\Theta(x) = a - i\omega \int_A^B K(x, y) \left(c(y) + \theta_0 \frac{\mu(y)^2}{\beta(y)}\right) \Theta(y)\,dy$$
$$+ i\omega\theta_0 b \int_A^B K(x, y) \frac{\mu(y)}{\beta(y)}\,dy.$$

In view of what has just been proved, the corresponding homogeneous equation

$$\Theta(x) = i\omega \int_A^B K(x, y) \left(c(y) + \theta_0 \frac{\mu(y)^2}{\beta(y)}\right) \Theta(y)\,dy,$$

which is obtained by setting $a = b = 0$, has no solution other than $\Theta(x) = 0$. Thus, as Fredholm's theory assures us, the integral equation has a unique solution for general values of a and b, and the same must be true of the boundary value problem.

The main step in the proof of the upper bound on the work done by the body in a period involves introducing a complex-valued kernel $G(x, y, \omega)$

associated with the operator

$$\frac{d}{dx}\left(k(x)\frac{d}{dx}\right) - i\omega\left(c(x) + \theta_0\frac{\mu(x)^2}{\beta(x)}\right).$$

The kernel $K(x, y)$, introduced earlier, coincides with $G(x, y, 0)$.

Let $\xi_A(x, \omega)$ and $\xi_B(x, \omega)$ be the complex-valued solutions of the initial value problems

$$(k\xi_A')' = i\omega\left(c + \theta_0\frac{\mu^2}{\beta}\right)\xi_A, \quad \xi_A(A, \omega) = 0, k(A)\,\xi_A'(A, \omega) = 1,$$

$$(k\xi_B')' = i\omega\left(c + \theta_0\frac{\mu^2}{\beta}\right)\xi_B, \quad \xi_B(B, \omega) = 0, k(B)\,\xi_B'(B, \omega) = -1.$$

The existence and uniqueness of ξ_A and ξ_B are guaranteed by standard arguments and, by virtue of our proof of uniqueness, ξ_A and ξ_B are linearly independent on $[A, B]$. Furthermore,

$$\Delta(\omega) = k(\xi_A'\xi_B - \xi_A\xi_B')$$

is independent of x and cannot vanish, and

$$\Delta(\omega) = \xi_B(A, \omega) = \xi_A(B, \omega).$$

At this point we define

$$G(x, y, \omega) = \begin{cases} \xi_A(x, \omega)\,\xi_B(y, \omega)/\Delta(\omega), & A \leqq x \leqq y \leqq B, \\ \xi_B(x, \omega)\,\xi_A(y, \omega)/\Delta(\omega), & A \leqq y \leqq x \leqq B, \end{cases}$$

and proceed to check that G has the following important property: *if ω is any positive number and $f(x)$ is any real-valued and continuous function which does not vanish identically on the interval $[A, B]$ then*

$$\operatorname{Re}\int_A^B\int_A^B G(x, y, \omega) f(x) f(y)\,dx\,dy > 0.$$

For, if we put

$$g(x) = \int_A^B G(x, y, \omega) f(y)\,dy,$$

g is the solution of the boundary value problem

$$(kg')' = i\omega\left(c + \theta_0\frac{\mu^2}{\beta}\right)g - f, \quad g(A) = g(B) = 0.$$

On multiplying the differential equation by $\bar{g}$, integrating with respect to x, and using the boundary conditions, we have

$$\int_A^B k\,|g'|^2\,dx + i\omega\int_A^B\left(c + \theta_0\frac{\mu^2}{\beta}\right)|g|^2\,dx - \int_A^B f\bar{g}\,dx = 0.$$

Thus, if we take the complex conjugate and remember that f is real-valued we see that

$$\int_A^B \int_A^B G(x, y, \omega) f(x) f(y)\, dx\, dy$$

$$= \int_A^B fg\, dx$$

$$= \int_A^B k\, |g'|^2\, dx - i\omega \int_A^B \left(c + \theta_0 \frac{\mu^2}{\beta}\right) |g|^2\, dx.$$

Hence

$$\mathrm{Re} \int_A^B \int_A^B G(x, y, \omega) f(x) f(y)\, dx\, dy = \int_A^B k\, |g'|^2\, dx \geqq 0.$$

Moreover, the inequality must be strict because the conditions

$$\int_A^B k\, |g'|^2\, dx = 0, \qquad g(A) = g(B) = 0,$$

force $g(x)$ to vanish identically and, by virtue of the differential equation that g satisfies, the same would have to be true of $f(x)$. Thus, the inequality is indeed a strict one.

It is now possible to verify that the work done by the body in a period is bounded above. To do so we observe that the boundary value problem which determines $\Theta(x)$ can be solved with the aid of the kernel $G(x, y, \omega)$: the solution is

$$\Theta(x) = \frac{a}{\Delta(\omega)} (\xi_A(x, \omega) + \xi_B(x, \omega)) + i\omega\theta_0 b \int_A^B G(x, y, \omega) \frac{\mu(y)}{\beta(y)}\, dy.$$

Therefore

$$\int_A^B \frac{\mu}{\beta} \Theta\, dx = \frac{a}{\Delta(\omega)} \int_A^B \frac{\mu(x)}{\beta(x)} (\xi_A(x, \omega) + \xi_B(x, \omega))\, dx$$

$$+ i\omega\theta_0 b \int_A^B \int_A^B G(x, y, \omega) \frac{\mu(x)}{\beta(x)} \frac{\mu(y)}{\beta(y)}\, dx\, dy$$

and the work done by the body in a period is given by the formula

$$-\pi\, \mathrm{Im}\left[\bar{b} \int_A^B \frac{\mu}{\beta} \Theta\, dx\right]$$

$$= -\pi\, \mathrm{Im}\left[\frac{a\bar{b}}{\Delta(\omega)} \int_A^B \frac{\mu(x)}{\beta(x)} (\xi_A(x, \omega) + \xi_B(x, \omega))\, dx\right]$$

$$- \pi\theta_0\omega\, |b|^2\, \mathrm{Re} \int_A^B \int_A^B G(x, y, \omega) \frac{\mu(x)}{\beta(x)} \frac{\mu(y)}{\beta(y)}\, dx\, dy$$

in which the coefficient of $|b|^2$ is known to be negative; it follows that the work done by the body in a period is bounded above as a function of b.

In fact, the maximum value of the work done by the body in a period occurs when b takes the value

$$\frac{\dfrac{ia}{\Delta(\omega)}\displaystyle\int_A^B \frac{\mu(x)}{\beta(x)}(\xi_A(x,\omega)+\xi_B(x,\omega))\,dx}{2\theta_0\omega\,\mathrm{Re}\displaystyle\int_A^B\int_A^B G(x,y,\omega)\frac{\mu(x)}{\beta(x)}\frac{\mu(y)}{\beta(y)}\,dx\,dy},$$

and the maximum is $\Omega(\omega)\,|a|^2$, where

$$\Omega(\omega)=\frac{\pi\left|\dfrac{1}{\Delta(\omega)}\displaystyle\int_A^B \frac{\mu(x)}{\beta(x)}(\xi_A(x,\omega)+\xi_B(x,\omega))\,dx\right|^2}{4\theta_0\omega\,\mathrm{Re}\displaystyle\int_A^B\int_A^B G(x,y,\omega)\frac{\mu(x)}{\beta(x)}\frac{\mu(y)}{\beta(y)}\,dx\,dy}.$$

The asymptotic behaviour of $\Omega(\omega)$ at low frequency is of interest. It is not difficult to show that, as $\omega\to 0$,

$$\frac{1}{\Delta(\omega)}\int_A^B \frac{\mu(x)}{\beta(x)}(\xi_A(x,\omega)+\xi_B(x,\omega))\,dx \to \int_A^B \frac{\mu(x)}{\beta(x)}\,dx,$$

$$\int_A^B\int_A^B G(x,y,\omega)\frac{\mu(x)}{\beta(x)}\frac{\mu(y)}{\beta(y)}\,dx\,dy \to \int_A^B\int_A^B K(x,y)\frac{\mu(x)}{\beta(x)}\frac{\mu(y)}{\beta(y)}\,dx\,dy,$$

and, hence,

$$\Omega(\omega)\sim\frac{\pi\left(\displaystyle\int_A^B \frac{\mu(x)}{\beta(x)}\,dx\right)^2}{4\theta_0\omega\displaystyle\int_A^B\int_A^B K(x,y)\frac{\mu(x)}{\beta(x)}\frac{\mu(y)}{\beta(y)}\,dx\,dy}.$$

It follows, in particular, that $\Omega(\omega)\to\infty$ as $\omega\to 0$ and, thus, the work done by the body in a period is unbounded in its dependence upon the frequency. In more exact terms: *the inertialess field theory predicts that, for a given nonzero complex a, it is possible to choose a positive* ω *and a complex b so that the work done by the body in a period exceeds any assigned value.*

This assertion differs from the corresponding assertion for the homogeneous field approximation in that the latter requires no adjustment of the frequency.

References

1. SERRIN, J., Lectures on thermodynamics. University of Naples (1979).
2. SERRIN, J., An outline of thermodynamical structure. In: New Perspectives in Thermodynamics, edited by J. SERRIN. Berlin Heidelberg New York, Springer-Verlag, 1986.
3. COLEMAN, B. D., D. R. OWEN, & J. SERRIN, The second law of thermodynamics for systems with approximate cycles. Arch. Rational Mech. Anal. **77**, 103–142 (1981)
4. CARLSON, D. E., Linear thermoelasticity. In Handbuch der Physik, vol. VIa/2, edited by C. TRUESDELL. Berlin Heidelberg New York, Springer-Verlag, 1972.
5. DAY, W. A., A Commentary on Thermodynamics. Springer Tracts in Natural Philosophy, Vol. 32. New York, Springer-Verlag, 1988.
6. DAY, W. A., Steady forced vibrations in coupled thermoelasticity. Arch. Rational Mech. Anal. **93**, 323–334 (1986).

Hertford College
Oxford

(Received November 2, 1987)

On the Stored Energy Functions of Hyperelastic Materials with Internal Constraints

H. COHEN & C.-C. WANG

Dedicated to B. D. Coleman on the Occasion of his Sixtieth Birthday

1. Introduction

In this paper we continue our study [1] on elastic materials with internal constraints by formulating the stored energy functions as functions defined only on the constraint surfaces of the materials. Since the constraint surfaces generally have dimensions lower than that of the proper linear group, differentiations and integrations with respect to the deformation gradient are not applicable and must be replaced by suitable operators associated with the constraint surfaces. It turns out that many results for hyperelastic materials free of constraints can be carried over to similar results for hyperelastic materials with internal constraints. However, there are also some constraint-free results whose counterparts with constraints are more complex.

For example, by virtue of the chain rule of differentiation the constraint-free stored energy function σ obeys the transformation rule

$$\boldsymbol{H}\,(\text{grad}\,\sigma)^T\,(\boldsymbol{FH}) = (\text{grad}\,\sigma)^T\,(\boldsymbol{F}) \tag{1.1}$$

for all deformation gradients $\boldsymbol{F}$ and all symmetry transformations $\boldsymbol{H}$ of σ. The direct counterpart of (1.1) for a function σ defined only on the constraint surface is

$$\boldsymbol{H}(d_{\boldsymbol{FH}}\,\sigma)^T = (d_{\boldsymbol{F}}\sigma)^T \tag{1.2}$$

where $d\sigma$ denotes the differential of the function σ. Unfortunately, the formula (1.2) is generally not true since the tangent space of the constraint surface at $\boldsymbol{FH}$ acted on the right by $\boldsymbol{H}^T$ generally does not coincide with the tangent space at $\boldsymbol{F}$. Since the formula (1.1) is an important step in the proof of the Truesdell theorem, we must be very careful in generalizing its implications.

The choice of a reference configuration may also be a problem when the material is subject to some internal constraints. Since we are mainly interested in developing the relation between the symmetry of the stored energy function and the symmetry of the distribution of the stress spaces on the constraint surface, an admissible (*i.e.* compatible with the constraint) configuration turns out to be most convenient as a reference configuration. However, there are materials for which not all undistorted (defined in some suitable way) configurations are admissible. As a result, an admissible reference configuration need not be undistorted. Because of such complications, the proofs of some constraint-free results must be modified before they can be extended to similar results for materials with constraints.

The concept of internal constraint as introduced in [1] is briefly reviewed in Section 2. Then the concept of a stored energy function is introduced in Section 3. We emphasize that only intrinsic entities are used in our formulation. Thus any extension of the stored energy function beyond its intrinsic domain – the constraint surface, plays no role at all in our formulation. We generalize the important Truesdell theorem to hyperelastic materials with internal constraints in Section 4. Then in Section 5 we obtain the general solution for the pair of material symmetry condition and material frame indifference condition in the context of internal constraints. Finally, in Section 6 we present general embedding theorems which describe the relation between representations for the stored energy functions and response functions of hyperelastic materials and elastic materials with internal constraints and the same for materials free of constraints. An open problem raised in [1], concerning the representation for the distribution of the stress spaces for elastic materials with internal constraints, is solved also in this section.

2. Internal Constraints

Let $\mathscr{E}$ and $\mathscr{V}$ denote the physical space and the physical translation space. Then the deformation gradient of a material point free of any internal constraints may be any proper linear isomorphism of $\mathscr{V}$. Let $\mathscr{M}$ denote the space of all linear transformations of $\mathscr{V}$. Then the proper linear group $\mathscr{L}$ is an open set in $\mathscr{M}$ defined by the condition

$$\boldsymbol{F} \in \mathscr{L} \quad \Leftrightarrow \quad \det \boldsymbol{F} > 0. \tag{2.1}$$

As usual, we identify the tangent space of $\mathscr{L}$ at any point $\boldsymbol{F} \in \mathscr{L}$ with the underlying space $\mathscr{M}$, and the space $\mathscr{M}$ is equipped with the standard inner product defined by

$$(\boldsymbol{A}, \boldsymbol{B}) = \operatorname{tr}(\boldsymbol{A}\boldsymbol{B}^T) \tag{2.2}$$

for all $\boldsymbol{A}, \boldsymbol{B} \in \mathscr{M}$. Of course, the dimension of $\mathscr{E}$ and $\mathscr{V}$ is three while that of $\mathscr{L}$ and $\mathscr{M}$ is nine.

As formulated in [1] an internal constraint of a material point may be defined by a connected smooth surface $\mathscr{C}$ in $\mathscr{L}$ such that, relative to a particular local reference configuration[1] $\varkappa$ of the material point, the deformation gradient $\boldsymbol{F}$ must be contained in $\mathscr{C}$. We have remarked in [1] that the abstract concept of

[1] We use a particular local reference configuration $\varkappa$ throughout this paper. Hence, for simplicity, we suppress the subscript $\varkappa$ from the notations $\mathscr{C}_\varkappa$, $\mathscr{N}_\varkappa$, $\mathscr{S}_\varkappa$, *etc.* introduced in [1].

an internal constraint is defined by a set of admissible local configurations; the surface $\mathscr{C}$ merely characterizes that set by means of its deformation gradients relative to the local reference configuration $\boldsymbol{\varkappa}$. Under a change of local reference configuration from $\boldsymbol{\varkappa}$ to $\boldsymbol{K}\boldsymbol{\varkappa}$, where $\boldsymbol{K} \in \mathscr{L}$, the constraint surface $\mathscr{C}$ associated with $\boldsymbol{\varkappa}$ is replaced by the constraint surface $\boldsymbol{K}^{-1}\mathscr{C}$ associated with $\boldsymbol{K}\boldsymbol{\varkappa}$. Hence without loss of generality we may require that $\mathscr{C}$ contains a particular tensor in $\mathscr{L}$, say

$$\mathscr{C} \ni \boldsymbol{I}, \tag{2.3}$$

where $\boldsymbol{I}$ denotes the identity transformation of $\mathscr{V}$. Indeed, the condition (2.3) is equivalent to the hypothesis that an admissible local configuration of the material point is chosen as the local reference configuration $\boldsymbol{\varkappa}$. We assume that (2.3) holds throughout this paper.

By virtue of consideration of the principle of material frame-indifference the constraint surface $\mathscr{C}$ is required to be closed under any superimposed rotation $\boldsymbol{Q}$ on an admissible deformation gradient $\boldsymbol{F} \in \mathscr{C}$. Thus

$$\boldsymbol{Q}\mathscr{C} = \mathscr{C} \tag{2.4}$$

for all rotations $\boldsymbol{Q}$ of $\mathscr{V}$. From (2.3) $\mathscr{C}$ contains the rotation group $\mathscr{SO}$ of $\mathscr{V}$. In fact, (2.4) implies that

$$\mathscr{C} \supset \mathscr{SO}\boldsymbol{F} \tag{2.5}$$

for each $\boldsymbol{F} \in \mathscr{C}$. Hence the dimension of $\mathscr{C}$ must be greater than three. (The extremal case $\mathscr{C} = \mathscr{SO}$ corresponds to the constraint of rigidity, which is not of interest in continuum mechanics.) The difference n between the dimension of $\mathscr{L}$ (namely nine) and that of $\mathscr{C}$ is called the degree of the constraint. Locally, $\mathscr{C}$ may be characterized by n independent smooth functions f_Γ, $\Gamma = 1, \ldots, n$, defined in $\mathscr{M}$ or in $\mathscr{L}$, such that

$$\boldsymbol{F} \in \mathscr{C} \Leftrightarrow f_\Gamma(\boldsymbol{F}) = 0 \quad \forall\, \Gamma = 1, \ldots, n. \tag{2.6}$$

The degree n is the co-dimension of $\mathscr{C}$ in $\mathscr{M}$ or in $\mathscr{L}$. At each point $\boldsymbol{F} \in \mathscr{C}$ the tangent space $\mathscr{T}(\boldsymbol{F})$ of $\mathscr{C}$ is a $(9 - n)$-dimensional subspace in $\mathscr{M}$ while its orthogonal complement $\mathscr{T}(\boldsymbol{F})^\perp$ is an n-dimensional subspace in $\mathscr{M}$. As in [1] we define the reaction space $\mathscr{N}(\boldsymbol{F})$ at $\boldsymbol{F} \in \mathscr{C}$ by

$$\mathscr{N}(\boldsymbol{F}) = \boldsymbol{F}(\mathscr{T}(\boldsymbol{F})^\perp)^T. \tag{2.7}$$

Then $\mathscr{N}(\boldsymbol{F})$ is the n-dimensional subspace in $\mathscr{M}$ such that

$$\boldsymbol{N} \in \mathscr{N}(\boldsymbol{F}) \quad \Leftrightarrow \quad \operatorname{tr}(\boldsymbol{N}\dot{\boldsymbol{F}}\boldsymbol{F}^{-1}) = 0 \tag{2.8}$$

for all $\dot{\boldsymbol{F}} \in \mathscr{T}(\boldsymbol{F})$. The right hand side of (2.8) is the condition that the stress power of the reaction vanishes in any admissible local process of the material point. The representation (2.7) is equivalent to the orthogonal direct sum decomposition

$$\mathscr{M} = \mathscr{N}(\boldsymbol{F})\,(\boldsymbol{F}^{-1})^T \oplus \mathscr{T}(\boldsymbol{F}), \tag{2.9}$$

where we have removed the transpose sign on the reaction space $\mathcal{N}(\boldsymbol{F})$ since all reactions are symmetric as we have remarked in [1]. In other words, $\mathcal{N}(\boldsymbol{F})$ may be represented also by

$$\mathcal{N}(\boldsymbol{F}) = \mathcal{T}(\boldsymbol{F})^{\perp} \boldsymbol{F}^T. \tag{2.10}$$

From (2.4) the tangent spaces of $\mathscr{C}$ obey the transformation rule

$$\mathcal{T}(\boldsymbol{QF}) = \boldsymbol{Q}\mathcal{T}(\boldsymbol{F}) \tag{2.11}$$

for all $\boldsymbol{F} \in \mathscr{C}$ and $\boldsymbol{Q} \in \mathscr{SO}$. From the representations (2.7) or (2.10) the reaction spaces on $\mathscr{C}$ obey the transformation rule

$$\mathcal{N}(\boldsymbol{QF}) = \boldsymbol{Q}\mathcal{N}(\boldsymbol{F})\,\boldsymbol{Q}^T \tag{2.12}$$

for all $\boldsymbol{F} \in \mathscr{C}$ and $\boldsymbol{Q} \in \mathscr{SO}$. In [1] we defined the symmetry group $\mathscr{k}$ of the constraint surface $\mathscr{C}$ in $\mathscr{L}$ by

$$\boldsymbol{K} \in \mathscr{k} \quad \Leftrightarrow \quad \mathscr{C}\boldsymbol{K} = \mathscr{C}. \tag{2.13}$$

Then for any $\boldsymbol{K} \in \mathscr{k}$ the tangent spaces of $\mathscr{C}$ obey the transformation rule

$$\mathcal{T}(\boldsymbol{FK}) = \mathcal{T}(\boldsymbol{F})\,\boldsymbol{K} \tag{2.14}$$

for all $\boldsymbol{F} \in \mathscr{C}$. From the representations (2.7) or (2.10) the reaction spaces obey the transformation rule

$$\mathcal{N}(\boldsymbol{FK}) = \mathcal{N}(\boldsymbol{F}) \tag{2.15}$$

for all $\boldsymbol{K} \in \mathscr{k}$ and $\boldsymbol{F} \in \mathscr{C}$.

It should be noted that the transformation rules (2.11), (2.12), (2.14), and (2.15) are properties of the constraint surface $\mathscr{C}$, so that they are entirely independent of the distribution of the stress spaces on $\mathscr{C}$. As a result, they are valid for an elastic material point or for a hyperelastic material point with internal constraint. The assumption of hyperelasticity concerns only the distribution of the stress spaces on $\mathscr{C}$.

We note here some special internal constraints considered in [1]. First, incompressibility is defined by the constraint surface $\mathscr{C} = \mathscr{SL}$, *viz.*,

$$\boldsymbol{F} \in \mathscr{C} \quad \Leftrightarrow \quad \det \boldsymbol{F} = 1, \tag{2.16}$$

where we have chosen the constant value of $\det \boldsymbol{F}$ as one in order to satisfy the condition (2.3). The symmetry group of this constraint surface is the special linear group $\mathscr{SL}$. Next, inextensibility in the direction of an unit vector $\boldsymbol{e}$ in the local reference configuration $\varkappa$ is defined by the constraint surface $\mathscr{C}$ such that

$$\boldsymbol{F} \in \mathscr{C} \quad \Leftrightarrow \quad \boldsymbol{Fe} \cdot \boldsymbol{Fe} = 1, \tag{2.17}$$

where we have chosen the constant magnitude of $\boldsymbol{Fe}$ as one so as to satisfy (2.3). We have shown in [1] that the symmetry group $\mathscr{k}$ of this constraint surface is given by

$$\boldsymbol{K} \in \mathscr{k} \quad \Leftrightarrow \quad \boldsymbol{Ke} = \pm \boldsymbol{e}. \tag{2.18}$$

Finally, total inextensibility is defined by

$$\boldsymbol{F} \in \mathscr{C} \quad \Leftrightarrow \quad \operatorname{tr}(\boldsymbol{F}^T\boldsymbol{F}) = 3 \tag{2.19}$$

which satisfies (2.3). In [1] we showed that the symmetry group $\mathscr{k}$ of this constraint is the rotation group $\mathscr{SO}$.

For the preceding three basic constraints the degree is one. When two of the three are imposed on a material point, the degree becomes two, and when all three are imposed, the degree is increased to three. Various material symmetries were considered in [1] in conjunction with the different types of internal constraints. We consider the same for hyperelastic material points in this paper.

3. Stored Energy Function

For a hyperelastic material point with internal constraint the stored energy function is a (scalar-valued) smooth function σ defined on the constraint surface $\mathscr{C}$ relative to the local reference configuration $\varkappa$. In any admissible local process $F = F(t)$ the stored energy density (per unit volume) of the point is given by the product $\varrho\sigma$, *viz.*,

$$\varrho\sigma = \varrho(t)\,\sigma(F(t)) \tag{3.1}$$

where ϱ denotes the mass density of the point in the process. The law of conservation of mass implies that

$$\varrho(t) = \varrho_\varkappa \det F(t), \tag{3.2}$$

where $\varrho_\varkappa$ denotes the mass density of the material point in the local configuration $\varkappa$.

In a purely mechanical theory, energy is present in the forms of kinetic energy, potential energy, and stored energy. Then the reduced form of the law of conservation of energy implies that

$$\varrho\dot{\sigma} = \operatorname{tr}(T\dot{F}F^{-1}), \tag{3.3}$$

where $T = T(t)$ denotes the stress tensor at the point in the local process. The right hand side of (3.3) is known as the stress power.

As remarked in [1] the stress tensor of a material point with internal constraint is determined in the local process only to within an additive reaction $N = N(t)$ which is contained in the reaction space $\mathscr{N}(F(t))$. By virtue of (2.8) the stress power of the reaction vanishes. Hence the stress power on the right hand side of (3.3) is a determinate function of t in each admissible process of the material point.

Since the stored energy function is a smooth function on $\mathscr{C}$, the rate of change $\dot{\sigma}$ on the left hand side of (3.3) may be calculated by

$$\dot{\sigma} = \frac{d}{dt}\sigma(F(t)) = \langle d_{F(t)}\sigma, \dot{F}(t)\rangle, \tag{3.4}$$

where $d_F\sigma$ denotes the differential of σ at the point $F \in \mathscr{C}$. The right hand side of (3.4) is a linear function on the tangent space $\mathscr{T}(F(t))$ which is an inner product space since it is a subspace of the inner product space $\mathscr{M}$. As a result, the differential $d_F\sigma$ may be represented by an element, which we denote by the same notation $d_F\sigma$, belonging to $\mathscr{T}(F)$ and such that

$$\langle d_F\sigma, \dot{F}\rangle = (d_F\sigma, \dot{F}) \tag{3.5}$$

for all $\dot{F} \in \mathcal{T}(F)$, where the right hand side of (3.5) denotes the inner product on $\mathcal{T}(F)$; *cf.* (2.2).

Using the representation (3.5), we can express the reduced energy equation (3.3) as

$$\operatorname{tr}\left([\varrho(t)\,(d_{F(t)}\,\sigma)^T - F^{-1}(t)\,T(t)]\,\dot{F}(t)\right) = 0. \tag{3.6}$$

Since at each point $F(t)$ in $\mathscr{C}$ the tangent $\dot{F}(t)$ is arbitrary in $\mathcal{T}(F(t))$, the condition (3.6) is possible if and only if the stress tensor T is given by

$$T = \varrho F (d_F \sigma)^T + N, \tag{3.7}$$

where $N \in \mathcal{N}(F)$. Using the notation $\mathcal{S}(F)$ introduced in [1] to denote the stress space, we can rewrite (3.7) as

$$\mathcal{S}(F) = \varrho F (d_F \sigma)^T + \mathcal{N}(F). \tag{3.8}$$

This is the constitutive equation of a hyperelastic material point relative to the local reference configuration.

In order that the right hand side of (3.8) be a valid representation for the stress space, the leading term must be a symmetric tensor in $\mathcal{M}$. It turns out that this important condition is equivalent to the requirement that σ be a frame-indifferent function on $\mathscr{C}$, *viz.*, it obeys the identity

$$\sigma(QF) = \sigma(F) \tag{3.9}$$

for all $F \in \mathscr{C}$ and $Q \in \mathcal{SO}$. Indeed, by virtue of (2.4) the curve

$$F(t) = Q(t)\,F \tag{3.10}$$

such that $Q(t) \in \mathcal{SO}$ with $Q(0) = I$ is contained in $\mathscr{C}$ at each $F \in \mathscr{C}$. Applying (3.4) to this curve at $t = 0$ and using the representation (3.5), we obtain

$$\operatorname{tr}\,[F(d_F\sigma)^T\,\dot{Q}(0)] = 0. \tag{3.11}$$

Since $\dot{Q}(0)$ may be any skew symmetric tensor, the condition (3.11) implies that $F(d_F\sigma)^T$ must be a symmetric tensor. Conversely, suppose that $F(d_F\sigma)^T$ is symmetric at all points $F \in \mathscr{C}$. Then using (3.4) for the curve (3.10) at an arbitrary t, we obtain

$$\begin{aligned} \frac{d}{dt}\sigma(F(t)) &= \frac{d}{dt}\sigma(Q(t)\,F) = \operatorname{tr}\,[F(d_{Q(t)F}\sigma)^T\,\dot{Q}(t)] \\ &= \operatorname{tr}\left([F(t)\,(d_{F(t)}\sigma)^T]\,[\dot{Q}(t)\,Q^T(t)]\right) = 0, \end{aligned} \tag{3.12}$$

because $\dot{Q}(t)\,Q^T(t)$ is skew symmetric since $Q(t) \in \mathcal{SO}$. As a result, σ takes constant value on all curves of the form (3.10). Since $\mathcal{SO}$ is connected, this condition is equivalent to the frame-indifference condition (3.9).

We shall now assume that σ is a frame-indifferent function on the constraint surface $\mathscr{C}$. Another important condition on the constitutive equation (3.8) is that the distribution of stress spaces on $\mathscr{C}$ must obey the transformation rule

$$\mathcal{S}(QF) = Q\,\mathcal{S}(F)\,Q^T \tag{3.13}$$

for all $F \in \mathscr{C}$ and $Q \in \mathscr{SO}$. For a hyperelastic material point with constitutive equation given by (3.8) the transformation rule (3.13) can be derived from the condition (3.9). Indeed, applying (3.4) to the curves $F(t)$ and $QF(t)$ and using (3.9), we obtain

$$\operatorname{tr}\left(\left[(d_{QF}\sigma)^T Q - (d_F\sigma)^T\right] \dot{F}\right) = 0, \tag{3.14}$$

where we have suppressed the argument t. Since $\dot{F}(t)$ is arbitrary in the tangent space $\mathscr{T}(F(t))$, the condition (3.14) is equivalent to

$$(d_{QF}\sigma)^T Q - (d_F\sigma)^T \in (\mathscr{T}(F)^{\perp})^T. \tag{3.15}$$

Using (2.7), we can rewrite this condition as

$$F(d_{QF}\sigma)^T Q - F(d_F\sigma)^T \in \mathscr{N}(F). \tag{3.16}$$

Since the leading term can be rewritten as

$$F(d_{QF}\sigma)^T Q = Q^T[QF(d_{QF}\sigma)^T]\, Q, \tag{3.17}$$

the condition (3.16) and the constitutive equation (3.8) imply that

$$QF(d_{QF}\sigma)^T \in Q\mathscr{S}(F)\, Q^T. \tag{3.18}$$

Combining (3.18) with the transformation rule (2.12) for the reaction spaces, we then obtain (3.13). Thus the distribution of stress spaces defined by (3.8) automatically obeys the transformation rule (3.13) provided that the stored energy function σ is frame-indifferent on the constraint surface $\mathscr{C}$.

We define the symmetry group $\mathscr{h}$ of the stored energy function as the subgroup of the symmetry group $\mathscr{k}$ of the constraint surface $\mathscr{C}$ such that

$$H \in \mathscr{h} \quad \Leftrightarrow \quad \sigma(FH) = \sigma(F) \tag{3.19}$$

for all $F \in \mathscr{C}$. This concept is similar to the concept of the symmetry group $\mathscr{g}$ of the distribution of the stress spaces on $\mathscr{C}$ defined by the conditions that $\mathscr{g} \subset \mathscr{k}$ and that

$$G \in \mathscr{g} \quad \Leftrightarrow \quad \mathscr{S}(FG) = \mathscr{S}(F) \tag{3.20}$$

for all $F \in \mathscr{C}$. Since the distribution $\mathscr{S}$ is determined by the function σ through (3.8), the symmetry group $\mathscr{h}$ and $\mathscr{g}$ are related. We show first that $\mathscr{h}$ is a subgroup of $\mathscr{g}$.

Consider the curves $F(t)$ and $F(t)\,H$ in $\mathscr{C}$ for any $H \in \mathscr{h}$. Applying (3.4) to the curves and using the right hand side of (3.19), we obtain

$$\operatorname{tr}\left(\left[H(d_{FH}\sigma)^T - (d_F\sigma)^T\right] \dot{F}\right) = 0. \tag{3.21}$$

Since $\dot{F}$ is arbitrary in the tangent space $\mathscr{T}(F)$, the condition (3.21) leads as before to

$$FH(d_{FH}\sigma)^T - F(d_F\sigma)^T \in \mathscr{N}(F). \tag{3.22}$$

Then by virtue of the transformation rule (2.15) we see that

$$\mathscr{S}(FH) = \mathscr{S}(F) \tag{3.23}$$

for all $F \in \mathscr{C}$ and $H \in \mathscr{h}$. Thus $\mathscr{h}$ is contained in $\mathscr{g}$.

For hyperelastic material points free of internal constraints some converse results relating the symmetry groups $\mathscr{h}$ and $\mathscr{g}$ were obtained by TRUESDELL [2]. We shall extend these results to hyperelastic material points with internal constraints in the next section. It should be noted that the results (3.18) and (3.23) are also generalizations of corresponding results for hyperelastic material points free of constraints. However, for hyperelastic material points free of constraints the condition of material frame-indifference is the requirement that

$$\boldsymbol{Q}\boldsymbol{F}(d_{\boldsymbol{Q}\boldsymbol{F}}\sigma)^T = \boldsymbol{Q}[\boldsymbol{F}(d_{\boldsymbol{F}}\sigma)^T]\,\boldsymbol{Q}^T \tag{3.24}$$

for all proper $\boldsymbol{F}$ and all rotations $\boldsymbol{Q}$. The condition (3.24) may be rewritten as

$$(d_{\boldsymbol{Q}\boldsymbol{F}}\sigma)^T\,\boldsymbol{Q} - (d_{\boldsymbol{F}}\sigma)^T = \boldsymbol{0}, \tag{3.25}$$

which is somewhat stronger than the condition (3.15). In fact, the stronger condition (3.25) is valid for hyperelastic material points with internal constraints, since from (2.11) the tensor $\boldsymbol{Q}^T\,d_{\boldsymbol{Q}\boldsymbol{F}}\sigma$ is contained in the tangent space $\mathscr{T}(\boldsymbol{F})$, *viz.*,

$$\boldsymbol{Q}^T\,d_{\boldsymbol{Q}\boldsymbol{F}}\sigma \in \boldsymbol{Q}^T\mathscr{T}(\boldsymbol{Q}\boldsymbol{F}) = \boldsymbol{Q}^T\boldsymbol{Q}\mathscr{T}(\boldsymbol{F}) = \mathscr{T}(\boldsymbol{F}). \tag{3.26}$$

Hence (3.14) implies not only (3.15) but actually the stronger condition (3.25) which is equivalent to (3.24).

For the condition of material symmetry the situation is not quite the same. For a hyperelastic material point free of constraint the stored energy function obeys the condition

$$\boldsymbol{F}\boldsymbol{H}(d_{\boldsymbol{F}\boldsymbol{H}}\sigma) = \boldsymbol{F}(d_{\boldsymbol{F}}\sigma)^T \tag{3.27}$$

for all proper $\boldsymbol{F}$ and all $\boldsymbol{H} \in \mathscr{h}$. The condition (3.27) may be rewritten as

$$\boldsymbol{H}(d_{\boldsymbol{F}\boldsymbol{H}}\sigma)^T - (d_{\boldsymbol{F}}\sigma)^T = \boldsymbol{0}, \tag{3.28}$$

which is stronger than the condition (3.22). Unlike the tensor $\boldsymbol{Q}^T\,d_{\boldsymbol{Q}\boldsymbol{F}}\sigma$, the tensor $(d_{\boldsymbol{F}\boldsymbol{H}}\sigma)\,\boldsymbol{H}^T$ need not belong to the tangent space $\mathscr{T}(\boldsymbol{F})$. Indeed, from (2.14), we have

$$(d_{\boldsymbol{F}\boldsymbol{H}}\sigma)\,\boldsymbol{H}^T \in \mathscr{T}(\boldsymbol{F}\boldsymbol{H})\,\boldsymbol{H}^T = \mathscr{T}(\boldsymbol{F})\,\boldsymbol{H}\boldsymbol{H}^T. \tag{3.29}$$

Since $\boldsymbol{H}$ need not be a rotation, $\mathscr{T}(\boldsymbol{F})\,\boldsymbol{H}\boldsymbol{H}^T$ need not coincide with $\mathscr{T}(\boldsymbol{F})$. As a result, the stronger conditions (3.27) or (3.28) generally do not follow from the condition (3.21).

4. Truesdell Theorem

Let $\boldsymbol{G}$ be an element in the symmetry group $\mathscr{g}$ of the distribution of stress spaces on $\mathscr{C}$. Then by virtue of (2.15) and (3.8) we have

$$\boldsymbol{F}\boldsymbol{G}(d_{\boldsymbol{F}\boldsymbol{G}}\sigma)^T - \boldsymbol{F}(d_{\boldsymbol{F}}\sigma)^T \in \mathscr{N}(\boldsymbol{F}) \tag{4.1}$$

for all $\boldsymbol{F} \in \mathscr{C}$. Using (2.7), we can rewrite this condition as

$$\operatorname{tr}\left([\boldsymbol{G}(d_{\boldsymbol{F}\boldsymbol{G}}\sigma)^T - (d_{\boldsymbol{F}}\sigma)^T]\,\dot{\boldsymbol{F}}\right) = 0 \tag{4.2}$$

for all $\dot{\boldsymbol{F}} \in \mathscr{T}(\boldsymbol{F})$. This is the restriction on σ due to any $\boldsymbol{G} \in \mathscr{g}$.

To analyze the restriction (4.2) we define a function α on $\mathscr{C}$ by

$$\alpha(F) = \sigma(FG), \tag{4.3}$$

where $G \in g$ is held fixed. Clearly, α is a smooth function. Applying (3.4) to the function α over any smooth curve $F(t)$, we obtain

$$\langle d_F\alpha, \dot{F}\rangle = \langle d_{FG}\sigma, \dot{F}G\rangle. \tag{4.4}$$

Now from (4.2) the right hand side of (4.4) is given by

$$\langle d_{FG}\sigma, \dot{F}G\rangle = \langle d_F\sigma, \dot{F}\rangle. \tag{4.5}$$

Combining (4.4) and (4.5), we see that the differentials of the functions α and σ coincide on the surface $\mathscr{C}$, *viz*.,

$$\langle d_F(\alpha - \sigma), \dot{F}\rangle = 0 \tag{4.6}$$

for all $\dot{F} \in \mathscr{T}(F)$ at all $F \in \mathscr{C}$. Since $\mathscr{C}$ is assumed to be connected, the condition (4.6) is equivalent to the requirement that

$$\alpha(F) - \sigma(F) = c, \tag{4.7}$$

where c is independent of $F \in \mathscr{C}$. Of course, the constant c generally depends on the tensor G, which is held fixed in g in order to define a particular function α on $\mathscr{C}$ by (4.3).

Conversely, let G be any element of h, the symmetry group of the constraint surface $\mathscr{C}$. We hold G fixed and define the function α on $\mathscr{C}$ by (4.3). Now suppose that the function α so defined differs from the function σ by a constant c as shown in (4.7). Then (4.6) holds. The condition (4.4) follows directly from the definition (4.3). Combining (4.6) and (4.4), we obtain (4.5), which is equivalent to (4.2) or (4.1). Hence by (3.8) G obeys the transformation rule

$$\mathscr{S}(FG) = \mathscr{S}(F) \tag{4.8}$$

for all $F \in \mathscr{C}$. Thus $G \in g$.

The preceding results may be summarized as follows:

Truesdell Theorem for Hyperelastic Material with Internal Constraint

An element G in the symmetry group k of the constraint surface $\mathscr{C}$ is contained in the symmetry group g of the distribution of stress spaces on $\mathscr{C}$ if and only if it obeys the condition

$$\sigma(FG) - \sigma(F) = c, \tag{4.9}$$

where c is a constant independent of $F \in \mathscr{C}$ but may depend on G.

As TRUESDELL observed, the condition (4.9) implies immediately that the symmetry group h of the stored energy function is contained in the symmetry group g of the distribution of stress spaces, since every $H \in h$ obeys the condition

$$\sigma(FH) - \sigma(F) = 0 \tag{4.10}$$

for all $F \in \mathscr{C}$. In general, the constant c need not vanish. If $c \neq 0$, then G belongs to $\mathscr{g}$ but not to $\mathscr{h}$.

As remarked in Section 2, we choose an admissible local configuration as the local reference configuration $\varkappa$. Then $\mathscr{C}$ contains the identity tensor I. Evaluating the constant c in (4.9) at $F = I$, we can rewrite (4.9) as

$$\sigma(FG) - \sigma(F) = \sigma(G) - \sigma(I). \tag{4.11}$$

This condition implies the corollary that a rotation Q belongs to $\mathscr{g}$ if and only if it belongs to $\mathscr{h}$. Indeed, replacing G in (4.11) by Q and using the frame-indifference condition (3.9) at $F = I$, we see that the constant c must vanish so that Q belongs to $\mathscr{h}$. As a result, if the material point is a hyperelastic solid such that there is an admissible undistorted local configuration, then $\mathscr{g}$ and $\mathscr{h}$ coincides.

It should be noted that the condition there be an admissible undistorted local configuration is important in the preceding argument. If the point has no admissible undistorted local configurations, then all admissible local configurations are distorted. Hence the symmetry group $\mathscr{g}$ relative to the admissible local reference configuration contains some non-rotation elements G. Then the argument for $\sigma(G) = \sigma(I)$ no longer follows from the frame-indifference condition (3.9) for σ.

In TRUESDELL'S paper [2], a second argument was presented to show that $\mathscr{g}$ coincides with $\mathscr{h}$ for all hyperelastic solids free of internal constraints. Fortunately, this argument prevails for hyperelastic solids with internal constraints, regardless whether there be any admissible undistorted local configurations.

TRUESDELL'S second argument is based on an invariant integral on a compact group. This argument may be modified to suit our situation here in the following way: Let the admissible local reference configuration differ from an undistorted local configuration $\varkappa$ by

$$\mu = A\varkappa, \tag{4.12}$$

where A belongs to $\mathscr{L}$ but not necessarily to $\mathscr{C}$. In [1] we have shown that the symmetry group obeys the transformation rule

$$\mathscr{g} = A^{-1}\,\mathscr{g}_\mu A \tag{4.13}$$

where $\mathscr{g}$ is $\mathscr{g}_\varkappa$ but the subscript $\varkappa$ is suppressed. Since μ is undistorted, the elements of $\mathscr{g}_\mu$ are rotations. Consider a particular element $\bar{Q} \in \mathscr{g}_\mu$ and its corresponding element

$$G = A^{-1}\bar{Q}A \tag{4.14}$$

in $\mathscr{g}$. Substituting (4.14) into (4.11), we obtain

$$\sigma(FA^{-1}\bar{Q}A) - \sigma(F) = \sigma(A^{-1}\bar{Q}A) - \sigma(I) \tag{4.15}$$

for all $F \in \mathscr{C}$. Since $\mathscr{g}$ is contained in $\mathscr{C}$, we can evaluate (4.15) for arbitrary F of the form

$$F = A^{-1}QA \tag{4.16}$$

where $Q \in \mathscr{g}_\mu$, thus obtaining

$$\sigma(A^{-1}Q\bar{Q}A) - \sigma(A^{-1}QA) = \sigma(A^{-1}\bar{Q}A) - \sigma(I). \tag{4.17}$$

This is the condition which may be integrated over the group $\mathscr{g}_\mu$ with respect to an arbitrary right-invariant measure ν.

More specifically, we define a function β on $\mathscr{g}_\mu$ by

$$\beta(Q) = \sigma(A^{-1}QA). \tag{4.18}$$

Then β obeys the condition

$$\beta(Q\bar{Q}) - \beta(Q) = \beta(\bar{Q}) - \beta(I). \tag{4.19}$$

for all Q, $\bar{Q} \in \mathscr{g}_\mu$. Clearly, β is smooth on the compact Lie group $\mathscr{g}_\mu$. Now the terms in (4.19) may be integrated over $\mathscr{g}_\mu$ as in [2], *viz.*,

$$\int \beta(Q\bar{Q})\, d\nu_Q - \int \beta(Q)\, d\nu_Q = (\beta(\bar{Q}) - \beta(I)) \int d\nu_Q \tag{4.20}$$

where Q is the integration variable while $\bar{Q}$ is held fixed. Since ν is right invariant, we may change the integration variable from Q to $Q\bar{Q}$ in the leading term of (4.20),

$$\int \beta(Q\bar{Q})\, d\nu_Q = \int \beta(Q\bar{Q})\, d\nu_{Q\bar{Q}}. \tag{4.21}$$

As a result, the left hand side of (4.20) vanishes. Thus

$$\beta(\bar{Q}) - \beta(I) = 0, \tag{4.22}$$

since the total measure of the compact group $\mathscr{g}_\mu$ is finite. Substituting (4.22) into (4.15) by means of the definition (4.18), we see that $G \in \mathscr{h}$.

The preceding result may be summarized as follows:

Truesdell Corollary for Hyperelastic Solids with Internal Constraint

The symmetry group $\mathscr{h}$ of the stored energy function coincides with the symmetry group $\mathscr{g}$ of the distribution of the stress spaces for all hyperelastic solids with internal constraint.

It should be noted that, although we have shown the preceding result for an admissible local reference configuration $\varkappa$ only, the result is valid for all reference configurations because the symmetry groups $\mathscr{h}$ and $\mathscr{g}$ obey exactly the same transformation rule under a change of reference configuration. Specifically, $\mathscr{g}$ obeys the transformation rule (4.13) while $\mathscr{h}$ obeys the same:

$$\mathscr{h} = A^{-1}\mathscr{h}_\mu A, \tag{4.23}$$

where μ is related to $\varkappa$ by (4.12). As a result, $\mathscr{g}_\mu$ coincides with $\mathscr{h}_\mu$ if and only if $\mathscr{g}$ coincides with $\mathscr{h}$.

Before closing this section we remark that the argument leading to the Truesdell Corollary may be modified easily to show that the condition of frame-indifference (3.13) for the distribution of the stress spaces is equivalent to the condition of frame-indifference (3.9) for the stored energy function for all hyperelastic material points with internal constraint. Indeed, in Section 3 we have shown that (3.9) implies (3.13) which is then shown to be equivalent to (3.14). We now show that (3.14) implies (3.9).

We define a function γ on $\mathscr{C}$ by

$$\gamma(F) = \sigma(\bar{Q}F) \tag{4.24}$$

for any particular rotation $\bar{Q}$. Then by following the argument leading from (4.2) to (4.7) we now start from (3.14) and obtain

$$\gamma(F) - \sigma(F) = c. \tag{4.25}$$

As before, the condition (4.24) may be rewritten as

$$\sigma(\bar{Q}F) - \sigma(F) = \sigma(\bar{Q}) - \sigma(I) \tag{4.26}$$

for all $F \in \mathscr{C}$. Choosing any rotation Q as F in (4.26), we obtain

$$\sigma(\bar{Q}Q) - \sigma(Q) = \sigma(\bar{Q}) - \sigma(I). \tag{4.27}$$

This condition may now be integrated over the compact group $\mathscr{S}\mathscr{O}$ with respect to a left invariant measure, and the result is

$$\sigma(\bar{Q}) - \sigma(I) = 0. \tag{4.28}$$

Hence (4.26) reduces to

$$\sigma(\bar{Q}F) - \sigma(F) = 0 \tag{4.29}$$

for all $F \in \mathscr{C}$. Since the fixed rotation $\bar{Q}$ is arbitrary, the condition (4.29) is the same as (3.9). Thus we have shown that when the distribution of stress spaces is related to the stored energy function by (3.8) the condition (3.9) holds if and only if so does the condition (3.13).

Although the preceding result is established under the assumption that the local reference configuration $\varkappa$ is admissible, the result is valid for all reference configurations because under a change of reference configuration of the form (4.12) the stored energy function σ_μ is given by

$$\sigma_\mu(F) = \sigma(FA) \tag{4.30}$$

while the stress space $\mathscr{S}_\mu(F)$ is given similarly by

$$\mathscr{S}_\mu(F) = \mathscr{S}(FA) \tag{4.31}$$

for all $F \in \mathscr{C}_\mu$, which is related to $\mathscr{C}$ by

$$\mathscr{C}_\mu = \mathscr{C}A^{-1}. \tag{4.32}$$

As a result, σ obeys (3.9) if and only if σ_μ obeys the same, *viz.*,

$$\sigma_\mu(QF) = \sigma_\mu(F) \tag{4.33}$$

for all $F \in \mathscr{C}_\mu$, while $\mathscr{S}$ obeys (3.13) if and only if $\mathscr{S}_\mu$ obeys the same, *viz.*,

$$\mathscr{S}_\mu(QF) = Q\mathscr{S}_\mu(F)\, Q^T \tag{4.34}$$

for all $F \in \mathscr{C}_\mu$. As a result, when $\mathscr{S}_\mu$ is related to σ_μ by

$$\mathscr{S}_\mu(F) = \varrho F(d_F\sigma_\mu)^T + \mathscr{N}_\mu(F) \tag{4.35}$$

for all $F \in \mathscr{C}_\mu$, the condition (4.33) holds if and only if so does the condition (4.34).

5. Further Implications of the Truesdell Theorem

The structure of a hyperelastic material point with internal constraint formulated so far remains closely parallel to that of a hyperelastic material point free of any constraint. This structure may now be developed further along the same line of thought as in constraint-free theory. By using an admissible configuration as a reference configuration, we have achieved the following inclusions:

$$\mathscr{h} \subset \mathscr{g} \subset \mathscr{k} \subset \mathscr{C}. \tag{5.1}$$

The relation between $\mathscr{h}$ and $\mathscr{g}$ is characterized by the condition (4.11), *viz.*,

$$\sigma(FG) - \sigma(F) = \sigma(G) - \sigma(I) \tag{5.2}$$

for all $F \in \mathscr{C}$ and $G \in \mathscr{g}$. Hence if $G \in \mathscr{g}$, then

$$G \in \mathscr{h} \Leftrightarrow \sigma(G) = \sigma(I). \tag{5.3}$$

The constraint-free structure is a special case of this general structure such that the constraint surface $\mathscr{C}$ and, therefore, its symmetry group $\mathscr{k}$ both coincide with the proper linear group $\mathscr{L}$.

The structure of the constraint-free special case of (5.2) has been developed by WANG [3]. We generalize the results to hyperelastic material points with internal constraints.

Theorem 1. *$\mathscr{h}$ is a normal subgroup of $\mathscr{g}$.*

This result means for any $H \in \mathscr{h}$ and $G \in \mathscr{g}$ the combination GHG^{-1} is contained in $\mathscr{h}$. By (5.3) this condition is equivalent to

$$\sigma(GHG^{-1}) = \sigma(I) \tag{5.4}$$

for all $H \in \mathscr{h}$ and $G \in \mathscr{g}$. The proof of (5.4) follows from repeated applications of (5.2) to the leading term in order to reduce its argument to a single tensor variable. Step by step the reduction is the same as in [3]. The assertion of the theorem may be expressed also by

$$G\mathscr{h}G^{-1} = \mathscr{h} \tag{5.5}$$

for all $G \in \mathscr{g}$.

Theorem 2. *The quotient group $\mathscr{g}/\mathscr{h}$ is Abelian.*

This result means for any G and $\bar{G}$ in $\mathscr{g}$ the combination $G\bar{G}G^{-1}\bar{G}^{-1}$ is contained in $\mathscr{h}$ (which corresponds to the identity element of the quotient group

$\mathscr{g}/\mathscr{h}$). By (5.3) this condition is equivalent to

$$\sigma(G\bar{G}G^{-1}\bar{G}^{-1}) = \sigma(I) \tag{5.6}$$

for all $G, \bar{G} \in \mathscr{g}$. The proof of (5.6) again follows from repeated applications (5.2) to the leading term in order to reduce its argument to a single tensor variable. Step by step the reduction is the same as in [3].

As in [3] we define a function $\bar{\sigma}$ on $\mathscr{g}$ by

$$\bar{\sigma}(G) = \sigma(G) - \sigma(I). \tag{5.7}$$

Then it obeys the transformation rule

$$\bar{\sigma}(GH) = \bar{\sigma}(HG) = \bar{\sigma}(G) \tag{5.8}$$

for all $G \in \mathscr{g}$ and $H \in \mathscr{h}$; furthermore, it satisfies the condition

$$\bar{\sigma}(H) = 0 \tag{5.9}$$

for all $H \in \mathscr{h}$. The transformation rule (5.8) means that $\bar{\sigma}$ takes on constant value on each coset of $\mathscr{h}$ in $\mathscr{g}$. Hence $\bar{\sigma}$ may be regarded as a function on the quotient group $\mathscr{g}/\mathscr{h}$.

Theorem 3. *$\bar{\sigma}$ is a additive function on the Abelian group $\mathscr{g}/\mathscr{h}$.*

The result means

$$\bar{\sigma}(\{G\}\{\bar{G}\}) = \bar{\sigma}(\{G\}) + \bar{\sigma}(\{\bar{G}\}) \tag{5.10}$$

for all $G, \bar{G} \in \mathscr{g}$, where for any $G \in \mathscr{g}$ the notation $\{G\}$ denotes the coset of $\mathscr{h}$ containing G. Since (5.8) implies that

$$\bar{\sigma}(\{G\}) = \bar{\sigma}(G), \tag{5.11}$$

the condition (5.10) amounts to

$$\bar{\sigma}(G\bar{G}) = \bar{\sigma}(G) + \bar{\sigma}(\bar{G}) \tag{5.12}$$

for all $G, \bar{G} \in \mathscr{g}$. Using (5.7), we can rewrite (5.12) as

$$\sigma(G\bar{G}) = \sigma(G) + \sigma(\bar{G}) - \sigma(I), \tag{5.13}$$

which is clearly valid by virtue of the basic condition (5.2) on σ.

The preceding three theorems show that the general solution of the condition (5.13), which is the restriction of the condition (4.11) to the symmetry group $\mathscr{g}$, may be described as follows: First, by taking the intersection of all normal subgroups $\mathscr{h}$ of $\mathscr{g}$ such that the quotient groups $\mathscr{g}/\mathscr{h}$ are Abelian, we obtain the smallest subgroup $\mathscr{h}^*$ of $\mathscr{g}$ among all such subgroups. Then $\mathscr{h}^*$ is a normal subgroup of $\mathscr{g}$ such that $\mathscr{g}/\mathscr{h}^*$ is Abelian and

$$\mathscr{h}^* \subset \mathscr{h} \subset \mathscr{g} \tag{5.14}$$

for all such normal subgroups $\mathscr{h}$. By virtue of this property every additive function on $\mathscr{g}/\mathscr{h}$ for any such normal subgroup $\mathscr{h}$ can be regarded as an additive function on $\mathscr{g}/\mathscr{h}^*$.

Theorem 4. *The general solution of* (5.13) *on* $\mathscr{g}$ *is given by*

$$\sigma(G) = \sigma^*(G) + c, \tag{5.15}$$

where c *is an arbitrary constant and where* σ^* *is an arbitrary additive function on* $\mathscr{g}/\mathscr{h}^*$.

Having determined the most general form of σ on the group $\mathscr{g}$, which is a subset of the domain $\mathscr{C}$ of σ, we can describe the form of σ on the rest of its domain in the following way: First, we define an equivalence relation $\sim$ on $\mathscr{C}$ by

$$\boldsymbol{F} \sim \bar{\boldsymbol{F}} \quad \Leftrightarrow \quad \boldsymbol{F}^{-1}\bar{\boldsymbol{F}} = \boldsymbol{G} \in \mathscr{g}. \tag{5.16}$$

Then $\mathscr{C}$ is decomposed into a disjoint union of equivalence classes, each of which is an orbit of the transformation group $\mathscr{g}$ in $\mathscr{C}$, *viz.*, a subset of the form

$$[\boldsymbol{F}_0] = \{\boldsymbol{F}_0\boldsymbol{G}_0, \boldsymbol{G}_0 \in \mathscr{g}\}. \tag{5.17}$$

In particular, the group $\mathscr{g}$ itself is the equivalence class

$$[\boldsymbol{I}] = \mathscr{g} = [\boldsymbol{G}_0], \tag{5.18}$$

where $\boldsymbol{G}_0$ may be any element in $\mathscr{g}$. We select a representative element, say $\boldsymbol{F}_0$, for each equivalence class $[\boldsymbol{F}_0]$ in $\mathscr{C}$, and we assign an arbitrary value $\sigma(\boldsymbol{F}_0)$ at $\boldsymbol{F}_0$. For the special equivalence class $\mathscr{g}$ we select $\boldsymbol{I}$ as its representative element; the constant c in (5.15) is then the value of σ at $\boldsymbol{I}$, *viz.*,

$$\sigma(\boldsymbol{I}) = \sigma^*(\boldsymbol{I}) + c = c, \tag{5.19}$$

where $\sigma^*(\boldsymbol{I})$ vanishes since σ^* is an additive function.

Theorem 5. *The general solution of* (5.2) *is given by*

$$\sigma(\boldsymbol{F}) = \sigma(\boldsymbol{F}_0\boldsymbol{G}_0) = \sigma^*(\boldsymbol{G}_0) + \sigma(\boldsymbol{F}_0) \tag{5.20}$$

for all $\boldsymbol{F} \in [\boldsymbol{F}_0]$ *and for all equivalence classes* $[\boldsymbol{F}_0]$ *in* $\mathscr{C}$.

Clearly, (5.20) is a general solution of (5.2), since it implies that

$$\sigma(\boldsymbol{F}\boldsymbol{G}) = \sigma(\boldsymbol{F}_0\boldsymbol{G}_0\boldsymbol{G}) = \sigma^*(\boldsymbol{G}_0\boldsymbol{G}) + \sigma(\boldsymbol{F}_0) \tag{5.21}$$

for any $\boldsymbol{G} \in \mathscr{g}$. Subtracting (5.20) from (5.21) and using the additivity of σ^*, we then get

$$\sigma(\boldsymbol{F}\boldsymbol{G}) - \sigma(\boldsymbol{F}) = \sigma^*(\boldsymbol{G}_0\boldsymbol{G}) - \sigma^*(\boldsymbol{G}_0) = \sigma^*(\boldsymbol{G}). \tag{5.22}$$

Substituting (5.15) into (5.22) and using (5.19), we finally obtain (5.2). Thus (5.20) is sufficient for (5.2). Conversely, (5.2) implies (5.20) since from (5.15) and (5.19) the value $\sigma^*(\boldsymbol{G}_0)$ is given by

$$\sigma^*(\boldsymbol{G}_0) = \sigma(\boldsymbol{G}_0) - \sigma(\boldsymbol{I}). \tag{5.23}$$

Thus (5.20) is the special case of (5.2) such that $\boldsymbol{G} = \boldsymbol{G}_0$ and $\boldsymbol{F} = \boldsymbol{F}_0$.

Among all functions σ given by (5.20) only those satisfying the frame-indifference condition (3.9) can be the stored energy function of a hyperelastic material with internal constraint. The general solution of the pair of conditions (5.2) and (3.9) may be described in the following way: First, we define the notion of an objective normal subgroup $\mathscr{h}$ of $\mathscr{g}$ relative to $\mathscr{C}$ by the conditions that $\mathscr{g}/\mathscr{h}$ be Abelian and that

$$(F\mathscr{h}F^{-1}) \cap \mathscr{S}\mathscr{O} = (F\mathscr{g}F^{-1}) \cap \mathscr{S}\mathscr{O} \tag{5.24}$$

for all $F \in \mathscr{C}$. The condition (5.24) is necessary for the symmetry groups $\mathscr{h}$ and $\mathscr{g}$, since according to the transformation rules (4.23) and (4.13)

$$\mathscr{h}_\gamma = F\mathscr{h}F^{-1}, \quad \mathscr{g}_\gamma = F\mathscr{g}F^{-1}, \tag{5.25}$$

where γ is the admissible local configuration $F\varkappa$. As a result, by the frame-indifference condition

$$\sigma_\gamma(QF) = \sigma_\gamma(F) \tag{5.26}$$

for all $Q \in \mathscr{S}\mathscr{O}$ and $F \in \mathscr{C}_\gamma$, which is equivalent to the condition (3.9), we must have

$$\sigma_\gamma(Q) = \sigma_\gamma(I). \tag{5.27}$$

Thus the groups $\mathscr{h}_\gamma$ and $\mathscr{g}_\gamma$ contain exactly the same rotations. By taking the intersection of all objective normal subgroups $\mathscr{h}$ of $\mathscr{g}$ relative to $\mathscr{C}$, we obtain the smallest subgroup $\mathscr{h}^\circ$ of $\mathscr{g}$ among all such subgroups.

Since we have imposed the additional conditions (5.24), the subgroup $\mathscr{h}^*$ defined previously by the condition (5.14) may be just a subgroup of $\mathscr{h}^\circ$. Because the symmetry group $\mathscr{h}$ of the stored energy function must contain the subgroup $\mathscr{h}^\circ$, we now define the function σ on $\mathscr{g}$ by

$$\sigma(G) = \sigma^\circ(G) + c, \tag{5.28}$$

where σ° is an arbitrary additive function on the Abelian quotient group $\mathscr{g}/\mathscr{h}^\circ$ and where c is an arbitrary constant as before. It should be noted that a function σ on $\mathscr{g}$ of the form (5.28) is a special case of the form (5.15) but not *vice versa*, since $\mathscr{h}^\circ$ contains $\mathscr{h}^*$ but not *vice versa*.

Next, we define an equivalence relation $\approx$ among the orbits of $\mathscr{g}$ in $\mathscr{C}$ by

$$[F] \approx [\bar{F}] \quad \Leftrightarrow \quad [\bar{F}] = Q[F], \tag{5.29}$$

where Q is a rotation. Notice that the rotation Q connecting $[F]$ to $[\bar{F}]$ on the right hand side is generally not unique. Indeed, $\bar{Q}[F]$ coincides with $Q[F]$ if and only if $\bar{Q}F$ and QF belong to the same orbit, *viz.*,

$$F^{-1}Q^{-1}\bar{Q}F \in \mathscr{g}. \tag{5.30}$$

Clearly, this condition is independent of the choice of F in the orbit $[F]$. Furthermore, we can replace the right hand side of (5.30) by $\mathscr{h}^\circ$, since that condition is equivalent to

$$Q^{-1}\bar{Q} \in (F\mathscr{g}F^{-1}) \cap \mathscr{S}\mathscr{O} = (F\mathscr{h}^\circ F^{-1}) \cap \mathscr{S}\mathscr{O} \tag{5.31}$$

by virtue of (5.24).

Among equivalent orbits with respect to the equivalence relation $\approx$ we require that their representative elements be connected to one another by rotations. Specifically, let $\boldsymbol{F}_0$ be the representative element of a particular orbit $[\boldsymbol{F}_0]$. Then we choose the representative element of the orbit $\boldsymbol{Q}[\boldsymbol{F}_0]$ of the form $\boldsymbol{Q}_0\boldsymbol{F}_0$, where $\boldsymbol{Q}_0$ is a particular rotation such that $\boldsymbol{Q}_0\boldsymbol{F}_0 \in \boldsymbol{Q}[\boldsymbol{F}_0]$, *viz.*,

$$\boldsymbol{F}_0^{-1}\boldsymbol{Q}^{-1}\boldsymbol{Q}_0\boldsymbol{F}_0 \in \mathscr{h}^\circ, \tag{5.32}$$

where the right hand side is taken to be $\mathscr{h}^\circ$ rather than $\mathscr{g}$ as we have explained before. Having chosen the representative elements of the orbits of $\mathscr{g}$ in $\mathscr{C}$ according to the preceding condition, we impose further the restriction that

$$\sigma(\boldsymbol{F}_0) = \sigma(\boldsymbol{Q}_0\boldsymbol{F}_0) \tag{5.33}$$

for all representative elements $\boldsymbol{Q}_0\boldsymbol{F}_0$ of orbits which are equivalent to the orbit $[\boldsymbol{F}_0]$ with respect to the equivalence relation $\approx$ defined by (5.29). Clearly, the condition (5.33) is necessary for (3.9).

The necessary conditions (5.24) and (5.33) are sufficient to characterize the general solution.

Theorem 6. *The general solution of the pair of conditions* (3.9) *and* (5.2) *is given by*

$$\sigma(\boldsymbol{F}) = \sigma(\boldsymbol{F}_0\boldsymbol{G}_0) = \sigma^\circ(\boldsymbol{G}_0) + \sigma(\boldsymbol{F}_0) \tag{5.34}$$

for all $\boldsymbol{F} \in [\boldsymbol{F}_0]$ *and for all orbits* $[\boldsymbol{F}_0]$ *in* $\mathscr{C}$.

Since (5.34) is a special case of (5.20), σ clearly satisfies the condition (5.2). To show that σ also satisfies the condition (3.9), we calculate the value $\sigma(\boldsymbol{QF})$ of σ at the point $\boldsymbol{QF}$, which belongs to the orbit $\boldsymbol{Q}[\boldsymbol{F}_0]$. According to the selection rule (5.32), the representative element of this orbit is of the form $\boldsymbol{Q}_0\boldsymbol{F}_0$. This means there is an element $\boldsymbol{G}_1 \in \mathscr{g}$ such that

$$\boldsymbol{QF} = \boldsymbol{Q}_0\boldsymbol{F}_0\boldsymbol{G}_1. \tag{5.35}$$

Then according to (5.34) and (5.33) the value $\sigma(\boldsymbol{QF})$ is given by

$$\sigma(\boldsymbol{QF}) = \sigma^\circ(\boldsymbol{G}_1) + \sigma(\boldsymbol{Q}_0\boldsymbol{F}_0) = \sigma^\circ(\boldsymbol{G}_1) + \sigma(\boldsymbol{F}_0). \tag{5.36}$$

Substituting the relation $\boldsymbol{F} = \boldsymbol{F}_0\boldsymbol{G}_0$ into (5.35), we obtain

$$\boldsymbol{G}_1\boldsymbol{G}_0^{-1} = \boldsymbol{F}_0^{-1}\boldsymbol{Q}_0^{-1}\boldsymbol{QF}_0. \tag{5.37}$$

Hence by (6.32) $\boldsymbol{G}_1\boldsymbol{G}_0^{-1} \in \mathscr{h}^\circ$. As a result,

$$\sigma^\circ(\boldsymbol{G}_1) - \sigma^\circ(\boldsymbol{G}_0) = \sigma^\circ(\boldsymbol{G}_1\boldsymbol{G}_0^{-1}) = 0, \tag{5.38}$$

since $\mathscr{h}^\circ$ is the identity element of the quotient group $\mathscr{g}/\mathscr{h}^\circ$ and σ° is an additive function on $\mathscr{g}/\mathscr{h}^\circ$. Combining (5.34), (5.36) and (5.38), we see that (3.9) holds. Thus (5.34) is sufficient for both (5.2) and (3.9). The necessity of (5.34) from (3.9) and (5.2) has already been remarked prior to the statement of the theorem, so that the proof is complete.

6. Embedding Theorems

The analysis leading to the general solution (5.34) for the pair of the conditions (3.9) and (5.2), presented in the preceding section, is almost same as that presented in [3] for the representation of the stored energy functions of a class of hyperelastic materials without any internal constraints. As remarked before, that case is included here by setting $\mathscr{C} = \mathscr{L}$ and, therefore, $\mathscr{k} = \mathscr{L}$ also. Hence the conditions (3.9) and (5.2) become

$$\sigma(\boldsymbol{QF}) = \sigma(\boldsymbol{F}) \tag{6.1}$$

and

$$\sigma(\boldsymbol{FG}) = \sigma(\boldsymbol{F}) + \sigma(\boldsymbol{G}) - \sigma(\boldsymbol{I}) \tag{6.2}$$

for all $\boldsymbol{Q} \in \mathscr{SO}$, $\boldsymbol{G} \in \mathscr{g}$, and $\boldsymbol{F} \in \mathscr{L}$.

There is a very subtle difference between the general solution for the pair of conditions (6.1) and (6.2) and the same for the pair of conditions (3.9) and (5.2), however. Specifically, for the pair of conditions (6.1) and (6.2) the symmetry group $\mathscr{h}$ must be an objective normal subgroup relative to $\mathscr{L}$. This means that $\mathscr{g}/\mathscr{h}$ must be Abelian and that

$$(\boldsymbol{F}\mathscr{h}\boldsymbol{F}^{-1}) \cap \mathscr{SO} = (\boldsymbol{F}\mathscr{g}\boldsymbol{F}^{-1}) \cap \mathscr{SO} \tag{6.3}$$

must be satisfied for all $\boldsymbol{F} \in \mathscr{L}$. Since $\mathscr{C}$ is contained in $\mathscr{L}$ but not *vice versa*, there may be more objective normal subgroups of $\mathscr{g}$ relative to $\mathscr{C}$ than those relative to $\mathscr{L}$. As a result, the intersection $\mathscr{h}^\circ$ of the former class of normal subgroups may be smaller than the intersection $\bar{\mathscr{h}}$ of the latter class, *viz.*,

$$\mathscr{h}^\circ \subset \bar{\mathscr{h}} \tag{6.4}$$

but not *vice versa.*

If the group $\mathscr{h}^\circ$ is actually a proper subgroup of $\bar{\mathscr{h}}$, then there may be additive functions σ° on $\mathscr{g}/\mathscr{h}^\circ$ which are not functions at all on $\mathscr{g}/\bar{\mathscr{h}}$. In that case, the restriction of the general solution for (6.1) and (6.2) to $\mathscr{C}$ constitutes only a class of particular solutions for (3.9) and (5.2). In other words, there may be some other particular solutions for (3.9) and (5.2) which cannot be extended to particular solutions for (6.1) and (6.2).

More importantly, however, if the core $\mathscr{h}^\circ$ of objective normal subgroups of $\mathscr{g}$ relative to $\mathscr{C}$ coincides with the core $\bar{\mathscr{h}}$ of the same relative to $\mathscr{L}$, then the restrictions of σ to $\mathscr{g}$ are exactly the same in the two cases. As a result, the variations of σ over an equivalence class with respect to the equivalence relation $\approx$ defined by (5.29) are also exactly the same in the two cases. Hence we have the following result:

Embedding Theorem for Hyperelastic Materials with Internal Constraints

Suppose that the core $\mathscr{h}^\circ$ of objective normal subgroups of $\mathscr{g}$ relative to $\mathscr{C}$ coincides with the core $\bar{\mathscr{h}}$ of the same relative to $\mathscr{L}$. Then the restriction of the general solution for (6.1) and (6.2) to $\mathscr{C}$ constitutes a general solution for (3.9) and (5.2).

Since we do not have any example in which $\mathscr{h}^\circ$ is actually a proper subgroup of $\bar{\mathscr{h}}$, it is not clear whether or not the hypothesis of the preceding theorem is necessary. For a hyperelastic material with internal constraint, the stronger condition (6.3) means that the symmetry group of the stored energy function and the symmetry group of the distribution of the stress spaces contain the same rotations relative to all local reference configurations, not just admissible local reference configurations as required by the weaker condition (5.24). We cannot find any argument to support the stronger condition (6.3) in the context of hyperelastic materials with internal constraints, however.

Suppose that $\mathscr{h}^\circ = \bar{\mathscr{h}}$. Then the restriction of the stored energy function σ on $\mathscr{g}$ is given by (5.28) regardless of constraints if any. As a result, the stored energy function takes the form (5.34) on each equivalence class of orbits of $\mathscr{g}$ with respect to the equivalence relation $\approx$ defined by (5.29), regardless of constraints if any. Indeed, the only difference between the stored energy functions of a class of hyperelastic materials with internal constraints and those of the same class (*i.e.*, with the same symmetry group $\mathscr{g}$) without any constraints is that the domain $\mathscr{C}$ of the former is the disjoint union of fewer equivalence classes of orbits of $\mathscr{g}$ than the domain $\mathscr{L}$ of the latter. However, on each common equivalence class of orbits of $\mathscr{g}$ in the domains (*i.e.*, any equivalence class in $\mathscr{C}$ since $\mathscr{C} \subset \mathscr{L}$) the form of the former is exactly the same as that of the latter. As a result, the former is just the restriction of the latter on $\mathscr{C}$, while the latter is the extension by the former to $\mathscr{L}$ according to (5.29). In some sense, this result means that the general solution for the pair of conditions (3.9) and (5.2) is embedded in that for the pair of conditions (6.1) and (6.2) just like the constraint surface $\mathscr{C}$ is embedded in the proper linear group $\mathscr{L}$ provided that the intersection $\mathscr{h}^\circ$ of all $\mathscr{h}$ satisfying (5.24) coincides with the intersection $\bar{\mathscr{h}}$ of all $\mathscr{h}$ satisfying (6.3).

Although we need the hypothesis $\mathscr{h}^\circ = \bar{\mathscr{h}}$ to prove the assertion of the embedding theorem, we do not have any example of $\mathscr{C}$ and $\mathscr{g}$ such that $\mathscr{h}^\circ$ is a proper subgroup of $\bar{\mathscr{h}}$ but we cannot prove that $\mathscr{h}^\circ$ coincides with $\bar{\mathscr{h}}$ in general either. A similar situation was noted in our preceding paper [1].

Specifically, we considered the distribution $\mathscr{S}$ of the stress spaces in the constraint surface $\mathscr{C}$, and we sought the general solution for the pair of conditions

$$\mathscr{S}(QF) = Q\mathscr{S}(F)\, Q^T \tag{6.5}$$

and

$$\mathscr{S}(FH) = \mathscr{S}(F) \tag{6.6}$$

for all $Q \in \mathscr{S}\mathscr{O}$, $H \in \mathscr{g}$, and $F \in \mathscr{C}$. For the case the elastic materials are free of constraints, the conditions (6.5) and (6.6) become

$$G(QF) = QG(F)\, Q^T \tag{6.7}$$

and

$$G(FH) = G(F) \tag{6.8}$$

for all $Q \in \mathscr{S}\mathscr{O}$, $H \in \mathscr{g}$, and $F \in \mathscr{L}$. We noted that the general solution for (6.5) and (6.6) may be embedded in that for (6.7) and (6.8) if the distribution $\mathscr{S}$ has a global cross-section G in $\mathscr{C}$ such that

$$\mathscr{S}(F) = G(F) + \mathscr{N}(F) \tag{6.9}$$

and that G obeys the conditions

$$G(QF) = QG(F)\,Q^T \tag{6.10}$$

and

$$G(FH) = G(F) \tag{6.11}$$

for all $Q \in \mathscr{SO}$, $H \in g$, and $F \in \mathscr{C}$.

In [1] we showed that such a global cross-section could be constructed provided that a certain condition (*cf.* [1, (5.33)] and we now use the notations of [1]) of the form

$$Q^*\mathscr{S}_\varkappa(F)\,Q^{*T} = \mathscr{S}_\varkappa(F) \tag{6.12}$$

for all Q^* belonging to a certain group $g(\mathscr{E}_\varkappa)$ defined by (*cf.* [1, (5.30)])

$$g(\mathscr{E}_\varkappa) = (F g_\varkappa F^{-1}) \cap \mathscr{SO} \tag{6.13}$$

could be satisfied by a particular tensor $G_\varkappa(F) \in \mathscr{S}_\varkappa(F)$ in the sense that

$$Q^* G_\varkappa(F)\,Q^{*T} = G_\varkappa(F) \tag{6.14}$$

for all $Q^* \in g(\mathscr{E}_\varkappa)$. Then the stress space $\mathscr{S}_\varkappa(F)$ may be represented by

$$\mathscr{S}_\varkappa(F) = G_\varkappa(F) + \mathscr{N}_\varkappa(F), \tag{6.15}$$

and by (6.14) $\mathscr{S}_\varkappa(F)$ obeys (6.12) since $\mathscr{N}_\varkappa(F)$ always satisfies the condition

$$Q^*\mathscr{N}_\varkappa(F)\,Q^{*T} = \mathscr{N}_\varkappa(F) \tag{6.16}$$

cf. [1, (5.39)]. Thus the condition (6.14) is the hypothesis for the embedding theorem. We noted in [1] some special cases for which the hypothesis (6.14) could be proved, but we were not able to prove the hypothesis in general.

We have given more thought to this hypothesis and now we are able to remove it completely. We establish first the following results:

Lemma 1. *Let $\mathscr{A}$ denote the space of symmetric tensors, $\mathscr{N}$ any subspace in $\mathscr{A}$, and $\mathscr{S}$ any hyperplane in $\mathscr{A}$ parallel to $\mathscr{N}$. Then there exists an unique tensor $G \in \mathscr{N}^\perp$, the orthogonal complement of $\mathscr{N}$ in $\mathscr{A}$, such that*

$$\mathscr{S} = G + \mathscr{N}. \tag{6.17}$$

Proof. Since $\mathscr{A}$ has the orthogonal sum decomposition

$$\mathscr{A} = \mathscr{N} \oplus \mathscr{N}^\perp,$$

any $S \in \mathscr{S}$ may be represented uniquely as the sum

$$S = N + G \tag{6.19}$$

where $N \in \mathscr{N}$ and $G \in \mathscr{N}^\perp$. Then $G \in \mathscr{S}$ since

$$G = S - N \in S + \mathscr{N} = \mathscr{S}. \tag{6.20}$$

As a result, $\mathscr{S}$ may be represented by (6.17). Thus the existence of G is proved. Suppose that there is another tensor $\bar{G} \in \mathscr{N}^{\perp}$ such that

$$\mathscr{S} = \bar{G} + \mathscr{N}. \tag{6.21}$$

Then $G - \bar{G}$ must belong to both $\mathscr{N}$ and $\mathscr{N}^{\perp}$. Hence

$$G - \bar{G} = \mathbf{0} \tag{6.22}$$

Thus the uniqueness of G is proved.

Lemma 2. *Suppose that the subspace $\mathscr{N}$ and hyperplane $\mathscr{S}$ in the preceding lemma satisfy the additional conditions*

$$Q^* \mathscr{N} Q^{*T} = \mathscr{N}, \ Q^* \mathscr{S} Q^{*T} = \mathscr{S} \tag{6.23}$$

for some particular rotation Q^*. Then the particular tensor $G \in \mathscr{S}$ given by the preceding lemma obeys the same condition, viz.,

$$Q^* G Q^{*T} = G. \tag{6.24}$$

Proof. Since the tensor G given by the preceding lemma is unique, it suffices to show that

$$Q^* G Q^{*T} \in \mathscr{N}^{\perp}. \tag{6.25}$$

By virtue of $(6.23)_1$ the elements of the subspace $\mathscr{N}$ are of the form $Q^* N Q^{*T}$ for all $N \in \mathscr{N}$. Then by (2.2) the inner product of $Q^* G Q^{*T}$ with a typical element $Q^* N Q^{*T}$ vanishes, *viz.*,

$$\operatorname{tr}(Q^* G Q^{*T} Q^* N Q^{*T}) = \operatorname{tr}(GN) = 0 \tag{6.26}$$

since $G \in \mathscr{N}^{\perp}$ and $N \in \mathscr{N}$. Thus (6.25) is proved.

By using the preceding lemmas, we see immediately that for any $\mathscr{S}_{\varkappa}(F)$ and $\mathscr{N}_{\varkappa}(F)$ obeying (6.12) and (6.16) there is a particular tensor $G_{\varkappa}(F) \in \mathscr{S}_{\varkappa}(F)$ satisfying (6.14). Thus the hypothesis for the embedding theorem (which we did not state formally in [1]) is superfluous.

Embedding Theorem for Elastic Materials with Internal Constraints

The restriction of the general solution for (6.7) and (6.8) to $\mathscr{C}$ constitutes the general solution for (6.10) and (6.11) and, therefore, it also yields the general solution for (6.5) and (6.7) through the representation (6.9).

We remark that the general solution for (6.7) and (6.8) is given originally in [4].

Acknowledgement. We acknowledge the support in part of the Natural Sciences and Engineering Research Council of Canada.

References

1. COHEN, H., & C.-C. WANG, On the Response and Symmetry of Elastic Materials with Internal Constraints. *Arch. Rational Mech. Anal.* **99**, 1–36 (1987).
2. TRUESDELL, C., A Theorem on the Isotropy Groups of a Hyperelastic Material. *Proc. Nat. Acad. Sci.* **52**, 1081–1083 (1964).
3. WANG, C.-C., On the Stored Energy Functions of Hyperelastic Materials. *Arch. Rational Mech. Anal.* **23**, 1–14 (1966).
4. WANG, C.-C., On the Response Functions of Elastic Materials, *Arch. Rational Mech. Anal.* **32**, 331–342 (1969).

Department of Civil Engineering
University of Manitoba
Winnipeg

and

Department of Mathematical Science
Rice University
Houston, Texas

(Received October 23, 1987)

The Bianchi Identities in an Explicit Form

A. W. MARRIS

Dedicated to Professor BERNARD D. COLEMAN

Introduction

Let $\boldsymbol{v}(\boldsymbol{x})$ be a given vector field in three-dimensional Euclidean space, and suppose that $\boldsymbol{s}(\boldsymbol{x})$ is the unit vector tangent to the vector-lines of $\boldsymbol{v}$; thus $\boldsymbol{v} = v\boldsymbol{s}$. The unit principal normal $\boldsymbol{n}$ to the vector-line of $\boldsymbol{v}$ is defined as the unit vector pointing in the direction of $\boldsymbol{s} \cdot \operatorname{grad} \boldsymbol{s}$, and the unit bi-normal $\boldsymbol{b}$ is defined by the cross product $\boldsymbol{b} = \boldsymbol{s} \times \boldsymbol{n}$. It is often convenient to study $\boldsymbol{v}$ and associated fields through the anholonomic space determined by the basis $\boldsymbol{s}$, $\boldsymbol{n}$, $\boldsymbol{b}$. As a case in point, $\boldsymbol{v}$ may be the velocity field for a steady motion; a simplication may then be afforded by the fact that, in the spatial representation, the acceleration $\boldsymbol{a} = \boldsymbol{v} \cdot \operatorname{grad} \boldsymbol{v}$ has no component in the bi-normal direction.

The components of $\operatorname{grad} \boldsymbol{s}$, $\operatorname{grad} \boldsymbol{n}$ and $\operatorname{grad} \boldsymbol{b}$ on the basis $\boldsymbol{s}$, $\boldsymbol{n}$ and $\boldsymbol{b}$, are connected by nine compatibility conditions which are contained in the relations $\operatorname{grad} \times \operatorname{grad} \boldsymbol{s} = \boldsymbol{0}$, $\operatorname{grad} \times \operatorname{grad} \boldsymbol{n} = \boldsymbol{0}$, $\operatorname{grad} \times \operatorname{grad} \boldsymbol{b} = \boldsymbol{0}$. The anholonomic space is embedded in Euclidean space, and these nine conditions imply the vanishing of the anholonomic components of the curvature tensor. For the three-dimensional space under consideration these compatibility conditions satisfy three Bianchi identities [BLUM (1947)].

In order to prove theorems concerning a particular vector field $\boldsymbol{v}$, one must involve the relations defining the field with the compatibility conditions. In this way we may generate further necessary conditions on the vector field.[1] In carrying out such an analysis we must clearly avoid the Bianchi identities.

In this short paper we present the three Bianchi identities in explicit form for the basis $\boldsymbol{s}$, $\boldsymbol{n}$, $\boldsymbol{b}$, and we reveal the three different paths leading to each identity.

---

[1] See, for example, MARRIS (1973).

Background Material

For the ortho-normal basis $\boldsymbol{s}$, $\boldsymbol{n}$, $\boldsymbol{b}$, already introduced, the curvature $\varkappa$ and torsion τ of the vector-lines of $\boldsymbol{s}$ are defined respectively by[2]

$$\boldsymbol{s} \cdot \operatorname{grad} \boldsymbol{s} \equiv \frac{\delta \boldsymbol{s}}{\delta s} = \varkappa \boldsymbol{n}, \tag{1.1}$$

and

$$\boldsymbol{s} \cdot \operatorname{grad} \boldsymbol{b} \equiv \frac{\delta \boldsymbol{b}}{\delta s} = -\tau \boldsymbol{n}. \tag{1.2}$$

By a Serret-Frenet formula

$$\boldsymbol{s} \cdot \operatorname{grad} \boldsymbol{n} \equiv \frac{\delta \boldsymbol{n}}{\delta s} = -\varkappa \boldsymbol{s} + \tau \boldsymbol{b}. \tag{1.3}$$

The abnormalities of the vector fields of $\boldsymbol{s}$, $\boldsymbol{n}$ and $\boldsymbol{b}$ are respectively

$$\Omega_s = \boldsymbol{s} \cdot \operatorname{curl} \boldsymbol{s}, \quad \Omega_n = \boldsymbol{n} \cdot \operatorname{curl} \boldsymbol{n}, \quad \text{and } \Omega_b = \boldsymbol{b} \cdot \operatorname{curl} \boldsymbol{b}, \tag{1.4}$$

and we introduce the two scalar functions

$$\psi = \boldsymbol{n} \cdot \operatorname{grad} \boldsymbol{s} \cdot \boldsymbol{n}, \tag{1.5}$$

and

$$\theta = \boldsymbol{b} \cdot \operatorname{grad} \boldsymbol{s} \cdot \boldsymbol{b}, \tag{1.6}$$

We have the following representations:

$$\begin{aligned}
\operatorname{grad} \boldsymbol{s} = \quad & && +\boldsymbol{sn}\varkappa && \\
& && +\boldsymbol{nn}\psi && +\boldsymbol{nb}[\Omega_s - (\Omega_n + \tau)] \\
& && -\boldsymbol{bn}(\Omega_n + \tau) && +\boldsymbol{bb}\theta,
\end{aligned} \tag{1.7}$$

$$\begin{aligned}
\operatorname{grad} \boldsymbol{n} = \quad & -\boldsymbol{ss}\varkappa && && +\boldsymbol{sb}\tau \\
& -\boldsymbol{ns}\psi && && -\boldsymbol{nb} \operatorname{div} \boldsymbol{b} \\
& +\boldsymbol{bs}(\Omega_n + \tau) && && +\boldsymbol{bb}(\varkappa + \operatorname{div} \boldsymbol{n})
\end{aligned} \tag{1.8}$$

$$\begin{aligned}
\operatorname{grad} \boldsymbol{b} = \quad & && -\boldsymbol{sn}\tau \\
& -\boldsymbol{ns}[\Omega_s - (\Omega_n + \tau)] && +\boldsymbol{nn} \operatorname{div} \boldsymbol{b} \\
& -\boldsymbol{bs}\theta && -\boldsymbol{bn}(\varkappa + \operatorname{div} \boldsymbol{n}).
\end{aligned} \tag{1.9}$$

---

[2] We use the symbol $\frac{\delta f}{\delta s}$ to denote $\boldsymbol{s} \cdot \operatorname{grad} f$ for a scalar valued or vector valued point function f. Then $\frac{\delta^2 f}{\delta s\, \delta n} = \boldsymbol{s} \cdot \operatorname{grad}(\boldsymbol{n} \cdot \operatorname{grad} f)$, and so on. We shall also refer to $\frac{\delta f}{\delta s}, \frac{\delta f}{\delta n}, \frac{\delta f}{\delta b}$, as directional derivatives with respect to $\boldsymbol{s}$, $\boldsymbol{n}$ and $\boldsymbol{b}$. Such wording is expedient but not entirely satisfactory, since derivative has the connotation of differential, and we are dealing with the components of a vector.

We sometimes refer to the components of these gradients generically as curvatures. In as much as they play an analogous part in the anholonomic space associated with the vector fields $\boldsymbol{s}$, $\boldsymbol{n}$ and $\boldsymbol{b}$, to the Christoffel symbols associated with the coordinates of a Riemannian space, we may see them as the anholonomic components of the Euclidean connection.

From (1.7), (1.8) and (1.9) we have

$$\operatorname{curl} \boldsymbol{s} = \operatorname{grad} \times \boldsymbol{s} = \Omega_s \boldsymbol{s} + \varkappa \boldsymbol{b}, \tag{1.10}$$

$$\operatorname{curl} \boldsymbol{n} = \operatorname{grad} \times \boldsymbol{n} = -\operatorname{div} \boldsymbol{b} \boldsymbol{s} + \Omega_n \boldsymbol{n} + \psi \boldsymbol{b}, \tag{1.11}$$

$$\operatorname{curl} \boldsymbol{b} = \operatorname{grad} \times \boldsymbol{b} = (\varkappa + \operatorname{div} \boldsymbol{n})\, \boldsymbol{s} - \theta \boldsymbol{n} + \Omega_b \boldsymbol{b}, \tag{1.12}$$

where

$$\Omega_b = \Omega_s - (\Omega_n + 2\tau). \tag{1.13}$$

The identity

$$\operatorname{grad} \times \operatorname{grad} F = \boldsymbol{0}, \tag{1.14}$$

where F may be taken as a scalar of vector valued function of the points $\boldsymbol{x}$ of the Euclidean space, yields the commutation formulae

$$\frac{\delta^2 F}{\delta b\, \delta n} - \frac{\delta^2 F}{\delta n\, \delta b} = \Omega_s \frac{\delta F}{\delta s} - \operatorname{div} \boldsymbol{b} \frac{\delta F}{\delta n} + (\varkappa + \operatorname{div} \boldsymbol{n}) \frac{\delta F}{\delta b}, \tag{1.15}$$

$$\frac{\delta^2 F}{\delta s\, \delta b} - \frac{\delta^2 F}{\delta b\, \delta s} = \Omega_n \frac{\delta F}{\delta n} - \theta \frac{\delta F}{\delta b}, \tag{1.16}$$

$$\frac{\delta^2 F}{\delta n\, \delta s} - \frac{\delta^2 F}{\delta s\, \delta n} = \varkappa \frac{\delta F}{\delta s} + \psi \frac{\delta F}{\delta n} + \Omega_b \frac{\delta F}{\delta b}. \tag{1.17}$$

where Ω_b is given by (1.13).

When these formulae are applied to the basis vectors $\boldsymbol{s}$, $\boldsymbol{n}$ and $\boldsymbol{b}$, they yield the nine compatibility conditions referred to in the Introduction.[3]

$$\frac{\delta}{\delta n}(\Omega_n + \tau) + \frac{\delta \psi}{\delta b} - (\Omega_s - 2(\Omega_n + \tau))(\varkappa + \operatorname{div} \boldsymbol{n}) - (\theta - \psi) \operatorname{div} \boldsymbol{b} - \Omega_s \varkappa = 0, \qquad \text{s–1}$$

$$\frac{\delta \theta}{\delta n} - \frac{\delta}{\delta b}(\Omega_s - (\Omega_n + \tau)) - (\Omega_s - 2(\Omega_n + \tau)) \operatorname{div} \boldsymbol{b} + (\theta - \psi)(\varkappa + \operatorname{div} \boldsymbol{n}) = 0, \qquad \text{s–2}$$

$$\frac{\delta}{\delta n}(\varkappa + \operatorname{div} \boldsymbol{n}) + \frac{\delta}{\delta b} \operatorname{div} \boldsymbol{b} + \theta \psi + (\operatorname{div} \boldsymbol{b})^2 + (\varkappa + \operatorname{div} \boldsymbol{n})^2 + \Omega_s \tau + (\Omega_n + \tau)(\Omega_s - (\Omega_n + \tau)) = 0, \qquad \text{s–3}$$

[3] This method avoids the repetitions inherent in calculating $\operatorname{grad} \times \operatorname{grad} \boldsymbol{s} = \boldsymbol{0}$, $\operatorname{grad} \times \operatorname{grad} \boldsymbol{n} = \boldsymbol{0}$, $\operatorname{grad} \times \operatorname{grad} \boldsymbol{b} = \boldsymbol{0}$.

$$\frac{\delta \varkappa}{\delta b} + \frac{\delta}{\delta s}(\Omega_n + \tau) + (\Omega_n + 2\tau)\,\theta + \Omega_n \psi = 0, \qquad \text{n–1}$$

$$\frac{\delta \tau}{\delta b} - \frac{\delta}{\delta s}(\varkappa + \operatorname{div} \boldsymbol{n}) - \Omega_n \operatorname{div} \boldsymbol{b} - \theta(2\varkappa + \operatorname{div} \boldsymbol{n}) = 0, \qquad \text{n–2} \qquad (1.18)$$

$$\frac{\delta \theta}{\delta s} + \theta^2 - \varkappa(\varkappa + \operatorname{div} \boldsymbol{n}) - \tau^2 - \Omega_n(\Omega_s - \Omega_n) = 0, \qquad \text{n–3}$$

$$\frac{\delta}{\delta s} \operatorname{div} \boldsymbol{b} + \frac{\delta \tau}{\delta n} - (\Omega_s - \Omega_n)\,\varkappa + \psi \operatorname{div} \boldsymbol{b}$$
$$- (\Omega_s - (\Omega_n + 2\tau))\,(\varkappa + \operatorname{div} \boldsymbol{n}) = 0, \qquad \text{b–1}$$

$$\frac{\delta}{\delta s}(\Omega_s - (\Omega_n + \tau)) + (\Omega_s - \Omega_n)\,\psi + \varkappa \operatorname{div} \boldsymbol{b}$$
$$+ (\Omega_s - (\Omega_n + 2\tau))\,\theta = 0, \qquad \text{b–2}$$

$$\frac{\delta \psi}{\delta s} - \frac{\delta \varkappa}{\delta n} + \varkappa^2 + \psi^2 - (2\Omega_s - 3\tau)\,\tau - \Omega_n(\Omega_s - \Omega_n - 4\tau) = 0. \qquad \text{b–3}$$

While these relations collectively will be labelled equation (1.18), we introduce the individual designations because of the following properties.

When $\Omega_s = \boldsymbol{s} \cdot \operatorname{curl} \boldsymbol{s} = 0$, the unit vector $\boldsymbol{s}$ is complex-lamellar, $\boldsymbol{s} = \xi \operatorname{grad} \eta$, and there is a one-parameter family of surfaces $\eta = \text{constant}$, orthogonal to the vector-lines of $\boldsymbol{s}$. Under this curcumstance the conditions s–1, s–2, and s–3 become respectively the two Mainardi-Codazzi equations and the Gauss equation for the surfaces orthogonal to the vector-lines of $\boldsymbol{s}$. For $\Omega_n = 0$ the relations n–1, n–2 and n–3 becomes respectively these equations for the surfaces orthogonal to the vector-lines of $\boldsymbol{n}$. In this case, the vector-lines of $\boldsymbol{s}$ and $\boldsymbol{b}$ are geodesics and geodesic parallels on the surface. For $\Omega_b = 0$, or by (1.13), $\Omega_s = \Omega_n + 2\tau$, the relations b–1, b–2, and b–3 are respectively these equations for the surfaces orthogonal to the vector-lines of $\boldsymbol{b}$. In this case, the vector-lines of $\boldsymbol{s}$ are asymptotic lines on the surfaces.[4]

The Bianchi Identities

First, we take the directional derivative with respect to $\boldsymbol{s}$ of the condition s–1, that is to say, we take the $\boldsymbol{s}$-component of the gradient of all the terms on the left hand side of the condition s–1.

---

[4] With this wording we really anticipate one of the conclusions of this paper. Normally we speak of the Mainardi-Codazzi and Gauss equations in the context of a single surface. We shall see in the following chapter that one interpretation of the Bianchi identities is that that these equations can be deduced for all members of the one-parameter family of surfaces.

The condition b–2 is obtained from $n-1$ by eliminating $\frac{\delta \varkappa}{\delta b}$ using the condition $\operatorname{div} \operatorname{curl} \boldsymbol{s} = 0$ or $\frac{\delta \Omega_s}{\delta s} + \frac{\delta \varkappa}{\delta b} + \Omega_s(\theta + \psi) + \varkappa \operatorname{div} \boldsymbol{b} = 0$. We use b–2 rather than the latter condition in the present discussion, since b–2 is a Mainardi-Codazzi equation (with asymptotic lines as one set of surface curves) when $\Omega_b = 0$.

We have

$$\frac{\delta}{\delta s}\{\text{s-1}\} = \frac{\delta^2}{\delta s\,\delta n}(\Omega_n + \tau) + \frac{\delta^2\psi}{\delta s\,\delta b} - (\varkappa + \operatorname{div}\boldsymbol{n})\frac{\delta}{\delta s}[\Omega_s - 2(\Omega_n + \tau)]$$

$$- [\Omega_s - 2(\Omega_n + \tau)]\frac{\delta}{\delta s}(\varkappa + \operatorname{div}\boldsymbol{n}) - \operatorname{div}\boldsymbol{b}\frac{\delta}{\delta s}(\theta - \psi) \tag{3.1}$$

$$- (\theta - \psi)\frac{\delta}{\delta s}\operatorname{div}\boldsymbol{b} - \varkappa\frac{\delta\Omega_s}{\delta s} - \Omega_s\frac{\delta\varkappa}{\delta s}.$$

For the condition s–1 to hold in a neighbourhood this expression must be zero. We show, however, that no new condition on the vector fields is obtained by equating this expression to zero. Rather, by virtue of the commutation formulae (1.15), (1.16) and (1.17) and the set of equations (1.18), the expression (3.1) is identically zero. We exhibit the structure of the identity by expressing (3.1) explicitly in terms of the directional derivative given by the conditions (1.18).

From (1.16) and (1.17) we have, noting (1.13),

$$\frac{\delta^2}{\delta s\,\delta n}(\Omega_n + \tau) + \frac{\delta^2\psi}{\delta s\,\delta b} = \frac{\delta^2}{\delta n\,\delta s}(\Omega_n + \tau) - \varkappa\frac{\delta}{\delta s}(\Omega_n + \tau)$$

$$- \psi\frac{\delta}{\delta n}(\Omega_n + \tau) - (\Omega_s - (\Omega_n + 2\tau))\frac{\delta}{\delta b}(\Omega_n + \tau)$$

$$+ \frac{\delta^2\psi}{\delta b\,\delta s} + \Omega_n\frac{\delta\psi}{\delta n} - \theta\frac{\delta\psi}{\delta b}. \tag{3.2}$$

The mixed gradients $\frac{\delta^2}{\delta n\,\delta s}(\Omega_n + \tau)$ and $\frac{\delta^2\psi}{\delta b\,\delta s}$ may be eliminated as follows. From n–1 and b–3 we have

$$\frac{\delta^2}{\delta n\,\delta s}(\Omega_n + \tau) + \frac{\delta^2\psi}{\delta b\,\delta s} = -\frac{\delta^2\varkappa}{\delta n\,\delta b} - \frac{\delta}{\delta n}[(\Omega_n + 2\tau) + \Omega_n\psi] + \frac{\delta^2\varkappa}{\delta b\,\delta n} \tag{3.3}$$

$$- \frac{\delta}{\delta b}[\varkappa^2 + \psi^2 - (2\Omega_s - 3\tau)\tau - \Omega_n(\Omega_s - \Omega_n - 4\tau)]$$

where by (1.15),

$$\frac{\delta^2\varkappa}{\delta b\,\delta n} - \frac{\delta^2\varkappa}{\delta n\,\delta b} = \Omega_s\frac{\delta\varkappa}{\delta s} - \operatorname{div}\boldsymbol{b}\frac{\delta\varkappa}{\delta n} + (\varkappa + \operatorname{div}\boldsymbol{n})\frac{\delta\varkappa}{\delta b}. \tag{3.4}$$

Expanding the terms in parenthesis in (3.3), substituting these expressions into (3.1) and collecting terms we get, formally,

$$
\begin{aligned}
\frac{\delta}{\delta s}\{\text{s-1}\} = &-(\theta + 2\psi)\left[\frac{\delta}{\delta n}(\Omega_n + \tau) + \frac{\delta\psi}{\delta b}\right] \\
&- (\Omega_n + 2\tau)\left[\frac{\delta\theta}{\delta n} - \frac{\delta}{\delta b}[\Omega_s - (\Omega_n + \tau)]\right] \\
&+ [-2\varkappa + (\varkappa + \operatorname{div} \boldsymbol{n})]\left[\frac{\delta\varkappa}{\delta b} + \frac{\delta}{\delta s}(\Omega_n + \tau)]\right] \\
&+ (\Omega_s - 2(\Omega_n + \tau))\left[\frac{\delta\tau}{\delta b} - \frac{\delta}{\delta s}(\varkappa + \operatorname{div} \boldsymbol{n})\right] \qquad (3.5) \\
&- \operatorname{div} \boldsymbol{b}\frac{\delta\theta}{\delta s} - (\theta - \psi)\left[\frac{\delta}{\delta s}\operatorname{div} \boldsymbol{b} + \frac{\delta\tau}{\delta n}\right] \\
&- [\varkappa + (\varkappa + \operatorname{div} \boldsymbol{n})]\frac{\delta}{\delta s}[\Omega_s - (\Omega_n + \tau)] \\
&+ \operatorname{div} \boldsymbol{b}\left[\frac{\delta\psi}{\delta s} - \frac{\delta\varkappa}{\delta n}\right].
\end{aligned}
$$

By substituting for the groups $\frac{\delta}{\delta n}(\Omega_n + \tau) + \frac{\delta\psi}{\delta b}, \frac{\delta\theta}{\delta n} - \frac{\delta}{\delta b}[\Omega_s - (\Omega_n + \tau)]$ and so on, the expressions given by the conditions (1.18), we verify that the right hand side of (3.5) vanishes identically.

If now we take the directional derivative with respect to s of the conditions s-2, this time eliminating $\frac{\delta^2\theta}{\delta s\,\delta n} - \frac{\delta^2}{\delta s\,\delta b}[\Omega_s - (\Omega_n + \tau)]$ in favor of $\frac{\delta^2\theta}{\delta n\,\delta s} - \frac{\delta^2}{\delta b\,\delta s}[\Omega_s - (\Omega_n + \tau)]$ by (1.16) and (1.17), and then using the expressions n-3, and b-2 for $\frac{\delta\theta}{\delta s}$ and $\frac{\delta}{\delta s}[\Omega_s - (\Omega_n + \tau)]$ we get,

$$
\begin{aligned}
\frac{\delta}{\delta s}\{\text{s-2}\} = &(\Omega_s - \Omega_n)\left[\frac{\delta}{\delta n}(\Omega_n + \tau) + \frac{\delta\psi}{\delta b}\right] \\
&- (\psi + 2\theta)\left[\frac{\delta\theta}{\delta n} - \frac{\delta}{\delta b}[\Omega_s - (\Omega_n + \tau)]\right] \\
&+ \varkappa\left[\frac{\delta}{\delta n}(\varkappa + \operatorname{div} \boldsymbol{n}) + \frac{\delta}{\delta b}\operatorname{div} \boldsymbol{b}\right] \\
&+ \operatorname{div} \boldsymbol{b}\left[\frac{\delta\varkappa}{\delta b} + \frac{\delta}{\delta s}(\Omega_n + \tau)\right] \\
&- (\theta - \psi)\left[\frac{\delta\tau}{\delta b} - \frac{\delta}{\delta s}(\varkappa + \operatorname{div} \boldsymbol{n})\right] \qquad (3.6)
\end{aligned}
$$

$$+ [-\varkappa + (\varkappa + \operatorname{div} \boldsymbol{n})] \frac{\delta\theta}{\delta s}$$

$$- (\Omega_s - 2(\Omega_n + \tau)) \left[\frac{\delta}{\delta s} \operatorname{div} \boldsymbol{b} + \frac{\delta\tau}{\delta n}\right]$$

$$- \operatorname{div} \boldsymbol{b} \frac{\delta}{\delta s} [\Omega_s - (\Omega_n + \tau)] - (\varkappa + \operatorname{div} \boldsymbol{n}) \left(\frac{\delta\psi}{\delta s} - \frac{\delta\varkappa}{\delta n}\right).$$

The right hand side of (3.6) vanishes identically when we substitute for $\frac{\delta}{\delta n}(\Omega_n + \tau) + \frac{\delta\psi}{\delta b}$ and so on, the expressions given by (1.18).

If we take the directional derivative with respect to $\boldsymbol{s}$ of the condition s–3 and then proceed in the same manner, we obtain

$$\frac{\delta}{\delta s}\{\text{s-3}\} = -\operatorname{div} \boldsymbol{b} \left[\frac{\delta}{\delta n}(\Omega_n + \tau) + \frac{\delta\psi}{\delta b}\right]$$

$$- [\varkappa + (\varkappa + \operatorname{div} \boldsymbol{n})] \left[\frac{\delta\theta}{\delta n} - \frac{\delta}{\delta b}[\Omega_s - (\Omega_n + \tau)]\right]$$

$$- (\theta + \psi) \left[\frac{\delta}{\delta n}(\varkappa + \operatorname{div} \boldsymbol{n}) + \frac{\delta}{\delta b} \operatorname{div} \boldsymbol{b}\right]$$

$$+ (\Omega_s - \Omega_n) \left[\frac{\delta\varkappa}{\delta b} + \frac{\delta}{\delta s}(\Omega_n + \tau)\right] \tag{3.7}$$

$$+ [\varkappa - 2(\varkappa + \operatorname{div} \boldsymbol{n})] \left[\frac{\delta\tau}{\delta b} - \frac{\delta}{\delta s}(\varkappa + \operatorname{div} \boldsymbol{n})\right]$$

$$+ \psi \frac{\delta\theta}{\delta s} + 2 \operatorname{div} \boldsymbol{b} \left[\frac{\delta}{\delta s} \operatorname{div} \boldsymbol{b} + \frac{\delta\tau}{\delta n}\right]$$

$$+ (\Omega_n + 2\tau) \frac{\delta}{\delta s} [\Omega_s - (\Omega_n + \tau)] + \theta \left[\frac{\delta\psi}{\delta s} - \frac{\delta\varkappa}{\delta n}\right].$$

The right hand side of (3.7) vanishes when the groups $\frac{\delta}{\delta n}(\Omega_n + \tau) + \frac{\delta\psi}{\delta b}$ and so on are given their expressions from (1.18).

The formulae (3.5), (3.6) and (3.7) represent the complete set of Bianchi identities. If we take the directional derivatives of n–1, n–2, and n–3 with respect to $\boldsymbol{n}$ we obtain respectively minus the expression (3.5), the expression (3.7), and minus the expression (3.6). Likewise, if we take the directional derivatives, of b–1, b–2, b–3 with respect to $\boldsymbol{b}$ we obtain respectively minus the expression (3.7), the expression (3.6), and minus the expression (3.5). Symbolically we may write

$$\begin{aligned}
&\frac{\delta}{\delta s}\{\text{s-1}\} = -\frac{\delta}{\delta n}\{\text{n-1}\} = -\frac{\delta}{\delta b}\{\text{b-3}\}\\
&\frac{\delta}{\delta s}\{\text{s-2}\} = -\frac{\delta}{\delta n}\{\text{n-3}\} = \frac{\delta}{\delta b}\{\text{b-2}\}\\
&\frac{\delta}{\delta s}\{\text{s-3}\} = \frac{\delta}{\delta n}\{\text{n-2}\} = -\frac{\delta}{\delta b}\{\text{b-1}\}.
\end{aligned} \tag{3.8}$$

We have already remarked that when the abnormality Ω_s vanishes there is a one-parameter family of surfaces $\eta = \text{constant}$ normal to the vector-lines of $\boldsymbol{s}$; thus $\boldsymbol{s} = \xi \operatorname{grad} \eta$. We noted that the relations s–1, s–2, s–3 then become respectively the Mainardi-Codazzi and Gauss equations for a representative surface $\eta = \text{constant}$. In the present context the Bianchi identities imply that the left hand sides of s–1, s–2, s–3 continue to be zero along the vector-line of $\boldsymbol{s}$ normal to the surface. The Mainardi-Codazzi and Gauss equations are shown for the family of surfaces. Again if $\Omega_n = 0$, or $\Omega_b = 0$ the same Bianchi identities fulfill this function for the family of surfaces orthogonal to the vector-lines of $\boldsymbol{n}$, or those orthogonal to the vector-lines of $\boldsymbol{b}$, as the case may be.

Finally, we may note the rather curious fact that in the present representation the Bianchi identity (3.5). involves only eight of the nine compatibility conditions (1.18); the condition s–3 does not appear in (3.5). One the other hand, the identities (3.6) and (3.7) exhibit all the nine conditions (1.18).

References

1947 R. BLUM, Sur les identités de Bianchi et Veblen, *C. R. Acad. Sci.* (Paris) **224**, 889–890.

1973 A. W. MARRIS, Hamel's Theorem, *Arch. Rational Mech. Anal.* **51**, 85–105.

3372 Embry Circle
Atlanta, Georgia 30341

(Received November 24, 1987)

On the Vorticity Numbers of Monotonous Motions

C. TRUESDELL

Dedicated to Bernard D. Coleman for his sixtieth birthday

1. Introduction

Among the phenomena best known in the mechanics of viscous fluids are boundary layers and the absence of them. In a typical flow past a stationary boundary there is a layer next to that boundary in which vorticity and stretching are both important, while at great distances from the boundary the velocity field grows more and more nearly irrotational. The effects of boundary layers are of main importance for fluids whose viscosity or viscosities are known. In contrast, when it comes to measuring the tangential and normal tractions exerted by a fluid body on confining walls—that is, to determine the several constitutive coefficients or functions—a flow throughout vorticose is usually laid down as basic and in experiment produced to within deviations such as end effects. Flows of this kind form the basis for design of instruments to measure shear tractions and normal tractions as functions of kinematical variables such as shearings.

In this note I address mainly phenomena of the latter kind, those in which spin balances or predominates stretching.

2. The vorticity number

A dimensionless measure of such relative importance is the field, generally time-dependent, of *vorticity numbers* $\mathfrak{W}$ [1953]:

$$\mathfrak{W} := \frac{|\boldsymbol{W}|}{|\boldsymbol{D}|}. \tag{2.1}$$

Here and subsequently I use the notations in my textbook [1977]: $\boldsymbol{W}$ is the spin tensor, $\boldsymbol{D}$ is the stretching tensor, *etc.* It has been noticed that $\mathfrak{W} = 1$ in simple shearing and in several other flows of the kind called "viscometric"; in examples of other kinds, the same is true; nonetheless, the steady simple vortices, given in cylindrical co-ordinates r, θ, z by the equations

$$\dot{r} = 0, \quad \dot{\theta} = \omega(r), \quad \dot{z} = 0, \tag{2.2}$$

are viscometric and have spatially dependent vorticity numbers taking on any and all values in $[0, \infty[$. Indeed, for (2.2)

$$\mathfrak{W}^2 = 1 + \frac{4\omega(r\omega)'}{r^2\omega'^2}, \quad \mathfrak{W} = \left|1 + \frac{2\omega}{r\omega'}\right|. \tag{2.3}$$

While $\mathfrak{W}$ generally varies with r, exceptions are the nth power vortices, $\omega \propto r^{-n}$, for which

$$\mathfrak{W} = \left|1 - \frac{2}{n}\right|. \tag{2.4}$$

For the potential vortex $n = 2$, and so $\mathfrak{W} = 0$. From (2.3), first taking $\omega > 0$ without loss of generality, we see that $\mathfrak{W} > 1$ in regions where the linear speed $r\omega$ is an increasing function of r, while $\mathfrak{W} < 1$ in regions where $r\omega$ decreases as r increases. Henceforth we assume that the motion is not rigid: $\boldsymbol{D} \neq \boldsymbol{0}$.

If $\mathfrak{W} \geqq \text{const.} > 0$, it is impossible that $\boldsymbol{W} \to \boldsymbol{0}$ at great distances from boundaries, and thus, unless also $\boldsymbol{D} \to \boldsymbol{0}$, such flows cannot conform with the old idea that outside some boundary layer next to a surface of adherence irrotational flow should prevail. The condition $\mathfrak{W} = 1$, which reflects perfect balance of spin and stretching, appears also in contexts other than viscometry. For example, $\mathfrak{W} = 1$ everywhere on a stationary wall to which a body in isochoric motion adheres [1970, 1, Equation (32)]. Indeed, the condition of adherence makes any isochoric flow essentially a simple shearing on and near a bounding surface. RAYLEIGH's criterion of purely inertial stability for a "Couette flow" $\omega = \Omega(1 - a^2/r^2)$ makes $\mathfrak{W} > 1$ necessary and sufficient.

3. General bound for $\mathfrak{W}$ in isochoric motions

For an isochoric motion having acceleration field $\ddot{\boldsymbol{x}}$ and spin field $\boldsymbol{W}$ it has long been known [1953] that

$$\frac{1}{\mathfrak{W}^2} = 1 + \frac{\operatorname{div} \ddot{\boldsymbol{x}}}{|\boldsymbol{W}|^2} \quad \text{if} \quad \mathfrak{W} \neq 0. \tag{3.1}$$

This statement is evident because $\operatorname{div} \ddot{\boldsymbol{x}} = |\boldsymbol{D}|^2 - |\boldsymbol{W}|^2$ if $\operatorname{div} \dot{\boldsymbol{x}} = 0$. In particular, for a rigid motion of spin $\boldsymbol{W}_r$ and acceleration $\ddot{\boldsymbol{x}}_r$

$$\operatorname{div} \ddot{\boldsymbol{x}}_r = -|\boldsymbol{W}_r|^2. \tag{3.2}$$

(If ω is the angular speed of the rotation, $\omega^2 = \frac{1}{2}|\boldsymbol{W}_r|^2$.)

Theorem 1. *Let a motion having spin* $\boldsymbol{W}$ *result from superposing a rigid motion having spin* $\boldsymbol{W}_r$ *upon an isochoric motion* $\dot{\boldsymbol{x}}_0$ *having spin* $\boldsymbol{W}_0$ *and vorticity number* $\mathfrak{W}_0$, *so that* $\boldsymbol{W} = \boldsymbol{W}_0 + \boldsymbol{W}_r$. *Then* $\mathfrak{W}$, *the vorticity number of the combined motion,* *is determined as follows: If* $\mathfrak{W} \neq 0$ *and* $\mathfrak{W}_0 \neq 0$, *then*

$$|W|^2\left(1-\frac{1}{\mathfrak{W}^2}\right)=|W_0|^2\left(1-\frac{1}{\mathfrak{W}_0^2}\right)+|W_r|^2+2W_r\cdot W_0. \tag{3.3}$$

Proof. If $\dot{x} := \dot{x}_0 + \dot{x}_r$, then

$$\operatorname{div}\ddot{x}=\operatorname{div}\ddot{x}_0-|W_r|^2-2W_r\cdot W_0. \tag{3.4}$$

Remark. Over a three-dimensional vector space a skew tensor S other than $\mathbf{0}$ has as its nullspace a unique line, which is called the *axis* of S. That line may be directed by convention, *e.g.* "the right-hand rule". Using the same convention for both W_r and W_0, and letting θ be the least non-negative angle between the oriented axes of W_r and W_0, we see that

$$W_r\cdot W_0=|W_r|\,|W_0|\cos\theta. \tag{3.5}$$

This statement holds trivially if $W_r = \mathbf{0}$. In a three-dimensional space, use of (3.5) lets (3.3) be written

$$|W|^2\left(1-\frac{1}{\mathfrak{W}^2}\right)=|W_0|^2\left(1-\frac{1}{\mathfrak{W}_0^2}\right)+|W_r|\left(|W_r|+2\,|W_0|\cos\theta\right). \tag{3.6}$$

If $W_r = -W_0$, then $W = \mathbf{0}$ and $\theta = \pi$, and so both (3.3) and (3.6) fail generally to make sense unless written as $|W|^2 = |W_0|^2 + |W_r|^2 - 2\,|W_r|\,|W_0|$.

Corollary. *Suppose that* $\mathfrak{W}_0 = 1$. *Then*

$$\mathfrak{W} \geqq 1 \Leftrightarrow \begin{cases} W_r = \mathbf{0} & \text{or} \\ |W_r| \geqq -2\,|W_0|\cos\theta. \end{cases} \tag{3.7}$$

In particular

$$\theta \leqq \tfrac{1}{2}\pi \Rightarrow \mathfrak{W} \geqq 1, \qquad \theta = \tfrac{1}{2}\pi \Rightarrow \mathfrak{W}^2 = 1 + \frac{|W_r|^2}{|W_0|^2}. \tag{3.8}$$

4. Monotonous motions

Coleman [1961] introduced and called "substantially stagnant" the class of motions in which the principal relative stretch histories at each body-point remain constant in time. Noll [1962], [1976] (*cf.* [1977, § IV.18]) characterized such motions in terms of a tensor N of magnitude 1 and a scalar $\varkappa$, both of which may depend upon time and place. The velocity gradient G of such a motion is the sum of $\varkappa N$ and the spin W_r of a rigid motion:

$$G=\varkappa N+W_r. \tag{4.1}$$

The scalar $\varkappa$, which has the dimensions of a time-rate, is called the *shearing*.

Noll classified these motions into three mutually exclusive types:

1. $N^2 = \mathbf{0}$; these flows are called "viscometric".
2. $N^3 = \mathbf{0}$ and $N^2 \neq \mathbf{0}$.
3. N is not nilpotent.

Use of the Hamilton-Cayley equation satisfied by N shows that in the first two types $\operatorname{tr} N = 0$, and so the motions are isochoric, and also $\operatorname{tr} N^2 = 0$. Flows of the third type need not be isochoric, and for them $\operatorname{tr} N^2$ may assume any value between 0 and 1.

NOLL [1976] designated these motions as *monotonous*. In them an observer stationed on a body-point can by suitable rotation see as he looks backward one single, never changing history of relative stretch.

If a monotonous flow is given, it determines $\varkappa$, N, and W_r uniquely. If, on the other hand, the fields $\varkappa$, N and W_r are assigned, and if they correspond to any flow at all, that flow is monotonous. The classification of monotonous flows is invariant under general change of $\varkappa$ and W_r.

From (4.1) we see that for a monotonous flow

$$W = \tfrac{1}{2}\varkappa(N - N^T) + W_r. \tag{4.2}$$

If $\varkappa$ and N are fixed, we can choose W_r in such a way as to make W have any value we like, for example $\mathbf{0}$. Thus the values of $\mathfrak{W}$ for a monotonous flow are neither bounded below nor in general independent of $\varkappa$, $\boldsymbol{x}$, and t. The monotonous flows for which $\mathfrak{W} = 1$ everywhere and always are very special.

Clearly (4.1) provides an example of the kind of motion mentioned in Theorem 1:

$$W_0 = \tfrac{1}{2}\varkappa(N - N^T), \quad D = \tfrac{1}{2}\varkappa(N + N^T), \tag{4.3}$$

and so

$$|W_0| = \frac{1}{\sqrt{2}}|\varkappa|\sqrt{1 - \operatorname{tr} N^2}, \quad \mathfrak{W}_0^2 = \frac{1 - \operatorname{tr} N^2}{1 + \operatorname{tr} N^2}. \tag{4.4}$$

Thus from (3.3) we conclude

Theorem 2. *In rotational monotonous motions of types 1 and 2*

$$|W|^2\left(1 - \frac{1}{\mathfrak{W}^2}\right) = |W_r|(|W_r| + \sqrt{2}\,|\varkappa|\cos\theta). \tag{4.5}$$

In particular,

$$\mathfrak{W} \geqq 1 \Leftrightarrow |W_r|\,(|W_r| + \sqrt{2}\,|\varkappa|\cos\theta) \geqq 0. \tag{4.6}$$

Corollary. *If $W_r \neq \mathbf{0}$ and $\theta > \frac{1}{2}\pi$, then $\mathfrak{W} > 1$ if $|\varkappa|$ is sufficiently small, while $\mathfrak{W} < 1$ if $|\varkappa|$ is sufficiently large.*

For an example of the effect of small values of the shearing we may consider a helical flow: in cylindrical co-ordinates

$$\dot{r} = 0, \quad \dot{\theta} = \omega(r), \quad \dot{z} = u(r).$$

For it (*cf.* the special instance (2.3))

$$\varkappa^2 = r^2\omega'^2 + u'^2, \quad \mathfrak{W}^2 = 1 + \frac{4\omega}{\varkappa^2}(r\omega)', \quad \mathfrak{W} = \left|1 + \frac{2\omega}{\varkappa}\right|, \tag{4.7}$$

and $\mathfrak{W} = 0$ if and only if $2\omega/\varkappa = -1$.

In some familiar examples of montonous flow the shearing $\varkappa$ does not affect $\mathfrak{W}$. From (4.5) we see at a glance that $\mathfrak{W}$ is independent of $\varkappa$ in exactly two cases:

1. $W_r = 0$. Then $\mathfrak{W} = 1$.
2. $W_r \neq 0$ and $\theta = \frac{1}{2}\pi$. Then $\mathfrak{W} > 1$.

The angle θ is easy to determine for monotonous flows of Types 1 and 2. For the first, namely viscometric flows, at each time and place a basis e_1, e_2, e_3 may be found such that

$$[N] = \begin{Vmatrix} 0 & 0 & 0 \\ 1 & 0 & 0 \\ 0 & 0 & 0 \end{Vmatrix}.$$

The axis of the spin W_0 is the proper direction of $N - N^T$ and is therefore parallel to e_3. YIN & PIPKIN [1970, 2] have shown that every viscometric flow can be regarded as generated by material surfaces, in general unsteady and deforming, which slide over each other isometrically. The basis vector e_2 is normal to them. The axis of W_r will usually be given or otherwise easily determinable.

For monotonous flows of Type 2 a basis e_1, e_2, e_3 may be found such that

$$[N] = \frac{1}{\sqrt{2}} \begin{Vmatrix} 0 & 1 & 0 \\ 0 & 0 & 1 \\ 0 & 0 & 0 \end{Vmatrix}.$$

Thus the proper line of W_0 subtends equal angles on e_3 and e_1.

Acknowledgment. I am indebted to Messrs. RAJAGOPAL, VIRGA, and W.-L. YIN for examples and discussion. The research reported here was partially supported by grants from the Applied Mathematics Program and the U.S.-Italy Co-operative Science Program of the U.S. National Science Foundation.

References

[1953] C. TRUESDELL, "Two measures of vorticity", *Journal of Rational Mechanics and Analysis* **2**: 173–217.

[1961] B. D. COLEMAN, "Kinematical concepts with applications in the mechanics and thermodynamics of incompressible fluids", *Archive for Rational Mechanics and Analysis* **9**: 273–300.

[1962] W. NOLL, "Motions with constant stretch history", *Archive for Rational Mechanics and Analysis* **11**: 97–105.

[1970, 1] C. TRUESDELL, "De pressionibus negativis in sinu et pariete regionis fluido viscoso moventi impletae schedula", *Annali di Matematica Pura ed Applicata* (4) **84**, 213–224.

[1970, 2] W.-L. YIN & A. C. PIPKIN, "Kinematics of viscometric flow", *Archive for Rational Mechanics and Analysis* **37**: 111–135.

[1976] W. NOLL, "The representation of monotonous processes by exponentials", *Indiana University Mathematics Journal* **25**: 209–214.

[1977] C. TRUESDELL, *A First Course in Rational Continuum Mechanics*, Volume 1, New York *etc.*, Academic Press.

The Johns Hopkins University
Baltimore

(Received April 1, 1988)

A Limiting "Viscosity" Approach to the Riemann Problem for Materials Exhibiting Change of Phase

M. SLEMROD

Dedicated to Bernard Coleman on the occasion of his sixtieth birthday

0. Introduction

The one-dimensional isothermal motion of a compressible elastic fluid or solid can be described in Lagrangian coordinates by the coupled system

$$u_t + p(w)_x = 0, \tag{0.1}$$

$$w_t - u_x = 0. \tag{0.2}$$

Here u denotes the velocity, w the specific volume for a fluid (or displacement gradient for a solid), and $-p$ is the stress which must be determined as a function of w by a constitutive relation. For many materials a natural condition placed on p is that $p'(w) < 0$ for all values of w (or all positive values of w) depending on the context of the problem. This makes (0.1), (0.2) a coupled system of hyperbolic conservation laws. In this paper, however, we shall consider the case where p has a graph illustrated by Figure 1. For convenience p will be globally defined, smooth, with

$$p' < 0, \quad w < \alpha, \quad w > \beta; \quad p' > 0, \quad \alpha < w < \beta;$$
$$p''(\alpha) > 0, \quad p''(\beta) < 0.$$

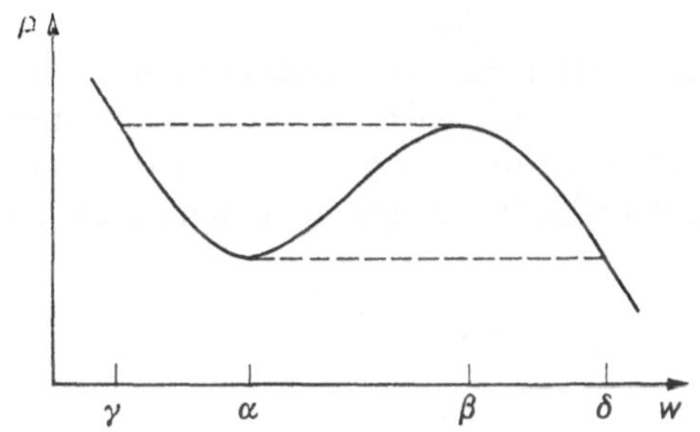

Fig. 1

This type of constitutive relation is usually associated with a van der Waals fluid where

$$p(w) = \frac{RT}{w-b} - \frac{a}{w^2}$$

and R, a, b are positive material constants, T is the temperature. Here we need nothing so specific as the van der Waals constitutive relation though our results will strongly depend at times on the global behavior of p as $|w| \to \infty$.

The reason for this non-standard choice of p is that it serves as a prototype problem for the dynamics of materials exhibiting changes of phase. For example in a van der Waals fluid the states $w < \alpha$ are viewed as liquid while states with $w > \beta$ are viewed as vapor. Because p is not monotone, liquid and vapor phases may co-exist.

The evolution of (0.1), (0.2) will be governed by initial data. Here we pose piecewise constant data

$$u(x, 0) = \begin{cases} u_- \\ u_+ \end{cases} \quad w(x, 0) = \begin{cases} w_- & x < 0 \\ w_+ & x \geqq 0 \end{cases} \tag{0.3}$$

which makes (0.1)–(0.3) into a *mixed hyperbolic-elliptic Riemann initial value problem* which we call *problem P*.

The classic method of solution of the Riemann problem is based on the construction of shock and wave curves for (0.1), (0.2). For materials like the van der Waals-like gas, discussions of this approach have been given first by R. JAMES [1] and later by M. SHEARER [2], [3]. The difficulty with this procedure is that even if shock admissibility conditions are known *a priori* it is not obvious in what manner the full solution which is a composite of shock and rarefaction waves is admissible. For example in [12], SHEARER proves existence of solutions to P when the data w_-, w_+ are in different phases but close to the well known Maxwell line. Each discontinuity in SHEARER's solution is admissible with respect to the viscosity-capillarity criterion discussed below.

The investigation here is based on a different approach. First we recall that earlier work [4], [5], [6] has suggested a reasonable admissibility criterion for (0.1), (0.2) to be the following *viscosity-capillarity criterion*. Namely weak solutions of (0.1), (0.2) will be admissible if they are limits boundedly a.e. of solutions u_ν, w_ν of the system

$$u_t + p(w)_x = \nu u_{xx} - \nu^2 A w_{xxx}, \tag{0.4}$$

$$w_t - u_x = 0, \tag{0.5}$$

as $\nu \to 0+$.

This system is derived from Korteweg's theory of capillarity where the total stress is written as the sum $-p(w) + \nu u_x - \nu^2 A w_{xx}$ comprising elastic, viscous, and capillary contributions.

As noted in [6] the substitutions

$$\begin{matrix} D_1 \\ D_2 \end{matrix} = \frac{\nu}{2} \pm \frac{\nu}{2}(1 - 4A)^{1/2}, \qquad v = u - D_2 w_x$$

bring ((0.4), (0.5)) into the parabolic form

$$v_t + p(w)_x = D_1 v_{xx}, \tag{0.6}$$

$$w_t - v_x = D_2 w_{xx}, \tag{0.7}$$

when $0 \leqq A \leqq 1/4$. In particular the choice of $A = 1/4$, $\nu = 2\varepsilon$ would say that admissible solutions would be limits as $\varepsilon \to 0+$ of solutions of (0.6), (0.7) with $D_1 = D_2 = \varepsilon$.

While (0.6), (0.7) is the common form for most artificial "viscosity" arguments, KALASNIKOV (1959 [13]], TUPCIEV (1964 [14], 1972 [15]), DAFERMOS (1973 [7]), 1974 [16], and DAFERMOS & DIPERNA (1976 [8]) suggested a variant advantageous for the study of Riemann initial value problems. Within the context of (0.1)–(0.2) the idea is to replace (0.6)–(0.7) by the system

$$u_t + p(w)_x = \varepsilon t u_{xx}, \tag{0.8}$$

$$w_t - u_x = \varepsilon t w_{xx}, \tag{0.9}$$

which is invariant under the transformation $(x, t) \to (ax, at)$, $a > 0$. (Here the letter v has been replaced by its former self u.) System (0.8), (0.9) has a decided advantage over (0.6), (0.7) in that it admits solutions that are functions of the single variable $\xi = \frac{x}{t}$. In fact a simple computation shows that $u\left(\frac{x}{t}\right)$, $w\left(\frac{x}{t}\right)$ is a solution of (0.8), (0.9), (0.3) if $u(\xi)$, $w(\xi)$ is a solution of the coupled system of non-autonomous ordinary differential equations

$$\varepsilon u'' = p(w)' - \xi u', \tag{0.10}$$

$$\varepsilon w'' = -u' - \xi w', \tag{0.11}$$

with boundary conditions

$$\begin{aligned} u(-\infty) &= u_-, \quad & w(-\infty) &= w_-, \\ u(+\infty) &= u_+, \quad & w(+\infty) &= w_+. \end{aligned} \tag{0.12}$$

Here $'$ denotes differentiation with respect to ξ. We will call the *boundary value problem* (0.10)–(0.12) *problem* P_ε.

Our program can now be broken into two parts. The first part carried out in Sections 1 and 2 establishes that if the data are in different phases there is a solution of P_ε which exhibits one change of phase. Also we give special conditions on the one-phase data which yield a one-phase solution of P_ε. The main feature of the proofs of these results is to note that DAFERMOS's arguments in [7] (which provided the successful resolution of P_ε in the case $p' < 0$) and those of DAFERMOS & DIPERNA in [8] do not directly apply. However a careful modification involving changes of underlying function spaces, application of the Leray-Schauder degree, and a new set of *a priori* estimates make P_ε solvable.

In Sections 3 and 4 we pursue the second part of our program, i.e. to give conditions on which solutions $u_\varepsilon(\xi)$, $w_\varepsilon(\xi)$ of P_ε possess limits as $\varepsilon \to 0+$ which solve the Riemann problem P. In the case $p' < 0$, DAFERMOS [7] and DAFERMOS & DIPERNA [8] succeeded by use of this method in solving P. Here we modify the ideas of [7], [8] to the case when $p' > 0$ in (α, β). In this case, when the above mentioned special data are in the same phase, assumptions on $p''(w)$ and behavior of p at infinity yield estimates on the total variation of u_ε, w_ε which combined with Helly's theorem show u_ε, w_ε do converge to a solution of P. For data in different phases similar estimates in the total variation may be obtained to yield solvability of P except in one case. The case in doubt is when there is a sequence $\tau^\varepsilon \to 0$ so that $|u_\varepsilon(\tau^\varepsilon)|$ becomes infinite as $\varepsilon \to 0+$. For this case we know u_ε, w_ε possesses a subsequence which converges a.e. to functions u, w as $\varepsilon \to 0+$. The limit functions u, w will be a solution of the Riemann problem if and only if the pressure p equilibrates across the stagnant phase boundary $\xi = 0$, *i.e.*, $\lim_{\xi \to 0+} p(w(\xi)) = \lim_{\xi \to 0-} p(w(\xi))$ (Theorem 4.13). Modulo this one case we see that the idea of artificial "viscosity" arguments, which play such a vital role in the existence theory of hyperbolic conservation laws can be extended to mixed hyperbolic-elliptic systems as well. (In this regard see also [9] for a study of a viscosity approach to a mixed hyperbolic-elliptic *boundary* value problem.)

1. Existence of connecting orbits assuming a priori estimates

In this section we will establish an existence theorem for the connecting orbit problem P_ε described in the introduction under the assumption of *a priori* estimates on (u, w). With this goal in mind, consider the two-parameter family of problems

$$\varepsilon u'' = \mu p(w)' - \xi u', \tag{1.1}$$

$$\varepsilon w'' = -\mu u' - \xi w', \tag{1.2}$$

$$u(-L) = u_-, \quad u(L) = u_+, \quad w(-L) = w_-, \quad w(L) = w_+, \tag{1.3}$$

where $\mu \in [0, 1]$ and $L \geqq 1$.

Theorem 1.1. *Assume* $w_- < \alpha$, $w_+ > \beta$ $(w_- > \beta, w_+ < \alpha)$ *and there is a constant* M_0 *such that every possible solution of* (1.1), (1.2), (1.3) *with* $w'(\xi) > 0$ $(w'(\xi) < 0)$ *when* $\alpha \leqq w(\xi) \leqq \beta$ *satisfies the* a priori *estimate*

$$\sup_{-L<\xi<L} (|u(\xi)| + |u'(\xi)| + |w(\xi)| + |w'(\xi)|) \leqq M_0. \tag{H0}$$

Here M_0 *can depend on* u_-, u_+, w_-, w_+, ε, p *but is independent of* μ *and* L. *Then there exist solutions of* P_ε *which satisfy the constraint that* $w'(\xi) > 0 (w'(\xi) < 0)$ *if* $\alpha \leqq w(\xi) \leqq \beta$, i.e. *the one phase change data connecting orbit problem possesses a one phase change solution.*

Proof. We consider the case $w_- < \alpha$, $w_+ > \beta$. The case $w_- > \beta$, $w_+ < \alpha$ is analogous. First notice that when $\mu = 0$ (1.1), (1.2), (1.3) possesses a unique solution

$$u_0(\xi) = \frac{(u_+ - u_-) \int\limits_{-L}^{\xi} \exp(-\zeta^2/2\varepsilon)\, d\zeta}{\int\limits_{-L}^{L} \exp(-\xi^2/2\varepsilon)\, d\xi} + u_-,$$

$$w_0(\xi) = \frac{(w_+ - w_-) \int\limits_{-L}^{\xi} \exp(-\zeta^2/2\varepsilon)\, d\zeta}{\int\limits_{-L}^{L} \exp(-\xi^2/2\varepsilon)\, d\xi} + w_-.$$

Also note that $w_0'(\xi) > 0$, $\xi \in [-L, L]$.

Now set $U(\xi) = u(\xi) - u_0(\xi)$, $W(\xi) = w(\xi) - w_0(\xi)$ and impose boundary conditions

$$U(-L) = U(L) = W(-L) = W(L) = 0. \tag{1.4}$$

If u, w are to solve (1.1), (1.2), (1.3), we see that U, W must satisfy (1.4) and

$$\begin{aligned} \varepsilon U'' &= \mu p(w_0 + W)' - \xi U', \\ \varepsilon W'' &= -\mu U' - \mu u_0' - \xi W'. \end{aligned} \tag{1.5}$$

Define the vectors

$$y(\xi) = \begin{pmatrix} U(\xi) \\ W(\xi) \end{pmatrix}, \quad f(\xi, y) = \begin{pmatrix} p(w_0 + W) \\ -U(\xi) - u_0 \end{pmatrix}.$$

Then the system (1.4), (1.5) takes the form

$$\varepsilon y''(\xi) = \mu f(\xi, y)' - \xi y'(\xi), \tag{1.6}$$

$$y(-L) = 0, \quad y(L) = 0. \tag{1.7}$$

Let $\boldsymbol{v} \in C^1([-L, L]; \mathbb{R}^2)$. Define T to be the solution map that carries $\boldsymbol{v}$ into $\boldsymbol{y}$ where $\boldsymbol{y}$ solves

$$\varepsilon y''(\xi) = f(\xi, v)' - \xi y'(\xi), \tag{1.8}$$

$$y(-L) = 0, \quad y(L) = 0. \tag{1.9}$$

A straightforward computation shows that $y(\xi)$ is given by the formula

$$\begin{aligned} y(\xi) = z \int\limits_{-L}^{\xi} \exp(-\zeta^2/2\varepsilon)\, d\zeta + \frac{1}{\varepsilon} \int\limits_{-L}^{\xi} f(\zeta, v(\zeta))\, d\zeta \\ - \frac{1}{\varepsilon^2} \int\limits_{-L}^{\xi} \int\limits_{0}^{\zeta} \tau f(\tau, v(\tau)) \exp\left(\frac{\tau^2 - \zeta^2}{2\varepsilon}\right) d\tau\, d\zeta \end{aligned} \tag{1.10}$$

where

$$\begin{aligned} z \int\limits_{-L}^{L} \exp(-\zeta^2/2\varepsilon)\, d\zeta = -\frac{1}{\varepsilon} \int\limits_{-L}^{L} f(\zeta, v(\zeta))\, d\zeta \\ + \frac{1}{\varepsilon^2} \int\limits_{-L}^{L} \int\limits_{0}^{\zeta} \tau f(\tau, v(\tau)) \exp\left(\frac{\tau^2 - \zeta^2}{2\varepsilon}\right) d\tau\, d\zeta. \end{aligned} \tag{1.11}$$

Notice the fixed points of μT are solutions of (1.6), (1.7) which in turn yield solutions of (1.1), (1.2), (1.3).

It is clear that T maps $C^0([-L, L]; \mathbb{R}^2)$ continuously into $C^0([-L, L]; \mathbb{R}^2)$. Of course this implies that T maps $C^1([-L, L]; \mathbb{R}^2)$ continuously into $C^0([-L, L]; \mathbb{R}^2)$. We now show T maps $C^1([-L, L]; \mathbb{R}^2)$ continuously into $C^1([-L, L]; \mathbb{R}^2)$.

For this purpose let $v_1, v_2 \in C^1([-L, L]; \mathbb{R}^2)$, $v_1 = (U_1, W_1)$, $v_2 = (U_2, W_2)$, and $y_1 = \mu T v_1$, $y_2 = \mu T v_2$. Differentiation of (1.10) shows

$$\begin{aligned} y_1'(\xi) - y_2'(\xi) &= (z_1 - z_2) \exp(-\xi^2/2\varepsilon) + \frac{f(\xi, {}_1v(\xi))}{\varepsilon} - \frac{f(\xi, v_2(\xi))}{\varepsilon} \\ &\quad - \frac{1}{\varepsilon^2} \int_0^\xi \tau(f(\tau, v_1(\tau)) - f(\tau, v_2(\tau)) \exp\left(\frac{\tau^2 - \xi^2}{2\varepsilon}\right) d\tau, \end{aligned} \tag{1.12}$$

where z_1, z_2 are defined in the obvious manner.

Now let v_1, v_2 be in a finite ball B in $C^1([-L, L;] \mathbb{R}^2)$. In particular for $v = (U, W)$ in B, $w_0 + W$ is uniformly bounded in $\mathbb{R}$ and hence p is a uniformly continuous function of the argument $w_0 + W$. But for $\delta > 0$ arbitrary we know from uniform continuity of p that there is $l(\delta) > 0$ such that $|p(w_0(\xi) + W_1(\xi)) - p(w_0(\xi) + W_2(\xi))| < \delta$ if $|(W_1(\xi) + w_0(\xi)) - (W_2(\xi) + w_0(\xi))| < l(\delta)$, i.e. if $|W_1(\xi) - W_2(\xi)| < l(\delta)$. Hence $\sup_{-L<\xi<L} |p(w_0(\xi) + W_1(\xi)) - p(w_0(\xi) + W_2(\xi))| < \delta$ if $\sup_{-L<\xi<L} |W_1(\xi) - W_2(\xi)| < l(\delta)$ and so $\sup_{-L<\xi<L} |p(w_0(\xi) + W_1(\xi)) - p(w_0(\xi) + W_2(\xi))| \to 0$ as $\sup_{-L<\xi<L} |W_1(\xi) - W_2(\xi)| \to 0$.
But this argument implies by the special nature of $f(\xi, v(\xi))$ that $\sup_{-L<\xi<L} |f(\tau, v_1(\tau)) - f(\tau, v_2(\tau))| \to 0$ as $\sup_{-L<\xi<L} |v_1(\tau) - v_2(\tau)| \to 0$. From (1.11), (1.12) we see then that $\sup_{-L<\xi<L} |y_1'(\xi) - y_2'(\xi)| \to 0$ as $\sup_{-L<\xi<L} |v_1(\xi) - v_2(\xi)| \to 0$, and so T is a continuous map of $C^1([-L, L]; \mathbb{R}^2)$ into itself.

Now note that (1.8) implies that if v is in a bounded set of $C^1([-L, L]; \mathbb{R}^2)$, y will be in a bounded set of $C^2([-L, L]; \mathbb{R}^2)$. This is because $f(\xi, v(\xi))'$ is uniformly bounded.

Hence T is a continuous compact map of $C^1([-L, L]; \mathbb{R}^2)$ into itself.

Now define $\Omega = \{U, W \in C^1([-L, L]; \mathbb{R}^2)$ such that $W(-L) + w_0(-L) < \alpha$, $W(L) + w_0(L) > \beta$; $W'(\xi) + w_0'(\xi) > 0$ if $\alpha \leqq W(\xi) + w_0(\xi) \leqq \beta$; and $\sup_{L<\xi<L} [|U(\xi) + u_0(\xi)| + |W(\xi) + w_0(\xi)| + |U'(\xi) + u_0'(\xi)| + |W'(\xi) + w_0'(\xi)|\}$ $< M + 1\}$. Ω is a *bounded set* in $C^1([-L, L]; \mathbb{R}^2)$.

In addition Ω is open. To see this let $U, W \in \Omega$. Note the definition of Ω implies the set $A \stackrel{\text{def}}{=} \{\xi \in |[-L, L]; \alpha \leqq w_0(\xi) + W(\xi) \leqq \beta\}$ is a closed interval $[\xi_1, \xi_2]$. For if $A \neq [\xi_1, \xi_2]$ for some $\xi_1, \xi_2 \in [-L, L]$ means by the monotonicity of $w_0 + W$ on $[\xi_1, \xi_2]$ that for some $\bar{\xi} \notin [\xi_1, \xi_2]$ either $w_0(\bar{\xi}) + W(\bar{\xi}) = \alpha$ or $w_0(\bar{\xi}) + W(\bar{\xi}) = \beta$, with $w_0'(\bar{\xi}) + W'(\bar{\xi}) \leqq 0$ in either case. Of course this would imply $U, W \notin \Omega$, contradiction.

Thus we have $A = [\xi_1, \xi_2]$, and we set $m = \min_{\xi \in A} (w_0'(\xi) + W'(\xi))$ which is positive. Since $w_0 + W \in C^1[-L, L]$ there is a larger interval $A_\delta \subsetneq [-L, L]$, $A_\delta \supset A$, $A_\delta = [\xi_1 - \delta, \xi_2 + \delta]$ for some small $\delta > 0$, so that $w_0'(\xi) + W'(\xi) \geqq \frac{1}{2} m$ for $\xi \in A_\delta$.

Let $D = \min \min\limits_{\xi_2 - \delta \leqq \xi \leqq L} (w_0(\xi) - W(\xi) - \beta), \min\limits_{-L \leqq \xi \leqq \xi_1 - \delta} (\alpha - w_0(\xi) - W(\xi))$. Since $w_0(\xi) - W(\xi) - \beta > 0$ on $[\xi_2 + \delta, L]$ and $\alpha - w_0(\xi) - W(\xi) > 0$ on $[-L, \xi_1 - \delta]$ we see that $D > 0$.

Now let $\overline{U}$, $\overline{W}$ be such that

$$\sup_{-L \leqq \xi \leqq L} (|\overline{U}(\xi)| + |\overline{U}'(\xi)| + |\overline{W}(\xi)| + |\overline{W}'(\xi)|) < \nu,$$

where $\nu = \min(\frac{1}{2} D, \frac{1}{4} m)$. Consider $\xi \in [-L, L]$ for which $\alpha \leqq w_0(\xi) + W(\xi) + \overline{W}(\xi) \leqq \beta$. If we can show that $w_0'(\xi) + \overline{W}'(\xi) + W'(\xi) > 0$, we will have proven Ω is open. But we see in this case that

$$w_0(\xi) + W(\xi) - \beta \leqq -\overline{W}(\xi), \qquad \alpha - w_0(\xi) - W(\xi) \leqq \overline{W}(\xi),$$

and hence $w_0(\xi) + W(\xi) - \beta \leqq D/2$, $\alpha - w_0(\xi) - W(\xi) \leqq D/2$. But this implies by the definition of D that $\xi \in (\xi_1 - \delta, \xi_2 + \delta)$. Thus we have shown that $\alpha \leqq w_0(\xi) + W(\xi) + \overline{W}(\xi) \leqq \beta$ implies $\xi \in A_\delta$. Now we compute at this value ξ:

$$w_0'(\xi) + W'(\xi) + \overline{W}'(\xi) \geqq \tfrac{1}{2} m + \overline{W}'(\xi) \geqq \tfrac{1}{4} m > 0.$$

Hence Ω is open.

Now we recall a well known theorem of Leray-Schauder type (see for example Mawhin [10], Corollary IV.7).

Proposition 1.2. *Let X be a real normed vector space, Ω an open bounded subset of X, and T a compact map of X into itself. If zero is an interior point of Ω and $\phi \neq \mu T \phi$ for all $\phi \in \partial\Omega$, $0 < \mu < 1$, then T has at least one fixed point in $\overline{\Omega}$.*

In our problem we take $X = C^1([-L, L]; \mathbb{R}^2)$ and T, Ω is as defined above. The origin is an interior point of Ω since the constraint $w_0'(\xi) + \overline{W}'(\xi) > 0$ is satisfied for all $\xi \in [-L, L]$ if $(\overline{U}, \overline{W})$ is a small $C^1([-L, L]; \mathbb{R}^2)$ perturbation. Note $\phi \in \partial\Omega$, $\phi = \mu T \phi$, $\mu \in (0, 1)$, means that there is a solution $(u(\xi), w(\xi))$ of (1.1), (1.2), (1.3) which satisfies $w'(\xi) \geqq 0$ if $\alpha \leqq w(\beta) \leqq \beta$ and either

(i) $w'(\xi_0) = 0$, $\alpha \leqq w(\xi_0) \leqq \beta$ for some $\xi_0 \in (-L, L)$

or

(ii) $\sup\limits_{-L < \xi < L} \{|u(\xi)| + |w(\xi)| + |u'(\xi)| + |w'(\xi)|\} = M_0 + 1$ or both (i) and (ii).

Let us first consider possibility (i). In this case either $\alpha < w(\xi_0) < \beta$, $w(\xi_0) = \alpha$, or $w(\xi_0) = \beta$. We consider these cases separately.

Case 1. $\alpha < w(\xi_0) < \beta$, $w'(\xi_0) = 0$. In this case there are three possibilities, either $w''(\xi_0) < 0$, $w''(\xi_0) > 0$, or $w''(\xi_0) = 0$. If $w''(\xi_0) < 0$ then $w(\xi_0)$ is a local maximum which implies $w'(\xi) < 0$ for some $\xi < \xi_0$, $|\xi - \xi_0|$ small. But this implies $\alpha < w(\xi) < \beta$ and violates the requirement that $w'(\xi) \geqq 0$.

An analogous statement holds if $w''(\xi_0) < 0$ and now $w(\xi_0)$ is a local minimum. The case $w''(\xi_0) = 0$ is excluded since $w''(\xi_0) = 0$, $w'(\xi_0) = 0$ implies because of (1.2) that $u'(\xi_0) = 0$. But in this case uniqueness of solutions for (1.1), (1.2) as an initial value problem (see [7], Lemma 4.1) $u'(\xi_0) = 0$, $w'(\xi_0) = 0$ implies $u(\xi) = u(\xi_0)$, $w(\xi) = w(\xi_0)$ for all $\xi \in [-L, L]$ and hence we cannot satisfy (1.3), $w_- < \alpha$, $w_+ > \beta$.

Case 2. $w(\xi_0) = \alpha$, $w'(\xi_0) = 0$. In this case there are again the three canonical possibilities, $w''(\xi_0) < 0$, $w''(\xi_0) > 0$, or $w''(\xi_0) = 0$. We can immediately dismiss $w''(\xi_0) > 0$ and $w''(\xi_0) = 0$ for the same reasons as in Case 1. So we need only consider $w''(\xi_0) < 0$. In this case $w(\xi_0) = \alpha$ is a local maximum. Hence if we are to satisfy $w(L) = w_+ > \beta$ we must proceed through a local minimum at $\xi_1 > \xi_0$, i.e. $w(\xi_1) < \alpha$, $w'(\xi_1) = 0$, $w''(\xi_1) \geqq 0$; $w(\xi) < \alpha$, $w'(\xi) < 0$, $\xi_0 < \xi \leqq \xi_1$. Again $w''(\xi_1) = 0$ is impossible since that forces $u'(\xi_1) = 0$ and the uniqueness theorem ([7], Lemma 4.1) is contradicted. Thus we need only consider $w''(\xi_1) > 0$. From (1.2) we see $u'(\xi_1) > 0$, $u'(\xi_0) < 0$ which implies u has a local maximum at a point $\xi_0 < \zeta < \xi_1$, $u'(\zeta) = 0$, $u''(\zeta) \leqq 0$, and again Lemma 4.1 of [7] tells us $u''(\xi) < 0$. Since $p'(w) < 0$ for $w < \alpha$ this implies by use of (1.1) that $w'(\zeta) > 0$ which contradicts the fact that w is decreasing on (ξ_0, ξ_1). Hence $w''(\xi_0) < 0$ is excluded as well.

Case 3. $w(\xi_0) = \beta$, $w'(\xi_0) = 0$. Here again we see we can exclude $w''(\xi_0) < 0$ and $w''(\xi_0) = 0$ immediately. If $w''(\xi_0) > 0$ it follows that $w(\xi_0) = \beta$ is a local minimum so to satisfy $w(-L) = w_- < \alpha$ there must be $\xi_1 < \xi_0$ where $w(\xi_1) > \beta$ and w has a local maximum, $w(\xi) > \beta$ on (ξ_1, ξ_0). But the same reasoning as in Case 2 yields a contradiction.

From Cases 1, 2, 3 of (i) we see there is no solution of (1.1), (1.2), (1.3), $\mu \in (0, 1)$, $(u(\xi) - u_0(\xi), w(\xi) - w_0(\xi))$ in Ω for which (i) can hold. Thus all solutions of (1.1), (1.2), (1.3), $\mu \in (0, 1)$ in $\bar{\Omega}$ must satisfy $w'(\xi) > 0$ in $\alpha \leqq w(\xi) \leqq \beta$. But now the hypothesis of our theorem says (ii) cannot hold either. Thus we conclude from Prop. 1.2 that (1.1), (1.2), (1.3) possesses a solution for which $u(\xi) - u_0(\xi)$, $w(\xi) - w_0(\xi)$ is in $\bar{\Omega}$.

To complete the proof we follow DAFERMOS [7] and extend the domains of u, w: Set

$$u(\xi; L) = u_+, \quad w(\xi; L) = w_+, \quad \xi > L,$$

$$u(\xi; L) = u_-, \quad w(\xi; L) = w_-, \quad \xi < -L.$$

The extended pair $\{u(\cdot, L), w(\cdot, L)\}$ form a sequence in $C^0((-\infty, \infty); \mathbb{R}^2)$ and by virtue of the hypothesis of theorem we know $\sup_{-L<\xi<L} \{|u'(\xi; L)| + |w'(\xi; L)|\} \leqq M$. Thus the sequence $\{(u(\cdot; L), w(\cdot; L)\}$ is precompact in $C^0((-\infty, \infty); \mathbb{R}^2))$ and so there is a subsequence $L_n \to \infty$ as $n \to \infty$ since that $(u(\xi; L), w(\xi; L)) \to (u(\xi), w(\xi))$ uniformly as $n \to \infty$ on $(-\infty, \infty)$. As in DAFERMOS [7] $u(\xi)$, $w(\xi)$ is a solution of P_ε and by its construction $w'(\xi) \geqq 0$ if $\alpha \leqq w(\xi) \leqq \beta$. But by the same reasoning used in Cases 2, 3, this connecting orbit must satisfy the more

restrictive requirement $w'(\xi) > 0$ if $\alpha \leqq w(\xi) \leqq \beta$. This completes the proof of Theorem 1.1. □

Having established Theorem 1.1 we can now proceed to weaken the hypothesis that the first derivatives of u, w be *a priori* bounded. This is done below.

Theorem 1.3. *The conclusion of Theorem 1.1 remains valid if* (H0) *is replaced by the* a priori *estimate*

$$\sup_{-L<\xi<L} (|u(\xi)| + |w(\xi)|) \leqq M_1 \tag{H1}$$

where again M_1 can depend on u_-, u_+, w_-, w_+, ε, p but is independent of μ and L.

Proof. All that we need to show is that if a solution of (1.1), (1.2), (1.3) satisfies (H1) it satisfies (H0). But this is precisely the nature of estimates (3.7), (3.8) given by Dafermos in [7]. For completeness we rederive these estimates.

Let $y(\xi) = \begin{pmatrix} u(\xi) \\ w(\xi) \end{pmatrix}$, $f(y) = \begin{pmatrix} p(w) \\ -u \end{pmatrix}$. Then (1.1), (1.2), (1.3) have the form

$$\varepsilon y''(\xi) = \mu f(y(\xi))' - \xi y'(\xi),$$

$$y(-L) = y_-, \quad y(L) = y_+,$$

where $y_- = (u_-, w_-)$, $y_+ = (u_+, w_+)$.

We see easily that

$$\frac{d}{d\xi}(\exp(\xi^2/2\varepsilon)\, y'(\xi)) = \frac{\mu}{\varepsilon}\left(\nabla f(y)\, y'(\xi) \exp\left(\frac{\xi^2}{2\varepsilon}\right)\right)$$

and hence by integrating from 0 to ξ we have

$$\exp(\varepsilon^2/2\varepsilon)\, y'(\xi) - y'(0) = \frac{\mu}{\varepsilon}\int_0^\xi (\nabla f(y)\, y'(\zeta) \exp(\zeta^2/2\varepsilon))\, d\zeta.$$

Since $\sup_{-L<\xi<L} (|u(\xi)|) + |w(\xi)| \leqq M_1$ we know that $\mu\,|\nabla f(y)| \leqq R$, R independent of L. Thus

$$|\exp(\xi^2/2\varepsilon)\, y'(\xi)| \leqq |y'(0)| + \frac{R}{\varepsilon}\int_0^\xi |y'(\zeta)| \exp\left(\frac{\zeta^2}{2\varepsilon}\right) d\zeta,$$

and using Gronwell's inequality we find

$$|y'(\xi)| \leqq |y'(0)| \exp\left(\frac{2R\,|\xi| - \xi^2}{2\varepsilon}\right).$$

But the function $\left(\frac{2R\,|\xi| - \xi^2}{2\varepsilon}\right)$ has a maximum value of $\left(\frac{R^2}{\varepsilon}\right)$ so we see

$$|y'(\xi)| \leqq |y'(0)| \exp\left(\frac{R^2}{\varepsilon}\right), \quad -L \leqq \xi \leqq L,$$

where R is independent of L. Thus $\sup_{L<\xi<L} |y'(\xi)|$ will be bounded independent of μ and L if $|y'(0)|$ is bounded independently of μ and L.

Now we derive the bound on $|y'(0)|$. First note that

$$y'(\xi) = z \exp(-\xi^2/2\varepsilon) + \frac{\mu}{\varepsilon} f(y) - \frac{\mu}{\varepsilon^2} \int_0^\xi \tau f(y(\tau)) \exp\left(\frac{\tau^2 - \xi^2}{2\varepsilon}\right) d\tau,$$

where

$$z \int_{-L}^{L} \exp(-\zeta^2/2\varepsilon) = y(L) - y(-L) - \frac{\mu}{\varepsilon} \int_{-L}^{L} f(y(\tau))\, d\tau$$
$$+ \frac{\mu}{\varepsilon^2} \int_{-L}^{L} \int_0^\zeta \tau f(y(\tau)) \exp\left(\frac{\tau^2 - \zeta^2}{2\varepsilon}\right) d\tau\, d\zeta.$$

Now set $\xi = 0$, $L = 1$ in the above expressions. Then

$$y'(0) = z + \frac{\mu}{\varepsilon} f(y(0)),$$

$$z \int_{-1}^{1} \exp(-\zeta^2/2\varepsilon)\, d\zeta = y(1) - y(-1) - \frac{\mu}{\varepsilon} \int_{-1}^{1} f(y(\tau))\, d\tau$$
$$+ \frac{\mu}{\varepsilon} \int_{-1}^{1} \int_0^\zeta \tau f(y(\tau)) \exp\left(\frac{\tau^2 - \zeta^2}{2\varepsilon}\right) d\tau\, d\zeta.$$

Since $\mu\, |f(y(\tau))| \leqq$ constant (independent of μ and L) we see that $|y'(0)|$ is bounded independently only of μ and L. This completes the proof. □

Theorem 1.3 gives a sufficient condition for P_ε to be solvable when the boundary values w_- and w_+ are in different phases. We now give a result which applies to the case when w_-, w_+ are in the same phase.

Theorem 1.4. *Assume* $u_- > u_+$ *and* $w_-, w_+ < \alpha$ ($u_- < u_+$ *and* $w_-, w_+ > \beta$) *and there is a constant* M_2 *such that every possible solution of* (1.1), (1.2), (1.3) *satisfies the* a priori *estimate*

$$\sup_{-L<\xi<L} (|u(\xi)| + |w(\xi)|) \leqq M_2. \tag{H2}$$

Here M_2 *can depend on* u_-, w_+, ε, p *but is independent of* μ *and* L. *Then there exist solutions of* P_ε *which satisfy the constraints* $w(\xi) < \alpha$ $(w(\xi) > \beta)$, i.e. *these special single phase data connecting orbit problems possesses single phase solutions.*

Proof. Theorem 1.4 is a special case of Theorem 3.1 of [7].

2. The *a priori* estimates

In this section we derive the *a priori* estimates needed to apply Theorems 1.3 and 1.4. Before doing this we give some useful lemmas. The first is a result from [7] Theorem 4.1 or [8] Theorem 2.2.

Lemma 2.1. *Let $(u(\xi), w(\xi))$ be a solution of(1.1), (1.2) on an interval $[-L, L]$, $\mu > 0$. Then on any subinterval (l_1, l_2) for which $p'(w(\xi)) < 0$ one the following holds:*
(i) *$u(\xi)$ and $w(\xi)$ are constant on (l_1, l_2);*
(ii) *$u(\xi)$ is a strictly increasing (or decreasing) function with no critical points in $l_1, l_2)$; $w(\xi)$ has, at most, one critical point in (l_1, l_2) that necessarily must be a maximum (or minimum);*
(iii) *$w(\xi)$ is a strictly increasing (or decreasing) function with no critical point in (l_1, l_2); $u(\xi)$ has, at most, one critical point in (l_1, l_2), that necessarily must be a maximum (or minimum).*

We will also need the following results.

Lemma 2.2. *Let $(u(\xi), w(\xi))$ be a solution of (1.1), (1.2) on an interval $[-L, L]$, $\mu > 0$. Then on any subinterval (l_1, l_2) for which $p'(w(\xi)) > 0$ the graph of $u(\xi)$ versus $w(\xi)$ is convex at points where $w'(\xi) > 0$ and concave at points where $w'(\xi) > 0$.*

Proof. We simply compute $\frac{d^2u}{dw^2}$ as follows:

$$\frac{du}{dw} = \frac{u'(\xi)}{w'(\xi)}; \quad \frac{d}{d\xi}\left(\frac{u'(\xi)}{w'(\xi)}\right) = \left(\frac{u''(\xi)\, w'(\xi) - u'(\xi)\, w''(\xi)}{w'(\xi)^2}\right).$$

Now we use (1.1), (1.2) to see that at $u(\xi)$, $w(\xi)$

$$\varepsilon \frac{d^2u}{dw^2} = \frac{\mu(p'(w(\xi))\, w'(\xi)^2 + u'(\xi)^2)}{w'(\xi)^3} = \mu\left(p'(w) + \left(\frac{du}{dw}\right)^2\right)\frac{1}{w'(\xi)}$$

which proves the result.

Lemma 2.3. *Let $u(\xi)$, $w(\xi)$ be a solution of (1.1), (1.2), (1.3) on an interval $[-L, L]$, $\mu > 0$ with $w'(\xi) > 0$ if $\alpha \leqq w(\xi) \leqq \beta$. Then u, w can have no local maxima or minima at points ξ for which $w(\xi) = \alpha$ or $w(\xi) = \beta$.*

Proof. Since $w'(\xi) > 0$ if $\alpha \leqq w(\xi) \leqq \beta$ certainly w has no local maxima or minima at points where $w(\xi) = \alpha$. On the other hand if $u(\xi)$ has a local maximum or minimum at such a point then $u'(\xi) = 0$ there and hence by (1.1) $u''(\xi) = 0$ as well. Differentiating (1.1) with respect to ξ, we see $p''(\alpha) > 0$ and $p''(\beta) < 0$ implies $u'''(\xi) \neq 0$ at such points so u could not have taken on a local maximum or minimum.

We can use Lemmas 2.1, 2.2, 2.3 to prove the following useful statement regarding possible connecting orbits in the two phase data case. (Notice that, extrema of u are denoted by τ's, extrema of w are denoted by σ's.)

Lemma 2.4. *Assume $w_- < \alpha$, $w_+ > \beta$ and let $u(\xi)$, $w(\xi)$ be a possible solution of (1.1), (1.2), (1.3) with $\mu > 0$ for which $w'(\xi) > 0$ when $\alpha \leqq w(\beta) \leqq \beta$. Then one of the following holds:*
(0) *No extremal points: $u(\xi)$, $w(\xi)$ have no local maxima or minima on $[-L, L]$. They are non-constant and monotone, w being monotone increasing.*

(i) *One extremal point*: (a) $w(\xi)$ *has a minimum at some* σ_-, $w(\sigma_-) < w_-$; $u(\xi)$ *is decreasing on* $[-L, L]$.
(b) $w(\xi)$ *has a maximum at some* σ_+, $w(\sigma_+) > w_+$; $u(\xi)$ *is increasing on* $[-L, L]$.
(c) $u(\xi)$ *has a maximum at some* τ_- (*or* τ_+); $w(\tau_-) < \alpha$ (*or* $w(\tau_+) > \beta$) *and* $w(\xi)$ *is increasing on* $[-L, L]$.
(d) $u(\xi)$ *has a minimum at some* τ; $\alpha < w(\tau) < \beta$ *and* $w(\xi)$ *is increasing on* $[-L, L]$.

(ii) *Two extremal points*: (a) $u(\xi)$ *has a local maximum at* τ_- (*or* τ_+) *and a local minimum at* τ; $w(\xi)$ *is increasing on* $[-L, L]$ *and* $w_- < w(\tau_-) < \alpha$ *or* $w_+ > w(\tau_+) > \beta$, $\alpha < w(\tau) < \beta$.
(b) $w(\xi)$ *has a minimum at* σ_-, $w(\sigma_-) < w$; $u(\xi)$ *has a local minimum at* τ, $\tau > \sigma_-$. $\alpha < w(\tau) < \beta$.
(c) $w(\xi)$ *has a maximum at* σ_+, $w(\sigma_+) > w_+$; $u(\xi)$ *has a local minimum at* τ, $\tau < \sigma_+, \alpha < w(\tau) < \beta$.

(iii) *Three extremal points*: (a) $u(\xi)$ *has local maxima at* τ_-, τ_+, *and a local minima at* τ, $\tau_- < \tau < \tau_+$; $w(\zeta)$ *is increasing with* $w_- < w(\tau_-) < \alpha$, $\alpha < w(\tau) < \beta$, $\beta < w(\tau_+) < w_+$.
(b) $w(\xi)$ *has a minimum at* σ_-, $w(\sigma_-) < w_-$ *and a maximum at* σ_+, $w(\sigma_+) > w_+$ *and* $u(\xi)$ *has a local minimum at* τ, $\sigma_- < \tau < \sigma_+$, $\alpha < w(\tau) < \beta$.
(c) $w(\xi)$ *has a minimum at* σ_-, $w(\sigma_-) < w_-$, $u(\xi)$ *has a local minimum at* τ, $\alpha < w(\tau) < \beta$ *and a local maximum at* τ_+, $\tau < w(\tau_+) < w_+$, $\sigma_- < \tau < \tau_+$.
(d) $w(\xi)$ *has a maximum at* σ_+, $w(\sigma_+) > w_+$, $u(\xi)$ *has a local maximum at* τ_-, $w_- < w(\tau_-) < \alpha$, *and a local minimum at* τ, $\alpha < w(\tau) < \beta$.

Proof.

(0) *No extremal points*: The non-constancy follows from Lemma 4.1 of [7] and $w_- \neq w_+$.

(i) *One extremal point*: Either $u(\xi)$ or $w(\xi)$ is monotone. (a) If $u(\xi)$ is decreasing then $w(\xi)$ can have either a maximum or minimum. But Lemma 2.1(ii) says it must be a minimum.
(b) If $u(\xi)$ is increasing the same reasoning as in (a) says $w(\xi)$ can possess only a maximum.
(c, d) On the other hand if $w(\xi)$ is monotone it must be monotone increasing since $w_- < w_+$. By Lemmas 2.1(iii), 2.2, 2.3 we see the only possibilities are a maximum for u at τ_- (or τ_+) with $w(\tau_-) < \alpha$ or $w(\tau_+) > \beta$ or a minimum at τ with $\alpha < w(\tau) < \beta$.

(ii) *Two extremal points*: First consider the case of one local maxima and one local minima for $u(\xi)$. (a) Since $w(\xi)$ must be monotone increasing Lemmas 2.1 (iii) and 2.2 say the local maximum occurs where $w < \alpha$ or $w > \beta$ and the local minimum occurs where $\alpha < w < \beta$.
One local maxima and one local minima for $w(\xi)$ is impossible with $u(\xi)$ montone. For if σ_1, σ_2 are such that w has a local maximum at σ_1 and a local minimum at σ_2 we must have $w'(\sigma_1) = 0$, $w''(\sigma_1) \leqq 0$, $w'(\sigma_2) = 0$, $w''(\sigma_2) \geqq 0$. Then (1.2) implies $u'(\sigma_1) \geqq 0$, $u'(\sigma_2) \leqq 0$. Thus monotonicity of u would yield u a constant. Lemma 4.1 of [7] would then give w a constant as well which contradicts $w_- \neq w_+$.
(b) If $w(\xi)$ has a minimum at σ_- and $u(\xi)$ has a local minimum at τ, then

certainly $w(\sigma_-) < w_-$. That means $\tau > \sigma_-$ and either $\alpha < w(\tau) < \beta$ or $w(\tau) > \beta$. But as $w'(\tau) > 0$ Lemma 2.1(iii) says $w(\tau) > \beta$ is impossible.
(c) If $w(\xi)$ has a maximum at σ_+ and $u(\xi)$ has a local minimum at τ, reasoning analogous to (b) above applies.
If $w(\xi)$ has a minimum at σ_-, $w(\sigma_-) < w_-$, u cannot have a maximum on $[-L, L]$. This is because such a maximum must occur at τ_1, $\tau_1 > \sigma_-$ implying $u'(\sigma_-) > 0$. This contradicts Lemma 2.1(ii). Similarly, if $w(\xi)$ has a maximum at σ_+, $w(\sigma_+) > w_+$, u cannot have a maximum on $[-L, L]$. Again this is because such a maximum occurs at τ_1, $\tau_1 < \sigma_+$ implying $u'(\sigma) < 0$ contradicting Lemma 2.1(ii).

(iii) *Three extremal points*: (a) First w cannot have three extreme points by Lemma 2.1(ii) but u can. By Lemmas 2.1(ii) and 2.2 we see they must go sequentially as a local maximum, local minimum, local maximum.
(b) If w has two extreme points one must be a minimum at σ_-, $w(\sigma_-) < w_-$ and the other a maximum at σ_+, $w(\sigma_+) > w_+$. Lemmas 2.1, 2.2 imply that the only possible extremal point for u is a minimum at τ, $\sigma_- < \tau < \tau_+$, $\alpha < w(\tau) < \beta$. If u has two extreme points and w has one then either
(c) w has a minimum at σ_-, $w(\sigma_-) < w_-$ or
(d) a maximum at σ_+, $w(\sigma_+) > w_+$. In (c) $w'(\xi) > 0$ for $\xi > \sigma_-$ so Lemma 2.1(iii) says u must have a local maximum at τ_+, $\beta < w(\tau_+) < w_+$ and a local minimum at $\alpha < w(\tau) < \beta$, $\sigma_- < \tau < \sigma_+$. In (d) $w'(\xi) > 0$, $\xi < \sigma_+$ so again Lemma 2.1(iii) says u must have a local maximum at τ_-, $w_- < w(\tau_-) < \alpha$ and a local minimum at τ, $\alpha < w(\tau) < \beta$, $\tau_- < \tau < \sigma_+$.
This completes all possible cases since extremal points at $w = \alpha$ or $w = \beta$ are excluded by Lemma 2.3.

Below are illustrated sketches in the u–w plane of the possible cases described in Lemma 2.4.

Theorem 2.5. *Assume* $w_- < \alpha$, $w_+ > \beta (w_- > \beta, w_+ < \alpha)$. *Then there is a constant* M_1 *such that every possible solution of* (1.1), (1.2), (1.3), $0 \leqq \mu \leqq 1$, *with* $w'(\xi) > 0 (w'(\xi) < 0)$ *when* $\alpha \leqq w(\xi) \leqq \beta$ *satisfies the estimate*

$$\sup_{-L<\xi<L} (|u(\xi)| + |w(\xi)|) \leqq M_1. \tag{H1}$$

M_1 *depends at most on* u_-, u_-, w_+, w_+, ε, p *and is independent of* μ *and* L.

Theorem 2.6. *Assume* $u_+ < u_-$ *and* w_-, $w_+ < \alpha (u_- < u_+$ *and* w_-, $w_+ > \beta)$. *Then there is a constant* M_2 *such that every possible solution of* (1.1), (1.2), (1.3), $0 \leqq \mu \leqq 1$, *satisfies the* a priori *estimate*

$$\sup_{-L<\xi<L} (|u(\xi)| + |w(\xi)|) \leqq M_2. \tag{H2}$$

M_2 *depends at most on* u_-, u_+, w_-, w_+, ε, p *and is independent of* μ *and* L.

Corollary 2.7. *If* $w_- < \alpha$, $w_+ > \beta (w_- > \beta, w_+ < \alpha)$ *there are solutions of* P_ε *which satisfy the constraints* $w'(\xi) > 0 (w'(\xi) < 0)$ *when* $\alpha \leqq w(\xi) \leqq \beta$, i.e.

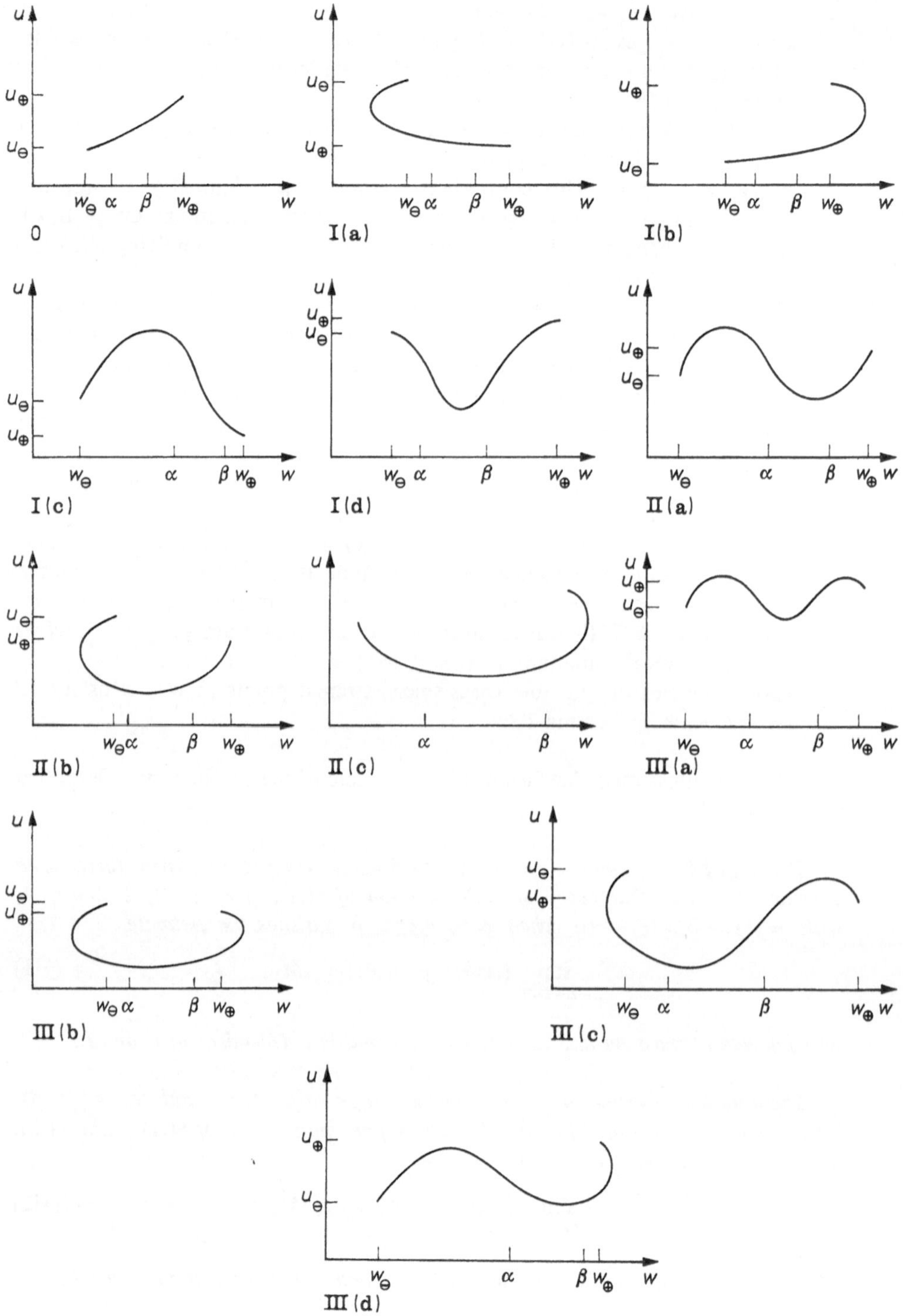

Fig. 2

the one phase change data connecting orbit problem possesses a one phase change solution.

If $u_+ < u$ *and* $w_-, w_+ < \alpha (u_- < u_+$ *and* $w_-, w_+ > \beta)$ *there are solutions of* P_ε *which satisfy the constraints* $w(\xi) < \alpha (w(\xi) > \beta)$, i.e. *the single phase data connecting orbit problem possesses single phase solutions.*

Corollary 2.7, our main result, follows directly from Theorems 1.3, 1.4, 2.5, 2.6, and we pass on to verifying Theorems 2.5, 2.6.

Proof of Theorem 2.5. We consider the cases listed in Lemma 2.4 for the choice $w_- < \alpha$, $w_+ > \beta$. The proof for $w_- > \beta$, $w_+ < \alpha$ is similar and is omitted.

(0) No extremal points certainly implies the assertion of the theorem.

(ia) Since u is decreasing we have $u_+ \leqq u(\xi) \leqq u_-$. Since w has a minimum at σ_- we need only bound w from below. To do this we follow the method given in [7], Theorem 4.2.

Assume $\sigma_- \geqq 0$ (similar arguments hold if $\sigma_- < 0$). Integrate (1.2) from σ_- to L and use $w'(\sigma_-) = 0$. Then

$$\varepsilon w'(L) + \int_{\sigma_-}^{L} \xi w'(\xi)\, d\xi = -\mu u(L) + \mu u(\sigma_-).$$

Since $w'(L) > 0$, we have

$$\int_{\sigma_-}^{L} \xi w'(\xi)\, d\xi \leqq -\mu u_+ + \mu u(\sigma_-). \tag{2.1}$$

If $\zeta \geqq \max\{1, \sigma_-\}$, then $w'(\xi) \leqq \xi w'(\xi)$ on (ζ, L), so that (2.1) implies

$$w(L) - w(\zeta) = \int_{\zeta}^{L} w'(\xi)\, d\xi \leqq \int_{\zeta}^{L} \xi w'(\xi)\, d\xi \leqq \int_{\sigma_-}^{L} \xi w'(\xi)\, d\xi \leqq -\mu u_+ + \mu u(\sigma_-),$$

and hence

$$w(\zeta) \geqq w_+ + \mu u_+ - \mu u(\sigma_-). \tag{2.2}$$

Since $u_+ \leqq u(\sigma_-) \leqq u_-$, $0 \leqq \mu \leqq 1$, we see that $w(\xi)$ is bounded below independently of μ and L if $1 \leqq \sigma_-$. If $0 \leqq \sigma_- < 1$, integrate (1.2) from σ_- to ξ where $\sigma_- < \xi < 1$. Then we see that

$$\varepsilon w'(\xi) + \int_{\sigma_-}^{\xi} \zeta w'(\zeta)\, d\zeta = -\mu u(\xi) + \mu u(\sigma_-). \tag{2.3}$$

Since $w'(\xi) > 0$ on (σ_-, L), we see that $\zeta w'(\zeta) > 0$ on (σ_-, ξ) and hence

$$\varepsilon w'(\xi) \leqq -\mu u(\xi) + \mu u(\sigma_-), \quad \sigma_- < \xi < 1. \tag{2.4}$$

Integrate (2.4) from σ_- to 1. We then see that

$$\varepsilon w(1) - \varepsilon w(\sigma_-) \leqq \mu \int_{\sigma_-}^{1} (u(\sigma_-) - u(\xi))\, d\xi$$

or

$$\varepsilon w(1) - \mu \int_0^1 (u(\sigma_-) - u(\xi))\, d\xi \leqq \varepsilon w(\sigma_-). \tag{2.5}$$

Since $u_+ \leqq u(\xi) \leqq u_-$ and $w(1)$ is bounded from below by (2.2), (2.5) provides the desired bound from below for $w(\sigma_-)$ when $0 \leqq \sigma_- < 1$.

Cases i(b), i(c) are proven similarly and in fact i(a–c) fall into the cases treated in Theorem 4.2 of [7]. Case i(d) was not possible in [7] because of the assumption of hyperbolicity. Nevertheless the above method still works as we show below.

In case i(d) $w(\xi)$ is increasing so $w_- \leqq w(\xi) \leqq w_+$. Again assume $\tau \geqq 0$ as the case $\tau < 0$ is similar. First integrate (1.1) from τ to L. This yields

$$\varepsilon u'(L) + \int_\tau^L \xi u'(\xi)\, d\xi = \mu p(w_+) - \mu p(w(\tau)). \tag{2.6}$$

Since $u'(L) > 0$ this implies

$$\int_\tau^L \xi u'(\xi)\, d\xi \leqq \mu p(w_+) - \mu p(w(\tau)). \tag{2.7}$$

If $\zeta \geqq \max\{1, \tau\}$, since $u'(\xi) > 0$ on (ζ, L) we find $u'(\xi) \leqq \xi u'(\xi)$ on (τ, L) and

$$u_+ - u(\zeta) = \int_\zeta^L u'(\xi)\, d\xi \leqq \int_\zeta^L \xi u'(\xi)\, d\xi \leqq \int_\tau^L \xi u'(\xi)\, d\xi \leqq \mu p(w_+) - \mu p(w(\tau)),$$

$$u(\zeta) \geqq u_+ - \mu p(w_+) + \mu p(w(\tau)). \tag{2.8}$$

Since $\alpha < w(\tau) < \beta$, we see for $1 \leqq \tau$, $u(\tau)$ is bounded from below independently of μ and L. Again if $0 \leqq \tau < 1$ integrate (1.1) from τ to ξ where $\tau < \xi < 1$. Then we see that

$$\varepsilon u'(\xi) + \int_\tau^\xi \zeta u'(\zeta)\, d\xi = \mu p(w(\xi)) - \mu p(w(\tau)).$$

Since $\zeta u'(\zeta) > 0$ on (τ, ξ), we find

$$\varepsilon u'(\xi) \leqq \mu p(w(\xi)) - \mu p(w(\tau)),$$

and integrating from τ to 1 we see that

$$\varepsilon u(1) - \varepsilon u(\tau) \leqq \mu \int_\tau^1 p(w(\xi)) - p(w(\tau))\, d\xi$$

and

$$\varepsilon u(1) - \mu \int_\tau^1 p(w(\xi)) - p(w(\tau))\, d\xi \leqq \varepsilon u(\tau). \tag{2.9}$$

We know $\max(u_-, u_+) \geqq u(\xi)$ and so u is bounded from above. Since $w(\xi)$ is bounded (2.8), (2.9) implies $u(\xi)$ is bounded from below on $[-L, L]$ independently of μ and L.

ii(a) Assume u has a local maximum at τ_-, $w(\tau_-) < \alpha$. (The case $w(\tau_+) > \beta$ is proved in a similar fashion.) Then the local minimum is at τ, $\tau_- < \tau$, $\alpha < w(\tau) < \beta$. For w we know $w_- \leqq w(\xi) \leqq w_+$. Trivially there are two possibilities we must consider:

$\tau \geqq 0$. In this case proceed exactly as in the proof of i(d) above and we find $u(\xi)$ bounded from below and certainly from above by u_+.

$\tau < 0$. If $\tau < 0$ then $\tau_- < 0$. We will show $u(\tau_-)$ is bounded from above. To do this consider the first case $\tau_- \leqq -1$ and then the case $-1 \leqq \tau_- \leqq 0$ in a manner similar to i(d). This proves $u(\tau_-)$ will be bounded from above while it is certainly bounded from below by u_-.

Thus we find either $u(\tau)$ is bounded from below or $u(\tau_-)$ is bounded from above where the bounds are independent of μ and L. In the first case we use i(c) on $-L \leqq \xi \leqq \tau$ to bound $u(\tau_-)$ from above; in the second case we use i(d) on $\tau_- \leqq \xi \leqq L$ to bound $u(\tau)$ from below. Again the bounds will be independent of μ and L.

ii(b) If $\tau \geqq 0$, we follow the argument of i(d) to note that $u(\tau)$ is bounded from below. Here we use the fact that $\alpha \leqq w(\xi) \leqq w_+$ for $\tau \leqq \xi \leqq L$. Since $u(\tau)$ is bounded from above by $\max(u_-, u_+)$ we know $u(\xi)$ is bounded from above and below. Finally we used result i(a) on $[-L, \tau]$ to see that w is also bounded from below at $\sigma_- \in (-L, \tau)$.

If $\tau < 0$ then using argument of i(d) again we find

$$u(\zeta) \geqq u_- - \mu p(w_-) + \mu p(w(\tau)) \tag{2.10}$$

if $\zeta \leqq \min\{-1, \tau\}$. But $\alpha < w(\tau) < \beta$ so (2.10) shows $u(\tau)$ is bounded from below if $\tau \leqq -1$. If $-1 < \tau \leqq 0$ then argument i(d) can be used again. We give the argument for completeness. First integrate (1.1) from τ to ξ where $\xi \in (-1, \tau)$. This yields

$$\varepsilon u'(\xi) + \int_\tau^\xi \zeta u'(\zeta)\, d\zeta = \mu p(w(\xi)) - \mu p(w(\tau)). \tag{2.11}$$

On (ξ, τ), $\zeta u'(\zeta) > 0$ so the integral in (2.11) is negative. Thus we see

$$\varepsilon u'(\xi) \geqq \mu(p(w(\xi)) - p(w(\tau))). \tag{2.12}$$

Now integrate (2.12) from -1 to τ. This implies that

$$\varepsilon u(\tau) \geqq \varepsilon u(-1) + \mu \int_{-1}^{\tau} p(w(\xi)) - p(w(\tau))\, d\xi. \tag{2.13}$$

Now $w(\xi) \leqq w(\tau)$ on $(-1, \tau)$ since $\alpha < w(\tau) < \beta$. Inspection of the graph of p (see Figure 3) shows that

$$p(w(\xi)) - p(w(\tau)) \geqq p(\alpha) - p(\beta). \tag{2.14}$$

(Notice $p(w(\xi)) - p(w(\tau))$ becomes positive if $w(\xi)$ decreases below γ.) Inserting (2.14) into (2.13) we find

$$\varepsilon u(\tau) \geqq \varepsilon u(-1) + \mu(\tau + 1)(p(\alpha) - p(\beta))$$

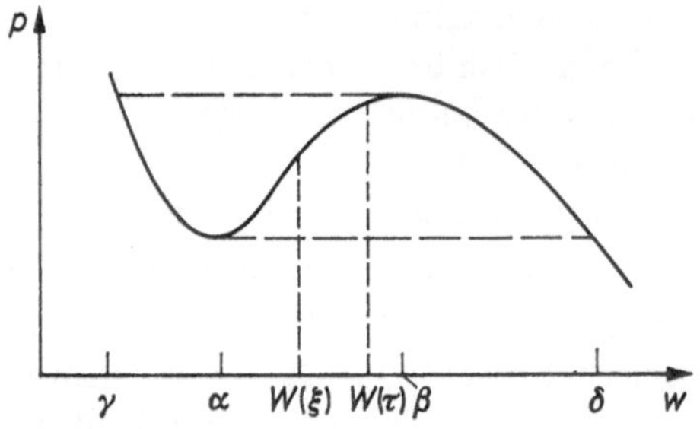

Fig. 3

and hence

$$\varepsilon u(\tau) \geqq \varepsilon u(-1) + \mu(p(\alpha) - p(\beta)).$$

Thus $u(\tau)$ is bounded from if $\tau \leqq 0$. Now use i(a) on $(-L, \tau)$ to see that $w(\sigma_-)$ is bounded from below.

ii(c) This case uses the same argument of ii(b). In this case we need

$$p(w(\tau)) - p(w(\xi)) \geqq p(\alpha) - p(\beta) \tag{2.15}$$

when $w(\xi) \geqq w(\tau)$. (It is *not* necessary that $p(w) \to \infty$ as $w \to \infty$.)

iii(a) In this case w is monotone increasing so $w_- \leqq w(\xi) \leqq w_+$ on $[-L, L]$. As to u, either $\tau_+ \geqq 0$ or it is not. If $\tau_+ \geqq 0$, we use an argument of type in i(c) to conclude $u(\tau_+)$ is bounded from above. If $\tau_+ < 0$, then $\tau_- < 0$ and again we use an argument of the type in i(c) to see $u(\tau_-)$ is bounded from above. Thus if $\tau_+ \geqq 0$, $u_+ \leqq u(\tau_+) \leqq$ bound from above; if $\tau_+ < 0$, then $u_- \leqq u(\tau_-) \leqq$ bound from above. But now we have reduced the problem to ii(a) which we have already considered.

iii(b) Either $\tau \geqq 0$ or it is not. If $\tau \geqq 0$ we use an argument as in ii(c) to find for $\tau \geqq 1$

$$u(\tau) \geqq u_+ - \mu p(w_+) + \mu(p(w(\tau))). \tag{2.16}$$

Since $\alpha \leqq w(\tau) \leqq \beta$, (2.16) shows $u(\tau)$ is bounded from below. If $0 \leqq \tau < 1$, we again proceed as in ii(c) to see

$$\varepsilon u(\tau) \geqq \varepsilon u(1) + \mu(p(\alpha) - p(\beta)).$$

Thus $u(\tau)$ is bounded from below. If $\tau < 0$, we proceed as in (ii)(a) to again show $u(\tau)$ is bounded from below. Thus in either case $u(\tau)$ is bounded from above and below. We now use i(a) on $[-L, \tau]$ and i(b) on $[\tau, L]$ to show $w(\sigma_-)$ is bounded from below and $w(\sigma_-)$ is bounded from above. Of course in each case we use the fact that the value of w at the endpoint σ is bounded from above and below since $\alpha \leqq w(\sigma) \leqq \beta$.

iii(c) If $\tau \leqq 0$ proceed as in ii(b). First note that if $\zeta \leqq \min\{-1, \tau\}$ then

$$u(\zeta) \geqq u_- + \mu(p(w_-) - p(w(\tau))).$$

Since $\alpha < w(\tau) < \beta$, $u(\tau)$ is bounded from below, if $\tau \leqq -1$. If $-1 < \tau \leqq 0$ we find

$$\varepsilon u(\tau) \geqq \varepsilon u(-1) + \mu \int_{-1}^{\tau} p(w(\xi)) - p(w(\tau))\, d\xi$$

where $w(\xi) \leqq w(\tau)$, $-1 \leqq \xi \leqq \tau$. In this case Figure 3 tells us

$$p(w(\xi)) - p(w(\tau)) \geqq p(\alpha) - p(\beta)$$

and so

$$\varepsilon u(\tau) \geqq \varepsilon u(-1) + \mu(p(\alpha) - p(\beta))$$

and $u(\tau)$ is bounded from below for $\tau \leqq 0$. If $\tau \geqq 0$ then $\tau_+ \geqq 0$. We then estimate $u(\tau_+)$ from above as in i(c). Thus if $\zeta \geqq \max\{\tau_+, 1\}$ we find

$$u(\zeta) \leqq u_+ + \mu p(w(\tau_+)) - \mu p(w_+). \tag{2.17}$$

Since $\beta \leqq w(\tau_+) \leqq w_+$, we know $u(\tau_+)$ is bounded from above if $\tau_+ \geqq 1$. If $0 \leqq \tau_+ < 1$, we find

$$\varepsilon u(\tau_+) \leqq \varepsilon u(1) - \mu \int_{\tau_+}^{1} p(w(\xi)) - p(w(\tau_+))\, d\xi. \tag{2.18}$$

But $\beta \leqq w(\xi) \leqq w_+$ for $\xi \in [\tau_+, 1]$ so the right hand side of (2.18) is bounded. Thus if $\tau \leqq 0$ we see $u(\tau)$ is bounded from above and below; if $\tau > 0$ then $u(\tau_+)$ is bounded from above and below. Notice that for the second possibility we have reduced the problem to Case ii(b) which we have already considered. For the first possibility we apply i(a) on $[-L, \tau]$ to bound $w(\sigma_-)$ from below remembering the endpoint $w(\tau)$ lies in (α, β). Finally apply i(c) on $[\tau, L]$ to bound $u(\tau_+)$ from above again using the fact that $w(\tau) \in (\alpha, \beta)$.

iii(d) The proof is analogous to iii(c). □

Proof of Theorem 2.6. In this case we never leave the hyperbolic regime $p'(w) < 0$. To see this consider the situation $w_- < w_+ < \alpha$. By Lemma 2.1 either $w(\xi)$ is monotone and hence we trivially have $w_- \leqq w(\xi) \leqq w_+$ or $u(\xi)$ is monotone decreasing and $w(\xi)$ possesses at most one critical point which must be a local minimum. Thus $w(\xi) < \alpha$ on $[-L, L]$. Now apply Theorem 4.1 of [7]. The other cases are done analogously. □

3. Existence of solutions of the Riemann problem: the case when $\{(u_\varepsilon(\xi), w_\varepsilon(\xi))\}$ are uniformly bounded

In this section we consider the applicability of the following result of Dafermos ([7], Theorem. 3.2) to prove existence of solutions to the Riemann problem.

Proposition 3.1. *For fixed* $\varepsilon > 0$, *let* $(u_\varepsilon(\xi), w_\varepsilon(\xi))$ *denote a solution of* P_ε. *Suppose the set* $\{(u_\varepsilon(\xi), w_\varepsilon(\xi)); 0 < \varepsilon < 1\}$ *is of uniformly bounded variation. Then* $\{(u_\varepsilon(\xi), w_\varepsilon(\xi))\}$ *possesses a subsequence which converges* a.e. *on* $(-\infty, \infty)$ *to* *function* $(u(\xi), w(\xi))$ *of bounded variation. The pair* $u\left(\frac{x}{t}\right)$, $w\left(\frac{x}{t}\right)$ *provide a weak solution of P.*

In order to apply Proposition 3.1 we need the desired estimates on $\{(u_\varepsilon(\xi), w_\varepsilon(\xi))\}$ on both the two phase and one phase data case. First, however, we state an assumption on $p(w)$.

Assumption 3.2
(a) *Assume* $p(w) \to +\infty$ *as* $w \to -\infty$.
(b) *Assume* $p(w) \to -\infty$ *as* $w \to +\infty$.

Now we can state an existence theorem for the case when the data refer only to one phase.

Theorem 3.3. *If* $u_- > u_+$ *and* $w_-, w_+ < \alpha$ *and Assumption 3.2(a) holds (or* $u_+ > u_-$ *and* $w_-, w_+ > \beta$ *and Assumption 3.2(b) holds) the sequence* $\{(u_\varepsilon(\xi), w_\varepsilon(\xi)); 0 < \varepsilon < 1\}$ *as given by Corollary 2.7 possesses a subsequence which converges* a.e. *on* $(-\infty, \infty)$ *to function* $(u(\xi), w(\xi))$ *of bounded variation. The pair* $u\left(\frac{x}{t}\right)$, $w\left(\frac{x}{t}\right)$ *provides a solution to the Riemann problem with* $w\left(\frac{x}{t}\right) < \alpha \left(w\left(\frac{x}{t}\right) > \beta\right)$.

Proof. Since the sequence $\{(u_\varepsilon(\xi), w_\varepsilon(\xi))\}$ is such that $p'(w_\varepsilon))(\xi < 0$ *and either* $w_\varepsilon(\xi) < \alpha$ *or* $w_\varepsilon(\xi) > \beta$ Theorem 4.2(ii) of [7] combined with Assumption 3.2(a) or (b) yields a uniform in ε, ξ bound for $\{u_\varepsilon(\xi), w_\varepsilon(\xi)\}$ on $(-\infty, \infty)$. Lemma 2.1 applied to the case $\mu = 1$, $-\infty < \xi < \infty$, shows $\{(u_\varepsilon(\xi), w_\varepsilon(\xi))\}$ is of uniformly bounded variation. Proposition 3.1 thus yields the result.

We now move on to the case when the data refer to two phases.
From now on unless otherwise stated *we assume Assumption 3.2(a), (b) holds as well as the following condition of genuine nonlinearity.*

Assunption 3.3. $p''(w) > 0$ if $w \leqq \alpha$; $p''(w) < 0$ *if* $w \geqq \beta$.

We now proceed to give a sequence of lemmas based on Lemma 4.1 [8] that will help us in our search for an estimate on $\{u_\varepsilon(\xi)\ w_\varepsilon(\xi)\}$. Here and *for the rest of this paper* $\{u_\varepsilon(\xi), w_\varepsilon(\xi); 0 < \varepsilon < 1\}$ *denotes the solution of* P_ε *given by Corollary 2.7 when* $w_- < \alpha, w_+ > \alpha$. (Results for the case $w_- > \beta, w_+ < \beta$ can always be obtained by analogous arguments.)

Lemma 3.4. *The list of possible graphs for* $(u_\varepsilon(\xi), w_\varepsilon(\xi))$ *given in Lemma 2.4 is still valid when* $L = \infty$.

Proof. The argument given in the finite L case still applies.

Notation. Points of minima, maxima of $u_\varepsilon(\xi)$, $w_\varepsilon(\xi)$ are denoted as in Lemma 2.4 with the addition of a superscript ε to emphasize the dependence on ε.

Lemma 3.5. *In cases* 0, *i(a, b, c) of Lemma* 2.4 $(u_\varepsilon(\xi), w_\varepsilon(\xi))$ *are uniformly bounded independenty of* ε *on* $(-\infty, \infty)$, i.e., *there is a constant N depending at most on* u_-, u_+, w_-, w_+, p *and independent of* ε, $0 < \varepsilon < 1$, *such that*

$$\sup_{-\infty<\xi<\infty} |u_\varepsilon(\xi)| + |w_\varepsilon(\xi)| \leqq N. \tag{3.1}$$

Proof.

0. Obvious.

i(a) In this case $u_\varepsilon(\xi)$ is monotone decreasing and hence $u_+ \leqq u(\xi) \leqq u_-$ on $(-\infty, \infty)$. Denote $\dfrac{dw_\varepsilon(\xi)}{du} = \dfrac{w'_\varepsilon(\xi)}{u'_\varepsilon(\xi)}$. We claim that $0 < \dfrac{dw_\varepsilon(\xi)}{du} < \left(\dfrac{-1}{p'(w_\varepsilon(\xi))}\right)^{1/2}$ on $(-\infty, \sigma^\varepsilon_-]$. For if not set

$$\xi_1 = \max\left\{\xi \in (-\infty, \sigma^\varepsilon_-], \quad \frac{dw_\varepsilon(\xi)}{du} \geqq +\left(\frac{-1}{p'(w_\varepsilon(\xi))}\right)^{1/2}\right\}.$$

Since w_ε has its minimum at σ^ε_-, $\dfrac{dw_\varepsilon}{du} = 0$ there and so $\xi_1 < \sigma^\varepsilon_-$ must exist. A direct computation shows that

$$\varepsilon \frac{d}{d\xi}\left(\frac{dw_\varepsilon}{du}(\xi)\right) = -1 - p'(w_\varepsilon(\xi))\left(\frac{dw_\varepsilon}{du}\right)^2 \tag{3.2}$$

and so $\dfrac{d}{d\xi}\left(\dfrac{dw_\varepsilon}{du}(\xi)\right) = 0$ at $\xi = \xi_1$. Furthermore, by the definition ξ_1 we have $0 < \dfrac{dw_\varepsilon}{du} < +\left(-\dfrac{1}{p'(w_\varepsilon(\xi))}\right)^{1/2}$ on $(\xi_1, \sigma^\varepsilon_-)$ and thus $\dfrac{d}{d\xi}\left(\dfrac{dw_\varepsilon}{du}(\xi)\right) < 0$ on $(\xi_1, \sigma^\varepsilon_-)$ and $\dfrac{d^2}{d\xi^2}\left(\dfrac{dw_\varepsilon}{du}(\xi)\right) < 0$ at $\xi = \xi_1$. On the other hand differentiation of (3.2) shows that

$$\varepsilon \frac{d^2}{d\xi^2}\left(\frac{dw_\varepsilon}{du}(\xi)\right) = -p''(w_\varepsilon(\xi))\, w'_\varepsilon(\xi)\left(\frac{dw_\varepsilon}{du}(\xi)\right)^2 \quad \text{at } \xi = \xi_1.$$

Since Assumption 3.3 implies $p''(w_\varepsilon(\xi_1)) > 0$ and we know $w'_\varepsilon(\xi_1) < 0$, we have $\varepsilon \dfrac{d^2}{d\xi^2}\left(\dfrac{dw_\varepsilon}{du}(\xi)\right) > 0$ at $\xi = \xi_1$, a contradiction. Thus we see $\dfrac{d}{d\xi}\left(\dfrac{dw_\varepsilon}{du}(\xi)\right) \leqq 0$ on $(-\infty, \sigma^\varepsilon_-]$. Hence for any $\xi \in (-\infty, \sigma^\varepsilon_-]$, $\dfrac{dw_\varepsilon}{du}(\xi) < \dfrac{dw_\varepsilon}{du}(-\infty) = +\left(-\dfrac{1}{p'(w^-)}\right)^{1/2}$. Now let us compute

$$w_\varepsilon(\sigma^\varepsilon_-) - w_- = \int_{u_-}^{u_\varepsilon(\sigma^\varepsilon_-)} \frac{dw_\varepsilon}{du}\, du$$

$$= \int_{u_\varepsilon(\sigma^\varepsilon_-)}^{u_-} \left(-\frac{dw_\varepsilon}{du}\right) du \geqq -\int_{u_\varepsilon(\sigma^\varepsilon_-)}^{u_-} \left(-\frac{1}{p'(w_-)}\right)^{1/2} du$$

and thus

$$w_\varepsilon(\sigma_-^\varepsilon) - w_- \geqq -\left(-\frac{1}{p'(w^-)}\right)^{1/2}(u_- - u_\varepsilon(\sigma_-^\varepsilon)). \tag{3.3}$$

Since $u_+ \leqq u_\varepsilon(\sigma_-^\varepsilon) \leqq u_-$, (3.3) shows $w_\varepsilon(\sigma_-^\varepsilon)$ is bounded from below independently of ε.

i(b) The proof is similar to i(a).

i(c) In this case $w_\varepsilon(\xi)$ is monotone increasing so $w_- \leqq w_\varepsilon(\xi) \leqq w_+$ on $(-\infty, \infty)$. Denote $\frac{du_\varepsilon}{dw}(\xi) = \frac{u_\varepsilon'(\xi)}{w_\varepsilon'(\xi)}$. We claim that $0 < \frac{du_\varepsilon}{dw} < +(-p'(w_\varepsilon(\xi))^{1/2}$ on $(-\infty, \tau_-^\varepsilon]$. For if not set

$$\xi_1 = \max\left\{\xi \in (-\infty, \tau_-^\varepsilon]; \frac{du_\varepsilon}{dw}(\xi) \geqq +(-p'(w_\varepsilon(\xi))^{1/2}\right\}$$

where $u_\varepsilon(\xi)$ has its maximum at τ_-^ε, $w_\varepsilon(\tau_-^\varepsilon) < \alpha$. (The case when the maximum is at τ_+^ε, $w_\varepsilon(\tau_+^\varepsilon) > \beta$ is done similarly). Since $u_\varepsilon(\xi)$ has a local maximum at τ_-^ε where $\frac{du_\varepsilon}{dw}(\xi) = 0$, ξ_1 must exist with $\xi_1 < \tau_-^\varepsilon$. A direct computation shows that

$$\varepsilon \frac{d}{d\xi}\left(\frac{du_\varepsilon}{dw}(\xi)\right) = p'(w(\xi)) + \left(\frac{du_\varepsilon}{dw}(\xi)\right)^2 \tag{3.4}$$

and so $\frac{d}{d\xi}\left(\frac{du_\varepsilon}{dw}(\xi)\right) = 0$ at $\xi = \xi_1$. Furthermore by the definition of ξ_1 we have $0 < \frac{du_\varepsilon}{dw}(\xi) < +(-p'(w_\varepsilon(\xi)))^{1/2}$ on $(\xi_1, \tau_-^\varepsilon)$ and thus $\frac{d}{d\xi}\left(\frac{du_\varepsilon}{dw}(\xi)\right) < 0$ on $(\xi_1, \tau_-^\varepsilon)$. Thus we have $\frac{d^2}{d\xi^2}\left(\frac{du_\varepsilon}{dw}(\xi)\right) < 0$ at $\xi = \xi_1$. On the other hand differentiation of (3.4) shows $\varepsilon \frac{d^2}{d\xi^2}\left(\frac{du_\varepsilon}{dw}(\xi)\right) = p''(w_\varepsilon(\xi))\, w_\varepsilon'(\xi) > 0$ at ξ_1 and we have $\varepsilon \frac{d^2}{d\xi^2}\left(\frac{du_\varepsilon}{dw}(\xi)\right) > 0$ at $\xi = \xi_1$, a contradiction. Thus we see that $\frac{d}{d\xi}\left(\frac{du_\varepsilon}{d\xi}(\xi)\right) < 0$ on $(-\infty, \tau_-^\varepsilon]$ and hence for any $\xi \in (-\infty, \tau_-^\varepsilon]$, $0 < \frac{du_\varepsilon}{dw}(\xi) < \frac{du_\varepsilon}{dw}(-\infty) = +(-p'(w^-))^{1/2}$. Now we compute

$$u_\varepsilon(\tau_-^\varepsilon) - u_- = \int_{w_-}^{w(\tau_-^\varepsilon)} \frac{du_\varepsilon}{dw}\, dw \leqq (-p'(w))^{1/2}(w_-(\tau_-^\varepsilon) - w_-).$$

Since $w_- \leqq w(\tau_-^\varepsilon) \leqq w_+$, we see that $u_\varepsilon(\tau_-^\varepsilon)$ is bounded from above independently of ε for $w(\tau_-^\varepsilon) < \alpha$. As noted above, analogous reasoning shows that if $w(\tau_+^\varepsilon) > \beta$ we have

$$u_\varepsilon(\tau_+^\varepsilon) \leqq u_+ + (-p'(w_+))^{1/2}\,(w_+ - w_\varepsilon(\tau_+^\varepsilon))$$

and since $w_- \leqq w_\varepsilon(\tau_+^\varepsilon) \leqq w_+$ a bound on $u_\varepsilon(\tau_+^\varepsilon)$ independent of ε is produced.

Lemma 3.6. *Let τ^ε denote the points where $u_\varepsilon(\xi)$ takes on its local minimum, $\alpha < w_\varepsilon(\tau^\varepsilon) < \beta$. If there is a subsequence $\{\tau^{\varepsilon_n}\}$ of $\{\tau^\varepsilon\}$, $\varepsilon_n \to 0+$ such that either*
(a) $\tau^{\varepsilon_n} \geqq m > 0$ or $\tau^{\varepsilon_n} \leqq -m < 0$, m a constant independent of ε, or
(b) $u_\varepsilon(\tau^{\varepsilon_n})$ is bounded from below independently of ε,
then for Case i(c) $\{(u_{\varepsilon_n}(\xi), w_{\varepsilon_n}(\xi))\}$ satisfies (3.1).

Proof. Assume $\tau^{\varepsilon_n} \leqq -m < 0$. Then $u'_{\varepsilon_n}(\xi) \leqq 0$ on $(-\infty, \tau^{\varepsilon_n}]$ and $\xi u'_{\varepsilon_n}(\xi) \geqq -mu'_{\varepsilon_n}(\xi)$ on $(-\infty, \tau^{\varepsilon_n}]$. Integrating (0.5) from $-\infty$ to τ^{ε_n}, we see that

$$-m(u_\varepsilon(\tau^{\varepsilon_n}) - u_-) \leqq \int_{-\infty}^{\tau^{\varepsilon_n}} \xi u'_{\varepsilon_n}(\xi)\, d\xi = p(w(\tau^{\varepsilon_n})) - p(w_-)$$

and hence

$$\frac{1}{m}(-p(w_\varepsilon(\tau^{\varepsilon_n}) + p(w_-)) - u_- \leqq u(\tau^{\varepsilon_n}). \tag{3.5}$$

Since $w_\varepsilon(\xi)$ is monotone, $w_- \leqq w_\varepsilon(\xi) \leqq w_+$, and we see that $u_\varepsilon(\tau^{\varepsilon_n})$ is bounded from below independently of ε. The case $\tau^{\varepsilon_n} \geqq m > 0$ is done similarly. Thus in (a) or (b) $u(\tau^{\varepsilon_n})$ is bounded from below and hence $\{(u_{\varepsilon_n}(\xi), w_{\varepsilon_n}(\xi)); 0 < \varepsilon_n < 1\}$ satisfies (3.1).

Lemma 3.7. *In Cases ii(a), (b), (c), iii(a), (b), (c), (d) assume $\{\tau^\varepsilon\}$ satisfies the hypothesis of Lemma 3.6. Then $\{(u_{\varepsilon_n}(\xi), w_{\varepsilon_n}(\xi)); 0 < \varepsilon_n < 1\}$ satisfies* (3.1).

Proof. ii(a). Let $\tau_-^{\varepsilon_n}$ denote the point where $u_{\varepsilon_n}(\xi)$ has its local maximum, $w_{\varepsilon_n}(\tau_-^{\varepsilon_n}) < \alpha$. The method of proof for i(c) in Lemma 3.5 shows $u_{\varepsilon_n}(\tau_-^{\varepsilon_n})$ is bounded from above independent of ε_n; $u_\varepsilon(\tau_-^\varepsilon)$ is bounded from below by u_-.

If $\tau^{\varepsilon_n} < -m < 0$ we know $\xi u'_{\varepsilon_n}(\xi) \geqq -mu'_{\varepsilon_n}(\xi)$ on $(\tau_-^{\varepsilon_n}, \tau^{\varepsilon_n})$. Integrate (0.5) from $\tau_-^{\varepsilon_n}$ to τ^{ε_n}. We see that

$$-m(u_{\varepsilon_n}(\tau_-^{\varepsilon_n}) - u_{\varepsilon_n}(\tau^{\varepsilon_n})) \leqq p(w_{\varepsilon_n}(\tau^{\varepsilon_n})) - p(w_{\varepsilon_n}(\tau_-^{\varepsilon_n})). \tag{3.6}$$

Inequality (3.6) combined with the monotonicity of $w_{\varepsilon_n}(\xi)$ gives the bound on $u_{\varepsilon_n}(\tau^{\varepsilon_n})$ from below. If $\tau^{\varepsilon_n} > m > 0$ integration of (0.6) from τ^{ε_n} to ∞ produces the bound from below on $u_{\varepsilon_n}(\tau^{\varepsilon_n})$. Thus, if the hypothesis of Lemma 2.6 holds, $\{(u_{\varepsilon_n}(\xi), w_{\varepsilon_n}(\xi)), 0 < \varepsilon_n < 1\}$ satisfies (3.1).

ii(b). First consider the case when $\tau^{\varepsilon_n} < -m < 0$. Proceed as in the proof of Lemma 3.6 to (3.5). Since $\alpha < w_{\varepsilon_n}(\tau^{\varepsilon_n}) < \beta$ (3.5) delivers a bound on $u_{\varepsilon_n}(\tau^{\varepsilon_n})$ from below. Now use the method of proof of Lemma 3.5 1(a) to bound $w_{\varepsilon_n}(\xi)$ from below. If $\tau^{\varepsilon_n} > m > 0$ (or $u_{\varepsilon_n}(\tau^{\varepsilon_n})$ is already bounded from below) an analogous argument works.

ii(c). Proceed as in the proof ii(b) above, only now use the argument of Lemma i(b) to bound $w_{\varepsilon_n}(\xi)$ from above.

iii(a). First bound the two local maxima of $u_{\varepsilon_n}(\xi)$ from above as in the proof of Lemma 3.5i(c). Now follow the proof of ii(a) above to bound $u_{\varepsilon_n}(\tau^{\varepsilon_n})$ from below.

iii(b). Proceed as in the proof of Lemma 3.6 to bound $u_{\varepsilon_n}(\tau^{\varepsilon_n})$ from below. Then bound $w_{\varepsilon_n}(\sigma_-^{\varepsilon_n})$ from below and $w_{\varepsilon_n}(\sigma_+^{\varepsilon})$ from above by using the proof of Lemma 3.5 i(a), (b).

iii(c). First bound the local maximum of $u_{\varepsilon_n}(\xi)$ from above by the method of Lemma 3.5i(c). Then bound the local minimum $u_{\varepsilon_n}(\tau^{\varepsilon_n})$ from below by the method of ii(a) above. Now use the method of Lemma 3.5i(a) to bound $w_{\varepsilon_n}(\sigma_-^{\varepsilon_n})$ from below.

iii(d). The proof is analogous to iii(c).

We are now in a position to state the main results of this section.

Theorem 3.8. *Assume* $w_- < \alpha$, $w_+ > \beta$ *(or* $w_- > \alpha$, $w_+ < \beta$*) and let* $(u_\varepsilon(\xi), w_\varepsilon(\xi))$ *denote the solution of* P_ε *given by Corollary 2.7. Let Assumptions 3.2, 3.3 and the hypothesis of Lemma 3.6 hold. Then* $\{u_{\varepsilon_n}(\xi), w_{\varepsilon_n}(\xi); 0 < \varepsilon_n < 1\}$ *possesses a subsequence which converges* a.e. *on* $(-\infty, \infty)$ *to a function* $u(\xi)$, $w(\xi)$ *of bounded variation. The pair* $u\left(\frac{x}{t}\right)$, $w\left(\frac{x}{t}\right)$ *provides a solution of the Riemann problem.*

Proof. If $w_- < \alpha$, $w_+ > \beta$ use Lemmas 3.5, 3.6, 3.7 and Prop. 3.1. If $w_- > \beta$, $w_+ < \alpha$ we can prove a set of lemmas similar to Lemma 3.5, 3.6, 3.7 and again use Prop. 3.1.

Remark 3.9. If the hypothesis of Lemmas 3.6 does *not* hold then $\tau^\varepsilon \to 0$, $u_\varepsilon(\tau^\varepsilon) \to -\infty$ as $\varepsilon \to 0+$.

Proof. If $\tau^\varepsilon \not\to 0$ as $\varepsilon \to 0+$ then there is a subsequence $\{\tau^{\varepsilon_k}\}$ so that $|\tau^{\varepsilon_k}| > m > 0$, m being a positive constant independent of ε_k. From this subsequence we can extract another subequence so that either $\tau^{\varepsilon_n} \geqq m > 0$ or $\tau^{\varepsilon_n} \leqq -m < 0$, a contradiction. On the other hand if $\tau^\varepsilon \to 0$ and $u_\varepsilon(\tau^\varepsilon) \not\to -\infty$ then of course the hypothesis of Lemma 3.6 holds.

From Remark 3.9 we see that the only situation that may cause difficulty in solving the Riemann problem (at least under Assumptions 3.2, 3.3) is when

$\tau^\varepsilon \to 0$, $u_\varepsilon(\tau^\varepsilon) \to -\infty$ as $\varepsilon \to 0+$. This possibility is the subject of the next section. Of course if we were to make the hypotheses that $u_\varepsilon(\xi)$, is uniformly bounded independent of ε, $0 < \varepsilon < 1$, when $\alpha \leqq w_\varepsilon(\xi) \leqq \beta$ then existence of a solution of the Riemann problem follows trivially from Lemma 3.6.

4. Existence of solutions to the Riemann problem: the case when $u_\varepsilon(\tau^\varepsilon) \to -\infty$ as $\tau^\varepsilon \to 0$

In this section we discuss the possible consequences of the case when $u_\varepsilon(\tau^\varepsilon) \to -\infty$ as $\tau^\varepsilon \to 0$. (We use the notation of Section 3.)

Our first goal is to show $u_\varepsilon(\xi)$, $w_\varepsilon(\xi)$ has a pointwise a.e. limit. To do this we need a sequence of lemmas. The first one is modelled on Theorem 2.3 of [8]. *We let Assumptions 3.2, 3.3 hold in this section.*

Lemma 4.1. *Let $(u_\varepsilon(\xi), w_\varepsilon(\xi))$ be a solution of P_ε as given by Corollary 2.7 when $w_- < \alpha$, $w_+ > \beta$. Let $\bar{u} = \min(u_-, u_+)$. Then if $u_\varepsilon(\xi)$ has a local minimum at τ^ε with $\alpha < w_\varepsilon(\tau^\varepsilon) < \beta$ (as in Cases i(d), ii(a), (b), (c), iii(a), (b), (c), (d) of Lemma 2.4 with $L = \infty$, $\mu = 1$) we have the estimates*

$$N_0(\sigma_2 - \sigma_1) \geqq \int_{\sigma_2}^{\sigma_1} u_\varepsilon(\xi)\, d\xi \geqq \bar{u}(\sigma_2 - \sigma_1) - (p(\beta) - p(\alpha)), \tag{4.1}$$

$$\bar{u} - \frac{(p(\beta) - p(\alpha))}{|\xi - \tau^\varepsilon|} \leqq u_\varepsilon(\xi) \leqq N_0, \qquad -\infty < \xi < \infty. \tag{4.2}$$

Here $(\sigma_1, \sigma_2) \subset (-\infty, \infty)$ and N_0 is a constant independent of ε.

Proof. The bound from above on $u_\varepsilon(\xi)$ in (4.1), (4.2) follows from the proof of Lemmas 3.5, 3.6, 3.7. Thus we now proceed to get the bounds from below. We first check i(d). Fix $l < \infty$ sufficiently large that $w_\varepsilon(-l) < \alpha$, $w_\varepsilon(l) > \beta$. Assume for the moment $u_\varepsilon(-l) \leqq u_\varepsilon(l)$, and let $\theta > -l$ be such that $u_\varepsilon(\theta) = u_\varepsilon(-l)$. Then as shown in Figure 4 we have $u_\varepsilon(\xi) \leqq u_\varepsilon(-l)$ on $(-l, \theta)$, $u_\varepsilon(\xi) \geqq u_\varepsilon(-l)$ on $\theta < \xi < l$ when $-l < \tau^\varepsilon < \theta < l$. From (0.10) we know that

$$\varepsilon(u_\varepsilon(\xi) - u_\varepsilon(-l))'' + \xi(u_\varepsilon(\xi) - u_\varepsilon(-l))' = p(w_\varepsilon)' \tag{4.3}$$

and integration of (4.3) from $-l$ to θ shows that

$$\varepsilon u_\varepsilon'(\theta) - \varepsilon u_\varepsilon'(-l) - \int_{-l}^{\theta} (u_\varepsilon(\xi) - u_\varepsilon(-l))\, d\xi = p(w_\varepsilon(\theta)) - p(w_\varepsilon(-l)). \tag{4.4}$$

But $u'(\theta) > 0$, $u'(-l) < 0$ and hence

$$\int_{-l}^{\theta} (u_\varepsilon(-l) - u_\varepsilon(\xi))\, d\xi \leqq p(w_\varepsilon(\theta)) - p(w_\varepsilon(-l)). \tag{4.5}$$

Now since $w_\varepsilon(\theta) > w_\varepsilon(-l)$ we know the right-hand side of (4.5) is bounded from above by $p(\beta) - p(\alpha)$. Thus for any $(\sigma_1, \sigma_2) \subset (-l, \theta)$ we have

$$\int_{\sigma_1}^{\sigma_2} (u_\varepsilon(-l) - u_\varepsilon(\xi))\, d\xi \leqq p(\beta) - p(\alpha) \tag{4.6}$$

and hence

$$u_\varepsilon(-l)(\sigma_2 - \sigma_1) - (p(\beta) - p(\alpha)) \leqq \int_{\sigma_1}^{\sigma_2} u_\varepsilon(\xi)\, d\xi .$$

Letting $l \to -\infty$ we find

$$\bar{u}(\sigma_2 - \sigma_1) - (p(\beta) - p(\alpha)) \leqq \int_{\sigma_1}^{\sigma_2} u_\varepsilon(\xi)\, d\xi . \tag{4.7}$$

If $(\sigma_1, \sigma_2) \subset (\theta, l)$, then $u_\varepsilon(\xi) \geqq u_\varepsilon(-l)$ and we see

$$\bar{u}(\sigma_2 - \sigma_1) \leqq \int_{\sigma_1}^{\sigma_2} u_\varepsilon(\xi)\, d\xi . \tag{4.8}$$

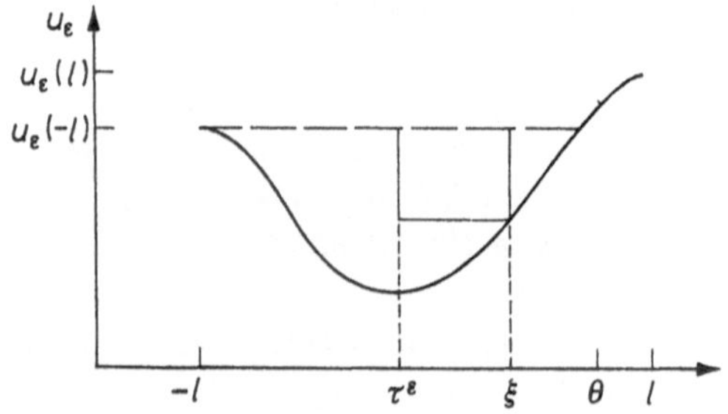

Fig. 4

Finally if $-l < \sigma_1 < \theta, \theta < \sigma_2 < l$, we write

$$\int_{\sigma_1}^{\sigma_2} u_\varepsilon(\xi)\, d\xi = \int_{\sigma_1}^{\theta} u_\varepsilon(\xi)\, d\xi + \int_{\theta}^{\sigma_2} u_\varepsilon(\xi)\, d\xi$$

and use (4.7), (4.8) to obtain (4.1) again.

To get the bound from below in (4.2) we observe from Figure 4 that when $\tau^\varepsilon < \xi < \theta$

$$(u_\varepsilon(-l) - u(\xi))(\xi - \tau^\varepsilon) \leqq \int_{-l}^{\theta} (u_\varepsilon(-l) - u_\varepsilon(\xi))\, d\xi . \tag{4.9}$$

From (4.9) and (4.6) we see that

$$(u_\varepsilon(-l) - u_\varepsilon(\xi))(\xi - \tau^\varepsilon) \leqq p(\beta) - p(\alpha). \tag{4.10}$$

Now letting $l \to \infty$ we obtain (4.2). If $-l < \xi < \tau^\varepsilon$ we again obtain (4.2) and of course if $\theta \leqq \xi \leqq l$ we trivially get (4.2). The proof for $u_\varepsilon(-l) > u_\varepsilon(l)$ is analogous.

Without going into details we sketch the appropriate constructions for ii(a), iii(a) if $u_\varepsilon(-l) \leqq u_\varepsilon(l)$ in Figure 5. Cases ii(b), ii(c) are done like i(a). Cases iii(b), (c), (d) are done like ii(a), In all cases crucial to the argument is the fact that $w_\varepsilon(\theta) > w_\varepsilon(\xi)$ so that $p(w_\varepsilon(\theta)) - p(w_\varepsilon(\zeta)) \leqq p(\beta) - p(\alpha)$ and inequality (4.6) is obtained without any monotonicity restrictions on $w_\varepsilon(\zeta)$ (in contrast to Theorem 2.3 of [8]).

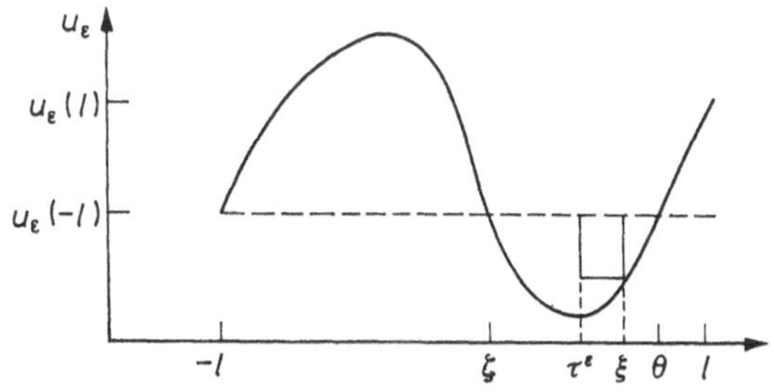

Fig. 5. Case ii(a). Integrate (4.3) from ζ to θ

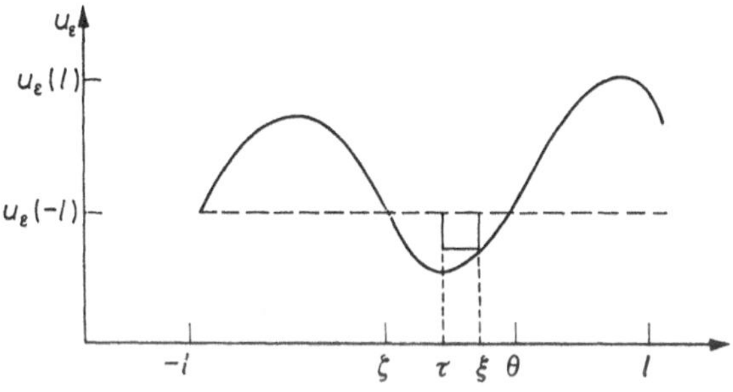

Fig. 6. Case iii(a). Integrate (4.3) from ζ to θ

This completes the proof of the lemma.

Lemma 4.2. *Let* $\{u_\varepsilon(\xi), w_\varepsilon(\xi); 0 < \varepsilon < 1\}$ *be a solution of* P_ε *as given by Corollary 2.7 when* $w_- < a$, $w_+ > \beta$. *Then for any given compact subset S of* $(-\infty, 0)$ *or* $(0, \infty)$ *there exist constants K and* ε_0 *(depending at most on* u_-, u_+, w_-, w_+, p, S*) such that*

$$\sup_{\xi\in S} (|u_\varepsilon(\xi)| + |w_\varepsilon(\xi)|) \leqq K \quad \text{for} \quad 0 < \varepsilon < \varepsilon_0 .$$

Proof. Let $S_+ \subset [a, b]$, $S_- \subset [-b, -a]$, $0 < a < b < \infty$. Then for ε sufficiently small $|\tau^\varepsilon| \leqq a/2$ and (4.2) yield $\sup_{\xi\in S_\pm} |u_\varepsilon(\xi)| \leqq K$. We now need to get a similar estimate on $w_\varepsilon(\xi)$. In Cases i(a), (b) of Lemma 2.4, the proof of Lemma 3.5, 3.6, 3.7 yields a uniform in ε and ξ, $(-\infty < \xi < \infty)$, bound on $w_\varepsilon(\xi)$ where as in Cases 0 i(c), ii(a), iii(a) $w_\varepsilon(\xi)$ is monotone so that trivially $w_- \leqq w_\varepsilon(\xi) \leqq w_+$ for $\xi \in (-\infty, \infty)$. Hence the only cases left to search are ii(b), ii(c), iii(b), iii(c), iii(d).

Case ii(b). On S_+, $w_\varepsilon(\xi)$ is un formly bounded in ε, ξ and so we need only verify S_-. Let $\eta \in S_-$, $\zeta \in S_+$. For ε sufficiently small $\eta < \tau_\varepsilon < \zeta$. Integrate (0.10) from η to ζ to obtain

$$\varepsilon u_\varepsilon'(\zeta) - \varepsilon u_\varepsilon'(\eta) + \int_\eta^\zeta \xi u_\varepsilon'(\xi)\, d\xi = p(w_\varepsilon(\zeta)) - p(w_\varepsilon(\eta)) . \tag{4.10}$$

Since $u'_\varepsilon(\zeta) > 0$, $u'_\varepsilon(\eta) < 0$ (4.10) implies

$$\int_\eta^\zeta \xi u'_\varepsilon(\xi)\, d\xi \leqq p(w_\varepsilon(\zeta)) - p(w_\varepsilon(\eta))$$

and integration by parts yields

$$\zeta u_\varepsilon(\zeta) - \eta u_\varepsilon(\eta) - \int_\eta^\zeta u_\varepsilon(\xi)\, d\xi \leqq p(w_\varepsilon(\zeta)) - p(w_\varepsilon(\eta)). \tag{4.11}$$

Now use (4.1), (4.2) to bound the left-hand side of (4.11) from below

$$\zeta\bar{u} - \frac{\zeta(p(\beta) - p(\alpha))}{|\zeta - \tau^\varepsilon|} - \eta N_0 - N_0(\zeta - \eta) \leqq p(w_\varepsilon(\zeta)) - p(w_\varepsilon(\eta)). \tag{4.12}$$

Since $\alpha \leqq w_\varepsilon(\zeta) \leqq w_+$, we see $p(w_\varepsilon(\zeta)) \leqq p(\beta)$. Hence this fact combined with $|\zeta - \tau^\varepsilon| \geqq a/2$ yields

$$-b\,|\bar{u}| - \frac{2b(p(\beta) - p(\alpha))}{a} - bN_0 - p(\beta) \leqq -p(w_\varepsilon(\eta)). \tag{4.13}$$

Since $w_\varepsilon(\eta) \leqq \beta$, (4.13) and the fact that $p(w) \to +\infty$ as $w \to -\infty$ show $w_\varepsilon(\eta)$ uniformly bounded in ε, η for ε sufficiently small, $\eta \in S_-$.

Cases ii(c), iii(b). Proceed as for Case ii(b).

Case iii(c). From the mean value theorem there is a $\zeta \in [1, 2]$ such that $u'_\varepsilon(\zeta) = u_\varepsilon(2) - u_\varepsilon(1)$ and so by (4.2) $\varepsilon u'_\varepsilon(\zeta)$ is uniformly bounded. Thus for this ζ and arbitrary $\eta \in S_-$ we again derive (4.10) and since $u'_\varepsilon(\eta) < 0$ we find that

$$\begin{aligned} \varepsilon u'_\varepsilon(\zeta) - \int_\eta^\zeta u_\varepsilon(\xi)\, d\xi &\leqq p(w_\varepsilon(\zeta)) - p(w_\varepsilon(\eta)) \\ &\leqq p(\beta) - p(w_\varepsilon(\eta)). \end{aligned}$$

The same argument as given above for case ii(b) shows $w_\varepsilon(\eta)$ uniformly bounded in ε, η for ε sufficiently small, $\eta \in S_-$.

Case iii(d). Proceed analogously as in Case iii(c).

Another set of bounds on $(u_\varepsilon(\xi), w_\varepsilon(\xi))$ is provided by the next lemma.

Lemma 4.3. *Let $\{u_\varepsilon(\xi), w_\varepsilon(\xi); 0 < \varepsilon < 1\}$ be a solution of P_ε as given by Corollary 2.7 when $w_- < a$, $w_+ > \beta$. Let σ^ε_-, σ^ε_+ denote the points of local minima and maxima for $w_\varepsilon(\xi)$ and τ^ε the point of local minima for $u_\varepsilon(\xi)$ (when they exist). Define* $\bar{u} = \min(u_-, u_+)$,

$$B_\varepsilon^- = \left(-\frac{1}{p'(w_-)}\right)^{\frac{1}{2}} \left(\bar{u} - \frac{(p(\beta) - p(\alpha))}{|\sigma^\varepsilon_- - \tau^\varepsilon|}\right) + w_- - \left(-\frac{1}{p'(w_-)}\right)^{\frac{1}{2}} u_-,$$

$$B_\varepsilon^+ = \left(-\frac{1}{p'(w_+)}\right)^{\frac{1}{2}} \left(\bar{u} + \frac{(p(\beta) - p(\alpha))}{|\sigma^\varepsilon_+ - \tau^\varepsilon|}\right) + w_+ + \left(-\frac{1}{p'(w_+)}\right)^{\frac{1}{2}} u_+.$$

Then in the cases of Lemma 2.4 (with $\mu = 1, L = \infty$) *we have the following estimates*:

In cases 0, i(a, b, c), (3.1) *holds.*
In the remaining cases $u_\varepsilon(\xi)$ *satisfies* (4.2) *and* $w_\varepsilon(\xi)$ *satisfies*

$w_- \leqq w_\varepsilon(\xi) \leqq w_+$ *in Cases i(d), ii(a), iii(a)*;

$B_\varepsilon^- \leqq w_\varepsilon(\xi) \leqq w_+$ *in Cases ii(b), iii(c)*;

$w_- \leqq w_\varepsilon(\xi) \leqq B_\varepsilon^+$ *in Cases ii(c), iii(d)*;

$B_\varepsilon^- \leqq w_\varepsilon(\xi) \leqq B_\varepsilon^+$ *in Case iii(b).*

Proof. For Cases 0, i(a, b, c) the result was given in Lemma 3.5 and Cases i(d), ii(a), iii(a) are trivial.

For Case ii(b) follow the method of proof of Case 1(a), Lemma 3.5. Upon reaching inequality (3.3) replace $u_\varepsilon(\sigma_-^\varepsilon)$ by the bound from below given by (4.2). Similarly in Case ii(c) obtain the bound for $w_\varepsilon(\sigma_+^\varepsilon)$:

$$w_+ - w_\varepsilon(\sigma_+^\varepsilon) \leqq \left(-\frac{1}{p'(w_+)}\right)(u_+ - u_\varepsilon(\sigma_+^\varepsilon))$$

and again bound $u_\varepsilon(\sigma_+^\varepsilon)$ using (4.2). Case iii(c) follows like ii(b), Case iii(d) follows as in ii(c), and Case iii(b) uses the arguments of both ii(b) and ii(c).

We now combine Lemmas 4.2 and 4.3 to get the following improved estimate on $\{u_\varepsilon(\xi), w_\varepsilon(\xi); 0 < \varepsilon < 1\}$.

Lemma 4.4. *Let* $\{u_\varepsilon(\xi), w_\varepsilon(\xi); 0 < \varepsilon < 1\}$ *be a solution of* P_ε *as given by Corollary 2.7 when* $w_- < \alpha$, $w_+ > \beta$. *Then on any semi-infinite interval* $(-\infty, -a]$ *or* $[a, \infty)$, $a > 0$ *there exist constants k and* ε_0 *(depending at most on* u_-, u_+, w_-, w_+, p, a) *such that*

$$\begin{aligned} &\sup_{(-\infty,a]} (|u_\varepsilon(\xi)| + |w_\varepsilon(\xi)|) \leqq k, \\ &\sup_{[a,\infty)} (|u_\varepsilon(\xi)| + |w_\varepsilon(\xi)|) \leqq k, \end{aligned} \tag{4.14}$$

for $0 < \varepsilon < \varepsilon_0$.

Proof. The bounds on $u_\varepsilon(\xi)$, $w_\varepsilon(\xi)$ in Cases 0, i(a, b, c) are known from inequality (3.1). For the rest of the cases the bounds on $u_\varepsilon(\xi)$ follow from inequality (4.2). Thus we only have to produce the relevant bounds on $w_\varepsilon(\xi)$. Here again Lemma 4.3 tells us Case i(d), ii(a), iii(a) are trivial and so let us move on to Case ii(b). In Case ii(b) we know from Lemma 4.3 that $B_\varepsilon^- \leqq w_\varepsilon(\xi) \leqq w^+$. Since $\tau^\varepsilon \to 0$ as $\varepsilon \to 0+$, B_ε^- will be bounded from below for sufficiently small ε if $\sigma^\varepsilon \not\to 0$ as $\varepsilon \to 0+$. On the other hand if $\sigma^\varepsilon \to 0$ as $\varepsilon \to 0+$ it means the minimum of w_ε on $(-\infty, \infty)$ is taken on at σ_-^ε where $|\sigma_-^\varepsilon| \leqq C$, a constant. If $-C \leqq \sigma_-^\varepsilon \leqq -a$ then Lemma 4.3 tells us $w_\varepsilon(\sigma_-^\varepsilon)$ is bounded from below. If $-a < \sigma_-^\varepsilon \leqq 0$ then the minimum of w_ε on $(-\infty, -a]$ is taken on at $\xi = -a$. By Lemma 4.2 we know that $w_\varepsilon(-a)$ is uniformly bounded from below, and so we see $w_\varepsilon(\xi)$ has a uniform (in ξ, ε) bound from below on $(-\infty, -a]$. The other cases, i.e. ii(c), iii(b), (c), (d) follow similarly. □

Lemma 4.5. *Let* $\{u_\varepsilon(\xi), w_\varepsilon(\xi), 0 < \varepsilon < 1\}$ *be a solution of* P_ε *as given by Corollary 2.7 when* $w_- < \alpha$, $w_+ > \beta$. *Then the sequence* $(u_\varepsilon(\xi), w_\varepsilon(\xi))$ *possesses a subsequence which converges* a.e. *on* $(-\infty, \infty)$ *to functions* $(u(\xi), w(\xi))$. *On compact subsets of* $(-\infty, 0) \cup (0, \infty)$ *the convergent subsequence is bounded uniformly in* ε *with uniformly bounded total variation. The limit functions have bounded variation on compact subsets of* $(-\infty, 0) \cup (0, \infty)$.

Proof. On the finite domain $[-1, -1/2] \cup [1/2, 2] (= R_2)$ Lemma 2.4 (with $\mu = 1, L = \infty$) and Lemma 4.4 combined with Helly's theorem ([10], p. 222) provides a convergent subsequence $\{(u_{\varepsilon_2}(\xi), w_{\varepsilon_2}(\xi))\}$ which converges boundedly to functions $u(\xi)$, $w(\xi)$ defined on R_2; $u(\xi)$, $w(\xi)$ have bounded variation R_2. Now consider $\{u_{\varepsilon_2}(\xi), w_{\varepsilon_2}(\xi)\}$ on $[-3, -1/3] \cup [1/3, 3] (= R_3)$. Again we extract a convergent subsequence which converges boundedly to the functions $u(\xi)$, $w(\xi)$ on R_2 and extensions of $u(\xi)$, $w(\xi)$ (also denoted by the same letters) on R_3. Continue this process on each R_n, $n = 2, 3, 4, \ldots$ Finally extract the diagonal element at each enumerated sequence. The sequence of diagonal elements is convergent at each $\xi \neq 0$ to functions $u(\xi)$, $w(\xi)$ defined on $(-\infty, 0) \cup (0, \infty)$. The remarks regarding compact subsets of $(-\infty, 0) \cup (0, \infty)$ follow directly from Helly's Theorem. □

Lemma 4.6. *The functions* $u(\xi)$, $w(\xi)$ *defined by Lemma 4.5 satisfy the boundary conditions*

$$u(-\infty) = u_-, \quad u(-\infty) = u_+,$$
$$w(-\infty) = w_-, \quad w(+\infty) = w_+.$$

Proof. We follow the proof of Theorem 3.2 of [7]. Let $y_\varepsilon(\xi) = \begin{pmatrix} u_\varepsilon(\xi) \\ w_\varepsilon(\xi) \end{pmatrix}$, $f(y_\varepsilon) = \begin{pmatrix} p(w_\varepsilon) \\ -u_\varepsilon \end{pmatrix}$. Then (0.10), (0.11) imply

$$\frac{d}{d\xi}(\exp(\xi^2/2\varepsilon)\, y_\varepsilon'(\xi)) = \frac{1}{\varepsilon}\left(\nabla f(y_\varepsilon)\, y_\varepsilon'(\xi) \exp\left(\frac{\xi^2}{2\varepsilon}\right)\right)$$

and integrating from 1 to ξ, $\xi > 1$, we find

$$\exp(\xi^2/2\varepsilon)\, y_\varepsilon'(\xi) - \exp\left(\frac{1}{2\varepsilon}\right) y_\varepsilon'(1) = \frac{1}{\varepsilon}\int_1^\xi \nabla f(y_\varepsilon y_\varepsilon'(\zeta) \exp\left(\frac{\zeta^2}{2\varepsilon}\right) d\zeta.$$

Since by Lemma 4.4 $|y(\xi_\varepsilon)|$ is uniformly bounded by k on $[1, \infty)$, we know $|\nabla f(y_\varepsilon)| \leq R$ for some constant $R > 0$. Thus

$$|\exp(\xi^2/2\varepsilon)\, y_\varepsilon'(\xi)| \leq \left|\exp\left(\frac{1}{2\varepsilon}\right) y_\varepsilon'(1)\right| + \frac{R}{\varepsilon}\int_1^\xi |y_\varepsilon'(\zeta)| \exp\left(\frac{\zeta^2}{2\varepsilon}\right) d\zeta$$

and using Gronwall's inequality we find that

$$|\exp(\xi^2/2\varepsilon)\, y_\varepsilon'(\xi)| \leqq \left|\exp\left(\frac{1}{2\varepsilon}\right) y_\varepsilon'(1)\right| \exp\left[\frac{R}{\varepsilon}\right](\xi - 1)$$

and hence

$$|y_\varepsilon'(\xi)| \leqq |y_\varepsilon'(1)| \exp\left(\frac{2R\xi - 2R + 1 - \xi^2}{2\varepsilon}\right). \tag{4.15}$$

Now note that

$$\begin{aligned}
\exp(\xi^2/2\varepsilon)\, y_\varepsilon'(\xi) &= z_1 + \frac{1}{\varepsilon}\int_1^\xi f(y_\varepsilon(\zeta))' \exp\left(\frac{\zeta^2}{2\varepsilon}\right) d\zeta \\
&= z_1 + \frac{1}{\varepsilon} f(y_\varepsilon(\xi)) \exp\left(\frac{\xi^2}{2\varepsilon}\right) - \frac{1}{\varepsilon} f(y_\varepsilon(1)) \exp\left(\frac{1}{2\varepsilon}\right) \\
&\quad - \frac{1}{\varepsilon^2}\int_1^\xi \zeta f(y_\varepsilon(\zeta)) \exp\left(\frac{\zeta^2}{2\varepsilon}\right) d\zeta \\
&= z_2 + \frac{1}{\varepsilon} f(y_\varepsilon(\xi)) \exp\left(\frac{\xi^2}{2\varepsilon}\right) - \frac{1}{\varepsilon^2}\int_1^\xi \zeta f(y_\varepsilon(\zeta)) \exp\left(\frac{\zeta^2}{2\varepsilon}\right) d\zeta
\end{aligned}$$

and hence

$$y_\varepsilon'(\xi) = z_2 \exp\left(\frac{-\xi^2}{2\varepsilon}\right) + \frac{1}{\varepsilon} f(y_\varepsilon(\xi)) - \frac{1}{\varepsilon^2}\int_1^\xi \zeta f(y_\varepsilon(\zeta)) \exp\left(\frac{\zeta^2}{2\varepsilon}\right) d\zeta. \tag{4.16}$$

Here

$$\begin{aligned}
z_2 \int_1^2 \exp(-\xi^2/2\varepsilon)\, d\xi &= y_\varepsilon(2) - y_\varepsilon(1) - \frac{1}{\varepsilon}\int_1^2 f(y_\varepsilon(\xi))\, d\xi \\
&\quad + \frac{1}{\varepsilon^2}\int_1^2 \zeta f(y_\varepsilon(\zeta)) \exp\left(\frac{\zeta^2}{2\varepsilon}\right) d\zeta.
\end{aligned} \tag{4.17}$$

Thus from (4.16) we see that

$$\begin{aligned}
|y_\varepsilon'(1)| &\leqq |z_2| \exp\left(-\frac{1}{2\varepsilon}\right) + \frac{1}{\varepsilon} |f(y_\varepsilon(1))| \\
&\leqq |z_2| \exp\left(-\frac{1}{2\varepsilon}\right) + \frac{\text{const.}}{\varepsilon}.
\end{aligned} \tag{4.18}$$

From (4.17) and the inequality

$$\int_1^2 \exp\left(-\frac{\xi^2}{2\varepsilon}\right) d\xi \geqq \exp\left(-\frac{2}{\varepsilon}\right)$$

we see that

$$|z_2| \leqq \left(\text{const.} + \frac{\text{const.}}{\varepsilon} + \frac{\text{const.}}{\varepsilon^2} \exp\left(\frac{2}{\varepsilon}\right)\right) \exp\left(\frac{2}{\varepsilon}\right)$$

and hence by (4.18) that

$$|y_\varepsilon'(1)| \leqq \frac{\text{const.}}{\varepsilon^2} \exp\left(\frac{7}{2\varepsilon}\right). \tag{4.19}$$

Now insert (4.19) into (4.15) to find that

$$|y_\varepsilon'(\xi)| \leqq \frac{\text{const.}}{\varepsilon^2} \left(\frac{2R\xi - 2R + 8 - \xi^2}{2\varepsilon}\right). \tag{4.20}$$

Thus for $\xi > R + (R^2 - 2R + 8)^{1/2}$ (4.20) shows that $|y_\varepsilon'(\xi)| \to 0$ as $\varepsilon \to 0+$. Recalling that $(u_\varepsilon(\xi), w_\varepsilon(\xi))$ converge pointwise to $u(\xi)$, $w(\xi)$ we see $u(\xi)$, $w(\xi)$ must be constants for $\xi > R + (R^2 - 2R + 8)^{1/2}$. Since for any $\varepsilon > 0$ $\lim\limits_{\xi \to \infty} u_\varepsilon(\xi) = u_+$, $\lim\limits_{\xi \to \sim} w_\varepsilon(\xi) = w_+$, these constants must be u_+ and w_+. A similar argument works for $\xi = -\infty$. □

Corollary 4.7. The functions $u(\xi)$, $w(\xi)$ defined by Lemma 4.5 satisfy the conditions

$$\begin{aligned} u(\xi) &= u_-, \\ w(\xi) &= w_-, \end{aligned} \quad \xi < -M, \qquad \begin{aligned} u(\xi) &= u_+ \\ w(\xi) &= w_+ \end{aligned} \quad \xi > M$$

for some positive constant M.

Lemma 4.8. *The functions $u(\xi)$, $w(\xi)$ defined by Lemma 4.5 satisfy*

$$\begin{aligned} p(w)' - \xi u' &= 0, \\ -u' - \xi w' &= 0 \end{aligned} \tag{4.21}$$

in the sense of distributions at any $\xi \neq 0$.

At any point $\xi_0 \neq 0$ of discontinuity of $u(\xi)$, $w(\xi)$ the Rankine-Hugoniot jump conditions are satisfied:

$$\begin{aligned} p(w(\xi_0+)) - p(w(\xi_0-)) - \xi_0(u(\xi_0+) - u(\xi_0-)) &= 0, \\ -(u(\xi_0+) - u(\xi_0-)) - \xi_0(w(\xi_0+) - w(\xi_0-)) &= 0. \end{aligned} \tag{4.22}$$

Proof. On any compact subset of $(0, \infty)$ or $(-\infty, 0)$ we have a sequence of solutions of (0.10), (0.11) which converges boundedly a.e. Hence if we multiply (0.10), (0.11) by C^∞ test functions with compact support excluding $\xi = 0$, integrate by parts, pass to the limits as the relevant sequence of ε's goes to zero, and use the Lebesgue dominated convergence theorem, we obtain (4.21). Equation (4.22) follows from (4.21) in the standard manner (*e.g.* [7], (3.14)). □

Corollary 4.7 and Lemma 4.8 take us very close to asserting that the Riemann problem is soluble. Unfortunately we still must deal with the behavior of u, w at the troublesome point $\xi = 0$, *i.e.* we have yet to show u, w solve (4.21) in a neighborhood of $\xi = 0$. In fact the derivation of the Rankine-Hugoniot jump condition for weak solutions [8] shows that u, w will be a distributional solution of

(4.21) at $\xi = 0$ if

$$\lim_{\xi\to 0-} p(w(\xi)) = \lim_{\xi\to 0+} p(w(\xi)),$$
$$\lim_{\xi\to 0-} u(\xi) = \lim_{\xi\to 0+} u(\xi), \tag{4.23}$$

where the indicated limits exist (finite).

Before pursuing the study of (4.23) we first show that $u(\xi)$, $w(\xi)$ are locally integrable. First we make

Assumption 4.9. Assume that

$$\frac{\left|\int\limits_{\beta}^{w} p(\zeta)\, d\zeta\right|}{|w|} \to \infty \quad \text{as } |w| \to \infty .$$

Lemma 4.10. *If Assumption 4.9 holds, then* $\{w_\varepsilon(\xi)\}$ *has absolutely equicontinuous integrals and the functions* $u(\xi)$, $w(\xi)$ *defined by Lemma 4.5 are locally integrable in* $(-\infty, \infty)$.

Proof. First we know from Lemma 4.1 (4.1) that $|u_\varepsilon(\xi)|$ is locally integrable. Since a subsequence of $u_\varepsilon(\xi)$ converges to $u(\xi)$, Fatou's theorem implies the limit function $u(\xi)$ is locally integrable. To show local integrability of $w(\xi)$ we proceed indirectly. A theorem of D. Vitali ([11], p. 152) tells us that if $\{w_\varepsilon(\xi)\}$ have absolutely equicontinuous integrals, then $w(\xi)$ is locally integrable and moreover

$$\lim_{\varepsilon_n\to 0} \int\limits_{\sigma_1}^{\sigma_2} w_{\varepsilon_n}(\xi)\, d\zeta = \int\limits_{\sigma_1}^{\sigma_2} w(\xi)\, d\xi . \tag{4.24}$$

Here the limit is taken on the a.e. convergent subsequence of $w_\varepsilon(\xi)$ denoted as $w_{\varepsilon_n}(\xi)$. Recall that to have an absolutely equicontinuous integral we need for every $\delta > 0$ an $l(\delta)$ such that if $0 < \sigma_2 - \sigma_1 \leqq l(\delta)$.

$$\left|\int\limits_{\sigma_1}^{\sigma_2} w_\varepsilon(\xi)\, d\xi\right| < \delta \quad \text{for all } \varepsilon > 0 .$$

We now establish absolute equicontinuity of the integral of $w_\varepsilon(\xi)$ by using the argument of Lemma 3.4 of [8] (which is itself a variant of the test of de la Vallée-Poussin [11], p. 154.)

First notice however, that in i(d) ii(a), iii(a) of Lemma 2.4 there is nothing to prove since $w_\varepsilon(\xi)$ is monotone and hence uniformly bounded in ξ, ε. Also recall that Cases 0, i(a), (b), (c) were covered by Theorem 3.8. Thus we need only consider Cases ii(b), (c), iii(b), (c), (d).

Let us consider Case ii(c). Given any interval (l_1, l_2) we either (I) divide it into the subintervals $(l_1, t_\varepsilon]$ where $w_- \leqq w_\varepsilon(\xi) \leqq \beta$ and $[t_\varepsilon, l_2)$ where $\beta \leqq w_\varepsilon(\xi)$, $w_\varepsilon(t_\varepsilon) = \beta$, (II) $w_\varepsilon(\xi) \geqq \beta$ on (l_1, l_2) or (III) $w_\varepsilon(\xi) \leqq \beta$ on (l_1, l_2).

First we consider possibility (I).

Multiply (0.10) by u and (0.11) by $-p(w)$ and add. If we define $\eta(u, w) = \frac{u^2}{2} - \int_\beta^w p(s)\, d\zeta$ and $\eta_\varepsilon(\xi) = \eta(u_\varepsilon(\xi), w_\varepsilon(\xi))$ we see that

$$\varepsilon\eta_\varepsilon''(\xi) + \xi\eta_\varepsilon'(\xi) - (u_\varepsilon(\xi))\, p(w_\varepsilon(\xi))' - \varepsilon u_\varepsilon'(\xi)^2 + ep'(w_\varepsilon(\xi))\, w_\varepsilon'(\xi)^2 = 0. \tag{4.25}$$

Let $\bar\eta = \max(\eta(u_-, w_-), \eta(u_+, w_+))$. On any sub-interval $(\sigma_1', \sigma_2') \subset [t_\varepsilon, l_2)$ if $\eta_\varepsilon(\sigma_1') > \bar\eta$ set $\zeta_\varepsilon = \sup\{\xi \in [t_\varepsilon, \sigma_1'); \eta_\varepsilon(\xi) \leqq \bar\eta\}$. If $\eta_\varepsilon(\sigma_1') \leqq \bar\eta$ set $\zeta_\varepsilon = \inf\{\zeta \in (\sigma_1', \sigma_2'); \eta_\varepsilon(\xi) \geqq \bar\eta\}$ if this set is not empty. Similarly if $\eta_\varepsilon(\sigma_2') > \bar\eta$ set $\theta_\varepsilon = \inf\{\xi \in (\sigma_2', l_2]; \eta_\varepsilon(\xi) \leqq \bar\eta\}$ while if $\eta_\varepsilon(\sigma_2') \leqq \bar\eta$ set $\theta_\varepsilon = \sup\{\xi \in (\sigma_1', \sigma_2'); \eta_\varepsilon(\xi) \geqq \bar\eta\}$. Observe that $\eta_\varepsilon'(\zeta_\varepsilon) \geqq 0$ and $\eta_\varepsilon'(\theta_\varepsilon) \leqq 0$, and

$$\int_{\sigma_1'}^{\sigma_2'} (\eta_\varepsilon(\xi) - \bar\eta)\, d\xi \leqq \int_{\zeta_\varepsilon}^{\theta_\varepsilon} (\eta_\varepsilon(\xi) - \bar\eta)\, d\xi = -\int_{\zeta_\varepsilon}^{\sigma_\varepsilon} \xi\eta_\varepsilon'(\xi)\, d\xi. \tag{4.26}$$

Thus if we integrate (4.25) over $(\zeta_\varepsilon, \sigma_\varepsilon)$ and use (4.26) we see that

$$\begin{aligned}\int_{\sigma_1'}^{\sigma_2'} (\eta_\varepsilon(\xi) - \bar\eta)\, d\xi + \varepsilon \int_{\zeta_\varepsilon}^{\theta_\varepsilon} (u_\varepsilon'(\xi)^2 - p'(w_\varepsilon(\xi))\, w_\varepsilon'(\xi)^2)\, d\xi \\ \leqq -u_\varepsilon(\theta_\varepsilon)\, p(w_\varepsilon(\theta_\varepsilon)) + u_\varepsilon(\zeta_\varepsilon)\, p(w_\varepsilon(\zeta_\varepsilon)).\end{aligned} \tag{4.27}$$

By the definitions of θ_ε, ζ_ε we see that $\eta(u_\varepsilon(\theta_\varepsilon), w_\varepsilon(\theta_\varepsilon))$ and $\eta(u_\varepsilon(\zeta_\varepsilon), w_\varepsilon(\zeta_\varepsilon))$ are uniformly bounded from above and since $w_\varepsilon(\theta_\varepsilon)$, $w_\varepsilon(\zeta_\varepsilon)$ are greater than β, η is convex at these values. This implies $u_\varepsilon(\theta_\varepsilon)$, $w_\varepsilon(\theta_\varepsilon)$, $u_\varepsilon(\zeta_\varepsilon)$, $w_\varepsilon(\zeta_\varepsilon)$ are uniformly bounded in ε. Hence the right-hand side of (4.27) is bounded by a constant K independent of ε.

Now since $-\frac{1}{w}\int_\beta^w p(s)\, ds \to \infty$ as $w \to \infty$ for arbitrary $\delta > 0$ there is a $w_0 \geqq \beta$ such that

$$\frac{w}{\eta(u, w)} < \frac{\delta}{2K} \quad \text{for all } w \geqq w_0.$$

We then set $l(\delta) = \dfrac{\delta}{\left(|w_-| + \beta + w_0 + \bar\eta\dfrac{\delta}{2K}\right)}$. Fix σ_1, σ_2, $0 < \sigma_2 - \sigma_1 < l(\delta)$.

We note that for any $\sigma_1, \sigma_2, \sigma_1 \in (l_1, t_\varepsilon], \sigma_2 \in (t_\varepsilon, l_2]$

$$\begin{aligned}\int_{\sigma_1}^{\sigma_2} w_\varepsilon(\xi)\, d\xi &= \int_{\sigma_1}^{t_\varepsilon} w_\varepsilon(\xi)\, d\xi + \int_{t_\varepsilon}^{\sigma_2} w_\varepsilon(\xi)\, d\xi \\ &\leqq \beta(t_\varepsilon - \sigma_1) + \int_{t_\varepsilon}^{\sigma_2} \left(w_0 + \frac{\delta}{2K}\eta(u_\varepsilon(\xi), w_\varepsilon(\xi))\right) d\xi \\ &\leqq \beta(t_\varepsilon - \sigma_1) + (\sigma_2 - t_\varepsilon)\, w_0 + \frac{\delta}{2K}\int_{t_\varepsilon}^{\sigma_2} \eta(u_\varepsilon(\xi), w_\varepsilon(\xi))\, d\xi.\end{aligned}$$

Now use (4.27) with $\sigma_2' = \sigma_2, \sigma_1' = t_\varepsilon$ and we see that

$$\begin{aligned}\int_{\sigma_1}^{\sigma_2} w_\varepsilon(\xi)\,d\xi &\leqq \beta(t_\varepsilon - \sigma_1) + (\sigma_2 - t_\varepsilon)\,w_0 + \frac{\delta}{2K}(K + \bar{\eta}(\sigma_2 - \sigma_1)) \\ &\leqq (\sigma_2 - \sigma_1)\left(\beta + w_0 + \frac{\bar{\eta}\delta}{2K}\right) + \frac{\delta}{2} \leqq \delta.\end{aligned} \tag{4.28}$$

If $\sigma_1, \sigma_2 \geqq t_\varepsilon$

$$\int_{\sigma_1}^{\sigma_2} w_\varepsilon(\xi)\,d\xi \leqq \int_{\sigma_1}^{\sigma_2} \left(w_0 + \frac{\delta}{2K}\eta(u_\varepsilon(\xi), w_\varepsilon(\xi))\right) d\xi \leqq \delta \tag{4.29}$$

and if $\sigma_1, \sigma_2 \leqq t_\varepsilon$

$$\int_{\sigma_1}^{\sigma_2} w_\varepsilon(\xi)\,d\xi \leqq \beta(\sigma_2 - \sigma_1) \leqq \delta. \tag{4.30}$$

Also since $w_\varepsilon(\xi) \geqq w_-$ we easily see that

$$\int_{\sigma_1}^{\sigma_2} w_\varepsilon(\xi)\,d\xi \geqq w_-(\sigma_2 - \sigma_1) \geqq -|w_-|\,(\sigma_2 - \sigma_1) \geqq -\delta. \tag{4.31}$$

Thus we see for (I) that (4.28)–(4.31) imply

$$\left|\int_{\sigma_1}^{\sigma_2} w_\varepsilon(\xi)\,d\xi\right| \leqq \delta \quad \text{if } 0 < \sigma_2 - \sigma_1 < l(\delta).$$

A similar argument of course works for (II) while (III) is trivial. Thus in ii(c) we know $w_\varepsilon(\xi)$ possesses absolutely equicontinuous integrals. Cases ii(b), iii(b), (c), (d) are done in a similar fashion. As noted above Vitali's theorem thus tells us that $w(\xi)$ is locally integrable and (4.24) holds. □

The next Lemma follows almost identically from Lemma 3.3 of [8].

Lemma 4.11. *The four limits which appear in* (4.23) *always exist (finite) and* $(4.23)_1$ *is always satisfied. Equation* (4.23_1) *is satisfied if the sequence* $\left\{\int_0^\xi u_\varepsilon(\xi)\,d\xi\right\}$ *(taken on the convergent subsequence of Lemma 4.5) is absolutely equicontinuous. Furthermore in general* $-(p(\beta) - p(\alpha)) \leqq \lim_{\xi\to 0-} - p(w(\xi)) + \lim_{\xi\to 0+} p(w(\xi)) \leqq 0.$

Proof. Let $\{u_\varepsilon(\xi), w_\varepsilon(\xi)\}$ denote the convergent subsequence of Lemma 4.5. Note that since $w_\varepsilon(\xi)$, $u_\varepsilon(\xi)$ are piecewise monotone as $(-\infty, \infty)$ (Lemma 2.4 with $L = \infty$) then the limit functions $u(\xi)$, $w(\xi)$ are also. Hence the set of points of continuity (in fact differentiability) of u, w is dense in any finite ξ interval.

Now let ζ and θ_- be points of continuity of $u(\xi)$, $w(\xi)$, $\zeta < 0 < \theta$. From the mean value theorem for every small $\varepsilon > 0$ we can find $\zeta_\varepsilon \in [\zeta - \varepsilon^{1/2}, \zeta]$, $\theta_\varepsilon \in [\sigma, \theta^{1/2} + \varepsilon^{1/2}]$ such that

$$\varepsilon^{\frac12} u_\varepsilon'(\zeta_\varepsilon) = u_\varepsilon(\zeta) - u_\varepsilon(\zeta - \varepsilon^{\frac12}), \quad \varepsilon^{\frac12} w_\varepsilon'(\zeta_\varepsilon) = w_\varepsilon(\zeta) - w_\varepsilon(\zeta - \varepsilon^{\frac12}),$$

$$\varepsilon^{\frac12} u_\varepsilon'(\theta_\varepsilon) = u_\varepsilon(\theta + \varepsilon^{\frac12}) - u_\varepsilon(\theta), \quad \varepsilon^{\frac12} w_\varepsilon'(\theta_\varepsilon) = w_\varepsilon(\theta_\varepsilon^{\frac12} + \varepsilon^{\frac12}) - w_\varepsilon(\theta).$$

By Lemma 2.4 there are constants K_θ, K_ζ such that

$$\begin{aligned} &|\varepsilon^{1/2} u_\varepsilon'(\zeta_\varepsilon)| \leqq K_\zeta, \quad |\varepsilon^{1/2} w_\varepsilon'(\zeta_\varepsilon)| \leqq K_\zeta, \\ &|\varepsilon^{1/2} w_\varepsilon'(\theta_\varepsilon)| \leqq K_\theta, \quad |\varepsilon^{1/2} w_\varepsilon'(\theta_\varepsilon)| \leqq K_\theta, \end{aligned} \tag{4.32}$$

for ε sufficiently small. (Of course K_θ, K_ζ may become unbounded as $\theta, \zeta \to 0$ but for the moment ζ, θ are fixed.)

Now we integrate (0.10). (0.11) on $(\zeta_\varepsilon, \theta_\varepsilon)$ obtaining

$$\begin{aligned} \varepsilon u_\varepsilon'(\theta_\varepsilon) - \varepsilon u_\varepsilon'(\zeta_\varepsilon) + \theta_\varepsilon u_\varepsilon(\theta_\varepsilon) - \zeta_\varepsilon u_\varepsilon(\zeta_\varepsilon) - \int_{\zeta_\varepsilon}^{\theta_\varepsilon} u_\varepsilon(\xi)\, d\xi \\ = p(w_\varepsilon(\theta_\varepsilon)) - p(w_\varepsilon(\zeta_\varepsilon)), \end{aligned} \tag{4.33}$$

$$\begin{aligned} \varepsilon w_\varepsilon'(\theta_\varepsilon) - \varepsilon w_\varepsilon'(\zeta_\varepsilon) + \theta_\varepsilon w_\varepsilon(\theta_\varepsilon) - \zeta_\varepsilon w_\varepsilon(\zeta_\varepsilon) - \int_{\zeta_\varepsilon}^{\theta_\varepsilon} w_\varepsilon(\xi)\, d\xi \\ = -u_\varepsilon(\theta_\varepsilon) + u_\varepsilon(\zeta_\varepsilon). \end{aligned} \tag{4.34}$$

Now let $\varepsilon \to 0+$ in (4.33), (4.34). Since θ, ζ are points of continuity of u, w we find by virtue of (4.32) that

$$\theta u(\theta) - \zeta u(\zeta) - p(w(\theta)) + p(w(\zeta)) = \lim_{\varepsilon \to 0+} \int_{\zeta_\varepsilon}^{\theta_\varepsilon} u_\varepsilon(\xi)\, d\xi, \tag{4.35}$$

$$\theta w(\theta) - \zeta w(\zeta) + u(\theta) - u(\zeta) = \int_\zeta^\theta w(\zeta)\, d\zeta, \tag{4.36}$$

where in (4.36) we have used (4.24) i.e. the absolute equicontinuity of the integrals of $w_\varepsilon(\xi)$. Notice the limit on the right hand side of (4.35) exists since the limits on the left hand side exist. From (4.1) we see that $\lim_{\varepsilon \to 0+} \int_{\zeta_\varepsilon}^{\theta_\varepsilon} u_\varepsilon(\xi)\, d\xi \overset{\text{def}}{=} S(\zeta, \theta)$ satisfies $\bar{u}(\zeta - \theta) - (p(\beta) - p(\alpha)) \leqq S(\zeta, \theta) \leqq N_0(\zeta - \theta)$. By Lemma 4.4 for fixed $\zeta < 0$, $S(\zeta, \theta)$ is continuous in θ, $\theta > 0$, $|\theta|$ small and for fixed $\theta > 0$, $S(\zeta, \theta)$ is continuous in ζ, $\zeta < 0$, $|\zeta|$ small.

Now since $|w(\xi)|$ may be infinite only at $\xi = 0$ (again by Lemma 4.4) pointwise limits of ii(b), (c), iii(b), (c), (d) of Lemma 2.4 shows that if $|w(0)| = \infty$, w must have one of three shapes shown in Figure 7.

In all three cases (I), (II), (III) we see that

$$|\zeta w(\zeta)| \leqq \int_\zeta^\theta |w(\xi)|\, d\xi,$$

$$|\theta w(\theta)| \leqq \int_\zeta^\theta |w(\xi)|\, d\xi.$$

But since $w(\xi)$ is locally integrable (Lemma 4.10),

$$\lim_{\theta \to 0+} \theta w(\theta) = \lim_{\zeta \to 0-} \zeta w(\zeta) = \lim_{\substack{\theta \to 0+ \\ \zeta \to 0-}} \int_\zeta^\theta w(\xi)\, d\xi = 0.$$

Since $u(\xi)$ has the shape of (I) near $\xi = 0$ and $u(\xi)$ is locally integrable (Lemma 4.10),

$$\lim_{\theta \to 0+} \theta u(\theta) = \lim_{\zeta \to 0+} \zeta u(\zeta) = 0.$$

Now let $\theta \to 0+$, $\zeta \to 0-$ along a sequence of points of continuity of u, w and possibly extracting a further subsequence such that $S(\zeta, \theta)$ converges we find that

$$\lim_{\theta \to 0+} -p(w(\theta)) + \lim_{\zeta \to 0-} p(w(\zeta)) = \lim_{\substack{\theta \to 0+ \\ \zeta \to 0-}} S(\zeta, \theta),$$

$$\lim_{\theta \to 0+} u(\theta) = \lim_{\zeta \to 0-} u(\zeta),$$

and $(9.23)_1$ is always satisfied. Moreover if $\int_0^\xi u_\varepsilon(\xi)\, d\xi$ is absolutely equicontinuous Vitali's theorem tells us we can pass the limit through the intcgral in (4.35) and hence show (4.23_2) holds as well. $\left(\text{In this case we have of course } S(\zeta, \theta) = \lim \int_\zeta^\theta u(\xi)\, d\xi.\right)$ Also the bounds on $S(\zeta, \theta)$ show that $p(\alpha) - p(\beta) \leqq -\lim_{\theta \to 0+} p(w(\theta)) + \lim_{\zeta \to 0-} p(w(\zeta)) \leqq 0$ holds in general. □

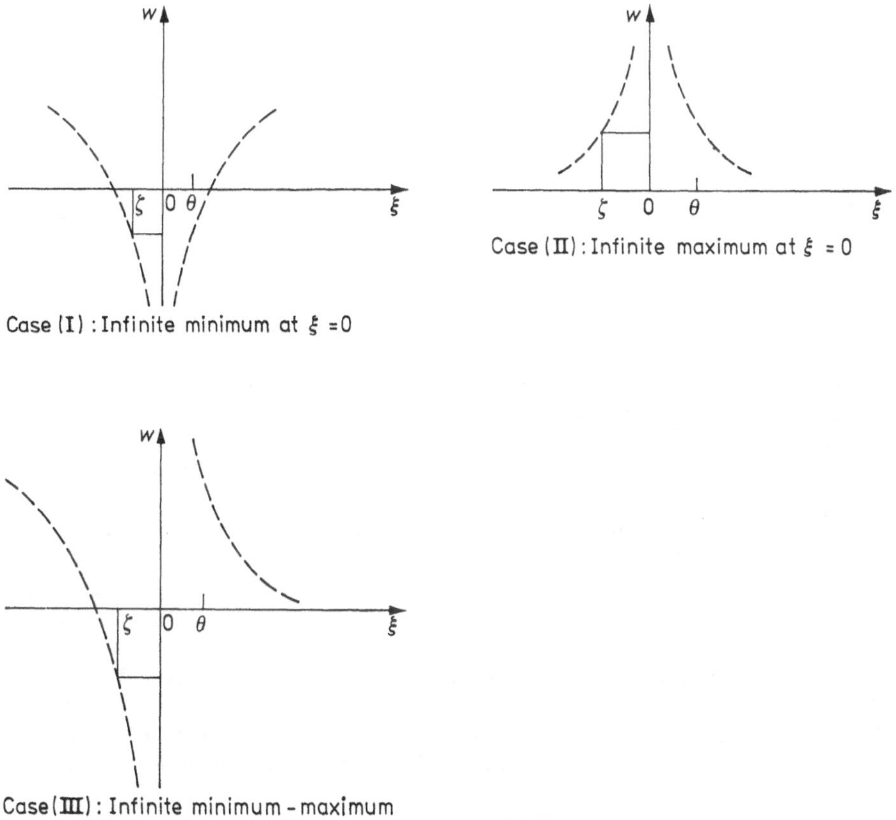

Case (I) : Infinite minimum at $\xi = 0$

Case (II) : Infinite maximum at $\xi = 0$

Case (III) : Infinite minimum - maximum

Fig. 7

Remark 4.12. As u_ε may have more than one critical point, the argument used in [8] to show that the absolute equicontinuity of $\int_0^\xi u_\varepsilon(\xi)\, d\xi$ is also *necessary* to have (4.32_2) hold does not seem to apply.

Theorem 4.13. *The functions $u(\xi)$, $w(\xi)$ defined by Lemma 4.5 provide a solution of the Riemann problem provided the pressure p equilibrates across the stagnant phase boundary at* $\xi = 0$, i.e.

$$\lim_{\xi\to 0-} p(w(\xi)) = \lim_{\xi\to 0+} p(w(\xi)).$$

Proof. Use Lemma 4.11.

Acknowledgement. I thank the Division of Applied Mathematics of Brown University and Departments of Theoretical and Applied Mathematics of the Weizmann Institute of Science for supporting me in part during the course of this work. Particular thanks go to the Michael family which provided the funds for the Michael Professorship which I held at the Weizmann Institute. Personal thanks go to my hosts: at Brown, Professors D. E. McCLURE and C. M. DAFERMOS; at the Weizmann Institute, Professors L. A. SEGEL and Z. ARTSTEIN. Finally an extra note of appreciation is extended to Professor Z. ARTSTEIN for his valuable suggestions and comments on the research presented here.

This research was supported in part by the Air Office of Scientific Research. Air Force Systems Command, USAF, under Contract/Grant No. AFOSR-85-0239 (R.P.I.) by the United States Army, Army Research Office under Contracts/Grants Nos. 5-28529 and 5-28317 (Brown University). The United States Government is authorized to reproduce and distribute reprints for Government purposes notwithstanding any copyright herein.

References

1. R. D. JAMES, The propagation of phase boundaries in elastic bars, *Archive for Rational Mechanics and Analysis.*
2. M. SHEARER, The Riemann problem for a class of conservation laws of mixed type, *J. Differential Equations* **46** (1982), 426–443.
3. M. SHEARER, Nonuniqueness of admissible solutions of Riemann initial value problems for a system of conservation laws of mixed type, *Archive for Rational Mechanics and Analysis* **93** (1986), 45–59.
4. M. SLEMROD, Admissibility criteria for propagating phase boundaries in a van der Waals fluid, *Archive for Rational Mechanics and Analysis* **81** (1983), 301–315.
5. R. HAGAN & M. SLEMROD, The viscosity-capillarity criterion for shocks and phase transitions, *Archive for Rational Mechanics and Analysis* **83** (1984), 333–361.
6. M. SLEMROD, Dynamics of first order phase transition, in *Phase Transformations and Material Instabilities in Solids*, ed. M. GURTIN, Academic Press: New York (1984), 163–203.
7. C. M. DAFERMOS, Solution of the Riemann problem for a class of hyperbolic systems of conservation laws by the viscosity method, *Archive for Rational Mechanics and Analysis* **52** (1973), 1–9.
8. C. M. DAFERMOS & R. J. DI PERNA, The Riemann problem for certain classes of hyperbolic systems of conservation laws, *J. Differential Equations* **20** (1976), 90–114.

9. C. S. MORAWETZ, On a weak solution for a transonic flow, *Comm. Pure and Applied Math.* **38** (1985), 797–818.
10. J. MAWHIN, Topological degree methods in nonlinear boundary value problems, Conference Board of Mathematical Sciences Regional Conference Series in Mathematics, No. 40, American Mathematical Society (1979).
11. I. P. NATANSON, Theory of functions of a real variable, Vol. 1, F. Ungar Publishing Co., New York (1955).
12. M. SHEARER, Dynamic phase transitions in a van der Waals gas, to appear *Quarterly of Applied Math.*
13. A. S. KALASNIKOV, Construction of generalized solutions of quasilinear equations of first order without convexity conditions as limits of solutions of parabolic equations with a small parameter, *Dokl. Akad. Nauk. SSR* **127** (1959), 27–30 (Russian).
14. V. A. TUPCIEV, The asymptotic behavior of the solution of the Cauchy problem for the equation $\varepsilon^2 t u_{xx} = u_t + [\varphi(u)]_x$ that degenerates for $\xi = 0$ into the problem of the decay of an arbitrary discontinuity for the case of a rarefraction wave. *Z. Vycisl. Mat. Fiz.* **12** (1972), 770–775; English translation in *USSR Comput. Math. and Phys.* **12**.
15. V. A. TUPCIEV, On the method of introducing viscosity in the study of problems involving decay of a discontinuity, *Dokl. Akad. Nauk. SSR* **211** (1973), 55–58; English translation in *Soviet Math. Dokl.* **14**.
16. C. M. DAFERMOS, Structure of solutions of the Riemann problem for hyperbolic systems of conservation laws, *Arch. for Rational Mechanics and Analysis* **53** (1974), 203–217.

Center for the Mathematical Sciences
University of Wisconsin
Madison

(Received April 1, 1988)

One Dimensional Infinite-horizon Variational Problems arising in Continuum Mechanics

ARIE LEIZAROWITZ & VICTOR J. MIZEL

Dedicated to Bernard D. Coleman in celebration of his sixtieth birthday

1. Introduction

In this paper we study a variational problem for real valued functions defined on an infinite semiaxis of the line. To wit, given $x \in \mathbb{R}^2$ we seek a "minimal solution" to the problem

Minimize the functional given by

$$I(w(\cdot)) = \int_0^\infty f(w(s), \dot{w}(s), \ddot{w}(s))\, ds, \tag{P$_\infty$}$$

$$w \in A_x = \{v \in W^{2,1}_{\text{loc}}(0, \infty) : (v(0), \dot{v}(0)) = x\}.$$

Here $W^{2,1}_{\text{loc}} \subset C^1$ denotes the Sobolev space of functions possessing a locally integrable second derivative, and $f = f(w, p, r)$ is a smooth function satisfying

$$\begin{gathered} f_{rr} \geqq 0, \ f(w, p, r) \geqq a\,|w|^\alpha - b\,|p|^\beta + c\,|r|^\gamma - d \quad (a, b, c, d > 0) \\ \text{where } \alpha, \gamma \in (1, \infty), \beta \in [1, \infty) \text{ satisfy } \alpha > \beta, \gamma \geqq \beta, \end{gathered} \tag{1.1}$$

as well as an upper growth condition to be described in § 2.

It can be appreciated that the notion of minimal solution for (P_∞) is a subtle one, since the infimum of I on A_x is typically either $+\infty$ or $-\infty$. The formulation which is best suited to our problem will be described and analyzed in § 3. It will also be shown in § 3 that the analysis given for (P_∞) applies to similar problems involving a functional identical to I except for the fact that integration is taken along the entire real line.

Our interest in variational problems of the form (P_∞) stems from a one-dimensional model recently proposed by BERNARD COLEMAN to describe the equilibrium behavior of a long slender bar of polymeric material under tension. It involves a fiber of material distributed along an infinite interval and possessing an equilibrium specific Helmholtz free energy function which, formally, is a higher order version of the VAN DER WAALS mean free energy for the density of a two phase fluid ([vdW], [C & H], see also [CGS]). This model goes beyond a model previously analyzed by COLEMAN in which, starting from a dynamical framework and a general nonlocal constitutive assumption for the stress in a slender rod of polymer, he arrived, by the use of quasistatic and retardation approximations in the limit of zero radius, at a constitutive relation of lower order for the equilibrium stress in a stressed one-dimensional fiber ([C1], [C2]). This relation of lower order includes, as an important special case, those constitutive formulas for the equilibrium stress in a finite fiber which arise from the minimization, under a fixed length constraint, of any one of a large class of free energies of the van der Waals type.

To describe the new, higher order, model we employ an unstressed reference configuration R for the material fiber, where $R = [Z_1, Z_2]$ is a long but finite interval and Z denotes the coordinate in R. The location z in the stressed fiber of the material point at Z in R is given in the form

$$z = \hat{z}(Z), \quad Z \in [Z_1, Z_2],$$

(time does not enter in the present equilibrium model), and the equilibrium *stretch ratio* (or "stretch") of the stressed fiber at the material point at Z in R is denoted by

$$\lambda(Z) = \hat{z}'(Z).$$

It is stipulated that, when the material is held under a fixed tension, the stress at the point Z will be that combination of the values of $\lambda(\cdot)$ and its derivatives at Z which is obtained by minimizing the free energy functional

$$I_{Z_1,Z_2}(\lambda(\cdot)) = \int_{Z_1}^{Z_2} f(\lambda(Z), \lambda'(Z), \lambda''(Z))\, dZ, \tag{1.2}$$

under the constraint that the fiber have a prescribed length:

$$\int_{Z_1}^{Z_2} \lambda(Z)\, dZ = l. \tag{1.3}$$

The form of free energy integrand proposed in this model is given by

$$f(w, p, r) = \Psi(w) - \tfrac{1}{2} b p^2 + \tfrac{1}{2} c r^2 \quad (b, c > 0), \tag{1.4}$$

where Ψ is any function possessing some of the basic features of the van der Waals potential, for instance

$$\Psi(w) = a(w - w_1)^2 (w - w_2)^2, \quad w \in \mathbb{R}, \quad \text{with } a > 0, w_2 > w_1. \tag{1.5}$$

Note that the function f given by (1.4), (1.5) obviously satisfies (1.1); in fact, much of our analysis permits a, b, c themselves to vary with w and p. We mention that the characterization of equilibrium states by means of (1.2), (1.3) is the one appropriate to a fiber held in a "hard device", one that maintains the fiber at length l.

It will be shown in § 2 that the functional I_{Z_1,Z_2} in (1.2) is bounded below. It then follows by a standard argument involving lower semicontinuity that there exists a stretch field $\lambda(\cdot)$ minimizing I_{Z_1,Z_2} subject to (1.3). Moreover, for f as in (1.4), (1.5), $\lambda(\cdot)$ is four times continuously differentiable and satisfies the Euler-

Lagrange equation:

$$\frac{d^2}{dZ^2}(c\lambda_{ZZ}) - \frac{d}{dZ}(b\lambda_Z) + \Psi'(\lambda) = T^0, \quad Z \in (Z_1, Z_2). \tag{1.6}$$

Furthermore, the tension T^0, which arises as a Lagrange multiplier associated with the constraint (1.3), is uniform over the fiber.

Since we are interested in very long physical fibers we are led to examine limiting cases in which $R = [0, \infty)$ or $R = (-\infty, \infty)$. In such cases the fixed length requirement (1.3) is useless, and we are instead led to postulate that the value T^0 of the tension is specified. This corresponds to the replacement of f in (1.4) by

$$f_0(w, p, r) = (\Psi(w) - T^0 w) - \frac{b}{2}p^2 + \frac{c}{2}r^2 \quad (b, c > 0). \tag{1.4$_0$}$$

It is easily verified that f_0 satisfies the conditions (1.1), whatever be the value of $T^0 \in \mathbb{R}$. Thus the first limiting case gives rise to problem (P_∞), the second limiting case to an analogous problem on $(-\infty, \infty)$. For convenience we restrict our attention for the remainder of this section to the integrand f in (1.4), (1.5). It will be shown in § 6 that if the parameter b is sufficiently large then the energy integral $I(\lambda(\cdot))$ in (P_∞) will have the value $-\infty$ for some choices $\lambda(\cdot) \in A_x$. Thus one cannot minimize (P_∞) in the usual sense. One way to overcome that difficulty is to consider the expression

$$J(\lambda(\cdot)) = \liminf_{L \to \infty} \frac{1}{L} \int_0^L f(\lambda(Z), \lambda'(Z), \lambda''(Z))\, dZ \tag{1.7}$$

and to look for a stretch field that minimizes J. In this paper we employ a more refined criterion to specify what is meant by a minimal solution for (P_∞), one which is a weakened version of that known in the control theory literature as the *overtaking optimality criterion* ([B & H], [Ca], [A & L]). The modification which we introduce is closely connected with the notion of *minimal energy configuration* employed by AUBRY & LE DAERON in the analysis of an infinite discrete model for crystals which undergo phase transitions in their ability to conduct electricity ([A & D]). This model, due to FRENKEL & KONTOROVA, is the object of current research by several investigators ([A & D], [G & C], [C & D], [Ma]).

The paper is organized as follows. In Section 2 we specify our notation and analyze the fixed endpoint variational problem, with f as in (1.1), corresponding to the integral in (P_∞) but taken over a bounded interval. In Section 3 we describe our criterion for a solution of (P_∞) to be mininal. In Section 4 we demonstrate the existence of a minimal energy solution, and in Section 5 we establish our main result: there always exists a *periodic* minimal solution for (P_∞). Then in Section 6 we prove that in the special case (1.4), (1.5) there is a threshold effect; for fixed a, c there is a value $b_0 > 0$, such that for $b \in (0, b_0)$ the periodic minimal solution mentioned above is constant, while for $b > b_0$ the infimum of I is $-\infty$ and the periodic solution whose existence was shown in Section 5 is nonconstant. Finally, in an appendix (Section 7) we establish an analytic result utilized in Section 4 which may be of independent interest.

2. The bounded interval problem

In pursuing our goal of analyzing the infinite semi-interval problem (P_∞) we begin by considering, for each $T > 0$ and $\boldsymbol{x}, \boldsymbol{y} \in \mathbb{R}^2$, the following variational problem for real valued functions on $[0, T]$:

Minimize the functional

$$I_T(w(\cdot)) = \int_0^T f(w(t), \dot{w}(t), \ddot{w}(t))\, dt,$$

$$w(\cdot) \in A_{x,y} = \{v \in W^{1,2}(0, T) : (v(0), \dot{v}(0)) = \boldsymbol{x}, (v(T), \dot{v}(T)) = \boldsymbol{y}\}. \qquad (P_T)$$

The function $f = f(w, p, r)$ is assumed to be smooth and to satisfy

$$\begin{aligned} &\text{(i)} \quad f_{,rr} \geqq 0, \\ &\text{(ii)} \quad f(w, p, r) \geqq a\,|x|^\alpha - b\,|p|^\beta + c\,|r|^\gamma - d, \qquad a, b, d, c > 0, \\ &\text{(iii)} \quad f(w, p, r) \leqq \varphi(w, p) + c'\,|r|^\gamma, \quad c' > 0, \end{aligned} \tag{2.1}$$

where $\alpha, \gamma > 1$, $\beta \geqq 1$ satisfy $\beta < \alpha, \beta \leqq \gamma$, and φ is continuous.

If we employ the Sobolev spaces $X = L^\beta(0, T)$, $Y = W^{1,\beta}(0, T)$, $Z = W^{2,\beta}(0, T)$, then it is an elementary consequence of the Arzelà-Ascoli theorem that these Banach spaces are compactly imbedded as follows:

$$Z \hookrightarrow Y \hookrightarrow X.$$

Hence by a result of LIONS & MAGENES [L & M, v. 1, p. 102] it follows that for each $\eta > 0$, there exists a $C(\eta) > 0$ such that

$$\|v\|_{W^{1,\beta}} \leqq \eta\,\|v\|_{W^{2,\beta}} + C(\eta)\,\|v\|_{L^\beta}, \qquad \forall\, v \in W^{2,\beta}(0, T). \tag{2.2}$$

It follows that

$$\|\dot{v}\|_{L^\beta} \leqq \eta\,\|\ddot{v}\|_{L^\beta} + \eta\,\|\dot{v}\|_{L^\beta} + (C(\eta) + \eta)\,\|v\|_{L^\beta},$$

so that for $\eta < 1$

$$(1 - \eta)^\beta\,\|\dot{v}\|_{L^\beta}^\beta \leqq (2\eta)^\beta\,\|\ddot{v}\|_{L^\beta}^\beta + [2(C(\eta) + \eta)]^\beta\,\|v\|_{L^\beta}^\beta \qquad \forall\, v \in W^{2,\beta}(0, T).$$

Putting $\eta' = [2\eta/(1 - \eta)]^\beta\, bc^{-1}$ we conclude that for each $\eta' > 0$ there exists $D(\eta') > 0$ satisfying

$$\int_0^T b\,|\dot{v}(t)|^\beta\, dt \leqq \eta' \int_0^T c\,|\ddot{v}(t)|^\beta\, dt + D(\eta') \int_0^T |v(t)|^\beta\, dt \qquad \forall\, v \in W^{2,\beta}(0, T). \tag{2.3}$$

Now for each $\delta > \beta$ and $a > 0$ there exists a $K = K(\eta', \delta, a) \geqq 1$ satisfying

$$ar^\delta + D(\eta')\,(\eta')^{-1}\,K^\beta \geqq D(\eta')\,(\eta')^{-1}\,r^\beta \qquad \forall\, r > 0.$$

Moreover, for each $\gamma \geqq \beta$ and $c > 0$ one has

$$cr^\gamma + c \geqq cr^\beta \qquad \forall\, r > 0.$$

Hence (2.3) ensures that for some $P = P(\eta') \geqq 0$ one has

$$(\eta')^{-1} \int_0^T b\,|\dot{v}(t)|^\beta\, dt \leqq \int_0^T (c\,|\ddot{v}(t)|^\gamma + a\,|v(t)|^\alpha)\, dt + P \qquad \forall\, v \in W^{2,\beta}(0, T), \tag{2.4}$$

(*e.g.*, $P(\eta') = D(\eta')(\eta')^{-1} K^\beta T + cT$ will do). By taking $\eta' \leqq 1/2$, say, we deduce that for some constants $R = R(\eta') > 0$, $Q = Q(\eta') > 0$ one has,

$$\int_0^T f(v(t), \dot{v}(t), \ddot{v}(t))\, dt \geqq \int_0^T R\, |\dot{v}(t)|^\beta\, dt - Q \qquad \forall\, v \in W^{2,\beta}(0, T). \tag{2.5}$$

Thus $I = I_T$ is bounded below on $A_{x,y}$ for each $T > 0$, $\boldsymbol{x}, \boldsymbol{y} \in \mathbb{R}^2$.

Remark 2.1. By using the existence for each $T' > T > 0$ of a bounded extension operator $E\colon W^{2,\beta}(0, T) \to W^{2,\beta}(0, T')$ (E can be chosen *uniformly* bounded for $T, T' \in [c, C]$ whenever $0 < c < C < \infty$) one readily concludes that $C(\eta)$ in (2.2) may be chosen uniformly for T varying in any compact subinterval of $(0, \infty)$. Thus the constants P, Q, R in (2.4) and (2.5) can be chosen uniformly for T varying in any such interval.

Moreover, we have the following result.

Theorem 2.1. *The function* $U_T\colon \mathbb{R}^2 \times \mathbb{R}^2 \to \mathbb{R}$ *defined by*

$$U_T(\boldsymbol{x}, \boldsymbol{y}) = \inf_{w(\cdot) \in A_{x,y}} I_T(w(\cdot)) \tag{2.6}$$

satisfies

$$\lim_{|x|+|y| \to \infty} U_T(\boldsymbol{x}, \boldsymbol{y}) = +\infty. \tag{2.7}$$

Note. Hereafter we omit the subscript T where no confusion will arise.

Proof. Given $M > 0$, it follows from (2.5) that

$$I(v(\cdot)) \geqq M \quad \text{whenever} \quad \int_0^T |\dot{v}(t)|^\beta\, dt \geqq \frac{1}{R}[M + Q].$$

Thus it will suffice to show that even for those $v(\cdot)$ satisfying

$$\int_0^T |\dot{v}(t)|^\beta\, dt < \frac{1}{R}[M + Q], \tag{2.8}$$

one has $I(v(\cdot)) \geqq M$ provided that $|\boldsymbol{x}| + |\boldsymbol{y}|$ is sufficiently large. Suppose first that $|x_1| = |v(0)|$ is sufficiently large that (2.8) implies

$$|v(t)| \geqq S, \quad 0 \leqq t \leqq T, \tag{2.9}$$

where S satisfies

$$S^\alpha \geqq \frac{1}{aT}\left[M + dT + \frac{b}{R}(Q + M)\right]. \tag{2.10}$$

Then by (2.1 ii) and (2.8)–(2.10),

$$I(v(\cdot)) \geqq \int_0^T a\, |v(t)|^\alpha\, dt - \int_0^T b\, |\dot{v}(t)|^\beta\, dt - dT \geqq M. \tag{2.11}$$

Similarly, if $|y_1| = |v(T)|$ is sufficiently large, (2.8) again implies that (2.9) holds and (2.11) follows.

Finally, suppose that (2.8) holds while $|x_1|, |y_1|$ are sufficiently small that the preceding argument does not apply. Note that (2.8) ensures that for some $t_0 \in (0, T)$

$$|\dot{v}(t_0)| \leqq \left[\frac{1}{RT}(M+Q)\right]^{1/\beta} =: \sigma.$$

Thus if $|x_2| = |\dot{v}(0)| \geqq S'$, where S' satisfies

$$S' \geqq \sigma + c^{-1/\gamma} T^{1/\gamma'} \left[M + \frac{b}{R}(M+Q) + dT\right]^{1/\gamma}, \tag{2.12}$$

then Hölder's inequality gives

$$t_0^{1/\gamma'} \left(\int_0^T |\ddot{v}(t)|^\gamma \, dt\right)^{1/\gamma} \geqq \int_0^{t_0} |\ddot{v}(t)| \, dt \geqq S' - \sigma, \qquad \left(\frac{1}{\gamma} + \frac{1}{\gamma'} = 1\right).$$

Hence (2.1ii), (2.8), and (2.12) imply

$$I(v(\cdot)) \geqq M. \tag{2.13}$$

A similar argument leads to (2.13) if $|y_2| = |\dot{v}(T)| \geqq S'$. This concludes the proof. □

Remark 2.2. A simple modification of this argument (basically by replacing T by $T/2$) reveals that (putting $m(w) := \max\{|w(t)| + |\dot{w}(t)|,\ t \in [0, T]\}$)

$$\lim_{m(w)\to\infty} I_T(w(\cdot)) = +\infty.$$

That is, for each $M > 0$ there exists a rectangle

$$Q_M = \{\boldsymbol{x} : |x_1| \leqq S, |x_2| \leqq S'\}$$

such that $I_T(w(\cdot)) \geqq M$ for any $w(\cdot)$ such that the corresponding trajectory $t \mapsto \boldsymbol{x}(t)$, $0 \leqq t \leqq T$, is not entirely contained in Q_M.

It is an elementary exercise to show that $U_T(\cdot, \cdot)$ is bounded on bounded sets, for instance by constructing polynomials belonging to $A_{x,y}$, for each $\boldsymbol{x}, \boldsymbol{y} \in \mathbb{R}^2$. We proceed to show that $U_T(\cdot, \cdot)$ is actually continuous.

Theorem 2.2. *For each* $T > 0$, *the function*

$$U_T : \mathbb{R}^2 \times \mathbb{R}^2 \to \mathbb{R}$$

defined in (2.6) *is continuous.*

Proof. 1. *Lower semicontinuity*. Given $\boldsymbol{x}, \boldsymbol{y} \in \mathbb{R}^2$, it follows from the convexity and growth conditions (2.1i), (2.1ii) [*cf.* (2.5)] (recall that $\gamma > 1$) that there exists a minimizer $w(\cdot) \in A_{x,y}$ for the functional I_T (*cf.* MORREY [M, Theorem 1.91] or GIAQUINTA [G, Theorem 3.1]]. Moreover by (2.4) one obtains the estimate

$$\begin{aligned} U_T(x, y) &= \int_0^T f(w(t), \dot{w}(t), \ddot{w}(t))\, dt \\ &\geqq (1-\eta') \int_0^T (c\,|\ddot{w}(t)|^\gamma + a\,|w(t)|^\alpha)\, dt - (1-\eta')\, P. \end{aligned} \tag{2.14}$$

Given any sequence $(x_k, y_k) \to (x, y)$, let us denote by $w_k(\cdot)$ a minimizer belonging to A_{x_k, y_k}, $k \geqq 1$. It follows from (2.14) and the local boundedness of U_T that the functions $\{w_k(\cdot)\}$ form a bounded subset of $W^{2,\gamma}(0, T)$. Since $\gamma > 1$ we can suppose, by extracting a subsequence and re-indexing, that for some $v(\cdot) \in W^{2,\gamma}$

$$w_k(\cdot) \to v(\cdot) \quad \text{weakly in } W^{2,\gamma}.$$

Thus

$$\begin{aligned} &w_k(t) \to v(t), \quad \dot{w}_k(t) \to \dot{v}(t), \text{ uniformly in } [0, T], \\ &\ddot{w}_k(\cdot) \to \ddot{v}(\cdot) \quad \text{weakly in } L^\gamma. \end{aligned} \tag{2.15}$$

These relations ensure by a lower semicontinuity result of TONELLI's (*e.g. cf.* GIAQUINTA [G, Ch. 1, Theorem 2.3]) that

$$\int_0^T f(v(t), \dot{v}(t), \ddot{v}(t))\, dt \leqq \liminf_{k\to\infty} U_T(x_k, y_k). \tag{2.16}$$

Moreover, by (2.15)

$$(v(0), \dot{v}(0)) = x, \qquad (v(T), \dot{v}(T)) = y,$$

so that (2.16) implies

$$U_T(x, y) \leqq \liminf_{k\to\infty} U_T(x_k, y_k).$$

This completes the proof of lower semicontinuity.

2. *Upper semicontinuity.* Let $w(\cdot) \in A_{x,y}$ denote as above a minimizer for I_T, and suppose $(x_k, y_k) \to (x, y)$ as $k \to \infty$. Put

$$u^k = x_k - x, \qquad v^k = y_k - y,$$

so that $u^k, v^k \to 0$ as $k \to \infty$, and define

$$\delta_k(t) = a_k + b_k t + c_k t^2 + d_k t^3,$$

where the coefficients are so chosen that

$$(\delta_k(0), \dot{\delta}_k(0)) = u^k, \quad (\delta_k(T), \dot{\delta}_k(T)) = v^k, \quad k \geqq 1. \tag{2.17}$$

(Explicitly

$$a_k = u_1^k,\ b_k = u_2^k,\ c_k = -\frac{1}{T^2}\,[3(u_1^k - v_1^k) + T(v_2^k + 2u_2^k)],$$

$$d_k = \frac{1}{T^3}\,[2(u_1^k - v_1^k) + T(u_2^k + v_2^k)].)$$

It is easy to see that

$$\delta_k(t),\ \dot{\delta}_k(t),\ \ddot{\delta}_k(t) \to 0 \quad \text{uniformly for } t \in [0, T] \text{ as } k \to \infty. \tag{2.18}$$

We define

$$z_k(t) = w(t) + \delta_k(t), \qquad k \geqq 1.$$

Then $z_k(\cdot) \in A_{x_k, y_k}$ by (2.17). so that (2.18) implies

$$|z_k(t)| \leqq |w(t)| + 1, \quad |\dot{z}_k(t)| \leqq |\dot{w}(t)| + 1, \quad |\ddot{z}_k(t)| \leqq |\ddot{w}(t)| + 1$$

for all $t \in [0, T]$ and all sufficiently large k. Consequently since $w(\cdot)$ and $\dot{w}$ take values in bounded sets (2.1 iii) implies that

$$\begin{aligned} |f(z_k(t), \dot{z}_k(t), \ddot{z}_k(t))| &\leqq \text{const} + c(|\ddot{w}(t)| + 1)^\gamma \\ &\leqq A + B\,|\ddot{w}(t)|^\gamma, \qquad t \in [0, T] \end{aligned}$$

for some constants A, B and all sufficiently large k. It now follows from the definition of $z_k(\cdot)$ and the dominated convergence theorem that

$$U_T(\boldsymbol{x}, \boldsymbol{y}) = I_T(w(\cdot)) = \lim_{k \to \infty} \int_0^T f(z_k(t), \dot{z}_k(t), \ddot{z}_k(t))\, dt,$$

and since

$$I_T(z_k) \geqq U_T(\boldsymbol{x}_k, \boldsymbol{y}_k)$$

this implies that

$$U_T(\boldsymbol{x}, \boldsymbol{y}) \geqq \limsup_{k \to \infty} U_T(\boldsymbol{x}_k, \boldsymbol{y}_k). \quad \square$$

By a simple modification of the above proof, utilizing the extension operators $E_{T,T'}$, from $W^{2,\beta}([0, T])$ to $W([0, T^{2,\beta'}])$, we can easily deduce

Corollary 2.3. *The mapping* $(T, \boldsymbol{x}, \boldsymbol{y}) \mapsto U_T(\boldsymbol{x}, \boldsymbol{y})$ *is continuous for* $T > 0$, $\boldsymbol{x}, \boldsymbol{y} \in \mathbb{R}^2$.

3. The optimality criterion

We will treat problem (P_∞) as a minimization, in the limit as $T \to \infty$, of the following functionals

$$\begin{aligned} I_T(w(\cdot)) &= \int_0^T f(w(s), \dot{w}(s), \ddot{w}(s))\, ds, \\ w(\cdot) \in A_x &:= \{v \in W^{2,1}_{\text{loc}} : (v(0), \dot{v}(0)) = \boldsymbol{x}\}. \end{aligned} \tag{3.1}$$

(Since this is an equilibrium problem, the use of T as parameter should cause no confusion.) However, in many cases it turns out that for *every* $w(\cdot) \in A_x$ one has $I_T(w(\cdot)) \to \infty$ as $T \to \infty$, in which case the minimization of $I_\infty(w(\cdot))$ has no meaning. Alternatively, it may turn out that there are functions $w(\cdot) \in A_x$ for which $I_T(w(\cdot)) \to -\infty$ as $T \to \infty$, in which case a straightforward minimization of $I_\infty(w(\cdot))$ again has no meaning.

As pointed out in the Introduction, one way to reduce this difficulty is to minimize the 'average energy over large intervals', that is to minimize the functional J defined by

$$J(w(\cdot)) = \liminf_{T\to\infty} \frac{1}{T} I_T(w(\cdot)), \qquad w(\cdot) \in A_x. \tag{3.2}$$

The infimum of the values assumed by J, namely

$$\mu = \inf J(w(\cdot)), \qquad w(\cdot) \in A_x, \tag{3.3}$$

is then called the *minimal growth rate* of the energy (it is easily seen that μ is independent of the initial vector $\boldsymbol{x}$). Unfortunately, this approach, too, suffers from a serious drawback, which we describe for the case in which the infimum in (3.3) is actually attained. Given an interval $[0, T_0]$, where T_0 may be arbitrarily large, let $w_0(\cdot)$ be *any* given element of A_x subject only to the condition

$$(w(T_0), \dot{w}(T_0)) = \boldsymbol{x}.$$

Now define $w_0(\cdot)$ on the half axis $[T_0, \infty)$ in such a way that $s \mapsto w_0(s + T_0)$ is in A_x, with

$$\liminf_{T\to\infty} \frac{1}{T - T_0} \int_{T_0}^{T} f(w_0(s), \dot{w}_0(s), \ddot{w}_0(s))\, ds = \mu.$$

Clearly the extension of $w_0(\cdot)$ to $[0, \infty)$ obtained in this way satisfies

$$J(w_0(\cdot)) = \mu, \qquad w_0(\cdot) \in A_x.$$

This is an unsatisfactory situation since for us the infinite horizon problem (P_∞) is merely a mathematical idealization for modelling problems on large intervals, while the above function $w_0(\cdot)$ is a very poor approximation on an interval of length T_0, where T_0 may be very large. This arbitrariness in the definition of $w(\cdot)$ on an initial interval can be removed by imposing some condition of stationarity. However, most conditions of this sort, such as periodicity, say, are rather artificial.

Another type of optimality criterion for infinite horizon problems was introduced in the economics literature by GALE [Ga] and VON WEIZSACKER [vW] and has been used in control theory by *e.g.* BROCK & HAURIE [B & H]. CARLSON [Ca] and ARTSTEIN & LEIZAROWITZ [A & L]. It is referred to as the "overtaking optimality criterion".

Definition 3.1. A function $w^*(\cdot) \in A_x$ will be called *overtaking minimal* relative to $\boldsymbol{x}$ if

$$\limsup_{T\to\infty} [I_T(w^*(\cdot)) - I_T(w(\cdot))] \leqq 0 \qquad \text{for all } w(\cdot) \in A_x.$$

Thus if $w^*(\cdot)$ is overtaking minimal then for each $\varepsilon > 0$ and $w(\cdot) \in W^{2,1}_{\text{loc}}$ with $\boldsymbol{x}(0) = \boldsymbol{x}^*(0)$., there is a T_0 such that $I_T(w^*(\cdot)) < I_T(w(\cdot)) + \varepsilon$, for every $T > T_0$. This implies, in particular, that $w^*(\cdot)$ is a minimizer of J in (3.2).

Just as the minimal growth rate criterion for minimizing $J(\cdot)$ is too loose, since there are infinitely many functions with a minimal 'average energy over large intervals', the overtaking optimality condition is too strict, and in general there will be no overtaking minimal functions. However, a closely related notion was considered by AUBRY & LE DAERON [A & D] in their study of the discrete Frenkel-Kontorova model describing one-dimensional crystals with phase transitions (*cf.* [N]). There they minimized an energy expression of the form

$$\sum_{k=-N}^{M} u(x_k, x_{k+1}) \quad \text{with } -\infty < \ldots < x_k < x_{k+1} < \ldots < \infty,$$

as $N, M \to \infty$ using the following criterion. A sequence $\{x_k^*\}_{k=-\infty}^{\infty}$ is called a *minimal energy configuration* if for each $M, N > 0$ the inequality

$$\sum_{k=-N}^{M} u(x_k^*, x_{k+1}^*) \leqq \sum_{k=-N}^{M} u(x_k, x_{k+1})$$

holds for every increasing sequence $\{x_k\}_{k=-N}^{M}$ satisfying

$$x_{-N} = x_{-N}^*, \qquad x_M = x_M^*.$$

An analogous criterion can be adapted to our framework, as follows.

Definition 3.2. A function $w^* \in W_{\text{loc}}^{2,1}$ is called a *locally minimal energy configuration* if

$$\int_{T_1}^{T_2} f(w^*(s), \dot{w}^*(s), \ddot{w}^*(s))\, ds \leqq \int_{T_1}^{T_2} f(w(s), \dot{w}(s), \ddot{w}(s))\, ds$$

for each T_1, T_2 such that $0 \leqq T_1 < T_2$ and each $w \in W^{2,1}([T_1, T_2])$ satisfying

$$(w(T_1), \dot{w}(T_1)) = (w^*(T_1), \dot{w}^*(T_1)), \quad (w(T_2), \dot{w}(T_2)) = (w^*(T_2), \dot{w}^*(T_2)).$$

If in addition to the above property $w^*(\cdot)$ also provides the minimal growth rate of energy, then $w^*(\cdot)$ is called a *minimal energy configuration.*

It is clear that if $w^*(\cdot)$ is overtaking minimal then it is also a minimal energy configuration. In the next section we will construct for each $\boldsymbol{x} \in \mathbb{R}^2$ a $w^*(\cdot) \in A_x$ which is a minimal energy configuration. The analysis given there will involve a reformulation of (P_∞) in discrete terms, but as will be seen, the reformulation is *not* an approximation to (P_∞).

The discrete problem to be analyzed in § 4, is of the following type. Consider expressions of the form

$$C_N(X) = \sum_{k=0}^{N-1} v(\boldsymbol{x}_k, \boldsymbol{x}_{k+1}), \tag{3.4}$$

for a given $\boldsymbol{x} \in \mathbb{R}^2$, where $X = \{\boldsymbol{x}_k\}_{k=0}^{\infty}$ is a sequence in $\mathbb{R}^2$ such that $\boldsymbol{x}_0 = \boldsymbol{x}$ and $v: \mathbb{R}^2 \times \mathbb{R}^2 \to \mathbb{R}$ is a continuous function satisfying

$$v(\boldsymbol{x}, \boldsymbol{y}) \to \infty \quad \text{as} \quad |\boldsymbol{x}| + |\boldsymbol{y}| \to \infty. \tag{3.5}$$

Remark. It will be seen that if $x(\cdot) = \begin{pmatrix} w(\cdot) \\ \dot{w}(\cdot) \end{pmatrix}$ is (globally) bounded for the locally minimal energy configuration $w(\cdot)$, then $w(\cdot)$ is automatically a minimal energy configuration.

It is desired to minimize $C_N(X)$ as $N \to \infty$, either in the overtaking sense or in the weaker sense of minimal energy configuration. A study of this problem was presented in [L]. There it was shown that when (3.5) holds one can restrict attention, insofar as optimality considerations are concerned, to sequences X lying inside some fixed ball:

$$|x_k| \leqq L \quad \forall k \geqq 0,$$

where $L > 0$ is a constant which does not depend on X (see [L], Theorem 8.1). Moreover the following result was proved ([L], Theorem 3.1).

Theorem 3.3. *Let* $v: \mathbb{R}^2 \times \mathbb{R}^2 \to \mathbb{R}$ *be a continuous function satisfying* (3.5). *Given an* $x \in \mathbb{R}^2$, *consider the expressions* $C_N(X)$ *where* $x_0 = x$. *Then there exist constants* μ *and* M *such that*

1. *For every* $X = \{x_k\}_{k=0}^{\infty}$ *the inequality*

$$\sum_{k=0}^{N} [v(x_k, x_{k+1}) - \mu] \geqq -M$$

holds for all $N \geqq 1$,

2. *There is a sequence* X^* *satisfying*

$$\left| \sum_{k=0}^{N} [v(x_k^*, x_{k+1}^*) - \mu] \right| \leqq M, \qquad \forall N \geqq 1.$$

The scalar μ describes the minimal growth rate for average energy of the energy expressions $C_N(X)$ in (3.4). By Theorem 3.3 every such expression is bounded below by a linear function of N whose slope is μ, while there is a sequence X^* for which $C_N(X^*)$ is bounded both from above and below by such functions.

Definition 3.4. A bounded sequence X^* will be referred to as a *minimal energy sequence* if for each $N_2 > N_1 > 0$ the inequality

$$\sum_{k=N_1}^{N_1-1} v(x_k^*, x_{k+1}^*) \leqq \sum_{k=N_1}^{N_2-1} v(x_k, x_{k+1})$$

holds for every sequence $\{x_k\}_{k=N_1}^{N_2}$ satisfying

$$x_{N_1} = x_{N_1}^*, \qquad x_{N_2} = x_{N_2}^*.$$

Now it has also been shown [L, Prop. 5.1] that Theorem 3.3 is equivalent to the following result.

Theorem 3.5. *Let* $v:\mathbb{R}^2\times\mathbb{R}^2\to\mathbb{R}$ *be continuous and satisfy* (3.5). *Then* $v(\cdot,\cdot)$ *can be decomposed in the form*

$$v(x, y) = \mu + \pi(x) - \pi(y) + \theta(x, y), \tag{3.6}$$

where μ *is a constant,* $\pi:\mathbb{R}^2\to\mathbb{R}$ *is continuous, and* $\theta:\mathbb{R}^2\times\mathbb{R}^2\to\mathbb{R}$ *is a continuous function satisfying*

$$\min_{y\in\mathbb{R}^2}\theta(x, y) = 0 \quad \textit{for every } x\in\mathbb{R}^2. \tag{3.7}$$

4. Existence of a minimal energy configuration

In this section we will prove the existence, for each $x\in\mathbb{R}^2$, of a minimal energy configuration in A_x. The construction will be given in two stages. First we consider a discrete reformulation of our problem and construct a minimal energy sequence $X^* = \{x_k^*\}_{k=0}^{\infty}$ for it (recall Definition 3.4). This sequence will determine the values of $(w^*(\cdot), \dot{w}^*(\cdot))$ at the points $\{kT\}_{k=0}^{\infty}$, for some fixed $T>0$ of a minimal energy configuration $w^*(\cdot)$ through the relation

$$(w^*(kT), \dot{w}^*(kT)) = x_k^* \quad k\geqq 0.$$

Then $w^*(\cdot)$ will be determined in each interval $[kT, (k+1)T]$ as a minimizer over $W^{2,1}([kT. (k+1)T])$ for

$$\int_{kT}^{(k+1)T} f(w(s), \dot{w}(s), \ddot{w}(s))\, ds, \quad \text{subject to}$$

$$(w(kT), \dot{w}(kT)) = x_k^* \quad (w((k+1)T), \dot{w}((k+1)T)) = x_{k+1}^*.$$

For a fixed $T>0$ consider the function U_T: $\mathbb{R}^2\times\mathbb{R}^2\to\mathbb{R}$ defined in (2.6). Examine the energy expressions associated with U_T, namely the quantities $C_N(X)$ defined for each sequence $X = \{x_k\}_{k=0}^{\infty}\subset\mathbb{R}^2$ by

$$C_N(X) = \sum_{k=0}^{N-1} U_T(x_k, x_{k+1}). \tag{4.1}$$

The following result will be proved.

Theorem 4.1. *For each fixed initial value* $x_0 = x\in\mathbb{R}^2$ *there is a bounded minimal energy sequence* X^*.

Proof. By Theorem 2.1 the function $U_T(\cdot,\cdot)$ satisfies

$$U_T(x, y)\to\infty \quad \text{as} \quad |x| + |y|\to\infty.$$

Then by Theorem 3.5 one can decompose U_T as follows:

$$U_T(x, y) = T\mu_T + \pi_T(x) - \pi_T(y) + \theta_T(x, y), \tag{4.2}$$

with μ_T a scalar, $\pi_T:\mathbb{R}^2\to\mathbb{R}$ continuous, and $\theta_T:\mathbb{R}^2\times\mathbb{R}^2\to\mathbb{R}$ a continuous function which satisfies

$$\min_{y\in\mathbb{R}^2}\theta_T(x, y) = 0 \quad \textit{for each } x\in\mathbb{R}^2. \tag{4.3}$$

Now define $\{x_k^*\}_{k=0}^{\infty}$ recursively as a sequence which satisfies

$$x_0^* = x_0, \quad \theta_T(x_k^*, x_{k+1}^*) = 0, \quad k = 0, 1, 2, \ldots \tag{4.4}$$

This scheme is applicable for each $x_0 \in \mathbb{R}^2$ by (4.3), and it results in a bounded sequence as follows easily from Theorem 3.3(1) and the fact that $\pi(y) \to \infty$ as $|y| \to \infty$. We claim that $X^* = \{x_k^*\}_{k=0}^{\infty}$ is a minimal energy sequence. For suppose that $1 \leqq M < N$ and that X is any sequence in $\mathbb{R}^2$ satisfying

$$x_M = x_M^*, \quad x_N = x_N^*. \tag{4.5}$$

Then (4.2) implies that

$$\sum_{k=M}^{N-1} U_T(x_k, x_{k+1}) = (N - M)\,\mu_T T + \pi_T(x_M) - \pi_T(x_N) + \sum_{k=M}^{N-1} \theta_T(x_k, x_{k+1}), \tag{4.6}$$

while by (4.4) one has

$$\sum_{k=M}^{N-1} U_T(x_k^*, x_{k+1}^*) = (N - M)\,\mu_T T + \pi_T(x_M^*) - \pi_T(x_N^*). \tag{4.7}$$

Comparing (4.6), (4.7) in the light of condition (4.5) and the nonnegativity of $\theta_T(\cdot, \cdot)$ yields

$$\sum_{k=M}^{N-1} U_T(x_k, x_{k+1}) \geqq \sum_{k=M}^{N-1} U_T(x_k^*, x_{k+1}^*),$$

which concludes the proof. □

We will now use the minimal energy sequence X^* to define a minimal energy configuration $w^*(\cdot)$.

Consider, for each integer $k \geqq 0$, problem (P_T) posed in the beginning of Section 2, with $(x, y) = (x_k^*, x_{k+1}^*)$. As indicated in the proof of lower semicontinuity in Theorem 2.2, there exists a minimizer $w_k(\cdot)$ for this problem. Now define $w^* : (0, \infty) \to \mathbb{R}$ as follows

$$w^*(t) = w_k(t - kT), \quad t \in [kT, (k+1)\,T), \quad k \geqq 0. \tag{4.8}$$

Thus $w^*|_{[kT,(k+1)T]}$ minimizes the expression

$$\int_{kT}^{(k+1)T} f(w(s), \dot{w}(s), \ddot{w}(s))\, ds, \quad w(\cdot) \in W^{2,1}([kT, (k+1)\,T)]),$$

subject to the conditions $x(kT) = x_k^*$, $x((k+1)\,T) = x_{k+1}^*$, where we denote $x(s) = \begin{pmatrix} w(s) \\ \dot{w}(s) \end{pmatrix}$ (thus by the proof of Theorem 2.1 $x(\cdot)$ is a bounded function from $(0, \infty)$ to $\mathbb{R}^2$). We proceed to demonstrate that this construction does provide a minimal energy configuration.

Theorem 4.2. *The function $w^*(\cdot)$ defined in* (4.8) *is a minimal energy configuration for problem* (P_∞).

In order to prove Theorem 4.2 we will need to compare, for each $w(\cdot) \in A_x$, the quantities

$$\int_{T_1}^{T_2} f(w(s), \dot{w}(s), \ddot{w}(s))\, ds \quad \text{and} \quad \int_{T_1}^{T_2} f(w^*(s), \dot{w}^*(s), \ddot{w}^*(s))\, ds$$

for every pair $T_2 > T_1 > 0$, not just integer multiples of some fixed $T > 0$. This will require the use of two results given below.

Lemma 4.3. *Let $\{a_k\}_{k=1}^{\infty}$ be an increasing sequence of positive numbers such that $a_k \to \infty$ as $k \to \infty$. Consider numbers $T > 0$ with the property that for every $m \geqq 1$,*

$$\inf_{k,n} \{a_k - nT : k \geqq m, n \geqq 0, a_k \geqq nT\} = 0. \tag{4.9}$$

Then there is a set $D \subset [0, \infty)$ with $m(D^c) = 0$ such that every $T \in D$ satisfies (4.9). (Here $m(ds)$ is Lebesgue measure and D^c denotes the complement of D in $[0, \infty)$).

The proof of this lemma will be given in Section 7.

Now set $S_x := \{X = \{x_k\}_{k=0}^{\infty} \subset \mathbb{R}^2 : x_0 = x\}$, let $\tilde{S}_x$ denote the set of *periodic* sequences in S_x, and consider the scalar μ_T and the function $\pi_T(\cdot)$ appearing in the decomposition (4.2) for $U_T(\cdot, \cdot)$. It has been shown (*cf.* [L] § 3, Prop. 5.1) that the following formulas define a μ_T and a $\pi_T(\cdot)$ for which (4.2) holds:

$$\mu_T = \inf_{S_x} \left[\liminf_{N\to\infty} \frac{1}{NT} C_N(X)\right] = \inf_{\tilde{S}_x} \left[\lim_{N\to\infty} \frac{1}{NT} C_N(X)\right], \tag{4.10}$$

$$\pi_T = \inf_{S_x} \left[\liminf_{N\to\infty} [C_N(X) - NT\mu_T]\right]. \tag{4.11}$$

Remark 4.3a. Here the lack of dependence of μ_T on x follows immediately from the *form* of (4.10), while its lack of dependence on T follows readily from Lemma 4.3, so that we obtain a $\mu \in \mathbb{R}$ such that

$$\mu_T = \mu \quad \text{for all } T > 0. \tag{4.12}$$

Moreover the function $\pi_T(\cdot)$ is (almost) independent of $T > 0$. In fact by using Lemma 4.3 we are able to prove the following.

Proposition 4.4. *There exists a continuous function $\pi(\cdot): \mathbb{R}^2 \to \mathbb{R}$ and a set $D \subset [0, \infty)$, with $m(D^c) = 0$, such that*

$$\pi_T(x) = \pi(x) \quad \text{for every } x \in \mathbb{R}^2 \quad \text{and} \quad T \in D. \tag{4.13}$$

Moreover, the decomposition (4.2) for $U_T(\cdot, \cdot)$ can be replaced by

$$U_T(x, y) = T\mu + \pi(x) - \pi(y) + \theta'_T(x, y), \quad \forall\, x, y \in \mathbb{R}^2 \quad \text{and} \quad T > 0 \tag{4.14}$$

where θ'_T is a continuous function which satisfies the condition

$$\min_{y\in\mathbb{R}^2} \theta'_T(x, y) = 0 \quad \text{for each } x \in \mathbb{R}^2. \tag{4.15}$$

Proof. By the proof of Theorem 2.1 it can be seen that the set $S_x^1 = S_x^1(T)$ of all sequences $\{X\} \in S_x$ for which the lim inf in (4.11) does not exceed $\pi_T(\boldsymbol{x}_0) + 1$ has the feature that

$$X \in S_{x_0}^1 \Rightarrow X = \{\boldsymbol{x}_k\}_{k=0}^{\infty} \subset K_0, \tag{4.16}$$

where K_0 is a compact subset of $\mathbb{R}^2$ which depends on the choice of $\boldsymbol{x}_0$ and $T > 0$. Moreover by Remark 2.1 it follows that for each interval $0 < \alpha \leqq T \leqq \beta < \infty$ the compact set K_0 can be chosen sufficiently large that (4.16) is valid for all $S_{x_0}^1(T)$, $\alpha \leqq T \leqq \beta$. Furthermore, the proof of Theorem 2.1 together with Remark 2.1 also implies that there is a compact set $K_1 \supset K_0$ with the following property. Given any $X \subset K_0$ and $T \in [\alpha, \beta]$, let $w(\cdot): [0, \infty) \to \mathbb{R}$ be defined as in (4.8):

$$w(t) = w_k(t - kT), \quad t \in [kT, (k+1)T), \quad k = 0, 1, 2, \ldots, \tag{4.17}$$

with $w_k(\cdot) \in W^{1,2}([0, T])$ a minimizer for problem (P_T) corresponding to $(\boldsymbol{x}, \boldsymbol{y}) = (\boldsymbol{x}_k, \boldsymbol{x}_{k+1})$. Then the associated function $\boldsymbol{x}: (0, \infty) \to \mathbb{R}^2$ defined by

$$\boldsymbol{x}(t) = \begin{pmatrix} w(t) \\ \dot{w}(t) \end{pmatrix}, \quad t \geqq 0$$

satisfies

$$\boldsymbol{x}(t) \in K_1, \quad \text{all } t \geqq 0. \tag{4.18}$$

Fix $T \in [\alpha, \beta]$. Given any $\varepsilon > 0$ let $X = X(\varepsilon) \in S_x^1$ be such that for this sequence the lim inf on the right of (4.11) is within ε of $\pi_T(\boldsymbol{x})$; let $w_\varepsilon(\cdot) \in A_x$ be the function associated with $X(\varepsilon)$ as in (4.17); and let $\boldsymbol{x}^\varepsilon(\cdot)$ be the corresponding $\mathbb{R}^2$-valued function. Thus $\boldsymbol{x}^\varepsilon(t) \in K_1$, for all $t \geqq 0$. Next let $N_j \to \infty$ be a sequence of integers satisfying

$$\liminf_{N\to\infty} [C_N(X(\varepsilon)) - NT\mu] = \lim_{j\to\infty} [C_{N_j}(X(\varepsilon)) - N_j T\mu]. \tag{4.19}$$

Set $a_k^\varepsilon = N_k T$, $k \geqq 0$. Then by Lemma 4.3 there is a set $D = D(\{a_k^\varepsilon\}_{k=0}^{\infty}) \subset (0, \infty)$ of full measure for which the condition (4.9) holds. Given any $T' \in D$ let the sequence $Y = (y_j\}_{j=0}^{\infty}$ be defined as

$$y_j = \boldsymbol{x}^\varepsilon(jT'), \quad j \geqq 0,$$

with $\boldsymbol{x}^\varepsilon(\cdot)$ as above. Clearly it follows that

$$U_T(y_j, y_{j+1}) \leqq \int_{jT'}^{(j+1)T'} f(w_\varepsilon(s), \dot{w}_\varepsilon(s), \ddot{w}_\varepsilon(s))\, ds, \quad j \geqq 0. \tag{4.20}$$

Now by (4.9) there exists a sequence $M_l \to \infty$, $l \geqq 1$, such that

$$\operatorname{dist}(M_l T', \{a_k\}) \to 0.$$

If we denote by a_{k_l} that element of $\{a_k\}$ satisfying

$$a_{k_l - 1} \leqq M_l T' < a_{k_l}, \quad l \geqq 1$$

then we obtain by use of (4.16)

$$C_{M_l}(Y) - \mu M_l T' - (C_{N_{K_l}}(X_\varepsilon) - \mu N_{k_l} T)$$

$$\leqq - \int_{M_l T'}^{a_{k_l}} f(w_\varepsilon(s)\, \dot{w}_\varepsilon(s), \ddot{w}_\varepsilon(s))\, ds - \mu(a_{k_l} - M_l T') \tag{4.21}$$

$$\leqq - \int_{M_l T'}^{a_{k_l}} (a\,|w_\varepsilon(s)|^\alpha - b\,|\dot{w}_\varepsilon(s)|^\beta - d)\, ds - \mu(a_{k_l} - M_l T'),$$

where we have used (2.1 ii) to obtain the last inequality. Now $x^\varepsilon(t) \in K_1$ for all $t \geqq 0$, so the integrand in the last integral in (4.21) is bounded uniformly, while by (4.9) $\liminf_{l\to\infty} \{a_{k_l} - M_l T'\} = 0$. Hence (4.21) implies that

$$\pi_{T'}(x_0) \leqq \pi_T(x_0) + \varepsilon \quad \forall\, T' \in D(\{a_k\}_{k=0}^\infty). \tag{4.22}$$

By taking a sequence $\varepsilon_m \to 0$, we deduce that the set

$$D^* = \bigcap_{m=0}^\infty D(\{a_k^{\varepsilon_m}\}_{k=0}^\infty)$$

has full measure. Moreover it follows by (4.22) that

$$\pi_{T'}(x_0) \leqq \pi_T(x_0) \quad \text{for all } T' \in D^*. \tag{4.22'}$$

Now set

$$\pi(x_0) = \inf\{\pi_T(x_0);\, T > 0\}. \tag{4.23}$$

It will be seen below that $\pi(x_0) \neq -\infty$. Given $\delta > 0$, select $T := T_\delta > 0$ such that

$$\pi_T(x_0) < \pi(x_0) + \delta.$$

Thus we obtain from (4.22') the existence of a set $D^* = D^*(\delta)$ such that $m((D^*)^c) = 0$ and

$$\pi_{T'}(x_0) \in [\pi(x_0), \pi(x_0) + \delta] \quad \text{for all } T' \in D^*(\delta).$$

By taking a sequence $\delta_l \to 0$ and setting $D^{**}(x_0) = \bigcap_{l=0}^\infty D^*(\delta_l)$ we deduce that

$$\pi_{T'}(x_0) = \pi(x_0) \quad \text{for all } T' \in D^{**}(x_0). \tag{4.24}$$

Note that if in (4.23) $\pi(x_0) = -\infty$, the same basic reasoning would show that $\pi_{T'}(x_0) = -\infty$ for all T' in a full set D^{**}, contradicting the fact that $\pi_{T'}(x_0) \in \mathbb{R}$.

The continuity of the function $\pi(\cdot)$ defined as in (4.23) now follows from the observation that for any sequence $y_{0,n} \to x_0$ there are values $T' \in \bigcap_{n=0}^\infty D^{**}(y_{0,n})$. Hence by the continuity of $\pi_{T'}(\cdot)$,

$$\pi(y_{0,n}) = \pi_{T'}(y_{0,n}) \Rightarrow \pi_{T'}(x_0) = \pi(x_0).$$

To demonstrate (4.13), it suffices to examine a countable dense set $\{z_i\} \subset \mathbb{R}^2$. Then setting $D = \bigcap_{i=0}^{\infty} D^{**}(z_i)$ we conclude that whenever $T' \in D$

$$\pi_{T'}(z_i) = \pi(z_i) \qquad i \geqq 0 .$$

It now follows by the continuity of $\pi_{T'}(\cdot)$ and $\pi(\cdot)$ that (4.13) is valid.

It remains to prove that (4.14), (4.15) hold. Examine for each $T > 0$ the function $\theta'_T(\cdot, \cdot)$ defined by

$$\theta'_T(x, y) := U_T(x, y) - T\mu + \pi(y) - \pi(x), \qquad x, y \in \mathbb{R}^2 . \tag{4.24}$$

Since $(T, x, y) \mapsto U_T(x, y)$ is continuous (see Corollary 2.3) it follows that the function

$$(T, x, y) \mapsto \theta'_T(x, y)$$

is continuous, and we know by (4.13) that

$$\theta'_T(\cdot, \cdot) = \theta_T(\cdot, \cdot) \qquad \text{for all } T \in D . \tag{4.25}$$

It follows from (4.24) that $\theta'_T(x, y) \to \infty$ as $|y| \to \infty$, uniformly on compact sets of the form

$$(T, x) \in [\alpha, \beta] \times S .$$

Thus there is a bounded set S_1 such that $\theta'_T(x, y) = 0$ with $(T, x) \in [\alpha, \beta] \times S \Rightarrow y \in S_1$. This together with the fact that for all $T \in D$

$$\min_{y \in \mathbb{R}^2} \theta'_T(x, y) = 0 \qquad \text{for} \qquad x \in S , \tag{4.26}$$

implies the validity of (4.26) for *all* $T > 0$. □ (Prop. 4.4.)

Proof of Theorem 4.2. Given $T_2 > T_1 > 0$, let $w(\cdot) \in W^{2,1}_{\text{loc}}$ satisfy

$$x(T_1) = x^*(T_1), \qquad x(T_2) = x^*(t_2)$$

where

$$x(\cdot) = \begin{pmatrix} w(\cdot) \\ \dot{w}(\cdot) \end{pmatrix}, \qquad x^*(\cdot) = \begin{pmatrix} w^*(\cdot) \\ \dot{w}^*(\cdot) \end{pmatrix},$$

with $w^*(\cdot)$ defined as in (4.8). We shall suppose that neither T_1 nor T_2 is a multiple of the fixed T relative to which $w^*(\cdot)$ was defined (the other cases are simpler). Denote by M, N the nonnegative integers determined by

$$T_1 \in ((M-1), T, MT), \qquad T_2 \in (NT, (N+1)\,T) .$$

Then we can estimate

$$\begin{aligned} \int_{T_1}^{T_2} f(w(s), \dot{w}(s), \ddot{w}(s))\, ds \geqq{} & U_{MT-T_1}(x(T_1), x(MT)) \\ & + \sum_{k=M}^{N-1} U_T(x(kT), x((k+1))\,T) \\ & + U_{T_2-NT}(x(NT), x(T_2)) \\ \geqq{} & (T_2 - T_1)\,\mu + \pi(x(T_1)) - \pi(x(T_2)) , \end{aligned}$$

where we have used (4.13).

We proceed to show that $w^*(\cdot)$ yields equality:

$$\int_{T_1}^{T_2} f(w^*(s), \dot{w}^*(s), \ddot{w}^*(s))\, ds = (T_2 - T_1)\,\mu + \pi(x(T_2)) - \pi(x(T_1)). \tag{4.28}$$

For this we examine the integral over $[(M-1)\,T, (N+1)\,T]$. By the definition of $w^*(\cdot)$ we have

$$\int_{(M-1)T}^{(N+1)T} f(w^*(s), \dot{w}^*(s), \ddot{w}^*(s))\, ds = (N - M + 2)\, T\mu$$
$$+ \pi(x^*((M-1)\,T)) - \pi(x^*((N+1)\,T)).$$

On the other hand, using the decomposition into integrals over the intervals $[(N-1)\,T, T_1]$, $[T_1, T_2]$ and $[T_2, N+1)\,T]$ we obtain by (4.14)

$$\int_{(M-1)T}^{(N+1)T} f(w^*(s), \dot{w}^*(s), \ddot{w}^*(s))\, ds \geqq U_{T_1-(M-1)T}(x^*((M-1)\,T), x^*(T_1))$$
$$+ U_{T_2-T_1}(x^*(T_1), x^*(T_2)) + U_{(N+1)T-T_2}(x^*(T_2), x^*((N+1)\,T))$$
$$\geqq (N - M + 2)\, T\mu + \pi(x^*((M-1)\,T)) - \pi(x^*((N+1)\,T))$$
$$+ \theta'_{T_1-(M-1)T}(x^*((M-1)\,T), x^*(T_1)) + \theta'_{T_2-T_1}(x^*(T_1), x^*(T_2))$$
$$+ \theta'_{(N+1)T-T_2}(x^*(T_2), x^*((N+1)\,T)).$$

By comparing these two decompositions we conclude that each of the θ' terms is 0. In particular

$$\theta'_{T_2-T_1}(x^*(T_1), x^*(T_2)) = 0,$$

whence (4.28) follows. □

5. Periodic minimal energy configuration

In this section we will demonstrate the existence of periodic minimal energy configurations. In order to state our result concisely we introduce the following notation for use with any integrand f of the sort described in (2.1):

$$m_f := \inf\{f(w, 0, s) : (w, s) \in \mathbb{R}^2\}. \tag{5.1}$$

We will prove the following assertion.

Theorem 5.1. *Suppose that the integrand f in (P_∞) is such that*

$$\mu < m_f, \tag{5.2}$$

where μ is the minimal growth rate for (P_∞). Then there exists a nonconstant periodic minimal energy configuration $w^*(\cdot)$ for (P_∞).

If f has the further property

$$\inf\{f(w, 0, 0) : w \in \mathbb{R}\} = m_f, \tag{5.1'}$$

then whenever (5.2) *fails there exists a constant minimal energy configuration*

$$w^*(s) = \bar{w} \quad \forall\, s \geqq 0,$$

where $\bar{w} \in \mathbb{R}$ *is any value for which*

$$m_f = f(\bar{w}, 0, 0).$$

Remark. It will be shown in § 6 that (5.2) and (5.1′) hold for a large and interesting class of problems.

The following result, which characterizes those periodic configurations which are minimal energy configurations, will be needed.

Lemma 5.2. *Let* $w(\cdot)$ *be a periodic configuration of period* $T > 0$:

$$w(t+T) = w(t) \quad \forall\, t \geqq 0.$$

Then $w(\cdot)$ *is a minimal energy configuration if and only if*

$$\frac{1}{T}\int_0^T f(w(s), \dot{w}(s), \ddot{w}(s))\, ds = \mu. \tag{5.3}$$

Proof. By the definition of μ [*cf.* (3.3)] one deduces the inequality

$$\mu \leqq \liminf_{k\to\infty} \frac{1}{kT}\int_0^{kT} f(w(t), \dot{w}(t), \ddot{w}(t))\, dt = \frac{1}{T}\int_0^T f(w(t), \dot{w}(t), \ddot{w}(t))\, dt.$$

Moreover if the inequality is strict then it is easily seen (using the periodicity) that $w(\cdot)$ cannot be a minimal energy configuration. Hence (5.3) is certainly necessary for $w(\cdot)$ to be a minimal energy configuration.

Now suppose (5.3) holds. Let $v(\cdot)$ be any configuration and let T_1, T_2 be a given pair of points in $[0, \infty)$, with $0 \leqq T_1 < T_2$. We proceed to show that if $v(T_1) = w(T_1)$, $\dot{v}(T_1) = \dot{w}(T_1)$, $v(T_2) = w(T_2)$, $\dot{v}(T_2) = \dot{w}(T_2)$ then

$$\int_{T_1}^{T_2} f(w(t), \dot{w}(t), \ddot{w}(t))\, dt \leqq \int_{T_1}^{T_2} f(v(t), \dot{v}(t), \ddot{v}(t))\, dt. \tag{5.4}$$

Let the integers m, n be determined by

$$(m-1)\,T \leqq T_1 < mT, \qquad nT \leqq T_2 < (n+1)\,T.$$

We now compute the integral I over $[(m-1)\,T, (n+1)\,T]$ in two ways, as follows. By (5.3) and the periodicity of $w(\cdot)$

$$I = \sum_{k=m-1}^{n} \int_{kT}^{(k+1)T} f(w(t), \dot{w}(t)), \ddot{w}(t))\, dt = \sum_{k=m-1}^{n} \mu T = (n-m+1)\,\mu T. \tag{5.5}$$

On the other hand, putting $x(t) = \begin{pmatrix} w(t) \\ \dot{w}(t) \end{pmatrix}$ we obtain

$$I \geqq U_{T_1-(m-1)T}(x((m-1)\,T), x(T_1)) + U_{T_2-T_1}(x(T_1), x(T_2)) \\ + U_{(n+1)T-T_2}(x(T_2), x((n+1)\,T)).$$

Using the decomposition (4.15) for $U(\cdot, \cdot)$ we obtain

$$I \geqq (T_1 - (m-1)T)\mu + \theta_{T_1-(m-1)T}(x((m-1)T), x(T_1)) + (T_2 - T_1)\mu$$
$$+ \theta_{T_2-T_1}(x(T_1), x(T_2)) + ((n+1)T - T_2)\mu + \theta_{(n+1)T-T_2}(x(T_2), x((n+1)T)).$$

Thus, because the functions $\theta(\cdot, \cdot)$ are nonnegative,

$$I \geqq (n - m + 2) T\mu + \theta_{T_2-T_1}(x(T_1), x(T_2)). \tag{5.6}$$

A comparison with (5.5) shows that equality holds in (5.5) and

$$\theta_{T_2-T_1}(x(T_1), x(T_2)) = 0.$$

Hence

$$\int_{T_1}^{T_2} f(w(t), \dot{w}(t), \ddot{w}(t))\, dt = \mu(T_2 - T_1) + \pi(x(T_1)) - \pi(x(T_2)). \tag{5.7}$$

On the other hand, since $v(\cdot)$ has the same end data as $w(\cdot)$,

$$\int_{T_1}^{T_2} f(v(t), \dot{v}(t), \ddot{v}(t))\, dt \geqq U_{T_2-T_1}(x(T_1), x(T_2))$$
$$\geqq \mu(T_1 - T_2) + \pi(x(T_1)) - \pi(x(T_2)).$$

Comparing this with (5.7) we conclude that (5.3) suffices for $w(\cdot)$ to be a minimal energy configuration. □

Proof of Theorem 5.1. Given $x_0 = x \in \mathbb{R}^2$ we consider the function $w^*(\cdot)$ defined in (4.8). By Theorem 4.2 this function is a minimal energy configuration. We will examine the phase-plane orbit

$$t \mapsto x^*(t) = \begin{pmatrix} w^*(t) \\ \dot{w}^*(t) \end{pmatrix}.$$

Recall that $w^*(\cdot)$ was constructed by using the minimal energy sequence (4.4). Such sequences are uniformly bounded if x_0 belongs to a bounded set in $\mathbb{R}^2$, as follows from Theorem 2.1 and Theorem 8.1 in [L]. It then follows that the orbits

$$t \mapsto x^*(t),$$

with $w^*(\cdot)$ as above and x_0 in a bounded set are uniformly bounded.

Suppose, for the sake of definiteness, that x_0 lies in the first quadrant of $\mathbb{R}^2$. Then $t \mapsto x_1^*(t)$, where $x_1^*(\cdot) = w(\cdot)$ is the first coordinate of $x^*(\cdot)$, is an increasing function so long as $x^*(\cdot)$ remains in the first quadrant. Moreover by (2.7), $x_1^*(\cdot)$ is bounded. Consequently $x^*(\cdot)$ either crosses the x_1-axis and enters the fourth quadrant in finite "time" or else $x^*(t)$ converges (essentially) to the point $(M_1, 0)$ as $t \to \infty$, where $M_1 = \lim_{t\to\infty} x_1^*(t) = \sup\{x_1^*(t) : t \geqq 0\}$. In the latter case, the finiteness of M_1 implies that for each $\varepsilon > 0$, the fraction of the time interval $[0, T]$ during which $x_2^*(t) \in (0, \varepsilon)$ and $x_1^*(t) \in (M_1 - \varepsilon, M_1)$ approaches 1 as $T \to \infty$. Hence when (5.2) holds this second possible behavior of $x^*(\cdot)$ contradicts the fact that $w^*(\cdot)$ is a minimal energy configuration, in light of (5.2) and the growth condition (2.1 ii) for f as a function of r. On the other hand, when (5.2) is false and (5.1′) holds, then the constant function $w^*(s) = \overline{w}$, $s \geqq 0$, is obviously a minimal energy configuration (so that (5.2) is replaced by $\mu = m_f$).

Hereafter we will suppose that (5.2) holds so that $\mathfrak{x}^*(\cdot)$ crosses the x_1-axis at some time $t_1 > 0$, and we have $x_2^*(t_1) = 0$, $x_2^*(t) < 0$ for $t > t_1$ sufficiently small. The same reasoning as above when applied to the decreasing function $t \to x_1^*(t)$ implies that there is a $t_2 > t_1$ at which another intersection of $\mathfrak{x}^*(\cdot)$ with the x_1-axis occurs. In this manner one obtains a sequence $\{t_k\}_{k \geqq 1}$ of successive times at which the orbit $\mathfrak{x}^*(\cdot)$ crosses the x_1-axis. We distinguish between two different cases:

First case: The orbit $\mathfrak{x}^*(\cdot)$ intersects itself; that is, for some $0 \leqq T_1 < T_2$ (5.8)

$$\mathfrak{x}^*(T_2) = \mathfrak{x}^*(T_1).$$

Second case: The orbit $\mathfrak{x}^*(\cdot)$ does not intersect itself. (5.9)

Proposition 5.3. *Assume that* (5.8) *holds. Then there exists a periodic minimal energy configuration.*

Proof. Let $0 \leqq T_1 < T_2$ be as in (5.8), set $T = T_2 - T_1$ and define for each t

$$w(t) = w^*(\tau),$$

where τ is the unique point satisfying

$$T_1 \leqq \tau < T_2, \quad t - \tau = nT \quad \text{for some integer } n.$$

Then the orbit $t \mapsto \begin{pmatrix} w(t) \\ \dot{w}(t) \end{pmatrix}$ associated with this function is obviously periodic of period T. In order to prove that $w(\cdot)$ is a minimal energy configuration it is enough by Lemma 5.2 to show that (5.3) holds, or equivalently, that

$$\int_{T_1}^{T_2} f(w^*(s), \dot{w}^*(s), \ddot{w}^*(s))\, ds = (T_2 - T_1)\mu. \tag{5.10}$$

But arguing as in the proof of Lemma 5.1 we conclude that

$$\begin{aligned}\int_{T_1}^{T_2} f(w^*(s), \dot{w}^*(s), \ddot{w}^*(s))\, ds &= U_{T_2-T_1}(\mathfrak{x}^*(T_1), \mathfrak{x}^*(T_2)) \\ &= (T_2 - T_1)\mu + \pi(\mathfrak{x}^*(T_1)) - \pi(\mathfrak{x}^*(T_2)) + \theta_{T_2-T_1}(\mathfrak{x}^*(T_1), \mathfrak{x}^*(T_2))\end{aligned}$$

and that the $\theta_{T_2-T_1}(\cdot, \cdot)$ term must vanish. Since $\mathfrak{x}^*(T_2) = \mathfrak{x}^*(T_1)$, the above equality reduces to (5.10), which completes the proof. □

Proposition 5.4. *Assume that* (5.9) *holds. Then there exists a periodic minimal energy configuration.*

Proof. Recall the increasing sequences $\{t_k\}_{k \geqq 1}$ of successive times where the orbit $t \mapsto \mathfrak{x}^*(t)$ intersects the x_1-axis. For every odd integer k we define

$$y_k(t) = \mathfrak{x}^*(t + t_k), \quad 0 \leqq t \leqq t_{k-1} - t_k, \quad k = 1, 3, 5, \ldots \tag{5.11}$$

We proceed to show that there is an interval $[0, \tau_1]$ such that $t_{k_i+1} - t_{k_i} \to \tau_1 > 0$ as $i \to \infty$ for some subsequence of odd numbers $k_i \to \infty$, and there is an orbit $y(\cdot)$ of the form $y(t) = \begin{pmatrix} v(t) \\ \dot{v}(t) \end{pmatrix}$ such that

$$\max_{0 \leq t \leq \tau_1} |y_{k_i}(t) - y(t)| \to 0 \quad \text{as} \quad i \to \infty. \tag{5.12}$$

In (5.12), when $t_{k_i+1} - t_{k_i} < \tau_1$ we extend $y_{k_i}(\cdot)$ to all of $[0, \tau_1]$ by setting

$$y_{k_i}(t) = y_{k_i}(t_{k_i+1} - t_{k_i}) \quad \text{for} \quad t \in [t_{k_i+1} - t_{k_i}, \tau].$$

For $y_k(\cdot)$ in (5.11) we denote $\alpha_k = t_{k+1} - t_k$. Thus $y_k(0)$ and $y_k(\alpha_k)$ both belong to the x_1-axis and $(y_k)_2(t) < 0$ for $0 < t < \alpha_k$. Moreover, since the orbit $y_k(\cdot)$ is part of the minimal energy configuration $x^*(\cdot)$, there is a bound

$$|y_k(t)| \leqq M, \quad 0 \leqq t \leqq \alpha_k \tag{5.13}$$

which is uniform for all k's. We show next that there is a bound of the form

$$\alpha_k \leqq T_0 \quad \text{for all} \quad k = 1, 3, 5, \ldots \tag{5.14}$$

To prove (5.14) let $\varepsilon > 0$ be small and consider the strip

$$S_\varepsilon = \{x \in \mathbb{R}^2 : -\varepsilon < x_2 < 0\}.$$

Observe that the boundedness of the first component of $y_k(\cdot)$ ensures that the total time spent outside s_ε cannot exceed $T_\varepsilon = \frac{2M'}{\varepsilon}$, where M' is a bound for $|w_k(\cdot)|$. But if α_k were very large compared to T_ε then the fraction of time spent by the orbit in S_ε would be arbitrarily close to one. For ε sufficiently small we would have by (5.2) that

$$\inf_{\substack{(x,p) \in S_\varepsilon \\ s > 0}} f(x, p, s) > \mu. \tag{5.15}$$

Thus by (5.15) we would have for some $\delta > 0$

$$\frac{1}{\alpha_k} \int_0^{\alpha_k} f(w_k(t), \dot{w}_k(t), \ddot{w}_k(t))\, dt > \mu + \delta, \tag{5.16}$$

(where $w_k(\cdot)$ is the first component of $y_k(\cdot)$), provided that α_k is large enough. On the other hand since $y_k(\cdot)$ corresponds to a minimal energy configuration,

$$\frac{1}{\alpha_k} \int_0^{\alpha_k} f(w_k(t), \dot{w}_k(t), \ddot{w}_k(t))\, dt = \frac{1}{\alpha_k} [\alpha_k \mu + \pi(y_k(0)) - \pi(y_k(\alpha_k))],$$

which is below $\mu + \delta$ for α_k sufficiently large. This contradiction implies the validity of (5.14).

We claim that there is a $C > 0$ such that

$$|y_k(0) - y_k(\alpha_k)| \geqq C, \quad k = 1, 3, 5, \ldots \tag{5.17}$$

For by the construction of the $\{y_k(\cdot)\}$ there are points x, z on the x_1-axis, such that monotonically (this uses (5.9))

$$y_k(0) \to x, \quad y_k(\alpha_k) \to z \quad \text{as } k \to \infty.$$

Now $x \neq z$, since otherwise $x^*(\cdot)$ would converge to a point on the x_1-axis in contradiction to (5.2), so (5.17) holds. Now (5.17) ensures that for some $t_0 > 0$ one has

$$\alpha_k > t_0, \quad k = 1, 3, 5, \ldots \tag{5.18}$$

Otherwise the quantities $\dot{w}_k(t)$ would be unbounded. It thus follows from (5.14) and (5.18) that there is a subsequence $\{k_i\}_{i \geqq 1}$ of the odd integers with

$$\alpha_{k_i} \to \tau_1 > 0 \quad \text{as } i \to \infty .$$

Since $\{y_{k_i}(\cdot))$ correspond to a minimal energy configuration it follows that the quantities

$$\|\ddot{w}_{k_i}(\cdot)\|_{L^\gamma(0,\alpha_k)} \text{ are uniformly bounded.}$$

Hence we may also suppose that

$$\{\dot{w}_{k_i}(\cdot)\}_{i \geqq 1}, \quad \{w_{k_i}(\cdot)\}_{i \geqq 1}$$

both converge uniformly on $[0, \tau_1]$ to limits $\dot{v}_1(\cdot)$ and $v_1(\cdot)$, respectively. Since each $y_{k_i}(\cdot)$ satisfies

$$\int_0^{\alpha_{k_i}} f(w_{k_i}(t), \dot{w}_{k_i}(t), \ddot{w}_{k_i}(t))\, dt = \alpha_{k_i}\mu + \pi(y_{k_i}(0)) - \pi(y_{k_i}(\alpha_{k_i})) \tag{5.19}$$

it follows, by letting $i \to \infty$ and using the lower semicontinuity, that

$$\int_0^{\tau_1} f(v_1(t), \dot{v}_1(t), \ddot{v}_1(t))\, dt \leqq \tau_1\mu + \pi(y(0)) - \pi(y(\tau_1)), \tag{5.20}$$

where $y(\cdot) = \begin{pmatrix} v_1(\cdot) \\ \dot{v}_1(\cdot) \end{pmatrix}$.

Now consider the functions $y_k(\cdot)$ defined for every *even* integer k by

$$y_k(t) = x^*(t + t_k), \quad 0 \leqq t \leqq t_{k+1} - t_k, \quad k = 2, 4, 6, \ldots$$

It follows by the same argument as above that there is an interval $[0, \tau_2]$, $\tau_2 > 0$ and a subsequence $\{k_i\}_{i \geqq 1}$ of the even integers such that $t_{k_i+1} - t_{k_i} \to \tau_2$ and an orbit $z(t) = \begin{pmatrix} v_2(t) \\ \dot{v}_2(t) \end{pmatrix}$ such that

$$\max_{0 \leqq t \leqq \tau_k} |y_{k_i}(t) - z(t)| \to 0 \quad \text{as} \quad i \to \infty .$$

Then we have

$$\int_0^{\tau_2} f(v_2(t), \dot{v}_2(t), \ddot{v}_2(t))\, dt \leqq \mu\tau_2 + \pi(z(0)) - \pi(z(\tau_2)). \tag{5.21}$$

Clearly

$$y(\tau_2) = z(0), \quad y(0) = z(\tau_1). \tag{5.22}$$

Now define a periodic configuration $v(\cdot)$ as follows

$$v(t) = \begin{cases} v_1(t) & 0 \leqq t \leqq \tau_1 \\ v_2(t - \tau_1) & \tau_1 < t \leqq \tau_1 + \tau_2, \end{cases}$$

while for $t > \tau_1 + \tau_2$ $v(t)$ is defined by $v(t) = v(t')$, where $t' \in (0, \tau_1 + \tau_2]$ is the unique point such that $t - t' = n(\tau_1 + \tau_2)$ for some integer n. In order to show that this is a minimal energy configuration it suffices by Lemma 5.1 to prove that

$$\int_0^{\tau_1+\tau_2} f(v(t), \dot{v}(t), \ddot{v}(t))\, dt = (\tau_1 + \tau_2)\, \mu. \tag{5.23}$$

But (5.20), (5.21) and (5.22) imply that

$$\int_0^{\tau_1+\tau_2} f(v(t), \dot{v}(t), \ddot{v}(t))\, dt \leqq (\tau_1 + \tau_2)\, \mu,$$

whence (5.23) holds, which completes the proof. □

6. A class of examples; single-well and double-well potentials

In this section we study (P_∞) for integrands of the form

$$f(w, p, r) = \psi(w) - bp^2 + cr^2, \quad b, c > 0, \tag{6.1}$$

where $\psi(\cdot)$ is a smooth function satisfying

$$\psi(w) \geqq a\,|w|^\alpha - d, \quad w \in \mathbb{R}, \quad \text{for some } \alpha > 2, \quad a, d > 0. \tag{6.2}$$

Thus f satisfies conditions (2.1). Furthermore, condition (5.1′) clearly holds:

$$m_f = \inf\{f(w, 0, r) : (w, r) \in \mathbb{R}^2\} = \inf\{f(w, 0, 0) : w \in \mathbb{R}\} = \min \psi(\cdot). \tag{6.3}$$

We have the following result.

Theorem 6.1. *Suppose that there are at most two absolute minimizers of* $\psi(\cdot)$: $\psi(w) = m_f \Leftrightarrow w \in M$, *where* $M = \{w_1\}$ *or* $M = \{w_1, w_2\}$. *Furthermore suppose that*

$$\psi''(w) > 0 \quad \forall\, w \in M. \tag{6.4}$$

Then for each fixed c, $\psi(\cdot)$ *as in* (6.1) *and* (6.2) *there is a scalar* $b_0 = b_0(c; \psi(\cdot)) > 0$ *such that the minimal energy growth rate for* f, $\mu = \mu(b; c, \psi(\cdot))$, *satisfies*

$$\begin{aligned} \mu &< m_f \quad \text{for } b \in (b_0, \infty) \\ \mu &= m_f \quad \text{for } b \in [0, b_0]. \end{aligned} \tag{6.5}$$

Proof. Consider for all $T > 0$ and all $w(\cdot) \in W^{2,1}_{\text{loc}}$ the Rayleigh quotient $R_T(w(\cdot)) = R_T(w(\cdot); \psi_0(\cdot))$, defined by

$$R_T(w(\cdot)) = \begin{cases} \dfrac{\int_0^T [\psi_0(w(t)) + c\ddot{w}^2(t)]\, dt}{\int_0^T \dot{w}^2(t)\, dt} & \text{if } \int_0^T \dot{w}^2(t)\, dt > 0, \\ +\infty & \text{otherwise,} \end{cases} \tag{6.6}$$

where $\psi_0(\cdot)$ is defined by

$$\psi_0(w) = \psi(w) - m_f \geqq 0 \quad \forall\, w \in \mathbb{R}; \tag{6.7}$$

thus $\psi_0(w) = 0 \Leftrightarrow w \in M$.

We will prove that $b_0 > 0$, where b_0 is defined by

$$b_0 := \inf_{\substack{T>0 \\ x(0)=x(T)}} R_T(w(\cdot)) \quad \text{with } x(\cdot) = \begin{pmatrix} w(\cdot) \\ \dot{w}(\cdot) \end{pmatrix}. \tag{6.8}$$

Relation (6.5a) follows directly because $b_0 > 0$; for each $b > b_0$ there exists by (6.8) a periodic function $w(\cdot)$ of period $T_0 > 0$ satisfying

$$\frac{1}{T_0} \int_0^{T_0} [\psi_0(w(t) - b\dot{w}^2(t) + c\ddot{w}^2(t)]\, dt < 0.$$

Thus (6.5a) follows from the definitions (3.3), (6.7) of μ and $\psi_0(\cdot)$.

Similarly, (6.5b) follows from Lemma 5.2 and the observation that when $b \leqq b_0$ then for all periodic $w(\cdot)$ of period $T_0 > 0$

$$\frac{1}{T_0} \int_0^{T_0} [\psi_0(w(t)) - b\dot{w}^2(t) + c\ddot{w}^2(t)]\, dt \geqq 0,$$

whereas for the functions $w(\cdot) \equiv w_i$, $w_i \in M$, equality holds.

If $M = \{w_1\}$ is a singleton, we note that by (6.2), (6.4) there is a constant e such that

$$0 < e \leqq \tfrac{1}{2}\psi''(w_1) \quad \text{and} \quad \psi(w) \geqq e(w - w_1)^2 \quad \forall\, w \in \mathbb{R}.$$

Hence it follows from (6.8) that

$$b_0 \geqq \inf_{\substack{T>0 \\ x(0)=x(T)}} R_T^e(w(\cdot) - w_1) = \inf_{\substack{T>0 \\ x(0)=x(T)}} R_T^e(w(\cdot)) =: b_1, \tag{6.9}$$

where

$$R_T^e(w(\cdot)) = \frac{\int_0^T [ew^2(t) + c\ddot{w}^2(t)]\, dt}{\int_0^T \dot{w}^2(t)\, dt}.$$

It will be seen later that $b_1 > 0$, whence $b_0 > 0$ as claimed.

To establish the positivity of b_0 when $M = \{w_1, w_2\}$ is a doubleton we begin by noting that (6.8) leads to an alternative recipe. We claim that for each fixed $y = \begin{pmatrix} y_1 \\ y_2 \end{pmatrix} \in \mathbb{R}^2$ we have

$$b_0 = \inf_{\substack{T>0 \\ x(0)=x(T)=y}} R_T(w(\cdot)). \tag{6.8'}$$

This follows from the observation that a function $w(\cdot)$ of period T is also of period kT for all positive integers k. Now for large k we can join the end values assumed by $x(\cdot)$ on an interval of length kT to the prescribed end values y (by extending $w(\cdot)$ onto the concentric interval of length $kT+2$, say) without making much change in the ratio (6.6). Thus (6.8′) holds.

Now take $y = \begin{pmatrix} w_* \\ 0 \end{pmatrix}$ where w_* is any interior point of the interval $[w_1, w_2]$. Observe that shifting each $w(\cdot) \in W^{2,1}_{\text{loc}}$ by an additive constant k corresponds to translating the function $\psi_0(\cdot)$ by the amount k. Thus we obtain the formula

$$b_0 = \inf_{\substack{T>0 \\ x(0)=x(T)=y}} R_T(w(\cdot)) = \inf_{\substack{T>0 \\ x(0)=x(T)=0}} R_T^*(w(\cdot)), \tag{6.10}$$

where $R_T^*(w(\cdot)) = R_T(w(\cdot); \psi_0(\cdot - w_*))$. Note that the zeros of this translate $\overline{\psi}_0(\cdot) = \psi_0(\cdot - w_*)$ are

$$\overline{w}_1 = w_1 - w_*, \quad \overline{w}_2 = w_2 - w_*.$$

Next, for functions $w(\cdot)$ satisfying $\mathbf{x}(0) = \mathbf{x}(T) = \mathbf{0}$ we shall examine the values assumed by the ratio in R_T^* over subintervals of $[0, T]$ where $w(\cdot)$ has constant sign. Given any such $w(\cdot)$ we decompose $[0, T]$ into three disjoint sets

$$A = \{t : w(t) > 0\}, B = \{t : w(t) < 0\}, C = \{t : w(t) = 0\}.$$

As is well known, $\dot{w}(t) = 0$ a.e. on C, whence $\ddot{w}(t) = 0$ a.e. on C as well. Hence

$$R_T^*(w(\cdot)) = \frac{\int_A [\overline{\psi}_0(w(t)) + c\ddot{w}^2(t)]\,dt + \int_B [\overline{\psi}_0(w(t)) + c\ddot{w}^2(t)]\,dt + \int_C \overline{\psi}_0(0)\,dt}{\int_A \dot{w}^2(t)\,dt + \int_B \dot{w}^2(t)\,dt}$$

$$\geqq \min\left\{ \frac{\int_A [\overline{\psi}_0(w(t)) + c\ddot{w}^2(t)]\,dt}{\int_A \dot{w}^2(t)\,dt}, \frac{\int_B [\overline{\psi}_0(w(t)) + c\ddot{w}^2(t)]\,dt}{\int_B \dot{w}^2(t)\,dt} \right\}. \tag{6.11}$$

Now denote by $\psi^L(\cdot)$ any smooth nonnegative extension of $\overline{\psi}_0(\cdot)$ from $[0, \infty)$ to $\mathbb{R}$ which possesses no zeros other than w_2 and which satisfies the growth condition (6.2) on $\mathbb{R}$. Likewise denote by $\psi^R(\cdot)$ any smooth nonnegative extension of $\overline{\psi}_0(\cdot)$ from $(-\infty, 0]$ to $\mathbb{R}$ which possesses no zeros other than $\overline{w}_1$ and which satisfies the growth condition (6.2). It then follows from (6.11), using the fact that the open sets A, B are disjoint unions of intervals, that

$$R_T^*(w(\cdot)) \geqq \min\left\{ \inf_{\substack{T_0 \in [0,T] \\ w(0)=w(T)=0}} R_{T_0}(w(\cdot); \psi^L(\cdot)), \inf_{\substack{T_0 \in [0,T] \\ w(0)=w(T)=0}} R_{T_0}(w(\cdot); \psi^R(\cdot)) \right\}. \tag{6.12}$$

Thus (6.10) implies

$$b_0 \geqq \min\left\{ \inf_{\substack{T>0 \\ w(0)=w(T)=0}} R_T(w(\cdot); \psi^L(\cdot)), \inf_{\substack{T>0 \\ w(0)=w(T)=0}} R_T(w(\cdot); \psi^R(\cdot)) \right\}. \tag{6.13}$$

We will proceed to show that both infima in (6.13) are positive. For the sake of brevity, we focus attention in what follows on the quantity

$$b_0^L := \inf_{\substack{T>0 \\ w(0)=w(T)=0}} R_T(w(\cdot); \psi^L(\cdot)),$$

but the treatment of the quantity

$$b_0^R := \inf_{\substack{T>0 \\ w(0)=w(T)=0}} R_T(w(\cdot); \psi^R(\cdot)),$$

is carried out in the same way.

Now by (6.4) and the construction of $\psi^L(\cdot)$ there is a constant $e' = e'(\psi^L(\cdot))$, $0 < e \leqq \frac{1}{2}\psi''(w_2)$, such that

$$\psi^L(w) \geqq e'(w - \overline{w}_2)^2 \quad \forall\, w \in \mathbb{R}. \tag{6.14}$$

Again shifting each $w(\cdot) \in W^{2,1}_{\text{loc}}$ by an additive constant so as to translate $\psi^L(\cdot)$ we obtain

$$b_0^L \geqq \inf_{\substack{T>0 \\ w(0)=w(T)=0}} R_T^{e'}(w(\cdot) - \overline{w}_2) = \inf_{\substack{T>0 \\ w(0)=w(T)=-\overline{w}_2}} R_T^{e'}(w(\cdot)) =: b_1' \tag{6.15}$$

where $R_T^{e'}$ denotes the Rayleigh quotient associated with $\psi(w) = e'w^2$, i.e.

$$R_T^{e'}(w(\cdot)) = \frac{\int_0^T [e'w^2(t) + c\ddot{w}^2(t)]\, dt}{\int_0^T \dot{w}^2(t)\, dt}.$$

Next we show that the infimum giving b_1' is not attained for small values of T. Since the end conditions in (6.15) imply

$$\int_0^T \dot{w}(t)\, dt = 0$$

it follows that for some $t_0 \in [0, T]$, $\dot{w}(t_0) = 0$. Hence by Schwarz's inequality

$$\int_0^T \dot{w}^2(t)\, dt \leqq \int_0^T \left| (t - t_0) \int_{t_0}^t \ddot{w}^2(s)\, ds \right| dt \leqq T^2/2 \int_0^T \ddot{w}^2(t)\, dt,$$

so that for each $w(\cdot)$ entering (6.15) one has

$$R_T^{e'}(w(\cdot)) \geqq \frac{2c}{T^2}.$$

Consequently for $\delta > 0$ sufficiently small we can give the following alternative formula for the right hand side of (6.15):

$$b_1' = \inf_{\substack{T \geqq \delta \\ w(0)=w(T)=-w_2}} R_T^{e'}(w(\cdot)). \tag{6.15'}$$

We now relax the conditions on $w(\cdot)$ under which the infimum in (6.15′) is taken; it will only be required that on $[0, T]$ $\dot{w}(\cdot) \not\equiv 0$. Clearly

$$b_1' \geqq \inf_{\substack{T \geqq \delta \\ \dot{w} \not\equiv 0}} R_T^{e'}(w(\cdot)) := b_2'. \tag{6.16}$$

Furthermore, we observe that b_2' is also given by the formula

$$b_2' = \inf_{\substack{T \in [\delta, 2\delta) \\ \dot{w} \not\equiv 0}} R_T^{e'}(w(\cdot)). \tag{6.16'}$$

This version of (6.16) holds because for each $T > \delta$ the interval $[0, T]$ can be decomposed into finitely many disjoint subintervals $I_j = [t_j, t_j + T_0)$ of common length $T_0 \in [\delta, 2\delta]$; hence

$$R_T^{e'}(w(\cdot)) = \frac{\Sigma \int_{I_j} [e' w^2(t) + c\ddot{w}^2(t)]\, dt}{\Sigma \int_{I_j} \dot{w}^2(t)\, dt} \geqq \min_j \frac{\int_{I_j} [e' w^2(t) + c\ddot{w}^2(t)]\, dt}{\int_{I_j} \dot{w}^2(t)\, dt}.$$

Finally, we use (6.16′) to demonstrate the positivity of b_2' (positivity of the analogous quantity associated with $\psi^R(\cdot)$ is proved in the same way), so that the positivity of b_0 will follow from (6.13)–(6.16). Let $\{(w_n(\cdot), T_n)\}_{n \geq 1}$ denote a minimizing sequence for (6.16′); *i.e.*

$$R_{T_n}^{e'}(w_n(\cdot)) \to b_2', \quad \text{with } T_n \in [\delta, 2\delta),\ \dot{w}_n \not\equiv 0 \quad \text{on } [0, T_n]. \tag{6.17}$$

Without loss of generality we can suppose that

$$T_n \to T_0 \in [\delta, 2\delta]. \tag{6.18}$$

Moreover by the homogeneity of $R_T^{e'}$ we can suppose that

$$\int_0^{T_n} \dot{w}_n^2(t)\, dt = 1 \quad \forall\, n \geqq 1. \tag{6.19}$$

For those values of n with $T_n < T_0$ we extend $w_n(\cdot)$ from $[0, T_n]$ onto $[T_n, T_0]$ as that (linear) function corresponding to the identically zero extension of $\ddot{w}_n(\cdot)$ onto $[T_n, T_0]$. Denote the resulting function in $W^{2,1}(0, T_0)$ by $\tilde{w}_n(\cdot)$. On the other hand, for values of n such that $T_n \geqq T_0$ let $\tilde{w}_n(\cdot)$ denote the restriction of $w_n(\cdot)$ to $[0, T_0]$. In general

$$\int_0^{T_0} \dot{\tilde{w}}_n^2(t)\, dt \neq 1,$$

but it is easy to see that (6.17)–(6.19) imply

$$\int_0^{T_0} \dot{\tilde{w}}_n^2(t)\, dt \to 1, \tag{6.20}$$

as well as

$$\int_0^{T_0} \tilde{w}_n^2(t)\, dt \leqq M, \quad \int_0^{T_0} \ddot{\tilde{w}}_n^2(t)\, dt \leqq M \quad \forall\, n \geqq 1, \text{ for some } M < \infty. \tag{6.21}$$

Hence without loss of generality we can suppose, by extracting a subsequence, that there is an element $\ddot{v}(\cdot) \in L^2(0, T_0)$ and continuous functions $v(\cdot)$, $\dot{v}(\cdot)$ for which

$$\begin{aligned} &\ddot{\tilde{w}}_n(\cdot) \to \ddot{v}(\cdot) \quad \text{weakly in } L^2(0, T_0), \\ &\tilde{w}_n(\cdot) \to v(\cdot), \quad \dot{\tilde{w}}_n(\cdot) \to \dot{v}(\cdot) \quad \text{uniformly in } C[0, T_0]. \end{aligned} \tag{6.22}$$

That is, $w_n(\cdot) \to v(\cdot)$ weakly in $W^{2,2}(0, T_0)$. It follows from (6.22) and the sequential weak lower semicontinuity of the L^2-norm that

$$\int_0^T \dot{v}^2(t)\,dt = 1, \quad R_{T_0}^{e'}(v(\cdot)) = \int_0^{T_0} (e'v^2(t) + c\ddot{v}^2(t)]\,dt = b_2'. \tag{6.23}$$

This obviously implies the asserted positivity of b_2', hence the positivity of b_1' in (6.15) (as well as the positivity of b_1 defined in (6.9)). □

Corollary 6.2. *Suppose that $\psi(\cdot)$ as in Theorem 1.6 has a single absolute minimizer (i.e., $M = \{w_1\}$) and in addition that $\psi(\cdot)$ satisfies*

$$\psi(w) \geqq e(w - w_1)^2, \quad w \in \mathbb{R}, \quad \text{where } e = \tfrac{1}{2}\psi''(w_1). \tag{6.24}$$

In this case the threshold value $b_0 = b_0(c; \psi(\cdot))$ is given by

$$b_0 = \sqrt{2c\psi''(w_1)}. \tag{6.25}$$

Proof. According to (6.9)

$$b_0 \geqq b_1 := \inf_{\substack{T>0\\ x(0)=x(T)}} R_T^e(w(\cdot)). \tag{6.26}$$

We proceed to appraise b_1 by making use of the arithmetic-geometric mean inequality, followed by integration by parts, for functions $w(\cdot)$ in (6.26):

$$\int_0^T [ew^2(t) + c\ddot{w}^2(t)] \geqq \int_0^T [-2\sqrt{ec}\,w(t)\,\ddot{w}(t)]\,dt$$

$$= 2\sqrt{ec}\left\{\left[\int_0^T \dot{w}^2(t)\,dt - w(t)\,\dot{w}(t)\right]_0^T\right\} = 2\sqrt{ec}\int_0^T \dot{w}^2(t)\,dt. \tag{6.27}$$

This yields the inequality

$$b_1 \geqq 2\sqrt{ec}. \tag{6.28}$$

Moreover equality holds in (6.27) if and only if

$$\sqrt{e}\,w(t) + \sqrt{c}\,\ddot{w}(t) = 0 \quad \text{a.e. } t \in [0, T].$$

It follows that

$$R_T^e(w(\cdot)) = 2\sqrt{ec}$$

whenever T is a multiple of $T^* = (c/e)^{1/4}\pi$ and $w(\cdot)$ has the form

$$w(t) = C\cos((e/c)^{1/4}t - \theta), \quad C, \theta \text{ constants}. \tag{6.29}$$

Hence (6.28) is actually an equality.

In order to verify (6.25) we now write (6.8) in the form

$$b_0 = \inf_{\substack{T>0\\ x(0)=x(T)}} R_T(w(\cdot) + w_1; \psi(\cdot)). \tag{6.8'}$$

Now when $w(\cdot)$ in (6.8′) is replaced by $\lambda w(\cdot)$, $\lambda \in (0, 1)$ then (6.24) implies that $R_T(\lambda w(\cdot) + w_1)$ satisfies

$$\frac{\int_0^T (\psi(\lambda w(t) + w_1) + c\lambda^2 \ddot{w}^2(t))\,dt}{\int_0^T \lambda^2 \dot{w}^2(t)\,dt} \geqq \frac{\int_0^T (ew^2(t) + c\ddot{w}^2(t))\,dt}{\int_0^T \dot{w}^2(t)\,dt} = R_T^e(w(\cdot)).$$

On the other hand by the formula $e = \frac{1}{2}\psi''(w_1)$ and the smoothness of $\psi(\cdot)$ we know that for each $\varepsilon > 0$ there is a $\delta > 0$ such that

$$ev^2 \leqq \psi(v + w_1) \leqq (1 + \varepsilon)\, ev^2, \quad |v| \leqq \delta.$$

Consequently for each fixed $w(\cdot)$ as in (6.8′) there exists $0 < \lambda \ll 1$ such that

$$R_T^e(w(\cdot)) \leqq R_T(\lambda w(\cdot) + w_1) \leqq \frac{\int_0^T [(1 + \varepsilon)\, ew^2(t) + c\ddot{w}^2(t)]\,dt}{\int_0^T \dot{w}^2(t)\,dt} = R_T^{(1+\varepsilon)e}(w(\cdot)). \tag{6.30}$$

It follows from (6.8′) and (6.30) that for each $\varepsilon > 0$

$$\inf_{\substack{T>0\\x(0)=x(T)}} R_T^e(w(\cdot)) \leqq b_0 \leqq \inf_{\substack{T>0\\x(0)=x(T)}} R_T^{(1+\varepsilon)e}(w(\cdot))$$

By (6.28) this is equivalent to

$$2\sqrt{ec} \leqq b_0 \leqq 2\sqrt{(1+\varepsilon)\, ec},$$

so that (6.25) follows. □

7. Proof of Lemma 4.3

We first prove the assertion of the Lemma for values of T in the interval (0, 1). Then for any $T_0 > 1$ the assertion follows for all T in the interval $(0, T_0)$ by considering the sequence $a_k' = a_k/T_0$ and applying the result for T in (0, 1) to the sequence $\{a_k'\}$. Thus the assertion follows for all T in $(0, \infty)$.

Let $\varepsilon > 0$ be fixed and let $I_k = (a_k - \varepsilon, a_k)$. For a fixed k we define

$$A_k = \bigcup_{n \geqq a_k} \frac{1}{n} I_k,$$

where for an interval I and a scalar $c \neq 0$

$$\frac{1}{c} I = \{x : cx \in I\}.$$

Then A_k is the set of all points in (0, 1) such that

$$a_k - \varepsilon < nx < a_k \quad \text{for some integer } n.$$

Thus $\bigcup_{k=m}^{\infty} A_k$ is the set of all points x in (0, 1) such that the relation $a_k - \varepsilon < nx < a_k$ holds for some $k \geqq m$, $n \geqq 1$, and

$$B_\varepsilon = \bigcap_{m=1}^{\infty} \bigcup_{k=m}^{\infty} A_k$$

is the set of all $x \in (0, 1)$ such that $a_k - \varepsilon < nx < a_k$ holds for infinitely many pairs of integers (k, n). Our goal is to prove that

$$m(B_\varepsilon) = 1 \quad \text{for each } \varepsilon > 0. \tag{7.1}$$

We then define $B = \bigcap_{n=2}^{\infty} B_{1/n}$, and (7.1) implies that $m(B) = 1$, so that the relation (4.9) holds for every $T \in B$. Since $B_\varepsilon = \bigcap_{m=1}^{\infty} \bigcup_{k=m}^{\infty} A_k$ it suffices, in order to establish (7.1), to show that

$$m\left(\bigcup_{k=m}^{\infty} A_k\right) = 1 \quad \text{for each } m \geqq 1. \tag{7.2}$$

This will conclude the proof of the Lemma, and the argument is given below.

We first examine the structure of the set A_k. It is a union of intervals $\frac{1}{n} I_k$ for $n \geqq a_k$. Put

$$n_0 = \left[\frac{a_k}{\varepsilon}\right], \tag{7.3}$$

(where $[x]$ is the largest integer not exceeding x), and consider any $n < n_0$. Then

$$\frac{a_k - \varepsilon}{n} \geqq \frac{a_k}{n+1}$$

so that the intervals $\frac{1}{n} I$ and $\frac{1}{n+1} I$ do not overlap. That is, the overlapping portions of A_k are composed of

$$\bigcup_{n \geqq n_0} \frac{1}{n} I_k$$

$\left(\text{if } \frac{a_k}{\varepsilon} \text{ is not an integer, which can be assumed without loss of generality by suitable choice of } \varepsilon\right)$, and thus lie in the interval $\left(0, \frac{a_k}{n_0}\right)$. This contains the interval $(0, \varepsilon)$ and is as close to it as we please for large k.

The measure of the nonoverlapping parts of A_k is $\varepsilon \sum_{a_k \leqq j < n_0} \frac{1}{j}$ and this can be approximated by $\varepsilon \log \frac{n_0}{a_k}$, which in turn can be approximated by the quantity

$$\delta := \varepsilon \log 1/\varepsilon.$$

The approximation is valid in the sense that the measure of the nonoverlapping parts of A_k lies between $(1-\theta)\,\delta$ and $(1+\theta)\,\delta$ for any prescribed small $\theta > 0$, provided that k is sufficiently large.

Now we take an interval (a, b) which is contained in $(\varepsilon, 1)$ and we estimate the measure of $(a, b) \cap A_k$. Since $(a, b) \subset (\varepsilon, 1)$, for large enough k only intervals in the nonoverlapping part of A_k will intersect (a, b), so

$$m[(a, b) \cap A_k] = \varepsilon \sum_{j \in S} \frac{1}{j},$$

S being the set of integers $\{j: [a_k/b] < j \leqq [a_k/a]\}$, which implies that

$$m[(a, b) \cap A_k] \geqq \varepsilon \tfrac{1}{2} \log \frac{b}{a} = \frac{\varepsilon}{2}\left(-\log \frac{a}{b}\right) \geqq \frac{\varepsilon}{2}\left(1 - \frac{a}{b}\right)$$

if k is large enough. Thus

$$m[(a, b) \cap A_k] \geqq \frac{\varepsilon}{2}(b - a), \tag{7.4}$$

provided that k is sufficiently large.

We will now select a sequence of integers $\{k_i\}_{i=1}^{\infty}$ increasing to infinity, and we will show that $m\left(\bigcup_{i=i_0}^{\infty} A_{k_i}\right) = 1$ for each i_0. Suppose $A_{k_1}, \ldots, A_{k_l}$ to have been already chosen. Then $\bigcup_{i=1}^{l} A_{k_i}$ is a finite union of intervals and so is its complement in $(0, 1)$, which we denote by Δ_l. We know that $\Delta_l \subset (\varepsilon, 1)$ and we write it as a finite union of intervals $\Delta_l = \bigcup_{p=1}^{m} J_p$. For each J_p we select a closed interval K_p satisfying

$$K_p \subset \operatorname{int} J_p, \tag{7.5}$$

$$m(K_p) > \tfrac{1}{2} m(J_p). \tag{7.6}$$

It follows from (7.5) that if k is sufficiently large then no interval in A_k will intersect both K_p and and $\bigcup_{i=1}^{l} A_{k_i}$. It also follows from (7.4) and (7.6) that if k is sufficiently large then

$$m[K_p \cap A_k] > \tfrac{1}{2} \varepsilon \tfrac{1}{2} m(J_p) = \frac{\varepsilon}{4} m(J_p). \tag{7.7}$$

Since we have only a finite number of intervals K_p we can find an integer k large enough so that (7.7) holds for every p, $1 \leqq p \leqq m$, and we choose k_{l+1} to be such an integer k. In fact by the same argument, k_{l+1} can be chosen large enough that (7.7) holds for every K_p which satisfies (7.5) and (7.6) where now $\bigcup_{p=1}^{m} J_p$ denotes the complement of $\bigcup_{i=i_0}^{l} A_{k_i}$ for some i_0, $1 \leqq i_0 \leqq l$.

Now let $i_0 \geqq 1$ be given and denote the measure of $\bigcup_{i=i_0}^{l} A_{k_i}$ by μ_l. We have the following relation

$$\mu_{l+1} = \mu_l + m\left\{\left[(0,1) \Big/ \bigcup_{i=i_0}^{l} A_{k_i}\right] \cap A_{k_{l+1}}\right\}$$
$$= \mu_l + m\left\{\left[(\varepsilon,1) \Big/ \bigcup_{i=i_0}^{l} A_{k_i}\right] \cap A_{k_{l+1}}\right\} \geqq \mu_l + \tfrac{1}{4}\sum_{p=1}^{m} \varepsilon\, m(J_p),$$

where the last inequality follows from (7.7) and, as above $\bigcup_{p=1}^{m} J_p$ denotes the complement of $\bigcup_{i=i_0}^{l} A_{k_i}$. We thus deduce

$$\mu_{l+1} \geqq \mu_l + \tfrac{1}{4}\,\varepsilon \sum_{p=1}^{m} m(J_p). \tag{7.8}$$

We know, however, that $\mu_l + \sum_1^m m(J_p) = 1$ (by definition) so that (7.8) implies

$$\mu_{l+1} \geqq \mu_l + \tfrac{1}{4}\,\varepsilon\,(1 - \mu_l). \tag{7.9}$$

The sequence $\{\mu_l\}_{l=i_0}^{\infty}$ is nondecreasing and bounded by 1, so it tends to a limit μ. Since for $\mu < 1$ (7.9) implies that $\mu_l \to \infty$ it follows that $\mu = 1$, so the measure of $\bigcup_{i=i_0}^{\infty} A_{k_i}$ is equal to 1 for each $i_0 \geqq 1$. □

Acknowledgements. We acknowledge our indebtedness to BERNARD COLEMAN for initially posing the problem considered here, as well as for his encouragement and enthusiasm throughout. We are grateful to WILLIAM HRUSA and MOSHE MARCUS for valuable comments. The research of V. J. MIZEL was partially supported by the National Science Foundation under Grant DMS 87040530.

Bibliography

[A & D] AUBRY, S., & P. Y. LE DAERON, The discrete Frenkel-Kontorova model and its extensions, *Physica* **8D** (1983), 381–422.

[A & L] ARTSTEIN, Z., & A. LEIZAROWITZ, Tracking periodic signals with overtaking criterion, *IEEE Trans. on Autom. Control* AC **30** (1985), 1122–1126.

[B & H] BROCK, W. A., A. HAURIE, On existence of overtaking optimal trajectories over an infinite time horizon, *Math. Op. Res.* **1** (1976), 337–346.

[Ca] CARLSON, D., On the existence of catching-up optimal solutions for Lagrange problems defined on unbounded intervals, *J. Optim. Thy. Appl.* **49** (1986), 207–225.

[Cl] COLEMAN, B. D., Necking and drawing in polymeric fibers under tension, *Arch. Rational Mech. Anal.* **83** (1983), 115–137.

[C2] Coleman, B. D., On the cold drawing of polymers, *Comp. & Maths. with Appl.* **11** (1985), 35–65.

[C & D] Chou, W., & R. J. Duffin, An additive eigenvalue problem of physics related to linear programming, *Adv. in Appl. Math.* **8** (1987), 486–498.

[C & H] Cahn, J. W., & J. E. Hilliard, Free energy of a nonuniform system. I, Interfacial free energy. *J. Chem. Physics* **28** (1958), 258–267.

[CGS] Carr, J., M. E., Gurtin & M. Slemrod, Structural phase transitions on a finite interval, *Arch. Rational Mech. Anal.* **86** (1984), 317–351.

[G] Giaquinta, M., *Multiple integrals in the calculus of variations and nonlinear elliptic systems*, Annals of Math. Studies # 105, Princeton U. Press, Princeton 1983.

[G & C] Griffiths, R. B., & W. Chou, Effective potentials: a new approach and new results for one-dimensional systems with competing length scales, *Phys. Rev. Letters* **56** (1986), 1929–1931.

[L] Leizarowitz, A., Infinite horizon autonomous systems with unbounded cost, *Appl. Math. and Opt.* **13** (1985), 19–43.

[L & M] Lions, J. L., & E. Magenes, *Non-homogeneous boundary value problems and Applications I*, Grundlehren # 181, Springer, Berlin, 1972.

[Ma] Mather, J. N., More Denjoy minimal sets for area preserving diffeomorphisms, *Comment. Math. Helvetici* **60** (1985), 508–557.

[M] Morrey, C. B., Jr., *Multiple integrals in the calculus of variations*, Grundlehren # 130, Springer, New York, 1966.

[N] Nabarro, F. R. N., *Theory of crystal dislocations*, Clarendon Press, Oxford, 1967.

[T] Tonelli, L., Sugli integrali del calcolo delle variazioni in forma ordinaria, *Ann. Scuola Norm. Pisa* **3** (1934), 401–450 (in L. Tonelli, *Opere Scelte* vol. III, # 105, Edizioni Cremonese, Roma, 1961).

[vdW] van der Waals, J. D., The thermodynamic theory of capillarity under the hypothesis of a continuous variation of density (in Dutch), *verhandel. Konink. Akad. Weten. Amsterdam* (*Sec.* 1) **1** (1893).

[vW] von Weizsacker, C. C., Existence of optimal programs of accumulation for an infinite horizon, *Rev. Econ. Studies* **32** (1965), 85–104.

Department of Mathematics
Carnegie Mellon University
Pittsburgh, Pennsylvania

(Received June 14, 1988)

Admissible Wave Fans in Nonlinear Hyperbolic Systems

C. M. DAFERMOS

Dedicated to Bernard Coleman on the occasion of his sixtieth birthday

1. Introduction

The field equations expressing the principles of balance for one-dimensional homogeneous continuous media with "elastic" response typically generate quasi-linear hyperbolic systems of the form

$$\partial_t U(x, t) + \partial_x F(U(x, t)) = 0; \tag{1.1}$$

here the *state vector* U takes values in R^n and F is a smooth mapping from R^n to R^n.

When F is nonlinear, solutions of the Cauchy problem for (1.1), starting out from smooth initial data, generally develop singularities in a finite time. Confronted with this difficulty in his study of isothermal flow of an ideal gas, STOKES [32] attempted to resolve it by introducing the notion of a generalized solution with shock waves. The present concept of such a solution is a bounded measurable function of class BV[1] which satisfies (1.1) in the sense of measures.

The domain of any BV solution U of (1.1) may be decomposed into the union of three, pairwise disjoint, subsets $\mathscr{C}$, $\mathscr{S}$ and $\mathscr{I}$, with the following properties (*cf.* VOLPERT [39]): $\mathscr{I}$ is a set of one-dimensional Hausdorff measure zero. $\mathscr{C}$ is the set of points of (Lebesgue) approximate continuity of U. $\mathscr{S}$ is the set of points of (Lebesgue) approximate jump discontinuity of U; it is contained in the union of a countable family of Lipschitz arcs. With each point $(\bar{x}, \bar{t})$ of $\mathscr{S}$ is associated a number s such that U attains distinct approximate limits U_- and U_+ on either side of the straight line $x = \bar{x} + s(t - \bar{t})$ at $(\bar{x}, \bar{t})$ and

$$F(U_+) - F(U_-) = s[U_+ - U_-]. \tag{1.2}$$

In the language of continuum physics, a shock wave propagating with speed s passes through the point $\bar{x}$ at time $\bar{t}$ and (1.2) is the Rankine-Hugoniot jump condition.

---

[1] That is, the distributional derivatives $\partial_t U$, $\partial_x U$ are locally finite Radon measures.

The issue of the admissibility of these weak solutions was raised soon after they were conceived. Any smooth isothermal flow of a gas that conserves mass and momentum also conserves mechanical energy. In contrast, the flows with shocks constructed by STOKES so as to conserve mass and momentum fail to balance mechanical energy; for that reason their admissibility was questioned by KELVIN and by RAYLEIGH [34; Sec. 253]. Responding to that criticism, STOKES revised his aforementioned paper (*cf.* [33]), abandoning the idea of a shock.

By the turn of the century, the objections raised by KELVIN and RAYLEIGH had been addressed and shocks were rehabilitated, albeit under conditions of admissibility. It was realized (STOKES [32], RANKINE [26], RAYLEIGH [36]) that internal dissipation generated by even minutely low viscosity and heat conductivity, which would have no detectable effect on smooth flows, exerts nevertheless considerable influence on flows with sharp transitions, by inducing the conversion of a substantial portion of the mechanical energy into heat. This provides the motivation for *viscosity admissibility criteria* which seek to characterize admissible solutions of (1.1) as the $\varepsilon \to 0$ limits of smooth solutions of "parabolic" systems

$$\partial_t U(x,t) + \partial_x F(U(x,t)) = \partial_x [A_\varepsilon \, \partial_x U(x,t)]. \tag{1.3}$$

Here A_ε is a properly selected family of $n \times n$ matrices, depending on the (generally vector-valued) parameter ε, such that $A_\varepsilon \to 0$ as $\varepsilon \to 0$. Thus, in the language of continuum physics, admissibility is to be decided by visualizing the "elastic" medium as a limiting case in an appropriate class of media with internal dissipation.

At this point it is natural to ask whether admissibility of solutions may be determined, to some extent, directly through the hyperbolic system (1.1), without making appeal to any argument that relies on the specific nature of internal dissipation. For systems expressing the principles of balance in continuum physics, admissible solutions should satisfy, as a minimal requirement, the Clausius-Duhem inequality (*cf.* TRUESDELL & TOUPIN [37; Sec. 258]), which is the proper statement of the Second Law of Thermodynamics. This will be compatible with any rational viscosity admissibility criterion, because, following the work of COLEMAN [2] and COLEMAN & NOLL [3], all constitutive classes of media with internal dissipation are now constructed so as to comply with the Clausius-Duhem inequality. Early advocates of this approach to admissibility were BURTON [1], WEBER [40; Sec. 179] and RAYLEIGH [35], who, in reevaluating the aforementioned work of STOKES and subsequent elaborations by RIEMANN [27], asserted that shocks in isothermal (or isentropic) flow should be admissible so long as they do not increase the mechanical energy.[2] It is easy to see that this assertion is equivalent to the Clausius-Duhem inequality in the situation where isothermal (or isentropic) flow is sustained by supplying externally, through radiation, the required amount of heat.

---

[2] "Mechanical energy" is to be understood as the sum of kinetic energy and strain energy, the latter being evaluated through a stored-energy function, which is the Helmholtz free energy restricted to constant temperature in the case of isothermal flow or the internal energy restricted to constant entropy in the case of isentropic flow.

In the context of the general system (1.1), the above considerations have motivated *entropy admissibility criteria* (Godunov [11], Lax [21]) according to which admissible solutions should satisfy the inequality

$$\partial_t \eta(U(x, t)) + \partial_x q(U(x, t)) \leqq 0, \tag{1.4}$$

for a designated pair of functions η, called *entropy*, and q, called *entropy flux*. In accordance with the interpretation of the Clausius-Duhem inequality in thermodynamics (*cf.* [2, 3]) it is natural to require that all smooth solutions of (1.1) should satisfy (1.4) automatically. This will be the case if and only if

$$\nabla q(U) = \nabla \eta(U) \nabla F(U). \tag{1.5}$$

If U is a solution of (1.1) in the space BV, it can be shown that, by virtue of (1.5), the measure on the left-hand side of (1.4) is concentrated in the set $\mathscr{S}$ of points of approximate jump discontinuity of U. In that case, (1.4) reduces to the local condition[3]

$$q(U_+) - q(U_-) - s[\eta(U_+) - \eta(U_-)] \leqq 0. \tag{1.6}$$

This suggests that admissible solutions of (1.1) may be characterized by means of *shock admissibility criteria* which impose on each point of $\mathscr{S}$ restrictions relating the local speed s of propagation with the local states U_+ and U_- ahead and behind the shock wave. Beyond (1.6), shock conditions may be induced by viscosity admissibility criteria[4] or may be motivated by stability arguments.

An early example of a shock admissibility criterion is the statement that only compressive shocks are admissible in classical gas dynamics. As observed by Riemann [27], this is equivalent to the requirement that the shock be "supersonic" relative to the state on its front and "subsonic" relative to the state on its rear. Lax [20] laid down this condition as a general shock criterion (*cf.* Section 3). A comprehensive shock admissibility criterion, which epitomizes the experience accumulated from detailed studies of special systems, was proposed by Liu [22] and will be discussed in Section 3.

Even though there is a consensus that admissibility of solutions of (1.1) may be characterized completely through local restrictions on points of singularity, it is being debated whether such restrictions need only be imposed on points of $\mathscr{S}$, as shock admissibility criteria profess, or whether more refined conditions, involving points of $\mathscr{I}$, should also be met.[5] There are strong indications that

---

[3] This condition was first introduced, in the context of gas dynamics, by Jouguet [18].

[4] There is voluminous literature on this subject, beginning with Rankine [26] and Rayleigh [36]. Interesting recent results and references to earlier work may be found in Majda & Pego [24].

[5] The question is even more relevant for systems in several spatial dimensions: It is not at all clear that, as is commonly assumed, admissibility may be decided completely at the level of manifolds of points of discontinuity of codimension one.

shock admissibility criteria suffice in the case of strictly hyperbolic systems but are inadequate for systems that are not strictly hyperbolic (*cf.* GLIMM [10]) or for systems that change type, modelling phase transitions (*cf.* SHEARER [29]). At this time, it is known for certain only that shock conditions suffice to characterize completely admissible solutions for the case $n = 1$, in one or several spatial dimensions (*cf.* VOLPERT [39]).

Reasonable admissibility criteria should comply with the principle of irreversibility of solutions, which pervades the entire theory, and thus may impose rules on wave sources but not on wave sinks. In addition, admissibility criteria should be compatible with translations and dilatations of coordinates, which leave the system (1.1) invariant. As we shall see in Section 4, the above requirements suggest that admissibility should be tested in the framework of the Riemann problem (RIEMANN [27], LAX [20], LIU [22]), *i.e.*, in the context of solutions of the form $U(x, t) = V(x/t)$, which represent wave fans emanating from the origin at time $t = 0$.

Wave fan admissibility criteria may be motivated through viscosity admissibility criteria by replacing (1.3) with the system

$$\partial_t U(x, t) + \partial_x F(U(x, t)) = t\, \partial_x [A_\varepsilon\, \partial_x U(x, t)], \tag{1.7}$$

which is invariant under dilatation of coordinates and thus admits solutions of the form $U(x, t) = V(x/t)$. This approach is pursued in TUPCIEV [38], DAFERMOS [6], DAFERMOS & DIPERNA [8], and SLEMROD [31].

The main object of this paper is to discuss the *entropy rate fan admissibility criterion*, introduced by DAFERMOS [5], which dubs admissible those solutions that maximize the rate of entropy production. There has been a limited investigation of the appropriateness of this criterion, particularly in the context of certain simple systems that change type, employed to model phase transitions (*cf.* HATTORI [12, 13, 14]).

Throughout this paper, we will be using, somewhat loosely, the term *wave of moderate strength* to describe waves whose amplitude does not exceed a value that depends solely on the maximum of second derivatives of F. These should be contrasted on the one hand with *weak waves*, whose amplitude is bounded by a constant that depends also on the maximum of third derivatives of F, and on the other hand with *strong waves*, whose amplitude is unrestricted. The main result, established in Section 5, is that in any strictly hyperbolic system wave fans made of rarefaction waves and shocks of moderate strength that satisfy the Liu admissibility criterion do maximize the rate of entropy production. In elastodynamics, this statement is valid for isothermal or isentropic motions with waves of arbitrarily large amplitude (*cf.* [5]) but is generally invalid for nonisothermal adiabatic motions with very strong waves, unless the medium is an ideal monatomic gas (*cf.* HSIAO [15]).

At the present time, the major challenge to the theory of admissibility of weak solutions in one space dimension is posed by hyperbolic systems that are not strictly hyperbolic (*cf.* KEYFITZ & KRANZER [19], ISAACSON, MARCHESIN, PLOHR & TEMPLE [16], SCHAEFFER & SHEARER [28]) and by systems of changing type (*cf.* JAMES [17], SLEMROD [30], SHEARER [29]).

2. Assumptions

The system (1.1) is strictly hyperbolic, that is, for each U in R^n the $n \times n$ matrix $\nabla F(U)$ has n real, distinct characteristic roots

$$\lambda_1(U) < \dots < \lambda_n(U) \tag{2.1}$$

and corresponding linearly independent right characteristic (column) vectors

$$r_1(U), \dots, r_n(U) \tag{2.2}$$

and left characteristic (row) vectors

$$l_1(U), \dots, l_n(U) \tag{2.3}$$

normalized so that

$$l_i(U)\, r_k(U) = \delta_{ik}. \tag{2.4}$$

Furthermore, (1.1) is equipped with a (locally) uniformly convex entropy function η and associated entropy flux q so that (1.5) holds. In particular, the system (1.1) is symmetrizable. If H denotes the symmetric and positive definite Hessian of η, it follows from (1.5) that

$$H(U)\, \nabla F(U) = \nabla F(U)^T H(U). \tag{2.5}$$

As observed by CONLON & LIU [4], (2.5) in turn implies

$$r_i(U)^T H(U)\, r_k(U) = 0, \quad i \neq k. \tag{2.6}$$

Throughout the paper we will be dealing exclusively with waves whose amplitude does not exceed a value that depends solely on the maximum of the second derivatives of F. Consequently, in the terminology enunciated in the Introduction, all waves here are of moderate strength.

3. Shock Admissibility Criteria

We fix U_- in R^n. By virtue of (2.1) and standard bifurcation theory, the set of points in R^n that may be joined to U_- by a shock of moderate strength is the union of n smooth arcs intersecting at U_-. Specifically, for $i = 1, \dots, n$, there is a smooth curve $U: (-a, a) \to R^n$ and a smooth function $s: (-a, a) \to R$ such that

$$F(U(\tau)) - F(U_-) = s(\tau)\, [U(\tau) - U_-], \quad -a < \tau < a, \tag{3.1}$$

$$s(0) = \lambda_i(U_-), \quad U(0) = U_-, \quad \dot{U}(0) = r_i(U_-). \tag{3.2}$$

The shock wave that joins $U(\tau)$ with U_- is called an *i-shock* and $U(\cdot)$ is the *i-Hugoniot curve* through U_-.

Upon rewriting (1.2) in the form

$$\left\{ \int_0^1 \nabla F(\sigma U_+ + (1 - \sigma)\, U_-)\, d\sigma - sI \right\} [U_+ - U_-] = 0, \tag{3.3}$$

we deduce that an i-shock of moderate strength with amplitude, say β, propagates with speed

$$s = \lambda_i(U_-) + O(\beta) \tag{3.4}$$

and the jump satisfies

$$U_+ - U_- = \{l_i(U_-)\,[U_+ - U_-]\}\, r_i(U_-) + O(\beta^2). \tag{3.5}$$

An i-shock that joins U_- with U_+ and propagates with speed s is called *a left contact discontinuity* if $s = \lambda_i(U_-)$ and a *right contact discontinuity* if $s = \lambda_i(U_+)$.

In order to distinguish admissible shocks of moderate strength, LIU [22] has proposed the following criterion, which was motivated by detailed studies of special systems (*cf.* OLEINIK [25], WENDROFF [41]): The i-shock that joins $U_+ := U(\bar{\tau})$ with U_- is *admissible* if

$$s(\bar{\tau}) \leqq s(\tau) \quad \text{for all } \tau \text{ between } 0 \text{ and } \bar{\tau}. \tag{3.6}$$

It is shown in CONLON & LIU [4] that any admissible i-shock of moderate strength that joins the states U_-, U_+ and propagates with speed s satisfies the entropy inequality (1.6) as well as (a slightly weakened version of) the admissibility condition of LAX [20]:

$$\lambda_i(U_+) \leqq s \leqq \lambda_i(U_-). \tag{3.7}$$

A partial converse of this proposition also holds: (3.7) and/or (1.6) imply (3.6) but, in general, only when the system (1.1) is *genuinely nonlinear* and the shock is weak (*cf.* LAX [21]). Consequently, the Liu admissibility condition is generally more discriminating than either the entropy criterion or the Lax condition. The relationship between (3.6) and the viscosity admissibility criterion is discussed in MAJDA & PEGO [24].

In special systems, for instance the equations of balance in one-dimensional elastodynamics, the portrait of Hugoniot curves described above extends globally in state space and so one may attempt to employ (3.6) in order to characterize even strong admissible shocks. In general, however, the global pattern is more complicated, as Hugoniot curves of one family may eventually bifurcate into several branches, Hugoniot curves of different families may merge, and disconnected branches may appear that cannot be assigned to any particular family of shocks. Moreover, in the presence of strong shocks the relation of the Liu condition to the entropy condition and the Lax condition is not *a priori* clear but has to be investigated case by case.

4. Wave Fan Admissibility Criteria

Assume $U(x, t)$ is a bounded measurable BV solution of (1.1) defined on the upper half-plane $\{(x, t): -\infty < x < \infty, 0 \leqq t < \infty\}$. Based on the premiss that admissibility criteria should be local and compatible with the principle of irreversibility, the admissibility of U should be tested individually at each point of its domain and the test at the point $(\bar{x}, \bar{t})$ of the upper half-plane should only involve the restriction of U to the semicircle $\{(x, t): (x - \bar{x})^2 + (t - \bar{t})^2 < \varrho^2, t \geqq \bar{t}\}$ where ϱ is an arbitrarily small positive number. Furthermore, conditions of admissibility should be invariant under translations and dilatations of the coordinates, *i.e.*, U should be admissible at $(\bar{x}, \bar{t})$ if and only if the new solution

$$U_\mu(x, t; \bar{x}, \bar{t}) := U(\bar{x} + \mu x, \bar{t} + \mu t) \tag{4.1}$$

is admissible at $(0, 0)$, for any $\mu > 0$.

We now make the assumption that for every fixed $(\bar{x}, \bar{t})$

$$U_\mu(x, t; \bar{x}, \bar{t}) \to U_0(x, t; \bar{x}, \bar{t}), \quad \mu \downarrow 0, \tag{4.2}$$

in measure, on the upper half-plane. This certainly holds when $(\bar{x}, \bar{t})$ belongs to the set $\mathscr{C}$ of points of approximate continuity of U, in which case U_0 is constant, equal to the approximate limit of U at $(\bar{x}, \bar{t})$. (4.2) also holds when $(\bar{x}, \bar{t})$ belongs to the set $\mathscr{S}$ of points of approximate jump discontinuity of U, in which case U_0 takes two values, namely the left approximate limit U_- for $x < st$ and the right approximate limit U_+ for $x > st$, s being the speed of propagation of the shock at $(\bar{x}, \bar{t})$. That (4.2) holds even when $(\bar{x}, \bar{t})$ belongs to the residual set $\mathscr{I}$ has been established so far only for solutions of genuinely nonlinear systems of two equations constructed by the random choice method (*cf.* DiPerna [9]). It is clear that U_0 is a solution of (1.1) which is constant along rays emanating from the origin. The basic premiss of wave fan admissibility criteria is that $U(x, t)$ is admissible at $(\bar{x}, \bar{t})$ if the solution $U_0(x, t; \bar{x}, \bar{t})$ is admissible at (0, 0).

Suppressing in the notation the dependence on $(\bar{x}, \bar{t})$, we may write

$$U_0(x, t) = V\left(\frac{x}{t}\right), \tag{4.3}$$

where V is a function of bounded variation defined on $(-\infty, \infty)$. We fix any C^∞ test function $\varphi(x, t)$ with compact support in $(-\infty, \infty) \times (0, \infty)$, define

$$\psi(\xi) := \int_0^\infty \varphi(\xi t, t)\, dt, \quad -\infty < \xi < \infty, \tag{4.4}$$

which is a C^∞ function with compact support in $(-\infty, \infty)$, and note the simple identity

$$\begin{aligned} &\int_0^\infty \int_{-\infty}^\infty \left\{V\left(\frac{x}{t}\right) \partial_t\varphi(x, t) + F\left(V\left(\frac{x}{t}\right)\right) \partial_x\varphi(x, t)\right\} dx\, dt \\ &\quad = \int_{-\infty}^\infty \{-V(\xi)\, \partial_\xi[\xi\psi(\xi)] + F(V(\xi))\, \partial_\xi\psi(\xi)\}\, d\xi. \end{aligned} \tag{4.5}$$

This implies that U_0 in (4.3) is a solution of (1.1) if and only if V satisfies the ordinary differential equation

$$\dot{F}(V) - \xi\dot{V} = 0, \quad -\infty < \xi < \infty, \tag{4.6}$$

in the sense of distributions and thereby in the sense of measures.

By the chain rule of Volpert [39] for BV functions, (4.6) may be rewritten in the form

$$\left\{\int_0^1 \nabla F(\sigma V(\xi+) + (1-\sigma)\, V(\xi-))\, d\sigma - \xi I\right\} \dot{V} = 0. \tag{4.7}$$

For ξ in the (at most countable) set of points of jump discontinuity of V, (4.7) and (3.3) yield the Rankine-Hugoniot jump condition

$$F(V(\xi+)) - F(V(\xi-)) = \xi[V(\xi+) - V(\xi-)]. \tag{4.8}$$

The set of points of continuity of V is the union of a set on which the measure $\dot{V}$ vanishes and a set on which

$$\lambda_i(V(\xi)) = \xi \tag{4.9}$$

for some $i = 1, \ldots, n$. The restriction of V on an interval where (4.9) holds, called an *i-rarefaction wave*, is differentiable almost everywhere and, in virtue of (4.7), (4.9), it satisfies

$$\dot{V}(\xi) = [\nabla\lambda_i(V(\xi))\, r_i(V(\xi))]^{-1}\, r_i(V(\xi)). \tag{4.10}$$

When the waves in the wave fan V are of moderate strength, the support of $\dot{V}$ is contained in the union of n disjoint intervals. The restriction of V on the ith of these intervals is generally composed of i-shocks, i-rarefaction waves and constant states; it is called an *i-wave*. We say an i-wave is *admissible* when all its shocks satisfy the Liu admissibility condition. So long as (3.7) holds, two i-shocks, two i-rarefaction waves or one i-shock and one i-rarefaction wave, contained in an i-wave, cannot be separated by a constant state. It follows that an admissible i-wave is composed of a (finite or infinite) sequence of i-shocks and i-rarefaction waves, placed in alternating order, and any shock adjacent on its left (or right) to a rarefaction wave is necessarily a left (or right) contact discontinuity.

As shown by LIU [22, 23], any two states U_-, U_+ with $|U_+ - U_-|$ sufficiently small may be connected by a unique wave fan consisting of $n+1$ constant states $U_- = U_1, U_2, \ldots, U_n, U_{n+1} = U_+$ and n waves, one from each family, so that, for $i = 1, \ldots, n$, U_i is joined with U_{i+1} by an admissible i-wave of moderate strength. This wave fan is a solution of the Riemann problem with initial data (U_-, U_+).

Despite the success of the Liu condition in yielding a satisfactory solution of the Riemann problem for strictly hyperbolic systems when the waves are of moderate strength, an admissibility criterion with global authority is still unknown. It is thus important, at this stage, to experiment with various criteria, especially those motivated by physics.

A seemingly promising approach is to attempt to characterize admissible wave fans by means of the viscosity criterion induced by (1.7): A solution V of (4.6) is admissible if it is the $\varepsilon \to 0$ limit of a family of (smooth) solutions V_ε of the system of ordinary differential equations

$$A_\varepsilon \ddot{V}_\varepsilon(\xi) = \dot{F}(V_\varepsilon(\xi)) - \xi \dot{V}_\varepsilon(\xi). \tag{4.11}$$

Only preliminary results in this direction have appeared in print (*cf.* TUPCIEV [38], DAFERMOS [7]) though several investigations are currently in progress.

The main objective here is to discuss the entropy rate criterion which stipulates that the solution U is admissible at $(\bar{x}, \bar{t})$ if

$$\limsup_{z \downarrow 0} \frac{D^+}{Dt} \int_{\bar{x}-z}^{\bar{x}+z} \{\eta(U(x,t)) - \eta(\bar{U}(x,t))\}\, dx \Bigg|_{t=\bar{t}} \leqq 0 \tag{4.12}$$

holds, where $\bar{U}$ is any other solution of (1.1) defined on some semicircle $\{(x,t): |x-\bar{x}|^2 + |t-\bar{t}|^2 < \varrho^2, t \geqq \bar{t}\}, \varrho > 0$, and satisfying $\bar{U}(x,\bar{t}) = U(x,\bar{t})$ for

$|x - \bar{x}| < \varrho$. Thus, according to this criterion, a solution is admissible if it dissipates entropy at a maximum rate or, in the language of thermodynamics, if it satisfies the Second Law with maximal efficiency.

It is clear that the entropy rate criterion complies with the principles laid down at the opening of this section. In particular, U is admissible at $(\bar{x}, \bar{t})$ if and only if U_μ, defined by (4.1), is admissible at (0, 0) for any $\mu > 0$. Thus, assuming (4.2) holds, it is reasonable to postulate an *entropy rate fan criterion* which stipulates that U be admissible at $(\bar{x}, \bar{t})$ if

$$\limsup_{z \downarrow 0} \frac{D^+}{Dt} \int_{-z}^{z} \{\eta((U_0(x,t)) - \eta(\bar{U}_0(x,t))\}\, dx \bigg|_{t=0} \leqq 0. \tag{4.13}$$

If $U_0(x,t) = V(x/t)$, $\bar{U}_0(x,t) = \bar{V}(x/t)$, with $V(\xi) = \bar{V}(\xi)$ for $|\xi|$ sufficiently large, a simple calculation shows that (4.13) may be rewritten in the form

$$\int_{-\infty}^{\infty} \{\eta(V(\xi)) - \eta(\bar{V}(\xi))\}\, d\xi \leqq 0. \tag{4.14}$$

We note the identity

$$\eta(V) = \frac{d}{d\xi}\{\xi\eta(V) - q(V)\} + \dot{q}(V) - \xi\dot{\eta}(V). \tag{4.15}$$

By the chain rule for BV functions,

$$\begin{aligned} \dot{q}(V) - \xi\dot{\eta}(V) = \bigg\{ & \int_0^1 \nabla q(\sigma V(\xi+) + (1-\sigma)\, V(\xi-))\, d\sigma \\ & - \xi \int_0^1 \nabla\eta(\sigma V(\xi+) + (1-\sigma)\, V(\xi-))\, d\sigma \bigg\}\, \dot{V}. \end{aligned} \tag{4.16}$$

By virtue of (1.5) and (4.7), the right-hand side of (4.16) vanishes at every point of continuity of V and so the measure $\dot{q}(V) - \xi\dot{\eta}(V)$ is concentrated on the set of points of jump discontinuity of V. Therefore, combining (4.14), (4.15) and (4.16), we conclude that V satisfies the entropy rate criterion if and only if it minimizes

$$\Phi := \sum_{\text{shocks}} \{q(V(\xi+)) - q(V(\xi-)) - \xi[\eta(V(\xi+)) - \eta(V(\xi-))]\} \tag{4.17}$$

over the set of wave fans with fixed end-states $V(-\infty)$ and $V(\infty)$.

In the following section we will discuss the relationship between the entropy rate criterion and the Liu shock admissibility condition.

5. Waves that Minimize the Entropy Rate Are Admissible

In this section we show that, for waves of moderate strength, the entropy rate criterion and the Liu admissibility condition are equivalent.

Theorem 5.1. *Let V be a wave fan, with waves of moderate strength, which minimizes the entropy rate Φ, defined by* (4.17). *Then every shock of V satisfies the Liu admissibility condition.*

Proof. To establish the assertion of the theorem, we shall show that if V contains at least one shock that violates the Liu condition then another wave fan may be constructed with the same end-states $V(\pm\infty)$ as V but with lower entropy rate Φ.

As noted in Section 4, V generally consists of $n+1$ constant states, say $U_1, \dots, U_{n+1}$, and n waves, one from each family, so that, for $i = 1, \dots, n$, U_i is joined to U_{i+1} by an i-wave. For concreteness and simplicity, we will make the following assumptions on V:

(a) For $i = 1, \dots, n$, the i-wave contains a finite number of i-shocks and/or i-rarefaction waves.

(b) For $i = 1, \dots, n$, if U is a state on the closure of the range of the i-wave, then

$$\nabla\lambda_i(U)\, r_i(U) \neq 0. \tag{5.1}$$

(c) For some index j, the j-wave contains one shock that violates the Liu admissibility condition. The remaining shocks in the j-wave and all shocks in the i-waves, $i \neq j$, satisfy the Liu condition (3.6), as a strict inequality.

The general situation, in which some i-waves may contain infinitely many shocks and rarefaction waves, the nondegeneracy condition (5.1) may fail at certain points of an i-wave, and more than one shock may violate the Liu admissibility condition, can be treated by a routine modification of the techniques that will be developed here.

Assume that the unique inadmissible shock in the j-wave joins the states U_-, U_+ and propagates with speed s_0:

$$F(U_+) - F(U_-) = s_0[U_+ - U_-]. \tag{5.2}$$

U_- and/or U_+ may be constant states of V; otherwise, the shock has to be a left and/or right contact discontinuity, in which case

$$\lambda_j(U_-) = s_0 \tag{5.3}$$

and/or

$$\lambda_j(U_+) = s_0. \tag{5.4}$$

The state U_+ lies on the j-Hugoniot curve $U(\cdot)$ through U_-, say $U_+ = U(\bar{\tau})$ for some $\bar{\tau} > 0$. Since (3.6) is violated for this shock, we can find τ_0 in $[0, \bar{\tau})$ such that $s(\tau_0) = s_0$ and $s(\cdot)$ is decreasing at τ_0. For definiteness, assume τ_0 lies in the open interval $(0, \bar{\tau})$ and

$$\dot{s}(\tau_0) < 0. \tag{5.5}$$

We set $U_0 = U(\tau_0)$. Since $s(\tau_0) = s_0$,

$$F(U_0) - F(U_-) = s_0[U_0 - U_-] \tag{5.6}$$

and hence, by account of (5.2),

$$F(U_+) - F(U_0) = s_0[U_+ - U_0]. \tag{5.7}$$

Motivated by (5.6), (5.7), we should view from now on the shock that joins U_- with U_+ as the superposition of two shocks, one joining U_- with U_0 the other joining U_0 with U_+, both travelling with the same speed s_0. Because of this observation and (5.5), it is to be expected that in the vicinity of V lie other wave fans in which the shock that joins U_- and U_+ has split into two shocks, one travelling with speed s_-, slightly lower than s_0, the other travelling with speed s_+, slightly higher than s_0. We shall demonstrate that this splitting increases locally the rate of entropy decay without affecting appreciably the rate of entropy change elsewhere.

We introduce a scalar parameter ε and seek a family V_ε of wave fans, defined for ε positive small, which collapse to V as $\varepsilon \downarrow 0$. V_ε consists of $n+1$ constant states, $\hat{U}_1(\varepsilon), \dots, \hat{U}_{n+1}(\varepsilon)$, and n waves, one from each family. For $i \neq j$, $\hat{U}_{i+1}(\varepsilon)$ is joined to $\hat{U}_i(\varepsilon)$ by an admissible i-wave. On the other hand, $\hat{U}_{j+1}(\varepsilon)$ is joined to $\hat{U}_j(\varepsilon)$ by a j-wave having the following parts:

(a) An admissible j-wave that joins $\hat{U}_j(\varepsilon)$ with a state $\hat{U}_-(\varepsilon)$, unless, of course, $U_- = U_j$ in which case we set $\hat{U}_-(\varepsilon) \equiv \hat{U}_j(\varepsilon)$.

(b) The constant state $\hat{U}_-(\varepsilon)$, provided U_- is a constant state of V other than U_j.

(c) A j-shock that joins $\hat{U}_-(\varepsilon)$ with a state $\hat{U}_0(\varepsilon)$:

$$F(\hat{U}_0(\varepsilon)) - F(\hat{U}_-(\varepsilon)) = s_-(\varepsilon)\,[\hat{U}_0(\varepsilon) - \hat{U}_-(\varepsilon)]. \tag{5.8}$$

(d) The constant state $\hat{U}_0(\varepsilon)$.

(e) A j-shock that joins $\hat{U}_0(\varepsilon)$ with a state $\hat{U}_+(\varepsilon)$:

$$F(\hat{U}_+(\varepsilon)) - F(\hat{U}_0(\varepsilon)) = s_+(\varepsilon)\,[\hat{U}_+(\varepsilon) - \hat{U}_0(\varepsilon)]. \tag{5.9}$$

(f) The constant state $\hat{U}_+(\varepsilon)$, provided U_+ is a constant state of V other than U_{j+1}.

(g) An admissible j-wave that joins $\hat{U}_+(\varepsilon)$ with $\hat{U}_{j+1}(\varepsilon)$, unless $U_+ = U_{j+1}$, in which case $\hat{U}_+(\varepsilon) \equiv \hat{U}_{j+1}(\varepsilon)$.

All functions of ε indicated above are continuously differentiable and $\hat{U}_1(\varepsilon) \equiv U_1$, $\hat{U}_{n+1}(\varepsilon) \equiv U_{n+1}$, $\hat{U}_i(0) = U_i$, $i = 1, \dots, n+1$, $\hat{U}_-(0) = U_-$, $\hat{U}_0(0) = U_0$, $\hat{U}_+(0) = U_+$, $s_-(\varepsilon) \leqq s_+(\varepsilon)$, $s(0) = s_0 = s_+(0)$. Furthermore, if U_- and/or U_+ are constant states of V other than U_j and/or U_{j+1}, then we require[6]

$$l_j(U_-)\,\hat{U}'_-(0) = 0 \tag{5.10}$$

and/or

$$l_j(U_+)\,\hat{U}'_+(0) = 0. \tag{5.11}$$

We let α denote the maximum of $|U_+ - U_0|$ and $|U_0 - U_-|$, say, for definiteness,

$$|U_+ - U_0| = \alpha, \qquad |U_0 - U_-| \leqq \alpha, \tag{5.12}$$

[6] Throughout this section a prime denotes differentiation with respect to ε.

and scale the parameter ε so that

$$s'_+(0) = 1. \tag{5.13}$$

Once the estimates that will be derived later in this section become available, the existence of the family V_ε of wave fans with the above properties may be established through standard arguments from bifurcation theory. We shall not get involved here with that lengthy but routine project.

Due to the assumed nondegeneracy conditions, there is a one-to-one correspondence between shocks and rarefaction waves of V_ε and shocks and rarefaction waves of V. Specifically, if the i-wave of V contains an i-shock joining the states V_-, V_+ and propagating with speed s, then the i-wave of V_ε contains an i-shock:

$$F(\hat{V}_+(\varepsilon)) - F(\hat{V}_-(\varepsilon)) = \hat{s}(\varepsilon)\,[\hat{V}_+(\varepsilon) - \hat{V}_-(\varepsilon)], \tag{5.14}$$

where $\hat{V}_-(\varepsilon)$, $\hat{V}_+(\varepsilon)$ and $\hat{s}(\varepsilon)$ are continuously differentiable functions with $\hat{V}_-(0) = V_-$, $\hat{V}_+(0) = V_+$ and $\hat{s}(0) = s$.

By virtue of (4.17), the entropy rate of V_ε is given by

$$\Phi(\varepsilon) = \sum_{\text{shocks}} \{q(\hat{V}_+(\varepsilon)) - q(\hat{V}_-(\varepsilon)) - \hat{s}(\varepsilon)\,[\eta(\hat{V}_+(\varepsilon)) - \eta(\hat{V}_-(\varepsilon))]\}. \tag{5.15}$$

The assertion of the theorem will be established if we show that $\Phi'(0) < 0$. We differentiate (5.15) and then set $\varepsilon = 0$; upon using (1.5), this yields

$$\begin{aligned}\Phi'(0) = \sum_{\text{shocks}} \{&\nabla\eta(V_+)\,[\nabla F(V_+) - sI]\,\hat{V}'_+(0) - \nabla\eta(V_-)\,[\nabla F(V_-) - sI]\,\hat{V}'_-(0)\\ &- \hat{s}'(0)\,[\eta(V_+) - \eta(V_-)]\}.\end{aligned} \tag{5.16}$$

Differentiating (5.14) with respect to ε and setting $\varepsilon = 0$ we get

$$[\nabla F(V_+) - sI]\,\hat{V}'_+(0) - [\nabla F(V_-) - sI]\,\hat{V}'_-(0) = \hat{s}'(0)\,[V_+ - V_-]. \tag{5.17}$$

Combining (5.16) with (5.17), we obtain

$$\begin{aligned}\Phi'(0) = &- \sum_{\hat{s}' \leq 0} \hat{s}'(0)\,\{\eta(V_+) - \eta(V_-) - \nabla\eta(V_+)\,[V_+ - V_-]\}\\ &- \sum_{\hat{s}' > 0} \hat{s}'(0)\,\{\eta(V_+) - \eta(V_-) - \nabla\eta(V_-)\,[V_+ - V_-]\}\\ &+ \sum_{\hat{s}' \leq 0} [\nabla\eta(V_+) - \nabla\eta(V_-)]\,[\nabla F(V_-) - sI]\,\hat{V}'_-(0)\\ &+ \sum_{\hat{s}' > 0} [\nabla\eta(V_+) - \nabla\eta(V_-)]\,[\nabla F(V_+) - sI]\,\hat{V}'_+(0).\end{aligned} \tag{5.18}$$

We examine the right-hand side of (5.18): As indicated by the notation, the first and third summations run over the set of shocks with $\hat{s}'(0) \leq 0$ while the second and fourth summations run over the set of shocks with $\hat{s}'(0) > 0$. Since η is uniformly convex, the first and second sum is positive. In particular, the second summation includes the term

$$s'_+(0)\,\{\eta(U_+) - \eta(U_0) - \nabla\eta(U_0)\,[U_+ - U_0]\}, \tag{5.19}$$

which is positive of order $O(\alpha^2)$, by virtue of (5.12) and (5.13). The remainder of the proof aims to establish that, when the strength of the wave fan V is sufficiently small, then the third and fourth sums are dominated by the term (5.19) and thus $\Phi'(0) < 0$.

From (5.17) it follows

$$\hat{s}'(0)\, l_i(V_+)\,[V_+ - V_-] = [\lambda_i(V_+) - s]\, l_i(V_+)\, \hat{V}'_+(0) - [\lambda_i(V_-) - s]\, l_i(V_-)\, \hat{V}'_-(0) + [l_i(V_-) - l_i(V_+)]\,[\nabla F(V_-) - sI]\, \hat{V}'_-(0). \tag{5.20}$$

Let β be the amplitude of the shock. Recalling (3.4), (3.5) and combining (5.17) with (5.20) we conclude that if either $\lambda_i(V_-) = s$ (left contact discontinuity) or $\hat{V}'_-(0)$ is nearly orthogonal to $l_i(V_-)$, then

$$\begin{aligned} |[\nabla F(V_-) - sI]\, \hat{V}'_-(0)| \leqq [1 + O(\beta)]\, \{ |[\nabla F(V_+) - sI]\, \hat{V}'_+(0)| \\ + |[\lambda_i(V_+) - s]\, l_i(V_+)\, \hat{V}'_+(0)| \}. \end{aligned} \tag{5.21}$$

Under the same conditions, if $[\nabla F(V_-) - sI]\, \hat{V}'_-(0)$ is nearly orthogonal to $l_k(V_-)$, for some $k \neq i$, then $[\nabla F(V_+) - sI]\, \hat{V}'_+(0)$ is nearly orthogonal to $l_k(V_+)$. In the above statements the roles of V_- and V_+ may be reversed.

An analogous estimate holds for rarefaction waves. Assume that the i-wave of V contains an i-rarefaction wave of amplitude γ, joining the states V_-, V_+. Then the i-wave of V_ε contains an i-rarefaction wave joining states $\hat{V}_-(\varepsilon)$, $\hat{V}_+(\varepsilon)$ where $\hat{V}_-(\varepsilon)$, $\hat{V}_+(\varepsilon)$ are continuously differentiable functions with $\hat{V}_-(0) = V_-$, $\hat{V}_+(0) = V_+$. By account of (4.10) and (5.1), we may introduce a convenient parametrization of rarefaction wave curves under which $\hat{V}_+(\varepsilon) = U(\hat{\tau}(\varepsilon), \varepsilon)$ where $U(\tau, \varepsilon)$ is the solution of the initial-value problem

$$\begin{aligned} \dot{U}(\tau, \varepsilon) &= r_i(U(\tau, \varepsilon)) \\ U(0, \varepsilon) &= \hat{V}_-(\varepsilon). \end{aligned} \tag{5.22}$$

Therefore,

$$\hat{V}'_+(0) = X(\hat{\tau}(0))\, \hat{V}'_-(0) + \hat{\tau}'(0)\, r_i(V_+), \tag{5.23}$$

where the $n \times n$ matrix-valued function $X(\tau)$ is determined by solving the variational equation

$$\begin{aligned} \dot{X}(\tau) &= \nabla r_i(U(\tau, 0))\, X(\tau) \\ X(0) &= I \end{aligned} \tag{5.24}$$

of (5.22). In particular,

$$X(\hat{\tau}(0))\, r_i(V_-) = r_i(V_+), \tag{5.25}$$

$$X(\hat{\tau}(0)) = I + O(\gamma). \tag{5.26}$$

Combining (5.23), (5.25) and (5.26), we deduce

$$[\nabla F(V_+) - \lambda_i(V_+)\, I]\, \hat{V}'_+(0) = [I + O(\gamma)]\, [\nabla F(V_-) - \lambda_i(V_-)\, I]\, V'_-(0) \tag{5.27}$$

whence we derive the estimate

$$|[\nabla F(V_+) - \lambda_i(V_+) I]\, \hat{V}'_+(0)| \leqq [1 + O(\gamma)]\, |[\nabla F(V_-) - \lambda_i(V_-) I]\, \hat{V}'_-(0)|. \quad (5.28)$$

From (5.27) also follows that if $[\nabla F(V_-) - \lambda_i(V_-) I]\, \hat{V}'_-(0)$ is nearly orthogonal to $l_k(V_-)$, for some $k \neq i$, then $[\nabla F(V_+) - \lambda_i(V_+) I]\, \hat{V}'_+(0)$ is nearly orthogonal to $l_k(V_+)$. In the above statements the roles of V_- and V_+ may be reversed.

Recall that any admissible i-wave is composed of a sequence of i-shocks and i-rarefaction waves so that any shock adjacent on its left (or right) to a rarefaction wave is necessarily a left (or right) contact discontinuity. Consequently, we may combine the results obtained above for individual shocks or rarefaction waves to arrive at the following conclusion: Assume that for $\varepsilon \geqq 0$ two states $\hat{V}_-(\varepsilon)$ and $\hat{V}_+(\varepsilon)$ are joined by an admissible i-wave, which is a part (or all) of the i-wave of the wave fan V_ε. $\hat{V}_-(\varepsilon)$ and $\hat{V}_+(\varepsilon)$ are continuously differentiable functions with $\hat{V}_-(0) = V_-$, $\hat{V}_+(0) = V_+$. In particular, V_- and V_+ are joined by an admissible i-wave W, of amplitude, say, δ. If W begins with a shock, we denote its speed by $\bar{s}_-$; otherwise we set $\bar{s}_- = \lambda_i(V_-)$. Similarly, if W ends with a shock, we let $\bar{s}_+$ denote its speed; otherwise we set $\bar{s}_+ = \lambda_i(V_+)$. Assume that either $\bar{s}_- = \lambda_i(V_-)$ or $\hat{V}'_-(0)$ is nearly orthogonal to $l_i(V_-)$. Then

$$|[\nabla F(V_-) - \bar{s}_- I]\, \hat{V}'_-(0)| \leqq [1 + O(\delta)]\, \{|[\nabla F(V_+) - \bar{s}_+ I]\, \hat{V}'_+(0)| + |[\lambda_i(V_+) - \bar{s}_+]\, l_i(V_+)\, \hat{V}'_+(0)|\}. \quad (5.29)$$

Furthermore, if $\hat{V}'_-(0)$ is nearly orthogonal to both $l_i(V_-)$ and $l_k(V_-)$, $k \neq i$, then $[\nabla F(V_+) - \bar{s}_+ I]\, \hat{V}'_+(0)$ is nearly orthogonal to $l_k(V_+)$ and hence, *a fortiori*, $\hat{V}'_+(0)$ is nearly orthogonal to $l_k(V_+)$. In the above statements, the roles of V_- and V_+ may be reversed.

We exploit this result as follows: Beginning with $\hat{U}'_1(0) = 0$, we show by induction:

$$\hat{U}'_i(0) \text{ is nearly orthogonal to } l_i(U_i), \dots, l_n(U_i);\ i = 1, \dots, j. \quad (5.30)$$

Next we prove that

$$[\nabla F(U_-) - s_0 I]\, \hat{U}'_-(0) \text{ is nearly orthogonal to } l_j(U_-), \dots, l_n(U_-). \quad (5.31)$$

Indeed, recalling that $\hat{U}_-(\varepsilon)$ is joined to $\hat{U}_j(\varepsilon)$ by an admissible j-wave, we deduce, by virtue of (5.30) and the result established above, that $[\nabla F(U_-) - s_0 I]$ $\hat{U}'_-(0)$ is nearly orthogonal to $l_{j+1}(U_-), \dots, l_n(U_-)$. However, this vector is also nearly orthogonal to $l_j(U_-)$, because, as noted earlier, it is either $\hat{U}_-(\varepsilon) \equiv \hat{U}_j(\varepsilon)$ or one of (5.3) and (5.10) holds.

In a similar manner, starting out from the other end of the wave fan V with $\hat{U}'_{n+1}(0) = 0$, we demonstrate by induction

$$\hat{U}'_i(0) \text{ is nearly orthogonal to } l_1(U_i), \dots, l_{i-1}(U_i);\quad i = j+1, \dots, n+1, \quad (5.32)$$

and then we show

$$[\nabla F(U_+) - s_0 I]\, \hat{U}'_+(0) \text{ is nearly orthogonal to } l_1(U_+), \dots, l_j(U_+). \tag{5.33}$$

In (5.8) and (5.9) we differentiate with respect to ε, then we put $\varepsilon = 0$ and add together the resulting equations thus obtaining

$$\begin{aligned} &[\nabla F(U_+) - s_0 I]\, \hat{U}'_+(0) - [\nabla F(U_-) - s_0 I]\, \hat{U}'_-(0) \\ &\quad = s'_+(0)\,[U_+ - U_0] + s'_-(0)\,[U_0 - U_-]. \end{aligned} \tag{5.34}$$

Both vectors $U_+ - U_0$ and $U_0 - U_-$ are nearly orthogonal to $l_1(U_-), \dots, \dots, l_{j-1}(U_-), l_{j+1}(U_-), \dots, l_n(U_-)$. Therefore, on account of (5.30), (5.34), the vectors $s'_+(0)\,[U_+ - U_0]$ and $s'_-(0)\,[U_0 - U_-]$ must cancel each other out to first order. In particular, recalling (5.12), (5.13), we infer

$$s'_-(0) < 0, \tag{5.35}$$

$$|[\nabla F(U_-) - s_0 I]\, \hat{U}'_-(0)| = O(\alpha^2), \tag{5.36}$$

$$|[\nabla F(U_+) - s_0 I]\, \hat{U}'_+(0)| = O(\alpha^2). \tag{5.37}$$

Note that (5.35) induces the required separation $s_-(\varepsilon) < s_0 < s_+(\varepsilon)$, $\varepsilon > 0$.

We apply (5.29) to the admissible j-wave that joins $\hat{U}_j(\varepsilon)$ with $\hat{U}_-(\varepsilon)$. Combining (5.26) with (5.3) or (5.10), whichever is applicable, and recalling that by (5.30) $\hat{U}'_j(0)$ is nearly orthogonal to $l_j(U_j)$, we conclude $|\hat{U}'_j(0)| = O(\alpha^2)$. Similarly, we show $|\hat{U}'_{j+1}(0)| = O(\alpha^2)$. Returning to (5.29) and using (5.30), (5.32), we easily deduce, by induction,

$$|\hat{U}'_i(0)| = O(\alpha^2), \qquad i = 1, \dots, n+1. \tag{5.38}$$

Finally we show that if the i-wave of V contains a shock joining V_- to V_+ and propagating with speed s, then

$$|[\nabla F(V_-) - sI]\, \hat{V}'_-(0)| = O(\alpha^2), \tag{5.39}$$

provided that $V_- \neq U_0$, and

$$|[\nabla F(V_+) - sI]\, \hat{V}'_+(0)| = O(\alpha^2), \tag{5.40}$$

provided that $V_+ \neq U_0$. For $i \neq j$, we verify (5.39) by applying (5.29) to the admissible i-wave that joins V_- with U_{i+1}, taking account of (5.38). We employ the same argument when $i = j$ but V_- lies to the right of U_0. On the other hand, when $i = j$ and V_- lies on the left of U_0, we get (5.39) by applying (5.29), with the roles of the endpoints reversed, to the admissible j-wave that joins U_i with V_-. The proof of (5.40) is, of course, similar.

We have now laid the preparation for estimating the last two sums on the right-hand side of (5.18). For definiteness, let us consider the typical term in the third summation. Assume that the i-wave of V contains an i-shock with amplitude β, which joins states V_-, V_+, propagates with speed s and $\hat{s}'(0) \leqq 0$.

In particular, on account of (5.13), $V_- \neq U_0$. Then

$$\nabla\eta(V_+) - \nabla\eta(V_-) = [V_+ - V_-]^T \int_0^1 H(\sigma V_+ + (1-\sigma) V_-)\, d\sigma$$
$$= [V_+ - V_-]^T [H(V_-) + O(\beta)], \tag{5.41}$$

where, as in Section 2, H denotes the Hessian of η. By virtue of (3.5),

$$V_+ - V_- = O(\beta)\, r_i(V_-) + O(\beta^2). \tag{5.42}$$

Finally,

$$[\nabla F(V_-) - sI]\, \hat{V}'_-(0) = [\lambda_i(V_-) - s]\, [l_i(V_-)\, \hat{V}'_-(0)]\, r_i(V_-)$$
$$+ \{[\nabla F(V_-) - sI]\, \hat{V}'_-(0) - [\lambda_i(V_-) - s]\, [l_i(V_-)\, \hat{V}'_-(0)]\, r_i(V_-)\}. \tag{5.43}$$

The first term on the right-hand side of (5.43); (a) vanishes if the shock is a left contact discontinuity; (b) also vanishes when $i = j$, $V_- = U_-$ and $U_- \neq U_j$, because then either (5.3) or (5.10) holds; (c) is of order $O(\beta\alpha^2)$ in the remaining case $V_- = U_i$, as follows by combining (3.4) with (5.38). The second term on the right-hand side of (5.43) lies on the subspace spanned by $r_1(V_-), \ldots, r_{i-1}(V_-)$, $r_{i+1}(V_-), \ldots, r_n(V_-)$ and is of order $O(\alpha^2)$, by virtue of (5.39). Combining this with (5.41), (5.42) and using (2.6) gives

$$[\nabla\eta(V_+) - \nabla\eta(V_-)]\, [\nabla F(V_-) - sI]\, \hat{V}'_-(0) = O(\beta^2\alpha^2). \tag{5.44}$$

A similar estimate holds for the typical term of the fourth summation on the right-hand side of (5.18). Therefore, the last two sums on the right-hand side of (5.18) are of order $O(\gamma\, \delta\alpha^2)$, where γ is an upper bound for the amplitude of each shock in V and δ is the sum of amplitudes of all shocks in V. It follows that if $\gamma\, \delta \ll 1$, the term (5.19), of order $O(\alpha^2)$, dominates and so $\Phi'(0) < 0$. This completes the proof of the theorem.

Acknowledgment. This work was supported in part by the National Science Foundation through Grant DMS-8504051, in part by the U.S. Army under Contract No. DAAL-3-86-0074, and in part by the U.S. Air Force URI F49620-86-C-0111.

References

1. Burton, C. V., On plane and spherical sound-waves of finite amplitude. *Philosophical Magazine*, Ser. 5, **35**, 317–333 (1893).
2. Coleman, B. D., Thermodynamics of materials with memory. *Arch. Rational Mech. Analysis* **17**, 1–46 (1964).
3. Coleman, B. D., & W. Noll, The thermodynamics of elastic materials with heat conduction and viscosity. *Arch. Rational Mech. Analysis* **13**, 167–177 (1963).
4. Conlon, J., & T.-P. Liu, Admissibility criteria for hyperbolic conservation laws. *Indiana U. Math. J.* **30**, 641–652 (1981).
5. Dafermos, C. M., The entropy rate admissibility criterion for solutions of hyperbolic conservation laws. *J. Diff. Eqs.* **14**, 202–212 (1973).

6. DAFERMOS, C. M., Solution of the Riemann problem for a class of hyperbolic conservation laws by the viscosity method. *Arch. Rational Mech. Analysis* **52**, 1–9 (1973).
7. DAFERMOS, C. M., Structure of solutions of the Riemann problem for hyperbolic systems of conservation laws. *Arch. Rational Mech. Analysis* **53**, 203–217 (1974).
8. DAFERMOS, C. M., & R. J. DIPERNA, The Riemann problem for certain classes of hyperbolic systems of conservation laws. *J. Diff. Eqs.* **20**, 90–114 (1976).
9. DIPERNA, R. J., Singularities of solutions of nonlinear hyperbolic systems of conservation laws. *Arch. Rational Mech. Analysis* **60**, 75–100 (1975).
10. GLIMM, J., The interactions of nonlinear hyperbolic waves. *Comm. Pure Appl. Math.* **41**, 569–590 (1988).
11. GODUNOV, S. K., An interesting class of quasilinear systems. *Dokl. Akad. Nauk SSSR* **139**, 521–523 (1961). English translation: *Soviet Math.* **2**, 947–949 (1961).
12. HATTORI, H., The Riemann problem for a van der Waals fluid with entropy rate admissibility criterion. Isothermal case. *Arch. Rational Mech. Analysis* **92**, 247–263 (1986).
13. HATTORI, H., The Riemann problem for a van der Waals fluid with entropy rate admissibility criterion. Nonisothermal case. *J. Diff. Eqs.* **65**, 158–174 (1986).
14. HATTORI, H., The entropy rate admissibility criterion and the double phase boundary problem. *Contemporary Math.* **60**, 51–65 (1987).
15. HSIAO, L., The entropy rate admissibility criterion in gas dynamics. *J. Diff. Eqs.* **38**, 226–238 (1980).
16. ISAACSON, E., MARCHESIN, D., PLOHR, B., & B. TEMPLE, The classification of solutions of quadratic Riemann problems. *SIAM J. Appl. Math.* **48**, 1009–1032 (1988).
17. JAMES, R. D., The propagation of phase boundaries in elastic bars. *Arch. Rational Mech. Anal.* **73**, 125–158 (1980).
18. JOUGUET, E., Sur la propagation des discontinuités dans les fluides. *C. R. Acad. Sci. Paris* **132**, 673–676 (1901).
19. KEYFITZ, B. L., & H. C. KRANZER, A system of non-strictly hyperbolic conservation laws arising in elasticity theory. *Arch. Rational Mech. Anal.* **72**, 219–241 (1980).
20. LAX, P. D., Hyperbolic systems of conservation laws. *Comm. Pure Appl. Math.* **10**, 537–566 (1957).
21. LAX, P. D., Shock waves and entropy. *Contributions to Nonlinear Functional Analysis* (E. H. ZARANTONELLO ed.), pp. 603–634. New York: Academic Press, 1971.
22. LIU, T.-P., The Riemann problem for general systems of conservation laws. *J. Diff. Eqs.* **18**, 218–234 (1975).
23. LIU, T.-P., Admissible solutions of hyperbolic conservation laws. *Memoirs Am. Math. Soc.* **240**, 1–78 (1981).
24. MAJDA, A., & R. PEGO, Stable viscosity matrices for systems of conservation laws. *J. Diff. Eqs.* **56**, 229–262 (1985).
25. OLEINIK, O. A., Uniqueness and stability of the generalized solution of the Cauchy problem for a quasilinear equation. *Uspehi Mat. Nauk* (N.S.), **14**, 165–170 (1959). English translation: *Am. Math. Soc. Transl.* **33**, 285–290.
26. RANKINE, W. J. M., On the thermodynamic theory of waves of finite longitudinal disturbance. *Phil. Trans. Roy. Soc. Lond.* **160**, 277–288 (1870).
27. RIEMANN, B., Ueber die Fortpflanzung ebener Luftwellen von endlicher Schwingungsweite. *Gött. Abh. Math. Cl.* **8**, 43–65 (1860).
28. SCHAEFFER, D. G., & M. SHEARER, The classification of 2×2 systems of non-strictly hyperbolic conservation laws with application to oil recovery. *Comm. Pure Appl. Math.* **40**, 141–178 (1987).
29. SHEARER, M., Nonuniqueness of admissible solutions of Riemann initial value prob-

lems for a system of conservation laws of mixed type. *Arch. Rational Mech. Anal.* **93**, 45–59 (1986).
30. SLEMROD, M., Admissibility criteria for propagating phase boundaries in a van der Waals fluid. *Arch. Rational Mech. Anal.* **81**, 301–315 (1983).
31. SLEMROD, M., A limiting "viscosity" approach to the Riemann problem for materials exhibiting change of phase. *Arch. Rational Mech. Anal.* **105**, No. 4 (1989).
32. STOKES, G. G., On a difficulty in the theory of sound. *Philosophical Magazine* **33**, 349–356 (1848).
33. STOKES, G. G., On a difficulty in the theory of sound. *Mathematical and Physical Papers*, Vol. II, pp. 51–55. Cambridge University Press, 1883.
34. STRUTT, J. W. (Lord RAYLEIGH), *The Theory of Sound*, Vol. II, London: Macmillan 1878.
35. STRUTT, J. W. (Lord RAYLEIGH), Note on tidal bores. *Proc. Roy. Soc. London* **A 81**, 448–449 (1908).
36. STRUTT, J. W. (Lord RAYLEIGH), Aerial plane waves of finite amplitude. *Proc. Roy. Soc. London* A **84**, 247–284 (1911).
37. TRUESDELL, C., & R. TOUPIN, *The Classical Field Theories. Handbuch der Physik* (S. FLÜGGE, Ed.), Vol. III/1, pp. 226–793. Berlin-Göttingen-Heidelberg: Springer-Verlag, 1960.
38. TUPCIEV, V. A., On the method for introducing viscosity in the study of problems involving the decay of a discontinuity. *Dokl. Akad. Nauk SSSR*, **211**, 55–58 (1973). English translation: *Soviet Math.* **14**, 978–982 (1973).
39. VOLPERT, A. I., The space BV and quasilinear equations. *Mat. Sbornik*, **73 (115)**, 255–302 (1967). English translation: *Math. USSR Sbornik* **2**, 225–267 (1967).
40. WEBER, H., *Die Partiellen Differential-Gleichungen der Mathematischen Physik*, Zweiter Band. Vierte Auflage. Braunschweig: Friedrich Vieweg und Sohn, 1901.
41. WENDROFF, B., The Riemann problem for materials with nonconvex equation of state. Part I: Isentropic flow. Part II: General flow. *J. Math. Anal. Appl.* **38**, 454–466; 640–658 (1972).

Division of Applied Mathematics
Brown University
Providence, Rhode Island

(Received August 26, 1988)

Weak Martensitic Transformations in Bravais Lattices

J. L. ERICKSEN

Dedicated to B. D. Coleman, on the Occasion of His Sixtieth Birthday

1. Introduction

Nonlinear thermoelasticity theory is being used, with some success, to analyze phenomena associated with phase transitions in some crystals, involving a change in crystal symmetry, what are often called Martensitic transformations. Roughly, these are the crystals which are, or at least behave as if they were Bravais lattices. For such lattices, molecular theories of thermoelasticity imply that the Helmholtz free energy should be invariant under an infinite discrete group. However, workers often use constitutive equations which are invariant only under a finite subgroup, to analyze behavior near transitions, Physicists are likely to use polynomials of as low degree as is feasible, what is sometimes called "Landau Theory", to treat second-order or "weak" first-order transitions. I don't know how to give a precise meaning to the notion of a weak transition. By one rather pragmatic interpretation a transition is weak if behavior near the transition of interest can be analyzed, satisfactorily, with a free energy function which is invariant only under some finite subgroup. Using this idea, one can deduce some properties which transitions must have, to be considered weak. My purpose is to elaborate this, to try to get some better understanding of what limits the ranges of applicability of what are, really, two versions of thermoelasticity theory.

Briefly, if an unloaded crystal is subjected to different temperatures, it is likely to experience phase transitions at isolated values of the temperature. Commonly, these are of first-order, involving sudden changes in deformation, the "transition strains", which can be large or small. Twins are likely to appear or disappear if some change in symmetry occurs, among other things. Application of loads is likely to shift the transition temperature and produce other unusual effects even when the loads are quite small. The idea is to use one constitutive equation to cover such deformations and a temperature interval. Physically, it is reasonable enough to think of its domain as bounded, but it does need to include the phases of interest. Roughly, I will argue that, in assuming that a finite invariance group suffices, we are, effectively, restricting the constitutive equation to a rather special kind of domain. To be weak, a transition must connect two states in one such domain and, as we shall see, it is fairly easy to deduce some necessary conditions for this.

2. Elastic crystals

First, we review some ideas about nonlinear thermoelasticity theory which are implied by various versions of molecular theory, for Bravais lattices. Think of what is, intuitively, a homogeneous configuration of a crystal filling all of space. According to the classical definition of a crystal, it must admit a translation group generated by three linearly independent lattice vectors e_a $(a = 1, 2, 3)$. That is, any vector with integer components relative to this basis translates any point to a physically indistinguishable point. The simplest kinds of configurations are the Bravais lattices. These consist of identical atoms, positioned so that all of their positions are given by applying such a translation group to one position. Molecular theories of elasticity for these tend to be simpler and involve fewer hypotheses than do those used for more complex crystals. Roughly, they deal with relatively short-range interactions.

Now, in crystallography, it has long been appreciated that different lattice vectors can generate the same lattice: e_a and $\bar{e}_a$ do provided

$$\bar{e}_a = m_a^b e_b, \tag{2.1}$$

where

$$m = \|m_a^b\| \tag{2.2}$$

is any matrix of integers with

$$\det m = \pm 1. \tag{2.3}$$

We employ the usual summation convention. Such matrices form a representation of the infinite group $GL(3, Z)$, the general linear transformations on the integers, in three-dimensions, a group I denote by G. Now, for the Bravais lattices, prescribing a set of lattice vectors fixes the atomic positions to within an unimportant translation of the crystal. From this, it is rather clear that molecular theories of thermoelasticity will imply a constitutive equation for the Helmholtz free energy density of the form.

$$\varphi = \hat{\varphi}(e_a, \theta) = \hat{\varphi}(m_a^b e_a, \theta), \quad m \in G, \tag{2.4}$$

G being the infinite group mentioned in the introduction. Another rather obvious implication is that rigidly rotating the crystal won't change φ, or

$$\hat{\varphi}(e_a, \theta) = \hat{\varphi}(\mathrm{R}e_a, \theta), \quad R \in SO(3), \tag{2.5}$$

$SO(3)$ being the group of rotations.

Tacitly assumed here is that only the atomic positions matter. Paramagnetic-ferromagnetic transitions do occur in Bravais lattices and, to treat them, one needs to introduce other "internal variables". With more complex crystals, there are more possibilities for encountering other kinds of variables which can play an important role in transitions. JAMES [1] and RIVLIN [2] discuss cases of this kind, associated with certain transformations in quartz.

In molecular theory, one strategy is to try to eliminate such variables by minimizing φ with respect to them, holding the lattice vectors fixed. If we use absolute minimizers, and all goes well with the elimination, we should be left with a function

$\hat{\varphi}(e_a, \theta)$ satisfying (2.4) so, to this extent, the crystal behaves as if it were a Bravais lattice. This does exclude some kinds of metastable configurations which can be important, physically. Or, if one allows them, $\hat{\varphi}$ can become multivalued, a possibility discussed by ERICKSEN [3]. The previous references illustrate one kind of difficulty which occurs.

To get to thermoelasticity theory, workers use the Cauchy-Born hypothesis. Pick some configuration, with lattice vectors E_a, as a reference. The assumption is that, if F is the macroscopic deformation gradient, then the vectors e_a given by

$$e_a = FE_a \tag{2.6}$$

are a possible set of lattice vectors in the deformed configuration. Then, we can introduce a group $\bar{G}$, conjugate to G, consisting of the linear transformations H given by

$$H \in G \Leftrightarrow HE_a = m_b^a E_b, \quad m \in G. \tag{2.7}$$

Then, we have, from above

$$\varphi = \hat{\varphi}(FE_a, \theta) = \hat{\varphi}(RFHE_a, \theta), \tag{2.8}$$

with fixed values of E_a, the kind of constitutive equation which is associated with thermoelasticity, incorporating some general features suggested by molecular theory.

One aim is to try to use such theory to analyze near-transition phenomena, which can involve twinning, in particular. ZANZOTTO [4] has explored the experimental evidence concerning the applicability of the above theory of twinning, writing that "... some experimental observations suggest that it is *not* safe to use thermoelasticity theory to describe the behavior of bodies whose structure is not given by a simple Bravais lattice. On the contrary, it does seem rather safe to use it for the latter. The reasons are rather unclear." Among other things, he discusses clear evidence that (2.6) fails to apply to deformations occurring in mechanical twinning of zirconium, which is a rather simple hexagonal close-packed crystal, but not a Bravais lattice. For it, there is no clear evidence that (2.4) fails. In some more complex crystals, for example calcite, (2.6) seems to apply to twinning deformations, so it seems hard to spot any clear pattern in this. In cases I have checked, transition strains in Bravais lattices also fit (2.6). Without belaboring the matter, ZANZOTTO's statement seems to me the best available rule of thumb. So, it seems that our theory is pretty good for Bravais lattices. Occasionally it seems to apply to a crystal not of this kind, but one better look hard at the available experimental evidence before gambling much on this. Primarily, we have used ideas of molecular theory to estimate the group $\bar{G}$. Conceivably, thermoelasticity theory could apply, when some of these ideas fail but, here, I will ignore this possibility. Workers involved in molecular theories of crystal elasticity have not considered the possibility that (2.6) fails to apply, as far as I know.

To a large extent, the classical theory of crystallographic groups aims at describing the symmetry of particular configurations. Such theory, as it applies to Bravais lattices, is briefly summarized by SCHWARZENBERGER [5], who also covers some topological considerations to be discussed later. I won't repeat the references he gives to group-theoretic results mentioned below.

Associated with any set of lattice vectors e_a is a finite subgroup $L(e_a)$ of G which I like to call the *lattice group*, defined by

$$L(e_a) = \{m \in G \mid Q = m^a_b e_a \otimes e^b \in O(3)\}. \tag{2.9}$$

Here $O(3)$ denotes the three-dimensional orthogonal group, e^a the reciprocal lattice vectors, the dual basis satisfying

$$e^a \cdot e_b = \delta^a_b, \quad e^a \otimes e_a = e_a \otimes e^a = 1. \tag{2.10}$$

Then, for example, the expression for Q in (2.9) is equivalent to the formula

$$Qe_a = m^b_a e_b. \tag{2.11}$$

As is clear from (2.1), there are infinitely many choices of lattice vectors for a given configuration. Applying (2.1) changes the lattice group, by a similarity transformation involving some $m \in G$. We account for this by considering all of these to be in an equivalence class, as is briefly indicated by

$$L(m^b_a e_b) \cong L(e_a), \quad m \in G. \tag{2.12}$$

We accept the classical view that two Bravais lattices have the same symmetry provided these equivalence classes coincide or, equivalently, if it is possible to choose lattice vectors so that their lattice groups coincide. SCHWARZENBERGER [5] calls the set of lattices having the same symmetry a Bravais lattice. For us, this might be confusing, since the different ones can be related by non-trivial deformations. Instead, I will say that they have the same *symmetry type*.

Associated with $L(e_a)$ is a conjugate group $P(e_a)$ consisting of the orthogonal transformations indicated by (2.9), the point group. Applying a rotation to e_a changes the point group to a conjugate group of $O(3)$, by the obvious similarity transformation. Since this is commonly regarded as a trivial operation, we regard these as being in an equivalence class indicated by

$$P(\mathrm{Re}_a) \cong P(e_a), \quad R \in SO(3). \tag{2.13}$$

Also, it is easy to verify that

$$P(m^b_a e_b) = P(e_a), \quad m \in G, \tag{2.14}$$

i.e. the point group is independent of the choice of lattice vectors. Following SCHWARZENBERGER [5], we say that two Bravais lattices belong to the same *crystal system* provided that their point groups are in the same equivalence class. It is known that there are 7 crystal systems. The corresponding point groups are described in Table 1 in the *Appendix*. The reader might be more familiar with the number 32, which is obtained by permitting proper subgroups of the 7. These can be realized in some crystal structures, but not Bravais lattices. Elasticians might be familiar with the number 11. The number 32 effectively reduces to 11 because elastic strain energies are automatically invariant under central inversions. Clearly, 4 of these can only be realized in structures more complex than Bravais lattices. From remarks made earlier one should be cautious, in using elasticity theory for them.

As is familiar to group theorists, and is discussed in some detail by ERICKSEN [6], any finite subgroup of G is a lattice group for some choices of lattice vectors and, independent of the choice, it determines a point group, to within the equivalence indicated by (2.13). Said differently, configurations having the same symmetry type are in the same crystal system. However, crystals in the same system can have different symmetry types, it being known that there are 14 symmetry types. The breakdown is described in Table 2, in the Appendix. Thus, the distinction between lattice and point groups is of some import. Elasticians are in the habit of glossing it and, sometimes, this is justifiable. One of our aims is to elaborate this.

It simplifies matters a bit to replace the groups by subgroups, discarding the elements with negative determinants, which involves no real loss in generality. We denote by G^+ the subgroup of G with $\det m = 1$, and similarly define $L^+(e_a)$ and $P^+(e_a)$.

Also, it follows from (2.6) that φ is expressible as a function of the scalar products forming the matrix

$$D = \|e_a \cdot e_b\| = D^T > 0 \tag{2.15}$$

or, with a slight abuse of notation,

$$\varphi = \varphi(D, \theta). \tag{2.16}$$

This is still subject to the invariance requirement

$$\varphi(m^T Dm, \theta) = \varphi(D, \theta), \qquad m \in G, \tag{2.17}$$

the same for all Bravais lattices. Configurations of particular interest are the natural states, minimizers of φ, values $\bar{D}$ such that

$$\varphi(D, \theta) \geqq \varphi(\bar{D}, \theta). \tag{2.18}$$

One of these will generate the infinite set $m^T \bar{D}m$, for $m \in G^+$. As is perhaps obvious, and easy to verify, these all have the same symmetry type or, if you like, they are symmetry-related minimizers. Reasonably, φ can have other minimizers, with a different symmetry type, at isolated values of θ, transition temperatures associated with first-order Martensitic transformations. When we say that a crystal has a certain symmetry, we generally mean that this is the symmetry of a natural state, as I interpret it.

Clearly, the assumption that (2.6) holds gives

$$e_a \cdot e_b = FE_a \cdot FE_b = E_a \cdot CE_b \tag{2.19}$$

where

$$C = F^T F \tag{2.20}$$

is the Cauchy-Green tensor commonly used in nonlinear elasticity theory. With (2.8) and (2.9), we then have

$$\varphi(E_a \cdot CE_b, \theta) = \varphi(E_a \cdot H^T CHE_b, \theta), \qquad H \in \bar{G}^+. \tag{2.21}$$

What is more conventional is to use (2.21) with E_a taken as lattice vectors for a natural state at some value of θ, and $\bar{G}^+$ replaced by the finite group $P^+(E_a)$.

When we allow the possibility of Martensitic transformations, more than one point group can fit this description. If one exercises some good judgement in making the choice, theory of this kind can produce good predictions for some kinds of transitions, those we loosely defined as weak. This is well illustrated by the work of BALL & JAMES [7], for example.

3. Weak transitions

Rather tacit in the previous discussion is the assumption that the constitutive equation is defined for all kinematically possible configurations, or at least some related by very large deformations. Physically, the theory cannot be expected to apply to such a wide range, so there is some motivation for replacing the previous function by a restriction of it to some smaller domain. This will affect the invariance, since the invariance group must map the domain to itself. What is most likely is that the restriction will be invariant under some subgroup of G^+ For our purposes, it would be nice to have it be a finite subgroup. There is a natural way of doing this. It stems from the following fact, discussed briefly by SCHWARZENBERGER [5], in more detail by PITTERI [8]. Bear in mind that, given any symmetric, positive definite matrix D, one can find vectors e_a satisfying (2.15) and they are determined to within an orthogonal transformation. Thus, D determines a lattice group as indicated by

$$L^+(D) = \{m \in G^+ \mid m^T Dm = D\}, \tag{3.1}$$

as follows from (2.10). Then, the fact is that, *given* $\hat{D} = \hat{D}^T > 0$, *there is a neighborhood of it*, $N(\hat{D})$, *such that, for* $m \in G^+$,

$$D \subset N(\hat{D}) \Rightarrow m^T Dm \not\subset N(\hat{D}) \quad \textit{if} \quad m \notin L^+(\hat{D}) \tag{3.2}$$

and

$$D \subset N(\hat{D}) \Rightarrow m^T Dm \subset N(\hat{D}) \quad \textit{if} \quad m \in L^+(\hat{D}). \tag{3.3}$$

Clearly, if φ is invariant under G^+, its restriction to $N(\hat{D})$ will be invariant under the finite group $L^+(\hat{D})$ and not under any larger subgroup of G^+. From (3.1)–(3.3), it follow s immediately that

$$D \subset N(\hat{D}) \Rightarrow L^+(D) \subset L^+(\hat{D}). \tag{3.4}$$

As I see it, assuming that φ is invariant under a finite group is tantamount to assuming that its domain is restricted to be one of these neighborhoods. Typically, we are interested in temperatures within a few degrees from the transition temperature, a relatively small temperature interval. So, configurations of most interest should lie in some such neighborhood. Physically, this will include some of the natural states which are correlated by the Cauchy-Born hypothesis, as it applies to deformations produced by thermal expansion and the phase transition. In this, we have at least a rough description of what it means for a transition to be weak.

One point is that, within $N(\hat{D})$, we can set up a one-to-one correspondence between lattice and point groups, so there is no important distinction between them. To see this, take as a reference lattice vectors E_a such that

$$\| E_a \cdot E_b \| = \hat{D}. \tag{3.5}$$

For any other $D \subset N(\hat{D})$, we can, by normalizing the choice of an orthogonal transformation, assume that corresponding lattice vectors e_a satisfy a special version of the Cauchy-Born rule,

$$e_a = U E_a, \quad U = U^T > 0. \tag{3.6}$$

Then,

$$R \in P^+(e_a) \Rightarrow \mathrm{Re}_a = m_a^b e_b, \quad m \in L^+(D). \tag{3.7}$$

From (3.4), it then follows from (3.4) that, for the same m, there is a rotation $\bar{R}$ such that

$$m \in L^+(\hat{D}) \Rightarrow \bar{R} E_a = m_a^b E_b. \tag{3.8}$$

Putting these together, we have

$$\mathrm{Re}_a = R U E_a = m_a^b e_b = U(m_a^b E_b) = U \bar{R} E_a$$

or

$$RU = U\bar{R}. \tag{3.9}$$

By an elementary exercise in linear algebra, (3.9) is satisfied if and only if

$$R = \bar{R} \Rightarrow UR = RU \quad \text{and} \quad R \in P^+(e_a). \tag{3.10}$$

Clearly, this makes $P^+(e_a)$ a subgroup of $P^+(E_a)$, and attaches a unique rotation to each $m \in L^+(E_a)$. So, to answer a question raised earlier, it is in this sense justifiable to gloss the difference between point and lattice groups, within one of our neighborhoods.

It is worth noting that, in deducing (3.10), we only used the inclusion of lattice groups indicated in (3.4), a condition which is only necessary that D be contained in $N(\hat{D})$. Also, the conclusion does not depend much on $\hat{D}$. Read differently, it says that, *if e_a is given and E_a is any other set of similarly oriented lattice vectors with $L^+(e_a) \subset L^+(E_a)$, then, to within a rotation, $E_a = U^{-1} e_a$, where U^{-1} commutes with all $R \in P^+(e_a)$.*

To (3.10), there is a kind of converse, easily proved by similar arguments. *Suppose* (3.6) *and* (3.8) *hold. Then*

$$\bar{R} U = U \bar{R} \Rightarrow \bar{R} \in P^+(e_a) \quad \textit{and} \quad m \in L^+(e_a). \tag{3.11}$$

If one considers e_a as a natural state, (3.10) describes the restrictions on U which are often considered to apply to stretches produced by thermal expansion, or application of an hydrostatic pressure. Essentially, this is based on assumptions of local uniqueness and smooth dependence of U on θ or pressure, likely to be good except at phase transitions. The possible forms of U for the various point groups are neatly deduced by COLEMAN & NOLL [9], for example. The shorter list for Bravais lattices is in Table 3, in the Appendix.

In, say, the* BCC-BCT transitions in Indium-Thallium alloys discussed by BALL & JAMES [7], one does not have a crystal, in the strict sense. As with other alloys, a small amount of the alloying element is more randomly spaced. However, the bulk of the atoms do fit a Bravais lattice. The observed transition strains do fit (3.10) with e_a corresponding to BCT natural states, E_a as a BCC natural state. So, the BCT lattice group is contained in the BCC lattice group. The latter is maximal, not contained in any other lattice group, so, from (3.4), we are pretty well forced into identifying $\hat{D}$ with the cubic phase, as is implied above. It is then a problem in kinematics to determine whether the indicated BCT phases lie in such a $N(\hat{D})$. Later, we will discuss some difficulties which could arise here. If the transition strain is small enough, it will be in $N(\hat{D})$ and, in the case at hand, it is quite small. I won't pursue this question. However, PITTERI's [8] analysis of (3.1) *etc.* is rather constructive, involving estimates which might by useful in settling some questions of this kind. Assuming the answer is affirmative in the case at hand, one has a fairly good justification for assuming that φ is invariant under the cubic point group, as BALL & JAMES do. Many workers would go a step further, assuming that it suffices to use a polynomial function of fairly low degree, the first few terms in a Taylor expansion of an unknown function, if you like. Again, this comes close to the idea that we are approximating a restriction of a function, perhaps to neighborhood which is a ball.

Now let us return to (3.6) with $L^+(e_a)$ a proper subgroup of $L^+(E_a)$ or, what is here the same, that $P^+(e_a)$ is a proper subgroup of $P^+(E_a)$. Consider the linear transformations given by

$$U_R = R^T U R, \qquad R \in P^+(E_a), \qquad U_1 = U \tag{3.12}$$

producing lattice vectors

$$(e_R)_a = U_R E_a, \tag{3.13}$$

configurations giving the same value to φ, if only it is invariant under $P^+(E_a)$. In dealing with Martensitic transformations, workers are likely to identify E_a with Austenite, a higher temperature natural state, the $(e_R)_a$ with lower-temperature, symmetry-related Martensitic natural states. For analysis of twins, likely to form when the Austenite is cooled enough to transform to Martensite, it is important to have such symmetry-related natural states.

Suppose $P^+(e_a)$ is a group of order N, with elements R_i $(i = 1, \ldots, N)$. Then, from (3.10), all these should leave U invariant

$$R_i^T U R_i = U \quad \text{(no sum on } i\text{)}. \tag{3.14}$$

Thus, one will get the same U_R for all rotations of the form

$$R_i R, \quad i = 1, \ldots, N, \tag{3.15}$$

that is, for all rotations in the same right coset of $P^+(e_a)$ relative to $P^+(E_a)$. Also, using (3.11), one can show that, for two rotations in $P^+(E_a)$ to give the same value to U_R, they must lie in the same right coset. The number n of symmetry related

* We use the abbreviations of symmetry types given in Table 2 of the Appendix.

phases is then also the number of right cosets and from the theory of finite groups this is

$$n = \bar{N}/N, \quad \bar{N} = \text{order } P^+(E_a). \tag{3.16}$$

If $L^+(E_a)$ is, in turn, a subgroup of a larger lattice group then, in some cases, another choice of E_a could give a larger value of n, a fact which might be helpful in attempting to analyze complex arrangements of twins, *etc.* Of course, we can get infinitely many, with φ invariant under G^+.

Given the two point groups, as laid out in Table 1 of the Appendix, it is rather easy to find the most commonly observed twins, those commonly called** type I or type II. Briefly, configurations given by U and U_R are so related, if R is not in $P^+(e_a)$ but it, or another member of the same right coset is a 180 degree rotation R^π, so that

$$U_R = R^\pi U R^\pi .$$

Then, one can always find a rotation $\bar{R}$ and vectors a and n satisfying the basic twinning equation for these types

$$\bar{R} U R^\pi = (1 + a \otimes n) U . \tag{3.17}$$

Commonly, this is used to describe twinned Martensite, considered as a piecewise homogeneous natural Martensite state, arising from homogeneous Austenite, by a continuous displacement. Different values of R^π will give different solutions provided they are in different cosets or, equivalently, if their product is not an element of the smaller point group $P^+(e_a)$. My experience is that it is easier to use the last criterion. With Table 1, it isn't very hard to sort this out, for any particular choice of group and subgroup.

Clearly, we can apply similar reasoning to get the possible forms of deformations taking any lattice vector in $N(\hat{D})$ to any other. Adding in arbitrary rotations, we will have, say

$$e_a^{(1)} = R^{(1)} U^{(1)} E_a, \quad e_a^{(2)} = R^{(2)} U^{(2)} E_a, \tag{3.18}$$

with $U^{(1)}$ and $U^{(2)}$ restricted as in (3.10). Then, we will have

$$e_a^{(2)} = F e_a^{(1)},$$

with

$$F = R^{(2)} U^{(2)} (U^{(1)})^{-1} R^{(1)T}. \tag{3.19}$$

Possibly, such relations would be useful, when possible choices of $\hat{D}$ exist, but are not intuitively obvious choices. With these observations, we have found some guidelines for deciding whether a phase transition can reasonably be considered to be weak, covered by a finite group. It does emphasize the need to better understand the structure of neighborhoods $N(\hat{D})$.

** As is noted by Zanzotto [4], two different definitions of these occur in the literature and are alleged to be equivalent, but are not mathematically equivalent. I disregard that involving conditions that certain directions be rational.

4. Examples

First, we consider an example of a Martensitic transformation which is certainly not weak. It is of the BCC-FCC type, which occurs in Iron, for example. These have the same point group, so one might be tempted to try to use a constitution equation invariant under it. I won't belabor the difficulties encountered if one tries this. Although they are in the same crystal system, they have different symmetry types, so there is no way of choosing lattice vectors so the two lattice groups coincide. These two groups must then generate a larger group. However, there is no larger lattice group, so the larger group must be infinite. It might be a proper subgroup of G^+, but it won't simplify matters much to use it instead of G^+ as the invariance group. Fonseca [10] discusses possible ways of constructing constitutive functions invariant under G^+, but I don't know of an example of a constitutive equation appropriate for a transition of this kind. Further, there are ramifications concerning bifurcation theory. Were equilibrium paths of the BCC and FCC phases to intersect, a continuity argument indicates that, at the intersection, the lattice group would contain both lattice groups, which is impossible.

A rather similar situation occurs with transitions of the TRIG-H type. A TRIG point group can be a proper subgroup of the H point group, and the latter is not a proper subgroup of any larger group. Thus, if any point group is appropriate, it must be the H group. From Tables 1 and 2 in the Appendix, allowing the possibility that the two orthonormal bases might differ, we see that at least the two vectors labelled as k must be parallel. Then, from (3.10) and Table 22, it follows that the linear transformation relating them must be of the form

$$U = \alpha 1 + \beta k \otimes k. \tag{4.1}$$

From Table 1, the larger point group contains the rotation $R^{\pi/3}(k)$. It commutes with U so, from (3.11) it should also be in the smaller point group. However it is not. So the TRIG lattice group is not contained in the H lattice group. Again, because the latter is maximal, the two lattice groups generate an infinite group, so one might as well use G^+ as the invariance group. Also, by an argument like that used before, equilibrium paths of these kinds cannot intersect. Another observation is that the deformation associated with a transition of this kind will not be of the form (4.1). Briefly, as above, one can use (4.1) and (3.4) to get elements which should be in the TRIG-point group, but aren't.

Now consider configurations with lattice vectors of the form

$$e_1 = ai, \quad e_2 = aj, \quad e_3 = \frac{a}{2}(i+j) + bk \tag{4.2}$$

with (i, j, k) an orthonormal basis, a and b positive scalars. For most values of a and b, these are BCT configurations, and linear transformations of the form (4.1) map one into any other. It isn't hard to show that (3.10) is then satisfied, with $P^+(e_a)$ interpreted as this tetragonal point group. So, it might seem that one might use this as the basic invariance group, for transitions leaving the symmetry unchanged. A quirk arises because some cubic configurations are included. We have

$$\begin{aligned} &\text{BCC} \quad \text{when} \quad b = a/2, \\ &\text{FCC} \quad \text{when} \quad b = a/\sqrt{2}. \end{aligned} \tag{4.3}$$

This divides the BCT configurations into 3 disjoint sets, according as $b > a/2$, $a/2 < b < a/\sqrt{2}$ or $b > a/\sqrt{2}$, what SCHWARZENBERGER [5] calls the short, medium and long body-centered tretagonals. It follows from (3.4) that, if $\hat{D}$ is tetragonal, $N(\hat{D})$ cannot contain cubic configurations and, from above, there is no $\hat{D}$, with $N(\hat{D})$ containing both BCC and FCC configurations. This doesn't prove that it is impossible to treat as weak such things as short-long BCT, BCC-long BCT or FCC-short BCT transitions but, with the obvious complications, it is not so clear that it would be advantageous to do so. Actually, FONSECA [10] has done some analysis of cubic-tetragonal transitions, with φ invariant under G^+, so it is not hopeless to use this approach. With, say, a short-medium BCT transition, the two configurations can be very close to each other, which might have tempted us to take one as $\hat{D}$ and use the tetragonal group. Now, it is not so sure that this can be done and even if it can, it might be better to use as $\hat{D}$ a nearby cubic configuration, and the associated group. Here, we have discussed classification of the BCT symmetry type into 3 subtypes. SCHWARZENBERGER [5] gives a more satisfactory and comprehensive analysis, using topological ideas to classify all symmetry types into subtypes. Clearly, this is of some help in understanding the structure of $N(\hat{D})$, although it is not a panacea. It is perhaps worth noting that (4.1) can obviously relate the BCC and FCC configurations given by (4.3). It is then what is called "Bain strain" in the metallurgical literature, this being in agreement with the observed deformation associated with such transitions in at least some crystals.

One point seems worth mentioning. For SCHWARZENBERGER and other mathematical crystallographers, one set of lattice vectors is as good as any other. Objectively, the X-ray crystallographer is in this position. HARTSHORNE & STUART [11, p. 1] note that different workers have, in fact, opted for distinguishably different choices. For us, this is not quite true, since the choice is linked to macroscopic deformation. For us, all similarity oriented sets of lattice vectors are topologically equivalent, in the sense of being related by linear transformations with positive determinant, and with this is associated, in a natural way, a set of neighborhoods, including those we have called $N(\hat{D})$. SCHWARZENBERGER has a somewhat different view of matters topological, shared by other mathematical crystallographers. From the viewpoint of the crystallographer, the usual picture of a face-centered cubic is nice, making clear what is the point group, and you can figure out where all the atoms are. There is the difficulty that it suggests, wrongly, that the edges of the cube are lattice-vectors. There is a similar situation with the body-centered cubic. Try to apply the Cauchy-Born hypothesis to such "lattice vectors", taking the Bain strain as correct, and you will see a pitfall. Enough said.

From one point of view, we have compared two kinds of thermoelasticity theory. One, more conventional and thereby more familiar, assumes φ invariant under some point group, and it can cope with some kinds of Martensitic transitions. With the newer, employing G^+, we have made a step toward putting this in perspective. The earlier version is covered, briefly in the textbook by TRUESDELL [12, Ch. IV], for example. The newer version is also mentioned here and, as far as I know, in no other text. Both have spurred the development of some rather deep mathematical theory, relevant to the subject at hand, as can be seen from the surveys by KINDERLEHRER [13, 14], for example. We should, I think learn to live with the idea that a constitutive equation in one theory is to be regarded as a restriction of one in the other. With the rather simple reasoning used, I have done no more than indicate that this has some rather definite implications.

Appendix

Here, for easy reference, I include some information on point groups, *etc.*, for Bravais lattices. If v is a unit vector, θ an angle expressed in radians, $R^{\theta}(v)$ denotes the rotation with v as axis, θ as an angle. Also, i, j and k represent vectors forming an orthonormal basis, v_n denoting the four unit vectors given by

$$\frac{1}{\sqrt{3}}(i \pm j \pm k).$$

In Table 1, "point groups" means the groups denoted by P^+ earlier, P being the point group generated by P^+ and the central inversion $Q = -1$. The groups are numbered according to their order.

Table 1. Point groups for Bravais lattices

Number	*Elements*	*Order*
1	1	1
2	$1,\ R^{\pi}(i)$	2
3	$1,\ R^{\pi}(i),\ R^{\pi}(j),\ R^{\pi}(k)$	4
4	$1,\ R^{\pi}(i),\ R^{\pi}\left(\frac{i \pm \sqrt{3}j}{2}\right)$ $R^{2\pi/3}(k),\ R^{4\pi/3}(k)$	6
5	$1,\ R^{\pi}(i),\ R^{\pi}(j),\ R^{\pi}(k),\ R^{\pi}\left(\frac{i \pm j}{\sqrt{2}}\right)$ $R^{\pi/2}(k),\ R^{3\pi/2}(k)$	8
6	$1,\ R^{\pi}(i),\ R^{\pi}(j),\ R^{\pi}(k),$ $R^{\pi}\left(\frac{i \pm \sqrt{3}j}{2}\right),\ R^{\pi}\left(\frac{\sqrt{3}i \pm j}{2}\right),\ R^{\pi/3}(k)$ $R^{2\pi/3}(k),\ R^{4\pi/3}(k),\ R^{5\pi/3}(k)$	12
7	$1,\ R^{\pi/2}(i),\ R^{\pi/2}(j),\ R^{\pi/2}(k),$ $R^{\pi}\left(\frac{i \pm j}{\sqrt{2}}\right),\ R^{\pi}\left(\frac{j \pm k}{\sqrt{2}}\right),\ R^{\pi}\left(\frac{i \pm k}{\sqrt{2}}\right),$ $R^{\pi}(i),\ R^{\pi}(j),\ R^{\pi}(k),\ R^{3\pi/2}(i),\ R^{3\pi/2}(j),\ R^{3\pi/2}(k)$ $R^{2\pi/3}(v_n),\ R^{4\pi/3}(v_n)$	24

Different workers do attach different names and symbols to these. With the same numbering and names which are, I think, somewhat familiar to elasticians, we have

Table 2. Systems and symmetry types

Number	*System*	*Symmetry types*
1	Triclinic	Triclinic (TRI)
2	Monoclinic	Simple Monoclinic (SM) Centered Monoclinic (CM)
3	Orthorhombic	Simple Orthorhombic (SO) Centered Orthorhombic (CO) Body-centered Orthorhombic (BCO) Face-centered Orthorhombic (FCO)
4	Trigonal	Trigonal (TRIG)
5	Tetragonal	Simple Tetragonal (ST) Body-centered Tetragonal (BCT)
6	Hexagonal	Hexagonal (H)
7	Cubic	Simple cubic (SC) Body-centered cubic (BCC) Face-centered cubic (FCC)

where letters in parentheses denote abbreviations used here. Lattice groups for the different symmetry types depend on the choice of lattice vectors, so it is impossible to list all of them. One can use some rule for picking particular sets of lattice vectors, as SCHWARZENBERGER [5] does, use the point group P^+ which fits and calculate $m \in L^+$, using

$$m^a_b = e^a \cdot \mathrm{R} e_b, \qquad R \in P^+.$$

The group generated by L^+ and $m^a_b = -\delta^a_b$ is L.

Again, with the same numbering and i, j, k as in Table 1, we have α and β being any scalars such that $\alpha > 0, \alpha + \beta > 0$.

Table 3. Forms of U satisfying (3.10)

Number	*Restriction on U*
1	None
2	U has i as eigenvector
3	U has i, j, k as eigenvectors
4	$U = \alpha 1 + \beta k \otimes k$
5	$U = \alpha 1 + \beta k \otimes k$
6	$U = \alpha 1 + \beta k \otimes k$
7	$U = \alpha 1$

Acknowledgement. I thank MARJORIE SENECHAL for bringing to my attention the paper by SCHWARZENBERGER, helping me to get a somewhat better understanding of those neighborhoods. This research was partially supported by AFOSR/XOP and NSF, under NSF grant DMS-8842048.

References

1. JAMES, R. D., The stability and metastability of quartz, in *Metastability and Incompletely Posed Problems*, IMA Vol. 3 (ed. S. ANTMAN, J. L. ERICKSEN, D. KINDERLEHRER & I. MÜLLER), Springer-Verlag, 1987, 147–176.
2. RIVLIN, R. S., Some thoughts on material stability, *Proc. IUTAM Symp. Finite Elasticity* (ed. D. E. CARLSON & R. T. SHIELD), Martinus Nijhoff, 1982, 105–122.
3. ERICKSEN, J. L., Multi-valued strain energy functions for crystals, *Int. J. Solids Structures* **18**, 913–916, 1982.
4. ZANZOTTO, G., Twinning in minerals and metals: remarks on the comparison of a thermoelastic theory with some available experimental results, to appear in *Rend. dell'accad. Lincei.*
5. SCHWARZENBERGER, R. L. E., Classification of crystal lattices, *Proc. Camb. Phil. Soc.* **72**, 325–349, 1972.
6. ERICKSEN, J. L., On the symmetry of deformable crystals, *Arch. Rational Mech. Anal.* **72**, 1–13, 1979.
7. BALL, J. M., & JAMES, R. D., Fine phase mixtures as minimizers of energy, *Arch. Rational Mech. Anal.* **100**, 13–52, 1987.
8. PITTERI, M., Reconciliation of local and global symmetries of crystals, *J. Elasticity* **14**, 175–190, 1984.
9. COLEMAN, B. D., & NOLL, W., Material symmetry and thermostatic inequalities in finite elastic deformations, *Arch. Rational Mech. Anal.* **15**, 87–111, 1964.
10. FONSECA, I., Variational methods for elastic crystals, *Ph. D. Thesis*, University of Minnesota, 1985.
11. HARTSHORNE, N. H., & STUART, A., *Practical Optical Crystallography*, American Elsevier Publishing Co., 1964.
12. TRUESDELL, C., *A First Course in Rational Continuum Mechanics*, Vol. I, Academic Press, 1977.
13. KINDERLEHRER, D., Remarks about equilibrium configurations of crystals, in *Material Instabilities in Continuum Mechanics* (ed. J. M. BALL), Oxford University Press, 1988.
14. KINDERLEHRER, D., Phase transitions in crystals: the analysis of microstructure, to appear in *Proc. Int. Colloquium in Free Boundary Problems: Theory and Applications.*

Department of Aerospace
Engineering & Mechanics, and School
of Mathematics
University of Minnesota,
Minneapolis

(Received September 9, 1988)

Necessary Conditions at the Boundary for Minimizers in Finite Elasticity

HENRY C. SIMPSON & SCOTT J. SPECTOR

Dedicated to Bernard Coleman on the occasion of his 60th birthday

1. Introduction

In this paper we derive pointwise algebraic conditions on the elasticity tensor $\boldsymbol{C}(\boldsymbol{x}, \nabla \boldsymbol{f}(\boldsymbol{x}))$ that are necessary for a deformation $\boldsymbol{f}$ to be a local minimizer of the energy of an elastic body that is subjected to dead loads. In particular we show that the Legendre-Hadamard condition, Agmon's condition, and the new condition: if, for some vector $\boldsymbol{e}$,

$$\boldsymbol{e} \otimes \boldsymbol{n}_0 \cdot \boldsymbol{C}_0[\boldsymbol{e} \otimes \boldsymbol{n}_0] = 0$$

then (1.1)

$$\boldsymbol{C}_0[\boldsymbol{e} \otimes \boldsymbol{n}_0] = \boldsymbol{0}$$

are necessary conditions. Here $\boldsymbol{n}_0 = \boldsymbol{n}(\boldsymbol{x}_0)$ is the outward unit normal to the boundary at any point $\boldsymbol{x}_0$ where the deformation is *not* prescribed and $\boldsymbol{C}_0 = \boldsymbol{C}(\boldsymbol{x}_0, \nabla \boldsymbol{f}(\boldsymbol{x}_0))$ where $\boldsymbol{C} = \partial^2 W / \partial(\nabla \boldsymbol{f})^2$; the second derivative of the stored energy W.

We also show that the three aforementioned conditions are sufficient for the nonnegativity of the second variation of the energy of a homogeneous deformation of a homogeneous body that, in the reference configuration, has the shape[1] of a half-ball with the deformation prescribed on the curved surface of the body.

A long standing problem[2] in the calculus of variations is that of finding pointwise algebraic conditions on $\boldsymbol{C}$ that are necessary and sufficient for the *nonnegativity of the second variation, i.e.,*

$$\int_{\Omega} \nabla \boldsymbol{u}(\boldsymbol{x}) \cdot \boldsymbol{C}(\boldsymbol{x}, \nabla \boldsymbol{f}(\boldsymbol{x})) [\nabla \boldsymbol{u}(\boldsymbol{x})] \, d\boldsymbol{x} \geqq 0 \tag{1.2}$$

[1] The crucial geometric assumption is that the boundary of the body contains a portion of a hyperplane that is the only surface upon which the deformation is not prescribed.

[2] See HADAMARD [11, p. 252] who referred to (1.2) as "la condition de stabilité de l'équilibre interne."

for all variations $\boldsymbol{u}$. It is well known that a necessary condition for (1.2) to be satisfied is the Legendre-Hadamard condition:

$$\boldsymbol{a} \otimes \boldsymbol{b} \cdot \mathsf{C}_0[\boldsymbol{a} \otimes \boldsymbol{b}] \geqq 0 \tag{1.3}$$

for all points $\boldsymbol{x}_0 \in \bar{\Omega}$ and all vectors $\boldsymbol{a}$ and $\boldsymbol{b}$. A well known condition that is sufficient for (1.2) is that C_0 be positive semi-definite:

$$\boldsymbol{M} \cdot \mathsf{C}_0[\boldsymbol{M}] \geqq 0 \tag{1.4}$$

for all points $\boldsymbol{x}_0 \in \bar{\Omega}$ and matrices $\boldsymbol{M}$.

In the special case when $\mathsf{C}(\boldsymbol{x}, \nabla \boldsymbol{f}(\boldsymbol{x})) \equiv \mathsf{C}_0$ is independent of $\boldsymbol{x}$ more precise results are known. In particular VAN HOVE [22] proved that (1.3) is necessary and sufficient for the nonnegativity of the second variation provided that the variations are required to satisfy $\boldsymbol{u} = \boldsymbol{0}$ on the boundary of Ω. If all smooth functions are admitted as variations then the choice $\boldsymbol{u}(\boldsymbol{x}) = \boldsymbol{M}\boldsymbol{x}$ shows that (1.4) is necessary and sufficient for (1.2). Thus the boundary conditions imposed upon the variations are a crucial part of this problem.

We note that our results are related to the existing work on surface instabilities in a half-space.[3] In such problems one is interested in finding algebraic conditions on the elasticity tensor such that the complementing (Lopatinsky-Shapiro) condition[4] fails. Thus the linearized equilibrium equations in a half-space $\mathscr{H}$ admit a nontrivial bounded exponential solution, *i.e.*,

$$\begin{aligned} \operatorname{div} \mathsf{C}_0[\nabla \boldsymbol{w}] &= \alpha^2 \boldsymbol{w} \quad \text{in } \mathscr{H}, \\ \mathsf{C}_0[\nabla \boldsymbol{w}]\, \boldsymbol{n} &= \boldsymbol{0} \quad \text{on } \partial\mathscr{H}, \end{aligned} \tag{1.5}$$

where $\alpha = 0$, $\boldsymbol{n}$ is the outward unit normal to $\partial\mathscr{H}$, and

$$\boldsymbol{w}(\boldsymbol{x}) = \boldsymbol{z}(-\boldsymbol{x} \cdot \boldsymbol{n}) \exp(i\boldsymbol{x} \cdot \boldsymbol{t})$$

with $\boldsymbol{z}$ bounded and $\boldsymbol{t}$ tangent to $\partial\mathscr{H}$.

One of the necessary conditions that we derive, Agmon's condition, is the requirement that (1.5) not admit a nontrivial bounded exponential solution for any $\alpha \neq 0$. The failure of this condition induces a more severe instability than the failure of the complementing condition since it is possible for the complementing condition to fail at the unique global minimizer of the energy (see Section 8) while, as a consequence of our results, the failure of Agmon's condition implies that the underlying deformation is not even a local minimizer.

2. Notations

We let

$$\mathrm{Lin}^n := \text{space of all linear transformations from } \mathbb{R}^n \text{ into } \mathbb{R}^n$$

with inner product and norm

$$\boldsymbol{G} \cdot \boldsymbol{H} := \operatorname{trace}(\boldsymbol{G}\boldsymbol{H}^T), \qquad |\boldsymbol{G}|^2 := \boldsymbol{G} \cdot \boldsymbol{G},$$

[3] See, *e.g.*, BIOT [6, pp. 204–216], NOWINSKI [16], and REDDY [17]. HUTCHINSON & TVERGAARD [13] conjecture that such solutions explain the onset of surface wrinkling in aluminum.

[4] *Cf.* AGMON, DOUGLIS & NIRENBERG [2], SIMPSON & SPECTOR [18, 19], and THOMPSON [20].

where H^T denotes the transpose of H. We write

$$\mathrm{Lin}^n_+ := \{H \in \mathrm{Lin}^n : \det H > 0\},$$

$$\mathrm{Sym}^n := \{E \in \mathrm{Lin}^n : E = E^T\},$$

where det denotes the determinant. Given any $M \in \mathrm{Lin}^n$ we extend its domain of definition to $\mathbb{C}^n$ and write $Mz := Mx + iMy$ where $x := \mathrm{Re}\{z\}$ and $y := \mathrm{Im}\{z\}$ are, respectively, the real and imaginary parts of $z \in \mathbb{C}^n$. We denote by $a \otimes b$ the tensor product of any two vectors $a, b \in \mathbb{C}^n$; in components

$$(a \otimes b)_{ij} = a_i b_j .$$

We write ∇ and div for the gradient and divergence operators in $\mathbb{R}^n$: for a vector field u, ∇u is the tensor field with components $(\nabla u)_{ij} = \partial u_i/\partial x_j$; for a tensor field S, div S is the vector field with components $\sum_j \partial S_{ij}/\partial x_j$. Given any function $\Phi(a, b, \ldots, c)$ with vector or tensor arguments we write, *e.g.*, $\partial\Phi/\partial a$ for the partial Fréchet derivative with respect to a holding the remaining arguments fixed.

We denote by $L^2((0, \infty), \mathbb{C}^n)$ the space (of equivalence classes) of square-integrable functions $z : (0, \infty) \to \mathbb{C}^n$. For $z \in C^1([0, \infty), \mathbb{C}^n)$ with $z, \dot{z} \in L^2((0, \infty), \mathbb{C}^n)$ we write $\|z\|_1$ for the H^1 norm of z, *i.e.*,

$$\|z\|_1^2 := \int_0^\infty |z(s)|^2 \, ds + \int_0^\infty |\dot{z}(s)|^2 \, ds. \tag{2.1}$$

Throughout this paper $\Omega \subset \mathbb{R}^n$ will denote a bounded open region with C^1 boundary, $\partial\Omega$. Further, we assume that

$$\partial\Omega = \mathscr{D} \cup \mathscr{S},$$

where $\mathscr{D}$ and $\mathscr{S}$ are disjoint and $\mathscr{S}$ is relatively open.

3. The Constitutive Relation

We consider a body that, for convenience, we identify with the region $\bar{\Omega}$ that it occupies in a fixed reference configuration. A **deformation** f of the body is a member of the space

$$\mathrm{Def} := \{f \in C^1(\bar{\Omega}, \mathbb{R}^n) : \det \nabla f > 0\}.$$

We assume that the body is hyperelastic with continuous response function $W: \bar{\Omega} \times \mathrm{Lin}^n \to \mathbb{R} \cup \{+\infty\}$. W gives the **stored energy**

$$W(x, \nabla f(x))$$

at any point $x \in \bar{\Omega}$ when the body is deformed by $f \in \mathrm{Def}$.

We further assume that W restricted to $\bar{\Omega} \times \mathrm{Lin}^n_+$ is C^2 and finite-valued. The derivatives

$$S(x, F) := \frac{\partial}{\partial F} W(x, F), \quad C(x, F) := \frac{\partial^2}{\partial F^2} W(x, F)$$

are called the (Piola-Kirchhoff) **stress** and the **elasticity tensor**, respectively.

Given $\boldsymbol{x}_0 \in \bar{\Omega}$ and $\boldsymbol{f} \in \mathrm{Def}$ we write $\mathsf{C}_0 := \mathsf{C}(\boldsymbol{x}_0, \nabla\boldsymbol{f}(\boldsymbol{x}_0))$. We say that C_0 satisfies the **Legendre-Hadamard condition** provided that

$$\boldsymbol{a} \otimes \boldsymbol{b} \cdot \mathsf{C}_0[\boldsymbol{a} \otimes \boldsymbol{b}] \geqq 0$$

whenever $\boldsymbol{a}, \boldsymbol{b} \in \mathbb{R}^n$.

For $\boldsymbol{x}_0 \in \partial\Omega$ we write $\boldsymbol{n}_0 = \boldsymbol{n}(\boldsymbol{x}_0)$ for the outward unit normal to $\partial\Omega$ and $\mathscr{H}$ for the half-space

$$\mathscr{H} := \{\boldsymbol{y} \in \mathbb{R}^n : (\boldsymbol{y} - \boldsymbol{x}_0) \cdot \boldsymbol{n}_0 < 0\}.$$

We consider the problem: Find $\boldsymbol{w}: \bar{\mathscr{H}} \to \mathbb{C}^n$ that satisfies

$$\begin{aligned} \operatorname{div} \mathsf{C}_0[\nabla \boldsymbol{w}] &= \alpha^2 \boldsymbol{w} \quad &\text{in } \mathscr{H}, \\ \mathsf{C}_0[\nabla \boldsymbol{w}]\, \boldsymbol{n}_0 &= \boldsymbol{0} \quad &\text{on } \partial\mathscr{H}, \end{aligned} \tag{3.1}$$

where $\alpha > 0$.
We are specifically interested in whether or not there are solutions of (3.1) that are *bounded exponentials*, *i.e.*, functions

$$\boldsymbol{w}(\boldsymbol{y}) = \boldsymbol{z}(-(\boldsymbol{y} - \boldsymbol{x}_0) \cdot \boldsymbol{n}_0) \exp(i(\boldsymbol{y} - \boldsymbol{x}_0) \cdot \boldsymbol{t}) \tag{3.2}$$

for some $\boldsymbol{t} \in \mathbb{R}^n$ that is *tangent* to $\partial\mathscr{H} (\boldsymbol{t} \cdot \boldsymbol{n}_0 = 0)$ and some $\boldsymbol{z} \in C^2([0, \infty), \mathbb{C}^n)$ satisfying $\sup_s |\boldsymbol{z}(s)| < \infty$.

We say that the pair $(\mathsf{C}_0, \boldsymbol{n}_0)$ satisfies **Agmon's condition**[5] provided that for every real $\alpha > 0$ the only bounded exponential solution of (3.1) is $\boldsymbol{w} = \boldsymbol{0}$.

Remark. Suppose that (3.1) has a bounded exponential solution $\boldsymbol{w}_\alpha$ for some $\alpha > 0$. Then (3.1) has such a solution for *all* $\alpha > 0$. The required solution is given by

$$\boldsymbol{w}_\beta(\boldsymbol{x}) = \boldsymbol{w}_\alpha\left(\frac{\beta}{\alpha}(\boldsymbol{x} - \boldsymbol{x}_0) + \boldsymbol{x}_0\right).$$

Remark. Agmon's condition and the Legendre-Hadamard condition can each be viewed as a dynamic stability[6] requirement for linear elastodynamics. The failure of Agmon's condition yields an $\alpha > 0$ and a bounded exponential $\boldsymbol{w}$ that satisfies (3.1). The failure of the Legendre-Hadamard condition yields, with the aid of the spectral theorem, $\boldsymbol{a}, \boldsymbol{b} \in \mathbb{R}^n$ and $\alpha > 0$ such that

$$\mathsf{C}_0[\boldsymbol{a} \otimes \boldsymbol{b}]\, \boldsymbol{b} = -\alpha^2 \boldsymbol{a}.$$

Thus

$$\begin{aligned} \boldsymbol{v}^{\mathrm{A}}(\boldsymbol{x}, t) &= \operatorname{Re}\{\boldsymbol{w}(\boldsymbol{x})\} \exp(\alpha t), \\ \boldsymbol{v}^{\mathrm{LH}}(\boldsymbol{x}, t) &= \boldsymbol{a} \sin(\boldsymbol{b} \cdot \boldsymbol{x}) \exp(\alpha t), \end{aligned}$$

---

[5] *Cf.* AGMON [1, p. 123].

[6] See HAYES & RIVLIN [12, Section 5] and SIMPSON & SPECTOR [19, p. 10].

are solutions of

$$\operatorname{div} \boldsymbol{C}_0[\nabla \boldsymbol{v}] = \boldsymbol{v}_{tt}.$$

Therefore if either condition fails the equilibrium solution, $\boldsymbol{v} = \boldsymbol{0}$, is unstable since there are initial data, arbitrarily close to $\boldsymbol{0}$, that give rise to a solution that grows exponentially in time.

For future reference we define $\boldsymbol{M}, \boldsymbol{N}_t, \boldsymbol{P}_t^\alpha \in \operatorname{Lin}^n$ by

$$\begin{gathered} \boldsymbol{M}\boldsymbol{a} := \boldsymbol{C}_0[\boldsymbol{a} \otimes \boldsymbol{n}_0]\,\boldsymbol{n}_0, \qquad \boldsymbol{N}_t\boldsymbol{a} := \boldsymbol{C}_0[\boldsymbol{a} \otimes \boldsymbol{t}]\,\boldsymbol{n}_0, \\ \boldsymbol{P}_t^\alpha \boldsymbol{a} := \boldsymbol{C}_0[\boldsymbol{a} \otimes \boldsymbol{t}]\,\boldsymbol{t} + \alpha^2 \boldsymbol{a}, \end{gathered} \tag{3.3}$$

and note that (3.1) and (3.2) reduce to a system of ordinary differential equations,

$$-\boldsymbol{M}\ddot{\boldsymbol{z}} + i(\boldsymbol{N}_t + \boldsymbol{N}_t^T)\,\dot{\boldsymbol{z}} + \boldsymbol{P}_t^\alpha \boldsymbol{z} = \boldsymbol{0} \quad \text{on } (0, \infty), \tag{3.4}$$

with boundary conditions,

$$\boldsymbol{M}\dot{\boldsymbol{z}}(0) - i\boldsymbol{N}_t\boldsymbol{z}(0) = \boldsymbol{0}. \tag{3.5}$$

Remark. The existence of a bounded solution of (3.4) and (3.5) is completely determined by $\boldsymbol{M}$, $\boldsymbol{P}_t^0$, and $\boldsymbol{N}_t$. Thus Agmon's condition is an *algebraic* condition on the pair $(\boldsymbol{C}_0, \boldsymbol{n}_0)$ just as the Legendre-Hadamard condition is an algebraic condition on $\boldsymbol{C}_0$ (see Section 8 for a simple example).

4. The Energy. Minimizers

We suppose that $\boldsymbol{b} \in C^0(\Omega, \mathbb{R}^n)$ is a given body force field, that $\boldsymbol{s} \in C^0(\mathcal{S}, \mathbb{R}^n)$ is a given surface force field, and let

$$E(\boldsymbol{f}) := \int_\Omega [W(\boldsymbol{x}, \nabla \boldsymbol{f}(\boldsymbol{x})) - \boldsymbol{b}(\boldsymbol{x}) \cdot \boldsymbol{f}(\boldsymbol{x})]\, d\boldsymbol{x} - \int_{\mathcal{S}} \boldsymbol{s}(\boldsymbol{x}) \cdot \boldsymbol{f}(\boldsymbol{x})\, d\mathcal{S}_{\boldsymbol{x}} \tag{4.1}$$

denote the **total energy** when the body is deformed by $\boldsymbol{f}$.

We are interested in a deformation $\boldsymbol{f}$ that minimizes the total energy among certain classes of competing deformations; *i.e.*,

$$E(\boldsymbol{f}) \leqq E(\boldsymbol{f} + \boldsymbol{u}) \tag{4.2}$$

for certain variations

$$\boldsymbol{u} \in \operatorname{Var} := \{\boldsymbol{u} \in C^1(\bar{\Omega}, \mathbb{R}^n) : \boldsymbol{u} = \boldsymbol{0} \text{ on } \mathcal{D}\}.$$

Definition. Let $\boldsymbol{f} \in C^1(\bar{\Omega}, \mathbb{R}^n)$ satisfy $E(\boldsymbol{f}) < +\infty$. We say that $\boldsymbol{f}$ is a

(i) **strong local minimizer** provided that there is an $\varepsilon > 0$ such that (4.2) is satisfied for all $\boldsymbol{u} \in \operatorname{Var}$ such that

$$\sup_{\boldsymbol{x} \in \Omega} |\boldsymbol{u}(\boldsymbol{x})| < \varepsilon;$$

(ii) **weak local minimizer** provided that there is an $\varepsilon > 0$ such that (4.2) is satisfied for all $\boldsymbol{u} \in \mathrm{Var}$ such that

$$\sup_{\boldsymbol{x}\in\Omega} |\boldsymbol{u}(\boldsymbol{x})| + \sup_{\boldsymbol{x}\in\Omega} |\nabla \boldsymbol{u}(\boldsymbol{x})| < \varepsilon.$$

Clearly, if $\boldsymbol{f}$ is a strong local minimizer then it is also a weak local minimizer. All of our results will be consequences of $\boldsymbol{f}$ being a weak local minimizer. Thus the conditions we develop will *also* be necessary conditions for $\boldsymbol{f}$ to be a strong local minimizer. Before proceeding we first recall some results concerning strong local minimizers.

Given any $x_0 \in \bar{\Omega}$, $\boldsymbol{f} \in C^1(\bar{\Omega}, \mathbb{R}^n)$, and $r > 0$ we define $\boldsymbol{F}_0 := \nabla \boldsymbol{f}(\boldsymbol{x}_0)$,

$$\boldsymbol{f}_0(\boldsymbol{y}) := \boldsymbol{F}_0 \boldsymbol{y}, \qquad W_0(\boldsymbol{G}) := W(\boldsymbol{x}_0, \boldsymbol{G})$$

for any $\boldsymbol{G} \in \mathrm{Lin}^n$ and

$$\boldsymbol{y} \in \mathscr{B} := \{\boldsymbol{y} \in \mathbb{R}^n \colon |\boldsymbol{y} - \boldsymbol{x}_0| < r\}.$$

If, in addition, $\boldsymbol{x}_0 \in \partial\Omega$ we write $\boldsymbol{n}_0 := \boldsymbol{n}(\boldsymbol{x}_0)$, the outward unit normal to $\partial\Omega$, and

$$\mathscr{H}\mathscr{B} := \{\boldsymbol{y} \in \mathscr{B} \colon (\boldsymbol{y} - \boldsymbol{x}_0) \cdot \boldsymbol{n}_0 < 0\}.$$

We interpret $\boldsymbol{f}_0$ as a homogeneous deformation of a homogeneous body that occupies the region $\mathscr{B}$ or $\mathscr{H}\mathscr{B}$ in a homogeneous reference configuration.

Proposition 4.1 (BALL & MARSDEN [5]). *Let $\boldsymbol{f}$ be a strong local minimizer. Then,*[7] *for every $\boldsymbol{x}_0 \in \Omega$,*

$$\int_{\mathscr{B}} W_0(\boldsymbol{F}_0)\, dy \leq \int_{\mathscr{B}} W_0(\boldsymbol{F}_0 + \nabla \boldsymbol{u}(\boldsymbol{y}))\, dy \tag{4.3}$$
[8]

for all $\boldsymbol{u} \in C_0^1(\mathscr{B}, \mathbb{R}^n)$ such that $W_0(\boldsymbol{F}_0 + \nabla \boldsymbol{u}(\cdot))$ is uniformly bounded on $\bar{\mathscr{B}}$. Moreover,[9] *for every $\boldsymbol{x}_0 \in \mathscr{S}$,*

$$\int_{\mathscr{H}\mathscr{B}} W_0(\boldsymbol{F}_0)\, dy \leq \int_{\mathscr{H}\mathscr{B}} W_0(\boldsymbol{F}_0 + \nabla \boldsymbol{u}(\boldsymbol{y}))\, dy - \int_{\partial(\mathscr{H}\mathscr{B}) \cap \partial\mathscr{H}} \boldsymbol{s}(\boldsymbol{x}_0) \cdot \boldsymbol{u}(\boldsymbol{y})\, d\mathscr{S}_y \tag{4.4}$$

for all $\boldsymbol{u} \in C^1(\overline{\mathscr{H}\mathscr{B}}, \mathbb{R}^n)$ vanishing in a neighborhood of $\partial\mathscr{B} \cap \partial(\mathscr{H}\mathscr{B})$ and such that $W_0(\boldsymbol{F}_0 + \nabla \boldsymbol{u}(\cdot))$ is uniformly bounded on $\overline{\mathscr{H}\mathscr{B}}$.

In other words, in order for $\boldsymbol{f}$ to be a strong local minimizer each of the homogeneous deformations $\boldsymbol{f}_0(\boldsymbol{y}) = [\nabla \boldsymbol{f}(\boldsymbol{x}_0)]\, \boldsymbol{y}$ must be a *global* minimizer of the energy of a ball (half-ball) composed of the material at $\boldsymbol{x}_0$.

We now suppose that $\boldsymbol{f} \in \mathrm{Def}$ is a *weak* local minimizer. Then, for every $\boldsymbol{u} \in \mathrm{Var}$,

$$0 \leq \varphi(t) := E(\boldsymbol{f} + t\boldsymbol{u}) - E(\boldsymbol{f})$$

---

[7] This result is essentially due to MEYERS [14, pp. 128–131]. See BALL [3, Theorem 2.1] or GURTIN [10, pp. 9–10] for a direct proof under the additional hypothesis that W is finite-valued.

[8] This inequality is MORREY's [15] *quasiconvexity* condition.

[9] This result is due to BALL & MARSDEN [5], who refer to the inequality as *quasiconvexity at the boundary.*

provided t is sufficiently small. Thus $\dot{\varphi}(0) = 0$ and $\ddot{\varphi}(0) \geqq 0$. If f is C^2 one can then show that f satisfies the *equation of equilibrium* and the *natural boundary condition*, that is,

$$\operatorname{div} S(\nabla f) + b = 0 \quad \text{in } \Omega,$$
$$S(\nabla f)\, n = s \quad \text{on } \mathcal{S}.$$

In addition (whether or not f is C^2) the *second variation* of E at f must be non-negative, *i.e.*,

$$0 \leqq \int_\Omega \nabla u(x) \cdot C(x, \nabla f(x))\, [\nabla u(x)]\, dx \tag{4.5}$$

for all $u \in \mathrm{Var}$.

If we now define

$$\hat{E}(u) := \int_\Omega \hat{W}(x, \nabla u(x))\, dx, \qquad \hat{W}(x, U) := U \cdot C(x, \nabla f(x))\, [U],$$

we find that (4.5) is equivalent to $\hat{E}(0) \leqq \hat{E}(u)$ for all $u \in \mathrm{Var}$. Thus the function 0 is a strong local minimizer of $\hat{E}$ and hence we can apply Proposition 4.1 to get the following result.

Proposition 4.2. *Let* $f \in \mathrm{Def}$ *be a weak local minimizer. Then, for every* $x_0 \in \Omega$,

$$0 \leqq \int_{\mathcal{B}} \nabla u(y) \cdot C_0[\nabla u(y)]\, dy \tag{4.6}$$

for all $u \in \mathrm{Var}(\mathcal{B})$. *Moreover, for every* $x_0 \in \mathcal{S}$,

$$0 \leqq \int_{\mathcal{H}\mathcal{B}} \nabla u(y) \cdot C_0[\nabla u(y)]\, dy$$

for all $u \in \mathrm{Var}(\mathcal{H}\mathcal{B})$. *Here* $C_0 := C(x_0, \nabla f(x_0))$ *and*

$$\mathrm{Var}(\mathcal{B}) := \{u \in C^1(\bar{\mathcal{B}}, \mathbb{R}^n) : u = 0 \text{ on } \partial\mathcal{B}\},$$
$$\mathrm{Var}(\mathcal{H}\mathcal{B}) := \{u \in C^1(\overline{\mathcal{H}\mathcal{B}}, \mathbb{R}^n) : u = 0 \text{ on } \partial\mathcal{B} \cap \partial(\mathcal{H}\mathcal{B})\}. \tag{4.7}$$

Remark. A deformation that satisfies (4.5) is usually said to be *infinitesimally stable.*[10] Thus the above result states that in order for a deformation f to be infinitesimally stable each of the homogeneous deformations $f_0(y) = [\nabla f(x_0)]\, y$ (of a ball or half-ball) must be infinitesimally stable.

Remark. Actually Proposition 4.1 implies that (4.6) is satisfied only for $u \in C_0^1(\mathcal{B}, \mathbb{R}^n)$. However, since $C_0^1(\mathcal{B}, \mathbb{R}^n)$ is dense in $H_0^1(\mathcal{B}, \mathbb{R}^n)$ and the quadratic form in (4.6) is continuous on $H_0^1(\mathcal{B}, \mathbb{R}^n)$ we deduce that (4.6) is valid for u in the larger space $\mathrm{Var}(\mathcal{B})$. In a similar way one extends the class of variations in the second part of this proposition.

---

[10] *Cf., e.g.,* TRUESDELL & NOLL [21, Section 68 bis].

5. Main Results

We suppose that f_0 is a homogeneous deformation of a homogeneous body composed of material with stored energy function W_0. Further, we suppose that the body occupies the region $\mathscr{H}\mathscr{B}$ in a homogeneous reference configuration. We then consider the following problem: Find pointwise algebraic conditions on the elasticity tensor C_0 that are necessary and sufficient for the nonnegativity of the second variation of the total energy, *i.e.*,

$$0 \leq \int_{\mathscr{H}\mathscr{B}} \nabla u(y) \cdot C_0[\nabla u(y)]\, dy \tag{5.1}$$

for all $u \in \operatorname{Var}(\mathscr{H}\mathscr{B})$ (*cf.* (4.7)).

The following result solves this problem.

Theorem 1. *Necessary and sufficient conditions for* (5.1) *to be satisfied for all* $u \in \operatorname{Var}(\mathscr{H}\mathscr{B})$ *are that:*

(i) C_0 *satisfies the Legendre-Hadamard condition;*

(ii) (C_0, n_0) *satisfies Agmon's condition;*

(iii) *If* $a \otimes n_0 \cdot C_0[a \otimes n_0] = 0$ *for some* $a \in \mathbb{R}^n$ *then* $C_0[a \otimes n_0] = \mathbf{0}$.

Here n_0 *is the outward unit normal to* $\partial\mathscr{H}$.

The following result, which is a direct consequence of Proposition 4.2 and Theorem 1, gives conditions that a general deformation of a general body must satisfy in order to be a local minimizer of the energy.

Theorem 2. *Let* $f \in \operatorname{Def}$ *be a weak local minimizer of*

$$E(f) = \int_{\Omega} [W(x, \nabla f(x)) - b(x) \cdot f(x)]\, dx - \int_{\mathscr{S}} s(x) \cdot f(x)\, d\mathscr{S}_x .$$

Then, for every $x_0 \in \mathscr{S}$, *conditions* (i), (ii) *and* (iii) *of Theorem 1 are satisfied. Here* $C_0 := C(x_0, \nabla f(x_0))$ $(C := \partial^2 W/\partial(\nabla f)^2)$ *and* $n_0 := n(x_0)$, *the outward unit normal to* $\partial\Omega$ *at* x_0.

Remark. Theorem 2 is valid for a slightly more general class of loadings. In particular the body force can take the form

$$b(x) = \hat{b}(x, f(x)) = \frac{\partial}{\partial f} \hat{\varphi}(x, f(x))$$

where $\hat{\varphi} \in C^3(\overline{\Omega} \times \mathbb{R}^n, \mathbb{R})$.

Proof of Theorem 2. *Proof of Theorem 1* (*Necessity*).[11] The necessity of (i) is well known.[12] We show that (ii) and (iii) are direct consequences of (5.1). In order to simplify the notation we first change coordinates so that $x_0 = \mathbf{0}$.

---

[11] The sufficiency portion of the proof of Theorem 1 has been divided into three parts. The first is at the end of this section and the other two parts are at the end of Sections 6 and 7.

[12] *Cf.*, *e.g.*, TRUESDELL & NOLL [21, Section 68 bis].

(5.1) $\Rightarrow$ (ii). Let $b \in C^\infty(\mathbb{R}, \mathbb{R})$ satisfy $b(s) = 1$ for $s \leqq r/3$ and $b(s) = 0$ for $s \geqq 2r/3$. Define $\varphi(\boldsymbol{x}) = b(|\boldsymbol{x}|)$. Then given any $\boldsymbol{w} \in C^2(\overline{\mathscr{H}}, \mathbb{C}^n)$, $\boldsymbol{u} := \operatorname{Re}\{\varphi \boldsymbol{w}\}$ and $\boldsymbol{v} := \operatorname{Im}\{\varphi \boldsymbol{w}\}$ are contained in Var $(\mathscr{H}\mathscr{B})$ and a straightforward computation shows that

$$\begin{aligned}\int_{\mathscr{H}\mathscr{B}} (\nabla \boldsymbol{u} \cdot \boldsymbol{C}_0[\nabla \boldsymbol{u}] + \nabla \boldsymbol{v} \cdot \boldsymbol{C}_0[\nabla \boldsymbol{v}]) &= \int_{\mathscr{H}\mathscr{B}} \nabla(\varphi \overline{\boldsymbol{w}}) \cdot \boldsymbol{C}_0[\nabla(\varphi \boldsymbol{w})] \\ &= \int_{\mathscr{H}\mathscr{B}} (\overline{\boldsymbol{w}} \otimes \nabla\varphi \cdot \boldsymbol{C}_0[\boldsymbol{w} \otimes \nabla\varphi] + \operatorname{Re}\{\nabla(\varphi^2 \overline{\boldsymbol{w}}) \cdot \boldsymbol{C}_0[\nabla \boldsymbol{w}]\})\end{aligned} \tag{5.2}$$

and, by the divergence theorem,

$$\int_{\mathscr{H}\mathscr{B}} \nabla(\varphi^2 \overline{\boldsymbol{w}}) \cdot \boldsymbol{C}_0[\nabla \boldsymbol{w}] = -\int_{\mathscr{H}\mathscr{B}} \varphi^2 \overline{\boldsymbol{w}} \cdot \operatorname{div} \boldsymbol{C}_0[\nabla \boldsymbol{w}] + \int_{\partial(\mathscr{H}\mathscr{B})} \varphi^2 \overline{\boldsymbol{w}} \cdot \boldsymbol{C}_0[\nabla \boldsymbol{w}]\, \boldsymbol{n}. \tag{5.3}$$

We now suppose that Agmon's condition is not satisfied and we let $\boldsymbol{w}_1$ be a bounded exponential solution of (3.1) with $\alpha = 1$ (see the remark following the definition of Agmon's condition). Then $\boldsymbol{w}_\alpha(\boldsymbol{x}) := \boldsymbol{w}_1(\alpha \boldsymbol{x})$ will be a bounded exponential solution of (3.1) for any $\alpha > 0$. Moreover, $|\boldsymbol{w}_\alpha(\boldsymbol{x})|$ is uniformly bounded independently of α.

By (3.1), (5.2), and (5.3) we find that there is a constant $k > 0$ such that

$$\int_{\mathscr{H}\mathscr{B}} \{\nabla \boldsymbol{u}_\alpha \cdot \boldsymbol{C}_0[\nabla \boldsymbol{u}_\alpha] + \nabla \boldsymbol{v}_\alpha \cdot \boldsymbol{C}_0[\nabla \boldsymbol{v}_\alpha]\} \leqq k - \alpha^2 \int_{\mathscr{H}\mathscr{B}} \varphi^2 |\boldsymbol{w}_\alpha|^2 \leqq k - \alpha^2 \int_{\mathscr{H}\mathscr{B}_{r/3}} |\boldsymbol{w}_\alpha|^2, \tag{5.4}$$

where $\mathscr{H}\mathscr{B}_\varrho$ is the intersection of $\mathscr{H}$ with the ball of radius ϱ centered at $\boldsymbol{x}_0$.

We next note that $\boldsymbol{w}_\alpha$ is of the form (3.2) and hence $|\boldsymbol{w}_\alpha(s, y)| = |\boldsymbol{z}_\alpha(s)|$. Thus

$$\int_{\mathscr{H}\mathscr{B}_{r/3}} |\boldsymbol{w}_\alpha(\boldsymbol{x})|^2 \, d\boldsymbol{x} \geqq (2r/3n)^{n-1} \int_0^{r/3n} |\boldsymbol{z}_\alpha(s)|^2 \, ds = \alpha^{-1}(2r/3n)^{n-1} \int_0^{\alpha r/3n} |\boldsymbol{z}_1(\sigma)|^2 \, d\sigma.$$

If we combine the last inequality with (5.4) we conclude that, for $\alpha \geqq 3n$,

$$\int_{\mathscr{H}\mathscr{B}} \{\nabla \boldsymbol{u}_\alpha \cdot \boldsymbol{C}_0[\nabla \boldsymbol{u}_\alpha] + \nabla \boldsymbol{v}_\alpha \cdot \boldsymbol{C}_0[\nabla \boldsymbol{v}_\alpha]\} \leqq k - \alpha(2r/3n)^{n-1} \int_0^r |\boldsymbol{z}_1(\sigma)|^2 \, d\sigma.$$

Therefore, by choosing α sufficiently large, we can find a $\boldsymbol{u}_\alpha \in$ Var $(\mathscr{H}\mathscr{B})$ (or a $\boldsymbol{v}_\alpha \in$ Var $(\mathscr{H}\mathscr{B})$) for which (5.1) is not satisfied.

(5.1) $\Rightarrow$ (iii). Suppose that $\boldsymbol{a} \in \mathbb{R}^n$ satisfies

$$\boldsymbol{a} \cdot \boldsymbol{M}\boldsymbol{a} = \boldsymbol{a} \otimes \boldsymbol{n}_0 \cdot \boldsymbol{C}_0[\boldsymbol{a} \otimes \boldsymbol{n}_0] = 0.$$

Then, since $\boldsymbol{C}_0$ is symmetric and satisfies the Legendre-Hadamard condition, $\boldsymbol{M}$ is symmetric and positive semi-definite. Thus

$$\boldsymbol{M}\boldsymbol{a} = \boldsymbol{C}_0[\boldsymbol{a} \otimes \boldsymbol{n}_0]\, \boldsymbol{n}_0 = \boldsymbol{0}. \tag{5.5}$$

Next, let $\boldsymbol{b}, \boldsymbol{t}_0 \in \mathbb{R}^n$ with $\boldsymbol{t}_0 \perp \boldsymbol{n}_0$ and $|\boldsymbol{t}_0| = 1$. Choose coordinates so that $\boldsymbol{n}_0 = (-1, 0, 0, \ldots, 0)$, $\boldsymbol{t}_0 = (0, 1, 0, \ldots, 0, \ldots, 0)$ and write $\boldsymbol{x} = (s, t, \boldsymbol{z})$, where

$s, t \in \mathbb{R}$ and $z \in \mathbb{R}^{n-2}$. Let $\varphi, \psi \in C^\infty(\mathbb{R}, \mathbb{R})$ and $\varrho \in C^\infty(\mathbb{R}^{n-2}, \mathbb{R})$ satisfy

$$\begin{aligned} \int_{-1}^{1} |\dot\psi(t)|^2\, dt = 1, &\quad \psi(t) = 0 \quad \text{for } |t| \geqq 1, \\ \varphi(0) = 1, &\quad \varphi(s) = 0 \quad \text{for } |s| \geqq 1, \\ \int_{|z| \leqq r} |\varrho(z)|^2\, dz = 1, &\quad \varrho(z) = 0 \quad \text{for } |z| \geqq r/3. \end{aligned} \tag{5.6}$$

For $\varepsilon > 0, \delta > 0$, and $\beta > 0$ define

$$u^\beta_{\delta,\varepsilon}(s, t, z) = \varrho(z)\left[a\,\delta^{-1}\,\varphi\left(\frac{s}{\varepsilon\beta}\right)\psi\left(\frac{t}{\beta}\right) + b\,\delta\varphi\left(\frac{s}{\beta}\right)\dot\psi\left(\frac{t}{\beta}\right)\right].$$

Then $u^\beta_{\delta,\varepsilon} \in \operatorname{Var}(\mathscr{H}\mathscr{B})$, provided ε and β are sufficiently small, and

$$\nabla u^\beta_{\delta,\varepsilon} = \begin{bmatrix} \beta^{-1}\,\varrho(a \otimes n_0\,\delta^{-1}\,\varepsilon^{-1}\,\dot\varphi\psi + b \otimes n_0\,\delta\dot\varphi\dot\psi) \\ +\,\beta^{-1}\,\varrho(a \otimes t_0\,\delta^{-1}\,\varphi\dot\psi + b \otimes t_0\,\delta\varphi\ddot\psi) \\ +\,(a\,\delta^{-1}\,\varphi\psi + b\,\delta\varphi\dot\psi) \otimes \nabla\varrho \end{bmatrix}. \tag{5.7}$$

Now, by (5.1),

$$\int_{|z| \leqq r} \int_{-\beta}^{\beta} \int_{0}^{\beta} \nabla u^\beta_{\delta,\varepsilon} \cdot C_0[\nabla u^\beta_{\delta,\varepsilon}]\, ds\, dt\, dz \geqq 0.$$

We make the change of variables $\sigma = s/\beta$ and $\tau = t/\beta$ (and hence $ds\,dt = \beta^2\, d\sigma\, d\tau$) and then let $\beta \to 0^+$ in the last inequality to conclude, with the aid of (5.5), (5.6), (5.7), and the bounded convergence theorem that

$$\begin{aligned} &b \otimes n_0 \cdot C_0[b \otimes n_0]\,\delta^2 \int_0^1 \dot\varphi^2 + a \otimes t_0 \cdot C_0[a \otimes t_0]\,\delta^{-2} \int_0^\varepsilon \left|\varphi\left(\frac{\sigma}{\varepsilon}\right)\right|^2 d\sigma \\ &+ b \otimes t_0 \cdot C_0[b \otimes t_0]\,\delta^2 \int_0^1 \varphi^2 \int_{-1}^{1} \ddot\psi^2 + 2a \otimes t_0 \cdot C[b \otimes n_0] \int_0^\varepsilon \dot\varphi(\sigma)\varphi\left(\frac{\sigma}{\varepsilon}\right) d\sigma \\ &- 2b \otimes t_0 \cdot C_0[a \otimes n_0]\,\varepsilon^{-1} \int_0^\varepsilon \dot\varphi\left(\frac{\sigma}{\varepsilon}\right)\varphi(\sigma)\, d\sigma \geqq 0. \end{aligned} \tag{5.8}$$

In deriving the above inequality we have used the identities

$$\int_{-1}^{1} \psi(\tau)\,\dot\psi(\tau)\, d\tau = \int_{-1}^{1} \dot\psi(\tau)\,\ddot\psi(\tau)\, d\tau = 0, \quad \int_{-1}^{1} \psi(\tau)\,\ddot\psi(\tau)\, d\tau = -1,$$

which are consequences of $(5.6)_1$.

We note that the change of variables $y = \sigma/\varepsilon$ yields

$$\varepsilon^{-1} \int_0^\varepsilon \dot\varphi\left(\frac{\sigma}{\varepsilon}\right)\varphi(\sigma)\, d\sigma = \int_0^1 \dot\varphi(y)\,\varphi(\varepsilon y)\, dy$$

and hence by the bounded convergence theorem and $(5.6)_2$ we find that

$$\lim_{\varepsilon \to 0^+} \varepsilon^{-1} \int_0^\varepsilon \dot\varphi\left(\frac{\sigma}{\varepsilon}\right)\varphi(\sigma)\, d\sigma = \varphi(0) \int_0^1 \dot\varphi(y)\, dy = -1. \tag{5.9}$$

Finally, we note that φ and $\dot{\varphi}$ are bounded and let $\varepsilon \to 0^+$ followed by $\delta \to 0^+$ in (5.8) to conclude, with the aid of (5.9), that

$$0 \leqq 2\boldsymbol{b} \otimes \boldsymbol{t}_0 \cdot \boldsymbol{C}_0[\boldsymbol{a} \otimes \boldsymbol{n}_0] = 2\boldsymbol{b} \cdot \boldsymbol{C}_0[\boldsymbol{a} \otimes \boldsymbol{n}_0]\, \boldsymbol{t}_0.$$

Since $\boldsymbol{b}$ is arbitrary we find that

$$\boldsymbol{C}_0[\boldsymbol{a} \otimes \boldsymbol{n}_0]\, \boldsymbol{t} = \boldsymbol{0} \quad \text{for any } \boldsymbol{t} \perp \boldsymbol{n}_0. \tag{5.10}$$

The desired result now follows from (5.5) and (5.10). □

The first step in proving the sufficiency portion of Theorem 1 is to define a Fourier transform in the tangential variables. With this in mind we choose coordinates so that $\boldsymbol{x}_0 = \boldsymbol{0}$, $\boldsymbol{n} := \boldsymbol{n}_0 = (-1, 0, 0, 0, \ldots, 0)$ and write $\boldsymbol{x} = (s, \boldsymbol{y})$ where $s \in \mathbb{R}$ and $\boldsymbol{y} \in \mathbb{R}^{n-1}$.

For any $\xi \in \mathbb{R}^{n-1}$, $\alpha \in \mathbb{R}$, and $\boldsymbol{z} \in C_0^\infty([0, \infty), \mathbb{C}^n)$ we write

$$\begin{aligned} Q_\alpha^\xi(\boldsymbol{z}) &:= \int_0^\infty \{\overline{\boldsymbol{H}}(\boldsymbol{z}(s), \xi) \cdot \boldsymbol{C}_0[\boldsymbol{H}(\boldsymbol{z}(s), \xi)] + \alpha^2\, |\boldsymbol{z}(s)|^2\}\, ds, \\ \boldsymbol{H}(\boldsymbol{z}(s), \xi) &:= -\dot{\boldsymbol{z}}(s) \otimes \boldsymbol{n} + i\boldsymbol{z}(s) \otimes (0, \xi). \end{aligned} \tag{5.11}$$

The following result then uses VAN HOVE's [22] technique of reducing an analytic inequality to an algebraic inequality in the transform variables. It is essentially due[13] to AGMON [1, Theorem 5.1] and DE FIGUEIREDO [8, Theorem 3.1].

Lemma 5.3. *A necessary and sufficient condition for* (5.1) *to be satisfied for all* $\boldsymbol{u} \in \mathrm{Var}\,(\mathscr{H}\mathscr{B})$ *is that*

$$Q_\alpha^\xi(\boldsymbol{z}) \geqq 0 \tag{5.12}$$

for every $\boldsymbol{z} \in C_0^1([0, \infty), \mathbb{C}^n)$, $\alpha > 0$, *and* $\xi \in \mathbb{R}^{n-1}$.

Proof of Theorem 1 (*Sufficiency*). *Case I.* $\boldsymbol{M} = \boldsymbol{0}$. Then condition (iii) and (3.3) imply that $\boldsymbol{N}_t = \boldsymbol{0}$. Thus (3.3) and the Legendre-Hadamard condition imply that

$$Q_\alpha^\xi(\boldsymbol{z}) = \int_0^\infty \bar{\boldsymbol{z}} \cdot \boldsymbol{P}_t^\alpha \boldsymbol{z}\, ds \geqq \alpha^2 \int_0^\infty |\boldsymbol{z}(s)|^2\, ds,$$

where $\boldsymbol{t} = (0, \xi)$. Therefore Q_α^ξ is nonnegative and hence by Lemma 5.3 we find that (5.1) is satisfied for all $\boldsymbol{u} \in \mathrm{Var}\,(\mathscr{H}\mathscr{B})$. □

6. Quadratic Forms on a Half-line I. Regular Forms

In order to complete the sufficiency portion of the proof of Theorem 1 we will need to establish some auxiliary results concerning quadratic forms on a half-line. Motivated by (3.3) and (5.11) we let $\boldsymbol{A}: (0, \infty) \to \mathrm{Sym}^m$, $\boldsymbol{B}: (0, \infty) \to \mathrm{Sym}^m$, and $\boldsymbol{F}: (0, \infty) \to \mathrm{Lin}^m$ be continuous and consider the quadratic functional

$$q_\alpha(\boldsymbol{f}) := \int_0^\infty (\dot{\bar{\boldsymbol{f}}} \cdot \boldsymbol{A}(\alpha)\dot{\boldsymbol{f}} + \bar{\boldsymbol{f}} \cdot \boldsymbol{B}(\alpha)\boldsymbol{f} - 2\,\mathrm{Re}\,\{i\dot{\bar{\boldsymbol{f}}} \cdot \boldsymbol{F}(\alpha)\boldsymbol{f}\})ds.$$

[13] See also CHEN [7, Proposition 1].

For fixed $\alpha > 0$ we associate with q_α the tensor field $\boldsymbol{R}: \mathbb{R} \to \mathrm{Sym}^m$ defined by

$$\boldsymbol{R}(\lambda) := \lambda^2 \boldsymbol{A}(\alpha) + \lambda(\boldsymbol{F}(\alpha) + \boldsymbol{F}^T(\alpha)) + \boldsymbol{B}(\alpha), \tag{6.1}$$

the system of ordinary differential equations (the Euler equations corresponding to q_α)

$$-\boldsymbol{A}(\alpha)\ddot{\boldsymbol{f}} + i(\boldsymbol{F}(\alpha) + \boldsymbol{F}^T(\alpha))\dot{\boldsymbol{f}} + \boldsymbol{B}(\alpha)\boldsymbol{f} = \boldsymbol{0} \quad \text{on } (0, \infty), \tag{6.2}$$

and the natural boundary condition

$$\boldsymbol{A}(\alpha)\dot{\boldsymbol{f}}(0) - i\boldsymbol{F}(\alpha)\boldsymbol{f}(0) = \boldsymbol{0}. \tag{6.3}$$

Definition. Fix $\alpha > 0$. We say that q_α is:

(i) **regular** provided that there are constants $c_1 = c_1(\alpha) > 0$ and $c_2 = c_2(\alpha) > 0$ such that, for every $\boldsymbol{e} \in \mathbb{R}^m$,

$$\boldsymbol{e} \cdot \boldsymbol{R}(\lambda)\, \boldsymbol{e} \geqq (c_1 \lambda^2 + c_2)\, |\boldsymbol{e}|^2;$$

(ii) **coercive** provided that there is a constant $k = k(\alpha) > 0$ such that (*cf.* (2.1))

$$q_\alpha(\boldsymbol{f}) \geqq k\, \|\boldsymbol{f}\|_1^2$$

for all $\boldsymbol{f} \in C_0^1([0, \infty), \mathbb{C}^m)$;

(iii) **nonnegative** provided that, for all $\boldsymbol{f} \in C_0^1([0, \infty), \mathbb{C}^m)$,

$$q_\alpha(\boldsymbol{f}) \geqq 0.$$

We note that if q_α is regular and $\boldsymbol{z}$ is any bounded solution of (6.2) then, by a standard result from ordinary differential equations (*cf.*, *e.g.* DUFFIN [9]), $\boldsymbol{z}$ is C^∞ and $\boldsymbol{z}$ and all its derivatives are contained in L^2. Thus *q_α is well defined on bounded solutions of* (6.2).

Proposition 6.1. (DE FIGUEIREDO [8, Theorem 1.2]). *Let $\alpha > 0$ be given. Then necessary and sufficient conditions for q_α to be coercive are that q_α be regular and that $q_\alpha(\boldsymbol{f}) > 0$ whenever $\boldsymbol{f}$ is a nontrivial bounded solution of* (6.2).

The following result establishes the relation between Agmon's condition and De Figuiredo's condition.

Proposition 6.2. *Let q_α be regular for all $\alpha > 0$ and nonnegative for sufficiently large α. Suppose that there is an $\alpha_* > 0$ such that q_{α_*} is not nonnegative. Then there exist an $\alpha_0 > 0$ and a nontrivial $\boldsymbol{f}_0 \in C^\infty([0, \infty), \mathbb{C}^m)$, with $\boldsymbol{f}_0$ and all its derivatives contained in $L^2((0, \infty), \mathbb{C}^m)$, such that $\boldsymbol{f}_0$ satisfies* (6.2) *and* (6.3) *with $\alpha = \alpha_0$.*

Proof. Define

$$\alpha_0 := \inf \{\alpha > 0 \colon q_\beta \text{ is nonnegative for all } \beta > \alpha\}. \tag{6.4}$$

Then by hypothesis $0 < \alpha_0 < +\infty$. Also the continuity of the maps $\alpha \mapsto A(\alpha)$, $\alpha \mapsto B(\alpha)$, and $\alpha \mapsto F(\alpha)$ implies that q_{α_0} is nonnegative and hence (since $C_0^1([0,\infty), \mathbb{R}^n)$ is dense in H^1) that $q_{\alpha_0}(f) \geqq 0$ whenever f is a bounded solution of (6.2) with $\alpha = \alpha_0$.

Suppose, for the sake of contradiction, that $q_\alpha(f) > 0$ whenever f is a nontrivial bounded solution of (6.2) with $\alpha = \alpha_0$. Then by Proposition 6.1 q_{α_0} is coercive. However, the continuity of the maps $\alpha \mapsto A(\alpha)$, $\alpha \mapsto B(\alpha)$, and $\alpha \mapsto F(\alpha)$ implies that q_α is then coercive for all α in some neighborhood of α_0. This contradicts (6.4). Thus there must exist a nontrivial, bounded solution f_0 of (6.2), with $\alpha = \alpha_0$, that satisfies $q_{\alpha_0}(f_0) = 0$. Moreover $f_0 \in C^\infty([0,\infty), \mathbb{C}^m)$ and f_0 and all its derivatives are contained in $L^2((0,\infty), \mathbb{C}^m)$.

Finally, since $q_{\alpha_0}(f_0) = 0$ and q_{α_0} is nonnegative we find that the first variation of q_{α_0} at f_0 is zero and hence that f_0 satisfies (6.3). □

Our next result gives a sufficient condition for Q_α^ξ (as given by (5.11)) to be nonnegative, for sufficiently large α, and regular.

Lemma 6.3. *Let C_0 satisfy the Legendre-Hadamard condition. Suppose, in addition, that M is strictly positive definite. Then:*

(a) *For every $\xi \in \mathbb{R}^{n-1}$ there is an $\alpha_1 > 0$ such that Q_α^ξ is nonnegative whenever $\alpha \geqq \alpha_1$;*

(b) *For every $\xi \in \mathbb{R}^{n-1}$ and $\alpha > 0$, Q_α^ξ is regular.*

Proof. Given $\xi \in \mathbb{R}^{n-1}$ it is easy to show that the integrand in (5.11) is nonnegative for sufficiently large α. This proves (a). In order to prove (b) we let $\alpha > 0$, $\xi \in \mathbb{R}^{n-1}$ and $t = (0, \xi)$. Then the Legendre-Hadamard condition implies that, for every $\lambda \in \mathbb{R}$ and $a \in \mathbb{R}^n$,

$$a \cdot L(\lambda)\, a := a \cdot [\lambda^2 M + \lambda(N_t + N_t^T) + P_t^\alpha]\, a \geqq \alpha^2 |a|^2, \tag{6.5}$$

where M, N_t and P_t^α are given by (3.3).

Suppose, for the sake of contradiction, that for every positive integer i there is a $\lambda_i \in \mathbb{R}$ and an $a^i \in \mathbb{R}^n$, with $|a^i| = 1$, such that

$$a^i \cdot L(\lambda_i)\, a^i < \left(\frac{1}{i}\lambda_i^2 + \tfrac{1}{2}\alpha^2\right) |a^i|^2. \tag{6.6}$$

Since the unit ball is compact in $\mathbb{R}^n$ there is an $a \in \mathbb{R}^n$, with $|a| = 1$, and a subsequence (also denoted by a^i) such that $a^i \to a$ as $i \to \infty$. In addition, there is a sub-subsequence (also denoted by λ_i) for which either $|\lambda_i| \to +\infty$ or $\lambda_i \to \lambda$ (for some $\lambda \in \mathbb{R}$) as $i \to \infty$.

Case I. $\lambda_i \to \lambda$ as $i \to \infty$. We note that L is continuous in both λ and a and let $i \to \infty$ in (6.6) to conclude that

$$a \cdot L(\lambda)\, a \leqq \tfrac{1}{2}\alpha^2 |a|^2.$$

This contradicts (6.5).

Case II. $|\lambda_i| \to +\infty$ as $i \to +\infty$. We first divide (6.6) by λ_i^2 and then let $i \to +\infty$ to conclude, with the aid of (6.5), that

$$\boldsymbol{a} \cdot \boldsymbol{M}\boldsymbol{a} \leqq 0.$$

This contradicts the assumption that $\boldsymbol{M}$ is strictly positive definite. □

Proof of Theorem 1 (*Sufficiency*). *Case II.* $\boldsymbol{M}$ is strictly positive definite. Fix $\xi \in \mathbb{R}^{n-1}$. Then the Legendre-Hadamard condition and Lemma 6.3 imply that Q_α^ξ is regular for all $\alpha > 0$ and nonnegative for sufficiently large α.

Suppose, for the sake of contradiction, that there is an $\alpha^* > 0$ such that $Q_{\alpha^*}^\xi$ is not nonnegative. Then by Proposition 6.2 there is an $\alpha_0 > 0$ and a nontrivial $\boldsymbol{f}_0 \in C^\infty([0,\infty), \mathbb{C}^n)$, with $\boldsymbol{f}_0$ and all its derivatives contained in $L^2([0,\infty), \mathbb{C}^n)$, such that $\boldsymbol{f}_0$ satisfies (6.2) and (6.3) (with $\alpha = \alpha_0$) where $A(\alpha) = \boldsymbol{M}$, $B(\alpha) = \boldsymbol{P}_t^\alpha$, and $F(\alpha) = \boldsymbol{N}_t$ $(\boldsymbol{t} = (0, \xi))$. Therefore (*cf.* (3.4)–(3.5)) Agmon's condition is not satisfied, a contradiction. Thus Q_α^ξ is nonnegative for every $\alpha > 0$ and $\xi \in \mathbb{R}^{n-1}$ and hence by Lemma 5.3 we find that (5.1) is satisfied for all $\boldsymbol{u} \in \operatorname{Var}(\mathscr{H}\mathscr{B})$. □

7. Quadratic Forms on a Half-line II. Singular Forms

Thus far we have proven the sufficiently portion of Theorem 1 under the additional hypothesis that either $\boldsymbol{M} = \boldsymbol{0}$ or $\boldsymbol{M}$ is strictly positive definite. In this section we consider the final case where $\boldsymbol{M} \neq \boldsymbol{0}$ has a nontrivial kernel.

The Legendre-Hadamard condition implies that $\boldsymbol{M}$ is positive semi-definite and that $\boldsymbol{P}_t^\alpha$ is strictly positive definite for $\alpha > 0$. Since $\boldsymbol{M}$ and $\boldsymbol{P}_t^\alpha$ are also symmetric we apply the spectral theorem and write

$$\boldsymbol{M} = \left[\begin{array}{c|c} \boldsymbol{A}_* & \boldsymbol{0} \\ \hline \boldsymbol{0} & \boldsymbol{0} \end{array}\right], \quad \boldsymbol{P}_t^\alpha = \left[\begin{array}{c|c} \boldsymbol{B}_\alpha^* & \boldsymbol{E}^T \\ \hline \boldsymbol{E} & \boldsymbol{D}_\alpha \end{array}\right], \tag{7.1}$$

where $\boldsymbol{A}_* \in \operatorname{Sym}^m$, $\boldsymbol{B}_\alpha^* (= \boldsymbol{B}_0^* + \alpha^2 \boldsymbol{I}) \in \operatorname{Sym}^m$ and $\boldsymbol{D}_\alpha (= \boldsymbol{D}_0 + \alpha^2 \boldsymbol{I}) \in \operatorname{Sym}^{n-m}$ are strictly positive definite for all $\alpha > 0$ and some positive integer $m < n$. In addition, condition (iii) of Theorem 1 implies that certain blocks in such a representation for $\boldsymbol{N}_t$ must be zero. In particular

$$\boldsymbol{N}_t = \left[\begin{array}{c|c} \boldsymbol{F}_* & \boldsymbol{G}^T \\ \hline \boldsymbol{0} & \boldsymbol{0} \end{array}\right]. \tag{7.2}$$

Thus, if we write

$$\boldsymbol{z}(s) = \left[\begin{array}{c} \boldsymbol{f}(s) \\ \hline \boldsymbol{z}_0(s) \end{array}\right], \tag{7.3}$$

then (3.3), (5.11), and (7.1)–(7.3) imply that

$$Q_\alpha^\xi(\boldsymbol{z}) = \int_0^\infty \{p(\dot{\boldsymbol{f}}, \boldsymbol{f}) + r(\dot{\boldsymbol{f}}, \boldsymbol{f}, \boldsymbol{z}_0)\}\, ds,$$

$$p(\boldsymbol{a}, \boldsymbol{b}) := \bar{\boldsymbol{a}} \cdot \boldsymbol{A}_* \boldsymbol{a} + \bar{\boldsymbol{b}} \cdot \boldsymbol{B}_\alpha^* \boldsymbol{b} - 2 \operatorname{Re}\{i\bar{\boldsymbol{a}} \cdot \boldsymbol{F}_* \boldsymbol{b}\},$$

$$r(\boldsymbol{a}. \boldsymbol{b}, \boldsymbol{c}) := \bar{\boldsymbol{c}} \cdot \boldsymbol{D}_\alpha \boldsymbol{c} + 2 \operatorname{Re}\{\bar{\boldsymbol{c}} \cdot \boldsymbol{E}\boldsymbol{b} - i\bar{\boldsymbol{a}} \cdot \boldsymbol{G}^T \boldsymbol{c}\}. \tag{7.4}$$

If $a, b \in \mathbb{C}^m$ are fixed then (7.4) is quadratic in the vector c. Since D_α is strictly positive definite this quadratic has a minimum that is achieved by

$$c_M = -D_\alpha^{-1}(Eb + iGa)$$

and hence

$$r(a, b, c_M) = -(\overline{Eb + iGa}) \cdot D_\alpha^{-1}(Eb + iGa).$$

Since c_M minimizes (7.4) we arrive at the following result.

Proposition 7.1. *Let C_0 satisfy the Legendre-Hadamard condition and condition* (iii) *of Theorem 1. Suppose that $M \neq 0$ is singular. Then for every $z = [f \mid z_0]^T \in C_0^1([0, \infty), \mathbb{C}^n)$*

$$Q_\alpha^\xi(z) \geqq q_\alpha(f)$$

with equality if and only if

$$z_0(s) = -D_\alpha^{-1}(Ef(s) + iG\dot{f}(s))$$

for every $s \in [0, \infty)$. Here

$$q_\alpha(f) := \int_0^\infty (\dot{\bar{f}} \cdot A(\alpha)\dot{f} + \bar{f} \cdot B(\alpha) f - 2 \operatorname{Re}\{i\dot{\bar{f}} \cdot F(\alpha) f\})\, ds,$$

$$A(\alpha) := A_* - G^T D_\alpha^{-1} G, \quad B(\alpha) := B_\alpha^* - E^T D_\alpha^{-1} E, \tag{7.5}$$

$$F(\alpha) := F_* - G^T D_\alpha^{-1} E.$$

In order to apply Proposition 6.2 we show that q_α is nonnegative, for sufficiently large α, and regular.

Lemma 7.2. *Let C_0 satisfy the Legendre-Hadamard condition and condition* (iii) *of Theorem 1. Suppose that $M \neq 0$ is singular, Then*:

(a) *For every $\xi \in \mathbb{R}^{n-1}$ there is an $\alpha_1 > 0$ such that q_α is nonnegative whenever $\alpha \geqq \alpha_1$;*

(b) *For every $\xi \in \mathbb{R}^{n-1}$ and $\alpha > 0$, q_α is regular.*

Proof. We first prove (b). Given $\alpha > 0$ and $t = (0, \xi) \in \mathbb{R}^n$, the Legendre-Hadamard condition implies that (*cf.* (6.5)), for every $\lambda \in \mathbb{R}$ and $a \in \mathbb{R}^n$,

$$a \cdot L(\lambda)\, a \geqq \alpha^2 \, |a|^2.$$

If we let $a = [e \mid c]^T$ then by (7.1), (7.2), and (6.1) we find that

$$\begin{bmatrix} \lambda^2 e \cdot A_* e + \lambda e \cdot (F_* + F_*^T)\, e + e \cdot B_\alpha^* e \\ + c \cdot D_\alpha c = + 2c \cdot (Ee + \lambda Ge) \end{bmatrix} \geqq \begin{bmatrix} \alpha^2 \, |e|^2 \\ + \\ \alpha^2 \, |c|^2 \end{bmatrix}.$$

In particular we let $c = -D_\alpha^{-1}(Ee + \lambda Ge)$ to conclude, with the aid of (7.5), that

$$e \cdot R(\lambda)\, e \geqq \alpha^2 \, |e|^2 \tag{7.6}$$

for every $e \in \mathbb{R}^m$.

The proof of Lemma 6.3 now shows that (7.6) implies (b) *provided* $A(\alpha)$ is strictly positive definite for every $\alpha > 0$. In order to prove this we let $e_0 \in \mathbb{R}^n$ and $u \in \mathbb{R}^n$ be of the form

$$e_0 = \left[\frac{0}{u_0}\right], \quad u = \left[\frac{u_*}{0}\right].$$

Then (7.1), (7.2), and the Legendre-Hadamard condition imply that

$$|u_* \cdot G^T u_0|^2 \leqq (u_0 \cdot D_0 u_0)\,(u_* \cdot A_* u_*). \tag{7.7}$$

Define

$$T_\alpha := \left[\begin{array}{c|c} A_* & G^T \\ \hline G & D_\alpha \end{array}\right].$$

Then T_α is symmetric and

$$\left[\frac{u_*}{u_0}\right] \cdot T_\alpha \left[\frac{u_*}{u_0}\right] = \left\{\begin{matrix}(u_* \cdot A_* u_*) + (u_0 \cdot D_0 u_0) \\ + 2(u_* \cdot G^T u_0) + (\alpha^2 \,|u_0|^2)\end{matrix}\right\}. \tag{7.8}$$

Since A_* is strictly positive definite and D_0 is positive semi-definite (by the Legendre-Hadamard condition), equations (7.7) and (7.8) imply that T_α is strictly positive definite for every $\alpha > 0$.

If we let $u \in \mathbb{R}^n$ be of the form

$$u = \left[\begin{matrix} u_* \\ -D_\alpha^{-1} G u_* \end{matrix}\right]$$

we find that

$$u \cdot T_\alpha u = u_* \cdot (A_* - G^T D_\alpha^{-1} G)\, u_*.$$

Therefore $A(\alpha) = A_* - G^T D_\alpha^{-1} G$ is strictly positive definite. This proves (b).

Finally since $A(\alpha)$ is strictly positive definite it is easy to show that, for each $\xi \in \mathbb{R}^{n-1}$, the integrand in (7.5), is nonnegative for sufficiently large α. This proves (a). □

Proof of Theorem 1 (*Sufficiency*). *Final Case.* $M \neq 0$ is singular. Fix $\xi \in \mathbb{R}^{n-1}$. Then the Legendre-Hadamard condition, condition (iii) of Theorem 1, and Lemma 7.2 imply that q_α is regular for all $\alpha > 0$ and nonnegative for sufficiently large α.

Suppose, for the sake of contradiction that there is an $\alpha_* > 0$ such that $Q^\xi_{\alpha_*}$ is not nonnegative. Then by Proposition 7.1 q_{α_*} is not nonnegative. Thus, by Proposition 6.2 there is an $\alpha_0 > 0$ and a nontrivial $f_0 \in C^\infty([0, \infty), \mathbb{C}^m)$, with f_0 and all its derivatives contained in $L^2([0, \infty), \mathbb{C}^m)$, such that f_0 satisfies (6.2) and (6.3), with $\alpha = \alpha_0$.

Define

$$z_0 = \left[\begin{matrix} f_0 \\ -D_\alpha^{-1}(E f_0 + iG\dot{f}_0) \end{matrix}\right].$$

Then a straightforward computation shows that z_0 satisfies (3.4) and (3.5). Therefore Agmon's condition is not satisfied, a contradiction. Thus Q_α^ξ is nonnegative for every $\alpha > 0$ and $\xi \in \mathbb{R}^{n-1}$ and hence by Lemma 5.3 we find that (5.1) is satisfied for all $u \in \operatorname{Var}(\mathscr{H}\mathscr{B})$. □

8. Some Examples

Our first example shows that condition (iii) of Theorem 1 is not a consequence of the Legendre-Hadamard condition and Agmon's condition.[14] Let $e \in \mathbb{R}^n$, with $|e| = 1$, and

$$W(F) = \frac{\mu}{2} F \cdot F + h(\det F) + \frac{\omega}{2} |Fe|^2 + \frac{\omega}{2} |F^{-T} e|^2.$$

Then

$$S(F) = \begin{bmatrix} \mu F + (\det F)\, h'(\det F)\, F^{-T} \\ + \omega F e \otimes e - \omega F^{-T} e \otimes F^{-1} F^{-T} e \end{bmatrix},$$

$$C(I)[H] = 2\mu E + \lambda (E \cdot I) I + 2\omega E e \otimes e + 2\omega e \otimes E e,$$

where $\lambda = h''(1) - \mu$ and $E = \frac{1}{2}(H + H^T)$. We note that the choice $\mu = -h'(1)$ ensures that the reference configuration is a natural state.

We now let $a, b \in \mathbb{R}^n$ and define

$$\alpha := a \cdot e, \qquad A := a - \alpha e,$$

$$\beta := b \cdot e, \qquad B := b - \beta e.$$

Then

$$C(I)[a \otimes b] = \begin{bmatrix} \mu(a \otimes b + b \otimes a) + \lambda(a \cdot b) I \\ + \omega\beta(a \otimes e + e \otimes a) + \omega\alpha(b \otimes e + e \otimes b) \end{bmatrix} \tag{8.1}$$

and hence

$$a \otimes b \cdot C[a \otimes b] = \begin{bmatrix} \mu |A|^2 |B|^2 + (\mu + \lambda)(A \cdot B)^2 + (2\mu + \lambda + 4\omega)\alpha^2\beta^2 \\ + 2(\mu + \lambda + \omega)\alpha\beta A \cdot B + (\mu + \omega)(\alpha^2 |B|^2 + \beta^2 |A|^2) \end{bmatrix},$$

where we have written C for $C(I)$. We let

$$2\mu + \lambda + 4\omega = 0, \tag{8.2}$$

$$\mu \geqq 0, \quad 2\mu + \lambda \geqq 0, \quad \text{and} \quad \lambda < 0. \tag{8.3}$$

Then (8.2) and the arithmetic-geometric mean inequality imply that

$$a \otimes b \cdot C[a \otimes b] \geqq \mu |A|^2 |B|^2 + (\mu + \lambda)(A \cdot B)^2 + \theta(\alpha^2 |B|^2 + \beta^2 |A|^2),$$

---

[14] It is known (see, *e.g.*, SIMPSON & SPECTOR [19, Section 8]) that Agmon's condition is not a consequence of the Legendre-Hadamard condition and condition (iii) of Theorem 1.

where $\theta = -\lambda$ if $\mu + \lambda + \omega \geqq 0$ and $\theta = (2\mu + \lambda)/2$ otherwise. Thus, given (8.2) and (8.3), the Legendre-Hadamard condition will be satisfied.

Next, we combine (8.1) and (8.2) to conclude that

$$\boldsymbol{C}[\boldsymbol{e} \otimes \boldsymbol{e}] = \lambda[\boldsymbol{I} - \boldsymbol{e} \otimes \boldsymbol{e}]$$

and hence that

$$\boldsymbol{e} \otimes \boldsymbol{e} \cdot \boldsymbol{C}[\boldsymbol{e} \otimes \boldsymbol{e}] = 0.$$

Thus, given (8.2) and (8.3), condition (iii) of Theorem 1 will *not* be satisfied.

Finally, in order to verify that Agmon's condition is satisfied, we let $\boldsymbol{n} = \boldsymbol{e}$ and, without loss of generality, we choose coordinates so that $\boldsymbol{n} = (-1, 0, 0, \ldots, 0)$ and $\boldsymbol{t} = (0, \tau, 0, 0, \ldots, 0)$. Then by (3.3), (8.1), and (8.2)

$$\begin{aligned} \boldsymbol{M}\boldsymbol{a} &= (\mu + \omega)\,[\boldsymbol{a} - (\boldsymbol{a} \cdot \boldsymbol{n})\,\boldsymbol{n}], \\ \boldsymbol{N}_t\boldsymbol{a} &= (\mu + \omega)\,(\boldsymbol{a} \cdot \boldsymbol{n})\,\boldsymbol{t} + \lambda(\boldsymbol{a} \cdot \boldsymbol{t})\,\boldsymbol{n}, \\ \boldsymbol{P}_t^\alpha\boldsymbol{a} &= (\mu\tau^2 + \alpha^2)\,\boldsymbol{a} + (\mu + \lambda)\,(\boldsymbol{a} \cdot \boldsymbol{t})\,\boldsymbol{t} + \omega\tau^2(\boldsymbol{a} \cdot \boldsymbol{n})\,\boldsymbol{n}, \end{aligned}$$

and hence (3.4) and (3.5) reduce to

$$\begin{aligned} -i\tau(\mu + \lambda + \omega)\,\dot{z}_2 + [(\mu + \omega)\,\tau^2 + \alpha^2]\,z_1 &= 0, \\ -(\mu + \omega)\,\ddot{z}_2 - i\tau(\mu + \lambda + \omega)\,\dot{z}_1 + [(2\mu + \lambda)\,\tau^2 + \alpha^2]\,z_2 &= 0 \\ -(\mu + \omega)\,\ddot{z}_j + (\mu\tau^2 + \alpha^2)\,z_j &= 0 \end{aligned} \tag{8.4}$$

and

$$\begin{aligned} \lambda\tau z_2(0) = 0, \quad (\mu + \omega)\,\dot{z}_j(0) &= 0, \\ (\mu + \omega)\,(\dot{z}_2(0) + i\tau z_1(0)) &= 0 \end{aligned} \tag{8.5}$$

for $j = 3, 4, \ldots, n$.

If we assume that μ and λ satisfy (8.3) then, by (8.2),

$$\mu + \omega > 0 \tag{8.6}$$

and hence $(8.4)_3$, (8.5), and the fact that $|\boldsymbol{z}|$ is bounded imply that $z_j \equiv 0$ for $j = 3, 4, \ldots, n$.

If we differentiate $(8.4)_1$ and substitute the result into $(8.4)_2$ we find, with the aid of (8.3), $(8.4)_1$, (8.5), and (8.6), that

$$-\left[(\mu + \omega) - \frac{\tau^2(\mu + \lambda + \omega)^2}{(\mu + \omega)\,\tau^2 + \alpha^2}\right]\ddot{z}_2 + [(2\mu + \lambda)\,\tau^2 + \alpha^2]\,z_2 = 0 \quad \text{on } (0, \infty),$$

$$\tau z_2(0) = 0, \quad \left[1 - \frac{\tau^2(\mu + \lambda + \omega)}{(\mu + \omega)\,\tau^2 + \alpha^2}\right]\dot{z}_2(0) = 0.$$

It is now easy to show that (8.3) implies that the coefficients in the above differential equation have opposite signs for every $\alpha \neq 0$ and hence that $z_2 \equiv 0$ is the only bounded solution of this initial value problem. It then follows from $(8.4)_1$ that $z_1 \equiv 0$ and hence that $\boldsymbol{z} \equiv \boldsymbol{0}$ is the only bounded solution of (8.4) and (8.5).

Our next example shows that, unlike Agmon's condition, the complementing condition[15] can fail at the unique global minimizer of the energy. Following BALL & JAMES [4, Section 7b], we let $n = 2$ and

$$W(F) = \tfrac{1}{2} F \cdot F + h(\det F), \tag{8.7}$$

where $h \in C^2(\mathbb{R}^+, \mathbb{R})$ and satisfies

$$\begin{gathered} \lim_{t\to 0^+} h(t) = \lim_{t\to\infty} h(t) = +\infty, \\ h''(t) > 0 \quad \text{for } t \in (0,1) \cup (1,\infty), \\ h'(1) = -1, \quad h''(1) = 0. \end{gathered} \tag{8.8}$$

Then, by (8.7) and $(8.8)_3$,

$$\begin{gathered} S(F) = F + (\det F)\, h'(\det F)\, F^{-T}, \\ C(I)[H] = H + H^T - (H \cdot I)\, I. \end{gathered} \tag{8.9}$$

Thus (*cf.*, *e.g.*, SIMPSON & SPECTOR [18, 19]) the complementing condition fails at $(C(I), n)$ for any $n \in \mathbb{R}^2$.

We now consider the problem of minimizing the energy of a half-ball composed of material with constitutive relation (8.7). We claim that

$$\int_{\mathscr{H}\mathscr{B}} W(I)\, dx < \int_{\mathscr{H}\mathscr{B}} W(I + \nabla u(x))\, dx$$

for every nontrivial $u \in \operatorname{Var}(\mathscr{H}\mathscr{B})$ and hence that $f(x) = x$ is the unique global minimizer of the energy.

To verify the above claim we first minimize (8.7). By $(8.8)_1$ we find that a global minimizer to (8.7) must exist and satisfy $S = 0$. By (8.9) we then conclude that this minimizer must satisfy

$$FF^T = kI, \quad k = -(\det F)\, h'(\det F) \tag{8.10}$$

and hence

$$W_{\min} = k + h(k).$$

If we minimize the last expression with respect to k we find that $h'(k) = -1$ and therefore we conclude, with the aid of $(8.8)_2$ and $(8.8)_3$ that $k = 1$. Thus any global minimizer of the energy must satisfy $\nabla f(x)\, \nabla f(x)^T = I$. However, a standard result then implies that $\nabla f(x) \equiv Q_0$, a constant. Since $f(x) = x$ on $\partial(\mathscr{H}\mathscr{B}) \setminus \partial\mathscr{H}$ this proves our assertion.

Acknowledgement. We thank Professors J. L. ERICKSEN, R. D. JAMES, K. A. PERICAK-SPECTOR, and H. WEINBERGER for their many useful discussions and suggestions. This work was supported in part by the Institute for Mathematics and its Applications and

---

[15] The complementing condition is equivalent to the requirement that $w = 0$ be the only bounded exponential solution to (3.1) when $\alpha = 0$.

by Heriot-Watt University. It was also supported by the National Science Foundation (DMS 86-00281 and DMS 86-02201) and the Air Force Office of Scientific Research (USAFOSR 86-0184).

References

1. AGMON, S., On the eigenfunctions and on the eigenvalues of general elliptic boundary value problems. Comm. Pure Appl. Math. **15** (1962), 119–147.
2. AGMON, S., A. DOUGLIS, & L. NIRENBERG, Estimates near the boundary for solutions of elliptic partial differential equations satisfying general boundary conditions II. Comm. Pure Appl. Math. **17** (1964), 35–92.
3. BALL, J. M., Constitutive inequalities and existence theorems in nonlinear elastostatics. *Nonlinear Analysis and Mechanics Vol. I* (R. J. KNOPS, ed.). London: Pitman, 1977.
4. BALL, J. M., & R. D. JAMES, Fine phase mixtures as minimizers of energy. Arch. Rational Mech. Anal. **100** (1987), 13–52.
5. BALL, J. M., & J. E. MARSDEN, Quasiconvexity at the boundary, positivity of the second variation, and elastic stability. Arch. Rational Mech. Anal. **86** (1984), 251–277.
6. BIOT, M. A., *Mechanics of Incremental Deformations*. New York: Wiley, 1965.
7. CHEN, Y.-C., Stability of pure homogeneous deformations of an elastic plate with fixed edges. Q. J. Mech. Appl. Math. **41** (1988), 249–264.
8. DE FIGUEIREDO, D. G., The coerciveness problem for forms over vector valued functions. Comm. Pure Appl. Math. **16** (1963), 63–94.
9. DUFFIN, R. J., Crystal's theorem on differential equation systems. J. Math. Anal. Appl. **8** (1963), 325–331.
10. GURTIN, M. E., Two-phase deformations of elastic solids. Arch. Rational Mech. Anal. **84** (1983), 1–29.
11. HADAMARD, J., *Leçons sur la Propagation des Ondes et les Equations de l'Hydrodynamique*. Paris: Hermann, 1903.
12. HAYES, M., & R. S. RIVLIN, Propagation of a plane wave in an isotropic elastic material subjected to pure homogeneous deformation. Arch. Rational Mech. Anal. **8** (1961), 15–22.
13. HUTCHINSON, J. W.. & V. TVERGAARD, Surface instabilities on statically strained plastic solids. Int. J. Mech. Sci. **22** (1980), 339–354.
14. MEYERS, N. G., Quasi-convexity and lower semi-continuity of multiple variational integrals of any order. Trans. Amer. Math. Soc. **119** (1965), 125–149.
15. MORREY, C. B., Quasi-convexity and the lower semicontinuity of multiple integrals. Pacific J. Math. **2** (1952), 25–53.
16. NOWINSKI, J. L., Surface instability of a half-space under high two-dimensional compression. J. Franklin Inst. **288** (1969), 367–376.
17. REDDY, B. D., Surface instabilities on an equibiaxially stretched elastic half-space. Math. Proc. Camb. Phil. Soc. **91** (1982), 491–501.
18. SIMPSON, H. C., & S. J. SPECTOR, On failure of the complementing condition and nonuniqueness in linear elastostatics. J. Elasticity **15** (1985), 229–331.
19. SIMPSON, H. C., & S. J. SPECTOR, On the positivity of the second variation in finite elasticity. Arch. Rational Mech. Anal. **98** (1987), 1–30.
20. THOMPSON, J. L., Some existence theorems for the traction boundary value problem of linearized elastostatics. Arch. Rational Mech. Anal. **32** (1969), 369–399.

21. TRUESDELL, C., & W. NOLL, *The Non-linear Field Theories of Mechanics*. Handbuch der Physik III/3 (S. FLÜGGE, *ed.*). Berlin Heidelberg New York: Springer-Verlag, 1965.
22. VAN HOVE, L., Sur l'extension de la condition de Legendre du calcul des variations aux intégrales multiples à plusieurs fonctions inconnues. Proc. Koninklijke Nederlandsche Akademie Van Wetenschappen **50** # 1 (1947), 18–23.

Department of Mathematics
University of Tennessee
Knoxville

and

Department of Mathematics
Southern Illinois University
Carbondale

(Received January 14, 1989)

Phase Transitions of Elastic Solid Materials

IRENE FONSECA

Dedicated to Bernard Coleman on his sixtieth birthday

Table of Contents

1. Introduction

In recent years the analysis of phase transitions for mixtures of two or more non-interacting fluids has been successfully undertaken within the Van der Waals-Cahn-Hilliard gradient theory of phase transitions (see BALDO [1], FONSECA & TARTAR [5], GURTIN [9], KOHN & STERNBERG [11], MODICA [12], OWEN [13], STERNBERG [15]). If the nonnegative Gibbs free energy W vanishes only at two points a and b, this theory permits us to select among all the minimizers of

$$\int_{\Omega} W(v(x))\,dx$$

with prescribed total mass

$$m = \int_{\Omega} v(x)\,dx = \text{meas}\,(\Omega)\,(\theta a + (1-\theta)\,b), \qquad \text{with } \theta \in (0, 1),$$

those that have minimal interfacial area, *i.e.* it singles out those solutions $v \in \{a, b\}$ a.e. such that the set $\{v = a\}$ minimizes $\text{Per}_{\Omega}(\omega)$ among all subsets ω of Ω with $\text{meas}\,(\omega) = \theta = \text{meas}\,(\{v = a\})$.

We consider an analogous situation in the context of nonlinear elasticity. Here the stored energy density W is nonnegative and, due to frame indifference, W has two orbits of minima $\{\text{RA} \mid \text{R rotation}\}$ and $\{\text{RB} \mid \text{R rotation}\}$ where A

and B differ by a rank one matrix,

$$A = B + a \otimes n.$$

It is clear that the problem

$$(\mathrm{P}_0) \quad \text{minimize} \int_\Omega W(\nabla u(x))\, dx,$$

for u such that

$$\int_\Omega u(x)\, dx = m \quad \text{and} \quad \int_\Omega \nabla u(x)\, dx = \operatorname{meas}(\Omega)\,(\theta A + (1-\theta) B)$$

where $\theta \in (0, 1)$ and $m \in \mathbb{R}^3$ are fixed, admits infinitely many solutions. In particular, if E is a subset of Ω layered normally to n and if $\operatorname{meas}(E) = \theta \operatorname{meas}(\Omega)$ then there exists a solution u of (P_0) with

$$\nabla u = B + \chi_E a \otimes n. \tag{1.1}$$

We search for a model that will select among all the solutions of (P_0) those of the form (1.1) for which E is a solution of

$$(\mathrm{P}^*_0) \quad \text{minimize} \ \operatorname{Per}_\Omega (E'),$$

where $E' \subset \Omega$ is layered normally to n and $\operatorname{meas}(E') = \theta \operatorname{meas}(\Omega)$.

In order to apply the gradient theory of phase transitions to this setting, we have to add to the former problem the constraint $\operatorname{curl} v = 0$ which renders the analysis very difficult (see FONSECA & TARTAR [6]). In this paper we study a model in which the second deformation gradient is replaced by a Radon measure penalizing the formation of interfaces and where the spinodal region is removed as in GURTIN'S theory for phase transitions for fluids (see GURTIN [8]). In Section 2 we discuss briefly some notions and results of the theory of functions of bounded variation. In Proposition 2.16 we prove the converse of a result due to BALL & JAMES [3] characterizing the Lipschitz deformations satisfying (1.1) (see Theorem 2.14). In Section 3 we introduce a model accomodating the constraint $\operatorname{curl} v = 0$. If we disregard frame indifference, it can be shown that a sequence of minimizers of the regularized problems admits a subsequence converging weakly to a solution of the problem (P_0) of the form (1.1) with minimal interfacial area (see Theorem 3.2). In Section 4 we adapt a model proposed by GURTIN [8] for the analysis of fluid phase transitions. Here the spinodal region is removed and we penalize directly the interface. In Theorem 4.5 we obtain the analog of Theorem 3.2 for the sequence of penalized problems. In order to handle frame indifference, we combine the models of Sections 3 and 4. In Theorem 5.10 we show that a sequence of minimizers of the approximation problems admits a subsequence converging weakly to a solution of the problem (P_0) of the form

$$\nabla u = R(x)\,(B + \chi_E a \otimes n), \tag{1.2}$$

where $R(x)$ is a rotation, ∇u, R and $\chi_E \in \mathrm{BV}(\Omega) \cap L^\infty(\Omega)$. We conjecture that $R =$ identity a.e. and that E is a solution of (P^*_0). We prove that the conjecture is confirmed if the set E is reasonably smooth, precisely if E determines a partition of Ω into countably many open, strongly Lipschitz, connected domains (see Theorem 5.8). The rest of Section 5 is dedicated to finding results asserting the required smoothness of E. In Proposition 5.19 we show that if a is parallel to $B^{-T}n$ then the outward unit normal to $\partial E \cap \Omega$ is parallel to n. In this case the set E is layered normally to n and the conjecture is valid (see Corollary 5.20). If a is not parallel to $B^{-T}n$, we prove in Proposition 5.19 that the normal to $\partial E \cap \Omega$ is parallel either to n or to a known vector m, where m is a linear combination of a and n. Moreover, it is possible to prove that $\partial E \cap \Omega$ cannot have "corners" (see Lemma 5.5). I do not know if these properties imply the smoothness of E required by Theorem 5.8, in which case the conjecture would be confirmed.

2. Statement of the Problem and Preliminaries

We discuss briefly some results of the theory of functions of bounded variation (see EVANS & GARIEPY [4], GIUSTI [7]).

Let Ω be an open bounded strongly Lipschitz domain of $\mathbb{R}^n$.

Definition 2.1. A function $u \in L^1(\Omega)$ is said to be a *function of bounded variation* ($u \in \mathrm{BV}(\Omega)$) if

$$\int_\Omega |\nabla u(x)|\, dx := \sup\left\{\int_\Omega u(x) \cdot \operatorname{div} \varphi(x)\, dx \mid \varphi \in C_0^1(\Omega; \mathbb{R}^n),\ \|\varphi\|_\infty \leqq 1\right\} < +\infty .$$

It follows immediately that if $u_\varepsilon \to u$ in $L^1(\Omega)$ then

$$\int_\Omega |\nabla u(x)|\, dx \leqq \liminf_{\varepsilon \to 0} \int_\Omega |\nabla u_\varepsilon(x)|\, dx. \tag{2.2}$$

It can be shown that the sets

$$\left\{u \in L^1(\Omega) \mid \int_\Omega |u(x)| + |\nabla u(x)|\, dx \leqq C < +\infty\right\} \tag{2.3}$$

are compact in $L^1(\Omega)$.

Proposition 2.4. *There exists a constant* $C > 0$ *such that for all* $f \in \mathrm{BV}(\Omega)$

$$\int_\Omega f(x)\, dx = 0 \Rightarrow \left(\int_\Omega |f(x)|^{n/n-1}\, dx\right)^{\frac{n-1}{n}} \leqq C \int_\Omega |Df(x)|\, dx .$$

Definition 2.5. If A is a subset of $\mathbb{R}^n$ then the *perimeter of* A *in* Ω is defined by

$$\mathrm{Per}_\Omega(A) := \int_\Omega |\nabla \chi_A(x)|\, dx = \sup\left\{\int_A \operatorname{div} \varphi(x)\, dx \mid \varphi \in C_0^1(\Omega; \mathbb{R}^n),\ \|\varphi\|_\infty \leqq 1\right\},$$

where χ_A denotes the characteristic function of A.

Proposition 2.6. *If* $f, g \in L^\infty(\Omega) \cap \mathrm{BV}(\Omega)$ *then* $f \cdot g \in L^\infty(\Omega) \cap \mathrm{BV}(\Omega)$.

Clearly, if $A \subset \Omega$ and if

$$u(x) = \begin{cases} a & \text{if } x \in A \\ b & \text{if } x \in \Omega \setminus A \end{cases}$$

then $u \in \mathrm{BV}(\Omega)$ if and only if $\mathrm{Per}_\Omega(A) < +\infty$. Suppose that E is a set of finite perimeter in $\mathbb{R}^n$. There exist a Radon measure $\|\partial E\|$ and a $\|\partial E\|$-measurable function

$$\nu_E : \mathbb{R}^n \to \mathbb{R}^n, \quad \|\nu_E\| = 1 \quad \|\partial E\| \text{ a.e.},$$

such that

$$\int_E \operatorname{div} \varphi(x)\, dx = \int_{\mathbb{R}^n} \varphi(x) \cdot \nu_E(x)\, d\|\partial E\| \quad \text{for all } \varphi \in C_0^1(\mathbb{R}^n; \mathbb{R}^n).$$

Definition 2.7. Let $x \in \mathbb{R}^n$. We say that $x \in \partial^* E$, the *reduced boundary of E*, if

$$\|\partial E\|(B(x, r)) > 0 \quad \text{for all } r > 0,$$

$$\|\nu_E(x)\| = 1$$

and

$$\lim_{\varepsilon \to 0} \frac{1}{\|\partial E\|(B(x, \varepsilon))} \int_{B(x,\varepsilon)} \nu_E(y)\, d\|\partial E\| = \nu_E(x).$$

Theorem 2.8. (*Blow-up of the reduced boundary*) *If* $x \in \partial^* E$ *then*

$$\chi_{x + \frac{E-x}{\varepsilon}} \to \chi_{H^-(x)} \quad \text{in } L^1_{\text{loc}} \quad \text{as } \varepsilon \to 0^+,$$

where

$$H^-(x) := \{y \in \mathbb{R}^n \mid \nu_E(x) \cdot (y - x) < 0\}.$$

Theorem 2.9. (*Generalized Gauss-Green Theorem*)

$$\int_E \operatorname{div} \varphi\, dx = \int_{\partial^* E} \varphi \cdot \nu_E\, dH_{n-1}$$

for all $\varphi \in C_0^1(\mathbb{R}^n; \mathbb{R}^n)$.

Given $f \in \mathrm{BV}(\Omega)$, we define

$$\mu(x) := \operatorname{ap}\limsup_{y \to x} f(y) = \inf \left\{ t \mid \lim_{\varepsilon \to 0} \frac{\operatorname{meas}(B(x, \varepsilon) \cap \{f > t\})}{\varepsilon^n} = 0 \right\}$$

and

$$\lambda(x) := \operatorname{ap}\liminf_{y \to x} f(y) = \sup \left\{ t \mid \lim_{\varepsilon \to 0} \frac{\operatorname{meas}(B(x, \varepsilon) \cap \{f < t\})}{\varepsilon^n} = 0 \right\}.$$

Let

$$J := \{x \in \Omega \mid \lambda(x) < \mu(x)\}$$

denote the set of points at which f is not approximately continuous.

Theorem 2.10. *Assume that* $f \in \mathrm{BV}(\Omega)$. *Then* λ *and* μ *are Borel measurable and*

(i) $\operatorname{meas}(J) = 0$;

(ii) $-\infty \leqq \lambda(x) \leqq \mu(x) \leqq +\infty$ *in* Ω;

$$\text{(iii)} \lim_{\varepsilon \to 0^+} \frac{1}{\operatorname{meas}(B(x, \varepsilon))} \int_{B(x,\varepsilon)} \left| f(y) - \frac{\lambda(x) + \mu(x)}{2} \right|^{n/n-1} dy = 0 \qquad (2.11)$$

for H_{n-1} a.e. $x \in \Omega \setminus J$;

(iv) *for* H_{n-1} a.e. $x \in J$ *there exists a unit vector* ν *such that*

$$\lim_{\varepsilon \to 0^+} \frac{1}{\operatorname{meas}(B(x, \varepsilon) \cap H_\nu^+)} \int_{B(x,\varepsilon) \cap H_\nu^+} |f(y) - \mu(x)|^{n/n-1} dy = 0$$

and

$$\lim_{\varepsilon \to 0^+} \frac{1}{\operatorname{meas}(B(x, \varepsilon) \cap H_\nu^-)} \int_{B(x,\varepsilon) \cap H_\nu^-} |f(y) - \lambda(x)|^{n/n-1} dy = 0. \qquad (2.12)$$

Consider a hyperelastic body that occupies in a reference configuration a bounded, simply connected, strongly Lipschitz domain $\Omega \subset \mathbb{R}^3$, with $\operatorname{meas}(\Omega) = 1$. Let $W: M^{3\times 3} \to [0, +\infty]$ denote the stored energy density, and assume that

(H1) $W(F) = +\infty$ if and only if $\det F \leqq 0$;

(H2) $W(F) = 0$ if and only if $F \in \{RA, RB \mid R \in O^+(\mathbb{R}^3)\}$, where $A = B + a \otimes n$, $\|n\| = 1$, $a \neq 0$.

Here, and in what follows, $M^{3\times 3}$ is the set of real 3×3 matrices and $O^+(\mathbb{R}^3)$ denotes the set of rotations of $\mathbb{R}^3$. Let $\theta \in (0, 1)$ and $m \in \mathbb{R}^3$ be fixed, and define the class of admissible deformations

$$\mathscr{A}_0 := \left\{ u \in W^{1,1}(\Omega; \mathbb{R}^3) \mid \int_\Omega u(x)\, dx = m,\ \int_\Omega \nabla u(x)\, dx = \theta A + (1 - \theta) B \right\}.$$

In this paper, we study the variational problem
(P_0) Minimize

$$\int_\Omega W(\nabla u(x))\, dx,$$

for $u \in \mathscr{A}_0$.

Remark 2.13. (P_0) admits infinitely many solutions u such that $\nabla u \in \mathrm{BV}(\Omega)$. In fact, let α be such that

$$\operatorname{meas}\{x \in \Omega \mid x \cdot n > \alpha\} = \theta,$$

and define

$$C := m - \int_{\{x \in \Omega \mid x \cdot n > \alpha\}} Ax\, dx - \int_{\{x \in \Omega \mid x \cdot n < \alpha\}} (Bx + \alpha a)\, dx.$$

Setting

$$u^*(x) := \begin{cases} Ax + C & \text{if } x \cdot n > \alpha \\ Bx + \alpha a + C & \text{if } x \cdot n < \alpha, \end{cases}$$

we see immediately that u^* is a solution of (P_0). Similarly, an infinite set of Lipschitz-continuous solutions with gradients taking only the values A and B can be found by layering Ω by finitely many parallel planes with normal n, in such a way that meas $(\{x \in \Omega \mid \nabla u(x) = A\}) = \theta$ (see Fig. 1).

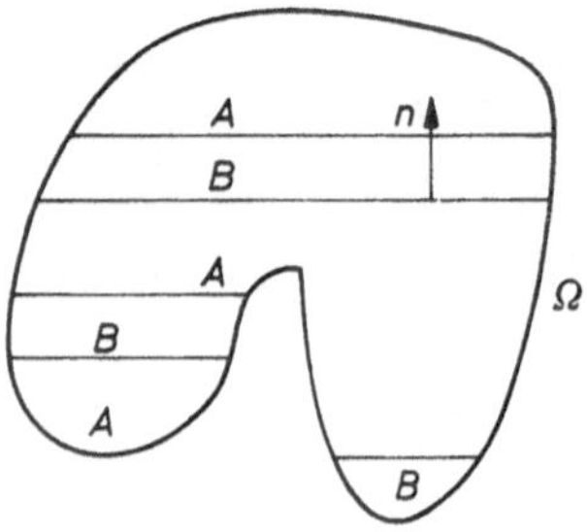

Fig. 1

The following result is due to BALL & JAMES [3].

Theorem 2.14. *Let* $E \subset \Omega$, $0 < \text{meas}(E) < \text{meas}(\Omega)$, *and let* χ_E *denote the characteristic function of E. If* $u \in W^{1,\infty}(\Omega; \mathbb{R}^3)$ *satisfies*

$$\nabla u = \chi_E C + (1 - \chi_E) D,$$

then

$$C = D + b \otimes m$$

for some $b \in \mathbb{R}^3$, $m \in \mathbb{R}^3$, $\|m\| = 1$. *Moreover, for all convex set* $\Gamma \subset \Omega$, $E \cap \Gamma$ *and* $(\Omega \setminus E) \cap \Gamma$ *consist of parallel layers normal to m; precisely, there exists a Lipschitz function f with* $f' \in \{0, 1\}$ a.e. *such that*

$$E \cap \Gamma = \{x \in \Gamma \mid f'(x \cdot n) = 1\}.$$

We define the set
$\mathcal{S} := \{E \subset \Omega \mid \text{meas}(E) = \theta$ and for all convex set $\Gamma \subset \Omega$ there exists a Lipschitz function f with $f' \in \{0, 1\}$ a.e. such that $E \cap \Gamma = \{x \in \Gamma \mid f'(x \cdot n) = 1\}\}$.

Remark 2.15. It is clear that the solutions u exhibited in Remark 2.13 satisfy

$$\nabla u = B + \chi_E a \otimes n, \quad \text{with } E \in \mathcal{S} \quad \text{and} \quad \nabla u \in BV(\Omega). \tag{2.15}$$

We prove that the converse of Theorem 2.14 is also true.

Proposition 2.16. *If* $E \in \mathcal{S}$ *then there exists* $u \in W^{1,\infty}(\Omega; \mathbb{R}^3) \cap \mathcal{A}_0$ *such that*

$$\nabla u = B + \chi_E a \otimes n.$$

Proof. Suppose for simplicity that $n = e_3$. Let $x_0 \in \Omega$ be fixed, and define for $x \in \Omega$ the function

$$\theta(x) := \int_{\gamma_x} \chi_E \, dx_3 = \int_0^1 \chi_E(\alpha(s)) \, \alpha_3'(s) \, ds$$

where $\gamma_x = \{\alpha(t) \mid t \in [0, 1], \alpha(0) = x_0, \alpha(1) = x\}$ is a piecewise C^1 curve in Ω joining x_0 to x.

(i) θ does not depend on the choice of the curve γ_x. Let $\alpha, \beta\colon [0\ 1] \to \Omega$ be piecewise C^1 functions such that $\alpha(0) = \beta(0) = x_0$, $\alpha(1) = \beta(1) = x$. As Ω is a simply connected domain, there exists a C^1 function $H\colon [0, 1] \times [0, 1] \to \Omega$ such that $H(0, t) = \alpha(t)$, $H(1, t) = \beta(t)$. Define

$$F(s) := \int_0^1 \chi_E(H(s, t)) \frac{\partial}{\partial t} H_3(s, t)\, dt.$$

We wish to show that $F(0) = F(1)$. Clearly, it suffices to prove that for all $s_0 \in [0, 1]$ there exists $\varepsilon > 0$ such that $F(s) = F(s_0)$ for all $s \in [0, 1]$, $|s - s_0| < \varepsilon$. Let $s_0 \in [0, 1]$ and let

$$H(s_0, t) \in \bigcup_{i=0}^{N} B(x_i, \varepsilon_i) \subset \Omega \quad \text{for all } t \in [0, 1], \quad \text{where } x_N \equiv x.$$

As H is smooth, there exists $\varepsilon > 0$ such that

$$H(s, t) \in \bigcup_{i=0}^{N} B(x_i, \varepsilon_i) \quad \text{for } |s - s_0| < \varepsilon \text{ and for all } t \in [0, 1].$$

Let f_i be such that

$$\chi'_E(x) = f'_i(x \cdot n) \quad \text{for } x \in B(x_i, \varepsilon_i). \tag{2.17}$$

Define the open sets

$$\omega_i := \{t \in \mathbb{R} \mid \exists\, x \in B(x_i, \varepsilon_i) \cap B(x_{i+1}, \varepsilon_{i+1}) \text{ such that } x \cdot n = t\}, \; i = 0, \ldots, N-1.$$

As

$$f'_i = f'_{i+1} \quad \text{in } \omega_i,$$

we can assume without loss of generality that

$$f_i = f_{i+1} \quad \text{in } \omega_i. \tag{2.18}$$

Choose points $\sigma(t_i)$, $\varrho(t_i)$ such that $0 < t_1 < \ldots < t_N < 1$ and

$$\sigma(t_i), \beta(t_i) \in B(x_{i-1}, \varepsilon_{i-1}) \cap B(x_i, \varepsilon_i), \quad \text{for } i = 1, \ldots, N,$$

where $\sigma(t) := H(s_0, t)$ and $\varrho(t) := H(s, t)$. By (2.17) and (2.18) we have

$$\begin{aligned}
F(s_0) &= \int_0^1 \chi_E(H(s_0, t)) \frac{\partial}{\partial t} H_3(s_0, t)\, dt \\
&= \int_0^{t_1} f'_0(\sigma(t) \cdot n)\, \sigma'(t) \cdot n\, dt + \sum_{i=1}^{N-1} \int_{t_i}^{t_{i+1}} f'_i(\sigma(t) \cdot n)\, \sigma'(t) \cdot n\, dt \\
&\quad + \int_{t_N}^1 f'_N(\sigma(t) \cdot n)\, \sigma'(t) \cdot n\, dt \\
&= f_0(\sigma(t_1) \cdot n) - f_0(x_0 \cdot n) + \sum_{i=1}^{N-1} [f_i(\sigma(t_{i+1}) \cdot n) \\
&\quad - f_i(\sigma(t_i) \cdot n)] + f_N(x \cdot n) - f_N(\sigma(t_N) \cdot n) \\
&= f_N(x \cdot n) - f_0(x_0 \cdot n).
\end{aligned}$$

In a similar way, we obtain

$$F(s) = f_N(x \cdot n) - f_0(x_0 \cdot n).$$

(ii) θ is a locally Lipschitz-continuous function. In fact, let $B(x, \varepsilon) \subset \Omega$ and let $y \in B(x, \varepsilon)$. By (i) we have

$$|\theta(x) - \theta(y)| \leqq \left| \int_0^1 \chi_E((1-t)\, x + ty)\, (y - x) \cdot n\, dt \right|$$
$$\leqq \|x - y\|.$$

(iii) We prove that $\nabla\theta = \chi_E n$ in $\mathscr{D}'(\Omega)$. Let $B(x, \varepsilon) \subset\subset \Omega$ and let

$$E \cap B(x, \varepsilon) = \{y \in B(x, \varepsilon) \mid f'(y \cdot n) = 1\}.$$

By (i) we deduce that

$$\theta(y) = \int_0^1 \chi_E((1-t)\, x + ty)\, (y - x) \cdot n\, dt$$
$$= \int_0^1 f'((1-t)\, x + ty)\, (y - x) \cdot n\, dt$$
$$= f(y \cdot n) - f(x \cdot n).$$

Therefore, if $\phi \in \mathscr{D}(B(x, \varepsilon))$ and if $i = 1, 2$ we have

$$\left\langle \frac{\partial\theta}{\partial x_i}, \varphi \right\rangle = \int_{B(x,\varepsilon)} [f(x \cdot n) - f(y \cdot n)] \frac{\partial\varphi}{\partial y_i}(y)\, dy$$
$$= 0$$

and

$$\left\langle \frac{\partial\theta}{\partial x_3}, \varphi \right\rangle = \int_{B(x,\varepsilon)} [f(x \cdot n) - f(y \cdot n)] \frac{\partial\varphi}{\partial y_3}(y)\, dy$$
$$= \int_{B(x,\varepsilon)} f'(y \cdot n)\, \varphi(y)\, dy$$
$$= \int_E \varphi(y)\, dy.$$

(iv) Finally, set

$u(x) := u_0 + Bx + \theta(x)\, a$, with u_0 such that

$$\int_\Omega u(x)\, dx = m.$$

Clearly, $u \in W^{1,\infty}(\Omega; \mathbb{R}^3) \cap \mathscr{A}_0$ and

$$\nabla u = B + \chi_E a \otimes n.$$

We search for a model that will select among all solutions of (P_0) those of the form (2.15) with minimal interfacial area, *i.e.* where E is a solution of the geometric variational problem:

(P_0^*) Minimize $\operatorname{Per}_\Omega(E')$ with $E' \in \mathscr{S}$.

In Sections 3 and 4 we consider simplified models for which W is not frame indifferent.

3. Regularization

Assume that W is continuous and verifies (H1) and

(H2′) $W(F) = 0$ if and only if $F \in \{A, B\}$.

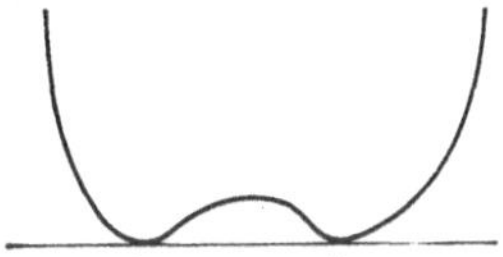

Fig. 2

In view of Remark 2.15, we define the class of admissible deformations

$$\mathscr{A}_1 := \left\{ u \in W^{1,1}(\Omega; \mathbb{R}^3) \mid \nabla u \in \mathrm{BV}(\Omega), \int_\Omega u(x)\, dx = m, \right.$$

$$\left. \int_\Omega \nabla u(x)\, dx = \theta A + (1-\theta) B \right\}$$

and for $\varepsilon > 0$ we introduce the "regularized" problem:

(P_ε) Minimize in $\mathscr{A}_1$

$$E_\varepsilon(u) := \int_\Omega W(\nabla u(x))\, dx + \varepsilon \int_\Omega |D \nabla u(x)|\, dx.$$

Here and in what follows we use the notation

$$\|D \nabla u\| = \int_\Omega |D \nabla u(x)|\, dx := \sum_{i,j=1}^{3} \int_\Omega \left| D \frac{\partial u_i}{\partial x_j}(x) \right| dx.$$

Proposition 3.1. *For all* $\varepsilon > 0$, (P_ε) *admits a solution.*

Proof. Let u^* be as defined in Remark 2.13. Then

$$E_\varepsilon(u^*) = \varepsilon C^* \quad \text{where} \quad C^* := \text{area}\,\{x \in \Omega \mid x \cdot n = \alpha\}.$$

Let $\{u_j\}$ be a minimizing sequence with j large enough that

$$E_\varepsilon(u_j) \leqq \varepsilon C^* + 1.$$

Thus

$$\{\|D \nabla u_j\|\} \text{ is a bounded sequence}$$

and so, by Proposition 2.4 and Poincaré's inequality, we deduce that

$$\{u_j\} \quad \text{is a bounded sequence in } W^{1,3/2}(\Omega)$$

and

$$\{\nabla u_j\} \text{ is a bounded sequence in } \mathrm{BV}(\Omega).$$

By (2.3) we conclude that there exist $u \in \mathscr{A}_1 \cap W^{1,3/2}(\Omega)$ and a subsequence $\{u_m\}$ such that

$$u_m \to u \quad \text{weakly in } W^{1,3/2}(\Omega)$$

and

$$\nabla u_m \to \nabla u \quad \text{strongly in } L^1(\Omega).$$

By Fatou's Lemma and (2.2) it follows that

$$E_\varepsilon(u) \leqq \liminf_{m\to\infty} E_\varepsilon(u_m) = \inf_{\mathscr{A}} E_\varepsilon(.).$$

Theorem 3.2. *Let v_ε be a sequence of minimizers of E_ε in $\mathscr{A}_1$. There exist a solution u of* (P_0) *and a subsequence u_ε such that*

$$u_\varepsilon \to u \quad \textit{in } W^{1,3/2}(\Omega; \mathbb{R}^3) \textit{ weak}$$

and

$$\nabla u = B + \chi_E a \otimes n \quad \text{a.e. } \textit{in } \Omega,$$

where $E \in \mathscr{S}$ is a solution of (P_0^*).

Proof. Considering u^* as in the proof of Proposition 3.1, we have

$$E_\varepsilon(v_\varepsilon) \leqq E_\varepsilon(u^*) = \varepsilon C^*$$

and so

$$\int_\Omega |D \nabla v_\varepsilon(x)| \, dx \leqq C^*.$$

Thus, by Proposition 2.4 and Poincaré's inequality, we conclude that

$$\{v_\varepsilon\} \quad \text{is a bounded sequence in } W^{1,2/3}(\Omega)$$

and

$$\{\nabla v_\varepsilon\} \text{ is a bounded sequence in } \mathrm{BV}(\Omega).$$

Hence, by (2.3) there exist $u \in \mathscr{A}_1 \cap W^{1,n/n-1}(\Omega)$ and a subsequence $\{u_\varepsilon\}$ such that

$$u_\varepsilon \to u \quad \text{weakly in } W^{1,3/2}(\Omega)$$

and

$$\nabla u_\varepsilon \to \nabla u \quad \text{strongly in } L^1(\Omega).$$

By Fatou's Lemma it follows that

$$\begin{aligned} \int_\Omega W(\nabla u(x)) \, dx &\leqq \liminf_{\varepsilon\to 0} \int_\Omega W(\nabla u_\varepsilon(x)) \, dx \\ &\leqq \lim \varepsilon C^* \\ &= 0. \end{aligned}$$

Therefore

$$\nabla u(x) \in \{A, B\} \quad \text{a.e. } x \in \Omega$$

and so, by Theorem 2.14, we deduce that

$$\nabla u(x) = B + \chi_E(x) \, a \otimes n, \quad \text{where } E \in \mathscr{S}.$$

Let $E' \in \mathcal{S}$ and, by Proposition 2.16, let $v \in W^{1,\infty}(\Omega; \mathbb{R}^3) \cap \mathcal{A}_0$ be such that

$$\nabla v = B + \chi_{E'} a \otimes n.$$

If $\mathrm{Per}_\Omega(E') < +\infty$ then $v \in \mathcal{A}_1$ and so

$$\begin{aligned} E_\varepsilon(v) &= \varepsilon \int_\Omega |D \nabla v(x)| \, dx \\ &\geqq E_\varepsilon(u_\varepsilon) \\ &\geqq \varepsilon \int_\Omega |D \nabla u_\varepsilon(x)| \, dx. \end{aligned}$$

Hence, by (2.2) we obtain

$$\begin{aligned} \int_\Omega |D \nabla v(x)| \, dx &\geqq \liminf_{\varepsilon \to 0} \int_\Omega |D \nabla u_\varepsilon(x)| \, dx \\ &\geqq \int_\Omega |D \nabla u(x)| \, dx, \end{aligned}$$

and so we conclude that

$$C \, \mathrm{Per}_\Omega(E') \geqq C \, \mathrm{Per}_\Omega(E)$$

where

$$C = \sum_{i,j=1}^{3} |a_i| \, |b_j|.$$

4. Direct Penalization of the Interface. Removal of the Spinodal Region

Here we adapt to our present setting a model proposed by GURTIN [8] for phase transitions in the case of a mixture of two fluids. Let D_1 and D_2 be closed, convex, bounded subsets of $\{F \in M^{3\times 3} \mid \det F \geqq 0\}$, with $A \in \mathrm{int}\, D_1$, $B \in \mathrm{int}\, D_2$ and $D_1 \cap D_2 = \emptyset$. Assume that W satisfies (H1), (H2') and

(H3) (polyconvexity) there exist convex functions $G_i: M^{3\times 3} \times M^{3\times 3} \times (0, +\infty) \to [0, +\infty)$, $i = 1, 2$, such that

$$W(F) = G_i(F, \mathrm{adj}\, F, \det F) \quad \text{for all } F \in D_i, \ \det F > 0, \ i = 1, 2.$$

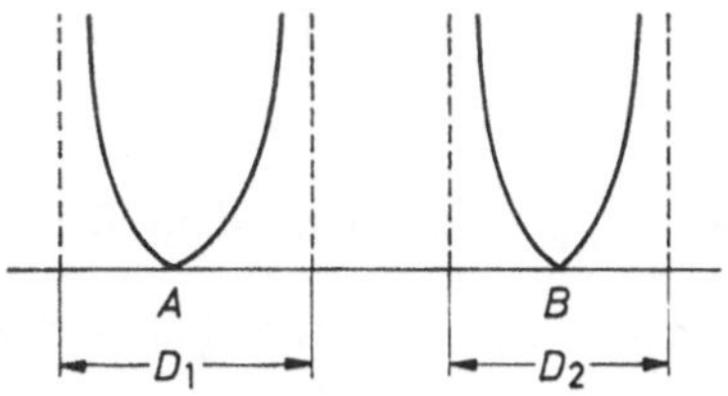

Fig. 3

Let

$$\mathscr{A}_2 := \Big\{ u \in W^{1,1}(\Omega; \mathbb{R}^3) \mid \nabla u \in D_1 \cap D_2 \text{ a.e.}, \int_\Omega u(x)\, dx = m,$$

$$\int_\Omega \nabla u(x)\, dx = \theta A + (1-\theta) B \Big\},$$

and consider the problem:
(P_0) Minimize in $\mathscr{A}_2$

$$\int_\Omega W(\nabla u(x))\, dx.$$

If $u \in \mathscr{A}_2$ we define

$$\Omega_1(u) := \{x \in \Omega \mid \nabla u(x) \in D_1\}$$

and

$$I(u) := \mathrm{Per}_\Omega\, (\Omega_1(u)).$$

We introduce the family of penalized problems:

(P_ε) Minimize in $\mathscr{A}_2$

$$E_\varepsilon(u) := \int_\Omega W(\nabla u(x))\, dx + \varepsilon I(u).$$

Proposition 4.1. *For all* $\varepsilon > 0$, (P_ε) *admits a solution.*

Proof. Let $\{u_j\}$ be a minimizing sequence. By Poincaré's inequality we have

$$\| u_j - m \|_{L^2(\Omega)} \leqq \text{Const.}\, \| \nabla u_j \|_{L^2(\Omega)} \leqq \text{Const.}\, \| \nabla u_j \|_\infty,$$

and so, as $\{\| \nabla u_j \|_\infty\}$ is a bounded sequence, there exist $u \in H^1(\Omega)$ and a subsequence $\{u_m\}$ such that

$$u_m \rightharpoonup u \quad \text{weakly in } H^1(\Omega),$$

$$u_m \to u \quad \text{strongly in } L^2(\Omega)$$

and (see BALL [2])

$$(\nabla u_m, \operatorname{adj} \nabla u_m, \det \nabla u_m) \rightharpoonup (\nabla u, \operatorname{adj} \nabla u, \det \nabla u) \quad \text{in } L^\infty(\Omega) \text{ weak*}.$$

Moreover, with u^* as in Proposition 3.1, for m sufficiently large we have

$$\varepsilon I(u_m) \leqq E_\varepsilon(u^*) + 1 = \varepsilon\, \mathrm{Per}_\Omega\, (\{x \in \Omega \mid x \cdot n \geqq \alpha\}) + 1,$$

and so the sequence

$$\{\chi_{\Omega_1(u_m)}\} \quad \text{is bounded in BV}\,(\Omega).$$

By (2.2) and (2.3) there exists a subsequence $\{u_k\}$ and a subset ω of Ω such that

$$\chi_{\Omega_1(u_k)} \to \chi_\omega \quad \text{in } L^1(\Omega) \text{ strong}$$

and

$$\mathrm{Per}_{\Omega}(\omega) \leqq \liminf_{k\to\infty} I(u_k). \tag{4.2}$$

Hence, as D_1 is a closed convex set, and since

$$\chi_{\Omega_1(u_k)} \nabla u_k \in D_1 \text{ a.e.}$$

and

$$\chi_{\Omega_1(u_k)} \nabla u_k \to \chi_\omega \nabla u \quad \text{in } L^\infty(\Omega) \text{ weak*},$$

it follows that

$$\chi_\omega \nabla u \in D_1 \text{ a.e.}$$

and, in a similar way,

$$(1 - \chi_\omega) \nabla u \in D_2 \text{ a.e.}$$

Therefore we conclude that

$$u \in \mathscr{A}_2.$$

Let

$$g_1(F, H, \delta) := \begin{cases} G_1(F, H, \delta) & \text{if } (F, H, \delta) \in M^{3\times 3} \times M^{3\times 3} \times (0, +\infty) \\ +\infty, & \text{otherwise.} \end{cases}$$

As g_1 is a nonnegative, convex and lower semicontinuous function, there exists an increasing sequence of piecewise affine convex nonnegative functions $g_{1,n}$ such that

$$g_1 = \sup_n g_{1,n}.$$

Since

$$0 \leqq \chi_{\Omega_1(u_k)} g_{1,n}(\nabla u_k, \operatorname{adj} \nabla u_k, \det \nabla u_k) \leqq C_n < +\infty \text{ a.e.}$$

there is a subsequence $\{u_{k,n}\}$ such that

$$\chi_{\Omega_1(u_{k,n})} g_{1,n}(\nabla u_{k,n}, \operatorname{adj} \nabla u_{k,n}, \det \nabla u_{k,n}) \to h_n \quad \text{in } L^\infty(\Omega) \text{ weak*}$$

and so

$$\chi_\omega g_{1,n}(\nabla u_{k,n}, \operatorname{adj} \nabla u_{k,n}, \det \nabla u_{k,n}) \to h_n \quad \text{in } L^\infty(\Omega) \text{ weak*}.$$

On the other hand, since

$$(\nabla u_{k,n}, \operatorname{adj} \nabla u_{k,n}, \det \nabla u_{k,n}) \to (\nabla u, \operatorname{adj} \nabla u, \det \nabla u) \quad \text{in } L^\infty(\Omega) \text{ weak*}$$

and $g_{1,n}$ is a convex nonnegative function, we have that, after extracting a subsequence of $\{u_{k,n}\}$,

$$g_{1,n}(\nabla u_{k,n}, \operatorname{adj} \nabla u_{k,n}, \det \nabla u_{k,n}) \to L_n \geqq g_{1,n}(\nabla u, \operatorname{adj} \nabla u, \det \nabla u) \quad \text{in } L^\infty(\Omega) \text{ weak*}.$$

Therefore

$$h_n = \chi_\omega L_n \geqq \chi_\omega g_{1,n}(\nabla u, \operatorname{adj} \nabla u, \det \nabla u) \quad \text{a.e. in } \Omega.$$

We conclude that

$$\int_{\Omega} \chi_\omega g_{1,n}(\nabla u, \operatorname{adj} \nabla u, \det \nabla u)\, dx \leqq \liminf \int_{\Omega_1(u_k)} W(\nabla u_k)\, dx$$

and so, by Lebesgue's monotone convergence theorem,

$$\int_{\omega} g_1(\nabla u, \operatorname{adj} \nabla u, \det \nabla u)\, dx \leqq \liminf \int_{\Omega_1(u_k)} W(\nabla u_k)\, dx \tag{4.3}$$

and, in a similar way,

$$\int_{\Omega \setminus \omega} g_2(\nabla u, \operatorname{adj} \nabla u, \det \nabla u)\, dx \leqq \liminf \int_{\Omega \setminus \Omega_1(u_k)} W(\nabla u_k)\, dx. \tag{4.4}$$

Therefore, $\det \nabla u > 0$ a.e. in Ω and (4.2), (4.3) and (4.4) yield

$$E_\varepsilon(u) \leqq \liminf E_\varepsilon(u_k).$$

Theorem 4.5. *Let v_ε be a sequence of minimizers of E_ε in $\mathscr{A}_2$. There exist a solution u of* (P_0) *and a subsequence u_ε such that*

$$u_\varepsilon \to u \quad \text{in } W^{1,\infty}(\Omega; \mathbb{R}^3) \text{ weak}^*,$$

and

$$\nabla u = B + \chi_E a \otimes n \quad \text{a.e. in } \Omega,$$

where $E \in \mathscr{S}$ is a solution of (P_0^*).

Proof. Let u^* be as in Remark 2.13. Since $\nabla v_\varepsilon \in D_1 \cup D_2$ a.e. and

$$\varepsilon I(v_\varepsilon) \leqq E_\varepsilon(v_\varepsilon) \leqq E_\varepsilon(u^*) = \varepsilon \text{ Const.}, \tag{4.6}$$

as in the proof of Proposition 4.1 we deduce that there is a function $u \in H^1(\Omega)$ a subsequence $\{u_\varepsilon\}$ such that

$$\begin{aligned} u_\varepsilon &\to u && \text{weakly in } H^1(\Omega), \\ u_\varepsilon &\to u && \text{strongly in } L^2(\Omega), \\ \nabla u_\varepsilon &\to \nabla u && \text{weak* in } L^\infty(\Omega), \\ \chi_{\Omega_1(u_\varepsilon)} &\to \chi_\omega && \text{in } L^1(\Omega) \text{ strong}, \end{aligned}$$

$$\operatorname{Per}_\Omega(\omega) \leqq \liminf I(u_\varepsilon) \tag{4.7}$$

and

$$\int_{\Omega} W(\nabla u(x))\, dx \leqq \liminf \int_{\Omega} W(\nabla u_\varepsilon(x))\, dx, \tag{4.8}$$

where

$$u \in \mathscr{A}_2, \quad \chi_\omega \nabla u \in D_1 \text{ a.e.} \quad \text{and} \quad (1 - \chi_\omega) \nabla u \in D_2 \text{ a.e.}$$

By (4.6) and (4.8) we have

$$W(\nabla u(x)) = 0 \quad \text{a.e. in } \Omega$$

and so

$$\nabla u(x) = B + \chi_\omega(x)\, a \otimes n \quad \text{a.e. } x \in \Omega$$

and, by Theorem 2.14, $\omega \in \mathscr{S}$. Let $E' \in \mathscr{S}$ and, by Proposition 2.16, let $v \in W^{1,\infty}(\Omega; \mathbb{R}^3) \cap \mathscr{A}_0$ be such that

$$\nabla v = B + \chi_{E'} a \otimes n.$$

If $\mathrm{Per}_\Omega(E') < +\infty$ then $v \in \mathscr{A}_2$ and

$$E_\varepsilon(v) = \varepsilon \,\mathrm{Per}_\Omega(E') \geqq E_\varepsilon(u_\varepsilon) \geqq \varepsilon I(u_\varepsilon).$$

Therefore, by (4.7) we conclude that

$$\mathrm{Per}_\Omega(E') \geqq \liminf I(u_\varepsilon) \geqq \mathrm{Per}_\Omega(\omega).$$

5. Regularization and Direct Penalization of the Interfaces

Suppose that W satisfies the hypotheses (H1) and (H2) and consider (P_0) and (P_ε) as in Section 3. In a similar way, we obtain existence of solutions v_ε of the problem (P_ε), with

$$u_\varepsilon \to u \quad \text{in } W^{1,3/2}(\Omega; \mathbb{R}^3) \text{ weak},$$

where $\{u_\varepsilon\}$ is a subsequence of $\{v_\varepsilon\}$ and u is a solution (P_0). Thus, $u \in \mathscr{A}_1, \nabla u \in \mathrm{BV}(\Omega)$ and, by (H2),

$$\nabla u(x) = R(x)\,(B + \chi_E(x)\, a \otimes n) \quad \text{a.e. in } \Omega, \text{ where } R(x) \in O^+(\mathbb{R}^3).$$

We search for a model implying that

$$\nabla u \in \{A, B\} \quad \text{a.e. in } \Omega \text{ and } E \text{ is a solution of } (\mathrm{P}_0^*). \tag{5.1}$$

We show that (5.1) is valid if E is a sufficiently smooth set. Assume that E determines a partition of Ω into countably many strongly Lipschitz-connected subdomains Ω_i. Then

$$\nabla u = \sum_i R_i(B + \chi_{\Omega_i \cap E} a \otimes n), \text{ with } R_i \text{ constant rotations}.$$

Here, we used the following result (see Reshetnyak [14] and Appendix Proposition A.1).

Proposition 5.2. *Let Ω be an open bounded connected domain of $\mathbb{R}^n$ and let $u \in W^{1,\infty}(\Omega; \mathbb{R}^n)$ be such that $\nabla u \in O^+(\mathbb{R}^n)$ a.e. in Ω. Then u is an affine function.*

By Theorem 2.14, if Ω_i and Ω_j are such that $\mathrm{int}\,(\Omega_i \cup \Omega_j \cup (\partial\Omega_i \cap \partial\Omega_j))$ is a connected set, then

$$\nabla u = \chi_{\Omega i} R_i B + \chi_{\Omega j} R_j (B + a \otimes n) \quad \text{in } \Omega_i \cup \Omega$$

with the interface between Ω_i and Ω_j planar and

$$R_j(B + a \otimes n) = R_i B + b \otimes m, \tag{5.3}$$

for some $b \in \mathbb{R}^3$, $m \in \mathbb{R}^3$, $\|m\| = 1$.

JAMES [10] studied the condition (5.3) in the context of elastic crystals. As a consequence of his analysis, we have the following result (see Appendix Lemma A.5).

Lemma 5.4 (JAMES [10]). *Given* a *and* $n \in \mathbb{R}^3$, $a \cdot n > -1, a \neq 0$ *and* $\|n\| = 1$, *let* $R \in O^+(\mathbb{R}^3)$, $b \in \mathbb{R}^3$, $m \in \mathbb{R}^3$, $\|m\| = 1$, *satisfy the equation*

$$R(\mathbb{1} + a \otimes n) = \mathbb{1} + b \otimes m.$$

Then

(i) $R = \mathbb{1}$ *and* m *is parallel to* n *if* a *is parallel to* n;

(ii) *If* a *is not parallel to* n, *either* $R = \mathbb{1}$ *and* m *is parallel to* n, *or* $R^2 \neq \mathbb{1}$ *and* $R = (\mathbb{1} + b \otimes m)(\mathbb{1} + a \otimes n)^{-1}$ *where* $b \otimes m$ *is uniquely defined,* $b \in$ Span $\{a, n\}$ *and*

$$m = \pm \frac{2a + \|a\|^2 n}{\|2a + \|a\|^2 n\|}.$$

Clearly, (5.3) is equivalent to

$$R_i^T R_j(\mathbb{1} + a \otimes B^{-T}n) = \mathbb{1} + R_i^T b \otimes B^{-T}m.$$

First case: a is parallel to $B^{-T}n$.

By Lemma 5.4 we conclude that m is parallel to n and $R_i = R_j$. Thus there is a fixed rotation R such that

$$\nabla u(x) = R(B + \chi_E(x)\, a \otimes n) \quad \text{a.e. in } \Omega,$$

and so, by Theorem 2.14 and due to the constraint

$$\int_\Omega \nabla u(x)\, dx = \theta A + (1 - \theta) B,$$

we conclude that $E \in \mathcal{S}$ and $R = \mathbb{1}$. Hence

$$\nabla u \in \{A, B\} \quad \text{a.e. in } \Omega$$

and (see Theorem 3.2) E is a solution of (P_0^*).

Second case: a is not parallel to $B^{-T}n$.

By Lemma 5.4, either $R_i = R_j$ and m is parallel to n, or

$$R_j = R_i Q,$$

where

$$Q := (\mathbb{1} + b' \otimes B^{-T}m)(\mathbb{1} + a \otimes B^{-T}n)^{-1}$$

with $b' \in$ Span $\{a, B^{-T}n\}$ uniquely defined and

$$m = \frac{2B^T a + \|a\|^2 n}{\|2B^T a + \|a\|^2 n\|}.$$

Hence, locally either E is layered normally to n or E is layered normally to m. In the next lemma we prove that the boundary of E in Ω cannot have "corners".

Lemma 5.5.

$$\partial E \cap \Omega = \bigcup_{i=1}^{\infty} \pi_i \cap \pi'_i,$$

where, for some ϱ_i, $\tau_i \in \mathbb{R}$, π_i *is a connected component of* $\Omega \cap \{x \cdot n = \varrho_i\}$, π'_i *is a connected component of* $\Omega \cap \{x \cdot m = \tau_i\}$ *and* $H_1(\pi_i \cap \pi'_i) = 0$.

Proof. Suppose that there exist $x_0 \in \Omega$, $\varepsilon > 0$, ϱ and τ such that either

(i) $B(x_0, \varepsilon) \cap E = \{x \in B(x_0, \varepsilon) \mid (x - x_0) \cdot n < \varrho \text{ and } (x - x_0) \cdot m < \tau\}$ (see Fig. 4)

or

(ii) $B(x_0, \varepsilon) \cap E = \{x \in B(x_0, \varepsilon) \mid [(x - x_0) \cdot n - \varrho] \cdot [(x - x_0) \cdot m - \tau] > 0\}$ (see Fig. 5).

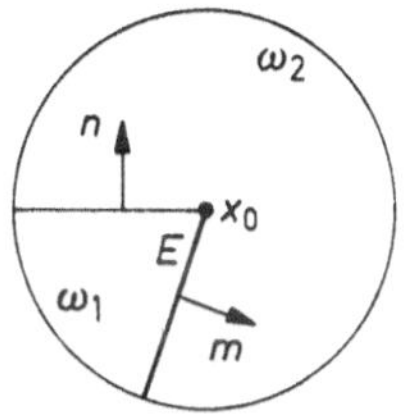

Fig. 4

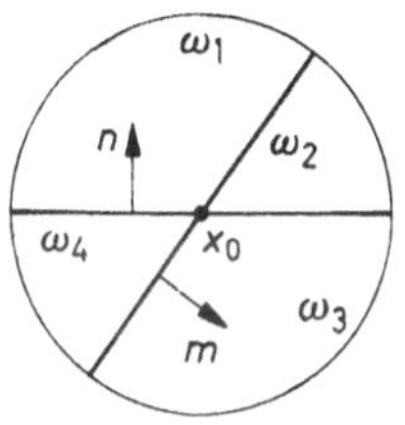

Fig. 5

In case (i) we have, using the notation of Fig. 4,

$$\nabla u = \begin{cases} R_1(B + a \otimes n) & \text{in } \omega_1 \\ R_2 B & \text{in } \omega_2 \end{cases}$$

and so there exist b and $b' \in \mathbb{R}^3$ such that

$$R_2 + b \otimes B^{-T} n = R_1(\mathbb{1} + a \otimes B^{-T} n) \tag{5.6$_1$}$$

and

$$R_2 + b' \otimes B^{-T} m = R_1(\mathbb{1} + a \otimes B^{-T} n). \tag{5.6$_2$}$$

Since two rotations cannot differ by a nonzero rank one matrix, $(5.6)_1$ implies that $R_1 = R_2$ and so by $(5.6)_2$ we have m parallel to n, in which case there are no corners. If (ii) holds then

$$\nabla u = \begin{cases} R_1 B & \text{in } \omega_1 \\ R_2(B + a \otimes n) & \text{in } \omega_2 \\ R_3 B & \text{in } \omega_3 \\ R_4(B + a \otimes n) & \text{in } \omega_4. \end{cases}$$

The necessary compatibility conditions imply that there exist b, b', c, $c' \in \mathbb{R}^3$ such that

$$R_1 + b \otimes B^{-T} m = R_2(\mathbb{1} + a \otimes B^{-T} n), \tag{5.7$_1$}$$

$$R_3 + c \otimes B^{-T} n = R_2(\mathbb{1} + a \otimes B^{-T} n), \tag{5.7$_2$}$$

$$R_3 + b' \otimes B^{-T} m = R_4(\mathbb{1} + a \otimes B^{-T} n) \tag{5.7$_3$}$$

and

$$R_1 + c' \otimes B^{-T}n = R_4(\mathbb{1} + a \otimes B^{-T}n). \qquad (5.7)_4$$

From $(5.7)_2$ and $(5.7)_4$ we have

$$R_2 = R_3 \quad \text{and} \quad R_1 = R_4,$$

which, together with $(5.7)_1$ and $(5.7)_3$ and Lemma 5.4, yield

$$R_1^T R_2 \quad \text{and} \quad R_2^T R_1 \in \{\mathbb{1}, Q\}.$$

If m is not parallel to n then $R_1 \neq R_2$ and so

$$R_1^T R_2 = R_2^T R_1 = Q$$

which implies that

$$Q^2 = \mathbb{1},$$

contradicting Lemma 5.4 (ii). We deduce that, as in case (i), m and n are parallel and $\partial E \cap \Omega$ does not have any "corners".

We conclude the proof of the second case (a is not parallel to $B^{-T}n$). By Lemma 5.5 we deduce that there is a constant rotation R such that

$$\nabla u \in \{RQ^k B, RQ^k(B + a \otimes n), RQ^{kT}B, RQ^{kT}(B + a \otimes n) \mid k \in \mathbb{N}_0\},$$

with

$$\sum_{k=0}^{\infty} \alpha_k RQ^k B + \beta_k RQ^k(B + a \otimes n) + \gamma_k RQ^{kT}B + \eta_k RQ^{kT}(B + a \otimes n)$$

$$= B + \theta a \otimes n,$$

where

$$\alpha_k, \beta_k, \gamma_k, \eta_k \geqq 0, \quad \sum_{k=0}^{\infty} (\alpha_k + \beta_k + \gamma_k + \eta_k) = 1,$$

$$\alpha_i + \beta_i = 0 \Rightarrow \alpha_j + \beta_j = 0 \quad \text{for all } j \geqq i$$

and

$$\gamma_i + \eta_i = 0 \Rightarrow \gamma_j + \eta_j = 0 \quad \text{for all } j \geqq i.$$

Suppose that $\alpha_1 + \beta_1 > 0$ and let $x \cdot n = 0$, $x \neq 0$. Then

$$\|Bx\| = \left\| \sum_{k=0}^{\infty} (\alpha_k + \beta_k) Q^k Bx + (\gamma_k + \eta_k) Q^{kT} Bx \right\|$$

$$\leqq \sum_{k=0}^{\infty} (\alpha_k + \beta_k + \gamma_k + \eta_k) \|Bx\| = \|Bx\|$$

and so, as $\alpha_0 + \beta_0$, $\alpha_1 + \beta_1 > 0$, we have

$$QBx = Bx \quad \text{for all } x \text{ such that } x \cdot n = 0.$$

As $\det Q = 1$ we conclude that $Q = \mathbb{1}$ which contradicts Lemma 5.4. Therefore we have

$$\alpha_i + \beta_i = 0 \quad \text{for all } i \geqq 1$$

and, in a similar way,

$$\gamma_i + \eta_i = 0 \quad \text{for all } i \geqq 1.$$

Finally, we conclude that

$$\nabla u(x) = R(B + \chi_E(x)\, a \otimes n) \text{ a.e. in } \Omega,$$

reducing this case to the first case.

The result that we have proved can be stated as follows.

Theorem 5.8. *Let* $u \in \mathscr{A}_1$ *be such that*

$$\nabla u(x) = R(x)\,(B + \chi_E(x)\, a \otimes n) \quad \text{a.e. } \textit{in } \Omega, \textit{ where } R(x) \in O^+(\mathbb{R}^3).$$

If the set E determines a partition of Ω *into countably many strongly Lipschitz-connected subdomains then*

$$\nabla u \in \{A, B\} \quad \text{a.e. } \textit{in } \Omega \quad \textit{and} \quad E \in \mathscr{S} \textit{ is a solution of } (\mathrm{P}_0^*).$$

Next, we try the same approach as in Section 4, with the hypothesis (H2) replaced by
(H2″):

if $F \in D_1$, then $W(F) = 0$ if and only if $F \in \{RA \mid R \in O^+(\mathbb{R}^3)\} \cap D_1$,

if $F \in D_2$, then $W(F) = 0$ if and only if $F \in \{RB \mid R \in \cap^+(\mathbb{R}^3)\} \cap D_2$.

As in Theorem 4.5 we obtain the existence of a subsequence $\{u_\varepsilon\}$ of solutions of (P_ε) such that

$$u_\varepsilon \to u \text{ in } W^{1,\infty}(\Omega; \mathbb{R}^3) \text{ weak*}$$

and

$$I(u) \leqq \liminf_{\varepsilon \to 0^+} I(u_\varepsilon).$$

where u is a solution (P_0). By (H2″) we have

$$\nabla u(x) = R(x)\,(B + \chi_E(x)\, a \otimes n) \text{ a.e. in } \Omega,$$

where $R(x) \in O^+(\mathbb{R}^3)$ and $\chi_E \in \mathrm{BV}(\Omega)$. In order to be able to conclude (5.1) using Theorem 5.8, we need to obtain more information regarding the set E. As we will see, this is possible if we consider a model that combines the approaches undertaken in Sections 3 and 4.

Let D_1 and D_2 be closed subsets of $\{F \in M^{3\times 3} \mid \det F \geqq 0\}$, with $A \in \operatorname{int} D_1$, $B \in \operatorname{int} D_2$ and $D_1 \cap D_2 = \emptyset$. Assume that W is continuous and satisfies (H1) and (H2″). We define the class of admissible deformations

$$\mathscr{A} := \Big\{ u \in W^{1,1}(\Omega; \mathbb{R}^3) \mid \nabla u \in D_1 \cup D_2 \text{ a.e.}, \nabla u \in \mathrm{BV}(\Omega), \int_\Omega u\, dx = m,$$

$$\int_\Omega \nabla u\, dx = B + \theta a \otimes n \Big\}.$$

Consider the approximating problems:
(P_ε) minimize in $\mathscr{A}$

$$E_\varepsilon(u) := \int_\Omega W(\nabla u(x))\,dx + \varepsilon \int_\Omega |D\nabla u(x)|\,dx + \varepsilon I(u).$$

Proposition 5.9. (P_ε) *admits a solution.*

Proof. Let $\{u_j\}$ be a minimizing sequence such that

$$E_\varepsilon(u_j) \leqq E_\varepsilon(u^*) + 1 = \varepsilon C^* + 1 \quad \text{for all } j,$$

where u^* is the particular deformation introduced in Remark 2.13. As

$$\int_\Omega |D\nabla u_j|\,dx \quad \text{and} \quad \mathrm{Per}_\Omega\,(\Omega_1(u_j)) \quad \text{are bounded}$$

by Proposition 2.4 we have

$$\{u_j\} \text{ is bounded in } W^{1,\frac{3}{2}},$$

$$\{\nabla u_j\} \text{ is bounded in } \mathrm{BV}\,(\Omega)$$

and

$$\{\chi_{\Omega_1(u_j)}\} \text{ is bounded in } \mathrm{BV}\,(\Omega).$$

Hence, there is a subsequence $\{u_m\}$ and there exists $u \in W^{1,3/2}$ such that

$$u_m \to u \text{ weakly in } W^{1,\frac{3}{2}},$$

$$u_m \to u \text{ strongly in } L^{\frac{3}{2}},$$

$$\nabla u_m \to \nabla u \text{ strongly in } L^1, \quad \nabla u \in \mathrm{BV}\,(\Omega)$$

and

$$\chi_{\Omega_1(u_m)} \to \chi_\omega \text{ strongly in } L^1, \text{ with } \chi_\omega \in \mathrm{BV}\,(\Omega).$$

Therefore, and since

$$\int_\Omega \chi_{\Omega_1(u_m)} |\nabla u_m - \chi_\omega \nabla u|\,dx \leqq \int_\Omega \chi_\omega\,|\nabla u_m - \nabla u|\,dx + \int_\Omega |\chi_{\Omega_1(u_m)} - \chi_\omega|\,|\nabla u_m|\,dx$$

$$\leqq \int_\Omega |\nabla u_m - \nabla u|\,dx + \left(\int_\Omega |\chi_{\Omega_1(u_m)} - \chi_\omega|\,dx\right)^{\frac{1}{3}} \left(\int_\Omega |\nabla u_m|^{3/2}\,dx\right)^{\frac{2}{3}},$$

it follows that

$$\chi_{\Omega_1(u_m)} \nabla u_j \to \chi_\omega \nabla u \text{ strongly in } L^1$$

which implies that

$$\chi_\omega \nabla u \in D_1 \quad \text{a.e.}$$

and, in a similar way,

$$(1 - \chi_\omega)\nabla u \in D_2 \quad \text{a.e.}$$

We conclude that $u \in \mathscr{A}$ and, by (2.2) and Fatou's Lemma,

$$E_\varepsilon(u) \leqq \liminf E_\varepsilon(u_m).$$

Theorem 5.10. *Let v_ε be a sequence of minimizers of E_ε in $\mathcal{A}$. There is a subsequence u_ε converging weakly in $W^{1,3/2}(\Omega;\mathbb{R}^3)$ to a solution u of* (P_0) *such that*

$$\nabla u(x) = R(x)\,(B + \chi_E(x)\, a \otimes n) \quad \text{a.e. in } \Omega.$$

Moreover, $u \in W^{1,\infty}(\Omega;\mathbb{R}^3)$ *and* $\nabla u, R, \chi_E \in \mathrm{BV}(\Omega)$.

Proof. Since

$$E_\varepsilon(v_\varepsilon) \leq E_\varepsilon(u^*) = \varepsilon C^*,$$

as in the proof of Proposition 5.9 we can extract a subsequence $\{u_\varepsilon\}$ and there exists $u \in \mathcal{A}$ such that

$$u_\varepsilon \to u \text{ weakly in } W^{1,\frac{3}{2}},$$

$$u_\varepsilon \to u \text{ strongly in } L^{\frac{3}{2}},$$

$$\nabla u_\varepsilon \to \nabla u \text{ strongly in } L^1,\ \nabla u \in \mathrm{BV}(\Omega),$$

$$\chi_{\Omega_1(u_\varepsilon)} \to \chi_E \text{ strongly in } L^1, \text{ with } \chi_E \in \mathrm{BV}(\Omega)$$

$$\chi_E \nabla u \in D_1 \text{ a.e. and } (1-\chi_E)\nabla u \in D_2 \quad \text{a.e.}$$

Since, by Fatou's Lemma,

$$\int_\Omega W(\nabla u(x))\,dx \leq \liminf \int_\Omega W(\nabla u_\varepsilon(x))\,dx = 0,$$

we deduce that

$$\nabla u(x) = R(x)\,(B + \chi_E(x)\, a \otimes n) \quad \text{a.e. in } \Omega.$$

Moreover, as

$$\nabla u \text{ and } \chi_E \in L^\infty \cap \mathrm{BV}(\Omega),$$

by Proposition 2.6 we conclude that

$$R \in L^\infty \cap \mathrm{BV}(\Omega).$$

Note that, since Ω is a strongly Lipschitz domain and as $\mathrm{Per}_\Omega(E) < +\infty$, E is a set of finite perimeter in $\mathbb{R}^3$. Also, by Theorem 2.10, to each entry R_{ij} of the rotation R correspond a set J_{ij} and functions $\lambda_{ij}, \mu_{ij}, \nu_{ij}$ with $\|\nu_{ij}\| = 1$, for which (2.11) and (2.12) hold.

Proposition 5.11. *For H_2 a.e.* $x \in \left(\partial^* E \setminus \bigcup_{i,j=1}^{3} J_{ij}\right) \cap \Omega$, *$\nu_E(x)$ is parallel to n.*

Proof. By (2.11), for H_2 a.e. $x_0 \in \left(\partial^* E \setminus \bigcup_{i,j=1}^{3} J_{ij}\right) \cap \Omega$ we have

$$\lim_{\varepsilon\to 0} \frac{1}{\mathrm{meas}\,(B(x_0,\varepsilon))} \int_{B(x_0,\varepsilon)} |R(x) - F(x_0)|^{\frac{3}{2}}\,dx = 0 \tag{5.12}$$

where $F(x_0)$ is the matrix with entries

$$F(x_0)_{ij} = \frac{\lambda_{ij} + \mu_{ij}}{2}.$$

For $\varepsilon > 0$ sufficiently small, define the difference quotient

$$v_\varepsilon(x) := \frac{u(x_0 + \varepsilon(x - x_0)) - u(x_0)}{\varepsilon} \quad \text{for } y \in B(x_0, 1).$$

As

$$\nabla v_\varepsilon(x) = R(x_0 + \varepsilon(x - x_0))\,(B + \chi_{x_0 + \frac{E - x_0}{\varepsilon}}(x)\, a \otimes n),$$

it follows that $\{v_\varepsilon\}$ is bounded in $W^{1,\infty}(B(x_0, 1))$, and so it admits a subsequence converging in $W^{1,\infty}(B(x_0, 1))$ weak* to a function v. By (5.12) and Theorem 2.8, we have

$$\nabla v(x) = F(x_0)\,(B + \chi_{H^-(x_0)}(x)\, a \otimes n) \tag{5.13}$$

where

$$H^-(x_0) = \{x \in \mathbb{R}^3 \mid \nu_E(x_0) \cdot (x - x_0) < 0\}.$$

We claim that

$$F(x_0) \text{ is a rotation.} \tag{5.14}$$

Indeed, by (5.12)

$$0 = \lim_{\varepsilon \to 0} \int_{B(x_0,1)} |R(x_0 + \varepsilon(x - x_0)) - F(x_0)|\, dx$$
$$= \lim_{\varepsilon \to 0} \int_{B(x_0,1)} |\mathbb{1} - R(x_0 + \varepsilon(x - x_0))^T F(x_0)|\, dx$$

and so

$$\left(\lim_{\varepsilon \to 0} \int_{B(x_0,1)} R(x_0 + \varepsilon(x - x_0))\, dx\right)^T F(x_0) = \text{meas}\,(B(x_0, 1))\, \mathbb{1}$$

which implies that

$$F^T(x_0)\, F(x_0) = \mathbb{1}.$$

Finally, by (5.13) and (5.14) we conclude that n is parallel to $\nu_E(x_0)$.

Proposition 5.15. *For* H_2 a.e. $x \in \partial^* E \cap J_{ij}$, $\nu_{ij}(x)$ *is parallel to* $\nu_E(x)$.

Proof. By (2.12), for H_2 a.e. $x_0 \in \partial^* E \cap J_{ij}$ and for all k, $l \in \{1, 2, 3\}$ we have

$$R_{kl}(x_0 + \varepsilon(x - x_0)) \to \chi_{H^+_{kl}(x_0)}(x)\, \mu_{kl}(x_0) + \chi_{H^-_{kl}(x_0)}(x)\, \lambda_{kl}(x_0) \text{ strongly in } L^1$$

where

$$H^+_{kl}(x_0) = \{x \in \mathbb{R}^3 \mid \nu_{kl}(x_0) \cdot (x - x_0) > 0\}$$

and

$$H^-_{kl}(x_0) = \{x \in \mathbb{R}^3 \mid \nu_{kl}(x_0) \cdot (x - x_0) < 0\}.$$

Suppose that $\nu_{ij}(x_0)$ is not parallel to $\nu_E(x_0)$. As $B(x_0, 1) \cap H^+(x_0)$ is sliced by the planes through x_0 with normal $\nu_{kl}(x_0)$, let ω be the union of the two adjacent slices ω^+ and ω^- with common boundary the portion of the plane with normal $\nu_{ij}(x_0)$. For $\varepsilon > 0$ sufficiently small consider the difference quotient

$$v_\varepsilon(x) := \frac{u(x_0 + \varepsilon(x - x_0)) - u(x_0)}{\varepsilon} \quad \text{for } x \in B(x_0, 1).$$

As in the proof of Proposition 5.11, $\{v_\varepsilon\}$ admits a subsequence converging in $W^{1,\infty}(B(x_0, 1))$ weak* to a function v, where, by Theorem 2.8 and (2.12),

$$\nabla v(x) = S(x)\,(B + \chi_{H^-(x_0)}(x)\, a \otimes n) \quad \text{in } B(x_0, 1)$$

and

$$S_{kl}(x) = \chi_{H^+_{kl}(x_0)}(x)\, \mu_{kl}(x_0) + \chi_{H^-_{kl}(x_0)}(x)\, \lambda_{kl}(x_0).$$

Therefore

$$\nabla v(x) = \begin{cases} S^+B & \text{in } \omega^+ \\ S^-B & \text{in } \omega^- \end{cases} \tag{5.16}$$

where

$$S^+_{kl} = \mu_{kl}(x_0), \quad S^-_{kl} = \lambda_{kl}(x_0) \quad \text{and } S^-_{ij} < S^+_{ij}. \tag{5.17}$$

We claim that

$$S^+ \text{ and } S^- \text{ are rotations.} \tag{5.18}$$

Indeed

$$\begin{aligned} 0 &= \lim_{\varepsilon \to 0} \int_{B(x_0,1) \cap \omega^+} |R(x_0 + \varepsilon(x - x_0)) - S^+|\, dx \\ &= \lim_{\varepsilon \to 0} \int_{B(x_0,1) \cap \omega^+} |\mathbb{1} - R^T(x_0 + \varepsilon(x - x_0))\, S^+|\, dx \end{aligned}$$

and so

$$\left(\lim_{\varepsilon \to 0} \int_{B(x_0,1) \cap \omega^+} R(x_0 + \varepsilon(x - x_0))\, dx\right)^T S^+ = \text{meas}\,(B(x_0, 1 \cap \omega^+))\, \mathbb{1}$$

i.e.

$$(S^+)^T S^+ = \mathbb{1} \quad \text{and, in a similar way} \quad (S^-)^T S^- = \mathbb{1}.$$

Finally, by (5.16) and (5.17) we conclude that S^+ and S^- differ by a nonzero rank one matrix, which contradicts (5.18).

Proposition 5.19.
(i) *If a is parallel to $B^{-T}n$ then $\nu_E(x)$ is parallel to n for H_2* a.e. $x \in \partial^* E \cap \Omega$.
(ii) *If a is not parallel to $B^{-T}n$, then $\nu_E(x)$ is either parallel to n or to m for H_2* a.e. $x \in \partial^* E \cap \Omega$.

Proof. By Proposition 5.11, for H_2 a.e. $x_0 \in \partial^* E \setminus \left(\bigcup_{i,j=1}^{3} J_{ij}\right) \cap \Omega$ we have $\nu_E(x_0)$ parallel to n. On the other hand, by Proposition 5.15 and (5.18), for H_2

a.e. $x_0 \in \partial^* E \cap \left(\bigcup_{i,j=1}^{3} J_{ij} \right)$ there exist rotations S^+ and S^- such that $S^+ \neq S^-$ and

$$R(x_0 + \varepsilon(x - x_0)) \to S^+ \text{ strongly in } L^1(B(x_0, 1) \cap H^+(x_0))$$

and

$$R(x_0 + \varepsilon(x - x_0)) \to S^- \text{ strongly in } L^1(B(x_0, 1) \cap H^-(x_0)).$$

Therefore, setting

$$v_\varepsilon(x) := \frac{u(x_0 + \varepsilon(x - x_0)) - u(x_0)}{\varepsilon} \quad \text{for } x \in B(x_0, 1),$$

$\{v_\varepsilon\}$ admits a subsequence converging in $W^{1,\infty}(B(x_0, 1))$ weak* to a function v such that

$$\nabla v(x) = \begin{cases} S^+ B & \text{if } \nu_E(x_0) \cdot (x - x_0) > 0 \\ S^-(B + a \otimes n) & \text{if } \nu_E(x_0) \cdot (x - x_0) < 0. \end{cases}$$

Thus there exists $b \in \mathbb{R}^3$ such that

$$S^-(\mathbb{1} + a \otimes B^{-T} n) = S^+ + b \otimes B^{-T} \nu_E(x_0)$$

and so, by Lemma 5.4 either a is parallel to $B^{-T}n$ and then $\nu_E(x)$ is parallel to n for H_2 a.e. $x \in \partial^* E \cap \Omega$ and $S^+ = S^-$ (*i.e.* H_2 a.e. $x \in \partial^* E$ is a point of approximate continuity of R) or a is not parallel to $B^{-T}n$ and then $\nu_E(x)$ is either parallel to n or to m for H_2 a.e. $x \in \partial^* E \cap \Omega$.

Corollary 5.20. *If a is parallel to $B^{-T}n$, then*

$$\nabla u = B + \chi_E a \otimes n \quad \text{a.e. in } \Omega,$$

where $E \in \mathcal{S}$ is a solution of (P_0^*).

Proof. By Proposition 5.19, $\nu_E(x)$ is parallel to n for H_2 a.e. $x \in \partial^* E \cap \Omega$. We can suppose, without loss of generality, that $n = e_3$. Hence, by the Generalized Gauss-Green Theorem 2.9,

$$\frac{\partial \chi_E}{\partial x_1} = 0 = \frac{\partial \chi_E}{\partial x_2} \quad \text{in } \mathcal{D}'(\Omega)$$

and so,

$$\frac{\partial}{\partial x_j}(\delta_{ki} + \chi_E a_k n_i) = \frac{\partial}{\partial x_i}(\delta_{kj} + \chi_E a_k n_j) \quad \text{for all } k, i, j \in \{1, 2, 3\}.$$

Therefore, there exists $v \in W^{1,\infty}(\Omega; \mathbb{R}^3)$ such that

$$\nabla v = \mathbb{1} + \chi_E a \otimes n \quad \text{a.e. in } \Omega$$

and, by Theorem 2.14, E is layered normally to n. As $\operatorname{Per}_\Omega(E) < +\infty$ and Ω is a strongly Lipschitz domain, if we set

$$\Omega_k := \{x \in \Omega \mid \operatorname{dist}(x, \partial\Omega) > 1/k\},$$

then $E \cap \Omega_k$ is formed by finitely many slices normal to n. By Proposition 5.2 in each one of these slices either

$$\nabla u = RB$$

or

$$\nabla u = R(B + a \otimes n),$$

with R a constant rotation. Due to the necessary jump condition of ∇u across an interface, for any two adjoint slices we have

$$R(B + a \otimes n) = R'B + b \otimes n$$

for some $b \in \mathbb{R}^3$. As a is parallel to $B^{-T}n$, from Lemma 5.4 we obtain

$$R = R'.$$

Using induction in k, with $k \to +\infty$, we obtain

$$\nabla u(x) = R(B + \chi_E(x)\, a \otimes n) \quad \text{a.e. in } \Omega,$$

for some fixed rotation R, and so (see proof of Theorem 5.8, first case) we conclude that $E \in \mathcal{S}$ is a solution of (P_0^*) and $R = \mathbb{1}$.

Final comments. I conjecture that the solutions obtained in Theorem 5.10 satisfy (5.1). However, I was able to confirm the conjecture only in the case where a is parallel to $B^{-T}n$ (see Corollary 5.20). If a is not parallel to $B^{-T}n$, by Theorem 5.8 the conjecture remains valid if E is sufficiently smooth. By Lemma 5.5, Theorem 5.10 and Proposition 5.19 we know that the set E has finite perimeter, the direction of the normal to $\partial^* E$ is either n or m, and, due to kinematic compatibility conditions, there cannot be either "corners" or "intersections" (see Fig. 3 and 4). Is it possible to infer that E is under the hypothesis of Theorem 5.8, namely that E determines a partition of Ω into countably many strongly Lipschitz connected domains?

Acknowledgment. This work was partially supported by the National Science Foundation under Grant No. DMS-8803315

Appendix

For completeness, in this appendix we prove Proposition 5.2 (see also RESHETNYAK [14]) and Lemma 5.4 (see JAMES [10]).

Proposition A.1. *Let Ω be an open bounded connected domain of $\mathbb{R}^n$ and let $u \in W^{1,\infty}(\Omega; \mathbb{R}^n)$ be such that $\nabla u \in O^+(\mathbb{R}^n)$ a.e. in Ω. Then u is an affine function.*

Proof. Let

$$R := \nabla u \in O^+(\mathbb{R}^n) \quad \text{a.e. in } \Omega \text{ and } \quad w_i := \nabla u_i, \quad i = 1, \ldots, n.$$

Since

$$R = \operatorname{adj} R$$

we have

$$\begin{aligned}\nabla u_i &= \operatorname{div} w_i \\ &= (\operatorname{adj} \nabla u)_{ij,j} \\ &= 0 \text{ in } \mathscr{D}'(\Omega).\end{aligned}$$

Therefore $u \in C^\infty(\Omega; \mathbb{R}^n)$ and so

$$R \in C^\infty(\Omega; \mathbb{R}^n). \tag{A.2}$$

Set

$$C^k := R^T \frac{\partial R}{\partial x_k} \quad \text{for } k = 1, \dots, n.$$

As

$$R^T R = \mathbb{1},$$

by (A.2) we have

$$C^{k^T} = C^k = 0. \tag{A.3}$$

Since

$$\frac{\partial^2 u_i}{\partial x_j\, \partial x_k} = \frac{\partial^2 u_i}{\partial x_k\, \partial x_j},$$

we obtain

$$R_{i,jk} = R_{i,kj}$$

i.e.

$$R_{im} C^k_{mj} = R_{is} C^j_{sk} \quad \text{for all } i, j, k \in \{1, \dots, n\}$$

and so

$$C^k_{mj} = C^j_{mk}. \tag{A.4}$$

Finally, (A.3) and (A.4) yield

$$C^k_{mj} = C^j_{mk} = -C^j_{km} = -C^m_{kj} = C^m_{jk} = C^k_{jm} = -C^k_{mj};$$

therefore

$$C^k = 0 \quad \text{for all } k \in \{1, \dots, n\}$$

and we conclude that

$$R(x) \text{ is constant in } \Omega.$$

Lemma A.5 (JAMES [10]). *Given a and $n \in \mathbb{R}^3$, $a \cdot n > -1$, $a \neq 0$ and $\|n\| = 1$, let $R \in O^+(\mathbb{R}^3)$, $b \in \mathbb{R}^3$, $m \in \mathbb{R}^3$, $\|m\| = 1$, satisfy the equation*

$$R(\mathbb{1} + a \otimes n) = \mathbb{1} + b \otimes m.$$

Then

(i) *$R = \mathbb{1}$ and m is parallel to n if a is parallel to n;*

(ii) *If a is not parallel to n, either $R = \mathbb{1}$ and m is parallel to n, or $R^2 \neq \mathbb{1}$ and $R = (\mathbb{1} + b \otimes m)(\mathbb{1} + a \otimes n)^{-1}$ where $b \otimes m$ is uniquely defined, $b \in$ Span$\{a, n\}$ and*

$$m = \pm \frac{2a + \|a\|^2 n}{\|2a + \|a\|^2 n\|}.$$

Proof. Suppose that

$$R(\mathbb{1} + a \otimes n) = \mathbb{1} + b \otimes m. \tag{A.6}$$

(i) If $a = \alpha n$ for some $\alpha \in \mathbb{R}$, by (A.6) we have

$$(2\alpha + \alpha^2)\, n \otimes n = b \otimes m + m \otimes b + \|b\|^2\, m \otimes m.$$

Therefore

$$m \text{ is parallel to } n$$

and since a rotation cannot differ from the identity by a nonzero rank one matrix, we conclude that

$$R = \mathbb{1}.$$

(ii) Assume that $R \neq \mathbb{1}$. Then

$$m \text{ is not parallel to } n \text{ and } Ra \text{ is not parallel to } b, \tag{A.7}$$

and so, if e is a unit vector on the axis of rotation of R, *i.e.*

$$Re = e = R^T e,$$

we have

$$Ra(n \cdot e) - b(m \cdot e) = 0$$

and

$$n(a \cdot e) - m(b \cdot e) = 0,$$

which, together with (A.7) imply

$$m \cdot e = 0, \quad n \cdot e = 0, \quad a \cdot e = 0 \quad \text{and} \quad b \cdot e = 0. \tag{A.8}$$

By part (i), a is not parallel to n and b is not parallel to m; thus we may set

$$e := \frac{a \wedge n}{\|a \wedge n\|}. \tag{A.9}$$

Since

$$a \otimes n + n \otimes a + \|a\|^2\, n \otimes n = b \otimes m + m \otimes b + \|b\|^2\, m \otimes m,$$

we obtain

$$a + (a \cdot n)\, n + \|a\|^2\, n = b(m \cdot n) + m(b \cdot n) + \|b\|^2\, (m \cdot n)\, m$$

and

$$n(a \cdot n \wedge e) = b(m \cdot n \wedge e) + m(b \cdot n \wedge e) + \|b\|^2\, m(m \cdot n \wedge e).$$

Thus we deduce that

$$2a \cdot n + \|a\|^2 = 2(b \cdot n)\,(m \cdot n) + \|b\|^2\, (m \cdot n)^2, \tag{A.10$_1$}$$

$$a \cdot n \wedge e = (b \cdot n \wedge e)\,(m \cdot n) + (m \cdot n \wedge e)\,(b \cdot n) + \|b\|^2\, (m \cdot n)\,(m \cdot n \wedge e) \tag{A.10$_2$}$$

and

$$0 = 2(b \cdot n \wedge e)\,(m \cdot n \wedge e) + \|b\|^2\, (m \cdot n \wedge e)^2. \tag{A.10$_3$}$$

By (A.7) and (A.8),

$$m \cdot n \wedge e \neq 0$$

and so, by (A.10$_1$–(A.10)$_3$ we have

$$m \cdot n = (2a \cdot n + \|a\|^2) \frac{(m \cdot n \wedge e)}{2a \cdot n \wedge e}$$

which, together with (A.8), implies that m has the direction of the vector

$$\frac{2a \cdot n + \|a\|^2}{2a \cdot n \wedge e} n + \frac{n \wedge e}{\|n \wedge e\|}.$$

Finally, by (A.9) we conclude that

$$m = \pm \frac{2a + \|a\|^2 n}{\|2a + \|a\|^2 n\|}.$$

Suppose that $R^2 = \mathbb{1}$ and let R' be the restriction of R to Span $\{n, n \wedge e\}$. Then R' is a rotation on a two-dimensional vector space and $R'^2 = \mathbb{1}$; therefore

$$R' = \pm \mathbb{1}.$$

As we are assuming that $R \neq \mathbb{1}$, we have

$$R' = -\mathbb{1}.$$

By (A.7) and (A.8) we have

$$n \notin \text{Span}\,\{m, e\};$$

thus it is possible to choose $x \in \mathbb{R}^3$ such that

$$m \cdot x = 0, \quad x \cdot e = 0 \quad \text{and} \quad n \cdot x \neq 0.$$

As

$$R = -n \otimes n - n \wedge e \otimes n \wedge e + e \otimes e,$$

by (A.6) and (A.8) we obtain

$$-x - (n \cdot x)\, a = x$$

which implies that

$$m \cdot a = 0$$

or

$$a \cdot n = -2,$$

contradicting the hypothesis

$$a \cdot n > -1.$$

We conclude that

$$R^2 \neq \mathbb{1}.$$

References

1. BALDO, S., "Minimal interface criterion for phase transitions in mixtures of Cahn-Hilliard fluids," to appear.
2. BALL, J. M., "Convexity conditions and existence theorems in nonlinear elastostatics," Arch. Rational Mech. Anal **63** (1977), 337–403.
3. BALL, J. M., & JAMES, R. D., "Fine phase mixtures as minimizers of energy," Arch. Rational Mech. Anal., **100** (1987), 13–52.
4. EVANS, L. C., & GARIEPY, R. F. *Lecture Notes on Measure Theory and Fine Properties of Functions.* Kentucky EPSCoR Preprint Series.
5. FONSECA, I., & TARTAR, L., "The gradient theory of phase transitions for systems with two potential wells," Proc. Royal Soc. Edin. **111 A** (1989), 89–102.
6. FONSECA, I., & TARTAR, L., in preparation.
7. GIUSTI, E. *Minimal Surfaces and Functions of Bounded Variation.* Birkhäuser, 1984.
8. GURTIN, M. E., "On phase transitions with bulk, interfacial, and boundary energy," Arch. Rational Mech. Anal. **87** (1986), 243–264.
9. GURTIN, M. E., "Some results and conjectures in the gradient theory of phase transitions," in Metastability and Incompletely Posed Problems. S. ANTMAN and others, eds., Springer-Verlag, 1987, 135–146.
10. JAMES, R. D., "Displacive phase transformations in solids," J. Mech. and Phys. Solids **34** (1986), 359–384.
11. KOHN, R., & STERNBERG, P., "Local minimizers and singular perturbations," Proc. Royal Soc. Edin. to appear.
12. MODICA, L., "Gradient theory of phase transitions and minimal interface criterion," Arch. Rational Mech. Anal. **98** (1987), 123–142.
13. OWEN, N. C., "Nonconvex variational problems with general singular perturbations," to appear in Trans. A. M. S.
14. RESHETNYAK, YU. G., "On the stability of conformal mappings in multidimensional spaces," Siberian Math. J. **8** (1967), 69–85.
15. STERNBERG, P., "The effect of a singular perturbation on nonconvex variational problems," Arch. Rational Mech. Anal. **101** (1988), 209–261.

Department of Mathematics
Carnegie Mellon University
Pittsburgh

(Received December 13, 1988)

Drops of Nematic Liquid Crystals

EPIFANIO G. VIRGA

Dedicated to Bernard D. Coleman on the occasion of his sixtieth birthday

Table of Contents

Introduction

The problem of finding the equilibrium shapes of a drop of nematic liquid crystal in contact with an isotropic fluid is fascinating, but hard to solve. I attempt here to crack this problem.

Following the pioneers, in Sect. 1 I write the energy functional. Sect. 2 is devoted to the equations which govern the equilibrium configurations of the drop in a class of admissible shapes whose boundaries may have sharp edges.

In Sect. 3 a heuristic argument leads to the statement of two limiting problems which apply, respectively, when the drop is *small* or *large*, in a sense which does not pertain to the geometric size only. The former problem was first considered by CHANDRASEKHAR (1966), while the latter has been recently proposed by ERICKSEN (1988).

In Sect. 4 a standard surface potential is adopted which depends on a dimensionless parameter ω (*cf.* eq. (1.12) below). The equilibrium shapes of a *small* drop are then determined with the aid of a geometric construction due to WULFF (1901). A criterion is worked out which predicts some qualitative properties of the set delivered by WULFF's construction. Use of such a criterion shows that the shapes that have been observed so far are like those the theory predicts when $\omega > 1$.

In Sect. 5 I consider a *large* drop floating on a dense isotropic liquid and in contact with an isotropic fluid all around the emerging boundary. The surface potential in Sect. 4 is employed on the submerged boundary. Under this and further assumptions, the equilibrium shapes of the floating drop are determined for all values of ω. Once again, the shapes the theory predicts when $\omega > 1$ agree with those that have been observed.

I believe that there is enough evidence in this paper to foresee that phase transitions involving the shape are likely to be observed in drops of nematic liquid crystals.

1. Energy Functional

Nematic liquid crystals are incompressible liquids with a preferred direction for the propagation of light. Such a direction, often called the *optical axis*, is usually described as a unit vector, called the *director*. If $\mathscr{B}$ is the region of the three-dimensional Euclidean space $\mathscr{E}^3$ occupied by a nematic liquid crystal and $\mathscr{S}^2$ denotes the unit sphere of $\mathscr{V}^3 := \mathscr{E}^3 - \mathscr{E}^3$, then $u\colon \mathscr{B} \to \mathscr{S}^2$ is the *director field* of the crystal. In the theory of OSEEN (1933) and FRANK (1958) the bulk energy associated with the director field is the functional

$$\mathscr{E}_b[\mathscr{B}, u] := \int_{\mathscr{B}} \sigma(u, \nabla u),$$

where σ, the energy per unit volume, is a scalar-valued function satisfying the requirements:

$$\sigma(u_0, 0) = 0 \quad \text{for all } u_0 \in \mathscr{S}^2, \tag{1.2}$$

$$\sigma(-u, -\nabla u) = \sigma(u, \nabla u), \tag{1.3}$$

$$\sigma(Qu, Q\,\nabla u Q^T) = \sigma(u, \nabla u) \quad \text{for all } Q \in \mathscr{O}^3, \tag{1.4}$$

$\mathscr{O}^3$ being the full orthogonal group of $\mathscr{V}^3$.

Let the mappings $u \mapsto K(u)$ and $u \mapsto \mathbb{K}(u)$ deliver, respectively, a second-order and a fourth-order tensor for all $u \in \mathscr{S}^2$. FRANK (1958) proved that among all functions σ of the form

$$\sigma(u, \nabla u) = K(u) \cdot \nabla u + \nabla u \cdot \mathbb{K}(u)\,[\nabla u], \tag{1.5}$$

the only one that satisfies requirements (2)–(4) is

$$\sigma_F(u, \nabla u) := \varkappa_1(\operatorname{div} u)^2 + \varkappa_2(u \cdot \operatorname{curl} u)^2 + \varkappa_3\,|u \times \operatorname{curl} u|^2$$
$$+ (\varkappa_2 + \varkappa_4)\,(\operatorname{tr}(\nabla u)^2 - (\operatorname{div} u)^2). \tag{1.6}$$

In (6) the coefficients $\varkappa_1$, $\varkappa_2$, $\varkappa_3$, and $\varkappa_4$ are material moduli depending on temperature (which here is constant). If $\varkappa_1 = \varkappa_2 = \varkappa_3 = \varkappa$ and $\varkappa_4 = 0$, then (6) becomes

$$\sigma_F = \varkappa\,|\nabla u|^2. \tag{1.7}$$

Although formula (6) is widely accepted, we do not confine attention to it. *We assume that* (i) σ *satisfies* (2)–(4), (ii) *there is a real number* $v > 1$ *such that*

$$\sigma(u, \mu \nabla u) = \mu^v \sigma(u, \nabla u) \quad \textit{for all } \mu \in R, \tag{1.8}$$

and that (iii)

$$\sigma(u, \nabla u) > 0 \quad \textit{unless } u \equiv u_0. \tag{1.9}$$

Of course, σ_F satisfies (8) with $v = 2$; σ_F satisfies also (9) if (see ERICKSEN (1966))

$$\varkappa_1 > 0, \quad \varkappa_2 > |\varkappa_4|, \quad \varkappa_3 > 0, \quad 2\varkappa_1 - \varkappa_2 - \varkappa_4 > 0.$$

When a liquid crystal is surrounded by an isotropic fluid at uniform pressure, following the pioneers (see FRIEDEL (1922) and OSEEN (1931, 1933)) we take the energy of the interface between the two fluids as

$$\mathscr{E}_s[\mathscr{B}, u] := \int_{\partial \mathscr{B}} \tau(u \cdot n), \tag{1.10}$$

where n is the outer normal to $\mathscr{B}$ and $\tau: [-1, 1] \to (0, +\infty)$ is an even function of class C^2. *We assume that* τ *is minimum at only one point of* $[0, 1]$:

$$\tau(\xi) > \tau(\xi_0) \quad \text{for all } \xi \in [0, 1] \setminus \{\xi_0\}. \tag{1.11}$$

The function defined by

$$\tau(\xi) := \tau_0(1 + \omega \xi^2) \quad \text{for all } \xi \in [-1, 1], \tag{1.12}$$

with $\tau_0 > 0$ and $\omega > -1$, is often employed in applications (*cf. e.g.* ERICKSEN (1976)). The values of τ_0 and ω depend on both the temperature and the nature of the fluids in contact.

Neglecting the work of body forces, we write the total energy of a drop of nematic liquid crystal surrounded by an isotropic fluid as

$$\mathscr{E}[\mathscr{B}, u] := \mathscr{E}_b[\mathscr{B}, u] + \mathscr{E}_s[\mathscr{B}, u]. \tag{1.13}$$

Each pair $(\mathscr{B}, u)$ is a *configuration* of the drop, and each set $\mathscr{B}$ is a *shape*. Since liquid crystals are incompressible, all shapes satisfy

$$\text{vol}(\mathscr{B}) = \beta, \tag{1.14}$$

where β is a positive given number and vol denotes Lebesgue measure. An *equilibrium configuration* is a configuration at which $\mathscr{E}$ attains an extremum; an equilibrium configuration is *stable* if it minimizes $\mathscr{E}$. When $(\mathscr{B}, u)$ is an equilibrium configuration we say that $\mathscr{B}$ is an *equilibrium shape*; when $(\mathscr{B}, u)$ is a stable equilibrium configuration we say that $\mathscr{B}$ is a *stable shape.*

Problem. *Minimize* $\mathscr{E}[\mathscr{B}, u]$ *subject to* (1.14).

This problem is still unsolved. Valuable progress towards its solution has recently been made by DE GIORGI & AMBROSIO (1987), AMBROSIO (1988a & b), and AMBROSIO, MORTOLA & TORTORELLI (1988).

2. Equilibrium Equations

We call Reg ($\mathscr{E}^3$) the collection of all bounded open regions of $\mathscr{E}^3$ whose boundaries are piecewise of class C^2. If $\mathscr{B} \in$ Reg ($\mathscr{E}^3$), we say that $\mathscr{B}$ is a *regular region* and denote by $\partial_s\mathscr{B}$ the union of all smooth surfaces of $\partial\mathscr{B}$. In this section we consider drops of nematic liquid crystals whose shapes are regular regions and whose director fields are C^2-mappings. *We assume that* $(u, \nabla u) \mapsto \sigma(u, \nabla u)$ *is a C^2-mapping satisfying requirements* (1.2)–(1.4). Under these hypotheses, we deduce the equations that govern the equilibrium configurations of the drop.

Let $\mathscr{E}^*$ be the functional

$$\mathscr{E}^*[\mathscr{B}, u] = \mathscr{E}[\mathscr{B}, u] + \pi_0 \operatorname{vol}(\mathscr{B}) \tag{2.1}$$

where π_0 is a Lagrange multiplier. Let a regular region $\mathscr{B}$ and a C^2-mapping $u: \mathscr{B} \to \mathscr{S}^2$ be given. We construct two *paths* of configurations *close* to $(\mathscr{B}, u)$. For every $\varepsilon \in [0, \varepsilon_0]$ let $f_\varepsilon: \mathscr{B} \to \mathscr{E}^3$ be a C^2-diffeomorphism of $\mathscr{B}$ defined by

$$f_\varepsilon(x) := x + \varepsilon v(x) \quad \text{for all } x \in \mathscr{B} \tag{2.2}$$

with $v: \mathscr{B} \to \mathscr{V}^3$ a mapping of class C^2. We denote by F_ε the gradient of f_ε:

$$F_\varepsilon := I + \varepsilon \nabla v, \tag{2.3}$$

I being the identity tensor, and set

$$\mathscr{B}_\varepsilon := f_\varepsilon(\mathscr{B}). \tag{2.4}$$

Clearly, $\mathscr{B}_\varepsilon \in$ Reg ($\mathscr{E}^3$) for every $\varepsilon \in [0, \varepsilon_0]$. It easily follows from (3) that

$$\det F_\varepsilon = 1 + \varepsilon \operatorname{div} v + o(\varepsilon).$$

Hence

$$\operatorname{vol}(\mathscr{B}_\varepsilon) = \operatorname{vol}(\mathscr{B}) + \varepsilon \int_{\partial\mathscr{B}} v \cdot n + o(\varepsilon),$$

where the Divergence Theorem has been used. We denote by n_ε the outer normal field to $\mathscr{B}_\varepsilon$. An easy computation (see *e.g.* VIRGA (1988) or FOSDICK & VIRGA (1989)) shows that

$$n_\varepsilon(f_\varepsilon(x)) = n(x) - \varepsilon(\nabla_\sigma v(x))^T n(x) + o(\varepsilon) \quad \text{for all } x \in \partial_s\mathscr{B}. \tag{2.5}$$

In (5) $\nabla_\sigma v$ is the *surface gradient* of v:

$$\nabla_\sigma v(x) := (\nabla v(x)) N(x) \quad \text{for all } x \in \partial_s\mathscr{B}, \tag{2.6}$$

where

$$N(x) := I - n(x) \otimes n(x) \tag{2.7}$$

is the projection onto the plane tangent to $\partial_s\mathscr{B}$ at x. Here definition (6) is equivalent to the intrinsic definition of surface gradient (see *e.g.* GURTIN & MURDOCH (1974) or NOLL & VIRGA (1989)).

For every $\varepsilon \in [0, \varepsilon_0]$ let the mapping $u_\varepsilon: \mathscr{B}_\varepsilon \to \mathscr{S}^2$ be defined by

$$u_\varepsilon(y) := u(f_\varepsilon^{-1}(y)) \quad \text{for all } y \in \mathscr{B}_\varepsilon. \tag{2.8}$$

It follows from (8) that u_ε is of class C^2 and, also in view of (3),

$$\nabla u_\varepsilon(f_\varepsilon(x)) = \nabla u(x)\, F_\varepsilon^{-1}(x)$$
$$= \nabla u(x) - \varepsilon(\nabla u(x))\,(\nabla v(x)) + o(\varepsilon) \quad \text{for all } x \in \mathscr{B}. \tag{2.9}$$

As ε spans the interval $[0, \varepsilon_0]$, $(\mathscr{B}_\varepsilon, u_\varepsilon)$ traces a path of configurations. Clearly,

$$(\mathscr{B}_\varepsilon, u_\varepsilon)\,|_{\varepsilon=0} = (\mathscr{B}, u).$$

We now look for a path of configurations having the shape $\mathscr{B}$ in common. Let $w\colon \mathscr{B} \to \mathscr{V}^3$ be a mapping of class C^2. For every $\eta \in [0, \eta_0]$ we define the mapping $u_\eta : \mathscr{B} \to \mathscr{S}^2$ by

$$u_\eta(x) := \frac{u(x) + \eta w(x)}{|u(x) + \eta w(x)|} \quad \text{for all } x \in \mathscr{B}. \tag{2.10}$$

The mapping u_η is of class C^2 and satisfies

$$u_\eta(x) = u(x) + \eta U(x)\, w(x) + o(\eta) \quad \text{for all } x \in \mathscr{B} \tag{2.11}$$

where

$$U(x) := I - u(x) \otimes u(x) \tag{2.12}$$

is the projection onto the plane orthogonal to $u(x)$. As η spans the interval $[0, \eta_0]$, $(\mathscr{B}, u_\eta)$ traces a path of configurations, and again

$$(\mathscr{B}, u_\eta)\,|_{\eta=0} = (\mathscr{B}, u).$$

We call *partial variations* of $\mathscr{E}^*$ at $(\mathscr{B}, u)$ the linear functionals

$$\delta_1\mathscr{E}^*(\mathscr{B}, u)\,[v] := \frac{d}{d\varepsilon}\,\mathscr{E}^*[\mathscr{B}_\varepsilon, u_\varepsilon]\,|_{\varepsilon=0}, \tag{2.13}$$

$$\delta_2\mathscr{E}^*(\mathscr{B}, u)\,[w] := \frac{d}{d\eta}\,\mathscr{E}^*[\mathscr{B}, u_\eta]\,|_{\eta=0}. \tag{2.14}$$

If $(\mathscr{B}, u)$ is an equilibrium configuration, both $\delta_1\mathscr{E}^*(\mathscr{B}, u)$ and $\delta_2\mathscr{E}^*(\mathscr{B}, u)$ must vanish.

The following facts are essential to our development.

Surface-Divergence Theorem. *Let $\mathscr{S} \subset \mathscr{E}^3$ be an orientable surface of class C^2 with normal field $n : \mathscr{S} \to \mathscr{S}^2$. If $h : \mathscr{S} \to \mathscr{V}^3$ is a C^1-mapping such that*

$$h \cdot n = 0 \quad \text{on } \mathscr{S},$$

then

$$\int_{\mathscr{S}} \operatorname{div}_s h = \int_{\partial_s\mathscr{S}} h \cdot \nu, \tag{2.15}$$

where $\partial_s\mathscr{S}$ is the union of all smooth curves in (clo $\mathscr{S}$)$\setminus\mathscr{S}$,

$$\operatorname{div}_s h(x) := \operatorname{tr}\,(\nabla_s h(x)) \quad \textit{for all } x \in \mathscr{S}$$

and $\nu : \partial_s\mathscr{S} \to \mathscr{S}^2$ is the outer normal field to $\partial_s\mathscr{S}$ such that

$$\nu(x) \cdot n(x) = 0 \quad \textit{for all } x \in \partial_s\mathscr{S}.$$

Similarly, if $H: \mathscr{S} \to \mathscr{L}^3$ *is a tensor-valued mapping of class* C^1 *such that*

$$Hn = 0 \quad on\ \mathscr{S},$$

then

$$\int_{\mathscr{S}} \operatorname{div}_s H = \int_{\partial_s \mathscr{S}} H\nu, \tag{2.16}$$

where $\operatorname{div}_s H$ *is the vector-valued mapping on* $\mathscr{S}$ *defined by*

$$\operatorname{div}_s H(x) \cdot e = \operatorname{div}_s (H^T(x)\, e) \quad for\ all\ x \in \mathscr{S}\ and\ e \in \mathscr{V}^3.$$

Main Identities. *If* $u: \mathscr{B} \to \mathscr{S}^2$ *is of class* C^1, *then*

$$(\nabla u)^T u = 0 \quad in\ \mathscr{B}. \tag{2.17}$$

Furthermore, if the mapping $(u, \nabla u) \mapsto \sigma(u, \nabla u)$ *is of class* C^1, *then* (see Appendix)

$$\frac{\partial \sigma}{\partial u} \cdot u = 0 \quad and \quad \left(\frac{\partial \sigma}{\partial \nabla u}\right)^T u = 0 \quad in\ \mathscr{B}. \tag{2.18}$$

Use of these facts leads us to

$$\begin{aligned} \delta_1 \mathscr{E}^*(\mathscr{B}, u)\,[v] = & \int_{\mathscr{B}} \left[\operatorname{div}\left((\nabla u)^T \frac{\partial \sigma}{\partial \nabla u}\right) - \nabla \sigma\right] \cdot v \\ & + \int_{\partial_s \mathscr{B}} \left[\left((\sigma + \pi_0)\, I - (\nabla u)^T \frac{\partial \sigma}{\partial \nabla u}\right) n - \operatorname{div}_s\, [(\tau I - \tau' n \otimes u)\, N]\right] \cdot v \\ & + \int_{\partial_s^2 \mathscr{B}} [(\tau I - \tau' n \otimes u)\, \nu] \cdot v, \end{aligned} \tag{2.19}$$

$$\delta_2 \mathscr{E}^*(\mathscr{B}, u)\,[w] = \int_{\mathscr{B}} U \left[\frac{\partial \sigma}{\partial u} - \operatorname{div}\left(\frac{\partial \sigma}{\partial \nabla u}\right)\right] \cdot w + \int_{\partial_s \mathscr{B}} U \left[\left(\frac{\partial \sigma}{\partial \nabla u}\right) n + \tau' n\right] \cdot w. \tag{2.20}$$

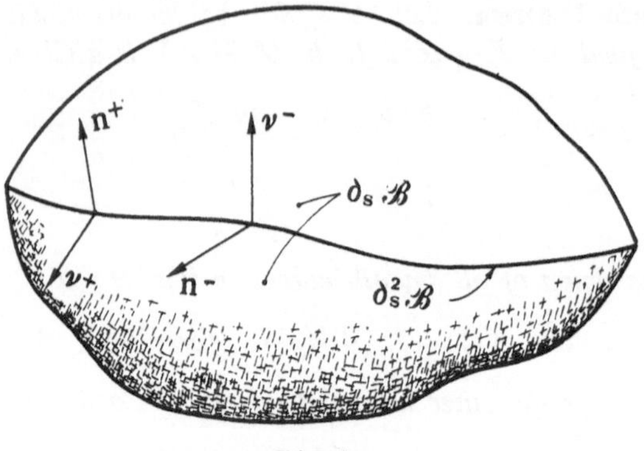

Fig. 1

In (19) the brackets [ ] denote the sum of the values taken by the enclosed field on the two sides adjacent to an edge of $\partial\mathcal{B}$ (see Figure 1):

$$[\Psi(n, \nu)] := \Psi(n^+, \nu^+) + \Psi(n^-, \nu^-),$$

where the astronomical symbol Ψ stands for any function depending on both n and ν. In both (19) and (20) $\partial_s^2\mathcal{B}$ is the union of all edges of $\partial\mathcal{B}$ and τ' denotes the first derivative of τ. We readily deduce from (19) and (20) the equilibrium equations and the natural boundary conditions of the drop:

$$\operatorname{div}\left(\sigma I - (\nabla u)^T \frac{\partial \sigma}{\partial \nabla u}\right) = 0 \tag{2.21}$$

and

$$\frac{\partial \sigma}{\partial u} - \operatorname{div}\left(\frac{\partial \sigma}{\partial \nabla u}\right) = \lambda u \quad \text{in } \mathcal{B}, \tag{2.22}$$

$$\left((\sigma + \pi_0)\, I - (\nabla u)^T \frac{\partial \sigma}{\partial \nabla u}\right) n = \operatorname{div}_s [(\tau I - \tau' n \otimes u)\, N] \tag{2.23}$$

and

$$\left(\frac{\partial \sigma}{\partial \nabla u} + \tau' I\right) n = \lambda u \quad \text{on } \partial_s\mathcal{B}, \tag{2.24}$$

and

$$[(\tau I - \tau' n \otimes u)\, \nu] = 0 \quad \text{on } \partial_s^2\mathcal{B}. \tag{2.25}$$

In (22) and (24) λ denotes an arbitrary scalar-valued function defined on the closure of $\mathcal{B}$; it has to be associated with the constraint on the length of u. Equation $(18)_2$ implies

$$\operatorname{div}\left(\frac{\partial \sigma}{\partial \nabla u}\right) \cdot u = -\frac{\partial \sigma}{\partial \nabla u} \cdot \nabla u.$$

Thus, by $(18)_1$ and (22), we get a formula for λ:

$$\lambda = \frac{\partial \sigma}{\partial \nabla u} \cdot \nabla u. \tag{2.26}$$

Equation (22) implies equation (21). A straightforward computation shows that (21) can be rewritten as

$$(\nabla u)^T \left[\frac{\partial \sigma}{\partial u} - \operatorname{div}\left(\frac{\partial \sigma}{\partial \nabla u}\right)\right] = 0$$

which by (17) is a consequence of (22). We do not discard (21) for it can be given a clearer mechanical interpretation than (22). Let T be the tensor-valued mapping on $\mathcal{B}$ defined by

$$T := -pI - (\nabla u)^T \frac{\partial \sigma}{\partial \nabla u} \qquad \text{with } p := -(\sigma + \pi_0). \tag{2.27}$$

Then (21) and (23) read, respectively,

$$\operatorname{div} T = 0 \quad \text{in } \mathscr{B}, \tag{2.28}$$

$$Tn = \operatorname{div}_s [(\tau I - \tau' n \otimes u) N] \quad \text{on } \partial_s \mathscr{B}, \tag{2.29}$$

and so T can be regarded as the stress tensor field of the drop (see ERICKSEN (1976), p. 261). Equations (29), (28), and (24) can also be found in § III.A of ERICKSEN (1976), while (25), the boundary condition along the edges of $\partial\mathscr{B}$, seems new.

3. Two Limiting Problems

Let ℓ_0 be the unit length in $\mathscr{E}^3$. We set $\beta := (\ell_0 \alpha)^3$ and define in $\mathscr{E}^3$ a change of variable $x \mapsto y(x)$ as follows

$$y(x) := \frac{x}{\alpha}. \tag{3.1}$$

Let a region $\mathscr{B}$ of volume β be given in Reg ($\mathscr{E}^3$). The region $\mathscr{B}_1 := y(\mathscr{B})$ is again in Reg ($\mathscr{E}^3$) and has unit volume. For every C^2-mapping $u: \mathscr{B} \to \mathscr{S}^2$ we define the mapping $\tilde{u}: \mathscr{B}_1 \to \mathscr{S}^2$ by

$$\tilde{u}(y) := u(\alpha y) \quad \textit{for all } y \in \mathscr{B}_1; \tag{3.2}$$

$\tilde{u}$ is again of class C^2 and

$$\nabla \tilde{u}(y) = \alpha \nabla u(\alpha y) \quad \textit{for all } y \in \mathscr{B}_1. \tag{3.3}$$

If σ satisfies (1.8), it is easily seen that the ratio of the bulk energy to the surface energy of a drop which occupies $\mathscr{B}$ is given by

$$\frac{\mathscr{E}_b[\mathscr{B}, u]}{\mathscr{E}_s[\mathscr{B}, u]} = \alpha^{(1-v)} \frac{\mathscr{E}_b[\mathscr{B}_1, \tilde{u}]}{\mathscr{E}_s[\mathscr{B}_1, \tilde{u}]}. \tag{3.4}$$

In the limit $\alpha \to 0^+$, that is when the drop is *small*, the bulk energy prevails over the surface energy, and so the total energy is minimum only if the bulk energy is. Since the bulk energy vanishes only if $u \equiv u_0$ (*cf.* (1.9) above), we are led to

Problem S. *Find* $\mathscr{B} \in$ Reg ($\mathscr{E}^3$) *such that the functional*

$$\mathscr{E}_S[\mathscr{B}] := \int_{\partial\mathscr{B}} \tau(u_0 \cdot n) \tag{3.5}$$

be minimum subject to

$$\operatorname{vol}(\mathscr{B}) = \beta. \tag{1.14$_r$}$$

Problem S is subsumed under a problem first considered by WULFF (1901). In the next section we shall examine Problem S in some detail.

Conversely, in the limit $\alpha \to \infty$, that is when the drop is *large*, the surface energy prevails over the bulk energy, and so the total energy is minimum only if the surface energy is. Thus hypothesis (1.11) on τ leads us to

Problem L. *Find* $\mathscr{B} \in \operatorname{Reg}(\mathscr{E}^3)$ *and* $u: \mathscr{B} \to \mathscr{S}^2$ *such that the functional*

$$\mathscr{E}_L[\mathscr{B}, u] := \int_{\mathscr{B}} \sigma(u, \nabla u) + \tau(\xi_0) \operatorname{area}(\partial \mathscr{B}) \tag{3.6}$$

be minimum subject to

$$u|_{\partial\mathscr{B}} \cdot n = \xi_0 \tag{3.7}$$

and to (1.14).

In (6) area denotes the surface measure, which here is taken as the two-dimensional Hausdorff measure. Problem L has been recently proposed by Ericksen (1988). In Sect. 5 the setting of Problem L will be employed to find the stable shapes of a drop floating on a dense liquid.

The heuristic argument which has led us to state both Problem S and Problem L gives rise to the question: Do *small* and *large* refer only to the geometric size of the drop? The answer is *no*. Here *small* and *large* have also a physical meaning. The drop cannot be either so small as to cause the continuum theory to fail or so large as not to be a drop anymore. The correct interpretation of the argument above relies on an accurate estimate of the ratio

$$\iota := \frac{\mathscr{E}_b[\mathscr{B}_1, \tilde{u}]}{\mathscr{E}_s[\mathscr{B}_1, \tilde{u}]}$$

occurring on the right-hand side of (4). If, for example, we employ formula (1.6) for σ and formula (1.12) for τ, then

$$\iota \sim \frac{\bar{k}}{\bar{\tau}} \ell_0,$$

where $\bar{k}$ is the order of magnitude of the moduli $\varkappa_i$s and $\bar{\tau}$ is the order of magnitude of τ_0. Using experimental data of Haller (1972) and Langevin (1972) for MBBA near 25°C (as reported by Stephen & Straley (1974)) we get $\iota \sim 10^{-8}$ when $\ell_0 = 1$ cm. Thus, the only conceivable drops of MBBA at room temperature are *large*. Nevertheless, we shall see in the next section that there are indeed nematic liquid crystals that give rise to *small* drops.

4. Small Drops

Wulff (1901) studied the growth of solid crystals. He assumed that the stable shapes of a solid crystal are the minimizers of the functional

$$\mathscr{E}_W[\mathscr{B}] := \int_{\partial\mathscr{B}} \chi(n) \tag{4.1}$$

subject to the isoperimetric constraint (1.14). In (1) $\chi: \mathscr{S}^2 \to R^+$ is a continuous positive function and n denotes the outer normal field to $\mathscr{B}$. Wulff conjectured that there was only one minimizer of $\mathscr{E}_W$ in $\operatorname{Reg}(\mathscr{E}^3)$: namely, the domain of volume β similar to

$$\mathscr{W}(\chi) := \{P \in \mathscr{E}^3 \mid (P - O) \cdot n \leqq \chi(n) \text{ for all } n \in \mathscr{S}^2\}, \tag{4.2}$$

where O is a point of $\mathscr{E}^3$. What WULFF conjectured is indeed a theorem. The major contributions to the proof were given by DINGHAS (1944), HERRING (1951) and TAYLOR (1974, 1978). A pretty and short proof can also be found in CHANDRASEKHAR (1966).

The set $\mathscr{W}(\chi)$ results from *Wulff's construction*. Select a point O in $\mathscr{E}^3$ and take a segment $OP(n)$ of length $\chi(n)$ along every direction $n \in \mathscr{S}^2$. Through every point $P(n)$ trace the plane $\pi(n)$ orthogonal to n and call $S(n)$ the half-space based on $\pi(n)$ that includes O. $\mathscr{W}(\chi)$ is then the intersection of all $S(n)$ as n spans $\mathscr{S}^2$. Clearly, WULFF's constructions delivers a convex set.

Since $\mathscr{E}_W = \mathscr{E}_S$ when $\chi(n) = \tau(u_0 \cdot n)$ for all $n \in \mathscr{S}^2$, Problem S can be solved by use of WULFF's construction. The corresponding set $\mathscr{W}(\tau)$ is axially symmetric about u_0 and symmetric with respect to the plane orthogonal to u_0 passing through O. Figure 2 illustrates a step of WULFF's construction on a plane containing u_0: the line $\ell(n)$ is the trace of $\pi(n)$ and the half-plane $\Sigma(n)$ is the cross-section of $S(n)$. The length of $OP(n)$ is $\tau(\cos\alpha)$ where α is the angle between n and u_0. The co-ordinates of the points where $\ell(n)$ intersects the direction of u_0 and the direction orthogonal to it passing through O are, respectively, $(0, \bar{y}(\alpha))$ and $(\bar{x}(\alpha), 0)$ (see Figure 2).

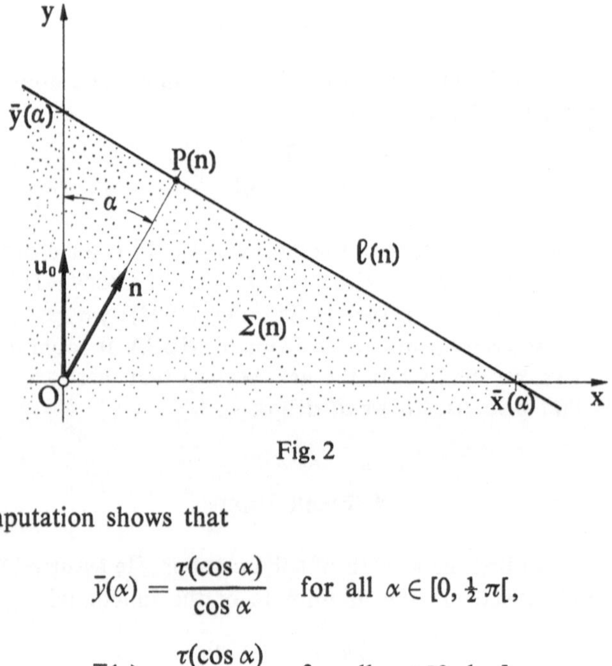

Fig. 2

An easy computation shows that

$$\bar{y}(\alpha) = \frac{\tau(\cos\alpha)}{\cos\alpha} \quad \text{for all } \alpha \in [0, \tfrac{1}{2}\pi[, \tag{4.3}$$

$$\bar{x}(\alpha) = \frac{\tau(\cos\alpha)}{\sin\alpha} \quad \text{for all } \alpha \in]0, \tfrac{1}{2}\pi]. \tag{4.4}$$

It follows from WULFF's construction that when the function $\bar{y}$ is decreasing near $\alpha = 0$, the normal field of $\mathscr{W}(\tau)$ is singular at poles: $\mathscr{W}(\tau)$ looks like a sharply pointed American football and is called a *tactoid*. If, on the contrary, $\bar{y}$ is not decreasing near $\alpha = 0$, then the poles of $\mathscr{W}(\tau)$ are regular points. Similarly, when $\bar{x}$ is increasing near $\alpha = \frac{1}{2}\pi$ the normal field of $\mathscr{W}(\tau)$ is singular along the equator: $\mathscr{W}(\tau)$ is *lens-shaped.* If $\bar{x}$ is not increasing near $\alpha = \frac{1}{2}\pi$, then the equator is not an edge of $\partial\mathscr{W}(\tau)$. Thus, if τ is of class C^2, we arrive at the following criterion:

If $\tau(1) < \tau'(1)$, *then* $\mathscr{W}(\tau)$ *is a tactoid*; *if* $\tau'(0) < 0$ *or* $\tau'(0) = 0$ *and* $\tau(0) < -\tau''(0)$, *then* $\mathscr{W}(\tau)$ *is lens-shaped. If the reverse inequalities hold, then* $\mathscr{W}(\tau)$ *is, respectively, regular at poles or along the equator.*

When τ is the function (1.12) this criterion ensures us that $\mathscr{W}(\tau)$ is lens-shaped for $-1 < \omega < -\frac{1}{2}$, is regular at poles and along the equator for $-\frac{1}{2} < \omega < 1$, and is a tactoid for $\omega > 1$. Figure 3 illustrates the cross-section of $\mathscr{W}(\tau)$ for various values of ω. We obtain the corresponding solution of Problem S by revolving these domains about u_0 and rescaling their volume.

Tactoids of nematic liquid crystals have indeed been observed by ZÖCHER (1925) in colloidal suspensions and by BERNAL & FANKUCHEN (1941) in preparations of plant virus. All the drops observed by both ZÖCHER and BERNAL & FANKUCHEN are much like the shape illustrated in Figure 3d. Thus formula (1.12) seems to be fit to serve as surface potential for nematic liquid crystals. It has also been shown in VIRGA (1987) that simple formulae for τ satisfying requirement (1.11), *but different from* (1.12), either do not lead to any tactoid or do lead to tactoids much different from that in Figure 3d. Using (1.12), we have predicted shapes that have not yet been observed (see Figures 3a–3c). Since ω depends on temperature, these shapes will indeed be observed only if the nematic phase persists in the range of temperature corresponding to $-1 < \omega < 1$.

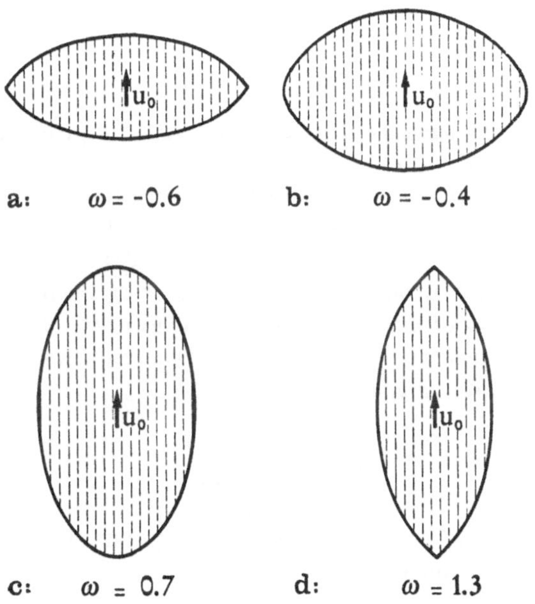

Fig. 3a–d

BERNAL & FANKUCHEN described the director field inside the drops they had seen; u was parallel to the axis of the drop in the core, but orthogonal to the normal at the boundary. Thus Figure 3d fails to represent the director field observed in the drops.

5. Floating Drops

Consider a drop of nematic liquid crystal floating on a dense isotropic liquid (say, gelatine) and in contact with an isotropic fluid (say, air). Suppose that it is a *large* drop according to the terminology of Sect. 3 and call $\mathscr{S}^+$ and $\mathscr{S}^-$, respectively, the emerged and submerged boundary of the drop. Suppose that $\mathscr{S}^+$ is so much larger than $\mathscr{S}^-$ as to make $\mathscr{S}^+$ and the whole boundary $\partial\mathscr{B}$ equivalent in the argument leading to Problem L in Sect. 3. Thus we rephrase Problem L as follows:

Problem L*. *Find* $\mathscr{B} \in \operatorname{Reg}(\mathscr{E}^3)$, $\mathscr{S}^+ \subset \mathscr{B}$ *and* $u\colon \mathscr{B} \to \mathscr{S}^2$ *such that the functional*

$$\mathscr{E}_L^*[\mathscr{B}, u] := \int_{\mathscr{B}} \sigma(u, \nabla u) + \tau^+(\xi_0) \operatorname{area}(\mathscr{S}^+) + \int_{\mathscr{S}^-} \tau^-(u \cdot n) \tag{5.1}$$

be minimum subject to

$$\partial\mathscr{B} = \mathscr{S}^+ \cup \mathscr{S}^-, \tag{5.2}$$

$$(u \cdot n)|_{\mathscr{S}^+} = \xi_0, \tag{5.3}$$

$$\operatorname{vol}(\mathscr{B}) = \beta, \tag*{(1.14)$_r$}$$

and

$$\operatorname{vol}(\mathscr{B}^-) = \beta^-, \tag{5.4}$$

where $\mathscr{B}^- \subset \mathscr{B}$ *is the submerged part of* $\mathscr{B}$, *and* β^- *a given positive number.*

In (1) τ^+ and τ^- are, respectively, the surface potential on $\mathscr{S}^+$ and $\mathscr{S}^-$.

Let $\xi_0 = 1$. There is much evidence that this is the case for many nematic liquid crystals in contact with air (*cf.* KLÉMAN (1983), p. 57).

We assume that $\mathscr{E}_L^*$ *is minimum when* (i) $\mathscr{B}$ *is symmetric about the axis* e_z *orthogonal to the plane of buoyancy*, (ii) $\mathscr{S}^+$ *belongs to the boundary of a ball* $\mathscr{B}_0$, (iii) *the submerged boundary of* $\mathscr{B}$ *is also the submerged boundary of* $\mathscr{B}^-$, (iv) *the director field in the drop is given by*

$$u_r(x) := \frac{x - O}{|x - O|}, \tag{5.5}$$

where O is the center of $\mathscr{B}_0$. Under these assumptions, Problem L* is reduced to minimizing the functional

$$\mathscr{E}_*^*[\mathscr{S}^-] := \int_{\mathscr{S}^-} \tau^-(u_r \cdot n) \tag{5.6}$$

among all surfaces $\mathscr{S}^-$ that are tangent to $\partial\mathscr{B}_0$ along a circle on the plane of buoyancy and satisfy (4).

We assume that the minimizer of $\mathscr{E}_\vartheta^*$ *does exist in the class of all piecewise smooth surfaces symmetric about* e_z. We represent the members of this class using polar co-ordinates (ϑ, ϱ), as is shown in Figure 4. The curve $\vartheta \mapsto \varrho(\vartheta)$ is the trace of $\mathscr{S}^-$ on each meridian plane of the drop. The unit vector e_r is the radial axis and n is the outer normal to $\mathscr{S}^-$. The co-ordinate ϑ varies in the interval $[0, \vartheta_0]$, $\frac{1}{2}\pi - \vartheta_0$ being the latitude of the circle where $\mathscr{S}^-$ is attached to $\partial\mathscr{B}_0$.

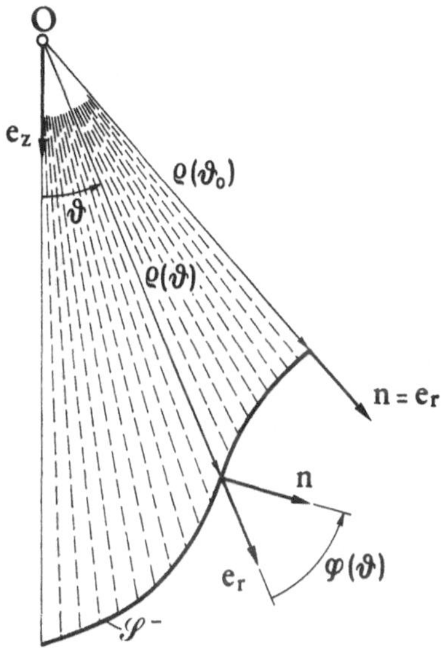

Fig. 4

Now both ϑ_0 and $\varrho_0 := \varrho(\vartheta_0)$ are regarded as prescribed; they are to be determined with the aid of the isoperimetric constraints (1.14) and (4). The function $\varphi\colon [0, \vartheta_0] \to [-\frac{1}{2}\pi, \frac{1}{2}\pi]$ delivers the angle between e_r and n (see Fig. 4). It is easy to represent the function $\vartheta \mapsto \varrho(\vartheta)$ in terms of the function $\vartheta \mapsto \operatorname{tg}\varphi(\vartheta)$ whenever the latter is integrable:

$$\varrho(\vartheta) = \varrho_0 e^{-\int_{\vartheta_0}^{\vartheta} \operatorname{tg}\varphi(\xi)d\xi}. \tag{5.7}$$

By (7) the energy $\mathscr{E}_\vartheta^*$ becomes a functional of φ:

$$\mathscr{I}[\varphi] := 2\pi\varrho_0^2 \int_0^{\vartheta_0} e^{-2\int_{\vartheta_0}^{\vartheta} \operatorname{tg}\varphi(\xi)d\xi} \sin\vartheta\psi(\cos\varphi(\vartheta))\, d\vartheta, \tag{5.8}$$

where the function $\psi\colon [-1, 1] \to R$ is defined by

$$\psi(\xi) := \frac{\tau^-(\xi)}{\xi}.$$

Here we employ formula (1.12) for τ^-, and so

$$\psi(\xi) = \tau_0 \frac{1 + \omega\xi^2}{\xi} \quad \text{with } \tau_0 > 0 \quad \text{and } \omega > -1. \tag{5.9}$$

Let $\vartheta_1 \in [0, \vartheta_0[$ be given, we define

$$\mathscr{C}(\vartheta_1) := \{\varphi\colon [0, \vartheta_0] \to [-\tfrac{1}{2}\pi, \tfrac{1}{2}\pi] \mid \varphi_{[\vartheta_1, \vartheta_0]} = 0, \quad \varphi|_{[0,\vartheta_1[} \in C^1([0, \vartheta_1[)\}.$$

$\mathscr{C}(0)$ is a singleton whose unique member is $\varphi = 0$; it represents a surface $\mathscr{S}^-$ which lies entirely on $\partial\mathscr{B}_0$, and so the corresponding shape of the drop is the ball $\mathscr{B}_0$ itself. Let $\varphi \in \mathscr{C}(\vartheta_1)$ with $\vartheta_1 \in]0, \vartheta_0[$. If $\varphi|_{[0,\vartheta_1[}$ is positive then $\mathscr{S}^-$ is a *bump* outside $\mathscr{B}_0$. If, on the contrary, $\varphi|_{[0,\vartheta_1[}$ is negative then $\mathscr{S}^-$ is a *crater* inside $\mathscr{B}_0$.

We assume that $\varphi \in \mathscr{C}(\vartheta_1)$ *for some* $\vartheta_1 \in [0, \vartheta_0[$. Let the function $\phi\colon [0, \vartheta_1] \to R$ be defined by

$$\phi(\vartheta) := \int_{\vartheta_1}^{\vartheta} \operatorname{tg} \varphi(\xi)\, d\xi \quad \text{for all } \vartheta \in [0, \vartheta_1]; \tag{5.10}$$

ϕ is of class C^2 where is finite, and it satisfies

$$\phi(\vartheta_1) = 0.$$

By inserting (10) and (9) into (8), we give $\mathscr{J}$ the form

$$\mathscr{J}[\varphi] = 2\pi\varrho_0^2\tau_0\mathscr{J}^*[\phi, \vartheta_1],$$

where

$$\mathscr{J}^*[\phi, \vartheta_1] := \int_0^{\vartheta_1} e^{-2\phi(\vartheta)} \frac{1 + \omega + (\phi'(\vartheta))^2}{[1 + (\phi'(\vartheta))^2]^{1/2}} \sin\vartheta\, d\vartheta + (1 + \omega)(\cos\vartheta_1 - \cos\vartheta_0).$$

The equilibrium shapes of the drops are described by the pairs (ϕ, ϑ_1) where $\mathscr{J}^*$ is stationary; an equilibrium shape is stable if it is described by a pair (ϕ, ϑ_1) that minimizes $\mathscr{J}^*$. It is easy to show that (ϕ, ϑ_1) describes an equilibrium shape if it satisfies

$$\frac{1 + \omega + (\phi'(\vartheta_1))^2}{[1 + (\phi'(\vartheta_1))^2]^{1/2}} = 1 + \omega, \tag{5.11}$$

$$\lim_{\vartheta \to 0^+} \left(e^{-2\phi(\vartheta)} \frac{1 - \omega + (\phi'(\vartheta))^2}{[1 + (\phi'(\vartheta))^2]^{3/2}} \phi'(\vartheta) \sin\vartheta \right) = 0, \tag{5.12}$$

and

$$\frac{d}{d\vartheta}\left(e^{-2\phi(\vartheta)} \frac{1 - \omega + (\phi'(\vartheta))^2}{[1 + (\phi'(\vartheta))^2]^{3/2}} \phi'(\vartheta) \sin\vartheta \right)$$
$$= -2e^{-2\phi(\vartheta)} \frac{1 + \omega + (\phi'(\vartheta))^2}{[1 + (\phi'(\vartheta))^2]^{1/2}} \sin\vartheta \quad \text{in } (0, \vartheta_1). \tag{5.13}$$

Equation (11) determines $\phi'(\vartheta_1)$. Let $h: R \to R$ be the function defined by

$$h(t) := \frac{1 + \omega + t^2}{(1 + t^2)^{1/2}} - (1 + \omega). \tag{5.14}$$

The graph of h is plotted in Figure 5. When $-1 < \omega \leqq 1$ h vanishes only at 0. When $\omega > 1$ h vanishes at $-\hat{t}$, 0, and $\hat{t}$, and it is minimum at $-t_0$ and t_0, being

$$\hat{t} := \sqrt{\omega^2 - 1} \quad \text{and} \quad t_0 := \sqrt{\omega - 1}. \tag{5.15}$$

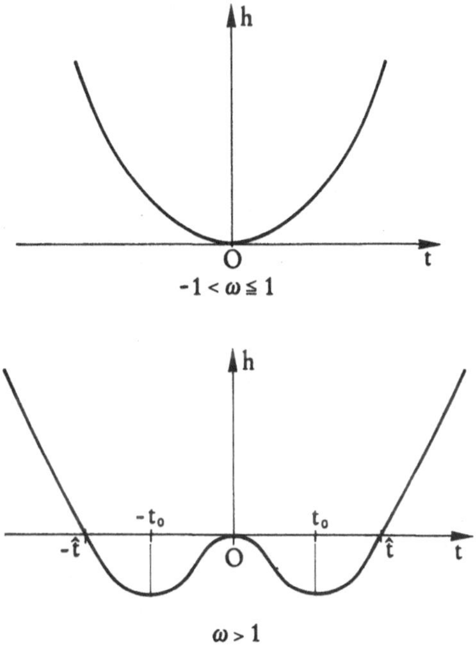

Fig. 5

Thus $\phi'(\vartheta_1)$ is uniquely determined when $-1 < \omega \leqq 1$, while it may take three different values when $\omega > 1$.

By use of (10) we see that (12) and

$$\lim_{\vartheta \to 0^+} e^{\int_{\vartheta}^{\vartheta_1} \operatorname{tg}\varphi(\xi)d\xi} (1 - \omega \cos^2 \varphi(\vartheta)) \sin \varphi(\vartheta) \sin \vartheta = 0, \tag{5.16}$$

are equivalent, while (13) can be written as

$$\begin{gathered} \varphi' \cos \varphi \, [\omega(1 - 3 \sin^2 \varphi) - 1] \sin \vartheta \\ - 2 \cos \varphi \, [1 + \omega(1 + \sin^2 \varphi)] \sin \vartheta + \sin \varphi \, (\omega \cos^2 \varphi - 1) \cos \vartheta = 0. \end{gathered} \tag{5.17}$$

Equation (17), subject to (16) and to

$$h(\operatorname{tg} \varphi(\vartheta_1)) = 0, \tag{5.18}$$

is the Euler-Lagrange equation of $\mathscr{J}$ in the class $\mathscr{C}(\vartheta_1)$.

It has been shown by ROCCATO & VIRGA (1989) that there is no solution of (17) and (18) in $\mathscr{C}(\vartheta_1)$ for all $\vartheta_1 \in]0, \vartheta_1[$ when $-1 < \omega < 1$ and that there is precisely one solution of (17) and (18) in $\mathscr{C}(\hat{\vartheta}_1)$ for precisely one $\hat{\vartheta}_1 \in [0, \vartheta_0[$ when $\omega > 1$. This solution, which we denote by $\hat{\varphi}$, describes a crater inside $\mathscr{B}_0$. It satisfies (16) and $\operatorname{tg} \hat{\varphi}(\hat{\vartheta}_1) = -\hat{t}$, $\hat{\varphi}|_{[0,\hat{\vartheta}_1[}$ is strictly decreasing, and

$$\operatorname{tg} \hat{\varphi}(0) = -t_0 \tag{5.19}$$

(*cf.* $(15)_2$). The angle $\hat{\vartheta}_1$ depends on ω, but does not depend on ϑ_0, which can be chosen in $]\hat{\vartheta}_1, \pi[$ so as to satisfy (4). Thus when $-1 < \omega \leqq 1$ there is only one equilibrium shape of the drop in the class we have employed here, that is the ball $\mathscr{B}_0$. When $\omega > 1$ there are two equilibrium shapes of the drop, $\mathscr{B}_0$ and $\hat{\mathscr{B}}$, the latter being the shape whose submerged boundary is described by $\hat{\varphi}$. As ω approaches 1, $\hat{\vartheta}_1$ approaches 0 (see ROCCATO & VIRGA (1989)), and so $\mathscr{B}_0$ and $\hat{\mathscr{B}}$ represent, in a sense, the branches of a bifurcation occurring at $\omega = 1$. To figure out which of the two shapes is stable we estimate the energy of both.

Let $\omega > 1$ and $\delta > 0$ be given. We define the function $\varphi_\delta : [0, \vartheta_0] \to [-\frac{1}{2}\pi, 0]$ by

$$\varphi_\delta(\vartheta) := \begin{cases} \hat{\varphi}(0) & \text{if } 0 \leqq \vartheta < \delta \\ 0 & \text{otherwise}, \end{cases} \tag{5.20}$$

(*cf.* $(15)_2$ and (19)). An easy computation shows that

$$\mathscr{I}[0] - \mathscr{I}[\varphi_\delta] = 2\pi\varrho_0^2\tau_0 \sqrt{\omega}\, \delta^2 \left(g(\omega) - 1\right) + o(\delta^2), \tag{5.21}$$

where the function $g: [1, \infty[\to R$ is defined by

$$g(\omega) := \frac{\omega + 1}{2\sqrt{\omega}} \tag{5.22}$$

The graph of g is plotted in Figure 6.

Since $g(\omega) > 1$ for all $\omega > 1$, we conclude that

$$\mathscr{I}[\varphi_\delta] < \mathscr{I}[0]$$

for all δ sufficiently small. Figure 7 illustrates qualitatively the stable shape of the drop when $\omega > 1$ and $\hat{\vartheta}_1$ is not too large (*cf.* ROCCATO & VIRGA (1989) for further details). By $(15)_2$ and (19), the angle γ is

$$\gamma := \sin^{-1} \frac{1}{\sqrt{\omega}}.$$

Floating drops having the shape illustrated in Fig. 7 have apparently been observed by P. PIÉRANSKI (*cf.* KLÉMAN (1983), p. xvi, plate III).

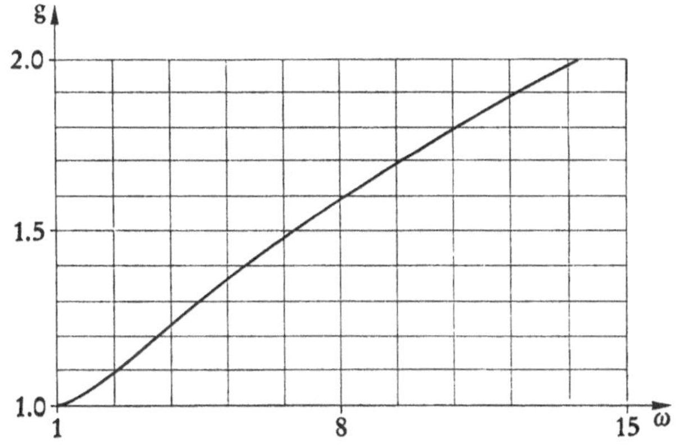

Fig. 6

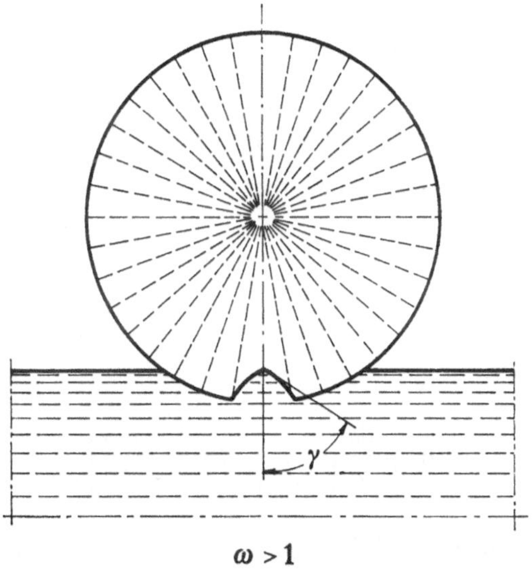

$\omega > 1$

Fig. 7

Acknowledgements. The research leading to this paper has been partially supported by a grant from the Italian Ministry of Education for the Research Project entitled: "Termomeccanica dei Continui". I am indebted to G. CAPRIZ, E. DE GIORGI, J. ERICKSEN, and D. ROCCATO for their helpful suggestions.

Appendix

Let $\mathscr{B}$ be an open and bounded subset of the three-dimensional Euclidean space $\mathscr{E}^3$. We denote by $\mathscr{V}^3$ the translation space of $\mathscr{E}^3$, by $\mathscr{S}^2$ the unit sphere of $\mathscr{V}^3$ and by $\mathscr{L}^3$ the space of all linear transformations of $\mathscr{V}^3$ into itself (second order tensors). For every $a \in \mathscr{S}^2$ we set

$$\mathscr{L}^3(a) := \{A \in \mathscr{L}^3 \mid A^T a = 0\},$$

where a superscript T denotes transposition, and define

$$\mathscr{L}_0 := \bigcup_{n \in \mathscr{S}^2} \mathscr{L}^3(n).$$

Let $\sigma: \mathscr{B} \times \mathscr{S}^2 \times \mathscr{L}_0 \to R$ be a C^1-mapping such that, for every $x \in \mathscr{B}$ and $u \in \mathscr{S}^2$, we have $\sigma(x, u, \cdot): \mathscr{L}^3(u) \to R$. Let $x \in \mathscr{B}$, $u \in \mathscr{S}^2$, and $D \in \mathscr{L}^3(u)$ be given and let $m: [0, 1] \to \mathscr{S}^2$, $M: [0, 1] \to \mathscr{L}_0$ be any pair of C^1-mappings such that

$$m(0) = u, \qquad M(0) = D, \tag{A.1}$$

$$(M(\eta))^T m(\eta) = 0 \quad \text{for all } \eta \in [0, 1]. \tag{A.2}$$

The partial derivatives $\frac{\partial \sigma}{\partial u}$ and $\frac{\partial \sigma}{\partial D}$ are the unique C^0-mappings of $\mathscr{B} \times \mathscr{S}^2 \times \mathscr{L}_0$ into $\mathscr{V}^3$ and $\mathscr{L}^3$, respectively, such that

$$\frac{d}{d\eta} \sigma(x, m(\eta), M(\eta)) \big|_{\eta=0} = \frac{\partial \sigma}{\partial u}(x, u, D) \cdot \dot{m}(0) + \frac{\partial \sigma}{\partial D}(x, u, D) \cdot \dot{M}(0), \tag{A.3}$$

where

$$\dot{m}(0) := \frac{dm}{d\eta}\bigg|_{\eta=0}, \qquad \dot{M}(0) := \frac{dM}{d\eta}\bigg|_{\eta=0}.$$

It easily follows from (1) and (2) that

$$\dot{m}(0) = U\dot{m}(0), \qquad \dot{M}(0) = U\dot{M}(0) - (u \otimes \dot{m}(0))\, D,$$

where

$$U := I - u \otimes u.$$

With the aid of these formulae we get from (3)

$$U \frac{\partial \sigma}{\partial u} = \frac{\partial \sigma}{\partial u}, \qquad U \frac{\partial \sigma}{\partial D} = \frac{\partial \sigma}{\partial D}. \tag{A.4}$$

Equations (4) and equations (2.18) in the text are equivalent.

References

AMBROSIO L. (1988a), A compactness theorem for a special class of functions of bounded variation, Publication no. 13, Scuola Normale Superiore, Pisa, *to appear in* Boll. Un. Mat. Ital.

AMBROSIO L. (1988b), Existence theory for a new class of variational problems, Publication no. 17, Scuola Normale Superiore, Pisa.

AMBROSIO L., S. MORTOLA & V. M. TORTORELLI (1988), Functionals with linear growth defined on vector-valued BV functions, Publication no. 14, Scuola Normale Superiore, Pisa.

BERNAL J. D. & I. FANKUCHEN (1941), X-ray and crystallografic studies of plant virus preparations, J. Gen. Physiol., **25**, 111–165.

CHANDRASEKHAR S. (1966), Surface tension of liquid crystals *in* Liquid crystals, BROWN, DINES & LABES *Eds.*, Gordon & Breach, New York, 331–340.

DE GIORGI E. & L. AMBROSIO (1987), Un nuovo tipo di funzionale del calcolo delle variazioni, *to appear in* Atti Acc. Lincei Rend. fis.

DINGHAS, A. (1944), Über einen geometrischen Satz von Wulff für die Gleichgewichtsform von Kristallen, Z. Kristallographie, **105**, 304–314.

ERICKSEN J. L. (1966), Inequalities in liquid crystal theory, Phys. Fluids, **9**, 1205–1207.

ERICKSEN J. L. (1976), Equilibrium theory of liquid crystals *in* Advances in liquid crystals, vol. II, GLENN & BROWN *Eds.*, Academic Press, New York, 233–298.

ERICKSEN J. L. (1988), Static theory of point defects in nematic liquid crystals, *forthcoming.*

FOSDICK R. L. & E. G. VIRGA (1989), A variational proof of the stress theorem of Cauchy, Arch. Rational Mech. Anal., **105**, 95–103.

FRANK F. C. (1958), On the theory of liquid crystals, Discussions Faraday Soc., **25**, 19–28.

FRIEDEL G. (1922), Les états mésomorphes de la matière, Ann. Phys. (Paris), **18**, 273–474.

GURTIN M. E. & A. I. MURDOCH (1974), A continuum theory of elastic material surfaces, Arch. Rational Mech. Anal., **57**, 291–323.

HALLER I. (1972), Elastic constants of nematic crystalline phase of MBBA, J. Chem. Phys., **57**, 1400–1405.

HERRING C. (1951), Some theorems on the free energy of crystals surface, Phys. Rev., **82**, 87–93.

KLÉMAN M. (1983), Points, lines and walls, J. Wiley & Sons, Chichester *etc.*

LANGEVIN D. (1972), Analyse spectrale de la lumière diffusée par la surface libre d'un crystal liquide nématique. Mesure de la tension superficielle et des coefficients de viscosité, J. Phys. (Paris), **33**, 249–256.

NOLL W. & E. G. VIRGA (1989), On edge interactions and surface tension, Arch. Rational Mech. Anal., *in press.*

OSEEN C. W. (1931), Probleme für die Theorie der anisotropen Flüssigkeiten, Z. Kristallographie, **79**, 173–185.

OSEEN C. W. (1933), The theory of liquid crystals, Trans. Faraday Soc., **29**, 883–899.

ROCCATO D. & E. G. VIRGA (1989), Drops of nematic liquid crystals floating on a fluid, *forthcoming.*

STEPHEN M. J. & J. P. STRALEY (1974), Physics of liquid crystals, Rev. Mod. Physics, **46**, 617–704.

TAYLOR J. E. (1974), Existence and structure of solutions to a class of nonelliptic variational problems, Symposia Mathematica, **14**, 499–508.

TAYLOR J. E. (1978), Crystalline variational problems, Bull. Amer. Math. Soc., **84**, 568–588.

Virga E. G. (1987), Forme di equilibrio di piccole gocce di cristallo liquido, Publication no. 562, Istituto Analisi Numerica, Pavia.

Virga E. G. (1988), Metastable equilibrium configurations of fluids with surface tension, Quart. Appl. Math., **46**, 217–228.

Wulff G. (1901), Zur Frage der Geschwindigkeit des Wachsthums und der Auflösung der Kristallflächen, Z. Kristallographie und Mineralogie, **34**, 449–530.

Zöcher H. (1925), Über freiwillige Strukturbildung in Solen, Z. Anorg. Chem., **147**, 91–110.

Dipartimento di Matematica
Università di Pavia
Pavia, Italy

(Received October 17, 1988)

Stability of Deformation of an Elastic Layer

Y. C. CHEN & K. R. RAJAGOPAL

Dedicated to Professor B. D. Coleman on his 60th Birthday

1. Introduction

ERICKSEN [1] proved all universal deformations of isotropic compressible elastic materials to be homogeneous. He also proved inhomogeneous universal deformations possible if the isotropic elastic material is incompressible (*cf.* ERICKSEN [2]). Recently, interest has grown in the study of inhomogeneous, not universal deformations of isotropic incompressible elastic materials. The importance of such deformations and the need to study is them discussed in some detail by RAJAGOPAL & CARROLL [3].

Recently, RAJAGOPAL & WINEMAN [4] have studied the inhomogeneous deformation of non-linear elastic layers of Neo-Hookean and Mooney-Rivlin materials. They found a one-parameter family of inhomogeneous exact solutions for the problem of uniaxial extension of the layer. They also obtained a similar result for the problem of the rotation of the layer about distinct axes. The classical solutions for homogeneous uniaxial extension and for torsion are embedded in these families of solutions, and it is natural to ask whether the inhomogeneous solutions exhibited by RAJAGOPAL & WINEMAN [4] are stable. Since the domain of deformation is unbounded, the analysis of the problem is not expected to be straightforward.

In this paper we study the stability, for Neo-Hookean materials, of the inhomogeneous deformations obtained by RAJAGOPAL & WINEMAN [4]. Since the domain is unbounded, we restrict our analysis to the class of disturbances which have compact support*. Thus, though the domain itself is unbounded, the perturbation is associated with a compact domain. Such a class of disturbances seems especially appropriate for problems involving displacments of fixed value on the boundaries, in this case at infinity.

---

* We could also study the problem within the context of disturbances which decay faster than $r^{-(2+\varepsilon)}$, $\varepsilon > 0$, as $r \to \infty$.

2. Preliminaries

Consider an infinite layer of thickness $2h$ that occupies the domain $\mathscr{B}$ in its reference configuration. Let $\boldsymbol{X}$ and $\boldsymbol{x}$ denote the positions of a particle in the reference configuration and the deformed configuration, respectively, and let $\boldsymbol{F}$ be the deformation gradient

$$\boldsymbol{F} = \frac{\partial \boldsymbol{x}}{\partial \boldsymbol{X}}. \tag{1}$$

We shall consider only homogeneous, incompressible bodies of hyperelastic material, *i.e.* material for which the stress is derivable from a stored energy function W.

For such bodies, in the absence of body forces, the equations of equilibrium take the form

$$-\nabla p + \boldsymbol{F}^T \operatorname{div} \frac{\partial W}{\partial \boldsymbol{F}} = \boldsymbol{0}. \tag{2}$$

Also, the constraint of incompressibility requires that

$$\det \boldsymbol{F} = 1. \tag{3}$$

For a Neo-Hookean material, the stored-energy function W has the structure

$$W(\boldsymbol{F}) = \tfrac{1}{2}\mu(\boldsymbol{F}\cdot\boldsymbol{F} - 3), \tag{4}$$

where μ is a positive constant and the scalar product of two tensors $\boldsymbol{A}$ and $\boldsymbol{B}$ is defined as usual: $\boldsymbol{A}\cdot\boldsymbol{B} = \operatorname{tr}(\boldsymbol{A}\boldsymbol{B}^T)$. Also, we shall denote by $|\boldsymbol{A}|$ the usual trace norm of $\boldsymbol{A}$, *i.e.*, $|\boldsymbol{A}|^2 = \operatorname{tr}(\boldsymbol{A}\boldsymbol{A}^T)$. Substituting (4) into (2), in the absence of body force we obtain the equilibrium equation

$$-\nabla p + \mu \boldsymbol{F}^T \operatorname{div} \boldsymbol{F} = \boldsymbol{0}. \tag{5}$$

RAJAGOPAL & WINEMAN [4] showed that there are inhomogeneous solutions of (5) subject to appropriate boundary conditions for the problems of uniaxial extension, and of rotations of a layer about distinct axes. In order to study the stability of these solutions, as mentioned in the introduction, we shall consider the following class of disturbances with compact support.

Let

$$\mathscr{U} \equiv \{\boldsymbol{u} \in C^2(\overline{\mathscr{B}}, \mathbb{R}^3) : \det(\boldsymbol{F} + \nabla \boldsymbol{u}) = 1 \text{ in } \mathscr{B},\ \boldsymbol{u} = \boldsymbol{0} \text{ on } \partial\mathscr{B},\ \boldsymbol{u} \text{ has compact support}\}. \tag{6}$$

Thus a perturbed deformation satisfies the constraint of incompressibility, and the perturbation $\boldsymbol{u}$ satisfies the null boundary condition at infinity. We shall define a C^1 seminorm on $\mathscr{U}$ by

$$\|\boldsymbol{u}\| \equiv \sup_{\mathscr{B}} |\nabla \boldsymbol{u}|. \tag{7}$$

It is natural to regard an equilibrium solution as being stable, by definition, if the energy associated with the equilibrium state is not greater than the energy associated with all the perturbed states when the perturbation belongs to the class of disturbances with compact support. Thus the following

Definition. *We say that an equilibrium solution* $\varkappa$ *is stable if*

$$\int [W(F + \nabla u) - W(F)]\, dv \geqq 0 \quad \forall\ u \in \mathscr{U}. \tag{8}$$

We say that the equilibrium solution $\varkappa$ *is locally stable if* (8) *holds for all* $u \in \mathscr{U}$ *with sufficiently small* $\|u\|$.

Here and henceforth the domain of integration is $\mathscr{B}$ unless stated otherwise.

Proposition 1. *If there is a constant* $\varkappa > 0$ *such that*

$$\int \left\{ u \cdot \nabla \left[\operatorname{div} \frac{\partial W}{\partial F} \right] F^{-1} u + \frac{\partial^2 W}{\partial F^2} [\nabla u] \cdot [\nabla u] - \varkappa |\nabla u|^2 \right\} dv > 0 \tag{9}$$

for all $u \in \mathscr{U}$ *with* $\|u\|$ *sufficiently small, then the equilibrium solution* $\varkappa$ *is locally stable.*

Proof. Since by (6)

$$1 = \det(F + \nabla u) = (\det F) \det(I + F^{-1} \nabla u),$$

by the Hamilton-Cayley theorem we immediately conclude that

$$(F^{-T}) \cdot \nabla u = \tfrac{1}{2} \operatorname{tr}(F^{-1} \nabla u)^2 - \tfrac{1}{2} [\operatorname{tr}(F^{-1} \nabla u)]^2 + o(|\nabla u|^2) \quad \text{as } |\nabla u| \to 0. \tag{10}$$

Next, assuming that W is sufficiently differentiable, by taking the Taylor series expansion of W and using (2), (6) and (10), we conclude that

$$\begin{aligned}
&\int [W(F + \nabla u) - W(F)]\, dv \\
&= \int \left\{ \frac{\partial W}{\partial F} \cdot \nabla u + \tfrac{1}{2} \frac{\partial W^2}{\partial F^2} [\nabla u] \cdot [\nabla u] + o(|\nabla u|^2) \right\} dv \\
&= \int \left\{ p(F^{-T}) \cdot \nabla u + \tfrac{1}{2} \frac{\partial^2 W}{\partial F^2} [\nabla u] \cdot [\nabla u] + o(|\nabla u|^2) \right\} dv \\
&= \tfrac{1}{2} \int \left\{ p \operatorname{tr}(F^{-1} \nabla u)^2 + \frac{\partial^2 W}{\partial F^2} [\nabla u] \cdot [\nabla u] + o(|\nabla u|^2) \right\} dv \\
&= \tfrac{1}{2} \int \left\{ u \cdot \nabla (F^{-T} \nabla p) F^{-1} u + \frac{\partial^2 W}{\partial F^2} [\nabla u] \cdot [\nabla u] + o(|\nabla u|^2) \right\} dv \\
&= \tfrac{1}{2} \int \left\{ u \cdot \nabla \left[\operatorname{div} \frac{\partial W}{\partial F} \right] F^{-1} u + \frac{\partial^2 W}{\partial F^2} [\nabla u] \cdot [\nabla u] + o(|\nabla u|^2) \right\} dv \quad \text{as } |\nabla u| \to 0.
\end{aligned} \tag{11}$$

Hence the proposition.

By proceeding along similar lines we can also establish the following

Proposition 2. *If the solution* $\boldsymbol{x}$ *is locally stable, then*

$$\int \left\{ \boldsymbol{u} \cdot \nabla \left[\operatorname{div} \frac{\partial W}{\partial \boldsymbol{F}} \right] \boldsymbol{F}^{-1} \boldsymbol{u} + \frac{\partial^2 W}{\partial \boldsymbol{F}^2} [\nabla \boldsymbol{u}] \cdot [\nabla \boldsymbol{u}] \right\} dv \geqq 0 \tag{12}$$

for all $\boldsymbol{u} \in \mathscr{U}$ *with* $\|\boldsymbol{u}\|$ *sufficiently small.*

Remark 1. The stability conditions (9) and (12) have been formulated for perturbations that satisfy the constraint of incompressibility. We may state them equivalently for perturbations that lie in the tangent space of the constraint, as is usually done in the calculus of variations. We thus define

$$\mathscr{U}' \equiv \{\boldsymbol{u} \in C^2(\overline{\mathscr{B}}, \mathbb{R}^3) : \boldsymbol{F}^{-T} \cdot \nabla \boldsymbol{u} = 0 \text{ in } \mathscr{B},\ \boldsymbol{u} = \boldsymbol{0} \text{ on } \partial\mathscr{B},$$
$$\boldsymbol{u} \text{ has compact support}\}. \tag{13}$$

Propositions 1 and 2 remain valid if $\mathscr{U}$ is replaced by $\mathscr{U}'$. A rigorous derivation of Propositon 2 in terms of $\mathscr{U}'$ has been given by FOSDICK & MACSITHIGH [5]*.

3. Non-uniform Uniaxial Extension

Let us consider the non-uniform uniaxial extension studied by RAJAGOPAL & WINEMAN [4]:

$$x = \frac{1}{\sqrt{\lambda'(Z)}} X, \quad y = \frac{1}{\sqrt{\lambda'(Z)}} Y, \quad z = \lambda(Z), \tag{14}$$

where $\lambda(Z)$ denotes the axial stretch, and the prime denotes the derivative with respect to Z. In a cylindrical co-ordinate system

$$r = \frac{R}{\sqrt{\lambda'(Z)}}, \quad \theta = \Theta, \quad z = \lambda(Z). \tag{15}$$

Then

$$(\boldsymbol{F}) = \begin{pmatrix} a^{-\frac{1}{2}} & 0 & R(a^{-\frac{1}{2}})' \\ 0 & a^{-\frac{1}{2}} & 0 \\ 0 & 0 & a \end{pmatrix}, \tag{16}$$

where

$$a \equiv \lambda'(Z).$$

* See inequality (3.24) of their paper, which is equivalent to inequality (12) of the present work.

On substituting (16) into (5), we obtain the appropriate equation that governs the equilibrium of the layer that is non-uniformly extended:

$$-a^{\frac{1}{2}}(a^{-\frac{1}{2}})'' = C, \tag{17}$$

where C is a constant. Equation (17) can be rewritten as

$$(a^{-\frac{1}{2}})'' + Ca^{-\frac{1}{2}} = 0, \tag{18}$$

which has the solution

$$a^{-\frac{1}{2}}(Z) = A \sin \sqrt{C}\, Z + B \cos \sqrt{C}\, Z \quad \text{when } C > 0 \tag{19}$$

and

$$a^{-\frac{1}{2}}(Z) = \bar{A}\, e^{\sqrt{-C}\, Z} + \bar{B}\, e^{-\sqrt{-C}\, Z} \quad \text{when } C < 0. \tag{20}$$

The case $C = 0$ includes the classical homogeneous solution

$$\lambda'(Z) = \text{constant}.$$

The constant C is a measure of the inhomogeneity of the solution.

It follows from (16) and (17) that

$$(\nabla \operatorname{div} \boldsymbol{F})\, \boldsymbol{F}^{-1} = \begin{pmatrix} -C & 0 & 0 \\ 0 & -C & 0 \\ 0 & 0 & 2C + \frac{3}{2} a^{-2} (a')^2 \end{pmatrix}. \tag{21}$$

Let us consider the following class of functions:

$$\mathscr{S} \equiv \{u \in C^2([-h, h], \mathbb{R}) : u \not\equiv 0,\ u(-h) = u(h) = 0\}. \tag{22}$$

We minimize the quotient

$$\frac{\int_{-h}^{h} [u'(Z)]^2\, dZ}{\int_{-h}^{h} [u(Z)]^2\, dZ}, \tag{23}$$

in $\mathscr{S}$, which leads to the eigenvalue problem

$$u''(Z) + \alpha u(Z) = 0, \quad -h < Z < h, \quad u(-h) = u(h) = 0.$$

The smallest eigenvalue α is found to be $\frac{\pi^2}{4h^2}$. Thus

$$\int_{-h}^{h} u^2(Z)\, dz \leq \frac{4h^2}{\pi^2} \int_{-h}^{h} [u'(Z)]^2\, dZ \qquad \forall u \in \mathscr{S}. \tag{24}$$

Hence we have the following particular version of the Poincaré inequality*:

$$\int |\boldsymbol{u}|^2\, dv \leq \frac{4h^2}{\pi^2} \int |\nabla \boldsymbol{u}|^2\, dv \qquad \forall \boldsymbol{u} \in \mathscr{U}. \tag{25}$$

---

* A discussion of the general form of the Poincaré inequality can be found, for example, in Morrey [6, Section 3.2].

We are now in a position to establish the following

Proposition 3. *The equilibrium solutions* (19) *and* (20) *are locally stable if*

$$-\frac{\pi^2}{8h^2} < C < \frac{\pi^2}{4h^2}. \tag{26}$$

Proof. Let us consider an equilibrium solution that satisfies (26). Choose a $\varkappa$ such that

$$0 < \varkappa \leqq \mu \min\left(1 - \frac{4h^2}{\pi^2} C, \quad 1 + \frac{8h^2}{\pi^2} C\right). \tag{27}$$

Suppose that $C \geqq 0$. It follows from (21), (25) and (27) that

$$\begin{aligned}\int \{\mu \boldsymbol{u} \cdot (\nabla \operatorname{div} \boldsymbol{F}) \boldsymbol{F}^{-1} \boldsymbol{u} + (\mu - \varkappa) |\nabla \boldsymbol{u}|^2\} \, dv \\ \geqq \int \{-\mu C |\boldsymbol{u}|^2 + (\mu - \varkappa) |\nabla \boldsymbol{u}|^2\} \, dv \\ \geqq \int \left\{-\frac{4h^2}{\pi^2} \mu C + \mu - \mu \left(1 - \frac{4h^2}{\pi^2} C\right)\right\} |\nabla \boldsymbol{u}|^2 \, dv = 0.\end{aligned} \tag{28}$$

The proposition follows from (4), (9) and Proposition 1.

Next, suppose $C < 0$. Then

$$\begin{aligned}\int \{\mu \boldsymbol{u} \cdot (\nabla \operatorname{div} \boldsymbol{F}) \boldsymbol{F}^{-1} \boldsymbol{u} + (\mu - \varkappa) |\nabla \boldsymbol{u}|^2\} \, dv \\ \geqq \int \{2\mu C |\boldsymbol{u}|^2 + (\mu - \varkappa) |\nabla \boldsymbol{u}|^2\} \, dv \\ \geqq \int \left\{\frac{8h^2}{\pi^2} \mu C + \mu - \mu \left(1 + \frac{8h^2}{\pi^2} C\right)\right\} |\nabla \boldsymbol{u}|^2 \, dv = 0,\end{aligned} \tag{29}$$

and thus the proposition again follows from (4), (9) and Proposition 1.

We next discuss an interesting instability associated with the above solution. We show that for a certain class of basic solutions the second variation becomes negative for an appropriately chosen perturbation.

Proposition 4. *An equilibrium solution of the form* (19) *is unstable if*

$$C > \frac{\pi^2}{4h^2}. \tag{30}$$

Proof. Consider a function $\boldsymbol{u}$ such that its components, with respect to the natural basis of the cylindrical co-ordinate system, are given respectively by

$$u_R = 0, \quad u_\Theta = \phi(R) \cos \frac{\pi Z}{2h}, \quad u_Z = 0, \tag{31}$$

where

$$\phi(R) \equiv \begin{cases} \dfrac{1}{M^5} R^2 (R - M)^2 & \text{when } 0 \leqq R \leqq M, \\ 0 & \text{when } R > M, \end{cases} \tag{32}$$

M being a positive number. It is an easy matter to verify that $\boldsymbol{u} \in \mathscr{U}'$. Moreover, it follows from (21), (31), (32) and a straightforward calculation that

$$\int [\boldsymbol{u} \cdot (\nabla \operatorname{div} \boldsymbol{F}) \boldsymbol{F}^{-1} \boldsymbol{u} + |\nabla \boldsymbol{u}|^2] \, dv$$

$$= \int \left[\left(-C\phi^2 + \phi'^2 + \frac{1}{R^2}\phi^2\right)\cos^2\frac{\pi Z}{2h} + \frac{\pi^2}{4h^2}\phi^2 \sin^2\frac{\pi Z}{2h}\right] dv$$

$$= 2\pi h \int_0^M \left[\left(-C + \frac{1}{R^2} + \frac{\pi^2}{4h^2}\right)\phi^2 + \phi'^2\right] R \, dR$$

$$= \pi h \left[\tfrac{1}{630}\left(-C + \frac{\pi^2}{4h^2}\right) + \frac{11}{420 M^2}\right]. \tag{33}$$

The condition (30) ensures that the right-hand side of (33) is negative for a sufficiently large M. The proposition then follows from (4), (12), Proposition 2 and Remark 1.

We next turn our attention to the problem of rotation of the layer about distinct axes.

4. Rotation About Distinct Axes

We shall consider a deformation of the form

$$\begin{aligned} x &= X\cos\Omega(Z) - Y\sin\Omega(Z) + f(Z), \\ y &= X\sin\Omega(Z) - Y\cos\Omega(Z) + g(Z), \\ z &= Z, \end{aligned} \tag{34}$$

where Ω, f and g are functions defined on the interval $[-h, h]$. The above mapping represents a deformation in which material points which lie in any plane parallel to the layers $Z = \pm h$ continue to remain in the plane, the planes rotating about a point. The x and y co-ordinates of this center of rotation vary from plane to plane. A more detailed description of such a deformation can be found in Rajagopal & Wineman [7].

A simple calculation yields

$$\boldsymbol{F} = \begin{pmatrix} c & -s & -Xs\Omega' - Yc\Omega' + f' \\ s & c & Xc\Omega' - Ys\Omega' + g' \\ 0 & 0 & 1 \end{pmatrix}, \tag{35}$$

where

$$c \equiv \cos\Omega(Z) \tag{36}$$

and

$$s \equiv \sin\Omega(Z). \tag{37}$$

Substituting (35) into (5) yields

$$\Omega'' = 0, \tag{38}$$

$$f''' + (\Omega')^2 f' = 0, \tag{39}$$

$$g''' + (\Omega')^2 g' = 0. \tag{40}$$

Equation (38) implies that

$$\Omega' = \psi = \text{constant}, \tag{41}$$

and thus the angle of rotation is a linear function of Z. Next, it follows from (35) and (39)–(41) that

$$(\nabla \operatorname{div} \boldsymbol{F})\, \boldsymbol{F}^{-1} = \begin{pmatrix} -\psi^2 & 0 & 0 \\ 0 & -\psi^2 & 0 \\ 0 & 0 & 0 \end{pmatrix}. \tag{42}$$

We are now in a position to prove the following

Proposition 5. *An equilibrium solution of the form* (34) *is locally stable if*

$$|\Omega(h) - \Omega(-h)| < \pi. \tag{43}$$

Proof. Consider an equilibrium solution satisfying (43). It follows from (41) that

$$|\psi| < \frac{\pi}{2h}.$$

We choose a $\varkappa$ such that

$$0 < \varkappa \leqq \mu \left(1 - \frac{4h^2}{\pi^2} \psi^2\right). \tag{44}$$

It immediately follows from (25), (42) and (44) that for all $\boldsymbol{u} \in \mathscr{U}$

$$\begin{aligned} \int [\mu \boldsymbol{u} \cdot (\nabla \operatorname{div} \boldsymbol{F})\, \boldsymbol{F}^{-1} \boldsymbol{u} + (\mu - \varkappa)\, |\nabla \boldsymbol{u}|^2]\, dv & \\ \geqq \int [-\mu\psi^2\, |\boldsymbol{u}|^2 + (\mu - \varkappa)\, |\nabla \boldsymbol{u}|^2]\, dv & \\ \geqq \int \left[-\frac{4h^2}{\pi^2} \mu\psi^2 + \mu - \varkappa\right] |\nabla \boldsymbol{u}|^2\, dv & \\ \geqq 0. & \end{aligned} \tag{45}$$

The proposition follows from (4), (9) and Proposition 1.

In the next proposition we prove a sufficient condition for instability.

Proposition 6. *An equilibrium solution of the form* (34) *is unstable if*

$$|\Omega(h) - \Omega(-h)| > 2\pi. \tag{46}$$

Proof. It follows from (41) and (46) that

$$|\psi| > \frac{\pi}{h}. \tag{47}$$

Consider a function $\boldsymbol{u}$ with components u_x, u_y and u_z given respectively by

$$u_x = 0, \quad u_y = \frac{\pi}{h}\,\phi(r)\sin\frac{\pi Z}{h}, \quad u_z = \frac{y}{r}\,\phi'(r)\left(\cos\frac{\pi Z}{h} + 1\right), \tag{48}$$

where

$$r \equiv \sqrt{x^2 + y^2}\,. \tag{49}$$

The variables x and y are related to X, Y and Z through (34), and the function ϕ is given again by (32). The function $\boldsymbol{u}$ defined by (48) belongs to $\mathcal{U}'$. Substituting (48) into (12), and using (4), (32), (34), (42) and (49), we find that

$$\int [\boldsymbol{u}\cdot(\nabla \operatorname{div} \boldsymbol{F})\,\boldsymbol{F}^{-1}\,\boldsymbol{u} + |\nabla\boldsymbol{u}|^2]\,dv = \frac{\pi^3}{630h}\left(\frac{\pi^2}{h^2} - \psi^2\right) + o\left(\frac{1}{M}\right) \qquad \text{as } M \to \infty.$$

By (47) the last expression is negative for sufficiently large M. The proposition immediately follows from (4), (12), Proposition 2 and Remark 1.

Acknowledgement. This work was partly supported by the U.S. Army Research Office through the Mathematical Sciences Institute of Cornell University. Y. C. Chen thanks Professors T. J. Healey and P. J. Holmes for their postdoctoral support under NSF Grants DMS-8519918, MSM-8402069 and AFOSR Grant 86-0185. K. R. Rajagopal thanks the National Science Foundation for its support.

Bibliography

1. Ericksen, J. L., Deformations possible in every compressible isotropic perfectly elastic material, J. Math. Phys. **34**, 126–128 (1955).
2. Ericksen, J. L., Deformations possible in every isotropic incompressible perfectly elastic body, Z. Angew. Math. Phys. **5**, 466–486 (1954).
3. Rajagopal, K. R., & Carroll, M. M., Non-homogeneous deformations of non-linear elastic wedges, submitted for publication.
4. Rajagopal, K. R., & Wineman, A. S., New exact solutions in non-linear elasticity, Intl. J. Eng. Science **23**, 216–234, (1985).
5. Fosdick, R. L., & MacSithigh, G. P., Minimization in incompressible nonlinear elasticity theory, J. Elasticity, **16**, 267–301 (1986).
6. Morrey, C. B., Jr., Multiple Integrals in the Calculus of Variations, Springer-Verlag, Berlin-Heidelberg-New York, 1966.
7. Rajagopal, K. R., & Wineman, A. S., On a class of deformations of a material with nonconvex stored energy function, J. Struct. Mech. **12**, 471–482 (1985).

Department of Theoretical Mechanics
Cornell University
Ithaca, New York

and

Department of Mechanical Engineering
University of Pittsburgh
Pittsburgh, Pennsylvania

(Received September 24, 1988)

Conditions for Mechanical Self-Annealing in Motions of Elastic-Plastic Oscillators

DAVID R. OWEN & JOHN P. THOMAS

Dedicated to Bernard D. Coleman
on the occasion of his sixtieth birthday

Abstract

In this study of the behavior for large times of a class of elastic–plastic oscillators under conservative loading, conditions on the relative stiffness of the oscillator and the loading agent are given that guarantee that permanent displacements arising during a motion automatically are removed in the limit of large times. When these conditions are not met, a sharp bound independent of initial data is obtained on the magnitude of the ultimate permanent displacement.

Contents

1. Introduction

The research described in this article is part of an ongoing study in mechanics that focuses on damping mechanisms that act intermittantly, in particular, on the damping that occurs when elastic–plastic materials undergo permanent deformation. In references [1979, 1], [1984, 1], [1985, 1], [1987, 1], [1988, 1], and [1988, 2], various types of elastic–plastic oscillators have been analyzed. These simple mechanical systems are useful model systems for the ongoing study, because they are governed by ordinary differential equations that mathematically are of independent interest and they have solutions displaying a rich variety of physically interesting characteristics. Knowledge of these systems provides a point of reference for studying the partial differential equations that arise in more detailed models of elastic–plastic behavior: the properties of solutions that one deduces from the ordinary differential equations governing elastic–plastic oscillators become conjectured properties of solutions of the partial differential equations of plasticity. Moreover, solutions of the ordinary differential equations describing coupled systems of elastic–plastic oscillators are candidates for lumped approximations to solutions of the partial differential equations. (In the lecture notes [1985, 1], MIYOSHI develops the idea of approximation in considerable detail for one-dimensional and two-dimensional problems in plasticity.) In a different setting, elastic–plastic oscillators also are useful tools in the design of structures that must withstand large ground motions due to earthquakes, because these oscillators are relatively simple to study numerically, and they display some of the observed features of structures that are severely deformed. (See the references in the article [1988, 1] for further details.)

The focus of this paper is the description of the behavior for large times of an elastic–perfectly plastic oscillator subject to a linear restoring force. This mechanical system also has been studied, particularly in the engineering literature, in an equivalent formulation as an unforced bilinear elastic–plastic oscillator. From either point of view, the principal issues that surround the long-range character of motions of this system are, first of all, those associated with solutions of every autonomous system of differential equations: simple stability, asymptotic stability, and the limiting character of solution trajectories; second of all, there are issues peculiar to elastic–plastic systems: the extent to which the system can be permanently deformed, and the amount of damping that can occur through permanent deformations. In this paper, we establish for the unforced, bilinear elastic–plastic oscillator the simple stability of equilibrium points, as well as their lack of asymptotic stability, we characterize the limiting behavior of the oscillator for all initial states, and we obtain detailed information about the permanent deformation that can occur. The main results presented have generalizations to cases when non-linear forces replace both the linearly elastic component of the response of the oscillator and the linear restoring force. We have attempted wherever possible to give proofs of results for the bilinear oscillators that carry over without substantial modification to the more general ones.

The most striking results that we obtain center on the discovery of a critical parameter that separates the motions in which oscillations are underdamped from those that are overdamped. Underdamped oscillations occur when that parameter, the stiffness ratio of the oscillator, is one-half or greater. The permanent deformations that arise during underdamped oscillations have the curious property of being automatically removed in the limit of large times, and we say that the oscillator is *self-annealing* in such motions. Overdamped oscillations occur when the stiffness ratio is less than one-half. In overdamped oscillations, permanent deformations generally do not disappear for large times, but we obtain a sharp bound, independent of the initial state, on the ultimate permanent deformation.

This paper contains, to a large extent, refinements and extensions of the results in the Ph.D. Thesis [1984] of THOMAS. Some of these results also are contained in the paper [1988, 1], where different, mathematically less detailed methods were used to describe the phenomenon of self-annealing in the context of earthquake engineering. In particular, unlike the bound obtained in this article, the bound on permanent deformations obtained in the paper [1988, 1] is not sharp.

While we were preparing this article for publication, we learned of the lecture notes of MIYOSHI [1985, 1], in which both ordinary differential equations and partial differential equations in the theory of plasticity are formulated and analyzed. The discussion of Chapter 2 in [1985, 1] covers the elastic–plastic oscillators discussed here in Sections 2–6 when the external force b in MIYOSHI's model with kinematical hardening is put equal to zero, and MIYOSHI's Theorems 2.3 and 2.9 are counterparts of our Theorems 3.1 and 5.1. MIYOSHI does not give results corresponding to our principal results, Theorems 6.1, 6.2 and the bound (6.9).

In Section 2, we formulate the autonomous system of ordinary differential equations in $\mathbb{R}^3$ that governs the motions of the unforced, bilinear elastic–plastic oscillator, and in Section 3 we show that the corresponding initial value problem is well posed. The main feature of solution trajectories that emerges in Sections 2 and 3 is their piecewise planar character, corresponding to the alternating elastic and plastic character of deformations of the oscillator. In contrast with the oscillators studied in [1979, 1] and [1987, 1], transition times that separate elastic from plastic segments of a trajectory for the oscillators studied here can accumulate only in the limit of large times. In Sections 4, 5, and 6, the large time behavior of unforced, bilinear elastic–plastic oscillators is discussed in detail, culminating in Section 6 with the principal results on self-annealing. In Section 7, we describe a more general class of elastic–plastic oscillators for which generalizations of the results in Sections 2–6 can be established.

2. Formulation of the Problem

The mechanical system that we study here is a rigid mass supported by a thin rod and acted upon by a conservative restoring force. We assume that the rod undergoes homogeneous, longitudinal motions and that the restoring force is applied in the direction of motion of the rod, so that the mass moves parallel to the longitudinal axis of the rod. In Sections 2 through 6 we consider the case where the rod is composed of a linearly elastic–perfectly plastic material and where the restoring force is linear. In Section 7 we discuss briefly more general cases that allow for a non-linearly elastic–perfectly plastic material in the rod and a non-linear restoring force.

Let X denote the position of a point in the rod in a reference configuration, and let X_u and X_ℓ denote the positions of the upper and lower ends of the rod in that configuration. The lower endpoint is attached to the rigid mass m, and the length $X_\ell - X_u$ of the rod in the reference configuration is taken to be one. The restriction to homogeneous deformations is obtained by considering only motions of the rod of the form

$$x(X, t) = X_u + \lambda(t)(X - X_u), \tag{2.1}$$

where $x(X, t)$ is the position at time t of the point in the rod at X in the reference configuration and $\lambda(t)$ is the deformation gradient in the rod at time t. The velocity of the mass at time t is given by

$$v(t) := \frac{\partial x}{\partial t}(X_\ell, t) = \dot{\lambda}(t)\,(X_\ell - X_u) = \dot{\lambda}(t), \tag{2.2}$$

where the superposed dot denotes differentiation with respect to time. We assume that the conservative restoring force $f(t)$ acting on the mass is proportional to the displacement of the mass,

$$u(t) := x(X_\ell, t) - X_\ell = \lambda(t) - 1, \tag{2.3}$$

so that there is a positive constant k such that

$$f(t) = -ku(t). \tag{2.4}$$

The force $\sigma(t)$ that the elastic–perfectly plastic rod exerts on the mass cannot exceed a *yield value* σ_y,

$$|\sigma(t)| \leqq \sigma_y, \tag{2.5}$$

and is related to the motion of the lower endpoint of the bar through the constitutive equation

$$\dot{\sigma}(t) = \begin{cases} 0, \text{ if } |\sigma(t)| = \sigma_y \text{ and} \\ \qquad \sigma(t)\,\dot{\lambda}(t) \geqq 0 \\ \mu\dot{\lambda}(t), \text{ otherwise} \end{cases} \tag{2.6}$$

in which μ is a positive number called the *elastic modulus* of the bar. The equation of motion of the mass

$$m\dot{v}(t) = f(t) - \sigma(t)$$

and the constitutive equation (2.6) now can be expressed in terms of the displacement $u(t)$ and the velocity $v(t) = \dot{u}(t)$ of the mass:

$$m\dot{v}(t) = -ku(t) - \sigma(t) \tag{2.7}$$

$$\dot{\sigma}(t) = \begin{cases} 0, \text{ if } |\sigma(t)| = \sigma_y \text{ and} \\ \qquad \sigma(t)\,v(t) \geqq 0 \\ \mu v(t), \text{ otherwise}. \end{cases} \tag{2.8}$$

It is useful to introduce the *elastic displacement*

$$u_e(t) := \frac{\sigma(t)}{\mu} \tag{2.9}$$

and the *plastic displacement*

$$u_p(t) := u(t) - u_e(t) \tag{2.10}$$

into the relations (2.5), (2.7) and (2.8) to obtain:

$$|u_e(t)| \leqq \alpha, \tag{2.11}$$

$$m\dot{v}(t) = -k(u_e(t) + u_p(t)) - \mu u_e(t), \tag{2.12}$$

$$\dot{u}_e(t) = \begin{cases} 0, \text{ if } |u_e(t)| = \alpha \text{ and} \\ \qquad u_e(t)\, v(t) \geqq 0 \\ v(t), \text{ otherwise,} \end{cases} \tag{2.13}$$

$$\dot{u}_p(t) = \begin{cases} v(t), \text{ if } |u_e(t)| = \alpha \text{ and} \\ \qquad u_e(t)\, v(t) \geqq 0 \\ 0, \text{ otherwise.} \end{cases} \tag{2.14}$$

The positive number $\alpha := \sigma_y/\mu$ is called the *yield displacement*. Relations (2.12)–(2.14) are the basic governing equations of the linearly elastic–perfectly plastic oscillator subject to a linear restoring force. Together with the constraint (2.11), the governing equations determine the evolution of the triple (v, u_e, u_p).

Because the velocity v is the derivative of the displacement u of the mass, specification of the displacement on a time interval in the relations (2.13) and (2.14) along with the bound (2.11) leads to a system of relations that should determine u_e and u_p as functionals of u. When the net force $F := (k + \mu)\, u_e + k u_p$ is plotted in a $u - F$ plane, with u_e and u_p the functionals of u determined by (2.11), (2.13) and (2.14), one obtains the characteristic force–displacement curves of a bilinear hysteretic element. Therefore, a solution (v, u_e, u_p) of (2.11)–(2.14) determines the displacement $u = u_e + u_p$ of a freely vibrating mass supported by a bilinear hysteretic element. It is in this formulation that one finds studies of the system (2.11)–(2.14) in the engineering literature. (See the article [1988, 1] for an analysis of some aspects of this system from the point of view of earthquake engineering.) The term "bilinear" follows from the fact that the slope $\frac{dF}{du}$ of the force–displacement curve is given by:

$$\frac{dF}{du} = \begin{cases} k, \text{ if } |u_e| = \alpha \quad \text{and} \quad u_e\dot{u} \geqq 0 \\ k + \mu, \text{ otherwise,} \end{cases} \tag{2.15}$$

so that the force–displacement curve is composed of line segments of slopes

$$k_1 := k + \mu \quad \text{and} \quad k_2 := k. \tag{2.16}$$

The *stiffnesses* k_1 and k_2 and the yield strain α characterize the bilinear, hysteretic element. As our discussion in Section 6 will reveal, the *stiffness ratio*

$$\varkappa := \frac{k_2}{k_1} = \frac{k}{k + \mu} \tag{2.17}$$

plays an important role in describing the long-term behavior of solutions (v, u_e, u_p) of (2.11)–(2.14).

We shall use either the phrase *linearly elastic–perfectly plastic oscillator under a linear restoring force* or the shorter phrase *unforced bilinear elastic–plastic oscillator* to describe the mechanical system embodied in equations (2.11)–(2.14). It is natural to try to decompose locally in time each motion of that system into finitely many *elastic segments, i.e.*, segments on which u_p is constant, and finitely many *plastic segments, i.e.*, segments on which u_e is constant. On an elastic segment the motion is governed by (2.11), (2.12), $(2.13)_2$, and $(2.14)_2$, while on a plastic segment the motion is governed by (2.11), (2.12), $(2.13)_1$ and $(2.14)_1$. As natural as this decomposition appears, there are examples of elastic–plastic systems where such a decomposition does not exist. For example, the linearly elastic–perfectly plastic oscillators under general forces studied in the article [1978, 1] can undergo motions during which elastic segments and plastic segments alternate and accumulate backwards in time to an initial time. The existence of such an initial time precludes a local decomposition of the motion into finitely many elastic and plastic segments; the mathematical consequence of this fact in the analysis of general forces is that solutions of the governing equations that are not necessarily locally piecewise smooth must be sought. Nevertheless, for the linear restoring force $-k(u_e(t) + u_p(t))$ studied here, as well as for the more general conservative forces discussed in Section 7, it will emerge that every motion can be decomposed locally in time into finitely many elastic and plastic segments, and this fact underlies the choice of smoothness requirements that we make in Section 3 on solutions of (2.11)–(2.14). These smoothness requirements translate into the geometrical restriction that the solution trajectories of (2.11)–(2.14) in $v - u_e - u_p$ space locally be piecewise planar, *i.e.*, each trajectory can be partitioned locally in time into finitely many segments in planes $u_p =$ constant or planes $u_e =$ constant. The possibility that these planar segments can accumulate as the time t tends to $+\infty$ forms the central theme of our discussion in Sections 5 and 6.

3. Qualitative Results for Initial Value Problems

Our discussion in Section 2 leads us to formulate initial value problems (IVP), $(\mathrm{IVP})_{\mathscr{E}}$, $(\mathrm{IVP})_+$, and $(\mathrm{IVP})_-$ associated with the unforced bilinear elastic–plastic oscillator. In stating these problems below, we let $\alpha, k, \mu \in \mathbb{R}^{++}$ be given, and we use the following additional notation and terminology:

$$\mathscr{S} := \{(v, u_e, u_p) \in \mathbb{R}^3 \mid |u_e| \leqq \alpha\}$$

... *state space*

$$\mathscr{Y}^+ := \{(v, u_e, u_p) \in \mathbb{R}^3 \mid u_e = \alpha,\ v \geqq 0\}$$

... *upper yield surface*

$$\mathscr{Y}^- := \{(v, u_e, u_p) \in \mathbb{R}^3 \mid u_e = -\alpha,\ v \leqq 0\}$$

... *lower yield surface*

$\mathscr{Y} := \mathscr{Y}^+ \cup \mathscr{Y}^-$ ... *space of plastic states*

$\mathscr{E} := \mathscr{S} \setminus \mathscr{Y}$ ... *space of elastic states*.

(IVP): *For each* $x^\circ \in \mathcal{S}$, *find a continuous, locally piecewise continuously differentiable function* $x = (v, u_e, u_p) : [0, \infty) \to \mathcal{S}$ *such that, for every t at which x is differentiable, x satisfies* (2.11)–(2.14), *and x satisfies the initial condition*

$$x(0) = x^\circ. \tag{3.1}$$

$(\text{IVP})_{\mathscr{E}}$: *For each* $x^\circ \in \mathcal{S}$ *and* $t_0 \in [0, \infty)$, *find a continuously differentiable function* $x = (v, u_e, u_p) : [t_0, \infty) \to \mathbb{R}^3$ *that satisfies* (2.12), $(2.13)_2$ *and* $(2.14)_2$ *at each* $t \in [t_0, \infty)$ *and that satisfies the initial condition*

$$x(t_0) = x^\circ. \tag{3.2}$$

$(\text{IVP})_\pm$: *For each* $x^\circ = (v^\circ, u_e^\circ, u_p^\circ) \in \mathcal{Y}^\pm$ *with* $v^\circ \neq 0$, *and for each* $t_0 \in [0, \infty)$, *find* $t_1 > t_0$ *and a continuously differentiable function* $x = (v, u_e, u_p) : [t_0, t_1) \to \mathcal{Y}^\pm$ *that satisfies* (2.12), $(2.13)_1$, *and* $(2.14)_1$ *at each* $t \in [t_0, t_1)$ *and that satisfies the initial condition* (3.2).

To save space, we have combined the statements of $(\text{IVP})_+$ and $(\text{IVP})_-$ into one statement in which either only the plus signs should be read or only the minus signs should be read. Note that the initial time is 0 for (IVP), but it is an arbitrary non-negative number t_0 for $(\text{IVP})_{\mathscr{E}}$, $(\text{IVP})_+$, and $(\text{IVP})_-$. Moreover, solutions of (IVP) are required only to be continuous and locally piecewise continuously differentiable, *i.e.*, continuous and, on every interval $[0, \tau]$ with $\tau > 0$, piecewise continuously differentiable, whereas solutions of the other initial value problems are required to be continuously differentiable. Solutions of (IVP) and of $(\text{IVP})_{\mathscr{E}}$ have domains an interval of the form $[t_0, +\infty)$, with $t_0 \geqq 0$, whereas solutions of $(\text{IVP})_+$ and of $(\text{IVP})_-$ can have domains that are bounded intervals. Finally, solutions of (IVP) take their values in the state space $\mathcal{S}$, solutions of $(\text{IVP})_+$ and $(\text{IVP})_-$ take their values in subsets of $\mathcal{Y}^+$ and $\mathcal{Y}^-$, respectively, and solutions of $(\text{IVP})_{\mathscr{E}}$ can take their values anywhere in $\mathbb{R}^3$.

Our goal in this section is to show that (IVP) is well posed, and that all solutions of (IVP) can be constructed by patching together solutions of the problems $(\text{IVP})_{\mathscr{E}}$, $(\text{IVP})_+$, and $(\text{IVP})_-$. We use explicit properties of solutions of the latter problems to show that the patching procedure never breaks down, *i.e.*, it always yields a solution of (2.11)–(2.14) that is continuous and locally piecewise continuously differentiable on the entire half-line $[0, \infty)$. We then use uniqueness of solutions of (IVP), established by means of a simple energy inequality, to show that the patching procedure yields all solutions of (IVP).

Before we discuss the details of the method of patching, we record the energy inequality for solutions $x = (v, u_e, u_p)$ and $\bar{x} = (\bar{v}, \bar{u}_e, \bar{u}_p)$ of (2.11)–(2.14) that implies uniqueness and continuous dependence of solutions of (IVP) on initial data:

$$\begin{aligned}&[\tfrac{1}{2} m(v - \bar{v})^2 + \tfrac{1}{2} k(u_e + u_p - \bar{u}_e - \bar{u}_p)^2 + \tfrac{1}{2} \mu(u_e - \bar{u}_e)^2]^{\cdot} \\ &\qquad \leqq \left(\frac{\mu}{m} + 1\right) [\tfrac{1}{2} m(v - \bar{v})^2 + \tfrac{1}{2} k(u_e + u_p - \bar{u}_e - \bar{u}_p)^2 + \tfrac{1}{2} \mu(u_e - \bar{u}_e)^2].\end{aligned} \tag{3.3}$$

This inequality is obtained by using the relations (2.11)–(2.14) at times when both x and $\bar{x}$ are differentiable. Similarly, one obtains for a single solution $x = (v, u_e, u_p)$ of (IVP):

$$[\tfrac{1}{2} mv^2 + \tfrac{1}{2} k(u_e + u_p)^2 + \tfrac{1}{2} \mu u_e^2]^\cdot \leqq 0, \tag{3.4}$$

and this inequality shows that the trajectory of each solution x lies in the compact subset of $\mathbb{R}^3$ consisting in all points (v, u_e, u_p) whose total energy

$$\Phi(v, u_e, u_p) := \tfrac{1}{2} mv^2 + \tfrac{1}{2} k(u_e + u_p)^2 + \tfrac{1}{2} \mu u_e^2 \tag{3.5}$$

does not exceed the total energy $\Phi(v^\circ, u_e^\circ, u_p^\circ)$ associated with the initial point $(v^\circ, u_e^\circ, u_p^\circ) = x(0)$.

The initial value problem $(\mathrm{IVP})_{\mathscr{E}}$ has a unique solution $x = (v, u_e, u_p)$ satisfying the linear system for (v, u_e)

$$\begin{aligned} m\dot{v} &= -(k + \mu)\, u_e - k u_p^\circ \\ \dot{u}_e &= v \end{aligned} \tag{3.6}$$

so that x is periodic of period $\left(\dfrac{k+\mu}{m}\right)^{1/2}$ and has as trajectory the ellipse

$$\tfrac{1}{2} mv^2 + \frac{1}{2(k+\mu)} [(k + \mu)\, u_e + k u_p^\circ]^2 = \tfrac{1}{2} m(v^\circ)^2 + \frac{1}{2(k+\mu)} [(k + \mu)\, u_e^\circ + k u_p^\circ]^2 \tag{3.7}$$

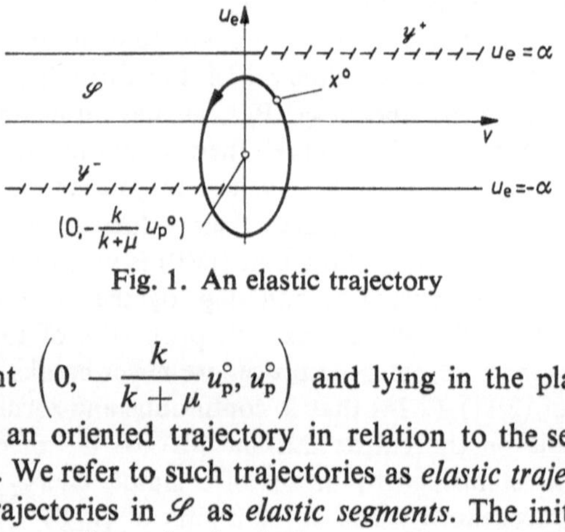

Fig. 1. An elastic trajectory

centered at the point $\left(0, -\dfrac{k}{k+\mu} u_p^\circ, u_p^\circ\right)$ and lying in the plane $u_p = u_p^\circ$. Figure 1 depicts such an oriented trajectory in relation to the sets $\mathscr{S}$, $\mathscr{Y}^+$, and $\mathscr{Y}^-$ when u_p° is positive. We refer to such trajectories as *elastic trajectories*, and to the portions of such trajectories in $\mathscr{S}$ as *elastic segments*. The initial value problem $(\mathrm{IVP})_+$ has a unique, graph-maximal solution $x = (v, u_e, u_p)$ satisfying the linear system for (v, u_p)

$$\begin{aligned} m\dot{v} &= -(k + \mu)\, \alpha - k u_p \\ \dot{u}_p &= v, \end{aligned} \tag{3.8}$$

so that x has trajectory a portion of the half-ellipse

$$\begin{gathered} \tfrac{1}{2} mv^2 + \frac{1}{2k} [k u_p + (k + \mu)\, \alpha]^2 = \tfrac{1}{2} m(v^\circ)^2 + \frac{1}{2k} [k u_p^\circ + (k + \mu)\, \alpha]^2, \\ v \geqq 0, \end{gathered} \tag{3.9}$$

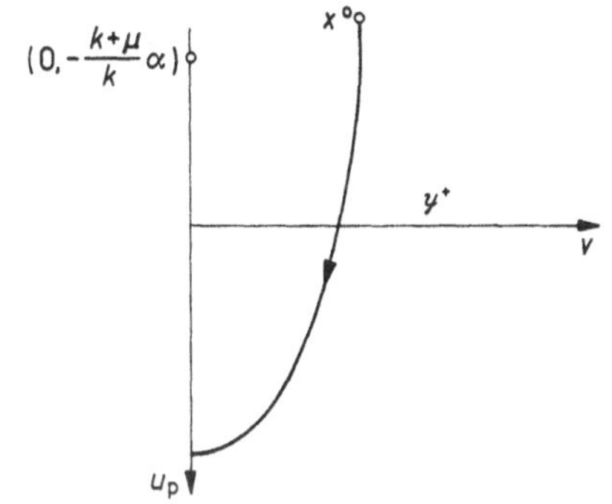

Fig. 2. A plastic segment in $\mathcal{Y}^+$

centered at $\left(0, \alpha, -\frac{(k+\mu)}{k}\alpha\right)$ and lying in the half-plane $\mathcal{Y}^+$. Figure 2 depicts such a trajectory. Similarly, $(\mathrm{IVP})_-$ has a unique graph-maximal solution $x = (v, u_e, u_p)$ satisfying

$$m\dot{v} = +(k+\mu)\alpha - ku_p, \qquad (3.10)$$
$$\dot{u}_p = v,$$

so that x has as trajectory in the half-ellipse

$$\tfrac{1}{2}mv^2 + \frac{1}{2k}[ku_p - (k+\mu)\alpha]^2 = \tfrac{1}{2}m(v^\circ)^2 + \frac{1}{2k}[ku_p^\circ - (k+\mu)\alpha]^2, \qquad (3.11)$$
$$v \leqq 0,$$

centered at $\left(0, -\alpha, \frac{k+\mu}{k}\alpha\right)$ and lying in the half-plane $\mathcal{Y}^-$, as depicted in Figure 3. We refer to trajectories for solutions of $(\mathrm{IVP})_+$ or of $(\mathrm{IVP})_-$ as *plastic segments.*

We describe now a procedure for patching together elastic and plastic segments to obtain solutions of (IVP). Consider first initial data x° for (IVP) lying in the set $\mathcal{E}$ of elastic states. Because $\mathcal{E}$ does not contain points of $\mathcal{Y}^+$ or of $\mathcal{Y}^-$ and

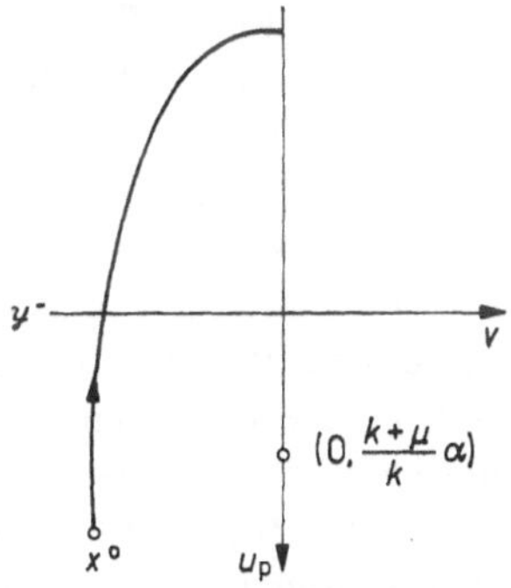

Fig. 3. A plastic segment in $\mathcal{Y}^-$

because of the orientation of elastic trajectories, we may follow the solution $x^{(0)} = (v^{(0)}, u_e^{(0)}, u_p^{(0)})$ of $(\mathrm{IVP})_{\mathscr{E}}$ satisfying $x^{(0)}(0) = x^\circ$ on a non-trivial time interval $[0, \tau_1)$ until the trajectory of $x^{(0)}$ leaves the state-space $\mathscr{S}$ by passing through $\mathscr{Y}^+$ or $\mathscr{Y}^-$ at time $\tau_1 > 0$. Suppose for definiteness that this occurs through $\mathscr{Y}^+$, so that $v^{(0)}(\tau_1) > 0$ and $u_e^{(0)}(\tau_1) = \alpha$. In order to continue along a solution of (IVP), *i.e.*, in order that a continuation of $x^{(0)}$ remain in $\mathscr{S}$, we put $x^\circ := (v^{(0)}(\tau_1), \alpha, u_p^{(0)}(\tau_1))$ and $t_0 := \tau_1$ in $(\mathrm{IVP})_+$ and follow the graph-maximal solution $x^{(1)} = (v^{(1)}, u_e^{(1)}, u_p^{(1)})$ of $(\mathrm{IVP})_+$ until it reaches the line $v = 0$ in $\mathscr{Y}^+$ at a time $\tau_2 > \tau_1$. In doing so, we have followed the elastic segment determined by $x^{(0)}$ with the plastic segment determined by $x^{(1)}$. The form of the plastic segment tells us that $v^{(1)}(\tau_2) = 0$ and

$$-(k + \mu)\, u_e^{(1)}(\tau_2) - k u_p^{(1)}(\tau_2) < 0. \tag{3.12}$$

If we put $t_0 := \tau_2$ and $x^\circ := (v^{(1)}(\tau_2), u_e^{(1)}(\tau_2), u_p^{(1)}(\tau_2)) = (0, \alpha, u_p^{(1)}(\tau_2))$ in $(\mathrm{IVP})_{\mathscr{E}}$, then we may follow the solution $x^{(2)} = (v^{(2)}, u_e^{(2)}, u_p^{(2)})$ of $(\mathrm{IVP})_{\mathscr{E}}$ and find that its trajectory remains in the state-space $\mathscr{S}$ during a non-trivial interval $[\tau_2, \tau_3]$ with $\tau_3 > \tau_2$. In fact, we have from (2.12) and (3.12),

$$\begin{aligned} v^{(2)\cdot}(\tau_2^+) &= \lim_{\tau \downarrow \tau_2} \frac{1}{m} (-(k + \mu)\, u_e^{(2)}(\tau) - k u_p^{(2)}(\tau)) \\ &= \frac{1}{m} (-(k + \mu)\, u_e^{(1)}(\tau_2) - k u_p^{(1)}(\tau_2)) < 0, \end{aligned}$$

and $v^{(2)}(\tau_2) = v^{(1)}(\tau_2) = 0$, so that $u_e^{(2)\cdot} = v^{(2)}$ is negative on an interval of the form $[\tau_2, \tau_3]$ with $\tau_3 > \tau_2$. Therefore, because $u_e^{(2)}(\tau_2) = u_e^{(1)}(\tau_2) = \alpha$, we find that $|u_e^{(2)}| \leqq \alpha$ on $[\tau_2, \tau_3]$, *i.e.*, the trajectory of $x^{(2)}$ remains in $\mathscr{S}$ during $[\tau_2, \tau_3]$.

This discussion shows how initial data x° in the set $\mathscr{E}$ of elastic states leads to a path consisting of an elastic, a plastic, and, again an elastic segment; the path starts at x° and lies in $\mathscr{S}$. The assumption that the first elastic segment ends in $\mathscr{Y}^+$ was not crucial for the construction of this path. Consequently, we have shown that every elastic trajectory that leaves $\mathscr{S}$ can be cut off at $\mathscr{Y}^+$ or at $\mathscr{Y}^-$ to form an elastic segment that can be followed by a plastic segment and then another elastic segment. This fact forms the basis of a recursive definition of a mapping $x: [0, t_*) \to \mathscr{S}$, with $t_* > 0$, that is continuous, locally piecewise continuously differentiable, satisfies (2.11) and, at its points of differentiability, (2.12)–(2.14), obeys the initial condition $x(0) = x^\circ \in \mathscr{E}$. We remark without proof that such a function x also can be constructed when the initial data x° is in $\mathscr{S} \setminus \mathscr{E} = \mathscr{Y}$.

We wish to indicate here why the function x constructed above is a solution of (IVP). It is clear that we need only show that $t_* = +\infty$, or, in other words, that the times separating successive elastic and plastic segments in the trajectory of x cannot accumulate at a finite time. We do so by considering only intermediate elastic segments of x, *i.e.*, only those elastic segments that occur between plastic segments. (If there are no intermediate segments, then it is easy to show that the trajectory of x ultimately is an elastic trajectory lying entirely within $\mathscr{S}$, so that the recursive construction of x terminates at a finite stage.) Each intermediate elastic segment has associated with it a traversal time, and we give now a geometrical argument that indicates why there is a positive number δ, depending only upon α, k, μ, m and the initial data x° for x, such that every traversal time is no less than δ. In fact, the inequality (3.4) tells us that the trajectory of x and, in particular, u_p is bounded. Therefore, all the intermediate elastic segments and all the centers of intermediate elastic segments lie in a bounded set. It is clear that an intermediate elastic segment must start in the plane $v = 0$ at $\mathcal{Y}^+$ or at $\mathcal{Y}^-$ and must end at $\mathcal{Y}^-$ or at $\mathcal{Y}^+$, respectively, so that each intermediate elastic segment must span the thickness 2α of the state-space $\mathcal{S}$ and must have it and its center lying in a predetermined bounded set. It follows that each angle subtended at its center by an intermediate elastic segment is bounded below by a positive number depending only upon α, k, μ, m and the initial data x°. Because every elastic trajectory is periodic with period $((k+\mu)/m)^{\frac{1}{2}}$, it follows that the traversal times of the elastic segments are also bounded below by a positive number depending only upon the same quantities.

The arguments given in this section can be used to produce a formal proof of the following result.

Theorem 3.1. *The initial value problem* (IVP) *for the unforced, bilinear elastic–plastic oscillator is well posed, and every solution of* (IVP) *can be obtained by patching together elastic and plastic segments.*

In closing this section, we emphasize the fact that the trajectories of solutions of (IVP) are piecewise planar; each planar segment forms a portion of an ellipse in a plane $u_p = \text{const.}$, $u_e = \alpha$ or $u_e = -\alpha$. The center of the ellipses in a plane $u_p = \overset{\circ}{u}_p$ is the point $v = 0$, $u_e = -\dfrac{k}{k+\mu}\overset{\circ}{u}_p$. If $\left|-\dfrac{k}{k+\mu}\overset{\circ}{u}_p\right| \leqq \alpha$, *i.e.*, if $|\overset{\circ}{u}_p| \leqq \dfrac{k+\mu}{k}\alpha$, then the point $\left(0, -\dfrac{k}{k+\mu}\overset{\circ}{u}_p, \overset{\circ}{u}_p\right)$ is an equilibrium state for the system (2.11)–(2.14), *i.e.*, the constant function with value $\left(0, -\dfrac{k}{k+\mu}\overset{\circ}{u}_p, \overset{\circ}{u}_p\right)$ is a solution of (IVP) with $x^\circ = \left(0, -\dfrac{k}{k+\mu}\overset{\circ}{u}_p, \overset{\circ}{u}_p\right)$. In fact, *all the equilibrium states of* (2.11)–(2.15) *are of the form*

$$\left(0, -\frac{k}{k+\mu}u_p^c, u_p^c\right), \quad \text{with } |u_p^c| \leqq \frac{k+\mu}{k}\alpha. \tag{3.13}$$

Thus, the equilibrium states form a line segment in the plane $v = 0$. The endpoint $\left(0, \alpha, -\dfrac{k+\mu}{k}\alpha\right) = \left(0, \alpha, -\dfrac{\alpha}{\varkappa}\right)$ of this line segment is the center of the plastic segments in the half-plane $\mathcal{Y}^+$, and the endpoint $\left(0, -\alpha, \dfrac{k+\mu}{k}\alpha\right) = \left(0, -\alpha, \dfrac{\alpha}{\varkappa}\right)$ is the center of the plastic segments in the half-plane $\mathcal{Y}^-$. Here $\varkappa$ is the stiffness ratio defined in (2.17).

Another feature associated with the piecewise planar character of solutions of (IVP) was already employed tacitly in the arguments leading to Theorem 3.1 and is recorded here for future use.

Remark 3.1. *Each intermediate elastic segment of a solution of* (IVP) *starts and ends in different yield planes. In other words, successive plastic segments alternate between* $\mathcal{Y}^+$ *and* $\mathcal{Y}^-$.

4. Stability of Equilibrium States

The equilibrium states for the mechanical system under consideration here are of interest because they represent solutions of the initial value problem (IVP) which do not evolve in time. Knowledge of the location of the equilibrium states also helps to characterize the behavior of the solutions which do evolve in time. As mentioned at the end of Section 3, the equilibrium state in each plane $u_p = \text{const.}$ is the center of the ellipses that contain the trajectories of the solutions of $(\text{IVP})_{\mathscr{E}}$. Similarly, the equilibrium state $\left(0, \alpha, -\frac{\alpha}{\varkappa}\right)$ $\left(\text{respectively } \left(0, -\alpha, \frac{\alpha}{\varkappa}\right)\right)$ is the center of the ellipses that contain the trajectories of the solutions of $(\text{IVP})_+$ (respectively of (IVP_-)). $\left(\text{Here, as above, } \varkappa \text{ is the stiffness ratio } \frac{k}{k+\mu}.\right)$

One issue of fundamental significance regarding the equilibrium states of the unforced bilinear elastic–plastic oscillator is the question of stability. We can ensure stability of the equilibrium states if we can construct an appropriate Liapunov function for each equilibrium state. Letting $(0, u_e^c, u_p^c)$ be an equilibrium state of the system (2.11)–(2.14), we consider

$$\hat{\Phi}(v, u_e, u_p) := \tfrac{1}{2} mv^2 + \tfrac{1}{2} \mu(u_e - u_e^c)^2 + \tfrac{1}{2} k[(u_e + u_p) - (u_e^c + u_p^c)]^2 \tag{4.1}$$

as a possible Liapunov function for this equilibrium state. This function has a strict minimum at $(0, u_e^c, u_p^c)$ and has the required smoothness. It remains to ensure that $\hat{\Phi}$ is non-increasing on solutions. To confirm this, we first differentiate $\hat{\Phi}$ along a solution $x = (v, u_e, u_p)$ to obtain

$$\hat{\Phi}^{\cdot} = mv\dot{v} + \mu(u_e - u_e^c)\,\dot{u}_e + k[(u_e + u_p) - (u_e^c + u_p^c)]\,(\dot{u}_e + \dot{u}_p). \tag{4.2}$$

Since we are considering the rate of change of $\hat{\Phi}$ along solutions, we may use equation (2.12) to substitute for $m\dot{v}$ in equation (4.2):

$$\begin{aligned} \hat{\Phi}^{\cdot} &= v[-k(u_e + u_p) - \mu u_e] + \mu(u_e - u_e^c)\,\dot{u}_e \\ &\quad + k[(u_e + u_p) - (u_e^c + u_p^c)]\,(\dot{u}_e + \dot{u}_p). \end{aligned} \tag{4.3}$$

Furthermore, since $v = \dot{u} = (u_e + u_p)^{\cdot}$, equation (4.3) becomes

$$\begin{aligned} \hat{\Phi}^{\cdot} &= v[-k(u_e + u_p) - \mu u_e] + \mu(u_e - u_e^c)\,\dot{u}_e \\ &\quad + k[(u_e + u_p) - (u_e^c + u_p^c)]\,v, \end{aligned} \tag{4.4}$$

or, equivalently,

$$\hat{\Phi}^{\cdot} = -\mu v u_e + \mu(u_e - u_e^c)\,\dot{u}_e - k(u_e^c + u_p^c)\,v. \tag{4.5}$$

Since $\varkappa = \dfrac{k}{k+\mu}$, we may eliminate u_p^c from equation (4.5) using (3.13) to obtain

$$\hat{\Phi}^{\cdot} = -\mu v u_e + \mu(u_e - u_e^c)\,\dot{u}_e - k\left[u_e^c - \frac{(k+\mu)}{k} u_e^c\right] v, \tag{4.6}$$

which becomes after rearrangement

$$\hat{\Phi}^{\cdot} = \mu(\dot{u}_e - v)\,(u_e - u_e^c). \tag{4.7}$$

When the solution is undergoing an elastic segment, we have from equation (2.13) $\dot{u}_e = v$, so that the right-hand side of equation (4.7) is identically zero on elastic segments. When the solution is undergoing a plastic segment in $\mathcal{Y}^+$, we have from equation (2.13), $\dot{u}_e = 0$, so that equation (4.7) becomes

$$\hat{\Phi}^{\cdot} = -\mu v(u_e - u_e^c). \tag{4.8}$$

Because the solution lies in $\mathcal{Y}^+$, $u_e = +\alpha$, so that equation (4.8) reduces to

$$\hat{\Phi}^{\cdot} = -\mu v(\alpha - u_e^c). \tag{4.9}$$

Since the solution is undergoing a plastic segment in $\mathcal{Y}^+$, we have $v \geqq 0$. Furthermore, because we are considering an equilibrium state $(v, u_e, u_p) = (0, u_e^c, u_p^c)$ of the initial value problem with governing equations (2.12)–(2.14) and constraint (2.11), we have $|u_e^c| \leqq \alpha$. Thus, we may conclude from (4.9) that $\hat{\Phi}^{\cdot} \leqq 0$ on solutions undergoing a plastic segment in $\mathcal{Y}^+$. The proof that $\hat{\Phi}^{\cdot} \leqq 0$ on solutions undergoing a plastic segment in $\mathcal{Y}^-$ is similar. Therefore, we may conclude that $\hat{\Phi}$ is non-increasing on solutions, that $\hat{\Phi}$ is a suitable Liapunov function for the equilibrium state, and we have

Theorem 4.1. *The equilibrium states of an unforced bilinear elastic–plastic oscillator are stable.*

Note that since $\hat{\Phi}$ is constant on elastic segments of solutions, we cannot claim asymptotic stability. In fact, recalling from Section 3 that the elastic trajectories of $(\mathrm{IVP})_{\mathscr{E}}$ in the plane $u_p = u_p^c$ are ellipses centered at $(0, u_e^c, u_p^c)$, we may conclude that when $|u_e^c| < \alpha$, the equilibrium state is a center for $(\mathrm{IVP})_{\mathscr{E}}$.

To make one additional observation, let us consider a plane $u_p = u_p^c$ with $|u_p^c| < \dfrac{\alpha}{\varkappa}$. In each such plane of constant plastic displacement, there are ellipses of the form (3.7) centered at the equilibrium state $(0, -\varkappa u_p^c, u_p^c)$ that do not intersect the planes $|u_e| = \alpha$, so that the solutions whose trajectories are contained in these ellipses remain elastic for all time. These equilibrium states therefore are isolated from one another in their respective planes of constant plastic displacement.

5. General Behavior of Solutions for Large Times

In this section we obtain a result which characterizes the asymptotic behavior of solutions of the unforced bilinear elastic–plastic oscillator. Of particular interest is the behavior of solutions which undergo an infinite number of plastic segments. The general framework provided in this section will be needed in the discussion of self-annealing in Section 6. We begin with some standard terminology.

Definition 5.1. *Let* $x^\circ \in \mathcal{S}$ *be given, and let* $x = (v, u_e, u_p)$ *be the corresponding solution of* (IVP). *The curve* $t \mapsto x(t)$ *is called the* **positive half-trajectory** *through* x° *and is denoted by* $C^+(x^\circ)$. *The* **ω-limit set** $L(C^+(x^\circ))$ *of* $C^+(x^\circ)$ *is the set of all points* $x^* \in \mathcal{S}$ *such that there is a sequence* $\{t_n\}$ *with limit* $+\infty$ *for which* $x(t_n) \to x^*$ *as* $n \to \infty$.

We need some preliminary results to obtain our characterization of ω-limit sets.

Remark 5.1. *A necessary and sufficient condition that a solution which has undergone a plastic segment in* $\mathcal{Y}^+$ *(respectively* $\mathcal{Y}^-$*) will have a subsequent plastic segment in* $\mathcal{Y}^-$ *(respectively* $\mathcal{Y}^+$*) after an intervening elastic segment is* $u_{p_1} > 0$ *(respectively* $u_{p_1} < 0$*), where* u_{p_1} *is the value of the plastic displacement as the solution departs from* $\mathcal{Y}^+$ *(respectively* $\mathcal{Y}^-$*).*

Proof. Let a solution which has undergone a plastic segment in $\mathcal{Y}^+$ be given, and let u_{p_1} denote the value of the plastic displacement as the solution departs $\mathcal{Y}^+$. Then the solution departs $\mathcal{Y}^+$ from the point $(0, \alpha, u_{p_1})$. Since, by (3.7), during the elastic segment the velocity v and elastic displacement u_e satisfy

$$\tfrac{1}{2} mv^2 = \frac{1}{2(k+\mu)} [(k+\mu)^2 (\alpha^2 - u_e^2) + 2k(k+\mu)(\alpha - u_e) u_{p_1}], \tag{5.1}$$

the elastic segment reaches the line $u_e = -\alpha$ to the left of the point $(0, -\alpha, u_{p_1})$ and thereby undergoes a segment in $\mathcal{Y}^-$ if and only if

$$\tfrac{1}{2} mv^2 = 2k\alpha u_{p_1} > 0. \tag{5.2}$$

(The possibility that the solution might cross the line $v = 0$ above the point $(0, -\alpha, u_{p_1})$ and subsequently enter $\mathcal{Y}^+$ with $v > 0$ is ruled out by the fact that the elastic unloading segment is contained in an ellipse symmetric about the line $v = 0$.) The proof of the analogous statement for a solution which has departed from $\mathcal{Y}^-$ is similar. □

The next result shows that the magnitudes of successive plastic displacements are decreasing.

Remark 5.2. *Let a solution undergo a plastic segment in* $\mathcal{Y}^+$ *(respectively* $\mathcal{Y}^-$*) and, after an intervening elastic segment, a plastic segment in* $\mathcal{Y}^-$ *(respectively* $\mathcal{Y}^+$*),* *subsequently departing from that plastic segment as well. Let* u_{p_1} *denote the plastic displacement as the solution departs from* $\mathcal{Y}^+$ *(respectively* $\mathcal{Y}^-$*), and let* u_{p_2} *denote the plastic displacement as the solution departs from* $\mathcal{Y}^-$ *(respectively* $\mathcal{Y}^+$*). Then* $|u_{p_2}| < |u_{p_1}|$.

Proof. When the solution departs from $\mathcal{Y}^+$, the total energy is $\frac{1}{2}\mu\alpha^2 + \frac{1}{2}k(\alpha + u_{p_1})^2$. When the solution departs from $\mathcal{Y}^-$, the total energy is $\frac{1}{2}\mu\alpha^2 + \frac{1}{2}k(-\alpha + u_{p_2})^2$. Since the total energy is constant on the intermediate elastic segment and since the change in total energy for a nontrivial plastic segment is negative (Remark 5.3), we have

$$\tfrac{1}{2}\mu\alpha^2 + \tfrac{1}{2}k(\alpha + u_{p_1})^2 > \tfrac{1}{2}\mu\alpha^2 + \tfrac{1}{2}k(-\alpha + u_{p_2})^2 . \tag{5.3}$$

By Remark 5.1, $u_{p_1} > 0$. Since $\dot{u}_p = v < 0$ while the solution lies in $\mathcal{Y}^-$ with $v \neq 0$, $u_{p_1} > u_{p_2}$. If $u_{p_2} \geqq 0$, then the conclusion is immediate. Otherwise, $u_{p_2} < 0$, in which case inequality (5.3) gives $\alpha + u_{p_1} > +\alpha - u_{p_2}$. Since $|u_{p_2}| = -u_{p_2}$ and $|u_{p_1}| = u_{p_1}$, the desired result follows. The proof with $\mathcal{Y}^+$ and $\mathcal{Y}^-$ switched in the statement is analogous. □

The following estimate on the energy of solutions also will be useful.

Remark 5.3. *If a solution of* (IVP) *enters a plastic segment in* $\mathcal{Y}^+$ *(respectively in* $\mathcal{Y}^-$*) at time* t_0*, and subsequently departs from the plastic segment in* $\mathcal{Y}^+$ *(respectively in* $\mathcal{Y}^-$*) at time* t_1*, with* $t_1 > t_0$*, then the net decrease in total energy* Φ *experienced during the segment is*

$$\Phi(x(t_1)) - \Phi(x(t_0)) = -\mu\alpha(u_p(t_1) - u_p(t_0)) < 0 \tag{5.4}$$

$$(\textit{respectively } \Phi(x(t_1)) - \Phi(x(t_0)) = \mu\alpha(u_p(t_1) - u_p(t_0)) < 0) . \tag{5.5}$$

Proof. Recall from equation (3.5) that $\Phi(v, u_e, u_p) = \frac{1}{2}mv^2 + \frac{1}{2}\mu u_e^2 + \frac{1}{2}k(u_e + u_p)^2$. Thus, during a plastic segment in $\mathcal{Y}^+$ (respectively $\mathcal{Y}^-$), we have

$$\begin{aligned}\dot{\Phi} &= mv\dot{v} + \mu u_e \dot{u}_e + k(u_e + u_p)(u_e + u_p)^{\cdot} \\ &= mv\dot{v} + k(\alpha + u_p)\, v\end{aligned} \tag{5.6}$$

$$(\text{respectively } \dot{\Phi} = mv\dot{v} + k(-\alpha + u_p)\, v.) \tag{5.7}$$

Substitution of equation (2.12) into equation (5.6) with $u_e = +\alpha$ (respectively into equation (5.7) with $u_e = -\alpha$) yields

$$\dot{\Phi} = -\mu\alpha v \tag{5.8}$$

$$(\text{respectively } \dot{\Phi} = +\mu\alpha v.) \tag{5.9}$$

Putting $v = \dot{u}_p$ into equations (5.8) and (5.9) and integrating yields

$$\Phi(x(t_1)) - \Phi(x(t_0)) = -\mu\alpha(u_p(t_1) - u_p(t_0)) \tag{5.10}$$

$$(\text{respectively } \Phi(x(t_1)) - \Phi(x(t_0)) = \mu\alpha(u_p(t_1) - u_p(t_0))) . \tag{5.11}$$

Since $v > 0$ (respectively $v < 0$) while the solution has $v \neq 0$ and remains in $\mathscr{Y}^+$ (respectively $\mathscr{Y}^-$), the function $t \mapsto u_p(t)$ is strictly increasing (respectively decreasing) during such a plastic segment and the energy difference $\Phi(x(t_1)) - \Phi(x(t_0))$ is negative. □

Note that equations (5.10) and (5.11) may be consolidated so that the net decrease W in energy experienced during a plastic segment can be written as

$$W = \mu\alpha\, |\Delta u_p|, \tag{5.12}$$

where Δu_p is the net change in plastic displacement experienced during the segment.

We can now prove the main result of this section:

Theorem 5.1. (*Characterization of ω-limit sets*). *Let $x^\circ \in \mathscr{S}$ be given. For the initial value problem corresponding to the unforced bilinear elastic–plastic oscillator the ω-limit set $L(C^+(x^\circ))$ of $C^+(x^\circ)$ consists in a single point in $\mathscr{S}$ (an equilibrium state) or an ellipse which lies in a plane $u_p =$ const. Furthermore, if $x^\circ \in \mathscr{S}$ is such that the resulting solution undergoes an infinite number of plastic segments, then the ω-limit set of $C^+(x_0)$ is the ellipse in the plane $u_p = 0$ given by*

$$\tfrac{1}{2} mv^2 + \tfrac{1}{2}(\mu + k)\, u_e^2 = \tfrac{1}{2}(\mu + k)\, \alpha^2. \tag{5.13}$$

Proof. Suppose first that $x^\circ \in \mathscr{S}$ is such that the resulting solution undergoes no nontrivial plastic segments. Then $\Phi(x(t)) = \Phi(x^\circ)$ and $u_p(t) = \overset{\circ}{u}_p$, for all $t \in [0, \infty)$. The curves $\Phi(x) =$ const. in the plane $u_p = \overset{\circ}{u}_p$ consist of a singleton (an equilibrium state) and ellipses. Hence the solution either is constant, in which case the singleton x° is the ω-limit set, or the solution is elastic and periodic, and the corresponding ellipse

$$\tfrac{1}{2} mv^2 + \tfrac{1}{2}\mu u_e^2 + \tfrac{1}{2} k(u_e + \overset{\circ}{u}_e)^2$$
$$= \tfrac{1}{2} mv^{\circ 2} + \tfrac{1}{2}\mu(\overset{\circ}{u}_e)^2 + \tfrac{1}{2} k(\overset{\circ}{u}_e + \overset{\circ}{u}_p)^2 \tag{5.14}$$

is the ω-limit set.

Now suppose that $x^\circ \in \mathscr{S}$ is such that the resulting solution undergoes one or more plastic segments, the number of such segments being finite. During its final plastic segment, the solution trajectory is contained in a half-ellipse of the type given by equation (3.7). One can explicitly solve the equations governing the evolution of the solution during the plastic segments, and using the point at which the solution entered $\mathscr{Y}^+$ as initial data, an explicit formula for the time required to traverse the plastic segment can be obtained. This formula confirms that the final plastic segment is left after a finite time, and the plastic displacement thereafter remains constant. Let u_{p_f} denote the ultimate plastic displacement, and suppose that the final plastic segment occurred in $\mathscr{Y}^+$ (respectively $\mathscr{Y}^-$). Then by Remark 5.1, $u_{p_f} \leq 0$ (respectively $u_{p_f} \geq 0$), and for all subsequent times the solution remains in the ellipse given by

$$\tfrac{1}{2} mv^2 + \tfrac{1}{2}\mu u_e^2 + \tfrac{1}{2} k(u_e + u_{p_f})^2 = \tfrac{1}{2}\mu\alpha^2 + \tfrac{1}{2} k(\alpha + u_{p_f})^2 \tag{5.15}$$

(respectively the ellipse given by

$$\tfrac{1}{2}mv^2 + \tfrac{1}{2}\mu u_e^2 + \tfrac{1}{2}k((u_e + u_{pf})^2 = \tfrac{1}{2}\mu\alpha^2 + \tfrac{1}{2}k(-\alpha + u_{pf})^2). \quad (5.16)$$

Since there are no additional plastic segments, this ellipse touches $\mathscr{Y}^+$ (respectively $\mathscr{Y}^-$) at the single point $(0, \alpha, u_{pf})$ (respectively $(0, -\alpha, u_{pf})$), and does not cross the line $u_e = -\alpha$ (respectively the line $u_e = +\alpha$) in the plane $u_p = u_{pf}$. Hence, after departing from its final plastic segment the solution is elastic and periodic, and the elliptical orbit in which it lies for all times thereafter is the ω-limit set of $C^+(x^\circ)$.

Finally, let $x^\circ \in \mathscr{S}$ be such that the resulting solution undergoes an infinite number of plastic segments. Let W_i be net the decrease in Φ associated with the ith plastic segment Remark 5.3 and equation (5.12) then imply that

$$W_i = \mu\alpha\,|(\Delta u_p)_i|, \quad (5.17)$$

where $(\Delta u_p)_i$ is the net change in plastic displacement for the ith plastic segment. Now let $(\Phi)_n$ be the value of Φ as the solution departs from the nth plastic segment. Then $(\Phi)_n$ is given by

$$(\Phi)_n = \Phi(x^\circ) - \sum_{i=1}^{n} \mu\alpha\,|(\Delta u_p)_i|. \quad (5.18)$$

Since $\Phi(x) \geqq 0$ for all $x \in \mathscr{S}$, equation (5.18) implies that the terms of the series $\sum_{i=1}^{\infty} \mu\alpha\,|(\Delta u_p)_i|$ are positive and its partial sums are bounded above by $\Phi(x^\circ)$, so that the series is convergent. We have $(u_p)_n = u_p^\circ + \sum_{i=1}^{n} (\Delta u_p)_i$. By Remark 3.1, the plastic segments alternate between the yield surfaces $\mathscr{Y}^+$ and $\mathscr{Y}^-$. By Remark 5.1, the $(u_p)_n$'s must then alternate in sign, and therefore, $(u_p)_n \to 0$ as $n \to \infty$.

Suppose for definiteness that the first plastic segment occurs in $\mathscr{Y}^+$ (the proof if the first plastic segment occurs in $\mathscr{Y}^-$ is analogous to what follows). Then the subsequence $\{(u_p)_{2n-1}\}$, $n = 1, 2, \ldots$ corresponds to the resultant plastic displacements as the solution leaves plastic segments in $\mathscr{Y}^+$, and the subsequence $\{(u_p)_{2n}\}$, $n = 1, 2, \ldots$ corresponds to the resultant plastic displacements as the solution leaves plastic segments in $\mathscr{Y}^-$.

When the solution has departed from $(0, \alpha, (u_p)_{2n-1})$ after a plastic segment in $\mathscr{Y}^+$, then during the subsequent elastic segment the solution lies in the ellipse given by

$$\tfrac{1}{2}mv^2 + \tfrac{1}{2}\mu u_e^2 + \tfrac{1}{2}k(u_e + (u_p)_{2n-1})^2 = \tfrac{1}{2}\mu\alpha^2 + \tfrac{1}{2}k(\alpha + (u_p)_{2n-1})^2. \quad (5.19)$$

Similarly, when a solution has departed from $(0, -\alpha, (u_p)_{2n})$ after a plastic segment in $\mathscr{Y}^-$, then during the subsequent elastic segment the solution lies in the ellipse given by

$$\tfrac{1}{2}mv^2 + \tfrac{1}{2}\mu u_e^2 + \tfrac{1}{2}k(u_e + (u_p)_{2n})^2 = \tfrac{1}{2}\mu\alpha^2 + \tfrac{1}{2}k(-\alpha + (u_p)_{2n})^2. \quad (5.20)$$

As a result of Remarks 5.1 and 5.2, the planes $u_p = \text{const.}$ on which these elastic unloading segments are located approach the plane $u_p = 0$ in the following manner: the planes $u_p = (u_p)_{2n-1}$ approach monotonically from the $u_p > 0$ direction, and the planes $u_p = (u_p)_{2n}$ approach monotonically from the $u_p < 0$ direction. Finally, because $(\Delta u_p)_n \to 0$ as $n \to \infty$ the curves containing the plastic segments in the half-planes $\mathscr{Y}^+$ and $\mathscr{Y}^-$ shrink down respectively to the points $(0, \alpha, 0)$ and $(0, -\alpha, 0)$.

Our final task is to confirm that the ellipse given by equation (5.13) is the ω-limit set of $C^+(x^\circ)$ for each $x^\circ \in \mathscr{S}$ whose solution undergoes an infinite number of plastic segments. Let t_{2n-1} be the time at which the solution occupies the point $(0, \alpha, (u_p)_{2n-1})$ (*i.e.*, the time when the solution leaves the $(2n-1)^{\text{st}}$ plastic segment). Note that the sequence $\{(0, \alpha, (u_p)_{2n-1})\}$ of points monotonically approaches the point $(0, \alpha, 0)$ as $n \to \infty$. Moreover, from the discussion preceding Theorem 3.1, $t_{2n-1} \to +\infty$ as $n \to \infty$. Thus, $(0, \alpha, 0) \in L(C^+(x^\circ))$ by the definition of ω-limit set. Similarly, it follows that $(0, -\alpha, 0) \in L(C^+(x^\circ))$. Thus, the ellipse E in (5.13) has two of its points in $L(C^+(x^\circ))$ and is itself the trajectory of the solution of (IVP) starting at $(0, \alpha, 0)$. Well known properties of ω-limit sets (see [1980, 1], Theorem 8.1, p. 47) then tell us that $L(C^+(x^\circ))$ includes the ellipse E in (5.13). To show that $L(C^+(x^\circ))$ equals E, we first note from earlier arguments that $L(C^+(x^\circ))$ is included in the plane $u_p = 0$. Let $(v, u_e, 0)$ be given in the exterior of E (with respect to the plane $u_p = 0$). The trajectory of the solution of (IVP) with $x^\circ = (v, u_e, 0)$ must contain a non-trivial plastic segment and, hence, points with $u_p \neq 0$. Because $L(C^+(x^\circ))$ is an invariant set for (2.12)–(2.15) and $u_p = 0$ at each point of $L(C^+(x^\circ))$, it follows that $(v, u_e, 0)$ is not in $L(C^+(x^\circ))$. Let a point $(v, u_e, 0)$ be given that is in the interior of E (with respect to the plane $u_p = 0$). The form of the elastic segments of $C^+(x^\circ)$ through the points $(0, \alpha, (u_p)_{2n-1})$ and the form of the elastic segments of $C^+(x^\circ)$ through the points $(0, -\alpha, (u_p)_{2n})$, for $n = 1, 2, 3, \ldots$, tell us that these segments, when projected on the plane $u_p = 0$, lie in the exterior of E. It follows that no point $(v, u_e, 0)$ interior to E can be in $L(C^+(x^\circ))$. Thus, $L(C^+(x^\circ)) = E$. □

6. Conditions for Self-Annealing

In this section we find non-trivial subsets of the state space $\mathscr{S}$ with the property that all initial data in a given subset lead to solution trajectories of (IVP) with the same character for large times. We already know that when the initial data is close to an equilibrium state other than the two endpoints of the line of equilibrium states, the solution remains elastic and periodic for all time. Here we shall delimit sets of initial data for which the resulting solutions undergo infinitely many plastic segments. From Theorem 5.1, we know that the plastic displacement associated with each such solution satisfies $u_p(t) \to 0$ as $t \to \infty$, and that the solution has the ellipse

$$\tfrac{1}{2} m v^2 + \tfrac{1}{2} (\mu + k) u_e^2 = \tfrac{1}{2} (\mu + k) \alpha^2 \tag{5.13}$$

as its ω-limit set. Specifically, we delimit here the data in the plane $u_p = 0$ that produce such solutions. In doing so, we find that the stiffness ratio $\varkappa = \dfrac{k}{k + \mu}$

is a critical parameter: when $\frac{1}{2} \leqq \varkappa < 1$, solutions of the (IVP) with an infinite number of plastic segments do indeed exist, and the choices of initial data whose solutions have this property can be completely characterized; on the other hand, when $0 < \varkappa < \frac{1}{2}$, solutions of (IVP) can have at most finitely many plastic segments. (See the thesis [1984, 1] for a detailed partition of initial data with $u_p \neq 0$ according to the long-term behavior of u_p.)

Let $x^\circ \in \mathcal{S}$ in the plane $u_p = 0$ lie outside of the ellipse given by equation (5.13). As we show below, when $\frac{1}{2} \leqq \varkappa < 1$, the solution of (IVP) starting at x° must undergo an infinite number of plastic segments. Furthermore, the plastic displacement of the solution satisfies $\lim_{t \to \infty} u_p(t) = 0$. Although the initial point x° in the plane $u_p = 0$ can be chosen so that the plastic displacement at the end of the first plastic segment has a magnitude that is arbitrarily large, the system nevertheless *completely removes* that plastic displacement from the solution in the limit of large times. We call this behavior *self-annealing, or mechanical self-annealing* to emphasize the absence of thermal devices to effect the removal of plastic displacements. The behavior of the unforced bilinear elastic–plastic oscillator when $\frac{1}{2} \leqq \varkappa \leqq 1$ also can be considered to be a type of *underdamped oscillation.* Indeed, the solution of (IVP) for initial $x^\circ \in \mathcal{S}$ in the plane $u_p = 0$ and outside the ellipse (5.13) has plastic displacements at the end of the plastic segments that alternate in sign and decrease monotonically in magnitude to zero. This behavior is analogous to the behavior of the displacements of a linear spring with linear viscosity in the underdamped case.

As mentioned above, we shall show that when $0 < \varkappa < \frac{1}{2}$, all solutions of (IVP) have only a finite number of plastic segments. Moreover, if x° is in the plane $u_p = 0$ and lies outside the ellipse (5.13), then the final plastic displacement is not necessarily zero, and that displacement is achieved in finite time. We also shall give a sharp bound, independent of the choice of initial data, on the magnitude of the final plastic displacement. The portion of Theorem 5.1 applicable to solutions with a finite number of plastic segments governs this case, so that we may conclude that after its final plastic segment, the solution remains in an elastic orbit contained in an ellipse which touches the yield surface in which the solution had its final plastic segment. Here again, until the final plastic segment is attained, the plastic displacements at the end of the plastic segments change sign and decrease monotonically in magnitude. Thus, for μ and k such that $0 < \varkappa < \frac{1}{2}$, the unforced bilinear elastic–plastic oscillator acts as if it were *overdamped,* because it allows only a finite number of plastic segments and, hence, only a finite number of oscillations in the sign of the resultant plastic displacements. Moreover, the value $\varkappa = \frac{1}{2}$ is a *critical stiffness ratio,* because it separates the underdamped and overdamped cases.

For the analysis that supports these results, two pairs of curves will be particularly useful: (1) the plastic segment in $\mathcal{Y}^+$ that passes through $(0, \alpha, 0)$ and its counterpart in $\mathcal{Y}^-$ which passes through $(0, -\alpha, 0)$, and (2) the locus of points (v, α, u_p) with $u_p < 0$ (respectively $(-v, -\alpha, u_p)$ with $u_p > 0$) where the elastic segment through $(0, -\alpha, u_p)$ (respectively $(0, \alpha, u_p)$) enters $\mathcal{Y}^+$ (respectively $\mathcal{Y}^-$).

The single curve, obtained from the pair of curves in (1) by joining at the origin $(0, 0)$ the projection on $u_e = 0$ of the curve in $\mathcal{Y}^+$ to the projection on $u_e = 0$ of its counterpart in $\mathcal{Y}^-$, will be denoted by $\mathcal{P}_0$. The portion of $\mathcal{P}_0$ with

$u_p \geqq 0$ is the half-ellipse

$$\tfrac{1}{2} mv^2 + \tfrac{1}{2} k(-\alpha + u_p)^2 - \mu\alpha u_p = \tfrac{1}{2} k\alpha^2, \quad v \leqq 0, \tag{6.1}$$

while the portion of $\mathscr{P}_0$ with $u_p \leqq 0$ is the half-ellipse

$$\tfrac{1}{2} mv^2 + \tfrac{1}{2} k(\alpha + u_p)^2 + \mu\alpha u_p = \tfrac{1}{2} k\alpha^2, \quad v \geqq 0. \tag{6.2}$$

Thus, on $\mathscr{P}_0$ with $u_p \geqq 0$, the kinetic energy is given in terms of u_p by rewriting (6.1) in the form

$$\tfrac{1}{2} mv^2 = \tfrac{1}{2} k\alpha^2 + \mu\alpha u_p - \tfrac{1}{2} k(-\alpha + u_p)^2, \tag{6.3}$$

with a similar equation on $\mathscr{P}_0$ holding when $u_p \leqq 0$. We note that the curve $\mathscr{P}_0$ is symmetric with respect to the origin in the plane $u_e = 0$.

The curve $\mathscr{P}_{\text{entry}}$ in the plane $u_e = 0$ is defined to be the union of the projections on $u_e = 0$ of the pair of curves in (2). For $u_p \geqq 0$, it is the half-parabola

$$\tfrac{1}{2} mv^2 + \tfrac{1}{2} k(-\alpha + u_p)^2 + \tfrac{1}{2} \mu\alpha^2 = \tfrac{1}{2} k(\alpha + u_p)^2 + \tfrac{1}{2} \mu\alpha^2, \quad v \leqq 0, \tag{6.4}$$

or

$$\tfrac{1}{2} mv^2 = 2k\alpha u_p, \quad v \leqq 0, \tag{6.5}$$

and for $u_p \leqq 0$, $\mathscr{P}_{\text{entry}}$ is the half-parabola

$$\tfrac{1}{2} mv^2 = -2k\alpha u_p, \quad v \geqq 0. \tag{6.6}$$

The curve $\mathscr{P}_{\text{entry}}$ is also symmetric with respect to the origin in the plane $u_e = 0$. Representative curves $\mathscr{P}_0$ and $\mathscr{P}_{\text{entry}}$ are shown in Figures 4 and 5. Subtracting the right-hand side of the equation in (6.3) from the right-hand side of the equation in (6.5) gives after some simplification

$$\tfrac{1}{2} m(v_{\text{entry}}^2 - v_0^2) = (k - \mu)\, \alpha u_p + \tfrac{1}{2} k u_p^2. \tag{6.7}$$

In (6.7) we have written v_{entry} and v_0 for the values of v on $\mathscr{P}_{\text{entry}}$ and on $\mathscr{P}_0$ corresponding to a given u_p.

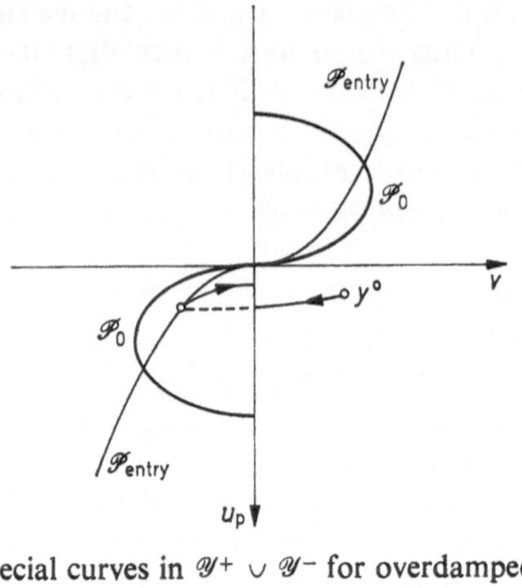

Fig. 4. Special curves in $\mathscr{Y}^+ \cup \mathscr{Y}^-$ for overdamped motions

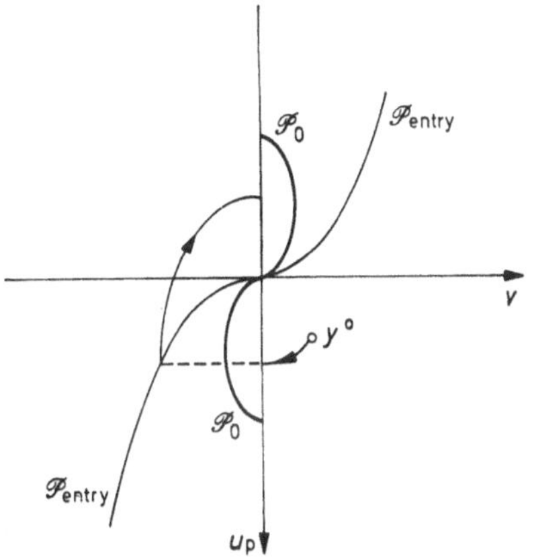

Fig. 5. Special curves in $\mathscr{Y}^+ \cup \mathscr{Y}^-$ for underdamped motions

Suppose a solution has just departed a plastic segment in $\mathscr{Y}^+$ (respectively $\mathscr{Y}^-$) with resultant plastic strain $u_{p_1} \in \mathbb{R}$. If $u_{p_1} \leqq 0$ (respectively $u_{p_1} \geqq 0$), then Remark 5.1 ensures that no further plastic segments occur. If $u_{p_1} > 0$, then Remark 5.1 guarantees that a subsequent plastic segment occurs in $\mathscr{Y}^-$ (respectively $\mathscr{Y}^+$) after an intermediate elastic segment. Knowledge of the curves $\mathscr{P}_0$ and $\mathscr{P}_{entry}$ enables one to determine whether yet another plastic segment in $\mathscr{Y}^+$ (respectively $\mathscr{Y}^-$) occurs. For example, a subsequent plastic segment will not take place if the point (v, u_{p_1}) on $\mathscr{P}_{entry}$ lies on $\mathscr{P}_0$ or in the interior of the region bounded by the curve $\mathscr{P}_0$ and the line $v = 0$; otherwise, a subsequent plastic segment will take place.

To illustrate more fully how the curves $\mathscr{P}_0$ and $\mathscr{P}_{entry}$ can be used to determine whether subsequent plastic segments occur, let $y^\circ = (v^\circ, \alpha, u_p^\circ) \in \mathscr{Y}^+$ be given such that $v^\circ > 0$ and $u_p^\circ > 0$. The solution of (IVP) for this initial data undergoes an initial plastic segment. The fact that $t \mapsto u_p(t)$ is strictly increasing on solutions in $\mathscr{Y}^+$ ensures that the solution departs $\mathscr{Y}^+$ with positive resultant plastic displacement. The solution then undergoes an intermediate elastic segment. Since the resultant plastic displacement at the end of the first plastic segment is positive, by Remark 5.1 the solution will undergo a subsequent plastic segment. If we let $u_p = u_p^1$ denote the resultant plastic strain at the end of the first plastic segment, then the intersection of the line $u_p = u_p^1$ with the curve $\mathscr{P}_{entry}$ in Figure 4 indicates where the solution enters the yield surface $\mathscr{Y}^-$. For the case indicated in Figure 4, the point of intersection lies between the curve $\mathscr{P}_0$ and the line $v = 0$. This ensures that the plastic strain at the end of the resulting plastic segment in $\mathscr{Y}^-$ is positive. We may then use Remark 5.1 to conclude that no further plastic segments occur.

Now consider Figure 5, where the same initial data are used, but the positions of $\mathscr{P}_0$ and $\mathscr{P}_{entry}$ are different. Once again, the solution undergoes an initial plastic segment with positive resultant plastic displacement $u_p^1 > 0$. After an intermediate elastic segment, the solution enters $\mathscr{Y}^-$ at the intersection of the line $u_p = u_p^1$ with the curve $\mathscr{P}_{entry}$. For the case indicated in Figure 5, the intersection point lies outside the region between $\mathscr{P}_0$ and $v = 0$. This ensures that the plastic strain at the end of the subsequent plastic segment in $\mathscr{Y}^-$ is negative. We may then conclude by Remark 5.1 that the solution will undergo a subsequent plastic segment in $\mathscr{Y}^+$.

It is now possible to use these observations to draw some conclusions about the asymptotic behavior of solutions for particular choices of μ and k. Consider first an unforced bilinear elastic–plastic oscillator for which $k \geqq \mu$, or, equivalently, $\frac{1}{2} \leqq \frac{k}{\mu + k} = \varkappa < 1$. Notice that this condition on k and μ ensures that the right-hand side of equation (6.7) is strictly positive for all $u_p > 0$. Hence, we may conclude that for all u_p not equal to zero, the curve $\mathscr{P}_0$ lies in the interior of the region bounded by the curve $\mathscr{P}_{entry}$ and the line $v = 0$. With this in mind, consider $x^\circ = (v^\circ, \overset{\circ}{u}_e, 0) \in \mathscr{S}$, *i.e.*, consider an initial condition in $\mathscr{S}$ with zero initial plastic displacement. For a choice of x° on or in the interior of the ellipse given by equation (5.13), the solution remains elastic and periodic for all time. Consider instead an $x^\circ = (v^\circ, \overset{\circ}{u}_e, 0) \in \mathscr{S}$ with $v^\circ > 0$ which lies in the plane $u_p = 0$ but in the exterior of the above ellipse. Since the initial elastic trajectory for x° eventually reaches $\mathscr{Y}^+$, the trajectory undergoes an initial plastic segment in $\mathscr{Y}^+$. Since the plastic displacement is strictly increasing along plastic segments in $\mathscr{Y}^+$, the trajectory for x° emerges from $\mathscr{Y}^+$ with positive resultant plastic displacement. By Remark 5.1, there will be a subsequent plastic segment in $\mathscr{Y}^-$. The above remark in the case $\frac{1}{2} \leqq \varkappa < 1$ concerning the curves $\mathscr{P}_0$ and $\mathscr{P}_{entry}$ ensures that the resultant plastic displacement at the end of this plastic segment in $\mathscr{Y}^-$ is negative. Again by Remark 5.1, the trajectory has another plastic segment in $\mathscr{Y}^+$, and the same remark about $\mathscr{P}_0$ and $\mathscr{P}_{entry}$ implies that the resultant plastic displacement at the end of the segment is positive. Let us define a *plastic cycle* to be a process in which a solution with elastic and plastic segments departs from $\mathscr{Y}^+$, undergoes an elastic segment and then a subsequent plastic segment in $\mathscr{Y}^-$, followed by another elastic segment and then a subsequent plastic segment in $\mathscr{Y}^+$. Then for $x^\circ \in \mathscr{S}$ as given, the following statement has been demonstrated:

(1) The solution undergoes a plastic cycle, and the resultant plastic displacement upon completion of this cycle is positive.

Label additional statements as follows:

(n) The solution undergoes n plastic cycles, and the resultant plastic displacement upon completion of cycle n is positive.

($n + 1$) The solution undergoes $n + 1$ plastic cycles, and the resultant plastic displacement upon completion of cycle $n + 1$ is positive.

The arguments used to prove that the solution corresponding to x° as given satisfies statement (1) may also be used to show that if statement (n) is assumed, then statement ($n + 1$) follows. Thus, it follows by induction that each $x^\circ = (v^\circ, \overset{\circ}{u}_e, 0) \in \mathscr{S}$ with $v^\circ > 0$ which lies in the exterior of the ellipse given by equation (5.13) has a solution which undergoes an infinite number of plastic segments. The proof for $x^\circ = (v^\circ, \overset{\circ}{u_e}, 0) \in \mathcal{S}$ with $v^\circ < 0$ lying in the exterior of the above ellipse is completely analogous. Thus, invoking Theorem 5.1, we have

Theorem 6.1. *Let* $\frac{1}{2} \leqq \varkappa < 1$, *and let* $x^\circ = (v^\circ, \overset{\circ}{u_e}, 0) \in \mathcal{S}$ *be given such that* x° *lies in the exterior of the ellipse given by*

$$\tfrac{1}{2} m v^2 + \tfrac{1}{2}(\mu + k) u_e^2 = \tfrac{1}{2}(\mu + k) \alpha^2. \tag{5.13}$$

The solution $t \mapsto (v(t), u_e(t), u_p(t))$ *then undergoes an infinite number of plastic segments,* $u_p(t) \to 0$ *as* $t \to \infty$, *and the* ω*-limit set of* $C^+(x^\circ)$ *is the ellipse given by the equation* (5.13).

One may analyze similarly the asymptotic behavior of solutions for initial data lying in planes $u_p = \text{const.}$, but $u_p \neq 0$. It turns out that in addition to the regions that correspond to solutions that remain elastic for all time, and the regions which correspond to solutions that undergo an infinite number of plastic segments, there is a third type of region when $\frac{1}{2} \leqq \varkappa < 1$. For the third region, the solution undergoes exactly one plastic segment, after which it is elastic and periodic for all time. The region lying in the half-planes $\mathcal{Y}^+$ and $\mathcal{Y}^-$ bounded by the curve $\mathcal{P}_0$ and the line $v = 0$ is an example of such a region.

We consider one additional case in this section. Suppose now that the unforced bilinear elastic-plastic oscillator is such that $0 < k < \mu$, which means that $0 < \varkappa < \frac{1}{2}$. The right-hand side of equation (6.7) then is negative for every positive but sufficiently small plastic displacement. Thus, for u_p positive and sufficiently small, points (v, u_p) on $\mathcal{P}_{\text{entry}}$ lie in the interior of the region enclosed by $\mathcal{P}_0$ and the line $v = 0$. However, as u_p is increased, the curves intersect, and for u_p sufficiently large, (v, u_p) in $\mathcal{P}_{\text{entry}}$ lies outside that region. The equations of the curves for $\mathcal{P}_0$ and $\mathcal{P}_{\text{entry}}$ are for $u_p \geqq 0$ respectively

$$\tfrac{1}{2} m v^2 + \frac{k}{2}\left(u_p - \frac{\alpha}{\varkappa}\right)^2 = \frac{k}{2}\left(\frac{\alpha}{\varkappa}\right)^2 \tag{6.8}$$

and

$$\tfrac{1}{2} m v^2 = 2k\alpha u_p. \tag{6.5}$$

Notice from equation (6.8) that the portion of $\mathcal{P}_0$ contained in the projection of $\mathcal{Y}^-$ into the plane $u_e = 0$ is a half-ellipse centered at $\left(0, \frac{\alpha}{\varkappa}\right)$, and that this half-ellipse crosses the line $v = 0$ at $u_p = 2\frac{\alpha}{\varkappa}$. On the other hand, as mentioned earlier, equation (6.5) describes the half-parabola representing that portion of $\mathcal{P}_{\text{entry}}$ contained in the projection of $\mathcal{Y}^-$ into the plane $u_e = 0$. This half-parabola is tangent to the line $v = 0$. The two curves described above intersect (in the projection of $\mathcal{Y}^-$) at the point $\left(-2\alpha\sqrt{\frac{2(\mu - k)}{m}}, \frac{2(\mu - k)}{k}\alpha\right)$.

Let us again examine the asymptotic behavior of a solution with initial data $x^\circ = (v^\circ, \overset{\circ}{u_e}, 0) \in \mathcal{S}$, *i.e.*, an initial condition in $\mathcal{S}$ with zero initial plastic displacement. Once again, if x° lies on or in the interior of the ellipse given by equation (5.13), the solution remains elastic and periodic for all time. Now consider $x^\circ = (v^\circ, \mathring{u}_e, 0) \in \mathscr{S}$ that lies in the exterior of this ellipse. Since the initial elastic trajectory for x° eventually reaches $\mathscr{Y}^+$, it will undergo an initial plastic segment in $\mathscr{Y}^+$. Since the plastic displacement is strictly increasing along plastic segments in $\mathscr{Y}^+$, the trajectory will emerge from $\mathscr{Y}^+$ with positive resultant plastic displacement. By Remark 5.1, the solution will have a subsequent plastic segment in $\mathscr{Y}^-$. There is a special segment of an ellipse $\mathscr{P}_2$ in $\mathscr{Y}^+$ (see Figure 6) which represents a plastic solution segment. It is that segment of an ellipse which crosses the line $v = 0$ at exactly the value of u_p at which $\mathscr{P}_0$ and $\mathscr{P}_{\text{entry}}$ intersect. The solution corresponding to this curve will undergo an intermediate elastic segment, enter $\mathscr{Y}^-$, and undergo a plastic segment in $\mathscr{Y}^-$ contained in the curve $\mathscr{P}_0$. Thus, this solution departs from its second plastic segment with zero resultant plastic displacement, and thereupon enters the ellipse given by equation (5.13) and remains elastic and periodic for all subsequent time. For $x^\circ = (v^\circ, \mathring{u}_e, 0) \in \mathscr{S}$ with $v^\circ > 0$ in the exterior of the ellipse given by equation (5.13) whose solution enters $\mathscr{Y}^+$ between $(0, \alpha, 0)$ and the point at which $\mathscr{P}_2$ crosses the line $u_p = 0$, the corresponding solution undergoes a plastic segment in $\mathscr{Y}^+$, an intermediate elastic segment, and then a plastic segment in $\mathscr{Y}^-$. Since this solution enters $\mathscr{Y}^-$ along $\mathscr{P}_{\text{entry}}$ at a point in the interior of $\mathscr{P}_0$, it emerges from its plastic segment in $\mathscr{Y}^-$ with positive resultant plastic displacement, and by Remark 5.1 then can undergo no further plastic segments. Thus, when $0 < \varkappa < \frac{1}{2}$, there is a region of the portion of $\mathscr{S}$ in the plane $u_p = 0$ with $v \geqq 0$ such that if x° lies in this region, then the corresponding solution undergoes exactly two plastic segments, and the solution is subsequently elastic and periodic in a plane $u_p = \text{const.} \geqq 0$.

There is an additional special segment of an ellipse $\mathscr{P}_3$ in $\mathscr{Y}^+$ (see Figure 6) which represents a plastic solution segment. A solution contained in that segment undergoes an initial plastic segment in $\mathscr{Y}^+$, followed by an intermediate elastic segment, followed by a plastic segment in $\mathscr{Y}^-$ such that, after the solution departs from that plastic segment and undergoes an additional intermediate elastic segment, the solution reenters $\mathscr{Y}^+$ at exactly the point where $\mathscr{P}_0$ and $\mathscr{P}_{\text{entry}}$ intersect in $\mathscr{Y}^+$. This solution will then be contained in a portion of $\mathscr{P}_0$, and will subsequently

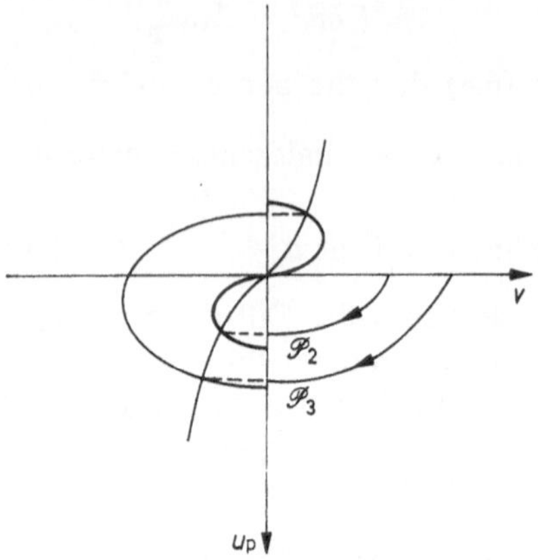

Fig. 6. Separating trajectories $\mathscr{P}_2$ and $\mathscr{P}_3$

emerge from its third and final plastic segment with zero resultant plastic displacement. It will subsequently remain in the ellipse given by equation (5.13) for all time, and the solution is then elastic and periodic. For each $x^\circ = (v^\circ, u_e^\circ, 0) \in \mathcal{S}$ with $v^\circ > 0$ whose solution enters its initial plastic segment in $\mathcal{Y}^+$ between the points at which $\mathcal{P}_2$ and $\mathcal{P}_3$ cross the line $u_p = 0$, the solution undergoes a plastic segment in $\mathcal{Y}^+$, an intermediate elastic segment, a plastic segment in $\mathcal{Y}^-$, an additional intermediate elastic segment, and then a plastic segment in $\mathcal{Y}^+$ with a resultant plastic displacement that is negative, so that by Remark 5.1, it cannot undergo a further plastic segment in $\mathcal{Y}^-$. Thus, when $0 < \varkappa < \frac{1}{2}$, there is a region of the portion of $\mathcal{S}$ in the plane $u_p = 0$ with $v \geqq 0$ such that if x° lies in this region, then the corresponding solution undergoes exactly three plastic segments, and the solution is subsequently elastic and periodic in a plane $u_p =$ const. $\leqq 0$.

It is clear that one can continue to construct such special elliptical segments forming regions of $\mathcal{Y}^+$ such that solutions corresponding to initial data in one region undergo one more plastic segment than do those for the previous region. The portion of $\mathcal{S}$ in the plane $u_p = 0$ lying in $v \leqq 0$ can be analyzed in a similar manner. Thus, one has the result:

Theorem 6.2. *Let* $0 < \varkappa < \frac{1}{2}$, *and let* $x^\circ = (v^\circ, u_e^\circ, 0) \in \mathcal{S}$ *be given such that* x° *lies in the exterior of the ellipse given by*

$$\tfrac{1}{2} m v^2 + \tfrac{1}{2}(\mu + k)\, u_e^2 = \tfrac{1}{2}(\mu + k)\, \alpha^2 . \tag{5.13}$$

For each natural number n, there is a region $\mathcal{D}_n$ *in the exterior of the ellipse such that if* $x^\circ \in \mathcal{D}_n$, *then the solution undergoes exactly n plastic segments, and is subsequently elastic and periodic. If* $v^\circ > 0$ *(respectively* $v^\circ < 0$*) and* $x^\circ \in \mathcal{D}_n$, *and if n is even, the solution undergoes its final plastic segment in* $\mathcal{Y}^-$ *(respectively* $\mathcal{Y}^+$*) and the final resultant plastic strain is nonnegative (respectively nonpositive); on the other hand, if n is odd the solution undergoes its final plastic segment in* $\mathcal{Y}^+$ *(respectively in* $\mathcal{Y}^-$*) and the final resultant plastic strain is nonpositive (respectively nonnegative).*

We note finally, when $0 < \varkappa < \frac{1}{2}$, there is an explicit bound for the maximum possible final resultant plastic displacement u_{p_f}. A solution which yields the extremal value is one whose final plastic segment lies in the region formed by $\mathcal{P}_{\text{entry}}$ and the line $v = 0$ and is tangent to $\mathcal{P}_{\text{entry}}$ at exactly one point. The maximum possible value of u_{p_f} is found by determining the smaller value at which the ellipse containing the final plastic segment intersects the line $v = 0$. The calculations to determine the point of tangency yield the bound

$$|u_{p_f}| \leqq \alpha \left[\frac{1}{\varkappa} - 2\sqrt{\frac{1}{\varkappa} - 1}\right]. \tag{6.9}$$

It should be observed that this estimate is sharp, and that it is independent of x°.

7. Generalizations of the Main Results

We consider now elastic–plastic oscillators for which the dependence of the force σ upon the elastic displacement u_e and the dependence of the restoring force f upon the total displacement $u = u_e + u_p$ both can be non-linear. Specifically, we replace the dynamical equation (2.12) by the relation

$$m\dot{v}(t) = -\phi'(u_e(t) + u_p(t)) - \Sigma'(u_e(t)), \tag{7.1}$$

where ϕ and Σ are smooth functions with further properties to be specified below, and where $'$ denotes differentiation. Of course, when

$$\phi(u) = \tfrac{1}{2} ku^2 \quad \text{and} \quad \Sigma(u) = \tfrac{1}{2} \mu u^2 \tag{7.2}$$

we obtain again (2.12), so that (7.1) together with (2.11), (2.13) and (2.14) does include the system (2.11)–(2.14) studied in Sections 2 through 6.

In this section we give conditions on the potentials ϕ and Σ that imply conclusions similar to those obtained when ϕ and Σ are quadratic. Specifically, the following conditions on ϕ and Σ yield generalizations of the main results in Sections 3, 5, and 6:

$$\Sigma, \phi \in C^2(\mathbb{R}, \mathbb{R}), \tag{7.3}$$

$$\phi(0) = 0, \quad \Sigma(0) = 0 \tag{7.4}$$

$$\Sigma(-u) - \Sigma(u) = \phi(-u) - \phi(u) = 0 \quad \text{for all } u \in \mathbb{R}, \tag{7.5}$$

$$\phi'(u) > 0 \quad \text{for all } u > 0, \tag{7.6}$$

$$\lim_{|u| \to \infty} \phi(u) = +\infty, \tag{7.7}$$

$$\Sigma''(u) > 0 \quad \text{for all } u \in \mathbb{R} \setminus \{0\}, \tag{7.8}$$

$$\phi''(u) > 0 \quad \text{for all } u \in \mathbb{R} \text{ and } \phi'(\alpha) \geqq \Sigma'(\alpha) \tag{7.9a}$$

$$\phi''(u) \leqq 0 \quad \text{for all } u > \alpha \text{ and } \sup_{u \in \mathbb{R}} \phi'(u) < \Sigma'(\alpha). \tag{7.9b}$$

When ϕ and Σ are the quadratics in (7.2), then conditions (7.3)–(7.8) are satisfied; if $k \geqq \mu$, then condition (7.9a) also holds. Examples of other functions that satisfy (7.3)–(7.9a) are the functions

$$\begin{aligned} \phi(u) &= au^2 + bu^4 \\ \Sigma(u) &= cu^4 \end{aligned} \tag{7.10}$$

with a, b, c positive and $2a\alpha + 4(b - c)\alpha^3$ positive. Examples of functions that satisfy (7.3)–(7.8), (7.9b) are the functions

$$\phi(u) = \begin{cases} 1 + \sin\dfrac{\pi(u-\alpha)}{2\alpha}, & 0 \leqq u \leqq \alpha \\ 1 + \dfrac{\pi}{2\alpha}(u-\alpha), & \alpha < u \\ \phi(-u), & u < 0 \end{cases} \tag{7.11}$$

$$\Sigma(u) = cu^4,$$

with $c > \pi/8\alpha^4$.

The distinction between the conditions (7.9a) and (7.9b) lies chiefly in the relative strengths of the restoring forces ϕ' and Σ'. In fact, when ϕ and Σ satisfy (7.2)–(7.8) and (7.9a), one can show that an analogue of Theorem 6.1 holds, so that all initial data in the plane $u_p = 0$ produce positive half-trajectories whose ω-limit sets are elastic trajectories in the plane $u_p = 0$. Thus, condition (7.9a), in which the restoring force $f = \phi'$ is stronger than the force $\sigma = \Sigma'$, leads to self-annealing for all initial data in the plane $u_p = 0$. In particular, if a solution of (2.11), (7.1), (2.13) and (2.14) with $u_p^\circ = 0$ undergoes at least one plastic segment, it undergoes infinitely many plastic segments and u_p tends to zero in the limit of large times. On the other hand, when ϕ and Σ satisfy (7.2)–(7.8), and (7.9b), then one can show that every solution of (2.11), (7.1), (2.13), and (2.14) undergoes at most two plastic segments, and, only for exceptional initial data is it true that u_p has limit zero for large times. Thus, condition (7.9b), in which the restoring force $f = \phi'$ is weaker than $\sigma = \Sigma'$, leads to self-annealing only in exceptional circumstances.

The methods used to obtain generalizations of the results of Sections 3, 5, and 6 to the case of non-quadratic potentials follow closely the ones presented in those sections for the case of quadratic potentials. Details for the special case when the potential Σ is quadratic can be found in the Ph.D. thesis of THOMAS, [1984, 1]. It should be noted that, when ϕ is not quadratic, there also are generalizations of Theorem 4.1 on stability of equilibrium states (see [1984, 1], Theorem 4.2), but the proofs of these generalizations are different from our proof of Theorem 4.1, because we have been unable to find Liapunov functions when the potential ϕ is not quadratic.

Acknowledgments. OWEN received support from the National Science Foundation during the period in which the research described in this article was carried out. We also thank ROBERT KOHN and KEMING WANG for pointing out and tracking down the interesting work of MIYOSHI [1985, 1].

References

1979 1. BUHITE, JACK L. & DAVID R. OWEN, An Ordinary Differential Equation from the Theory of Plasticity, *Arch. Rational Mech. Anal.* **71**, 357–383.

1980 1. HALE, JACK K., *Ordinary Differential Equations* (Second Edition), Krieger, New York.

1984 1. THOMAS, JOHN P., Ph.D. Thesis, Department of Mathematics, Carnegie Mellon University.

1985 1. MIYOSHI, TETSUHIKO, *Foundations of the Numerical Analysis of Plasticity*, Lecture Notes in Numerical and Applied Analysis, Vol. 7, North Holland: Amsterdam, New York, Oxford.

1987 1. OWEN, DAVID R., Weakly Decaying Energy Separation and Uniqueness of Motions of Elastic-Plastic Oscillators with Work Hardening, *Arch. Rational Mech. Anal.* **98**, 95–114.

1988 1. BIELAK, JACOBO, DAVID R. OWEN, & JOHN P. THOMAS, Permanent Drift of Bilinear Hysteretic Oscillators Subject to Impulsive Loading.

2. MIHAELESCU-SULICIU, M., I. SULICIU, & W. WILLIAMS, On Viscoplastic and Elastic-Plastic Oscillators, *Q. Applied Math.*, to appear.

Department of Mathematics
Carnegie Mellon University
Pittsburgh, Pennsylvania

and

Institut für Strukturmechanik
Deutsche Forschungs- und Versuchsanstalt
für Luft- und Raumfahrt e.V.
Braunschweig

(Received December 14, 1988)

On Saint-Venant's Principle in Finite Anti-Plane Shear: An Energy Approach

C. O. HORGAN & L. E. PAYNE

Dedicated to Bernard Coleman on the occasion of his 60th birthday

1. Introduction

Some time ago a version of Saint-Venant's principle was formulated and established for *finite* elastostatics [1]. As was discussed in [1], the issues of concern in connection with Saint-Venant's principle in the nonlinear theory of elasticity are considerably more involved then those arising in the linear theory. (For a survey of results on Saint-Venant's principle, primarily for linear theories, see *e.g.* [2–5].) One difficulty is that the appropriate Saint-Venant solutions need to be carefully characterized (see *e.g.* [6–13] and the references cited therein). Secondly, in the absence of superposition, consideration of *self-equilibrated* end loads is no longer sufficient. Furthermore, instabilities may have to be taken into account. Also the decay rate for end effects, even if exponential, might depend on the overall loading as well as on geometry and material characteristics. Several of these issues have been considered in recent studies in the nonlinear elasticity context [1, 14–23] as well as in investigations of spatial decay of solutions of nonlinear elliptic partial differential equations [24–32].

The setting considered in [1] is the simplest possible within the exact theory of finite elasticity: finite *anti-plane shear* of an infinitely long cylinder composed of homogeneous, isotropic, incompressible material. The constitutive law is assumed to belong to the special class of such laws that permit nontrivial states of finite anti-plane shear [33]. The undeformed cross-section of the cylinder is a semi-infinite strip, the long sides of which remain traction-free while the short side carries a prescribed shear traction directed parallel to the generators. This given traction is not necessarily uniformly distributed and the associated average shear stress p (and hence the resultant shear force) does not necessarily vanish. At infinity in the strip, the displacement is that of a simple shear with shear stress p. It is shown in [1] that, along the long sides of the strip, the difference between the nonvanishing component of shear stress and its average value p is bounded by an exponentially decaying function of the distance from the end. A lower bound is given for the rate of decay for materials for which the governing partial differential equation is uniformly elliptic and which either "harden" or "soften" in simple shear. This lower bound depends on the width of the strip, on the average stress p, on the strain-energy density of the material and on a measure of the departure from uniformity of the given end traction. The latter quantity enters into the estimates obtained in [1] as a result of the main analytical technique employed there, namely, a comparison (or maximum) principle for second-order quasi-linear elliptic partial differential equations.

The purpose of this paper is to investigate an analog of the problem considered in [1] using an energy approach. Such methods, establishing the decay of energy-like quadratic functionals with distance from the loaded boundary, have been extensively used in investigations of Saint-Venant's principle (see *e.g.* [2–5]). This approach is widely applicable in contrast to the arguments based on the maximum principle which are generally limited to second-order problems.

In Sections 2 and 3 we formulate the boundary-value problem to be studied. We consider an analog of the problem studied in [1] for a finite rectangular region of width h, with traction-free lateral sides, prescribed shear traction at the near end with resultant shear force ph, and subjected to a *uniformly distributed* shear stress p at the far end. Under rather general constitutive assumptions (which do not, for example, necessarily require that the governing second-order, quasilinear partial differential equation be elliptic, although of course that is the case of primary interest), our main results are established in Section 4. It is shown there that an energy-like quadratic functional, defined on the difference between the deformation field and a state of simple shear corresponding to p, decays exponentially with distance from the near end. The estimated decay rate (which is a lower bound for the actual rate of exponential decay) is characterized in terms of the load level p, the domain geometry and material properties. In Section 5, the special situation when the load is *self-equilibrated* (corresponding to $p = 0$) is considered and the results compared with previous work [1, 17]. Furthermore, a specific class of incompressible materials, namely those of power-law type, is examined for which the estimated decay rate can be written down explicitly, as was also possible in [1]. In Section 6 an alternative energy technique is described which is valid for a sub-class of the materials considered.

The energy method employed here has certain advantages and disadvantages when compared to the technique based on maximum principles used in [1]. In addition to the wider applicability of energy methods already mentioned, it is not necessary here to convert the basic Neumann problem to a Dirichlet problem as was the case in [1]; furthermore we obtain results on decay for a finite domain. In this paper, we shall not address the issue of obtaining pointwise estimates — this can often be technically complicated when using the energy approach (see *e.g.* [2]) while immediate from the maximum principle [29]. The estimated decay rates that we obtain here are in some cases better and in some cases not as good as those derived in [1].

2. Finite anti-plane shear deformations

Consider a body composed of an elastic material which in an undeformed state occupies an infinitely long cylinder with generators parallel to the x_3-axis of a rectangular Cartesian coordinate system. Let R be the open cross-section of the cylinder in the plane $x_3 = 0$. An *anti-plane shear* of the cylinder is a deformation with a single out-of-plane displacement component $u(x_1, x_2)$.

Suppose that the material is homogeneous, isotropic, incompressible and is characterized by an elastic potential W representing the strain-energy density per unit undeformed volume. Then W is a function of the first two fundamental scalar invariants I_1, I_2 of the Cauchy-Green deformation tensors, and so we have $W = \hat{W}(I_1, I_2)$. Not every such material can sustain a non-trivial state of anti-plane shear [33]. The entire subclass of these materials has been determined in [33] and it is only this class of materials that we consider here.

It may easily be shown that in all anti-plane deformations one has $I_1 = I_2 = 3 + q^2$, $q^2 = u_{,\alpha}u_{,\alpha}$. (The usual Cartesian tensor notation is used here and henceforth, with a repeated subscript indicating summation and a subscript preceded by a comma denoting partial differentiation with respect to the corresponding coordinate.) It is sufficient for our purposes to consider the restriction of $\hat{W}(I_1, I_2)$ to the line $I_1 = I_2$. Hence we define $W(I)$ by

$$W = W(I) = \hat{W}(I, I), \qquad I \geqq 3. \tag{2.1}$$

At infinitesimal deformations, the shear modulus is given by $\mu = 2W'(3) > 0$ where the prime denotes differentiation; $W(I)$ is assumed to be twice continuously differentiable for $I \geqq 3$.

The response of this material in simple shear is described by

$$\hat{\tau}(k) = 2kW'(3 + k^2) \qquad \text{for all } k, \tag{2.2}$$

where the odd function $\hat{\tau}(k)$ gives the shear stress associated with an amount of shear k. The (secant) modulus of shear, M, is taken to be positive:

$$M(k) \equiv \hat{\tau}(k)/k = 2W'(3 + k^2) > 0 \qquad \text{for all } k. \tag{2.3}$$

The components τ_{ij} of the true stress accompanying an anti-plane shear are given by (see [33, 34])

$$\tau_{\alpha 3} = \tau_{3\alpha} = 2W'(I)\, u_{,\alpha}, \qquad \tau_{\alpha\beta} = 0, \qquad \tau_{33} = 2W'(I)\, q^2, \tag{2.4}$$

where $I = 3 + q^2$. If the tractions on the lateral surface of the cylinder are independent of x_3, the field equations of finite elastostatics, in the absence of body forces, can be shown to reduce (see *e.g.* [33–35], to the following single quasilinear equation for u:

$$[\varrho(q^2)\, u_{,\alpha}]_{,\alpha} = 0 \qquad \text{on } R, \tag{2.5}$$

where

$$\varrho(q^2) = 2W'(3 + q^2), \qquad q^2 = u_{,\alpha}u_{,\alpha}. \tag{2.6}$$

3. Statement of the boundary-value problem

We are concerned with finite anti-plane shear deformation of the rectangular region $R = \{(x_1, x_2) \mid 0 < x_1 < l, 0 < x_2 < h\}$. The long sides of the rectangle are to be traction-free and so by (2.4)

$$u_{,2}(x_1, 0) = u_{,2}(x_1, h) = 0, \quad 0 < x_1 < l. \tag{3.1}$$

The end $x_1 = 0$ is subjected to a prescribed axial shear traction so that by (2.4) and (2.6)

$$\varrho u_{,1} = f(x_2) \quad \text{on } x_1 = 0, \quad 0 \leqq x_2 \leqq h, \tag{3.2}$$

where f is a given function taken to be continuous on $[0, h]$. The average shear stress on the end $x_1 = 0$ is denoted by

$$p = \frac{1}{h} \int_0^h f(x_2)\, dx_2 . \tag{3.3}$$

For simplicity in what follows, we shall assume that $p \geqq 0$. It will be seen subsequently that if p were negative, all our results would hold with p replaced by $|p|$. The far end of the rectangle $x_1 = l$ is subjected to a uniformly distributed shear stress p so that

$$\varrho u_{,1} = p \quad \text{at } x_1 = l, \quad 0 \leqq x_2 \leqq h. \tag{3.4}$$

If $f(x_2)$ were *constant* on $[0, h]$, then the function

$$u(x_1, x_2) = k x_1 \equiv v(x_1) \tag{3.5}$$

would satisfy the differential equation (2.5), the free surface conditions (3.1), and the end conditions (3.2) and (3.4) provided the constant k were chosen so that

$$\hat{\tau}(k) = k\varrho(k^2) = p. \tag{3.6}$$

If $\hat{\tau}(k)$ is invertible, (3.6) determines k uniquely as $k = \hat{k}(p)$. When f is *not* constant, one expects that solutions $u(x_1, x_2)$ of (2.5) and (3.1)–(3.4) might approach the state of simple shear (3.5) as x_1 increases. The principle of Saint-Venant type established here quantifies this expectation.

We assume the existence of solutions $u(x_1, x_2)$ of (2.5) and (3.1)–(3.4) which are continuously differentiable on R and twice continuously differentiable on the interior of R. The problem considered in [1] is analogous to the boundary-value problem described above when the rectangle R is replaced by a semi-infinite strip. The special case of *self-equilibrated* end loading occurs when $p = 0$. In this case one finds from (3.6) that $k = 0$ and so the state of simple shear (3.5) is also zero.

A sufficient condition to ensure the invertibility of (3.6) is given in [1]. Thus it is assumed in [1] that $\hat{\tau}(k)$ satisfies

$$k\hat{\tau}'(k) \geqq c\hat{\tau}(k) \quad \text{for all } k \geqq 0, \tag{3.7}$$

for some positive constant c. In view of (2.3) and (2.6), the condition (3.7) may also be written as

$$\varrho + 2\varrho' q^2 \geqq c\varrho, \quad c > 0, \tag{3.8}$$

for all solutions u of (2.5) and at all points of R, where $\varrho' = d\varrho/dq^2$. This, together with (2.3) and (2.6), implies that (2.5) is *uniformly elliptic* and ensures that (3.6) can be inverted to give k as an odd, monotone, strictly increasing function of $p: k = \hat{k}(p)$ with $\hat{k}(\infty) = \infty$.

It will be seen in the sequel that the results obtained here do not require that (2.5) be uniformly elliptic. If it is not, one is not assured of the invertibility of (3.6). However, in this case, one may consider the boundary-value problem as one in which the boundary condition (3.4) is replaced by requiring $u(x_1, x_2)$ to be given by (3.5) at $x_1 = l$, where the amount of shear k is prescribed. Then, corresponding to this k, a unique p can be determined by use of (3.6) and the decay estimates are expressed in terms of this computed value of p. Of course, it is well known that if (2.5) is *not* uniformly elliptic, the existence of a solution is not assured for arbitrary boundary data. For example, if (2.5) is the minimal surface equation, in which case $\varrho = (1 + q^2)^{-\frac{1}{2}}$, we see that the prescribed function $f(x_2)$ in (3.2) must be such that $|f(x_2)| < 1$ on $[0, h]$.

4. Energy decay estimates

In this section we establish our main energy decay estimates for solutions of (2.5) and (3.1)–(3.4). It is shown that the energy measure

$$E(z) = \int_{R_z} \varrho(q^2)\, w_{,\alpha} w_{,\alpha}\, dA, \qquad q^2 = u_{,\beta} u_{,\beta}, \tag{4.1}$$

contained in the subdomain $R_z = \{(x_1, x_2) \mid 0 \leqq z < x_1 < l,\ 0 < x_2 < h\}$ has exponential decay in z, where the function $w(x_1, x_2)$ is defined by

$$w(x_1, x_2) = u(x_1, x_2) - v(x_1), \tag{4.2}$$

and $v(x_1)$ is given by (3.5). Thus $E(z)$ can be viewed as a "weighted energy" associated with the difference (4.2) between $u(x_1, x_2)$ and the state of simple shear (3.5).

The hypotheses that we make concerning ϱ may be conveniently separated into two cases. It is asumed that positive constants m_α, d_α and positive functions $c_\alpha(t)$ $(\alpha = 1, 2)$ exist such that for all values of the arguments s, t we have either

Case 1.

$$\varrho \geqq m_1 > 0, \tag{4.3}$$

$$|\varrho(s^2) - \varrho(t^2)| \leqq c_1(t)\, |t - s|\, \varrho(s^2), \tag{4.4}$$

and

$$|\varrho(s^2) - \varrho(0)| \leqq d_1\, s^2 \varrho(s^2), \tag{4.5}$$

or

Case 2.

$$\varrho^{-1} \geqq m_2 > 0, \tag{4.6}$$

$$|\varrho^{-1}(s^2) - \varrho^{-1}(t^2)| \leqq c_2(t)\, |t - s|, \tag{4.7}$$

and

$$|\varrho^{-1}(s^2) - \varrho^{-1}(0)| \leqq d_2 s^2 \varrho(s^2), \tag{4.8}$$

respectively.

It should be noted that neither (4.3)–(4.5) nor (4.6)–(4.8) implies that equation (2.5) is elliptic.

We now proceed to establish our main conclusion, namely that

$$E(z) \leqq K_\alpha\, e^{-2(\pi/h)\nu_\alpha z}, \quad 0 \leqq z \leqq l, \tag{4.9}$$

where $\alpha = 1$ or 2 in Case 1 or 2, respectively. Here

$$\nu_\alpha = \nu_\alpha(p), \quad \alpha = 1, 2, \tag{4.10}$$

are functions of p (to be determined) which depend on the strain-energy density. The estimated rate of energy decay is thus $2\pi\nu_\alpha/h$. The constants

$$K_\alpha = K_\alpha(E_0, m_\alpha, c_\alpha, d_\alpha, p), \quad \alpha = 1, 2, \tag{4.11}$$

depend on the quantity $E_0 \equiv E(0)$, which is the total energy contained in the rectangle R. We show in Appendix B how an upper bound for E_0 in terms of geometry, boundary data and strain-energy density of the material may be obtained.

The result (4.9) is established in several stages. First we derive the differential inequality[1]

$$E'(z) + 2\varkappa_\alpha(z)\, E(z) \leqq 0, \quad z \geqq 0, \tag{4.12}$$

where

$$\varkappa_\alpha(z) = \frac{m_\alpha \pi (1 - s_\alpha)}{B_\alpha(z)\,(1 + 2s_\alpha)}, \quad \alpha = 1, 2. \tag{4.13}$$

Here

$$\begin{aligned} B_1(z) &= \int_0^h \varrho(q^2)\, dx_2 \qquad \text{(Case 1)}, \\ B_2(z) &= \int_0^h \varrho^{-1}(q^2)\, dx_2 \qquad \text{(Case 2)} \end{aligned} \tag{4.14}$$

and

$$s_1 = c_1 k \quad \text{(Case 1)}, \qquad s_2 = c_2 p \quad \text{(Case 2)}. \tag{4.15}$$

The inequality (4.12) may be immediately integrated to yield

$$E(z) \leqq E_0 \exp\left[-2 \int_0^z \varkappa_\alpha(s)\, ds\right], \quad z \geqq 0. \tag{4.16}$$

The second stage of the proof uses the different hypotheses (4.5) or (4.8) in Cases 1 or 2, respectively, to obtain bounds for the exponential term in (4.16) leading to the result (4.9).

---

[1] The prime denotes differentiation with respect to z.

To establish (4.12) we proceed as follows: If L_z denotes the line segment $x_1 = z,\ 0 \leqq x_2 \leqq h$, we find, on using the divergence theorem and (2.5), (3.1) and (3.4)–(3.6), that

$$E(z) = \int_{L_z} [\varrho(k^2)\, v_{,1} - \varrho(q^2)\, u_{,1}]\, w\, dx_2 + \int_{R_z} [\varrho(k^2) - \varrho(q^2)]\, v_{,\beta} w_{,\beta}\, dA. \quad (4.17)$$

Now it follows from the divergence theorem and (2.5), (3.1), (3.4) that

$$\int_{L_z} \varrho(q^2)\, u_{,1}\, dx_2 = p, \quad 0 \leqq z \leqq l, \quad (4.18)$$

so that, by virtue of (3.6), we have

$$\int_{L_z} [\varrho(k^2)\, v_{,1} - \varrho(q^2)\, u_{,1}]\, dx_2 = 0, \quad 0 \leqq z \leqq l. \quad (4.19)$$

Thus, on introducing the function $W_\alpha(x_1, x_2)$ ($\alpha = 1, 2$, in Cases 1, 2 respectively) by

$$W_1 = w - \overline{w}_1(x_1) \quad \text{or} \quad W_2 = w - \overline{w}_2(x_1), \quad (4.20)$$

where

$$\overline{w}_1(x_1) = \frac{1}{B_1} \int_0^h \varrho(q^2)\, w\, dx_2, \quad \overline{w}_2(x_1) = \frac{1}{B_2} \int_0^h \varrho^{-1}(q^2)\, w\, dx_2 \quad \text{(Cases 1, 2)}, \quad (4.21)$$

we may write (4.17) as

$$\begin{aligned} E(z) &= \int_{L_z} [\varrho(k^2)\, v_{,1} - \varrho(q^2)\, u_{,1}]\, W_\alpha\, dx_2 + \int_{R_z} [\varrho(k^2) - \varrho(q^2)]\, v_{,\beta} w_{,\beta}\, dA \\ &= -\int_{L_z} \varrho(q^2)\, w_{,1} W_\alpha\, dx_2 - \int_{L_z} [\varrho(q^2) - \varrho(k^2)]\, v_{,1} W_\alpha\, dx_2 \\ &\quad + \int_{R_z} [\varrho(k^2) - \varrho(q^2)]\, v_{,\beta} w_{,\beta}\, dA, \end{aligned} \quad (4.22)$$

where (4.2) has been used in obtaining the final form of (4.22).

At this stage, we treat Case 1 and Case 2 separately. Let I denote the last integral on the right-hand side of (4.22). In Case 1, on using (4.4) with $s = q,\ t = k$ we obtain

$$I \leqq c_1 \int_{R_z} |k - q|\, \varrho(q^2)\, |v_{,\beta}|\, |w_{,\beta}|\, dA \quad (4.23)$$

$$\leqq c_1 k \int_{R_z} \varrho(q^2)\, w_{,\beta} w_{,\beta}\, dA, \quad (4.24)$$

where we have made use of

$$|k - q| = (k^2 - 2kq + q^2)^{\frac{1}{2}} \leqq (v_{,\beta} v_{,\beta} - 2 v_{,\beta} u_{,\beta} + u_{,\beta} u_{,\beta})^{\frac{1}{2}} = (w_{,\beta} w_{,\beta})^{\frac{1}{2}}, \quad (4.25)$$

which in turn follows from

$$k^2 = v_{,1}^2 = v_{,\beta} v_{,\beta}, \quad kq = (v_{,\beta} v_{,\beta})^{\frac{1}{2}} (u_{,\alpha} u_{,\alpha})^{\frac{1}{2}} \geqq v_{,\beta} u_{,\beta}, \quad (4.26)$$

and the definition of w in (4.2). Thus, in view of (4.1) we may write (4.24) as

$$I \leqq c_1 k E(z), \tag{4.27}$$

and so (4.22) now yields, in Case 1,

$$(1 - c_1 k)\, E(z) \leqq -\int_{L_z} \varrho(q^2)\, w_{,1} W_1\, dx_2 - \int_{L_z} [\varrho(q^2) - \varrho(k^2)]\, v_{,1} W_1\, dx_2 \equiv J_1 + J_2. \tag{4.28}$$

Differentiation of (4.1) gives

$$E'(z) = -\int_{L_z} \varrho(q^2)\, w_{,\alpha} w_{,\alpha}\, dx_2. \tag{4.29}$$

We now show how the integrals J_1, J_2 may be bounded above in terms of $E'(z)$, which will lead to the desired inequality (4.12) in Case 1. We have

$$|J_1| \leqq \int_{L_z} \varrho(q^2)\, |w_{,1}|\, |W_1|\, dx_2 \leqq \left(\int_{L_z} \varrho(q^2)\, w_{,1}^2\, dx_2\right)^{\frac{1}{2}} \left(\int_{L_z} \varrho(q^2)\, W_1^2\, dx_2\right)^{\frac{1}{2}}, \tag{4.30}$$

on using Schwarz's inequality. Employing the hypothesis (4.4) with $s = q$, $t = k$ we obtain

$$\begin{aligned} |J_2| &\leqq c_1 k \int_{L_z} \varrho(q^2)\, |k - q|\, |W_1|\, dx_2 \\ &\leqq c_1 k \int_{L_z} \varrho(q^2)\, (w_{,\beta} w_{,\beta})^{\frac{1}{2}}\, |W_1|\, dx_2, \end{aligned} \tag{4.31}$$

on making use of (4.25). Finally, by Schwarz's inequality, we obtain

$$|J_2| \leqq c_1 k \left(\int_{L_z} \varrho(q^2)\, w_{,\beta} w_{,\beta}\, dx_2\right)^{\frac{1}{2}} \left(\int_{L_z} \varrho(q^2)\, W_1^2\, dx_2\right)^{\frac{1}{2}}. \tag{4.32}$$

It remains to find an upper bound for the integral

$$J_3 \equiv \int_{L_z} \varrho(q^2)\, W_1^2\, dx_2 \tag{4.33}$$

appearing in both (4.30) and (4.32). We recall from its definition in (4.20), (4.21) that

$$\int_0^h \varrho(q^2)\, W_1\, dx_2 = 0. \tag{4.34}$$

Following a scheme introduced in [24], we now make the change of variable, for fixed x_1,[2]

$$t = \int_0^{x_2} \varrho(x_1, \eta)\, d\eta, \tag{4.35}$$

[2] Here we regard ϱ evaluated on a solution u as a function of x_1 and x_2.

so that $dt = \varrho\, dx_2$. Since (4.34) can now be written

$$\int_0^{B_1} W_1\, dt = 0, \tag{4.36}$$

we invoke the Wirtinger inequality

$$\int_0^{B_1} W_1^2\, dt \leqq \frac{B_1^2}{\pi^2} \int_0^{B_1} \left(\frac{dW_1}{dt}\right)^2 dt, \tag{4.37}$$

which can be written as

$$J_3 = \int_0^h \varrho(q^2)\, W_1^2\, dx_2 \leqq \frac{B_1^2}{\pi^2} \int_0^h w_{,2}^2\, \varrho^{-1}(q^2)\, dx_2, \tag{4.38}$$

where we have used $W_{1,2} = w_{,2}$ as follows from (4.20). Since $\varrho(q^2) \geqq m_1$ in Case 1, we deduce from (4.38) that

$$J_3 \leqq \frac{B_1^2}{m_1^2 \pi^2} \int_0^h \varrho(q^2)\, w_{,2}^2\, dx_2. \tag{4.39}$$

On combining (4.39), (4.32), and (4.30) we have

$$|J_1 + J_2| \leqq \frac{B_1}{m_1 \pi} \left(\int_{L_z} \varrho(q^2)\, w_{,2}^2\, dx_2\right)^{\frac{1}{2}} \left[\left(\int_{L_z} \varrho(q^2)\, w_{,1}^2\, dx_2\right)^{\frac{1}{2}} + c_1 k \left(\int_{L_z} \varrho(q^2)\, w_{,\alpha} w_{,\alpha}\, dx_2\right)^{\frac{1}{2}}\right]. \tag{4.40}$$

Using the arithmetic-geometric mean inequality twice in (4.40), we thus obtain

$$|J_1 + J_2| \leqq \frac{B_1}{2 m_1 \pi} (1 + 2 c_1 k) \int_{L_z} \varrho(q^2)\, w_{,\beta} w_{,\beta}\, dx_2. \tag{4.41}$$

By virtue of (4.29) we may write (4.41) as

$$|J_1 + J_2| \leqq \frac{B_1}{2 m_1 \pi} (1 + 2 c_1 k)\, [-E'(z)]. \tag{4.42}$$

On employing (4.42) in (4.28), we arrive at the result (4.12) in Case 1, where

$$\varkappa_1(z) = \frac{m_1 \pi (1 - c_1 k)}{B_1(z)\, (1 + 2 c_1 k)}. \tag{4.43}$$

To ensure that $\varkappa_1(z) \geqq 0$, it is clear from (4.43) that the constant $c_1 = c_1(k)$ must be such that

$$c_1(k)\, k < 1. \tag{4.44}$$

The derivation of (4.12) in Case 2 proceeds along similar lines. We start with (4.22) with $\alpha = 2$ and consider the last integral on the right-hand side of (4.22), denoted again by I. On using the hypothesis (4.7) with $s=q, t=k$ we obtain

$$I \leqq c_2 \int_{R_z} |k-q|\, \varrho(k^2)\, \varrho(q^2)\, |v_{,\beta}|\, |w_{,\beta}|\, dA \tag{4.45}$$

$$\leqq c_2 k \varrho(k^2) \int_{R_z} \varrho(q^2)\, w_{,\beta} w_{,\beta}\, dA \tag{4.46}$$

where (4.25) and (3.5) have been employed to get (4.46). By virtue of (3.6), (4.1), we write (4.46) as

$$I \leqq c_2 p E(z), \tag{4.47}$$

and so (4.22) now yields, in Case 2,

$$(1-c_2 p)\, E(z) \leqq -\int_{L_z} \varrho(q^2)\, w_{,1} W_2\, dx_2 - \int_{L_z} [\varrho(q^2) - \varrho(k^2)]\, v_{,1} W_2\, dx_2 \equiv J_4 + J_5 . \tag{4.48}$$

As in (4.30), we immediately obtain

$$|J_4| \leqq \left(\int_{L_z} \varrho(q^2)\, w_{,1}^2\, dx_2\right)^{\frac{1}{2}} \left(\int_{L_z} \varrho(q^2)\, W_2^2\, dx_2\right)^{\frac{1}{2}}. \tag{4.49}$$

On using (4.7) with $s=q, t=k$ we obtain

$$|J_5| \leqq c_2 k \varrho(k^2) \int_{L_z} \varrho(q^2)\, |k-q|\, |W_2|\, dx_2 \leqq c_2 p \int_{L_z} \varrho(q^2)\, (w_{,\beta} w_{,\beta})^{\frac{1}{2}}\, |W_2|\, dx_2 , \tag{4.50}$$

the final inequality following from (4.25) and (3.6). Thus, we deduce from (4.50) that

$$|J_5| \leqq c_2 p \left(\int_{L_z} \varrho(q^2)\, w_{,\beta} w_{,\beta}\, dx_2\right)^{\frac{1}{2}} \left(\int_{L_z} \varrho(q^2)\, W_2^2\, dx_2\right)^{\frac{1}{2}}. \tag{4.51}$$

We now obtain an upper bound for the integral

$$J_6 \equiv \int_{L_z} \varrho(q^2)\, W_2^2\, dx_2 \tag{4.52}$$

appearing on the right-hand sides of (4.49) and (4.51). By virtue of its definition in (4.20) and (4.21)

$$\int_0^h \varrho^{-1}(q^2)\, W_2\, dx_2 = 0 . \tag{4.53}$$

We use the change of variable, for fixed x_1,

$$t = \int_0^{x_2} \varrho^{-1}(x_1, \eta)\, d\eta . \tag{4.54}$$

Since (4.53) can now be written

$$\int_0^{B_2} W_2\, dt = 0, \tag{4.55}$$

we again invoke the Wirtinger inequality

$$\int_0^{B_2} W_2^2\,dt \leqq \frac{B_2^2}{\pi^2}\int_0^{B_2}\left(\frac{dW_2}{dt}\right)^2 dt, \tag{4.56}$$

which can be written as

$$\int_0^h \varrho^{-1}(q^2)\,W_2^2\,dx_2 \leqq \frac{B_2^2}{\pi^2}\int_0^h \varrho(q^2)\,w_{,2}^2\,dx_2. \tag{4.57}$$

Since $\varrho^{-1}(q^2) \geqq m_2$ in Case 2, we deduce from (4.52) and (4.57) that

$$J_6 \leqq \frac{B_2^2}{m_2^2\pi^2}\int_0^h \varrho(q^2)\,w_{,2}^2\,dx_2. \tag{4.58}$$

On combining (4.58), (4.51) and (4.49) we obtain

$$|J_4 + J_5| \leqq \frac{B_2}{m_2\pi}\left(\int_{L_z}\varrho(q^2)\,w_{,2}^2\,dx_2\right)^{\frac{1}{2}}$$

$$\times\left[\left(\int_{L_z}\varrho(q^2)\,w_{,1}^2\,dx_2\right)^{\frac{1}{2}} + c_2p\left(\int_{L_z}\varrho(q^2)\,w_{,\alpha}w_{,\alpha}\,dx_2\right)^{\frac{1}{2}}\right], \tag{4.59}$$

$$\leqq \frac{B_2}{2m_2\pi}(1 + 2c_2p)\int_{L_z}\varrho(q^2)\,w_{,\beta}w_{,\beta}\,dx_2, \tag{4.60}$$

where the arithmetic–geometric mean inequality has been used to obtain (4.60). The result (4.60) has the same from as (4.41) with B_2, m_2, c_2p in place of B_1, m_1, c_1k. Thus, as in Case 1, we employ (4.60) in (4.48) to arrive at the result (4.12) in Case 2. Note that in this case, by analogy with (4.44), we require that $c_2p < 1$. This completes the verification of (4.12), and thus of (4.16), in both cases.

Next we proceed to show that the decay estimate (4.16) yields the following energy estimate:

$$E(z) \leqq E_0 \exp\left[-2\pi m_\alpha(1 - s_\alpha)(1 + 2s_\alpha)^{-1} z^2\Big/\int_0^z B_\alpha(s)\,ds\right], \quad \alpha = 1, 2. \tag{4.61}$$

Let

$$k_\alpha(z) = \frac{m_\alpha\pi}{B_\alpha(z)}, \quad \alpha = 1, 2, \tag{4.62}$$

so that

$$\varkappa_\alpha(z) = (1 - s_\alpha)(1 + 2s_\alpha)^{-1} k_\alpha(z). \tag{4.63}$$

On using Schwarz's inequality, it is easily shown that (see *e.g.*, pp. 315–316 of [24])

$$\frac{1}{m_\alpha\pi}\int_0^z k_\alpha(\eta)\,d\eta \geqq z^2\Big/\int_0^z B_\alpha(s)\,ds, \tag{4.64}$$

which, together with (4.63) and (4.16), yields the result (4.61).

The final step in establishing (4.9) is to use the hypotheses (4.3), (4.5) and (4.6), (4.8) in Cases 1 and 2, respectively, to obtain an upper bound for the quantity $\int_0^z B_\alpha(s)\,ds \equiv G_\alpha(z)$ $(\alpha = 1, 2)$ appearing in (4.61). We observe that it is only at this stage that we make use of the hypotheses (4.5) and (4.8), respectively.

In Case 1, we have, by virtue of (4.14),

$$G_1(z) = \int_0^z B_1(s)\,ds = \int_0^z \int_0^h \varrho(q^2)\,dA. \tag{4.65}$$

We write

$$G_1(z) = \int_0^z \int_0^h [\varrho(q^2) - \varrho(0)]\,dA + \varrho(0)\,hz, \tag{4.66}$$

so that, on using (4.5) with $s = q$, we obtain

$$G_1(z) \leqq d_1 \int_0^z \int_0^h q^2 \varrho(q^2)\,dA + \varrho(0)\,hz. \tag{4.67}$$

On writing

$$q^2 = u_{,\alpha} u_{,\alpha} = w_{,\alpha} w_{,\alpha} + 2k u_{,1} - k^2, \tag{4.68}$$

which makes use of (4.2) and (3.5), from (4.67) we have

$$\begin{aligned} G_1(z) \leqq d_1 \int_0^z \int_0^h \varrho(q^2)\, w_{,\alpha} w_{,\alpha}\,dA + 2 d_1 k \int_0^z \int_0^h u_{,1} \varrho(q^2)\,dA \\ - d_1 k^2 \int_0^z \int_0^h \varrho(q^2)\,dA + \varrho(0)\,hz. \end{aligned} \tag{4.69}$$

By virtue of (4.65), and making use of (4.19), from (4.69) we obtain

$$\begin{aligned} (1 + d_1 k^2)\, G_1(z) &\leqq d_1 \int_0^z \int_0^h \varrho(q^2)\, w_{,\alpha} w_{,\alpha}\,dA + [2 d_1 k^2 \varrho(k^2) + \varrho(0)]\,hz \\ &= d_1 [E_0 - E(z)] + [2 d_1 k^2 \varrho(k^2) + \varrho(0)]\,hz, \end{aligned} \tag{4.70}$$

the last equality following from the definition (4.1) of $E(z)$. Thus, from (4.70), we deduce the desired upper bound:

$$G_1(z) \leqq (1 + d_1 k^2)^{-1} \{d_1 E_0 + [2 d_1 k^2 \varrho(k^2) + \varrho(0)]\,hz\}. \tag{4.71}$$

To use (4.71) in (4.61), we write

$$\left(\int_0^z B_1(s)\,ds\right)^{-1} \equiv G_1^{-1}(z) \geqq (1 + d_1 k^2) \{d_1 E_0 + [2 d_1 k^2 \varrho(k^2) + \varrho(0)]\,hz\}^{-1} \tag{4.72}$$

so that

$$z^2 G^{-1}(z) \geqq (1 + d_1 k^2)\, z \left[[2d_1 k^2 \varrho(k^2) + \varrho(0)]\, h + \frac{d_1 E_0}{z} \right]^{-1}$$

$$= \frac{(1 + d_1 k^2)\, z}{[2d_1 k^2 \varrho(k^2) + \varrho(0)]\, h} \left[1 + \frac{d_1 E_0}{[2d_1 k^2 \varrho(k^2) + \varrho(0)]\, hz} \right]^{-1}$$

$$\geqq \frac{(1 + d_1 k^2)\, z}{[2d_1 k^2 \varrho(k^2) + \varrho(0)]\, h} - \frac{(1 + d_1 k^2)\, d_1 E_0}{[2d_1 k^2 \varrho(k^2) + \varrho(0)]^2\, h^2}. \tag{4.73}$$

On substitution from (4.73) into (4.61), we thus obtain the result

$$E(z) \leqq K_1\, e^{-2(\pi/h)\nu_1 z}, \quad 0 \leqq z \leqq l, \tag{4.74}$$

where

$$\nu_1 = \nu_1(p) = \frac{m_1(1 - s_1)(1 + d_1 k^2)}{[2d_1 k^2 \varrho(k^2) + \varrho(0)]\,(1 + 2s_1)}, \tag{4.75}$$

$$K_1 \equiv K_1(E_0, m_1, p) = E_0 \exp \left[\frac{2m_1 \pi E_0\, d_1(1 + d_1 k^2)(1 - s_1)}{\{2d_1 k^2 \varrho(k^2) + \varrho(0)\}^2\, (1 + 2s_1)\, h^2} \right]. \tag{4.76}$$

In (4.75), (4.76), $s_1 = c_1(k)\, k$, and $k = \hat{k}(p)$ is obtained by use of (3.6). This completes the proof of (4.9), (4.10) and (4.11) in Case 1.

In Case 2, we proceed in an analogous fashion. We have

$$G_2(z) = \int_0^z B_2(s)\, ds = \int_0^z \int_0^h \varrho^{-1}(q^2)\, dA, \tag{4.77}$$

which we write as

$$G_2(z) = \int_0^z \int_0^h [\varrho^{-1}(q^2) - \varrho^{-1}(0)]\, dA + \varrho^{-1}(0)\, hz. \tag{4.78}$$

On using (4.8) with $s = q$ we obtain

$$G_2(z) \leqq d_2 \int_0^z \int_0^h q^2 \varrho(q^2)\, dA + \varrho^{-1}(0)\, hz, \tag{4.79}$$

which, by (4.68), yields

$$G_2(z) \leqq d_2 \int_0^z \int_0^h \varrho(q^2)\, w_{,\alpha} w_{,\alpha}\, dA + 2d_2 k \int_0^z \int_0^h u_{,1} \varrho(q^2)\, dA$$

$$- d_2 k^2 \int_0^z \int_0^h \varrho(q^2)\, dA + \varrho(0)^{-1}\, hz. \tag{4.80}$$

Now using (4.19) in the second integral, dropping the third term and using the definition (4.1) of $E(z)$, we deduce from (4.80) the upper bound

$$G_2(z) \leqq d_2 E_0 + [2d_2 k^2 \varrho(k^2) + \varrho^{-1}(0)]\, hz. \tag{4.81}$$

Proceeding as we did in Case 1 to arrive at (4.73), we similarly find in Case 2 that

$$z^2 G_2^{-1}(z) \geqq \frac{z}{[2d_2 k^2 \varrho(k^2) + \varrho^{-1}(0)]\, h} - \frac{d_2 E_0}{[2d_2 k^2 \varrho(k^2) + \varrho^{-1}(0)]^2\, h^2}. \tag{4.82}$$

On substitution from (4.82) into (4.61), in Case 2 we finally obtain

$$E(z) \leqq K_2\, e^{-2(\pi/h)\nu_2 z}, \quad 0 \leqq z \leqq l, \tag{4.83}$$

where

$$\nu_2 = \nu_2(p) = \frac{m_2(1 - s_2)}{[2d_2 k^2 \varrho(k^2) + \varrho^{-1}(0)]\,(1 + 2s_2)}, \tag{4.84}$$

$$K_2 \equiv K_2(E_0, m_2, p) = E_0 \exp\left[\frac{2m_2 \pi E_0\, d_2(1 - s_2)}{\{2d_2 k^2 \varrho(k^2) + \varrho^{-1}(0)\}^2\,(1 + 2s_2)\, h^2}\right]. \tag{4.85}$$

In (4.84) and (4.85), $s_2 = c_2(k)\, p$, and $k = \hat{k}(p)$ is obtained by use of (3.6). This completes the proof of (4.9), (4.10) and (4.11) in Case 2.

5. Discussion

(i) Self-equilibrated loading

We begin by examining the form taken by the decay estimates (4.74)–(4.76) (Case 1) and (4.83)–(4.85) (Case 2) in the special case of *self-equilibrated* end loading where $p = 0$ in (3.3) and (3.4), and so, from (3.6), $k = 0$. Thus we are concerned with spatial evolution of solutions of (2.5), (3.1)–(3.4) to the *trivial state* of simple shear $v \equiv 0$. In Case 1, we have $s_1 \equiv c_1 k = 0$ and so from (4.75) we obtain

$$\nu_1 = \nu_1(0) = m_1/\varrho(0). \tag{5.1}$$

By the definition of m_1 in (4.3), we see that $\varrho(0) \geqq m_1$ and so

$$\nu_1(0) \leqq 1 \quad \text{in Case 1}. \tag{5.2}$$

Thus the estimated energy decay rate $2(\pi/h)\,\nu_1(0)$ *is never greater than that for Laplace's equation* $(2\pi/h)$ *subject to the same boundary conditions. Whenever* $\varrho(0) = m_1$, *the estimated decay rate is at least as fast as that for Laplace's equation.* (Note that the estimated decay rate given by (4.74) and (4.75) provides a *lower bound* for the actual decay rate.) In Case 1, from (4.76), with $k = 0$, we also obtain

$$K_1 = E_0 \exp\left[\frac{2\pi m_1\, d_1 E_0}{\varrho(0)^2\, h^2}\right]. \tag{5.3}$$

In Case 2 we have $s_2 = c_2(k)\,p = 0$ when $p = 0$. From (4.84) and (4.85) we thus obtain

$$\nu_2 = \nu_2(0) = m_2/\varrho^{-1}(0), \tag{5.4}$$

and

$$K_2 = E_0 \exp\left[\frac{2\pi m_2\, d_2 E_0}{[\varrho^{-1}(0)]^2\, h^2}\right], \tag{5.5}$$

respectively. Again, by virtue of the definition of m_2 in (4.6), we have $\varrho^{-1}(0) \geqq m_2$ and so

$$\nu_2(0) \leqq 1 \quad \text{in Case 2}. \tag{5.6}$$

Results analogous to those described above in Case 1, involving comparison between the estimated decay rates and that for Laplace's equation, are thus seen to hold also in Case 2.

In [24], which was concerned solely with the case of self-equilibrated end loading of a semi-infinite strip, results similar to those just described were obtained using hypotheses on ϱ somewhat similar to those assumed here. Thus, in [24], we also considered two cases in which either (4.3) and a simpler hypothesis than (4.5) (*cf.* (1.6) of [24]) *or* (4.6) and a hypothesis simpler than (4.8) (*cf.* (1.7) of [24]) were assumed to hold. In [1], in which arguments based on maximum principles were employed, a different set of hypotheses on ϱ were assumed to hold. More precisely, it was assumed in [1] that $\varrho(q^2) > 0$ for all q^2, and that

$$\varrho + 2\varrho' q^2 \geqq c\varrho, \qquad c\ (\text{constant}) > 0, \tag{5.7}$$

where ϱ' denotes $d\varrho/dq^2$, for all solutions u of (2.5) and at all points of R. The condition (5.7) is equivalent to assuming that (2.5) is *uniformly elliptic*. Furthermore, the materials considered in [1] were assumed to *harden* or *soften* in shear, *i.e.* either $\varrho'(s) > 0$ or $\varrho'(s) < 0$ for all s. For hardening materials, a pointwise decay estimate was obtained in [1] with estimated decay rate *coinciding* with that for Laplace's equation. For softening materials, the estimated decay rate obtained in [1] was always *less than* that for Laplace's equation. Similar results were obtained in [17] under less restrictive constitutive assumptions.

(*ii*) *Power-law materials*

The decay estimates (4.75) and (4.76) in Case 1, and (4.84) and (4.85) in Case 2, can be written down explicitly for particular classes of materials. We consider, as in [1] and [17], the class of power-law materials for which

$$\varrho = \mu(1 + bq^2/n)^{n-1}, \tag{5.8}$$

where μ, b and n are positive constants. The case $n = 1$ in (5.8) corresponds to the neo-Hookean materials for which ϱ is constant and (2.5) is Laplace's equation. When $n = \frac{1}{2}$ in (5.8), equation (2.5) is reducible, by a change of scale, to the minimal surface equation.

When $n > 1$, it is shown in Appendix A that ϱ, as given by (5.8), may be described by Case 1, while when $\frac{1}{2} \leq n < 1$, the hypotheses of Case 2 are seen to hold. [It is also easily seen from (5.8) that $\varrho' > 0$ or $\varrho' < 0$ according as $n > 1$ or $n < 1$, respectively.]

(a) $\frac{1}{2} \leq n < 1$; *Case 2*:

Since the results we obtain are simpler to describe in Case 2, we begin with this case. Thus suppose that $\frac{1}{2} \leq n < 1$. From (5.8), we see that (4.6) is satisfied with

$$m_2 = \mu^{-1}. \tag{5.9}$$

It is shown in Appendix A that, for the power-law materials (5.8), the hypotheses (4.7) and (4.8) of Case 2 hold with

$$c_2 = \mu^{-1}(b/n)^{\frac{1}{2}}, \quad \tfrac{1}{2} \leq n < 1,$$

$$d_2 = \mu^{-2}\, b/n, \quad \tfrac{1}{2} \leq n < \tfrac{2}{3}, \quad d_2 = \mu^{-2}\, b(1-n)/n, \quad \tfrac{2}{3} \leq n < 1. \tag{5.10}$$

Consider first the case of *self-equilibrated* end loading where $p = 0$, and so $k = 0$. Since $\varrho^{-1}(0) = \mu^{-1}$ for the materials (5.8), we thus conclude from (5.9) and (5.4) that, for self-equilibrated loading,

$$\nu_2 = \nu_2(0) = 1. \tag{5.11}$$

By virtue of (5.11), (5.9), (5.5), (4.85) we see that the exponential decay estimate (4.83) may be written as

$$E(z) \leq E_0 \exp\left[\frac{2\pi b E_0}{\mu n h^2}\right] e^{\frac{-2\pi z}{h}}, \qquad \tfrac{1}{2} \leq n < \tfrac{2}{3}, \tag{5.12a}$$

$$E(z) \leq E_0 \exp\left[\frac{2\pi(1-n)\, b E_0}{\mu n h^2}\right] e^{\frac{-2\pi z}{h}}, \quad \tfrac{2}{3} \leq n < 1, \tag{5.12b}$$

where $E(z)$ is given by

$$E(z) = \int_{R_z} \varrho(q^2)\, u_{,\alpha} u_{,\alpha}\, dA, \tag{5.13}$$

and $\varrho(q^2)$ is given by (5.8). Thus, for power-law materials with $\frac{1}{2} \leq n < 1$, we see that the estimated decay rate provided by (5.12) *coincides* with that for Laplace's equation. A result similar to (5.12) was also established in [24]. We note that the pointwise estimates obtained in [1], [17] when $\frac{1}{2} < n < 1$ have an estimated decay rate which is *slower* than that of (5.12).

We turn next to the general situation where the loading (3.12) is *not* self-equilibrated. In this case, the estimated decay rate $\nu_2 = \nu_2(p)$, given by (4.84), depends on the load level p. Using (5.9), (5.10), we write (4.84) as

$$\nu_2(p) = (1 - c_2 p)(1 + 2c_2 p)^{-1}\, [1 + 2\mu^{-1}(b/n)\, p\hat{k}(p)]^{-1}, \quad \tfrac{1}{2} < n < \tfrac{2}{3}, \tag{5.14}$$

$$\nu_2(p) = (1 - c_2 p)(1 + 2c_2 p)^{-1}\, [1 + 2\mu^{-1}(b(1-n)/n)\, p\hat{k}(p)]^{-1}, \quad \tfrac{2}{3} < n < 1, \tag{5.15}$$

where

$$c_2 = \mu^{-1}(b/n)^{\frac{1}{2}} \tag{5.16}$$

and $k = \hat{k}(p)$ is the unique value of k satisfying

$$\mu(1 + bk^2/n)^{n-1}\, k = p, \quad \tfrac{1}{2} < n < 1. \tag{5.17}$$

Alternatively, we can express $v_2 = v_2(k)$ explicitly in terms of k as

$$v_2(k) = \frac{[1 - (b/n)^{\frac{1}{2}}(1 + bk^2/n)^{n-1}\, k]\,[1 + 2(b/n)^{\frac{1}{2}}(1 + bk^2/n)^{n-1}\, k]^{-1}}{[1 + 2(b/n)(1 + bk^2/n)^{n-1}\, k^2]},$$

$$\tfrac{1}{2} < n < \tfrac{2}{3}, \tag{5.18}$$

with an analogous expression (using (5.15)) when $\frac{2}{3} < n < 1$.

An explicit expression for v_2 for the case of softening power-law materials (that is, (5.8) when $\frac{1}{2} < n < 1$) has also been obtained in [1] (see equations (8.7) and (8.8) of [1], valid for $\frac{3}{4} \leqq n < 1$). A direct comparison between the results of [1] and those established here is not possible, however, since the estimated decay rates obtained in [1] depend not only on the load level p but also on a parameter measuring a departure from uniformity of the prescribed end data $f(x_2)$ of (3.2). The introduction of this parameter in [1] arises naturally in an estimation procedure based on maximum principles.

As $p \to 0$ in (5.14) and (5.15), we recover the result (5.11) for the case of self-equilibrated loading. It is clear, however, from (5.14) and (5.15) that $v_2(p) > 0$ (thus providing an exponential *decay* in (4.85)) if and only if $1 - c_2 p > 0$, that is,

$$p < \mu(n/b)^{\frac{1}{2}}. \tag{5.19}$$

Thus the load level p (given by (3.3)), or equivalently, the boundary data (3.2) must be restricted to satisfy (5.19) in order for the decay estimate (4.83) to hold for the power-law materials (5.8) when $\frac{1}{2} < n < 1$.

(b) $n > 1$; *Case 1*:

Suppose now that $n > 1$. From (5.8) we see that the first hypothesis (4.3) of Case 1 is satisfied with

$$m_1 = \mu. \tag{5.20}$$

It is shown in Appendix A that the remaining hypotheses (4.4) and (4.5) of Case 1 hold with $c_1 = c_1(k)$ and d_1 given by

$$c_1(k) = \begin{cases} (b/n)^{\frac{1}{2}}(n-1)(1 + bk^2/n)^{\frac{n-1}{2}}, & 1 < n \leqq \frac{3}{2}, \quad (5.21)\\ (b/n)^{\frac{1}{2}}(n-1)(1 + bk^2/n)^{\frac{1}{2}}, & \frac{3}{2} < n \leqq 2, \quad (5.22)\\ (b/n)^{\frac{1}{2}}(n-1)\,[(1 + bk^2/n)^{\frac{1}{2}} + (b/n)^{\frac{1}{2}} k]\,(1 + bk^2/n)^{n-2}, & 2 < n < 3, \quad (5.23)\\ (b/n)^{\frac{1}{2}}(n-1)/4\,[(1 + bk^2/n)^{\frac{1}{2}} + (b/n)^{\frac{1}{2}} k]\,[1 + (1 + bk^2/n)^{n-2}], & n \geqq 3, \quad (5.24) \end{cases}$$

$$d_1 = (n-1)\, b/n, \quad n > 1, \tag{5.25}$$

respectively.

Consider first the case of self-equilibrated end loading where $p = 0$, and so $k = 0$. Since $\varrho(0) = \mu$ for the materials (5.8) when $n > 1$, we thus conclude from (5.20) and (5.1) that

$$\nu_1(0) = 1. \tag{5.26}$$

By virtue of (5.26), (5.20), (5.2), and (4.76) we see that the exponential decay estimate (4.74) may be written as

$$E(z) \leqq E_0 \exp\left[\frac{2\pi(n-1)\,bE_0}{\mu nh^2}\right] e^{-2\pi z/h}, \quad z \geqq 0. \tag{5.27}$$

Thus for power-law materials with $n > 1$, the estimated decay rate provided by (5.27) *coincides* with that for Laplace's equation. The pointwise decay estimates obtained in [1], [17] when $n > 1$ have an estimated decay rate the same as that of (5.27). A result similar to (5.27) was also established in [24].

We turn now to the situation where the loading (3.2) is not self-equilibrated. As in Case 2, we find that the estimated decay rate $\nu_1 = \nu_1(p)$, given by (4.75), depends on the load level p. Using (5.20), (5.25), we write (4.75) as

$$\nu_1(p) = [1 - c_1(\hat{k})\,\hat{k}]\,[1 + 2c_1(\hat{k})\,\hat{k}]^{-1}\,[1 + (b(n-1)/n)\,\hat{k}^2]$$
$$\times [1 + 2\mu^{-1}\,(b(n-1)/n)\,p\hat{k}]^{-1}, \quad n > 1, \tag{5.28}$$

where $c_1(\hat{k})$ is given by (5.21)–(5.24) and $k = \hat{k}(p)$ is the unique value of k satisfying

$$\mu(1 + bk^2/n)^{n-1}\,k = p, \quad n > 1. \tag{5.29}$$

Alternatively, one can express $\nu_1 = \nu_1(k)$ explicitly in terms of k on using (5.29), and the appropriate value of $c_1(k)$ from (5.21)–(5.24), in (5.28). We shall not write this result down here as it is somewhat cumbersome algebraically; moreover, an improved estimate for Case 1 is obtained in Section 6.

As $p \to 0$ in (5.28), we recover the result (5.26) for the case of self-equilibrated loading. It is also clear from (5.28) that $\nu_1(p) > 0$ (thus providing an exponential *decay* in (4.75)) if and only if

$$1 - c_1(\hat{k})\,\hat{k} > 0. \tag{5.30}$$

where $k = \hat{k}(p)$ is given by (5.29). In view of (5.21)–(5.24), this entails a restriction on $\hat{k}(p)$, oi equivalently, on the load level p, in order for (4.74) to hold for the power law materials (5.8) when $n > 1$. Since such an *a priori* restriction on the load level, or equivalently, on the boundary data, is an undesirable feature of the foregoing results, we turn now to a discussion of an alternative estimate in Case 1 which does not have this defect.

6. An alternative energy estimate in Case 1

As we have seen in Section 5 the results obtained in Case 1 when the loading is *not* self-equilibrated are only valid for a restricted range of the load p. The results obtained in [1] did not have this limitation. We now describe a modification of the preceding argument in Case 1 which may be used to provide an alternative estimate which, in general, holds for all p.

In addition to the hypotheses (4.3)–(4.5) made in Case 1, we now further assume that for all values of the arguments s and t

$$-[\varrho(s^2)-\varrho(t^2)](s^2-t^2) \leqq \tilde{c}_1[\varrho(s^2)+\varrho(t^2)](s-t)^2 \tag{6.1}$$

for some function $\tilde{c}_1(t)<1$, and we consider an alternative energy measure defined by

$$\tilde{E}(z)=\int_{R_z}[\varrho(q^2)+\varrho(k^2)]\,w_{,\alpha}w_{,\alpha}\,dA, \tag{6.2}$$

where w is still given by (4.2). We first derive the differential inequality

$$\tilde{E}'(z)+2\varkappa(z)\,\tilde{E}(z)\leqq 0, \tag{6.3}$$

where

$$\varkappa(z)=\frac{m_1\pi(1-\tilde{c}_1)}{(1+2c_1k)\,B(z)}, \tag{6.4}$$

$$B(z)=\int_0^h[\varrho(q^2)+\varrho(k^2)]\,dx_2. \tag{6.5}$$

The proof of (6.3)–(6.5) is similar to that used in establishing (4.12)–(4.14). On using the divergence theorem, and (2.5), (3.1), (3.4)–(3.6), we have

$$\tilde{E}(z)=\int_{L_z}[\varrho(k^2)\,v_{,1}-\varrho(q^2)\,u_{,1}]\,w\,dx_2+\int_{R_z}[\varrho(k^2)\,u_{,\alpha}-\varrho(q^2)\,v_{,\alpha}]\,w_{,\alpha}\,dA, \tag{6.6}$$

which may be written as

$$\begin{aligned}\tilde{E}(z)=&\int_{L_z}[\varrho(k^2)\,v_{,1}-\varrho(q^2)\,u_{,1}]\,w\,dx_2+\tfrac{1}{2}\int_{R_z}[\varrho(q^2)+\varrho(k^2)]\,w_{,\alpha}w_{,\alpha}\,dA\\&+\tfrac{1}{2}\int_{R_z}[\varrho(k^2)-\varrho(q^2)]\,(u_{,\alpha}+v_{,\alpha})\,(u_{,\alpha}-v_{,\alpha})\,dA\end{aligned} \tag{6.7}$$

on recalling from (4.2) that $w=u-v$. Since $v=kx_1$, we now employ the hypothesis (6.1) with $s=q$, $t=k$ to obtain an upper bound for the third integral on the right in (6.7) and so find

$$\begin{aligned}\tilde{E}(z)\leqq&\int_{L_z}[\varrho(k^2)\,v_{,1}-\varrho(q^2)\,u_{,1}]\,w\,dx_2+\tilde{E}(z)/2\\&+(\tilde{c}_1/2)\int_{R_z}[\varrho(q^2)+\varrho(k^2)]\,(q-k)^2\,dA,\end{aligned} \tag{6.8}$$

where the definition (6.2) has also been used. Making use of the ineqality (4.25) in the third integral on the right in (6.8), and recalling (6.2), we obtain

$$\tilde{E}(z)\leqq 2(1-\tilde{c}_1)^{-1}\int_{L_z}[\varrho(k^2)\,v_{,1}-\varrho(q^2)\,u_{,1}]\,w\,dx_2. \tag{6.9}$$

The line integral on the right in (6.9), denoted by J, will now be bounded in terms of $\tilde{E}'(z)$ to yield the desired result (6.3). To this end, we recall the conservation result (4.19), introduce the function $W(x_1, x_2)$ by

$$W = w - \overline{w}(x_1) \tag{6.10}$$

where

$$\overline{w}(x_1) = \frac{1}{B} \int_0^h [\varrho(q^2) + \varrho(k^2)]\, w\, dx_2 \tag{6.11}$$

(recall that B is defined in (6.5)), and obtain

$$J = \int_{L_z} [\varrho(k^2)\, v_{,1} - \varrho(q^2)\, u_{,1}]\, W\, dx_2 . \tag{6.12}$$

We write

$$J = \int_{L_z} -\varrho(q^2)\, w_{,1} W\, dx_2 - \int_{L_z} [\varrho(q^2) - \varrho(k^2)]\, v_{,1} W\, dx_2 = J^1 + J^2 . \tag{6.13}$$

Now

$$\begin{aligned} |J^1| &\leqq \int_{L_z} \varrho(q^2)\, |w_{,1}|\, |W|\, dx_2 \leqq \int_{L_z} [\varrho(q^2) + \varrho(k^2)]\, |w_{,1}|\, |W|\, dx_2 \\ &\leqq \left(\int_{L_z} [\varrho(q^2) + \varrho(k^2)]\, w_{,1}^2\, dx_2 \right)^{\frac{1}{2}} \left(\int_{L_z} [\varrho(q^2) + \varrho(k^2)]\, W^2\, dx_2 \right)^{\frac{1}{2}} . \end{aligned} \tag{6.14}$$

On using the hypothesis (4.4) with $s = q$, $t = k$ we obtain

$$\begin{aligned} |J^2| &\leqq c_1 k \int_{L_z} \varrho(q^2)\, |k - q|\, |W|\, dx_2 \leqq c_1 k \int_{L_z} \varrho(q^2)\, (w_{,\beta} w_{,\beta})^{\frac{1}{2}}\, |W|\, dx_2 , \\ &\leqq c_1 k \int_{L_z} [\varrho(q^2) + \varrho(k^2)]\, (w_{,\beta} w_{,\beta})^{\frac{1}{2}}\, |W|\, dx_2 , \end{aligned} \tag{6.15}$$

where we have made use of (4.25). Thus, by Schwarz's inequality

$$|J^2| \leqq c_1 k \left(\int_{L_z} [\varrho(q^2) + \varrho(k^2)]\, w_{,\beta} w_{,\beta}\, dx_2 \right)^{\frac{1}{2}} \left(\int_{L_z} [\varrho(q^2) + \varrho(k^2)]\, W^2\, dx_2 \right)^{\frac{1}{2}} . \tag{6.16}$$

It remains to find an upper bound for the integral

$$J^3 \equiv \int_{L_z} [\varrho(q^2) + \varrho(k^2)]\, W^2\, dx_2 \tag{6.17}$$

appearing in both (6.14) and (6.16). By virtue of its definition in (6.10), (6.11), we have

$$\int_{L_z} [\varrho(q^2) + \varrho(k^2)]\, W\, dx_2 = 0 . \tag{6.18}$$

Now, for fixed x_1, make the change of variable,

$$\sigma = \int_0^{x_2} [\varrho(x_1, \eta) + \varrho(k^2)]\, d\eta \tag{6.19}$$

so that $d\sigma = [\varrho(q^2) + \varrho(k^2)]\, dx_2$. Since (6.18) can be written as

$$\int_0^B W\, d\sigma = 0, \tag{6.20}$$

we have the Wirtinger inequality

$$\int_0^B W^2\, d\sigma \leqq \frac{B^2}{\pi^2} \int_0^B \left(\frac{dW}{d\sigma}\right)^2 d\sigma, \tag{6.21}$$

which can be written as

$$J^3 = \int_{L_z} [\varrho(q^2) + \varrho(k^2)]\, W^2\, dx_2 \leqq \frac{B^2}{\pi^2} \int_{L_z} [\varrho(q^2) + \varrho(k^2)]^{-1}\, w_{,2}^2\, dx_2, \tag{6.22}$$

where we have used $W_{,2} = w_{,2}$ as follows from (6.10). Since $\varrho \geqq m_1 > 0$, we deduce from (6.22) that

$$J^3 \leqq \frac{B^2}{4m_1^2\pi^2} \int_{L_z} [\varrho(q^2) + \varrho(k^2)]\, w_{,2}^2\, dx_2\,. \tag{6.23}$$

On combining (6.23), (6.16) and (6.14) we obtain

$$|J| \leqq \frac{B}{2m_1\pi} \left(\int_{L_z} [\varrho(q^2) + \varrho(k^2)]\, w_{,2}^2\, dx_2\right)^{\frac{1}{2}} \left[\left(\int_{L_z} [\varrho(q^2) + \varrho(k^2)]\, w_{,1}^2\, dx_2\right)^{\frac{1}{2}}\right.$$
$$\left. + c_1 k \left(\int_{L_z} [\varrho(q^2) + \varrho(k^2)]\, w_{,\beta} w_{,\beta}\, dx_2\right)^{\frac{1}{2}}\right]. \tag{6.24}$$

Thus employing the arithmetic-geometric mean inequality, we find that

$$J \leqq B(1 + 2c_1k)\,(4m_1\pi)^{-1} \int_{L_z} [\varrho(q^2) + \varrho(k^2)]\, w_{,\alpha} w_{,\alpha}\, dx_2\,, \tag{6.25}$$

which, in view of (6.4), can be written as

$$J \leqq -\tilde{E}'(1 + 2c_1k)\, B/(4m_1\pi)\,. \tag{6.26}$$

On inserting (6.26) into the right-hand side of (6.9), we obtain the desired result (6.3)–(6.5).

The differential inequality (6.3) may now be integrated as before, and an obvious analog of (4.64) used to deduce that

$$\tilde{E}(z) \leqq \tilde{E}_0 \exp\left(\frac{-2m_1\pi(1 - \tilde{c}_1)\, z^2}{(1 + 2c_1k)} \Big/ \int_0^z B(s)\, ds\right). \tag{6.27}$$

It remains to find an upper bound for

$$G(z) \equiv \int_0^z B(s)\,ds = \int_0^z \int_0^h [\varrho(q^2) + \varrho(k^2)]\,dA. \tag{6.28}$$

We write

$$G(z) = \int_0^z \int_0^h [\varrho(q^2) - \varrho(0)]\,dA + [\varrho(k^2) + \varrho(0)]\,hz, \tag{6.29}$$

so that, on using (4.5) with $s = q$ we have

$$G(z) \leqq d_1 \int_0^z \int_0^h q^2 \varrho(q^2)\,dA + [\varrho(k^2) + \varrho(0)]\,hz. \tag{6.30}$$

We now use (4.68) to obtain from (6.30)

$$\begin{aligned} G(z) \leqq d_1 \int_0^z \int_0^h \varrho(q^2)\, w_{,\alpha} w_{,\alpha}\,dA + 2k\, d_1 \int_0^z \int_0^h \varrho(q^2)\, u_{,1}\,dA \\ - d_1 k^2 \int_0^z \int_0^h \varrho(q^2)\,dA + [\varrho(k^2) + \varrho(0)]\,hz. \end{aligned} \tag{6.31}$$

Making use of (4.19), we thus obtain

$$\begin{aligned} G(z) \leqq d_1 \int_0^z \int_0^h \varrho(q^2)\, w_{,\alpha} w_{,\alpha}\,dA + [3k^2 \varrho(k^2)\, d_1 + \varrho(k^2) + \varrho(0)]\,hz \\ - d_1 k^2 \int_0^z \int_0^h [\varrho(q^2) + \varrho(k^2)]\,dA, \end{aligned} \tag{6.32}$$

which in turn yields

$$\begin{aligned} (1 + d_1 k^2)\, G(z) \leqq d_1 \int_0^z \int_0^h [\varrho(q^2) + \varrho(k^2)]\, w_{,\alpha} w_{,\alpha}\,dA \\ + [3k^2 \varrho(k^2)\, d_1 + \varrho(k^2) + \varrho(0)]\,hz \end{aligned} \tag{6.33}$$

$$= d_1 [\tilde{E}_0 - \tilde{E}(z)] + [3 d_1 k^2 \varrho(k^2) + \varrho(k^2) + \varrho(0)]\,hz, \tag{6.34}$$

the last equality following from the definition (6.2) of $\tilde{E}(z)$. Thus, from (6.34), we deduce the desired upper bound

$$G(z) \leqq (1 + d_1 k^2)^{-1} \{d_1 \tilde{E}_0 + [3 d_1 k^2 \varrho(k^2) + \varrho(k^2) + \varrho(0)]\,hz\}. \tag{6.35}$$

This result is the analog, in the present context, of the bound (4.71).

As in Section 4, to obtain the final decay estimate, we use (6.35) in (6.27) to yield

$$\tilde{E}(z) \leqq \tilde{K} e^{-2(\pi/h)\tilde{\nu} z}, \quad 0 \leqq z \leqq l \tag{6.36}$$

where

$$\tilde{\nu} = \nu(p) = \frac{m_1(1 - \tilde{c}_1)(1 + d_1 k^2)}{[3d_1 k^2 \varrho(k^2) + \varrho(k^2) + \varrho(0)](1 + 2c_1 k)}, \tag{6.37}$$

$$\tilde{K} = \tilde{K}(\tilde{E}_0, m_1, p) = \tilde{E}_0 \exp\left[\frac{2m_1 \pi \tilde{E}_0 \, d_1(1 + d_1 k^2)(1 - k_1)}{\{3d_1 k^2 \varrho(k^2) + \varrho(k^2) + \varrho(0)\}^2 (1 + 2c_1 k) h^2}\right], \tag{6.38}$$

and $k = \hat{k}(p)$. Thus the energy measure $\tilde{E}(z)$, *defined by* (6.2), *has been shown to satisfy the decay estimate* (6.36)–(6.38) *when the constitutive function* ϱ *is assumed to be such that* (4.3)–(4.5) *and* (6.1) *hold.*

We now compare the results (6.36)–(6.38) with the estimates (4.74)–(4.76) obtained in Section 4 for Case 1. In the case of *self-equilibrated* loading $p = 0$ and so $k = 0$. Thus (6.37) yields

$$\tilde{\nu} = \tilde{\nu}(0) = \frac{m_1(1 - \tilde{c}_1)}{2\varrho(0)}, \tag{6.39}$$

which, in general, is *not* as sharp as the result (5.1). However, as was pointed out at the beginning of this section, the estimates (6.36)–(6.38) resulted from a technique designed to extend the range of validity of the results of Case 1 and we turn now to this aspect.

From (6.37) it is clear that $\tilde{\nu}(p) > 0$ for all finite p whenever the function $\tilde{c}_1(t)$ in the hypothesis (6.1) can be taken to be zero. This is the case, for example, when ϱ is a monotone nondecreasing function of its argument since the left-hand side of (6.1) is then always nonpositive. Thus, in particular, for hardening materials which satisfy (4.4) and (4.5), the inequality (6.36) provides a decay estimate valid for all finite p.

Consider now the power-law materials (5.8) with $n > 1$. Since $\varrho' > 0$ in this case, we have $\tilde{c}_1 = 0$. The remaining quantities m_1, $c_1(k)$ and d_1 appearing in the estimated decay rate (6.37) are still given by (5.20) and (5.21)–(5.25). Thus, for the power-law materials (5.8) with $n > 1$, we obtain

$$\begin{aligned}\tilde{\nu} = \tilde{\nu}(p) &= [(1 + 2c_1(\hat{k})\,\hat{k})]^{-1}\,[1 + (b(n-1)/n)\,\hat{k}^2] \\ &\times [1 + \mu^{-1} p\hat{k}^{-1} + 3\mu^{-1}(b(n-1)/n)\,p\hat{k}]^{-1},\end{aligned} \tag{6.40}$$

where $c_1(\hat{k})$ is given by (5.21)–(5.24) and $k = \hat{k}(p)$ is the unique value of k satisfying

$$\mu(1 + bk^2/n)^{n-1}\,k = p, \quad n > 1. \tag{6.41}$$

The inequality (6.36), with estimated decay rate $\tilde{\nu}$ given by (6.40), provides a decay estimate for power-law materials with $n > 1$ valid for all finite p.

Alternatively, we can express $\tilde{\nu} = \tilde{\nu}(k)$ explicitly in terms of k and so obtain from (6.40)

$$\tilde{\nu} = \tilde{\nu}(k) = \frac{[1 + (b(n-1)/n)\,k^2]\,[1 + 2c_1(k)\,k]^{-1}}{[1 + (1 + bk^2/n)^{n-1}\,\{1 + 3(b(n-1)/n)\,k^2\}]}, \tag{6.42}$$

where $c_*(k)$ is given by (5.21)–(5.24). The result (6.42) provides an explicit expression for the estimated decay rate in (6.36) in terms of the amount of shear k for power-law materials with $n > 1$. An explicit expression for an estimated decay rate for hardening power-law materials has also been obtained in [1] (see equations (8.5), (8.6) of [1]). The estimate in [1] also involves a parameter measuring a departure from uniformity of the prescribed end data $f(x_2)$ of (3.2) and so a direct comparison between the results of [1] and (6.43) is not possible. In common with the results of [1], however, it can be shown from (6.42) that $\tilde{\nu}(k)$ is a *decreasing* function of k as k increases, or equivalently, that $\tilde{\nu}(p)$ is decreasing function of p as p increases. As $p \to \infty$ in (6.40), it is readily verified that $\tilde{\nu}(p) \to 0$. Thus (6.36)–(6.38) suggest a progressively slower decay of end effects with increasing load level for power-law materials with $n > 1$.

Appendix A: Verification of (5.10), (5.21)–(5.25)

When ϱ has the power-law form (5.8), we have seen that for $\frac{1}{2} \leqq n < 1$, the first hypothesis (4.6) of Case 2 is satisfied with $m_2 = \mu^{-1}$. We verify here that the remaining hypotheses of Case 2, namely (4.7), (4.8), are satisfied provided c_2, d_2 are chosen as in (5.10).

To verify that (4.7) holds with $s = q$, $t = k$ and with c_2 chosen as in (5.10), we must show that

$$J_2(q, k) \equiv \mu^{-1} \left|(1 + bq^2/n)^{1-n} - (1 + bk^2/n)^{1-n}\right| |k - q|^{-1} \leqq c_2. \tag{A.1}$$

We use the algebraic inequality

$$|c^\alpha - d^\alpha| \leqq \frac{|c - d|}{c^{1-\alpha} + d^{1-\alpha}}, \quad (0 < \alpha \leqq \tfrac{1}{2}) \tag{A.2}$$

for any real numbers $c, d > 0$, an inequality which may be readily verified on direct multiplication. Thus, since $0 < 1 - n \leqq \frac{1}{2}$, we may apply (A.2) with $c = 1 + bq^2/n$, $d = 1 + bk^2/n$, $\alpha = 1 - n$ on the left-hand side of (A.1) to get

$$J_2(q, k) \leqq \frac{\mu^{-1}(b/n)\,(q + k)}{(1 + bq^2/n)^n + (1 + bk^2/n)^n}. \tag{A.3}$$

Since $n \geqq \frac{1}{2}$, we find

$$J_2(q, k) \leqq \frac{\mu^{-1}(b/n)\,(q + k)}{(1 + bq^2/n)^{\frac{1}{2}} + (1 + bk^2/n)^{\frac{1}{2}}} \tag{A.4}$$

$$\leqq \mu^{-1}(b/n)^{\frac{1}{2}} \tag{A.5}$$

after use of the inequality

$$(b/n)^{\frac{1}{2}}\,(q + k) \leqq (1 + bq^2/n)^{\frac{1}{2}} + (1 + bk^2/n)^{\frac{1}{2}}. \tag{A.6}$$

Thus we have shown that (A.1) holds with c_2 given by

$$c_2 = \mu^{-1}(b/n)^{\frac{1}{2}}, \quad \tfrac{1}{2} \leqq n < 1. \tag{A.7}$$

This completes the verification of $(5.10)_1$.

To verify that (4.8) holds with $s = q$ and with d_2 chosen as in $(5.10)_2$, we must show that

$$\mu^{-1} \left|(1 + bq^2/n)^{1-n} - 1\right| \leqq \mu\, d_2 q^2 (1 + bq^2/n)^{n-1}, \tag{A.8}$$

that is,

$$P_2(q) \equiv \mu^{-2} \left|(1 + bq^2/n)^{1-n} - 1\right| (1 + bq^2/n)^{1-n} q^{-2} \leqq d_2. \tag{A.9}$$

Using (A.2) with $c = 1 + bq^2/n$, $d = 1$, $\alpha = 1 - n$ on the left-hand side of (A.9), we find that

$$P_2(q) \leqq \mu^{-2}(b/n)\,(1 + bq^2/n)^{1-n}/[(1 + bq^2/n)^{1-n} + 1] \leqq \mu^{-2}(b/n), \quad \tfrac{1}{2} \leqq n < 1. \tag{A.10}$$

Thus (A.9) has been shown to hold with

$$d_2 = \mu^{-2}(b/n), \quad \tfrac{1}{2} \leqq n < 1. \tag{A.11}$$

It turns out that the choice (A.11) for d_2 may be improved in the range $\frac{2}{3} \leqq n < 1$ by using an algebraic inequality different from (A.2). This inequality (stated in a different form in [36], p. 356) reads

$$a^{\alpha} - 1 \leqq \frac{\alpha(a-1)}{a^{(1-\alpha)/2}}, \quad 0 \leqq \alpha \leqq \tfrac{1}{2}, \tag{A.12}$$

for all real numbers $a > 0$. On applying (A.12) with $a = 1 + bq^2/n$, $\alpha = 1 - n$, on the left-hand side of (A.9) we obtain

$$P_2(q) \leqq \mu^{-2}\, b(1 - n)/n\, (1 + bq^2/n)^{1-\frac{3n}{2}}, \tag{A.13}$$

$$\leqq \mu^{-1}\, b(1 - n)/n \quad \text{if } n \geqq \tfrac{2}{3}. \tag{A.14}$$

Thus, if $\frac{2}{3} \leqq n < 1$, we have shown that (A.9) holds with

$$d_2 = \mu^{-2}\, b(1 - n)/n. \tag{A.15}$$

This completes the verification of $(5.10)_2$.

Turning now to the hypotheses (4.3)–(4.5) of Case 1 for the power-law materials described by (5.8), we have seen in Section 5 that when $n > 1$, the first hypothesis (4.3) is satisfied with $m_1 = \mu$. We verify here that (4.4), (4.5) are satisfied provided c_1 and d_1 are chosen as in (5.21)–(5.25).

To verify that (4.4) holds with $s = q$, $t = k$ we must show that for suitably chosen c_1

$$J_1(q, k) = \left|(1 + bq^2/n)^{n-1} - (1 + bk^2/n)^{n-1}\right| (1 + bq^2/n)^{1-n} |k - q|^{-1} \leqq c_1. \tag{A.16}$$

We employ the algebraic inequality (A.12) with $a = (1 + bq^2/n)/(1 + bk^2/n)$, $\alpha = n - 1$, on the left-hand side of (A.16). Since $0 \leqq \alpha \leqq \frac{1}{2}$, this restricts n to the range $1 \leqq n \leqq \frac{3}{2}$. Thus, for this range of n, we obtain

$$J_1(q, k) \leqq [b(n-1)/n]\,(1 + bk^2/n)^{\frac{n}{2}-1}\,(q + k)\,(1 + bq^2/n)^{-\frac{n}{2}}. \qquad \text{(A.17)}$$

Since $n \geqq 1$, we have

$$(q + k)\,(1 + bq^2/n)^{-n/2} \leqq (q + k)\,(1 + bq^2/n)^{-\frac{1}{2}} \equiv F(q, k). \qquad \text{(A.18)}$$

It is readily verified that

$$F(q. k) \leqq (n/b)^{\frac{1}{2}}\,(1 + bk^2/n)^{\frac{1}{2}}. \qquad \text{(A.19)}$$

Thus, by virtue of (A.19), (A.18) and (A.17), we find

$$J_1(q, k) \leqq (b/n)^{\frac{1}{2}}\,(n - 1)\,(1 + bk^2/n)^{\frac{n-1}{2}}, \quad 1 \leqq n \leqq \tfrac{3}{2}, \qquad \text{(A.20)}$$

and so (A.16) has been shown to hold for the choice

$$c_1 = c_1(k) = (b/n)^{\frac{1}{2}}\,(n - 1)\,(1 + bk^2/n)^{\frac{n-1}{2}}, \quad 1 \leqq n \leqq \tfrac{3}{2}. \qquad \text{(A.21)}$$

When $n > \frac{3}{2}$, we treat (A.16) in a slightly different way. Instead of (A.12), we use the inequality (stated in a different form in [36] p. 356),

$$a^\alpha - 1 \leqq \frac{2\alpha(a-1)}{1 + a^{1-\alpha}}, \quad \tfrac{1}{2} \leqq \alpha \leqq 1, \qquad \text{(A.22)}$$

for all real numbers $a > 0$. We employ (A.22) on the left-hand side of (A.16) with $a = (1 + bq^2/n)/(1 + bk^2/n)$, $\alpha = n - 1$. Since $\frac{1}{2} \leqq \alpha \leqq 1$ in (A 22), this restricts n to the range $3/2 \leqq n \leqq 2$. Thus, for this range of n, we obtain

$$\begin{aligned} J_1(q, k) \leqq\; & [2b(n-1)/n]\,(1 + bq^2/n)^{1-n}\,(q + k) \\ & \times [(1 + bq^2/n)^{2-n} + (1 + bk^2/n)^{2-n}]^{-1}. \end{aligned} \qquad \text{(A.23)}$$

Since $n \leqq 2$, we deduce from (A.23) that

$$J_1(q, k) \leqq [(n-1)\,b/n]\,(q + k)\,(1 + bq^2/n)^{1-n}, \qquad \text{(A.24)}$$

$$\leqq [(n-1)\,b/n]\,(q + k)\,(1 + bq^2/n)^{-\frac{1}{2}}, \qquad \text{(A.25)}$$

the last inequality following from the fact that $n \geqq \frac{3}{2}$. We now use (A.19) (with $F(q, k)$ defined in (A.18)) in (A.25) to find

$$J_1(q, k) \leqq (b/n)^{\frac{1}{2}}\,(n - 1)\,(1 + bk^2/n)^{\frac{1}{2}}, \quad \tfrac{3}{2} \leqq n \leqq 2, \qquad \text{(A.26)}$$

and so (A.16) has been shown to hold for the choice

$$c_1 = c_1(k) = (b/n)^{\frac{1}{2}}\,(n - 1)\,(1 + bk^2/n)^{\frac{1}{2}}, \quad \tfrac{3}{2} \leqq n \leqq 2. \qquad \text{(A.27)}$$

When $n \geqq 2$, we use another algebraic inequality (stated in a different form in [36], p. 356),

$$a^\alpha - 1 \leqq \alpha(a-1)\max(1, a^{\alpha-1}), \quad 1 \leqq \alpha \leqq 2, \qquad \text{(A.28)}$$

for all real $a > 0$. On using (A.28) on the left hand-side of (A.16) with $a = (1 + bq^2/n)/(1 + bk^2/n)$, $\alpha = n - 1$ (thus requiring n to lie in the range $2 \leqq n \leqq 3$), and carrying out the required maximization, we find that

$$J_1(q, k) \leqq (b/n)^{\frac{1}{2}} (n - 1)/2[(1 + bk^2/n)^{\frac{1}{2}} + (b/n)^{\frac{1}{2}} k] (1 + bk^2/n)^{n-2}, \quad \text{(A.29)}$$

and so (A.16) has been shown to hold for the choice

$$c_1 = c_1(k) = (b/n)^{\frac{1}{2}} (n - 1)/2[(1 + bk^2/n)^{\frac{1}{2}} + (b/n)^{\frac{1}{2}} k] (1 + bk^2/n)^{n-2}, \quad 2 \leqq n \leqq 3. \quad \text{(A.30)}$$

Finally, when $n \geqq 3$, we use yet another algebraic inequality (stated in a different form in [36], p. 356),

$$a^\alpha - 1 \leqq \frac{\alpha}{2} (a - 1)(1 + a^{\alpha-1}), \quad \alpha \geqq 2, \quad \text{(A.31)}$$

for all real $a > 0$. On using (A.31) on the left-hand side of (A.16) with $a = (1 + bq^2/n)/(1 + bk^2/n)$, $\alpha = n - 1$, we find that

$$J_1(q, k) \leqq [b(n - 1)/2n] \left[\frac{q + k}{1 + bk^2/n} + \frac{(q + k)(1 + bk^2/n)^{n-2}}{(1 + bq^2/n)^{n-1}}\right], \quad n \geqq 3. \quad \text{(A.32)}$$

One way of obtaining an upper bound for the right-hand side of (A.32) for all q is to use the inequality $(1 + bq^2/n)^{n-1} \geqq 1 + bq^2/n$ in the second term on the right in (A.32) and so deduce that

$$J_1(q, k) \leqq [b(n - 1)/2n] [1 + (1 + bk^2/n)^{n-2}] (q + k)(1 + bq^2/n)^{-1}. \quad \text{(A.33)}$$

It is readily verified that

$$(q + k)(1 + bq^2/n)^{-1} \leqq (n/b)^{\frac{1}{2}}/2[(1 + bk^2/n)^{\frac{1}{2}} + (b/n)^{\frac{1}{2}} k], \quad \text{(A.34)}$$

and so (A.33) yields

$$J_1(q, k) \leqq (b/n)^{\frac{1}{2}} (n - 1)/4[(1 + bk^2/n)^{\frac{1}{2}} + (b/n)^{\frac{1}{2}} k] [1 + (1 + bk^2/n)^{n-2}], \quad n \geqq 3. \quad \text{(A.35)}$$

Thus (A.16) has been shown to hold for the choice

$$c_1 = c_1(k) = (b/n)^{\frac{1}{2}} (n - 1)/4[(1 + bk^2/n)^{\frac{1}{2}} + (b/n)^{\frac{1}{2}} k] [1 + (1 + bk^2/n)^{n-2}], \quad n \geqq 3. \quad \text{(A.36)}$$

This completes the verification of (4.4), with c_1 given by (5.21)–(5.24).

It is readily verified that (4.5) holds, with d_1 given by (5.25). The proof of this is immediate on using the inequalities (A.12), (A.22), (A.28) and (A.31) with $a = 1 + bq^2/n$, $\alpha = n - 1$. One easily finds that (4.5) holds with

$$d_1 = (n - 1)\, b/n, \quad n > 1. \quad \text{(A.37)}$$

Appendix B: Total energy bounds

The total energy $\tilde{E}_0 = \tilde{E}(0)$ contained in the rectangle R appears in the multiplicative constant $\tilde{K}$ in the improved estimate (6.36) in Case 1. Here we show how an upper bound for $\tilde{E}_0$ in terms of the geometry, boundary data, ϱ and p can be obtained. Bounds of this type have also been discussed in our previous work [24].

We begin with the inequality (6.9) evaluated at $z = 0$. By virtue of the conservation result (4.19), w in the integral on the right in (6.9) may be replaced by $w - \hat{w}(x_1)$, where $\hat{w} = \int_{L_z} w \, dx_2/h$. Using the boundary conditions (3.2) and equations (3.5) and (3.6), we find that

$$\tilde{E}_0 \leqq 2(1 - \tilde{c}_1)^{-1} \int_0^h |p - f(x_2)| \, |w - \hat{w}| \, dx_2, \tag{B.1}$$

$$\leqq 2(1 - \tilde{c}_1)^{-1} \left(\int_0^h |p - f(x_2)|^2 \, dx_2 \right)^{\frac{1}{2}} \left(\int_0^h (w - \hat{w})^2 \, dx_2 \right)^{\frac{1}{2}}, \tag{B.2}$$

on using Schwarz's inequality. We now employ the inequality

$$\int_0^h (w - \hat{w})^2 \, dx_2 \leqq \frac{1}{\omega_1} \int_R \{w_{,2}^2 + (w_{,1} - \hat{w}_{,1})^2\} \, dA \tag{B.3}$$

where ω_1 is the smallest positive eigenvalue of the Stekloff problem

$$\Delta\phi = 0 \quad \text{on } R, \tag{B.4}$$

$$\phi_{,1} = 0 \quad \text{at } x_1 = l, \quad 0 \leqq x_2 \leqq h, \tag{B.5}$$

$$\phi_{,1} + \omega\phi = 0 \quad \text{at } x_1 = 0, \quad 0 \leqq x_2 \leqq h, \tag{B.6}$$

$$\phi_{,2} = 0 \quad \text{at } x_2 = 0, h, \quad 0 \leqq x_1 \leqq l, \tag{B.7}$$

$$\int_{L_0} \phi \, dx_2 = 0, \quad 0 \leqq z \leqq l. \tag{B.8}$$

For the rectangle R, it is easily shown that

$$\omega_1 = \frac{\pi}{h} \tanh\left(\frac{\pi l}{h}\right), \tag{B.9}$$

$$\simeq \frac{\pi}{h} \quad \text{for } l \gg h. \tag{B.10}$$

Since $\varrho \geqq m_1 > 0$ in Case 1, we have

$$\int_R \{w_{,2}^2 + (w_{,1} - \hat{w}_{,1})^2\} \, dA \leqq \int_R w_{,\alpha} w_{,\alpha} \, dA \tag{B.11}$$

$$\leqq \frac{1}{2m_1} \int_R [\varrho(q^2) + \varrho(k^2)] \, w_{,\alpha} w_{,\alpha} \, dA = \frac{\tilde{E}_0}{2m_1}. \tag{B.12}$$

On combining (B.12), (B.3) and (B.2) we deduce that

$$\tilde{E}_0 \leqq 2(1 - \tilde{c}_1)^{-2} (m_1\omega_1)^{-1} \left(\int_0^h |p - f(x_2)|^2 \, dx_2\right), \tag{B.13}$$

which is the desired upper bound for $\tilde{E}_0$.

By similar arguments, bounds on E_0 in Case 2 may also be obtained. In this case, an *a priori* restriction on the boundary data is, in general, necessary in order to obtain a result comparable to (B.13). Similar considerations arise in the analysis carried out in [24].

Acknowledgements. The work of C.O.H. was supported by the National Science Foundation under Grant No. MSM-89-04719 and by the Center for Advanced Studies, University of Virginia. The research of L. E. P. was carried out while he held an appointment as Visiting Professor, Department of Applied Mathematics, University of Virginia, Spring 1989, and was partially supported by the National Science Foundation under Grant No. DMS-86-00250.

References

1. HORGAN, C. O., & KNOWLES, J. K., The effect of nonlinearity on a principle of Saint-Venant type. Journal of Elasticity **11**, 271–291 (1981).
2. HORGAN, C. O., & KNOWLES, J. K., Recent developments concerning Saint-Venant's principle, Advances in Applied Mechanics, J. W. HUTCHINSON, ed., Vol. 23, pp. 179–269. Academic Press, New York, 1983.
3. HORGAN, C. O., Recent developments concerning Saint-Venant's principle: an update. Applied Mechanics Reviews **42** (1989) (in press).
4. FICHERA, G., Il principio di Saint-Venant: Intuizione dell'ingegnere e rigore del matematico. Rend. di Mat. Serie VI **10**, 1–24 (1977).
5. FICHERA, G., Remarks on Saint-Venant's principle. Rend. di Mat. Serie VI **12**, 181–200 (1979).
6. ERICKSEN, J. L., On the formulation of St. Venant's problem, in Nonlinear Analysis and Mechanics: Heriot-Watt Symposium (R. J. KNOPS, ed.) Vol. I, pp. 158–186. Pitman, London, 1977.
7. MUNCASTER, R. G., Saint-Venant's problem in nonlinear elasticity: a study of cross-sections, in Nonlinear Analysis and Mechanics: Heriot-Watt Symposium (R. J. KNOPS, ed.), Vol. IV, pp. 17–75. Pitman, London, 1979.
8. MUNCASTER, R. G., Saint-Venant's problem for slender prisms. Utilitas Math. **23**, 75–101 (1983).
9. ERICKSEN, J. L., Saint-Venant's problem for elastic prisms, in Systems of Nonlinear Partial Differential Equations (J. M. BALL, ed.), pp. 87–93. D. Reidel, Dordrecht, 1983.
10. ERICKSEN, J. L., Problems for infinite elastic prisms, in Systems of Nonlinear Partial Differential Equations (J. M. BALL, ed.), pp. 80–86. D. Reidel, Dordrecht, 1983.
11. KINDERLEHRER, D., A relation between semi-inverse and Saint-Venant solutions for prisms. SIAM J. Math. Anal. **17**, 626–640 (1986).

12. KINDERLEHRER, D., Remarks about Saint-Venant solutions in finite elasticity, in Nonlinear Functional Analysis and its Applications: Proceedings of Symposia in Pure Mathematics (F. E. BROWDER, ed.), Vol. 45, Part 2, pp. 37–50. American Mathematical Society, Providence, Rhode Island, 1986.
13. MIELKE, A., Saint-Venant's problem and semi-inverse solutions in nonlinear elasticity. Arch. Rational Mech. Anal. **102**, 205–229 (1988).
14. BREUER, S., & ROSEMAN, J. J., On Saint-Venant's principle in three-dimensional nonlinear elasticity. Arch. Rational Mech. Anal. **63**, 191–203 (1977).
15. BREUER, S., & ROSEMAN, J. J., Saint-Venant's principle in nonlinear plane elasticity with sufficiently small strains. Arch. Rational Mech. Anal. **80**, 19–37 (1982).
16. KNOPS, R. J., & PAYNE, L. E., A Saint-Venant principle for nonlinear elasticity. Arch. Rational Mech. Anal. **81**, 1–12 (1983).
17. HORGAN, C. O., & ABEYARATNE, R., Finite anti-plane shear of a semi-infinite strip subject to a self-equilibrated end traction. Q. Appl. Math. **40**, 407–417 (1983).
18. GALDI, G. P., KNOPS, R. J., & RIONERO, S., Asymptotic behavior in the nonlinear elastic beam. Arch. Rational Mech. Anal. **87**, 305–318 (1985).
19. ABEYARATNE, R., HORGAN, C. O., & CHUNG, D.-T., Saint-Venant end effects for incremental plane deformations of incompressible nonlinearly elastic materials. J. Appl. Mech. **52**, 847–852 (1985).
20. MIELKE, A., Normal hyperbolicity and Saint-Venant's principle, Preprint 10, Mathematical Problems in Nonlinear Mechanics, Univ. Stuttgart (1988).
21. DURBAN, D., & STRONGE, W. J., Diffusion of self-equilibrating end loads in elastic solids. J. Appl. Mech. **55**, 492–495 (1988).
22. VAFEADES, P., & HORGAN, C. O., Exponential decay estimates for solutions of the von Kármán equations on a semi-infinite strip. Arch. Rational Mech. Anal. **104**, 1–25 (1988).
23. QUINTANILLA, R., Asymptotic behavior of solutions in elasticity (preprint, Univ. Polit. de Catalunya, Barcelona, Spain, 1987).
24. HORGAN, C. O., & PAYNE, L. E., Decay estimates for second-order quasilinear partial differential equations. Advances in Appl. Math. **5**, 309–332 (1984).
25. HORGAN, C. O., & PAYNE, L. E., Decay estimates for a class of second-order quasilinear equations in three dimensions. Arch. Rational Mech. Anal. **86**, 279–289 (1984).
26. HORGAN, C. O., A note on the spatial decay of a three-dimensional minimal surface over a semi-infinite cylinder. J. Math. Anal. Appl. **107**, 285–290 (1985).
27. BREUER, S., & ROSEMAN, J. J., Phragmén-Lindelöf decay theorems for classes of nonlinear Dirichlet problems in a circular cylinder. J. Math. Anal. Appl. **113**, 59–77 (1986).
28. BREUER, S., & ROSEMAN, J. J., Decay theorems for nonlinear Dirichlet problems in semi-infinite cylinders. Arch. Rational Mech. Anal. **94**, 363–371 (1986).
29. HORGAN, C. O., Some applications of maximum principles in linear and nonlinear elasticity, in Maximum Principles and Eigenvalue Problems in Partial Differential Equations, pp. 49–67, ed. P. W. SCHAEFER, Pitman Research Notes in Mathematics Series, 175, Longman, New York, 1988.
30. HORGAN, C. O., & PAYNE, L. E., Decay estimates for a class of nonlinear boundary value problems in two dimensions. SIAM J. Math. Anal. **20**, 782–788 (1989).
31. HORGAN, C. O., & PAYNE, L. E., On the asymptotic behavior of solutions of inhomogeneous second-order quasilinear partial differential equations. Q. Appl. Math. **47** (1989).
32. HORGAN, C. O., & SIEGEL, D., On the asymptotic behavior of a minimal surface over a semi-infinite strip. J. Math. Anal. Appl. (in press).
33. KNOWLES, J. K., On finite anti-plane shear for incompressible elastic materials. J. Austral. Math. Soc. Series B **19**, 400–415 (1976).

34. Knowles, J. K., The finite anti-plane shear field near the tip of a crack for a class of incompressible elastic solids. Int. J. Fracture **13**, 611–639 (1977).
35. Gurtin, M. E., Topics in Finite Elasticity, SIAM Regional Conference Series in Applied Mathematics, No. 35, SIAM, Philadelphia, 1981.
36. Mitrinovic, D. S., Analytic Inequalities, Springer-Verlag, Berlin, 1970.

Department of Applied Mathematics
University of Virginia
Charlottesville

and

Department of Mathematics
Cornell University
Ithaca, New York

(Received April 7, 1989)

Mixture Invariance and its Applications

I. SAMOHÝL & M. ŠILHAVÝ

Dedicated to Professor BERNARD COLEMAN *on his* 60th *birthday*

Abstract

This paper shows that the balance equations and the entropy inequality for mixtures have the following property: some partial quantities can be changed in a prescribed way without affecting the form of the basic equations. This property, called here mixture invariance, is used to simplify the constitutive equations of reacting and non-reacting mixtures. Also agreement with the classical thermochemistry of mixtures can be achieved.

Contents

§ 1. Introduction

Soon after the pioneer work of COLEMAN & NOLL [1] and COLEMAN [2], [3], where the second law of thermodynamics was shown to place restrictions on the constitutive functions or functionals of single bodies, the ideas and methods of [1]–[3] were extended to mixtures (BOWEN [4], [5], MÜLLER [6], [7], WILLIAMS

[8], TRUESDELL [9, Lectures 5, 6], SAMOHÝL [10], [11]; see also Appendices 5A–5E in Ref. [9] and the review articles by ATKIN & CRAINE [12], [13]). It was shown in most of these works that the total thermodynamic quantities (*i e.*, the entropy, internal energy and free energy of the whole mixture) are independent of the non-equilibrium vaiiables (like the velocity and its gradient or the gradient of temperature). But it is an unpleasant fact that the partial thermodynamic quantities may depend on the non-equilibrium variables (*cf.* BOWEN [4], BOWEN & WIESE [14], SAMOHÝL [10]). This was a mysterious point of the theory for there was no thermodynamical restriction on the nature of this dependence. Moreover, the above theories of mixtures have never led to a complete agreement with the classical thermochemistry of mixtures (*cf.* the discussion of this point in [15], [10], [11] and in § 8, below).

Following SAMOHÝL [15], [16], [10], [11], we shall show here that these difficulties have their roots in a remarkable property of invariance of the balance equations and the entropy inequality of mixture theory. In what follows we use the term "mixture invariance" for this property.

Mitxure invariance states that two different sets of constitutive functions for the dependent quantities of the mixture may lead, when inserted into the balance equations, to identical evolution equations for measurable quantities such as concentrations, temperature, diffusion velocities *etc*. The fact that the evolution equations are not in one-to-one correspondence with the constitutive functions does not have a counterpart in the theory of single bodies and indicates that some of the quantities of the theory are not uniquely determined by the structure of the balance equations. This may cause certain problems of interpretation, but on the positive side the freedom in the constitutive equations may be used to pass to an equivalent theory with the constitutive equations having simpler properties.

The principal aim of the present paper is to demonstrate these points in the constitutive equations of a fairly general non-linear model of a mixture. We call it a reacting mixture with heat conduction, diffusion and viscosity. The constitutive functions of this mixture depend on the following quantities: the first and second deformation gradients, the densities of the reacting constituents and their gradients, the velocities of the constituents and their gradients, and on the temperature and its gradient. Moreover, the interaction force density is allowed to depend on the gradients of these quantities. A systematic exposition of this constitutive class will be the subject of a future paper [17]. Here we concentrate on how the invariance mentioned above can be used to simplify the constitutive equations. The constitutive class considered here is more general than those considered hitherto (BOWEN [4], [5], BOWEN & WIESE [14], and SAMOHÝL [10]), and hence our considerations apply and simplify even these particular cases.

The desired simplification is demonstrated here for three different cases. First, we shall show for the general reacting mixture with heat conduction, diffusion and viscosity, that the passage to a properly chosen equivalent model leads to the partial free energies, partial entropies and partial internal energies independent of all velocities, their gradients and of the gradient of the temperature. Secondly, we shall show that for a nonreacting mixture an equivalent theory exists which has partial free energies, partial internal energies and partial entropies independent of the second gradient of deformation. Finally, at the end of the paper we also recall from SAMOHÝL [15], [10], [11], that mixture invariance can be used to achieve a complete agreement of the partial thermodynamic quantities of the theory with those of the classical thermochemistry of mixtures.

§ 2. Basic Concepts and Kinematics

Following TRUESDELL & TOUPIN [18], TRUESDELL [9], and BOWEN [4], [5] we consider the mixture as a superposition of its constituents (*i.e.* components of the mixture). Namely, we consider the mixture as the superposition of the configurations of the constituent bodies. The mixture is composed of a fixed number n of constituents of which the first m are chemically reacting. These constituents are denoted by greek subscripts having values $\alpha, \gamma = 1, \ldots, n$ and $\beta, \delta = 1, \ldots, n-1$; for reacting constituents we use also $\varphi, \nu = 1, \ldots, m$, and for nonreacting ones $\omega = m+1, \ldots, n$.

Points $\boldsymbol{x}$ of Euclidean space are simultaneously occupied by all n constituents and the motion is described by n deformation functions $\boldsymbol{\chi}_\alpha$ (invertible and smooth)

$$\boldsymbol{x} = \boldsymbol{\chi}_\alpha(\boldsymbol{X}_\alpha, t), \tag{2.1}$$

where the scalar t is the time and the vector $\boldsymbol{\chi}_\alpha$ is the place of a particle of the body of the constituent α in certain reference configuration [4], [9].

For each constituent α we define
the velocity $\boldsymbol{v}_\alpha$,

$$\boldsymbol{v}_\alpha \doteq \frac{\partial \boldsymbol{\chi}_\alpha}{\partial t}, \tag{2.2}$$

the deformation gradient $\boldsymbol{F}_\alpha$,

$$\boldsymbol{F}_\alpha = \frac{\partial \boldsymbol{\chi}_\alpha}{\partial \boldsymbol{X}_\alpha}, \tag{2.3}$$

and the second deformation gradient $\boldsymbol{G}_\alpha$,

$$\boldsymbol{G}_\alpha = \frac{\partial \boldsymbol{F}_\alpha}{\partial \boldsymbol{X}_\alpha} = \operatorname{Grad} \boldsymbol{F}_\alpha, \tag{2.4}$$

where Grad denotes gradient in the reference configuration. This second gradient (G_α^{iJK} in components) is symmetric in the last two indices

$$G_\alpha^{iJK} = G_\alpha^{iKJ}. \tag{2.5}$$

We use only Cartesian components and the summation rule is implied. The components are denoted by Latin superscripts: lower case for the space coordinates and upper case for the reference coordinates.

While $\boldsymbol{v}_\alpha$ is independent of the reference configuration, both (2.3), (2.4) are not; here and in the remaining text we assume that the reference configuration is fixed and known. By the invertibility of (2.1), any field ψ may be expressed either as function of $\boldsymbol{X}_\alpha$, t or as function of $\boldsymbol{x}$, t. The material derivative $\overset{\backslash\alpha}{\psi}$ is

$$\overset{\backslash\alpha}{\psi} = \frac{\partial \psi(\boldsymbol{X}_\alpha, t)}{\partial t} = \frac{\partial \psi(\boldsymbol{x}, t)}{\partial t} + \boldsymbol{v}_\alpha \cdot \operatorname{grad} \psi, \tag{2.6}$$

where grad and $\partial/\partial t$ are the gradients and the time derivatives, respectively, in actual configuration (often we write $\grave{\psi}_\alpha$ instead of $\overset{\backslash\alpha}{\psi}_\alpha$). The velocity gradient $\boldsymbol{L}_\alpha$ is given by

$$\boldsymbol{L}_\alpha = \operatorname{grad} \boldsymbol{v}_\alpha = \grave{\boldsymbol{F}}_\alpha \boldsymbol{F}_\alpha^{-1}, \tag{2.7}$$

the determinant of $\boldsymbol{F}_\alpha$ is denoted by

$$J_\alpha = \det \boldsymbol{F}_\alpha > 0, \tag{2.8}$$

and Euler's relation is

$$\grave{J}_\alpha = J_\alpha \operatorname{div} \boldsymbol{v}_\alpha = J_\alpha \operatorname{tr} \boldsymbol{L}_\alpha . \tag{2.9}$$

The constitutive functions are distinguished from the corresponding values of fields given by them by a mark over the symbol, *e.g.*,

$$s_\alpha = \hat{s}_\alpha(\varrho_\gamma, T). \tag{2.10}$$

The shortened description denotes n functions of the variables $\varrho_1, \varrho_2, \dots, \varrho_n, T$; *e.g.* for $\alpha = 1$, $s_1 = \hat{s}_1(\varrho_1, \varrho_2, \dots, \varrho_n, T)$.

§ 3. Balance Equations and the Entropy Inequality

The balance equations of mass, momentum, moment of momentum, energy and entropy are essentially those formulated by TRUESDELL & TOUPIN [18] and TRUESDELL [19] for each constituent as well as for the mixture in accord with the concept of mixture mentioned in § 2. Here they are given in local forms which are the same (disregarding some definitions) as those given by BOWEN [4]. In this article we discuss a mixture all of whose constituents $\alpha = 1, \dots, n$ have the same temperature and therefore we formulate the balance equations of energy and entropy only for the mixture as a whole (more general case will be discussed below in this section):

$$\grave{\varrho}_\alpha + \varrho_\alpha \operatorname{div} \boldsymbol{v}_\alpha = r_\alpha, \tag{3.1}$$

$$\sum_{\alpha=1}^{n} r_\alpha = 0, \tag{3.2}$$

$$\varrho_\alpha \grave{\boldsymbol{v}}_\alpha = \operatorname{div} \boldsymbol{T}_\alpha + \varrho_\alpha \boldsymbol{b}_\alpha + \boldsymbol{k}_\alpha, \tag{3.3}$$

$$\sum_{\alpha=1}^{n} (\boldsymbol{k}_\alpha + r_\alpha \boldsymbol{v}_\alpha) = \boldsymbol{0}, \tag{3.4}$$

$$\boldsymbol{M}_\alpha = \boldsymbol{T}_\alpha - \boldsymbol{T}_\alpha^T, \tag{3.5}$$

$$\sum_{\alpha=1}^{n} \boldsymbol{M}_\alpha = \boldsymbol{0}, \tag{3.6}$$

$$\sum_{\alpha=1}^{n} \varrho_\alpha \grave{u}_\alpha + \sum_{\alpha=1}^{n} r_\alpha u_\alpha = \sum_{\alpha=1}^{n} \operatorname{tr} \boldsymbol{T}_\alpha \boldsymbol{L}_\alpha^T - \operatorname{div} \boldsymbol{q} + Q - \sum_{\alpha=1}^{n} \boldsymbol{k}_\alpha \cdot \boldsymbol{v}_\alpha - \tfrac{1}{2} \sum_{\alpha=1}^{n} r_\alpha v_\alpha^2, \tag{3.7}$$

$$\sigma = \sum_{\alpha=1}^{n} \varrho_\alpha \grave{s}_\alpha + \sum_{\alpha=1}^{n} s_\alpha r_\alpha + \operatorname{div}\left(\frac{\boldsymbol{q}}{T}\right) - \frac{Q}{T} \geqq 0. \tag{3.8}$$

In these balance equations for mass (3.1), (3.2), momentum (3.3), (3.4), moment of momentum (3.5), (3.6), energy (3.7) and entropy (3.8), ϱ_α is the density of the constituent α, r_α is the mass supply, $\boldsymbol{T}_\alpha$ is the partial stress tensor, $\boldsymbol{b}_\alpha$ is the body force, $\boldsymbol{k}_\alpha$ is the interaction force (the momentum supply to the constituent α from the other constituents), $\boldsymbol{M}_\alpha$ is the interaction couple, u_α is the partial internal energy, $\boldsymbol{q}$ is the heat flux and Q is the heat supply. Finally, s_α is the partial entropy, T is the absolute temperature (*i.e.* $T > 0$) and σ is the entropy production which is assumed to be non-negative; $(3.8)_2$ is the entropy inequality.

In accord with our assumption that the number of constituents is fixed, we stipulate that the densities ϱ_α are always positive. We also assume that the first m constituents are reacting and the remaining $n - m$ constituents are non-reacting:

$$r_\omega = 0, \qquad \omega = m + 1, \dots, n. \tag{3.9}$$

Then the balance equation of mass for the mixture may also be written

$$\sum_{\varphi=1}^{m} r_\varphi = 0. \tag{3.10}$$

The balance equations of mass of the reacting constituents (3.1) may be written (see (2.8), (2.9))

$$\grave{\gamma}_\varphi = r_\varphi J_\varphi, \qquad \varphi = 1, \dots, m, \tag{3.11}$$

where we define (*cf.* [4]):

$$\gamma_\varphi = \varrho_\varphi J_\varphi. \tag{3.12}$$

The balance equations of mass for the non-reacting constituents (3.9) have the form

$$\varrho_\omega J_\omega = \varrho_\omega^0, \qquad \omega = m + 1, \dots, n, \tag{3.13}$$

where ϱ_ω^0 is the density in the reference configuration. Taking the derivative of (3.13), we obtain the relation

$$h_\omega^i = J_\omega^{-1} \left[(\partial \varrho_\omega^0 / \partial X_\omega^J) - \varrho_\omega^0 \overset{-1}{F}{}_\omega^{Kj} G_\omega^{jKJ} \right] \overset{-1}{F}{}_\omega^{Ji}, \tag{3.14}$$

which relates the gradient of reference density $(\operatorname{Grad} \varrho_\omega^0)^J = \partial \varrho_\omega^0 / \partial X_\omega^J$ to the density gradient $\boldsymbol{h}_\omega$ defined by

$$\boldsymbol{h}_\alpha = \operatorname{grad} \varrho_\alpha, \qquad \alpha = 1, \dots, n. \tag{3.15}$$

In our applications (in §§ 5, 6) we use the reference configuration with constant reference density, *i.e.* $\operatorname{Grad} \varrho_\omega^0 = \boldsymbol{0}$. For this special case eqs. (3.14) give

$$h_\omega^i = -\varrho_\omega G_\omega^{jKJ} \overset{-1}{F}{}_\omega^{Kj} \overset{-1}{F}{}_\omega^{Ji}. \tag{3.16}$$

To use the Coleman & Noll method [1] we define the partial free energy f_α of the constituent α by

$$f_\alpha = u_\alpha - T s_\alpha, \qquad \alpha = 1, \dots, n. \tag{3.17}$$

We use it in the following combination of (3.7) and (3.8) known as the reduced dissipation inequality:

$$-T\sigma = \sum_{\alpha=1}^{n} \varrho_\alpha \overset{\backslash}{f}_\alpha + \sum_{\alpha=1}^{n} f_\alpha r_\alpha + \sum_{\alpha=1}^{n} \varrho_\alpha s_\alpha \overset{\backslash\alpha}{T} + T^{-1} \boldsymbol{q}\cdot\boldsymbol{g}$$

$$+ \sum_{\alpha=1}^{n} \boldsymbol{k}_\alpha \cdot \boldsymbol{v}_\alpha + \tfrac{1}{2} \sum_{\alpha=1}^{n} r_\alpha v_\alpha^2 - \sum_{\alpha=1}^{n} \operatorname{tr} \boldsymbol{T}_\alpha \boldsymbol{L}_\alpha^T \leqq 0, \tag{3.18}$$

where $\boldsymbol{g}$ is the temperature gradient,

$$\boldsymbol{g} = \operatorname{grad} T. \tag{3.19}$$

Further useful quantities are the density of the mixture ϱ, defined by

$$\varrho = \sum_{\alpha=1}^{n} \varrho_\alpha, \tag{3.20}$$

the mass fraction w_α,

$$w_\alpha = \varrho_\alpha/\varrho, \quad \alpha = 1, \dots, n, \tag{3.21}$$

with the property

$$\sum_{\alpha=1}^{n} w_\alpha = 1. \tag{3.22}$$

Using the partial thermodynamic quantities u_α, s_α, f_α, we define the total internal energy u, the total entropy s and the total free energy f of the mixture by

$$u = \sum_{\alpha=1}^{n} w_\alpha u_\alpha, \quad s = \sum_{\alpha=1}^{n} w_\alpha s_\alpha, \quad f = \sum_{\alpha=1}^{n} w_\alpha f_\alpha; \tag{3.23}$$

hence

$$f = u - Ts. \tag{3.24}$$

The total stress $\boldsymbol{T}$ is defined by

$$\boldsymbol{T} = \sum_{\alpha=1}^{n} \boldsymbol{T}_\alpha, \tag{3.25}$$

which is a symmetric tensor (*cf.* (3.5) and (3.6)).

At the end of this section we formulate the balance equations for mixtures with different temperatures $T_\alpha > 0$ of the constituents (see BOWEN [4]). The balance equations of mass, momentum, and moment of momentum are the same as in the case of a single temperature (see (3.1)–(3.6)). The equation of balance of energy for the constituent α is

$$\varrho_\alpha \overset{\backslash}{u}_\alpha = \operatorname{tr} \boldsymbol{T}_\alpha \boldsymbol{L}_\alpha^T - \operatorname{div} \boldsymbol{q}_\alpha + Q_\alpha + e_\alpha, \quad \alpha = 1, \dots, n, \tag{3.26}$$

where $\boldsymbol{q}_\alpha$ is the partial heat flux, Q_α is the partial energy supply and e_α is the interaction energy. The equation of balance of energy for the mixture is

$$\sum_{\alpha=1}^{n} (\boldsymbol{k}_\alpha \cdot \boldsymbol{v}_\alpha + r_\alpha(u_\alpha + \tfrac{1}{2} v_\alpha^2) + e_\alpha) = 0. \tag{3.27}$$

Eliminating the interaction energies, we obtain from (3.26), (3.27) the equation of balance energy of the mixture in the previous form (3.7) if we define the heat flux q and the heat supply Q in the mixture as

$$q = \sum_{\alpha=1}^{n} q_\alpha, \tag{3.28}$$

$$Q = \sum_{\alpha=1}^{n} Q_\alpha . \tag{3.29}$$

The local equation of balance of entropy for the constituents $\alpha = 1, \ldots, n$, is

$$\varrho_\alpha \grave{s}_\alpha + s_\alpha r_\alpha = -\operatorname{div}\left(\frac{q_\alpha}{T_\alpha}\right) + \frac{Q_\alpha}{T_\alpha} + \sigma_\alpha , \tag{3.30}$$

where σ_α is the entropy production for the constituent α. The local entropy balance for mixture is

$$\sum_{\alpha=1}^{n} \sigma_\alpha = \sigma \geqq 0, \tag{3.31}$$

with the additional postulate that the entropy production σ for the mixture is not negative. The entropy inequality for the mixture may be written as

$$\sigma = \sum_{\alpha=1}^{n} \varrho_\alpha \grave{s}_\alpha + \sum_{\alpha=1}^{n} s_\alpha r_\alpha + \operatorname{div}\left(\sum_{\alpha=1}^{n} \frac{q_\alpha}{T_\alpha}\right) - \sum_{\alpha=1}^{\alpha} \left(\frac{Q_\alpha}{T_\alpha}\right) \geqq 0 . \tag{3.32}$$

For the mixture with a single temperature T of all its constituents we obtain the entropy inequality (3.8) formulated above. In this case we can see that the reduced inequality (3.18) does not contain e_α, and q_α are present only through the sum q (3.28). Therefore the COLEMAN & NOLL method [1] based on inequality (3.31) gives no information about the constitutive equations for such quantities and the energy balance for the mixture (3.7) suffices in that case.

§ 4. Mixture Invariance of the Balance Equations

In this section we show that the balance equations (3.1)–(3.7) of the mixture and the entropy inequality (3.8) are invariant under specific changes of the quantities occurring in them. We first demonstrate this for the case when all the constituents have the same temperature (which is our main concern in this paper) and then we shall extend the invariance to the mixture whose constituents may have different temperatures.

Consider a process described by the quantities occurring in the balance equations and the entropy inequality (3.1)–(3.8) and let

$$\varepsilon_\alpha, \eta_\alpha, \quad \alpha = 1, \ldots, n, \tag{4.1}$$

be two sets of fields defined on the region occupied by the mixture, the ε_α having the physical dimensions of energy, and the η_α having the physical dimensions of entropy, such that

$$\sum_{\alpha=1}^{n} \varrho_\alpha \varepsilon_\alpha = 0, \quad \sum_{\alpha=1}^{n} \varrho_\alpha \eta_\alpha = 0, \tag{4.2}$$

identically. Define φ_α by

$$\varphi_\alpha = \varepsilon_\alpha - T\eta_\alpha, \qquad \alpha = 1, \ldots, n, \tag{4.3}$$

so that

$$\sum_{\alpha=1}^{n} \varrho_\alpha \varphi_\alpha = 0. \tag{4.4}$$

Now we replace u_α, s_α, $\boldsymbol{T}_\alpha$, $\boldsymbol{k}_\alpha$, $\boldsymbol{q}$ in (3.1)–(3.8) by the following primed quantities

$$u'_\alpha = u_\alpha + \varepsilon_\alpha, \tag{4.5}$$

$$s'_\alpha = s_\alpha + \eta_\alpha, \tag{4.6}$$

$$\boldsymbol{T}'_\alpha = \boldsymbol{T}_\alpha + \varrho_\alpha \varphi_\alpha \boldsymbol{1}, \tag{4.7}$$

$$\boldsymbol{k}'_\alpha = \boldsymbol{k}_\alpha - \operatorname{grad}(\varrho_\alpha \varphi_\alpha), \tag{4.8}$$

$$\boldsymbol{q}' = \boldsymbol{q} - T \sum_{\alpha=1}^{n} \varrho_\alpha \eta_\alpha \boldsymbol{v}_\alpha, \tag{4.9}$$

while the remaining quantities in (3.1)–(3.8) are left unchanged *i.e.*

$$\varrho'_\alpha = \varrho_\alpha, \quad \boldsymbol{v}'_\alpha = \boldsymbol{v}_\alpha, \quad r'_\alpha = r_\alpha, \quad \boldsymbol{b}'_\alpha = \boldsymbol{b}_\alpha,$$
$$\boldsymbol{L}'_\alpha = \boldsymbol{L}_\alpha, \quad Q' = Q, \quad T' = T, \quad \sigma' = \sigma. \tag{4.10}$$

We call the quantities with trivial transformations such as (4.10) mixture invariant quantities; from (4.5), (4.6) and from definition (3.17) it follows that

$$f'_\alpha = f_\alpha + \varphi_\alpha. \tag{4.11}$$

Hexe and in what follows we assume that the definitions themselves are mixture invariant.

In view of the relations (4.2), only $2(n-1)$ of the quantities (4.1) are independent. Hence, for example, Eq. (4.5) may also be written as

$$u'_\alpha = u_\alpha + \prod_{\beta=1}^{n-1} (\delta_{\alpha\beta} - w_\beta)(\varepsilon_\beta - \varepsilon_n), \qquad \alpha = 1, \ldots, n, \tag{4.12}$$

where $\delta_{\alpha\beta}$ is the Kronecker delta. This form was used in Refs. [15], [10], [11] together with the term "form invariance" instead of the "mixture invariance".

Theorem 4.1. (Mixture invariance of the balance equations). *Consider a process of the mixture described by the unprimed quantities in* (4.5)–(4.10) *that satisfy* (3.1)–(3.8) *and define the new primed quantities by* (4.5)–(4.10). *Then the primed quantities satisfy* (3.1)–(3.8).

The proof follows by a direct substitution of (4.5)–(4.10) into (3.1)–(3.8).

The possibility of changing systematically the values of certain quantities without breaking the validity of the balance equations indicates a certain degree of arbitrariness of the quantities in question. We shall see in the next section that when this invariance property is used on the constitutive level, *i.e.*, when the additional quantities (4.1) are given by constitutive equations similar to those for the main quantities, then the new mixture with the new constitutive functions given by (4.5)–(4.9) leads to the same evolution equations for temperature, densities, and motions of the constituents provided the same external fields of force and radiation are applied. Hence this new mixture is indistinguishable from, or, as we say, equivalent to the old one. In view of all these facts only the quantities invariant under these transformations have a direct physical significance, *e.g.* they are expected to be measurable. Such is the case, first of all, of the external fields of force and radiation (cf. $(4.10)_4$ and $(4.10)_6$). Also all of the kinematical quantities and the densities are invariant and, as a consequence of the definitions (3.23), (3.25), the total internal energy, the total entropy and the total free energy, and the total stress are invariant:

$$u' = u, \qquad s' = s, \qquad f' = f, \tag{4.13}$$

$$\boldsymbol{T}' = \boldsymbol{T}. \tag{4.14}$$

It is important to note that the total heat flux $\boldsymbol{q}$ in general is not mixture invariant. That reflects the different choices of the entropy flux in different classical and rational theories of mixtures (*cf.* DE GROOT & MAZUR [20], MÜLLER [4], [7]). It is also worth noting that the heat flux is invariant in the absence of diffusion (*i.e.* when the velocities of constituents are the same). Finally let us note that while the partial quantities (4.5), (4.6), (4.11), (4.7) themselves are not invariant, the total quantities are.

We conclude this section by extending this mixture invariance to the case of constituents having different temperatures. Then the balance equations (3.1)–(3.6), (3.26), (3.27), (3.30), (3.31) are appropriate and we again consider the quantities ε_α, η_α as in (4.1), (4.2). If we define φ_α by

$$\varphi_\alpha = \varepsilon_\alpha - T_\alpha \eta_\alpha, \tag{4.15}$$

then the transformation rules (4.4)–(4.8) and (4.11) remain valid, while (4.9) must be replaced with

$$\boldsymbol{q}'_\alpha = \boldsymbol{q}_\alpha - T_\alpha \varrho_\alpha \eta_\alpha \boldsymbol{v}_\alpha, \tag{4.16}$$

and the following transformations rules must be added:

$$e'_\alpha = e_\alpha + \varrho_\alpha \dot{\varepsilon}_\alpha - \operatorname{div}(\varrho_\alpha \varepsilon_\alpha \boldsymbol{v}_\alpha) + \boldsymbol{v}_\alpha \cdot \operatorname{grad}(\varrho_\alpha \varphi_\alpha), \tag{4.17}$$

$$\sigma'_\alpha = \sigma_\alpha + \partial(\varrho_\alpha \eta_\alpha)/\partial t. \tag{4.18}$$

Theorem 4.1 may now be extended:

Theorem 4.2. (Mixture invariance for a mixture whose constituents have different temperatures). *Consider a process of the mixture described by the unprimed quantities in* (4.5)–(4.8), $(4.10)_{1-5}$, (4.16)–(4.18) *that satisfy* (3.1)–(3.6), (3.26), (3.27), (3.30), (3.31) *and define the new, primed, quantities by* (4.5)–(4.8), $(4.10)_{1-5}$, (4.16)–(4.18). *Then the primed quantities also satisfy* (3.1)–(3.6), (3.26), (3.27), (3.30), (3.31).

The proof follows again by direct substitution.

§ 5. Reacting Mixtures with Heat Condution, Diffusion and Viscosity

In the rest of this paper we demonstrate the usefulness of mixture invariance by applying it to deduce the thermodynamic restrictions on the constitutive functions, limiting attention to mixtures with a single temperature. In this section we define a reacting mixture with heat conduction, diffusion and viscosity. Included as special cases are the reacting mixture without viscosity (BOWEN [4]), the non-reacting mixture without viscosity (BOWEN & WIESE [14]), and reacting mixture with viscosity and non-polar constituents (SAMOHÝL [10]). Similar approaches were poposed by MÜLLER [6], [7] and many other authors (*cf.* TRUESDELL [9], Lecture 5 and its Appendices and review articles by ATKIN & CRAINE [12], [13]).

We postulate the constitutive equations of the reacting mixture with heat conduction, diffusion and viscosity composed of n polar constituents of which first m are chemically reacting in the following form

$$u_\alpha = \bar{u}_\alpha(\boldsymbol{F}_\gamma, \boldsymbol{G}_\gamma, \varrho_\varphi, \boldsymbol{h}_\varphi, \boldsymbol{v}_\gamma, \boldsymbol{L}_\gamma, T, \boldsymbol{g}), \tag{5.1}$$

$$s_\alpha = \bar{s}_\alpha(\boldsymbol{F}_\gamma, \boldsymbol{G}_\gamma, \varrho_\varphi, \boldsymbol{h}_\varphi, \boldsymbol{v}_\gamma, \boldsymbol{L}_\gamma, T, \boldsymbol{g}), \tag{5.2}$$

$$\boldsymbol{T}_\alpha = \bar{\boldsymbol{T}}_\alpha(\boldsymbol{F}_\gamma, \boldsymbol{G}_\gamma, \varrho_\varphi, \boldsymbol{h}_\varphi, \boldsymbol{v}_\gamma, \boldsymbol{L}_\gamma, T, \boldsymbol{g}), \tag{5.3}$$

$$\boldsymbol{q} = \bar{\boldsymbol{q}}(\boldsymbol{F}_\gamma, \boldsymbol{G}_\gamma, \varrho_\varphi, \boldsymbol{h}_\varphi, \boldsymbol{v}_\gamma, \boldsymbol{L}_\gamma, T, \boldsymbol{g}), \tag{5.4}$$

$$r_\nu = \bar{r}_\nu(\boldsymbol{F}_\gamma, \boldsymbol{G}_\gamma, \varrho_\varphi, \boldsymbol{h}_\varphi, \boldsymbol{v}_\gamma, \boldsymbol{L}_\gamma, T, \boldsymbol{g}), \tag{5.5}$$

and

$$\boldsymbol{k}_\alpha = \bar{\boldsymbol{k}}_\alpha(\boldsymbol{F}_\gamma, \boldsymbol{G}_\gamma, \varrho_\varphi, \boldsymbol{h}_\varphi, \boldsymbol{v}_\gamma, \boldsymbol{L}_\gamma, T, \boldsymbol{g}, \operatorname{Grad} \boldsymbol{G}_\gamma, \operatorname{grad} \boldsymbol{h}_\varphi, \operatorname{grad} \boldsymbol{L}_\gamma, \operatorname{grad} \boldsymbol{g}), \tag{5.6}$$

$\alpha, \gamma = 1, \ldots, n$; $\nu, \varphi = 1, \ldots, m$. Using the definitions (3.17), (3.23) and the constitutive equations (5.1), (5.2), we see that the partial free energies f_α and the total internal energy, total entropy, and the total free energy are given by constitutive equations similar to (5.1)–(5.5):

$$f_\alpha = \bar{f}_\alpha(\boldsymbol{F}_\gamma, \boldsymbol{G}_\gamma, \varrho_\varphi, \boldsymbol{h}_\varphi, \boldsymbol{v}_\gamma, \boldsymbol{L}_\gamma, T, \boldsymbol{g}), \tag{5.7}$$

and

$$u = \bar{u}(\boldsymbol{F}_\gamma, \boldsymbol{G}_\gamma, \varrho_\varphi, \boldsymbol{h}_\varphi, \boldsymbol{v}_\gamma, \boldsymbol{L}_\gamma, T, \boldsymbol{g}), \tag{5.8}$$

$$s = \bar{s}(\boldsymbol{F}_\gamma, \boldsymbol{G}_\gamma, \varrho_\varphi, \boldsymbol{h}_\varphi, \boldsymbol{v}_\gamma, \boldsymbol{L}_\gamma, T, \boldsymbol{g}), \tag{5.9}$$

$$f = \bar{f}(\boldsymbol{F}_\gamma, \boldsymbol{G}_\gamma, \varrho_\varphi, \boldsymbol{h}_\varphi, \boldsymbol{v}_\gamma, \boldsymbol{L}_\gamma, T, \boldsymbol{g}). \tag{5.10}$$

The constitutive functions occurring on the right-hand sides of the constitutive equations are assumed to be as many times continuously differentiable as needed by the analysis. Notice that the constitutive equations above violate the principle of equipresence because of the presence of the higher gradients in (5.6). Our attitude towards the principle of equipresence is that there is no reason for insisting on it provided a departure from it is well motivated. And our reason for our assuming a more complicated constitutive equation for $\boldsymbol{k}_\alpha$ will become soon apparent.

Definition 5.1. (Admissible processes). *An admissible process for the reacting mixture with heat conduction, diffusion and viscosity is the following collection of fields (for* $\alpha = 1, \ldots, n;\ \varphi = 1, \ldots, m$)

$$\boldsymbol{x} = \boldsymbol{\chi}_\alpha(\boldsymbol{X}_\alpha, t), \qquad \varrho_\varphi = \varrho_\varphi(\boldsymbol{X}_\varphi, t), \qquad T = T(\boldsymbol{x}, t), \tag{5.11}$$

$$r_\varphi = r_\varphi(\boldsymbol{X}_\varphi, t), \qquad \boldsymbol{T}_\alpha = \boldsymbol{T}_\alpha(\boldsymbol{X}_\alpha, t), \qquad \boldsymbol{k}_\alpha = \boldsymbol{k}_\alpha(\boldsymbol{X}_\alpha, t),$$

$$\boldsymbol{q} = \boldsymbol{q}(\boldsymbol{x}, t), \qquad u_\alpha = u_\alpha(\boldsymbol{X}_\alpha, t), \qquad s_\alpha = s_\alpha(\boldsymbol{X}_\alpha, t), \tag{5.12}$$

$$\boldsymbol{b}_\alpha = \boldsymbol{b}_\alpha(\boldsymbol{X}_\alpha, t), \qquad Q = Q(\boldsymbol{x}, t), \tag{5.13}$$

provided they satisfy the balance equations (3.1)–(3.7) *and the constitutive equations* (5.1)–(5.6).

Definition 5.2. (Dissipation principle). *A reacting mixture with heat conduction, diffusion and viscosity is said to satisfy the dissipation principle of* Coleman & Noll *if the entropy inequality* (3.8) *is satisfied in every admissible process.*

As is well known, when interpreted in this way the entropy inequality places restrictions on the constitutive functions of the bodies in question. The constitutive equations are restricted also by several other constitutive principles but only the dissipation principle concerns in this paper.

We now introduce mixture invariance in to the constitutive theory.

Definition 5.3. (Equivalent constitutive functions and equivalent theories). *Consider a set of constitutive functions for the reacting mixture with heat conduction, diffusion and viscosity* (5.1)–(5.6). *An equivalent set of constitutive functions is the set*

$$\bar{u}'_\alpha, \bar{s}'_\alpha, \bar{\boldsymbol{q}}', \bar{\boldsymbol{T}}'_\alpha, \bar{r}'_\nu, \bar{\boldsymbol{k}}'_\alpha, \tag{5.14}$$

($\alpha = 1, \ldots, n; \nu = 1, \ldots, m$), *such that there are two collections of functions*

$$\varepsilon_\alpha = \bar{\varepsilon}_\alpha(\boldsymbol{F}_\gamma, \boldsymbol{G}_\gamma, \varrho_\varphi, \boldsymbol{h}_\varphi, \boldsymbol{v}_\gamma, \boldsymbol{L}_\gamma, T, \boldsymbol{g}), \tag{5.15}$$

$$\eta_\alpha = \bar{\eta}_\alpha(\boldsymbol{F}_\gamma, \boldsymbol{G}_\gamma, \varrho_\varphi, \boldsymbol{h}_\varphi, \boldsymbol{v}_\gamma, \boldsymbol{L}_\gamma, T, \boldsymbol{g}), \tag{5.16}$$

satisfying identically the relations

$$\sum_{\alpha=1}^{n} \varrho_\alpha \bar{\varepsilon}_\alpha = 0, \qquad \sum_{\alpha=1}^{n} \varrho_\alpha \bar{\eta}_\alpha = 0, \tag{5.17}$$

such that

$$\bar{u}'_\alpha = \bar{u}_\alpha + \bar{\varepsilon}_\alpha, \tag{5.18}$$

$$\bar{s}'_\alpha = \bar{s}_\alpha + \bar{\eta}_\alpha, \tag{5.19}$$

$$\bar{\boldsymbol{T}}'_\alpha = \bar{\boldsymbol{T}}_\alpha + \varrho_\alpha \bar{\varphi}_\alpha \boldsymbol{1}, \tag{5.20}$$

$$\bar{\boldsymbol{k}}'_\alpha = \bar{\boldsymbol{k}}_\alpha - \operatorname{grad}(\varrho_\alpha \bar{\varphi}_\alpha), \tag{5.21}$$

$$\bar{\boldsymbol{q}}' = \bar{\boldsymbol{q}} - T \sum_{\alpha=1}^{n} \varrho_\alpha \bar{\eta}_\alpha \boldsymbol{v}_\alpha, \tag{5.22}$$

where

$$\bar{\varphi}_\alpha = \bar{\varepsilon}_\alpha - T\bar{\eta}_\alpha. \tag{5.23}$$

The mixture described by the set of primed constitutive functions is also said to be equivalent to the mixture described by the original constitutive functions.

Observe that the constitutive functions of the equivalent theory are again of the forms (5.1)–(5.6). Indeed, this is clear for (5.1)–(5.5) immediately, while as for the constitutive equation (5.6) for $\boldsymbol{k}_\alpha$ is concerned, it is enough to realize that if f_α and φ_α depend on $\boldsymbol{F}_\gamma$, $\boldsymbol{G}_\gamma$, ϱ_φ, $\boldsymbol{h}_\varphi$, $\boldsymbol{v}_\gamma$, $\boldsymbol{L}_\gamma$, T, $\boldsymbol{g}$, then grad φ_α depends on the variables occurring on the right-hand side of (5.6), and even linearly on the additional higher gradients. It is exactly for this reason that we postulate a constitutive equation for $\boldsymbol{k}_\alpha$ that is more complicated than for the quantities in (5.1)–(5.5). When seeking the detailed restrictions on the constitutive functions, the linearity assumption simplifies things considerably (see [17]), but on the general level considered here, that is irrelevant and hence we here do not assume linearity.

It is clear that when the new constitutive functions are put into the balance equations, then the resulting differential equations for the motions of the constituents and their temperature are identical, *i.e.*, the contributions of the additional terms (5.15), (5.16) cancel provided the fields of external force and radiation are unchanged. This is our reason for calling the new theory equivalent to the original. On the other hand, it should be mentioned that if the boundary conditions are formulated in terms of the non-invariant quantities like $\boldsymbol{T}_\alpha$ or $\boldsymbol{q}$ (in the presence of diffusion), then the boundary conditions for the new theory do not coincide with the boundary conditions of the original theory.

We conclude this section with the following simple observations which will be useful in the sequel. Their obvious proofs are omitted.

Proposition 5.1. *If a set of constitutive functions* (5.1)–(5.6) *for a reacting mixture with heat conduction, diffusion and viscosity satisfies the dissipation principle, then any set of equivalent constitutive functions satisfies the dissipation principle.*

Lemma 5.1. *For any choice of fields in* (5.11) *satisfying the balance equations of mass for reacting constituents* (3.11) *there exists a uniquely determined admissible process of a mixture with given set of constitutive functions.*

Lemma 5.2. *The fields* (5.11), (5.12) *of an admissible process determine the fields of the partial free energies f_α and the fields of the total internal energy, entropy, and free energy.*

Lemma 5.3. *The dissipation principle is equivalent to requiring that the reduced dissipation inequality be satisfied in every set of fields* (5.11) *satisfying the balance equations of mass for reacting constituents* (3.11).

In the last lemma we assume, following COLEMAN & GURTIN [21] and BOWEN [4] that the solution of the differential equation (3.11) exists for any set of initial conditions.

§ 6. Reduction of the Partial Thermodynamic Quantities of the General Model by Mixture Invariance

In this section we use mixture invariance to simplify the consequences of the dissipation principle on the constitutive functions. Our main result is contained in Theorem 6.1 below. We start with the following result of the application of the dissipation principle to our reacting mixture.

Proposition 6.1. (Results of the dissipation principle). *Consider a reacting mixture with heat conduction, diffusion and viscosity determined by the constitutive functions* (5.1)–(5.6). *If this mixture satisfies the dissipation principle then the total internal energy, total entropy and total free energy are independent of* $\boldsymbol{v}_\gamma$, $\boldsymbol{L}_\gamma$, *and* $\boldsymbol{g}$, i.e.

$$u = \bar{u}(\boldsymbol{F}_\gamma, \boldsymbol{G}_\gamma, \varrho_\varphi, \boldsymbol{h}_\varphi, T), \tag{6.1}$$

$$s = \bar{s}(\boldsymbol{F}_\gamma, \boldsymbol{G}_\gamma, \varrho_\varphi, \boldsymbol{h}_\varphi, T), \tag{6.2}$$

$$f = \bar{f}(\boldsymbol{F}_\gamma, \boldsymbol{G}_\gamma, \varrho_\varphi, \boldsymbol{h}_\varphi, T), \tag{6.3}$$

and the entropy relation

$$s = -\frac{\partial \bar{f}}{\partial T} \tag{6.4}$$

holds.

Proof. By the constitutive equation (5.7) and the chain rule we obtain the following expression for $\overset{\backslash}{f}_\alpha$:

$$\begin{aligned} \overset{\backslash}{f}_\alpha = {} & \sum_{\gamma=1}^{n} \frac{\partial \bar{f}_\alpha}{\partial F_\gamma^{iJ}} \overset{\backslash\alpha}{F}{}_\gamma^{iJ} + \sum_{\gamma=1}^{n} \frac{\partial \bar{f}_\alpha}{\partial G_\gamma^{iJK}} \overset{\backslash\alpha}{G}{}_\gamma^{iJK} + \sum_{\gamma=1}^{n} \frac{\partial \bar{f}_\alpha}{\partial v_\gamma^{i}} \overset{\backslash\alpha}{v}{}_\gamma^{i} + \sum_{\gamma=1}^{n} \frac{\partial \bar{f}_\alpha}{\partial L_\gamma^{ij}} \overset{\backslash\alpha}{L}{}_\gamma^{ij} \\ & + \frac{\partial \bar{f}_\alpha}{\partial T} \overset{\backslash\alpha}{T} + \frac{\partial \bar{f}_\alpha}{\partial g^i} \overset{\backslash\alpha}{g}{}^{i} + \sum_{\varphi=1}^{m} \frac{\partial \bar{f}_\alpha}{\partial \varrho_\varphi} \overset{\backslash\alpha}{\varrho}_\varphi + \sum_{\varphi=1}^{m} \frac{\partial \bar{f}_\alpha}{\partial h_\varphi^i} \overset{\backslash\alpha}{h}{}_\varphi^{i}. \end{aligned} \tag{6.5}$$

This expression is now inserted into the reduced inequality (3.18), using simplified expressions for the material time derivatives like

$$\overset{\backslash\alpha}{L}{}_\gamma^{ij} = \overset{\backslash\backslash}{F}{}_\gamma^{iK} \overset{-1}{F}{}_\gamma^{Kj} - L_\gamma^{ik} L_\gamma^{kj} + (v_\alpha^k - v_\gamma^k)\, \overset{\backslash}{G}{}_\gamma^{iML} \overset{-1}{F}{}_\gamma^{Lj} \overset{-1}{F}{}_\gamma^{Mk} - (v_\alpha^k - v_\gamma^k)\, L_\gamma^{il} G_\gamma^{lJK} \overset{-1}{F}{}_\gamma^{Jj} \overset{-1}{F}{}_\gamma^{Kk}. \tag{6.6}$$

Even though the full calculation is laborious (see [17]), it may be seen that the following inequality is valid:

$$\varrho\left(\frac{\partial \bar{f}}{\partial T} + s\right)\frac{\partial T}{\partial t} + \varrho\,\frac{\partial \bar{f}}{\partial g^i}\frac{\partial g^i}{\partial t} + \sum_{\gamma=1}^{n}\left[\varrho\,\frac{\partial \bar{f}}{\partial L_\gamma^{ij}} \overset{-1}{F}{}_\gamma^{Kj}\right] \overset{\backslash\backslash}{F}{}_\gamma^{iK} + \sum_{\gamma=1}^{n} \varrho\,\frac{\partial \bar{f}}{\partial v_\gamma^i} \overset{\backslash}{v}{}_\gamma^{i} + A \leq 0 \tag{6.7}$$

where A is independent of $\dfrac{\partial T}{\partial t}, \dfrac{\partial \boldsymbol{g}}{\partial t}, \overset{\backslash\backslash}{\boldsymbol{F}}_\gamma$ and $\overset{\backslash}{\boldsymbol{v}}_\gamma$ (this last fact follows from inspection of (6.5), (3.18), (6.6)). As a consequence of the linearity in these quantities and Lemma 5.3 we obtain the result. *Q.E.D.*

Further consequences on the constitutive functions will be stated in our future paper [17].

Lemma 6.1. (Consequence of mixture invariance). *Consider n functions* $\bar{y}_\alpha$ $(\alpha = 1, \ldots, n)$ *of two sets of independent variables a, b and assume that the mass fractions* w_α *may be expressed through the variables a and the variables b may take on zero values,* $b = 0$. *If the function* $\bar{y}$, *defined by*

$$\bar{y} = \sum_{\alpha=1}^{n} w_\alpha \bar{y}_\alpha, \tag{6.8}$$

is independent of b, $y = \bar{y}(a)$, *then there are functions* $\bar{\zeta}_\alpha$ *of a, b satisfying*

$$\sum_{\alpha=1}^{n} w_\alpha \bar{\zeta}_\alpha = 0, \tag{6.9}$$

such that the function $\bar{y}'_\alpha$ *defined by*

$$\bar{y}'_\alpha = \bar{y}_\alpha + \bar{\zeta}_\alpha \tag{6.10}$$

are independent of b.

Proof. Define $\bar{\zeta}_\alpha$ by

$$\bar{\zeta}_\alpha(a, b) = \bar{y}_\alpha(a, 0) - \bar{y}_\alpha(a, b), \tag{6.11}$$

so that, by (6.8),

$$\sum_{\alpha=1}^{n} w_\alpha \bar{\zeta}_\alpha = \sum_{\alpha=1}^{n} (w_\alpha \bar{y}_\alpha(a, 0) - w_\alpha \bar{y}_\alpha(a, b)) = \bar{y}(a, 0) - \bar{y}(a, b).$$

By the hypothesis of the lemma, $\bar{y}(a, b)$ is independent of b and hence (6.9) follows. On the other hand, (6.10) and (6.11) imply that

$$\bar{y}'_\alpha(a, b) = \bar{y}_\alpha(a, 0),$$

and therefore $\bar{y}'_\alpha$ is independent of b. *Q.E.D.*

Theorem 6.1. (Independence of the partial thermodynamic quantities of the non-equilibrium variables). *For each reacting mixture* (5.1)–(5.6) *with heat conduction, diffusion and viscosity there is an equivalent set of constitutive functions* $\bar{u}_\alpha, \bar{s}_\alpha, \bar{f}_\alpha, \bar{\boldsymbol{q}}, \bar{\boldsymbol{k}}_\alpha, \bar{\boldsymbol{T}}_\alpha$, *such that the partial internal energies, entropies, and free energies are independent of* $\boldsymbol{v}_\gamma$, $\boldsymbol{L}_\gamma$, $\boldsymbol{g}$, i.e.,

$$u_\alpha = \bar{u}_\alpha(\boldsymbol{F}_\gamma, \boldsymbol{G}_\gamma, \varrho_\varphi, \boldsymbol{h}_\varphi, T), \tag{6.12}$$

$$s_\alpha = \bar{s}_\alpha(\boldsymbol{F}_\gamma, \boldsymbol{G}_\gamma, \varrho_\varphi, \boldsymbol{h}_\varphi, T), \tag{6.13}$$

$$f_\alpha = \bar{f}_\alpha(\boldsymbol{F}_\gamma, \boldsymbol{G}_\gamma, \varrho_\varphi, \boldsymbol{h}_\varphi, T). \tag{6.14}$$

Proof. In view of Proposition 6.1 we can apply Lemma 6.1 with the choices $\bar{y}_\alpha = \bar{u}_\alpha$, $\bar{s}_\alpha$, and $\bar{y} = \bar{u}, \bar{s}$, respectively, and with the identifications of the variables as follows: $a = (\boldsymbol{F}_\gamma, \boldsymbol{G}_\gamma, \varrho_\varphi, \boldsymbol{h}_\varphi, T)$, $b = (\boldsymbol{v}_\gamma, \boldsymbol{L}_\gamma, \boldsymbol{g})$. This way we obtain (6.12), (6.13); (6.14) is then a consequence of (6.12), (6.13) and the definition (3.17) of f_α. *Q.E.D.*

§ 7. Reduction of the Partial Thermodynamic Quantities for a Non-Reacting Mixture

If the constitutive functions of the mixture are simpler in one or another respect than the preceding ones, then the mixture invariance enables one to pass to an equivalent model to reach further simplifications. We shall demonstrate it on a non-reacting mixture.

A non-reacting mixture with heat conduction, diffusion and viscosity is a special case of the model described in § 5 for which the number of the reacting constituents is zero, i.e., $m = 0$ in the notation of that section; then

$$r_\alpha = 0, \qquad \alpha = 1, \ldots, n. \tag{7.1}$$

The equation $m = 0$ means that the variables ϱ_φ, $\boldsymbol{h}_\varphi$ in the constitutive functions (5.1)–(5.6) may be omitted and also that the constitutive variables grad $\boldsymbol{h}_\varphi$ may be omitted from (5.6).

Special non-reacting mixtures with heat conduction, diffusion and viscosity have been studied by Bowen & Wiese [14] (a non-reacting mixture of elastic materials with heat conduction and diffusion), Bowen [5] and Samohýl [10] (a non-reacting mixture with heat conduction, diffusion and viscosity), but with the constitutive functions $\bar{\boldsymbol{k}}_\alpha$ independent of the additional higher gradients. The particular cases of the following Proposition have been established in the works just cited.

Proposition 7.1. (Results of dissipation principle). *For a non-reacting mixture with heat conduction, diffusion and viscosity satisfying the dissipation principle the total free energy, the total entropy and the total internal energy are functions of $\boldsymbol{F}_\gamma$ and T only*, i.e.,

$$f = \bar{f}(\boldsymbol{F}_\gamma, T), \tag{7.2}$$

$$s = \bar{s}(\boldsymbol{F}_\gamma, T), \tag{7.3}$$

$$u = \bar{u}(\boldsymbol{F}_\gamma, T). \tag{7.4}$$

Proof. For a non-reacting mixture the total free energy is automatically independent of ϱ_φ, $\boldsymbol{h}_\varphi$. Moreover, by Proposition 6.1 it is also independent of $\boldsymbol{v}_\gamma$, $\boldsymbol{L}_\gamma$, $\boldsymbol{g}$, and hence

$$f = \bar{f}(\boldsymbol{F}_\gamma, \boldsymbol{G}_\gamma, T). \tag{7.5}$$

We now apply the COLEMAN & NOLL procedure to the situation when $\boldsymbol{v}_\alpha = \mathbf{0}$, $\alpha = 1, \ldots, n$ at a given point. In this case the derivatives in (6.5) $\overset{\backslash\alpha}{\boldsymbol{F}}_\gamma$, $\overset{\backslash\alpha}{\boldsymbol{G}}_\gamma, \ldots$ coincide with $\overset{\backslash}{\boldsymbol{F}}_\gamma$, $\overset{\backslash}{\boldsymbol{G}}_\gamma, \ldots$, respectively, and we further choose $\overset{\backslash}{\boldsymbol{F}}_\gamma = \mathbf{0}$, $\overset{\backslash}{\boldsymbol{v}}_\gamma = \mathbf{0}$, $\overset{\backslash}{\boldsymbol{L}}_\gamma = \mathbf{0}$, $\overset{\backslash}{\boldsymbol{g}} = \mathbf{0}$, $\boldsymbol{g} = \mathbf{0}$, $\boldsymbol{L}_\gamma = \mathbf{0}$, $\dfrac{\partial T}{\partial t} = 0$ at the point. The derivative $\overset{\backslash}{f}_\alpha$ in (6.5) then takes the form

$$\overset{\backslash}{f}_\alpha = \sum_{\gamma=1}^{n} \frac{\partial \bar{f}_\alpha}{\partial G_\gamma^{iJK}} \overset{\backslash}{G}{}_\gamma^{iJK},$$

where the partial derivatives are evaluated at $\boldsymbol{F}_\gamma$, $\boldsymbol{G}_\gamma$, $\boldsymbol{v}_\gamma = 0$, $\boldsymbol{L}_\gamma = \mathbf{0}$, T, $\boldsymbol{g} = \mathbf{0}$. This expression is now inserted in the reduced dissipation inequality (3.18). Taking into account the fact that for a non-reacting mixture one has $r_\alpha = 0$ and using the definition $(3.23)_3$ of f we see that (3.18) now reads as follows:

$$\sum_{\gamma=1}^{n} \frac{\partial \varrho \bar{f}}{\partial G_\gamma^{iJK}} \overset{\backslash}{G}{}_\gamma^{iJK} \leqq 0,$$

where now the partial derivatives are evaluated at $\boldsymbol{F}_\gamma$, $\boldsymbol{G}_\gamma$, T; see (7.5). In view of the arbitrariness of $\overset{\backslash}{\boldsymbol{G}}_\gamma$ this implies

$$\frac{\partial \bar{f}}{\partial \boldsymbol{G}_\gamma} = \mathbf{0},$$

which gives the result (7.2). Then, by (6.4) (valid also in this non-reacting case) and (3.24), we obtain (7.3), (7.4). *Q.E.D.*

Making use of mixture invariance, a further simplification of the constitutive equations may be achieved.

Theorem 7.1. (Independence of partial thermodynamic quantities of the second deformation gradients). *For a non-reacting mixture with heat conduction, diffusion and viscosity satisfying the dissipation principle there is a non-reacting mixture described by an equivalent set of constitutive functions such that its partial thermodynamic quantities f_α, s_α, and u_α, are functions of $\boldsymbol{F}_\gamma$, and T only*, i.e.,

$$f_\alpha = \bar{f}_\alpha(\boldsymbol{F}_\gamma, T), \tag{7.6}$$

$$s_\alpha = \bar{s}_\alpha(\boldsymbol{F}_\gamma, T), \tag{7.7}$$

$$u_\alpha = \bar{u}_\alpha(\boldsymbol{F}_\gamma, T). \tag{7.8}$$

Proof. This follows from Lemma 6.1 and Proposition 7.1. We apply Lemma 6.1 with the choices $y_\alpha = u_\alpha, s_\alpha$, and $y = u, s$, respectively, and with the identifications of the variables $a = (\boldsymbol{F}_\gamma, T), b = (\boldsymbol{G}_\gamma)$. This way we obtain (7.7), (7.8); (7.6) is then a consequence of (7.7) and (7.8) and of the definition (3.17) of f_α. *Q.E.D.*

§ 8. Partial Thermodynamic Quantities and Classical Thermochemistry

There is a discrepancy between the partial thermodynamic quantities of the rational thermodynamics of mixtures discussed above and the partial thermodynamic quantities used in the classical thermochemistry of fluid mixtures. Indeed, we shall show below (see (8.23), (8.24)) that the partial quantities of the rational theory do not satisfy all of the relations of classical thermochemistry. This is caused by the fact that in the rational approach the partial quantities are treated as primitives while in the classical theory they are defined from the total thermodynamic quantities of the mixture. This definition was adopted also by linear irreversible thermodynamics by use of the "hypothesis of local equilibrium" (see *e.g.* [20]) and this way the classical partial thermodynamic quantities are used as fields.

As the last application of mixture invariance we shall show here how to obtain complete agreement with the results of classical thermochemistry. We follow here SAMOHÝL [15], [10], [11].

In this section we shall deal with the rational theory of mixtures for which the constitutive equations for the fields of the partial thermodynamic quantities have the form

$$u_\alpha = \hat{u}_\alpha(\varrho_\gamma, T), \tag{8.1}$$

$$s_\alpha = \hat{s}_\alpha(\varrho_\gamma, T), \tag{8.2}$$

$$f_\alpha = \hat{f}_\alpha(\varrho_\gamma, T). \tag{8.3}$$

Therefore for total quantities we have *e.g.* (*cf.* (3.23))

$$f = \hat{f}(\varrho_\gamma, T). \tag{8.4}$$

Such is the case of the mixture of linear fluids (SAMOHÝL [15], [10], [11]) or the non-reacting mixture of § 7 specialized for fluids [16], because the ϱ_γ may be used in (7.6)–(7.8) instead of $\boldsymbol{F}_\gamma$ (*cf.* similar results of BOWEN [5], [4]), MÜLLER [6], [7] and others [9], [12], [13]).

It is useful to define the partial Gibbs energy, or chemical potential, g_α (for $\alpha = 1, \ldots, n$)

$$g_\alpha = \frac{\partial \varrho \hat{f}}{\partial \varrho_\alpha}, \tag{8.5}$$

partial thermodynamic pressure P_α

$$P_\alpha = \varrho_\alpha(g_\alpha - f_\alpha), \tag{8.6}$$

and total thermodynamic pressure P by

$$P = \sum_{\alpha=1}^{n} P_\alpha. \tag{8.7}$$

Further we define the partial volume v_α by

$$v_\alpha = \frac{P_\alpha}{\varrho_\alpha P}. \tag{8.8}$$

and the partial enthalpy h_α by

$$h_\alpha = u_\alpha + Pv_\alpha. \tag{8.9}$$

We denote by y_α any of the partial thermodynamic quantities u_α, s_α, f_α, g_α, v_α, h_α; the corresponding total thermodynamic quantities u, s, f, g, v, h are denoted by y and defined generally by

$$y = \sum_{\alpha=1}^{n} w_\alpha y_\alpha, \tag{8.10}$$

where w_α is the mass fraction (3.21), cf. (3.23).

As a consequence of these definitions, the following classical relations are valid.

Proposition 8.1. (Classical thermodynamic relations). *We have*

$$v = 1/\varrho, \tag{8.11}$$

$$\sum_{\alpha=1}^{n} \varrho_\alpha v_\alpha = 1, \tag{8.12}$$

$$g_\alpha = f_\alpha + Pv_\alpha, \tag{8.13}$$

$$f = u - Ts, \quad g = f + Pv, \quad h = u + Pv, \tag{8.14}$$

$$P_\alpha = \sum_{\gamma=1}^{n} \varrho_\alpha \varrho_\gamma \left(\frac{\partial \hat{f}_\gamma}{\partial \varrho_\alpha}\right). \tag{8.15}$$

Further, if the relation

$$\frac{\partial \hat{f}}{\partial T} = -s \tag{8.16}$$

is valid, then the Gibbs equations

$$d(\varrho f) = -\varrho s\, dT + \sum_{\alpha=1}^{n} g_\alpha\, d\varrho_\alpha, \tag{8.17}$$

$$du = T\, ds - P\, dv + \sum_{\beta=1}^{n-1} (g_\beta - g_n)\, dw_\beta, \tag{8.18}$$

$$dg = -s\, dT + v\, dP + \sum_{\beta=1}^{n-1} (g_\beta - g_n)\, dw_\beta, \tag{8.19}$$

and the Gibbs-Duhem equation (for the chemical potential)

$$-s\, dT + v\, dP - \sum_{\alpha=1}^{n} w_\alpha\, dg_\alpha = 0, \tag{8.20}$$

are valid.

Here the differential operator d may be replaced by the partial time derivative $\partial/\partial t$, the material derivative (2.6) or the gradients grad or Grad. Property (8.16) is standard, *e.g.* (6.4) is valid also for nonreacting fluid mixture.

Proposition 8.2. (Discrepancy from classical relations). *If the function (see* (8.7), (3.21), (3.22), (8.11))

$$P = \hat{P}(\varrho_\gamma, T) = \hat{P}\left((w_1/v, \ldots, w_{n-1}/v, \left(1 - \sum_{\beta=1}^{n-1} w_\beta\right)\Big/ v, T\right) \tag{8.21}$$

is invertible for v, then all thermodynamic quantities y_α, y may be also expressed as functions of T, P, $w_1, \ldots, w_{n-1}$:

$$y_\alpha = \tilde{y}_\alpha(T, P, w_\beta), \quad y = \tilde{y}(T, P, w_\beta), \tag{8.22}$$

and then the following relations hold:

$$\frac{\partial \tilde{y}}{\partial w_\beta} = y_\beta - y_n + \sum_{\alpha=1}^{n} w_\alpha \frac{\partial \tilde{y}_\alpha}{\partial w_\beta}, \tag{8.23}$$

$$0 = \frac{\partial \tilde{y}}{\partial T} dT + \frac{\partial \tilde{y}}{\partial P} dP - \sum_{\alpha=1}^{n} w_\alpha \, dy_\alpha + \sum_{\beta=1}^{n-1} \sum_{\alpha=1}^{n} w_\alpha \frac{\partial \tilde{y}_\alpha}{\partial w_\beta} dw_\beta . \tag{8.24}$$

These last relations (8.23), (8.24) *simplify only for the Gibbs energies g_α where*

$$\sum_{\alpha=1}^{n} w_\alpha \frac{\partial \tilde{g}_\alpha}{\partial w_\beta} = 0. \tag{8.25}$$

(In all these relations the free indices are $\alpha, \gamma = 1, \ldots, n$; $\beta = 1, \ldots, n-1$.)

The proofs of the above propositions are easy; for details see [15], [10], [11].

Proposition 8.1 shows that the thermodynamic quantities and pressures introduced above satisfy the classical equations. But Proposition 8.2 shows that the partial thermodynamic quantities y_α are not the same as the classical partial thermodynamic quantities of thermochemistry (in mass units; see *e.g.* [22]) because the latter satisfy

$$\sum_{\alpha=1}^{n} w_\alpha \frac{\partial \tilde{y}_\alpha}{\partial w_\beta} = 0, \quad \beta = 1, \ldots, n-1, \tag{8.26}$$

for $y_\alpha = u_\alpha, s_\alpha, f_\alpha, g_\alpha, v_\alpha, h_\alpha$ whereas here (8.26) is guaranteed only for $y_\alpha = g_\alpha$, see (8.25).

To remove this discrepancy, we now apply mixture invariance; in view of the form of the constitutive equations (8.1)–(8.3) we consider only the changes (5.18)–(5.19) in which the functions $\hat{\varepsilon}_\alpha$ and $\hat{\eta}_\alpha$ depend only on ϱ_γ, T,

$$\varepsilon_\alpha = \hat{\varepsilon}_\alpha(\varrho_\gamma, T), \tag{8.27}$$

$$\eta_\alpha = \hat{\eta}_\alpha(\varrho_\gamma, T), \tag{8.28}$$

with properties (5.17).

Proposition 8.3. (Mixture invariance of thermodynamic quantities). *Assuming that the primitives T, ϱ_γ and the definitions of the quantities* (3.17), (8.5) - (8.10) *themselves are mixture invariant, we have the following assertions: the Gibbs energy g_α, the total thermodynamic quantities y and the total thermodynamic pressure P are mixture invariant,* i.e.,

$$g'_\alpha = g_\alpha, \quad y' = y, \quad P' = P. \tag{8.29}$$

On the other hand, the remaining partial quantities y_α *and the partial thermodynamic pressures* P_α *are not mixture invariant, for they transform as*

$$y'_\alpha = y_\alpha + \zeta_\alpha, \tag{8.30}$$

$$P'_\alpha = P_\alpha - \varrho_\alpha \varphi_\alpha, \tag{8.31}$$

with the property

$$\sum_{\alpha=1}^{n} w_\alpha \zeta_\alpha = 0, \tag{8.32}$$

where

$$\zeta_\alpha = \varepsilon_\alpha, \eta_\alpha, \varphi_\alpha = \varepsilon_\alpha - T\eta_\alpha, -\varphi_\alpha/P, T\eta_\alpha, 0, \tag{8.33}$$

for $y_\alpha = u_\alpha, s_\alpha, f_\alpha, v_\alpha, h_\alpha, g_\alpha,$ *respectively.*

The proof is simple; see SAMOHÝL [15], [10], [11], for details.

It is remarkable that of all partial thermodynamic quantities y_α only the chemical potential g_α is mixture invariant; this corresponds with the fact that only g_α are measurable in principle.

Theorem 8.1. (Accord with classical theory). *For every theory satisfying the constitutive equations* (8.1)–(8.3) *and the hypotheses of Propositions* 8.1 *and* 8.2 *there is an equivalent theory achieved by the functions* $\hat{\varepsilon}_\alpha, \hat{\eta}_\alpha$, *of the form* (8.27), (8.28) *such that the partial quantities* y_α *satisfy* (8.26) *identically. Then instead of* (8.23), (8.24) *we have*

$$\frac{\partial \tilde{y}}{\partial w_\beta} = y_\beta - y_n, \tag{8.34}$$

$$\frac{\partial \tilde{y}}{\partial T} dT + \frac{\partial \tilde{y}}{\partial P} dP - \sum_{\alpha=1}^{n} w_\alpha \, dy_\alpha = 0; \tag{8.35}$$

the ranges of the indices are $\alpha = 1, \ldots, n,\ \beta = 1, \ldots, n-1$.

Proof. We pass to an equivalent theory by use of the functions $\hat{\varepsilon}_\alpha, \hat{\eta}_\alpha$ defined by

$$\begin{aligned} \zeta_\delta &= \sum_{\gamma=1}^{n} w_\gamma \frac{\partial \tilde{y}_\gamma}{\partial w_\delta} - \sum_{\beta=1}^{n-1} \sum_{\gamma=1}^{n} w_\beta w_\gamma \frac{\partial \tilde{y}_\gamma}{\partial w_\beta}, \quad \delta = 1, \ldots, n-1, \\ \zeta_n &= -\sum_{\beta=1}^{n-1} \sum_{\gamma=1}^{n} w_\beta w_\gamma \frac{\partial \tilde{y}_\gamma}{\partial w_\beta}, \end{aligned} \tag{8.36}$$

where $\zeta_\alpha = \varepsilon_\alpha, \eta_\alpha$ and $y_\alpha = u_\alpha, s_\alpha$ respectively (by Proposition 8.2, the ζ_α are also functions of T, P, w_β). It is easy to verify that this choice satisfies (8.32) and some algebra gives

$$\sum_{\alpha=1}^{n} w_\alpha \frac{\partial \tilde{y}'_\alpha}{\partial w_\beta} = 0, \tag{8.37}$$

$y'_\alpha = u'_\alpha, s'_\alpha$. With primes deleted this proves (8.26) for $y_\alpha = u_\alpha, s_\alpha$. With the choice (8.36), equation (8.37) holds also for the other quantities y_α (*i.e.* for $f_\alpha, v_\alpha, h_\alpha, g_\alpha$), because the corresponding ζ_α are combinations of $\varepsilon_\alpha, \eta_\alpha$, see (8.33) (for g_α trivially, *cf.* (8.25)). Eqs. (8.34), (8.35) now follow from (8.23), (8.24). *Q.E.D.*

The results (8.34), (8.35) of the last theorem show the complete agreement of the new quantities y_α with the partial thermodynamic quantities of classical thermochemistry. For example, the classical relations

$$\frac{\partial \tilde{g}_\alpha}{\partial T} = -s_\alpha, \quad \frac{\partial \tilde{g}_\alpha}{\partial P} = v_\alpha, \tag{8.38}$$

are valid only with this choice. (These relations (8.38) follow from the Gibbs equation (8.19) and the interchangeability of the second derivatives using (8.34), for the partial entropy and volume).

In the classical thermochemistry of mixtures [22] usually the uniform cases are studied (when the fields are independent of place) and in this case we start with the extensive thermodynamic quantities Y for the mixture (Y denotes the internal energy U, entropy S, the free energy F, the volume V, the enthalpy H and the Gibbs energy G, respectively),

$$Y = \bar{Y}(T, P, m_\alpha), \tag{8.39}$$

where m_α ($\alpha = 1, \ldots, n$) are the masses of the constituents. These quantities Y are related to the previous total thermodynamic quantities y by

$$\bar{Y}(T, P, m_\alpha) = m\check{y}(T, P, w_\beta), \quad \alpha = 1, \ldots, n, \quad \beta = 1, \ldots, n-1, \tag{8.40}$$

where $m = \sum_{\alpha=1}^{n} m_\alpha$ and the mass fraction is $w_\alpha = m_\alpha/m$. Relation (8.34) reads as

$$\frac{\partial \bar{Y}}{\partial m_\alpha} = y_\alpha, \tag{8.41}$$

and this is exactly the definition of classical partial thermodynamic quantities [22] in mass units.

The great advantage of these partial quantities y_α of Theorem 8.1 is that they are obtainable from measurable (and therefore mixture invariant) quantities of mixture y or Y (by (8.34) or (8.41)) or from chemical potentials by (8.38) (which are also measurable and mixture invariant). But the disadvantage of these new quantities y_α is that their definition is based on pressure P (used as an independent variable) and this is difficult to generalize to nonfluid mixtures. (Moreover, even in some linear models, the thermodynamic pressure coincides with the stress only in equilibrium; *cf.* such viscous models with chemical reactions [15], [10], [11]).

From (8.38), (8.22), (8.13), it follows that

$$-s_\alpha = \frac{\partial \hat{f}_\alpha}{\partial T} + P\frac{\partial \hat{v}_\alpha}{\partial T}, \tag{8.42}$$

which shows that the analogue of (8.16) for partial quantities

$$\frac{\partial \hat{f}_\alpha}{\partial T} = -s_\alpha \tag{8.43}$$

is not valid for these new partial quantities except in some special cases as *e.g.* a mixture of ideal gases (see [10], [11]). Therefore the partial thermodynamic quantities different from (8.41), although possible, are difficult to use because it is not clear how to obtain such quantities from measurable ones. Exceptions are some special cases as the simple mixtures [6], [7], where these quantities may be obtained from the properties of pure constituents before mixing.

References

1. COLEMAN, B. D., & NOLL, W., The thermodynamics of elastic materials with heat conduction and viscosity. Arch. Rational Mech. Anal. **13**, 167–178 (1963).
2. COLEMAN, B. D., Thermodynamics of materials with memory. Arch. Rational Mech. Anal. **17**, 1–46 (1964).
3. COLEMAN, B. D., On thermodynamics, strain impulses, and visco-elasticity. Arch. Rational Mech. Anal. **17**, 230–254 (1964).
4. BOWEN, R. M., The thermochemistry of a reacting mixture of elastic materials with diffusion. Arch. Rational Mech. Anal. **34**, 97–127 (1969).
5. BOWEN, R. M., Theory of mixtures. Continuum Physics, Vol. III, Mixtures and EM Field Theories (ed. ERINGEN, A. C.). Academic Press, New York 1976.
6. MÜLLER, I., A thermodynamic theory of mixtures of fluids. Arch. Rational Mech. Anal. **28**, 1–39 (1968).
7. MÜLLER, I., Thermodynamik. Grundlagen der Materialtheorie. Bertelsmann Universitätsverlag, Düsseldorf 1973.
8. WILLIAMS, W. O., On the theory of mixtures. Arch. Rational Mech. Anal. **51**, 239–260 (1973).
9. TRUESDELL, C., Rational Thermodynamics (2nd Ed.). Springer, New York 1984.
10. SAMOHÝL, I., Racionální termodynamika chemicky reagujících směsí (Rational Thermodynamics of Chemically Reacting Mixtures). Academia, Prague 1982.
11. SAMOHÝL, I., Thermodynamics of Irreversible Processes in Fluid Mixtures. Teubner, Leipzig 1987.
12. ATKIN, R. J., & CRAINE, J. E., Continuum theories of mixtures: Basic theory and historical development. Q. J. Mech. appl. Math. **29**, 209–244 (1976).
13. ATKIN, R. J., & CRAINE, R. E., Continuum theories of mixtures: Applications. J. Inst. Maths. Applics. **17**, 153–207 (1976).
14. BOWEN, R. M., & WIESE, J. C., Diffusion in mixtures of elastic materials. Int. J. Engng. Sci. **7**, 689–722 (1969).
15. SAMOHÝL, I., Comparison of classical and rational thermodynamics of reacting fluid mixtures with linear transport properties. Coll. Czechoslov. Chem. Commun. **40**, 3421–3435 (1975).
16. SAMOHÝL, I., Reduction of thermodynamic constitutive equations for fluid mixtures using form invariance. Coll. Czechoslov. Chem. Commun. **54**, 277–283 (1989).
17. SAMOHÝL, I., & ŠILHAVÝ, M., The thermodynamics of reacting mixtures with heat conduction, diffusion and viscosity. In preparation.
18. TRUESDELL, C., & TOUPIN, R., The Classical Field Theories. Handbuch der Physik III/1 (ed. FLÜGGE, S.). Springer, Berlin, 1960.
19. TRUESDELL, C., Sulle basi della termodinamica delle miscele. Rend. Accad. Naz. Lincei (8) **44**, 381–383 (1968).

20. de Groot, S. R., & Mazur, P., Nonequilibrium Thermodynamics. North-Holland, Amsterdam 1962.
21. Coleman, B. D., & Gurtin, M. E., Thermodynamics with internal state variables. J. Chem. Phys. **47**, 597–613 (1967).
22. Guggenheim, E. A., Thermodynamics. North-Holland, Amsterdam 1957.

Department of Physical Chemistry
Institute of Chemical Technology
Prague

and

Mathematical Institute
Czechoslovak Academy of Sciences
Prague

(Received June 15, 1989)

Interactions in General Continua with Microstructure

GIANFRANCO CAPRIZ & EPIFANIO G. VIRGA

Dedicated to Bernard D. Coleman on the occasion of his 60th birthday

Table of Contents

1. Introduction

A satisfactory axiomatic treatment of the interaction among bodies modelled as standard continua was achieved by NOLL almost thirty years ago and was presented in the seminal paper [1], variously reprinted. A full review of the matter is also available in the first chapters of the treatise [2].

Adaptations of NOLL's axioms and theorems for more general continua were suggested by BEATTY [3] and WILLIAMS [4]. However, they did not involve some essential features of actions present in continua with microstructure. As is well known from special theories (*e.g.*, liquid crystals, Cosserat continua) self-action occurs in general, and hence it should be allowed by the basic axioms. Also, interactions are, generally, much more complex than those finally envisaged in [3] and [4]; it is not only a question of admitting torques which are not moments of forces. In fact, the choice of the proper setting for the description of microstructural actions and events in general continua is usually much more delicate than is admitted even in the very comprehensive initial hypotheses of [4]: a manifold and its fibre bundle should be involved. A naive approach towards satisfying this need is proposed in the tract [5], together with concrete examples that justify such complex setting.

In substance it is shown in [5] that in a body with microstructure specification of the state of each material element requires assignment of a finite number of *order parameters* $\nu^1, \nu^2, \dots, \nu^m$ and these parameters are best interpreted as the coordinates, in a local chart, of an element $\boldsymbol{\nu}$ of a differentiable manifold $\boldsymbol{M}$ of finite dimension m which need not be a linear space. Correspondingly it is proposed that the local mechanical equations of balance be: (i) CAUCHY's classic equation of momentum balance and (ii) an equation of balance of micro-momentum which involves, as variables, elements of the cotangent bundle $\mathcal{T}^*\boldsymbol{M}$.

In [1] CAUCHY's equation of balance is *deduced* from general properties of interactions, but for continua where microactions are present, when we look at the matter, as it were, from the bottom upward, it is not clear what kind of objects the global quantities that measure interactions could be. The difficulty is particularly acute when one has to deal with bodies endowed with different types of microstructure (*e.g.*, a liquid crystal in its container). In this paper we consider exclusively interactions among bodies with an identical type of microstructure (as for the subbodies of a given homogeneous body). Then a way out is to imagine $\boldsymbol{M}$ embedded in a Euclidean space $\boldsymbol{E}$ of a dimension n sufficiently greater than m: by WHITNEY's theorem a diffeomorphism can be found that sends $\boldsymbol{M}$ into a closed set of $\boldsymbol{E}$. Each term in the equation of balance of micromomentum can be interpreted as an element of a linear subspace of the translation space of $\boldsymbol{E}$ and it becomes possible to introduce totals over a whole body. At the same time each element of $\boldsymbol{M}$ can be specified by a point $\boldsymbol{x}$ in $\boldsymbol{E}$. Freeing temporarily $\boldsymbol{x}$ from the constraint of belonging to $\boldsymbol{M}$, one can imagine a new 'virtual' model of the physical situation where the microstructure is now specified by a point in $\boldsymbol{E}$. Within this framework one can develop a theory of interactions; as we shall see, the 'real' physical situation can be recovered later through reintroduction of the constraint.

The plan of the present paper is as follows.

In Section 2 we explore interactions within a setting which is formally very near to NOLL's but allows for self-action and permits also consideration of actions between non-disjoint parts of a body. In particular we put in evidence the severity of a set of hypotheses advanced by RIZZO (see [2], Sect. I.5).

In Section 3 we take over from [6] a concept of quasi-balanced interaction and adapt it to our purposes. leading to a local equation of balance which in·volves the density of self-action.

That such self-action has no part in the natural boundaiy conditions to be associated with the local equation of balance is shown in Section 4.

In Section 5 we introduce an event world which is rich enough to comprise all manner of microstructural phenomena. The setting is that of an extended linear space, as in the paper by WILLIAMS [4]; however, an essential additional ingredient is introduced here: a generalized concept of rigidly related frames.

In parallel with the assumptions of Section 5, we specify, in Section 6, what a motion (and in particular a rigid motion) should be in a continuum with virtual microstructure. and then distinguish. in Section 7, macrointeractions and micro-interactions and deduce the corresponding local equations of balance.

In Section 8 we exploit NOLL's axiom of frame-indifference of the power of all actions exerted on a body. From that axiom and from the assumed indifference of the microstate to translations follows the removal of a self-action term in the equation of balance of momentum. More delicate, in general, is the condition of rotational frame-indifference; it implies an expression of the skew part of the Cauchy stress tensor in terms of the density of microstructural actions, as was shown in [5].

In a final Section 9 we reintroduce the constraint that $\varkappa$ belong to $\boldsymbol{M}$, make use of the *extended principle of determinism* and deduce 'pure' equations of balance. These equations are exactly those proposed in [5] for the study of the mechanics of bodies with general microstructure.

2. Interactions

Let $\mathscr{E}$ be the 3-dimensional Euclidean space with translation space $\mathscr{V} := \mathscr{E} - \mathscr{E}$. Let $\mathscr{B} \subset \mathscr{E}$ be a fit region in the sense of NOLL & VIRGA [7]. We denote by $\Omega_{\mathscr{B}}$ the collection of all fit regions of $\mathscr{E}$ that are included in $\mathscr{B}$. It is shown in [7] that $\Omega_{\mathscr{B}}$ is a *material universe* if inclusion in the sense of set theory is taken as the relation "*part of*" and *meet, join,* and *exterior* are defined thus:

$$\mathscr{C} \wedge \mathscr{D} := \mathscr{C} \cap \mathscr{D},$$

$$\mathscr{C} \vee \mathscr{D} := \operatorname{intclo}(\mathscr{C} \cup \mathscr{D}),$$

$$\mathscr{C}^e := \operatorname{int}(\mathscr{B} \setminus \mathscr{C}), \qquad \text{for all } \mathscr{C}, \mathscr{D} \in \Omega_{\mathscr{B}}$$

(*cf.* [8], [9], and Sect. I.2 of [2] for the definition and a detailed treatment of material universes in continuum mechanics).

Following [7] we define the *difference region* of two fit regions by

$$\mathscr{C} \setminus\!\!\!\triangle\, \mathscr{D} := \operatorname{int}(\mathscr{C} \setminus \mathscr{D}).$$

We call *parts* of $\mathscr{B}$ all members of $\Omega_{\mathscr{B}}$. The members of

$$(\Omega_{\mathscr{B}})_{\text{int}} := \{\mathscr{P} \in \Omega_{\mathscr{B}} \mid \operatorname{clo} \mathscr{P} \subset \mathscr{B}\}$$

are the *interior parts* of $\mathscr{B}$. We use the notation

$$(\Omega^2_{\mathscr{B}})_{\text{int}} := \{(\mathscr{P}, \mathscr{Q}) \in \Omega_{\mathscr{B}} \times \Omega_{\mathscr{B}} \mid \mathscr{P}, \mathscr{Q} \in (\Omega_{\mathscr{B}})_{\text{int}}\}.$$

If $\mathscr{Q} \in \Omega_{\mathscr{B}}$,

$$\Omega_{\mathscr{Q}} := \{\mathscr{P} \in \Omega_{\mathscr{B}} \mid \mathscr{P} \subset \mathscr{Q}\}$$

is the collection of all parts of $\mathscr{Q}$.

Two parts of $\mathscr{B}$ are *disjoint* if their meet is empty. All pairs of either disjoint or equal parts of $\mathscr{B}$ are members of the collection

$$(\Omega^2_{\mathscr{B}})_{\text{deq}} := \{(\mathscr{P}, \mathscr{Q}) \in \Omega_{\mathscr{B}} \times \Omega_{\mathscr{B}} \mid \mathscr{P} \wedge \mathscr{Q} = \emptyset \text{ or } \mathscr{P} = \mathscr{Q}\}.$$

We define the subcollections $(\Omega^2_{\mathscr{B}})_{\text{dis}}$ and $(\Omega^2_{\mathscr{B}})_{\text{eq}}$ of $(\Omega^2_{\mathscr{B}})_{\text{deq}}$ by

$$(\Omega^2_{\mathscr{B}})_{\text{dis}} := \{(\mathscr{P}, \mathscr{Q}) \in \Omega_{\mathscr{B}} \times \Omega_{\mathscr{B}} \mid \mathscr{P} \wedge \mathscr{Q} = \emptyset\}$$

and

$$(\Omega^2_{\mathscr{B}})_{\mathrm{eq}} := \{(\mathscr{P}, \mathscr{P}) \mid \mathscr{P} \in \Omega_{\mathscr{B}}\},$$

respectively. Hence $(\Omega^2_{\mathscr{B}})_{\mathrm{deq}} = (\Omega^2_{\mathscr{B}})_{\mathrm{dis}} \cup (\Omega^2_{\mathscr{B}})_{\mathrm{eq}}$. If $\mathscr{Q} \in \Omega_{\mathscr{B}}$ we use the notation

$$(\Omega^2_{\mathscr{Q}})_{\mathrm{deq}} := \{(\mathscr{P}_1, \mathscr{P}_2) \in (\Omega^2_{\mathscr{B}})_{\mathrm{deq}} \mid \mathscr{P}_1, \mathscr{P}_2 \in \Omega_{\mathscr{Q}}\},$$

$$(\Omega^2_{\mathscr{Q}})_{\mathrm{dis}} := \{(\mathscr{P}_1, \mathscr{P}_2) \in (\Omega^2_{\mathscr{B}})_{\mathrm{dis}} \mid \mathscr{P}_1, \mathscr{P}_2 \in \Omega_{\mathscr{Q}}\},$$

$$(\Omega^2_{\mathscr{Q}})_{\mathrm{eq}} := \{(\mathscr{P}_1, \mathscr{P}_2) \in (\Omega^2_{\mathscr{B}})_{\mathrm{eq}} \mid \mathscr{P}_1, \mathscr{P}_2 \in \Omega_{\mathscr{Q}}\}.$$

If $\mathscr{P} \in \Omega_{\mathscr{B}}$ we denote by $\partial^*\mathscr{P}$ the *reduced boundary* of $\mathscr{P}$ (*cf. e.g.* [7]). Let $(\mathscr{P}, \mathscr{Q}) \in (\Omega^2_{\mathscr{B}})_{\mathrm{dis}}$; we call *contact of $\mathscr{P}$ and $\mathscr{Q}$* the set

$$\operatorname{Cct}(\mathscr{P}, \mathscr{Q}) := \partial^*\mathscr{P} \cap \partial^*\mathscr{Q}.$$

The contact of two parts of $\mathscr{B}$ always has finite area if the latter is taken as the 2-dimensional Hausdorff measure (see [7]).

Let $\mathscr{I}$ be a normed linear space. We say that a mapping $F: \Omega_{\mathscr{B}} \to \mathscr{I}$ is *additive* if

$$F(\mathscr{P}_1 \vee \mathscr{P}_2) = F(\mathscr{P}_1) + F(\mathscr{P}_2)$$

for all $(\mathscr{P}_1, \mathscr{P}_2) \in (\Omega^2_{\mathscr{B}})_{\mathrm{dis}}$. Of coulse, if $\mathscr{Q} \in \Omega_{\mathscr{B}}$ and $F: \Omega_{\mathscr{B}} \to \mathscr{I}$ is additive, then also $F|_{\Omega_{\mathscr{Q}}}$ is additive. An *interaction* is a mapping $I: (\Omega^2_{\mathscr{B}})_{\mathrm{deq}} \to \mathscr{I}$ such that

$$I(\mathscr{P}_1 \vee \mathscr{P}_2, \mathscr{Q}) = I(\mathscr{P}_1, \mathscr{Q}) + I(\mathscr{P}_2, \mathscr{Q}), \tag{2.1}$$

$$I(\mathscr{Q}, \mathscr{P}_1 \vee \mathscr{P}_2) = I(\mathscr{Q}, \mathscr{P}_1) + I(\mathscr{Q}, \mathscr{P}_2), \tag{2.2}$$

for all $\mathscr{Q} \in \Omega_{\mathscr{B}}$ and $(\mathscr{P}_1, \mathscr{P}_2) \in (\Omega^2_{\mathscr{Q}^e})_{\mathrm{dis}}$.

Axioms (1) and (2) are NOLL's (see [8] or Sect. I.5 of [2]). We refer to (1) and (2) by saying that I is *biadditive* on $(\Omega^2_{\mathscr{B}})_{\mathrm{dis}}$. $I(\mathscr{P}, \mathscr{Q})$ is the *action* of $\mathscr{Q}$ on $\mathscr{P}$ when $\mathscr{P}$ and $\mathscr{Q}$ are disjoint parts of $\mathscr{B}$ and $I(\mathscr{P}, \mathscr{P})$ is the *self-action of* $\mathscr{P}$.

An interaction I is *skew* if

$$I(\mathscr{P}, \mathscr{Q}) + I(\mathscr{Q}, \mathscr{P}) = 0 \quad \text{for all } (\mathscr{P}, \mathscr{Q}) \in (\Omega^2_{\mathscr{B}})_{\mathrm{dis}}.$$

I is a *contact interaction* if $I(\mathscr{Q}, \mathscr{P}) = 0$ whenever $\operatorname{Cct}(\mathscr{P}, \mathscr{Q}) = \emptyset$.

Proposition 1. *If I is an interaction then*

$$I(\mathscr{P}_1 \vee \mathscr{P}_2, \mathscr{Q}) = I(\mathscr{P}_1, \mathscr{Q}) + I(\mathscr{P}_2, \mathscr{Q}) - I(\mathscr{P}_1 \wedge \mathscr{P}_2, \mathscr{Q}), \tag{2.3}$$

$$I(\mathscr{Q}, \mathscr{P}_1 \vee \mathscr{P}_2) = I(\mathscr{Q}, \mathscr{P}_1) + I(\mathscr{Q}, \mathscr{P}_2) - I(\mathscr{Q}, \mathscr{P}_1 \wedge \mathscr{P}_2), \tag{2.4}$$

for all $\mathscr{Q} \in \Omega_{\mathscr{B}}$ *and* $\mathscr{P}_1, \mathscr{P}_2 \in \Omega_{\mathscr{Q}^e}$.

Proof. If $\mathscr{P}_1$ and $\mathscr{P}_2$ are members of $\Omega_{\mathscr{B}}$, $\mathscr{P}_1 \vee \mathscr{P}_2$ is also the join of three pairwise disjoint members of $\Omega_{\mathscr{B}}$:

$$\mathscr{P}_1 \vee \mathscr{P}_2 = (\mathscr{P}_1 \setminus \mathscr{P}_2) \vee (\mathscr{P}_1 \wedge \mathscr{P}_2) \vee (\mathscr{P}_2 \setminus \mathscr{P}_1).$$

Then by (1)

$$I(\mathscr{P}_1 \vee \mathscr{P}_2, \mathscr{Q}) = I(\mathscr{P}_1 \setminus \mathscr{P}_2, \mathscr{Q}) + I(\mathscr{P}_1 \wedge \mathscr{P}_2, \mathscr{Q}) + I(\mathscr{P}_2 \setminus \mathscr{P}_1, \mathscr{Q}). \tag{2.5}$$

Since $\mathscr{P}_1 = (\mathscr{P}_1 \setminus \mathscr{P}_2) \vee (\mathscr{P}_1 \wedge \mathscr{P}_2)$ and $\mathscr{P}_2 = (\mathscr{P}_2 \setminus \mathscr{P}_1) \vee (\mathscr{P}_1 \wedge \mathscr{P}_2)$, use of (1) in (5) proves (3). The same argument with the use of (2) proves (4). □

Definition 1. *Let an interaction* $I: (\Omega^2_{\mathscr{B}})_{\text{deq}} \to \mathscr{I}$ *be given and let* $\mathscr{Q} \in \Omega_{\mathscr{B}}$. *We define the mappings* $I^{\mathscr{Q}}: \Omega_{\mathscr{Q}} \to \mathscr{I}$ *and* $I_{\mathscr{Q}}: \Omega_{\mathscr{Q}} \to \mathscr{I}$ *by*

$$I^{\mathscr{Q}}(\mathscr{P}) := I(\mathscr{P}, \mathscr{P}) + I(\mathscr{P}, \mathscr{Q} \setminus \mathscr{P}), \tag{2.6}$$

$$I_{\mathscr{Q}}(\mathscr{P}) := I(\mathscr{P}, \mathscr{P}) + I(\mathscr{Q} \setminus \mathscr{P}, \mathscr{P}), \tag{2.7}$$

for all $\mathscr{P} \in \Omega_{\mathscr{Q}}$.

$I^{\mathscr{Q}}$ is the *action of* $\mathscr{Q}$ *on its parts* and $I_{\mathscr{Q}}$ is the *action of the parts of* $\mathscr{Q}$ *on* $\mathscr{Q}$.

Proposition 2. *Let* $\mathscr{Q} \in \Omega_{\mathscr{B}}$. *For all* $(\mathscr{P}_1, \mathscr{P}_2) \in (\Omega^2_{\mathscr{B}})_{\text{dis}}$, $I^{\mathscr{Q}}$ and $I_{\mathscr{Q}}$ *satisfy*

$$I^{\mathscr{Q}}(\mathscr{P}_1 \vee \mathscr{P}_2) = I^{\mathscr{Q}}(\mathscr{P}_1) + I^{\mathscr{Q}}(\mathscr{P}_2) + D(\mathscr{P}_1, \mathscr{P}_2), \tag{2.8}$$

$$I_{\mathscr{Q}}(\mathscr{P}_1 \vee \mathscr{P}_2) = I_{\mathscr{Q}}(\mathscr{P}_1) + I_{\mathscr{Q}}(\mathscr{P}_2) + D(\mathscr{P}_1, \mathscr{P}_2), \tag{2.9}$$

where $D: (\Omega^2_{\mathscr{B}})_{\text{dis}} \to \mathscr{I}$ *is defined by*

$$\begin{aligned} D(\mathscr{P}_1, \mathscr{P}_2) := {} & I(\mathscr{P}_1 \vee \mathscr{P}_2, \mathscr{P}_1 \vee \mathscr{P}_2) - I(\mathscr{P}_1, \mathscr{P}_1) - I(\mathscr{P}_2, \mathscr{P}_2) \\ & - I(\mathscr{P}_1, \mathscr{P}_2) - I(\mathscr{P}_2, \mathscr{P}_1). \end{aligned} \tag{2.10}$$

Proof. By (6) and (1)

$$\begin{aligned} I^{\mathscr{Q}}(\mathscr{P}_1 \vee \mathscr{P}_2) = {} & I(\mathscr{P}_1 \vee \mathscr{P}_2, \mathscr{P}_1 \vee \mathscr{P}_2) + I(\mathscr{P}_1, \mathscr{Q} \setminus (\mathscr{P}_1 \vee \mathscr{P}_2)) \\ & + I(\mathscr{P}_2, \mathscr{Q} \setminus (\mathscr{P}_1 \vee \mathscr{P}_2)). \end{aligned} \tag{2.11}$$

Since $\mathscr{Q} \setminus \mathscr{P}_1 = (\mathscr{Q} \setminus (\mathscr{P}_1 \vee \mathscr{P}_2)) \vee \mathscr{P}_2$ and $\mathscr{Q} \setminus \mathscr{P}_2 = (\mathscr{Q} \setminus (\mathscr{P}_1 \vee \mathscr{P}_2)) \vee \mathscr{P}_1$, by (1) we get

$$I(\mathscr{P}_1, \mathscr{Q} \setminus (\mathscr{P}_1 \vee \mathscr{P}_2)) = I(\mathscr{P}_1, \mathscr{Q} \setminus \mathscr{P}_1) - I(\mathscr{P}_1, \mathscr{P}_2), \tag{2.12}$$

$$I(\mathscr{P}_2, \mathscr{Q} \setminus (\mathscr{P}_1 \vee \mathscr{P}_2)) = I(\mathscr{P}_2, \mathscr{Q} \setminus \mathscr{P}_2) - I(\mathscr{P}_2, \mathscr{P}_1). \tag{2.13}$$

Introducing (12) and (13) into (11) and using (6), we arrive at (8). The same argument applied to $I_{\mathscr{Q}}$ completes the proof. □

Corollary 1. *The mapping* $F: \Omega_{\mathscr{Q}} \to \mathscr{I}$ *defined by*

$$F(\mathscr{P}) := I^{\mathscr{Q}}(\mathscr{P}) - I_{\mathscr{Q}}(\mathscr{P}) \quad \text{for all } \mathscr{P} \in \Omega_{\mathscr{Q}} \tag{2.14}$$

is additive.

Corollary 2. *The mappings* $I^{\mathscr{Q}}$ *and* $I_{\mathscr{Q}}$ *are additive if and only if* $D(\mathscr{P}_1, \mathscr{P}_2) = 0$ *for all* $(\mathscr{P}_1, \mathscr{P}_2) \in (\Omega^2_{\mathscr{Q}})_{\text{dis}}$.

Remark 1. If the interaction I is skew and the mapping $\mathcal{P} \mapsto I(\mathcal{P}, \mathcal{P})$ is additive, then $D(\mathcal{P}_1, \mathcal{P}_2) = 0$ for all $(\mathcal{P}_1, \mathcal{P}_2) \in (\Omega^2_{\mathcal{B}})_{\mathrm{dis}}$.

The following example shows that I need not be skew when D vanishes on $(\Omega^2_{\mathcal{B}})_{\mathrm{dis}}$.

Example. Let the mapping $i: \mathcal{B} \times \mathcal{B} \to \mathcal{I}$ be Lebesgue-integrable and let $\hat{I}: (\Omega^2_{\mathcal{B}})_{\mathrm{deq}} \to \mathcal{I}$ be a skew interaction. The mapping $I: (\Omega^2_{\mathcal{B}})_{\mathrm{deq}} \to \mathcal{I}$ defined by

$$I(\mathcal{P}, \mathcal{Q}) = \int_{\mathcal{P}} dv_x \int_{\mathcal{Q}} dv_y \, i(x, y) + \hat{I}(\mathcal{P}, \mathcal{Q}) \quad \text{for all } (\mathcal{P}, \mathcal{Q}) \in (\Omega^2_{\mathcal{B}})_{\mathrm{dis}},$$

and

$$I(\mathcal{P}, \mathcal{P}) = \int_{\mathcal{P} \times \mathcal{P}} i \quad \text{for all } \mathcal{P} \in \Omega_{\mathcal{B}},$$

is an interaction. It satisfies $D(\mathcal{P}_1, \mathcal{P}_2) = 0$ for all $(\mathcal{P}_1, \mathcal{P}_2) \in (\Omega^2_{\mathcal{B}})_{\mathrm{dis}}$, but it is skew only for very special choices of i.

The following Proposition is an easy consequence of Props. 1 and 2.

Proposition 3. *Let* $\mathcal{Q} \in \Omega_{\mathcal{B}}$. *For all* $\mathcal{P}_1, \mathcal{P}_2 \in \Omega_{\mathcal{Q}}$, $I^{\mathcal{Q}}$ *and* $I_{\mathcal{Q}}$ *satisfy*

$$I^{\mathcal{Q}}(\mathcal{P}_1 \vee \mathcal{P}_2) = I^{\mathcal{Q}}(\mathcal{P}_1) + I^{\mathcal{Q}}(\mathcal{P}_2) - I^{\mathcal{Q}}(\mathcal{P}_1 \wedge \mathcal{P}_2) + \tilde{D}(\mathcal{P}_1, \mathcal{P}_2), \tag{2.15}$$

$$I_{\mathcal{Q}}(\mathcal{P}_1 \vee \mathcal{P}_2) = I_{\mathcal{Q}}(\mathcal{P}_1) + I_{\mathcal{Q}}(\mathcal{P}_2) - I_{\mathcal{Q}}(\mathcal{P}_1 \wedge \mathcal{P}_2) + \tilde{D}(\mathcal{P}_1, \mathcal{P}_2), \tag{2.16}$$

where $\tilde{D}: \Omega_{\mathcal{B}} \times \Omega_{\mathcal{B}} \to \mathcal{I}$ *is defined by*

$$\begin{aligned} \tilde{D}(\mathcal{P}_1, \mathcal{P}_2) := {} & I(\mathcal{P}_1 \vee \mathcal{P}_2, \mathcal{P}_1 \vee \mathcal{P}_2) + I(\mathcal{P}_1 \wedge \mathcal{P}_2, \mathcal{P}_1 \wedge \mathcal{P}_2) - I(\mathcal{P}_1, \mathcal{P}_1) - I(\mathcal{P}_2, \mathcal{P}_2) \\ & - I(\mathcal{P}_1 \setminus \mathcal{P}_2, \mathcal{P}_2 \setminus \mathcal{P}_1) - I(\mathcal{P}_2 \setminus \mathcal{P}_1, \mathcal{P}_1 \setminus \mathcal{P}_2). \end{aligned} \tag{2.17}$$

Corollary. *The mapping* $F: \Omega_{\mathcal{Q}} \to \mathcal{I}$ *defined by* (14) *satisfies*

$$F(\mathcal{P}_1 \vee \mathcal{P}_2) = F(\mathcal{P}_1) + F(\mathcal{P}_2) - F(\mathcal{P}_1 \wedge \mathcal{P}_2) \quad \text{for all } \mathcal{P}_1, \mathcal{P}_2 \in \Omega_{\mathcal{Q}}.$$

Remark 2. If $I(\emptyset, \emptyset) = 0$ then $\tilde{D}\big|_{(\Omega^2_{\mathcal{B}})_{\mathrm{dis}}} = D$.

When $\tilde{D}$ vanishes, (15) and (16) read as the axioms posited by RIZZO for interactions between parts of $\mathcal{B}$ not necessarily joint (*cf.* Sect. I.5 of [2]) at least within the restricted context of actions of regions on their parts and *vice versa.* Indeed RIZZO's axioms taken in their full strength are too strong for our purposes.

3. Quasi-balanced interactions

Let $\mathcal{E}$ be the 3-dimensional Euclidean space and $\mathcal{V}$ its translation space. We denote by $\mathcal{S}$ the unit sphere of $\mathcal{V}$. Let an interaction $I: (\Omega^2_{\mathcal{B}})_{\mathrm{deq}} \to \mathcal{I}$ as described in Sect. 2 be given. We denote by Lin $(\mathcal{V}, \mathcal{I})$ the space of all linear transformations of $\mathcal{V}$ into $\mathcal{I}$. Let the mapping $I^{\mathcal{B}}: \Omega_{\mathcal{B}} \to \mathcal{I}$ be defined according to Def. 1 of Sect. 2.

The following definition comes from [6]:

Definition 1. *The interaction I is quasi-balanced if there is a continuous function $f\colon \operatorname{clo}\mathscr{B} \to \mathscr{I}$ such that*

$$I^{\mathscr{B}}(\mathscr{P}) = \int_{\mathscr{P}} f\, dv \quad \text{for all } \mathscr{P} \in (\Omega_{\mathscr{B}})_{\text{int}}. \tag{3.1}$$

Proposition 1. *If I is a quasi-balanced interaction then*

$$I(\mathscr{P}_1 \vee \mathscr{P}_2, \mathscr{P}_1 \vee \mathscr{P}_2) - I(\mathscr{P}_1, \mathscr{P}_1) - I(\mathscr{P}_2, \mathscr{P}_2) = I(\mathscr{P}_1, \mathscr{P}_2) + I(\mathscr{P}_2, \mathscr{P}_1) \tag{3.2}$$

for all $(\mathscr{P}_1, \mathscr{P}_2) \in (\Omega^2_{\mathscr{B}})_{\text{dis}} \cap (\Omega^2_{\mathscr{B}})_{\text{int}}$.

The proof follows immediately from Prop. 2 of Sect. 2 and the fact that the mapping $\mathscr{P} \mapsto \int_{\mathscr{P}} f\, dv$ is additive.

Assumption 1. *There are two positive constants k_v and k_a such that*

$$|I(\mathscr{P}, \mathscr{P})| \leqq k_v v(\mathscr{P}) \quad \text{for all } \mathscr{P} \in \Omega_{\mathscr{B}} \tag{3.3}$$

and

$$|I(\mathscr{P}, \mathscr{Q})| \leqq k_a a(\operatorname{Cct}(\mathscr{P}, \mathscr{Q})) \quad \text{for all } (\mathscr{P}, \mathscr{Q}) \in (\Omega^2_{\mathscr{B}})_{\text{dis}}, \tag{3.4}$$

where v and a denote the Lebesgue measure and the 2-dimensional Hausdorff measure, respectively.

Remark 1. Inequality (4) guarantees that I is a contact interaction.

Theorem 1. *If I is a quasi-balanced interaction that satisfies Assumption 1, then there is a bounded mapping $c\colon \mathscr{P} \times \mathscr{S} \to \mathscr{I}$ such that for all $(\mathscr{P}, \mathscr{Q}) \in (\Omega^2_{\mathscr{B}})_{\text{dis}} \cap (\Omega^2_{\mathscr{B}})_{\text{int}}$*

$$I(\mathscr{P}, \mathscr{Q}) = \int_{\mathscr{C}} c(x, n_{\mathscr{P}}(x))\, da_x \tag{3.5}$$

where $\mathscr{C} := \operatorname{Cct}(\mathscr{P}, \mathscr{Q})$ and $n_{\mathscr{P}}\colon \partial^* \mathscr{P} \to \mathscr{S}$ is the outer normal field to $\mathscr{P}$.

This theorem is NOLL's [1]. A proof suited for fit regions can be found in [10] and, under rather weaker hypotheses, also in [11]; an exhaustive account is also given in the forthcoming second edition of [2].

Proposition 2. *If I is a quasi-balanced interaction that satisfies Assumption 1, then I is skew on $(\Omega^2_{\mathscr{B}})_{\text{dis}} \cap (\Omega^2_{\mathscr{B}})_{\text{int}}$ and the mapping $\mathscr{P} \mapsto I(\mathscr{P}, \mathscr{P})$ is additive on $(\Omega^2_{\mathscr{B}})_{\text{int}}$.*

Proof. Since I is quasi-balanced, Prop. 1 applies. For any $(\mathscr{P}_1^1, \mathscr{P}_2^1) \in (\Omega^2_{\mathscr{B}})_{\text{dis}} \cap (\Omega^2_{\mathscr{B}})_{\text{int}}$ such that $\operatorname{Cct}(\mathscr{P}_1^1, \mathscr{P}_2^1) \neq \emptyset$ construct a sequence of pairs $\{(\mathscr{P}_1^n, \mathscr{P}_2^n)\}_{n \in \mathbb{N}}$ in $(\Omega^2_{\mathscr{B}})_{\text{dis}} \cap (\Omega^2_{\mathscr{B}})_{\text{int}}$ such that $\mathscr{P}_1^{n+1} \subset \mathscr{P}_1^n$, $\mathscr{P}_2^{n+1} \subset \mathscr{P}_2^n$ for all $n \in \mathbb{N}$ and $\operatorname{Cct}(\mathscr{P}_1^n, \mathscr{P}_2^n) = \operatorname{Cct}(\mathscr{P}_1^1, \mathscr{P}_2^1)$ for all $n \in \mathbb{N}$. Then, by Theorem 1, $I(\mathscr{P}_1^n, \mathscr{P}_2^n) + I(\mathscr{P}_2^n, \mathscr{P}_1^n)$ is independent of n. Suppose that $\lim_{n\to\infty} v(\mathscr{P}_1^n) = \lim_{n\to\infty} v(\mathscr{P}_2^n) = 0$. If we write (2) for all pairs $(\mathscr{P}_1^n, \mathscr{P}_2^n)$ and take the limit as $n \to \infty$, by (3) we get

$$I(\mathscr{P}_1^n, \mathscr{P}_2^n) + I(\mathscr{P}_2^n, \mathscr{P}_1^n) = 0 \quad \text{for all } n \in \mathbb{N}.$$

Hence I is skew on $(\Omega_{\mathscr{B}}^2)_{\text{dis}} \cap (\Omega_{\mathscr{B}}^2)_{\text{int}}$. The additivity on $(\Omega_{\mathscr{B}})_{\text{int}}$ of the self-action then follows from (2). □

Corollary. *For all* $x \in \mathscr{B}$ *the mapping* $c(x, \cdot): \mathscr{S} \to \mathscr{I}$ *has the property*

$$c(x, -n) = -c(x, n) \quad \text{for all } n \in \mathscr{S}. \tag{3.6}$$

Since the mapping $\mathscr{P} \mapsto I(\mathscr{P}, \mathscr{P})$ is additive on $(\Omega_{\mathscr{B}})_{\text{int}}$ by Prop. 2 and absolutely continuous with respect to Lebesgue measure by Assumption 1, a standard argument (see *e.g.* the Appendix of [12] or [13]) shows that $I(\mathscr{P}, \mathscr{P})$ can be represented as an integral over $\mathscr{P}$ for all $\mathscr{P} \in (\Omega_{\mathscr{B}})_{\text{int}}$. The Proposition below then follows from the fact that the topological boundary of a fit region has Lebesgue measure zero (*cf.* [7]).

Proposition 3. *There is an integrable mapping* $s: \mathscr{B} \to \mathscr{I}$ *such that*

$$I(\mathscr{P}, \mathscr{P}) = \int_{\mathscr{P}} s \, dv \quad \text{for all } \mathscr{P} \in \Omega_{\mathscr{B}}. \tag{3.7}$$

The following Assumption is crucial to establish the existence of the stress tensor.

Assumption 2. *For all* $n \in \mathscr{S}$ *the mapping* $c(\cdot, n): \mathscr{B} \to \mathscr{I}$ *is continuous.*

Theorem 2. *If* I *is a quasi-balanced interaction and Assumptions 1 and 2 are satisfied, then there is a continuous mapping* $C: \mathscr{B} \to \operatorname{Lin}(\mathscr{V}, \mathscr{I})$ *such that*

$$c(x, n) = C(x)\, n \quad \text{for all } x \in \mathscr{B} \text{ and } n \in \mathscr{S}. \tag{3.8}$$

This theorem is CAUCHY's. Weaker versions of it have been proved under assumptions weaker than Assumption 2 by GURTIN, MIZEL & WILLIAMS [14], GURTIN & MARTINS [15], ZIEMER [10], and ŠILHAVÝ [11]. Among these authors only ŠILHAVÝ weakens also Assumption 1. FOSDICK & VIRGA [16] have proved Theorem 2 for regions of $\mathscr{E}$ far more regular than fit regions, under the assumption that $I(\mathscr{P}, \mathscr{Q})$, for disjoint $\mathscr{P}$ and $\mathscr{Q}$, be the integral over $\operatorname{Cct}(\mathscr{P}, \mathscr{Q})$ of a mapping of class C^1 that depends on the place in $\operatorname{Cct}(\mathscr{P}, \mathscr{Q})$, the outer unit normal to $\mathscr{P}$, and its surface gradient. The latter assumption is consistent with ŠILHAVÝ's, but it is more general than (4) of Assumption 1.

Assumption 3. *The mapping* $s: \mathscr{B} \to \mathscr{I}$ *of* (7) *is continuous.*

The following Proposition is an easy consequence of Theorem 2 and the Divergence Theorem.

Proposition 4. *Let I be a quasi-balanced interaction and let Assumptions 1, 2, and 3 be satisfied. If the mapping* $C\colon \mathscr{B} \to \operatorname{Lin}(\mathscr{V}, \mathscr{I})$ *of* (8) *is of class* C^1, *then*

$$s + \operatorname{div} C = f \quad \text{in } \mathscr{B}, \tag{3.9}$$

where f is the mapping that appears in (1).

Remark 2. Apart from the density of self-action s, (9) has the familiar form of a balance equation if f is interpreted as the action per unit volume of $\mathscr{B}$ exerted by the "external world" and is intended to comprise the action of inertia.

4. Boundary conditions

Let a quasi-balanced interaction $I\colon (\Omega^2_{\mathscr{B}})_{\text{deq}} \to \mathscr{I}$ as described in Sect. 3 be given. The following assumption comes from [6].

Assumption 4. *There is a continuous mapping* $h\colon \partial^*\mathscr{B} \to \mathscr{I}$ *such that*

$$I^{\mathscr{B}}(\mathscr{P}) = \int_{\mathscr{P}} f\,dv + \int_{\partial^*\mathscr{P}\cap\partial^*\mathscr{B}} h\,da \quad \text{for all } \mathscr{P} \in \Omega_{\mathscr{B}}, \tag{4.1}$$

where f is the same as in (3.1).

Remark 1. For all $\mathscr{P} \in (\Omega_{\mathscr{B}})_{\text{int}}$ (1) is nothing but (3.1).

Theorem. *If Assumptions 1, 2, and 4 are satisfied, if* $\partial\mathscr{B}$ *is piecewise of class* C^1, *and if* $\lim_{y\to x} C(y)\, n_{\mathscr{B}}(x)$ *exists for all* $x \in \partial^*\mathscr{B}$, *then*

$$\lim_{y\to x} (C(y)\, n_{\mathscr{B}}(x)) + h(x) = 0 \quad \text{for all } x \in \partial^*\mathscr{B}, \tag{4.2}$$

where $n_{\mathscr{B}}\colon \partial^*\mathscr{B} \to \mathscr{S}$ *is the unit outer normal field of* $\mathscr{B}$.

Proof. Let $\mathscr{C} \subset \partial^*\mathscr{B}$ and $e \in \mathscr{S}$ be such that $n_{\mathscr{B}}(x) \cdot e \leqq 0$ for all $x \in \mathscr{C}$. For all positive and sufficiently small δ we define the subset of $\mathscr{B}$

$$\mathscr{C}_\delta := \{x + \delta e \mid x \in \mathscr{C}\}.$$

Let $\mathscr{P}_\delta := \{x + \alpha e \mid x \in \mathscr{C}, \alpha \in]0, \delta[\}$. There are two disjoint parts of $\mathscr{P}_\delta$, say $\mathscr{P}'_\delta$ and $\mathscr{P}''_\delta$, such that

$$\mathscr{P}_\delta = \mathscr{P}'_\delta \vee \mathscr{P}''_\delta, \quad \mathscr{P}'_\delta \in (\Omega_{\mathscr{B}})_{\text{int}}, \quad \mathscr{C}_\delta = \operatorname{Cct}(\mathscr{P}'_\delta, \mathscr{P}^e_\delta).$$

The sketch in Figure 1, overleaf, illustrates the situation envisaged here.

It follows from Assumption 1 that there are two positive constants, k_1 and k_2, such that

$$|I(\mathscr{P}_\delta, \mathscr{P}_\delta)| \leqq k_1\, \delta, \tag{4.3}$$

$$|I(\mathscr{P}''_\delta, \mathscr{P}^e_\delta)| \leqq k_2\, \delta. \tag{4.4}$$

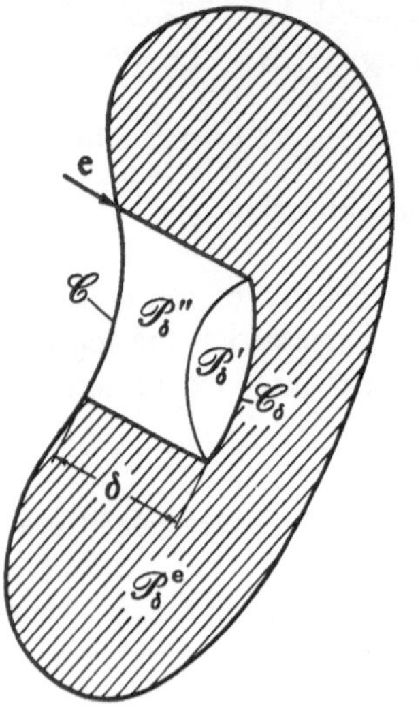

Fig. 1

Let $n_{\mathscr{P}'_\delta}$ denote the unit outer normal field of $\mathscr{P}'_\delta$. Since

$$n_{\mathscr{P}'_\delta}(x) = -n_{\mathscr{B}}(x - \delta e) \quad \text{for all } x \in \mathscr{C}_\delta,$$

by Theorems 1 and 2 of Sect. 3 we get

$$I(\mathscr{P}'_\delta, \mathscr{P}^e_\delta) = - \int_{\mathscr{C}_\delta} C(x)\, n_{\mathscr{B}}(x - \delta e)\, da_x. \tag{4.5}$$

Hence, by Assumption 4,

$$\int_{\mathscr{C}_\delta} C(x)\, n_{\mathscr{B}}(x - \delta e)\, da_x + \int_{\mathscr{C}} h\, da = I(\mathscr{P}_\delta, \mathscr{P}_\delta) + I(\mathscr{P}''_\delta, \mathscr{P}^e_\delta) - \int_{\mathscr{P}_\delta} f\, dv. \tag{4.6}$$

Thus (3) and (4) lead us to

$$\lim_{\delta \to 0} \left(\int_{\mathscr{C}_\delta} C(x)\, n_{\mathscr{B}}(x - \delta e)\, da_x \right) + \int_{\mathscr{C}} h\, da = 0. \tag{4.7}$$

The desired conclusion then follows from (7), since $\mathscr{C}$ can be chosen of arbitiarily small area. □

Remark 2. If besides Assumptions 1, 2, and 4 also Assumption 3 is satisfied, (2) reads as the *natural* boundary condition for the balance equation (3.9).

Remark 3. By saying that $\partial\mathscr{B}$ is piecewise of class C^1 we mean that $\partial\mathscr{B}$ consists of a *finite* number of C^1-surfaces. This hypothesis cannot be removed from the theorem; our proof above as it now stands does not apply to a body with countably many holes accumulating on a subset of its reduced boundary. An example of such a body can be obtained by taking the exterior relative to the rectangle $\{(s, t) \in \mathbb{R}^2 \mid s \in]0, 1[, t \in]0, d[\}$ of the fit region $\mathscr{D}_d$ defined on p. 18 of [7].

Remark 4. Should we trace now (*i.e.*, on the strength also of Assumption 4) the path that leads to the proof of Theorem 1 of Sect. 3 (*cf.* Sect. III.1 of [2]), we could actually prove that $\lim_{y \to x} C(y)\, n_{\mathscr{B}}(x)$ exists (*cf.* Sect. 8 of [6]), thus avoiding the explicit hypothesis of the Theorem.

5. Event world

In Classical Mechanics a basic primitive entity is the *event world*, whose elements are the *events*, "atoms of experience" in NOLL's words. Each model of physical reality is based on a particular event world. Though this subject has been explored profoundly by NOLL in [17] (see also [8]), here we do not take over his treatment. Rather, we adjust to our purposes the simplified treatment in Section I.6 of the forthcoming second edition of [2].

The mathematical structure of the event world $\mathscr{W}$ employed here aims to fill the needs expressed in general terms in the Introduction; the structure is made precise by introducing a *frame*, which is a bijection $\mathbf{f}$ of $\mathscr{W}$ onto $\mathscr{E}\times \boldsymbol{E}\times\mathscr{T}$, where $\mathscr{E}$, $\boldsymbol{E}$ and $\mathscr{T}$ are Euclidean spaces with dimension 3, n, and 1, respectively:

$$\mathbf{f}: \mathscr{W} \to \mathscr{E}\times\boldsymbol{E}\times\mathscr{T}.$$

We call $\mathscr{E}\times\boldsymbol{E}\times\mathscr{T}$ the *space of reference*. We denote by $\mathscr{V}$ and V the translation spaces of $\mathscr{E}$ and $\boldsymbol{E}$, respectively. We use the notation

$$\varepsilon \to \mathbf{f}(\varepsilon) =: (x, \boldsymbol{x}, \tau),$$

where x is the *place* where the event ε occurs, $\boldsymbol{x}$ is the *virtual microstructure* related to ε, and τ is the *instant* at which ε occurs. As mentioned in the Introduction, real microstructures related to the events may actually be described by the elements of a differentiable manifold $\boldsymbol{M}$ embedded in $\boldsymbol{E}$, but we leave aside this constraint for the moment; that is the reason why we have used the term virtual microstructure.

Two events are said to be simultaneous if they occur at the same instant. *We assume that events simultaneous in one frame are simultaneous in all.* We denote by $\mathscr{W}_\tau$ the collection of all events that occur at the instant τ. Let a frame $\mathbf{f}$ be given; we define the mapping $\mathbf{f}_\tau: \mathscr{W}_\tau \to \mathscr{E}\times\boldsymbol{E}$ as the bijection that assigns to all events occurring at the instant τ the places where they occur and the related virtual microstructures. Formally,

$$\mathbf{f}_\tau(\varepsilon) = (x, \boldsymbol{x}) \quad \text{if} \quad \mathbf{f}(\varepsilon) = (x, \boldsymbol{x}, \tau).$$

We assume that any other frame $\mathbf{f}^*$ *assigns to the events the same instants as* $\mathbf{f}$.

The event world described so far will be called *complete* to emphasize that it is different from the *classical* event world, whose space of reference is merely $\mathscr{E}\times\mathscr{T}$. For the classical event world the concept of rigid relation between frames is crucial: the frames $\mathbf{f}$ and $\mathbf{f}^*$ are said then to be *rigidly related* if the bijections $\mathbf{f}^*_\tau \circ \mathbf{f}_\tau^{-1}$ and $\mathbf{f}_\tau \circ (\mathbf{f}^*_\tau)^{-1}$ are isometries of $\mathscr{E}$ for all $\tau \in \mathscr{T}$. Let an event ε be given and let $\mathbf{f}(\varepsilon) = (x, \tau)$, $\mathbf{f}^*(\varepsilon) = (x^*, \tau)$. A well known theorem about the representation of isometries of Euclidean spaces (preserving orientation) assures us that there are a place $x_0 \in \mathscr{E}$ and two mappings $x_0^*: \mathscr{T} \to \mathscr{E}$, $Q: \mathscr{T} \to \text{Orth}^+(\mathscr{V})$ such that

$$x^* = x_0^*(\tau) + Q(\tau)(x - x_0). \tag{5.1}$$

We wish to render meaningful the concept of rigid relation between frames also for the complete event woıld $\mathscr{W}$. *We assume that there is a continuous group* $\mathbf{G}$ *of transformations of* $\mathbf{E}$ *and a mapping* $\Gamma: \text{Orth}^+(\mathscr{V}) \to \mathbf{G}$ *which is a differentiable homomorphism.* For all $Q \in \text{Orth}^+(\mathscr{V})$ we put $\Gamma_Q := \Gamma(Q)$. We interpret

$$\varkappa_Q := \Gamma_Q(\varkappa) \tag{5.2}$$

as the result of applying Q *to* $\varkappa$.

We say that the frames $\mathbf{f}$ and $\mathbf{f}^*$ of $\mathscr{W}$ are *rigidly related* if for all $\tau \in \mathscr{T}$

$$(\mathbf{f}_\tau \circ (\mathbf{f}_\tau^*)^{-1})(x^*, \varkappa^*) = (x, \varkappa) \quad \text{and} \quad (\mathbf{f}_\tau^* \circ \mathbf{f}_\tau^{-1})(x, \varkappa) = (x^*, \varkappa^*), \tag{5.3}$$

in which x^* is related to x as in (1) and

$$\varkappa^* := \varkappa_{Q(\tau)}, \tag{5.4}$$

$Q(\tau)$ being the same as in (1). All frames rigidly related to the frame $\mathbf{f}$ constitute an equivalence class that we call the *rigid class of* $\mathbf{f}$. *We shall assume that it is possible to choose a preferred rigid class of frames within which the axioms of mechanics apply.* All developments below refer to such a class.

Remark. It follows from (4) that translations do not affect the value of $\varkappa$.

6. Bodies and motions

The mathematical structures introduced in previous sections (material universe, interactions, events) are the basic tools to describe the behavior of a class of real bodies which can be modelled as continua with microstructure. The connection between physical experience and mathematical description is established through a set of informal statements, which are listed below without comment, as they are simply the adaptation to ouı needs of those which are advanced within the classical context in the papers of NOLL (see, in particular, [8]) and in TRUESDELL's treatise [2].

A real body is imagined as made up of material points ξ. These points experience events: in a given frame $\mathbf{f}$ at time τ each point occupies some one place $x(\xi, \tau)$ and has a microstructure $\varkappa(\xi, \tau)$. The set of places occupied by all material points of the body at a given instant τ (*i.e.*, the placement of the body at that instant) is a fit region $\mathscr{B}_\tau$.

Any two placements induce a smooth bijection between the corresponding fit regions. Thus each material point may be specified through the place x_* it occupies in a fixed placement $\mathscr{B}_*$ of the body; as a consequence x and $\varkappa$ can be thought as functions of x_* $(x_* \in \mathscr{B}_*)$ and τ (within an appropriate interval $[0, \bar{\tau}]$ of $\mathbb{R}$).

The set of events $\{(x(x_*, \hat{\tau}), \varkappa(x_*, \hat{\tau}), \hat{\tau}) \mid x_* \in \mathscr{B}_*, \hat{\tau} \in [0, \bar{\tau}]\}$ experienced by the material points of a body at a given instant $\hat{\tau}$ is called the *complete placement* of the body at that instant. Thus a complete placement generates over $\mathscr{B}_*$ (or indifferently over $\mathscr{B}_\tau$) also a smooth field with values in $\boldsymbol{E}$.

The complete motion of a body point is a smooth mapping of times into $\mathscr{E}\times\boldsymbol{E}$; the complete motion of a body is engendered by the motions of its points:

$$(x_*, \tau) \mapsto (x(x_*, \tau), \varkappa(x_*, \tau)).$$

Velocity and rate of change of the microstructure (microvelocity) are given by

$$\dot{x}(x_*, \tau) := \frac{\partial x(x_*, \tau)}{\partial \tau}, \qquad \dot{\varkappa}(x_*, \tau) := \frac{\partial_\varkappa(x_*, \tau)}{\partial \tau}. \tag{6.1}$$

The notion of rigid complete motion of a body is in line with the notion of rigidly related frames introduced in Sect. 5. Given a reference placement $\mathscr{B}_*$ and a frame $\mathbf{f}$, a complete motion of the body is rigid if there is a frame in which the body appears at complete rest, *i.e.*, a frame in which velocity and microvelocity both vanish (*cf.* [2], Sect. I.10). Let $\mathbf{f}_*$ be the frame where the complete motion of the body is described by the mappings

$$(x_*, \tau) \mapsto x_*, \quad (x_*, \tau) \mapsto \varkappa_*(x_*) \quad \text{for all } \tau \in [0, \bar{\tau}],$$

where $\varkappa_*: \mathscr{B}_* \to \boldsymbol{E}$ is a given mapping. Thus, by (5.1) and (5.4), a complete rigid motion in the frame $\mathbf{f}$ is the mapping

$$(x_*, \tau) \mapsto (x_R(x_*, \tau), \varkappa_R(x_*, \tau)),$$

where $x_R: \mathscr{B}_* \times [0, \bar{\tau}] \to \mathscr{E}$ and $\varkappa_R: \mathscr{B}_* \times [0, \bar{\tau}] \to \boldsymbol{E}$ are defined by

$$x_R(x_*, \tau) = x_0^*(\tau) + Q(\tau)(x_* - x_0),$$
$$\varkappa_R(x_*, \tau) = \Gamma_{Q(\tau)}(\varkappa_*). \tag{6.2}$$

In $(2)_1$ $x_0^*: [0, \bar{\tau}] \to \mathscr{E}$ and $Q: [0, \bar{\tau}] \to \mathrm{Orth}^+$ are smooth mappings such that

$$x_0^*(0) = x_0 \quad \text{and} \quad Q(0) = 1,$$

x_0 being a point of $\mathscr{E}$ and 1 the identity tensor.

It follows easily from $(2)_1$ that in a rigid motion the velocity reads

$$\dot{x}_R = \dot{x}_0^* + W(x_R - x_0^*), \tag{6.3}$$

where W is the mapping of $[0, \bar{\tau}]$ into Skw $(\mathscr{V})$, the space of all skew tensors of Lin $(\mathscr{V})$, defined by

$$W(\tau) := \dot{Q}(\tau)\, Q^T(\tau) \quad \text{for all } \tau \in [0, \bar{\tau}]. \tag{6.4}$$

To calculate the microvelocity of a rigid motion we first remark that

$$\varkappa_R(\tau + \delta) = \Gamma_{Q(\tau+\delta)Q^T(\tau)}(\varkappa_R(\tau)) \quad \text{for all} \quad \tau \in [0, \bar{\tau}] \quad \text{and} \quad \delta > 0, \tag{6.5}$$

because Γ is a group homomorphism. Then we arrive at

$$\dot{\varkappa}_R(\tau) = \Gamma'(\varkappa_R(\tau))\, W(\tau), \tag{6.6}$$

where W is the mapping defined by (4) and $\Gamma': \boldsymbol{E} \to \mathrm{Lin}\,(\mathrm{Skw}(\mathscr{V}), V)$. Since $W = -\mathsf{e}w$, where e is the Ricci permutation tensor and $w: [0, \bar{\tau}] \to \mathscr{V}$, (3) can be rewritten as

$$\dot{x}_R = \dot{x}_0^* + w \times (x_R - x_0^*) \tag{6.7}$$

and w can be interpreted as the angular velocity of the motion. Accordingly, if $\boldsymbol{A}: \boldsymbol{E} \to \operatorname{Lin}(\mathscr{V}, \boldsymbol{V})$ is the mapping defined by

$$\boldsymbol{A}(\boldsymbol{x})\, w := -\Gamma'(\boldsymbol{x})\,(\mathrm{e}w) \qquad \text{for all } \boldsymbol{x} \in \boldsymbol{E},$$

then (6) reads

$$\dot{\boldsymbol{x}}_R = \boldsymbol{A}(\boldsymbol{x}_R)\, w. \tag{6.9}$$

The values of velocity and microvelocity are generally not the same in different frames. The differences in values read on two rigidly related frames have expressions formally identical with the right-hand sides of (7) and (9), provided that $\dot{x}_0^*$ and w are now interpreted as the relative translatory and rotational speeds of one frame with respect to the other.

Acceleration and microacceleration have obvious definitions; the difference of their values in different frames could easily be expressed and the related questions of inertia and microinertia could be discussed. However, we do not deal here with these matters because they are irrelevant to our present purposes; inertial actions will be imagined as simply included in the total action of the external world as announced in Remark 2 of Section 3.

7. Macrointeractions and microinteractions

We proceed now to a specialization of the objects introduced in Sections 2, 3, 4 so as to fit the event world proposed in Section 5.

Let $\mathscr{V}$ and $\boldsymbol{V}$ again be, respectively, the translation spaces of the Euclidean spaces $\mathscr{E}$ and $\boldsymbol{E}$ as defined in Sect. 5. Let a fit region $\mathscr{B}$ be given in $\mathscr{E}$. $\mathscr{B}$ is to be identified with anyone of the regions $\mathscr{B}_\tau$ defined in the preceding section. *We assume from now on that $\mathscr{I}$ coincide with $\mathscr{V} \times \boldsymbol{V}$.* It is easily seen that $\mathscr{I}$ is a normed linear space. We call *projections* the mappings $P: \mathscr{I} \to \mathscr{V}$ and $\boldsymbol{P}: \mathscr{I} \to \boldsymbol{V}$ defined by

$$P(v, \boldsymbol{v}) := v, \qquad \boldsymbol{P}(v, \boldsymbol{v}) := \boldsymbol{v} \qquad \text{for all } (v, \boldsymbol{v}) \in \mathscr{I}.$$

We consider an interaction $I: (\Omega^2_{\mathscr{B}})_{\text{deq}} \to \mathscr{I}$ as defined in Sect. 2. We define the mappings $I_M: (\Omega^2_{\mathscr{B}})_{\text{deq}} \to \mathscr{V}$ and $\boldsymbol{I}_M: (\Omega^2_{\mathscr{B}})_{\text{deq}} \to \boldsymbol{V}$ as follows:

$$I_M(\mathscr{P}, \mathscr{Q}) := P(I(\mathscr{P}, \mathscr{Q})), \qquad \boldsymbol{I}_M(\mathscr{P}, \mathscr{Q}) = \boldsymbol{P}(I(\mathscr{P}, \mathscr{Q}))$$

for all $(\mathscr{P}, \mathscr{Q}) \in (\Omega^2_{\mathscr{B}})_{\text{deq}}$. It is easily seen that both I_M and $\boldsymbol{I}_M$ are interactions. We call I_M a *macrointeraction* and $\boldsymbol{I}_M$ a *microinteraction*. If one of the Assumptions 1–4 is valid for I, it is also valid for both I_M and $\boldsymbol{I}_M$. If I is quasi-balanced according to Def. 1 of Sect. 3, also I_M and $\boldsymbol{I}_M$ are quasi-balanced.

Here we accept as valid Assumptions 1–4 for I and assume also that I is quasi-balanced.

We define the continuous mappings $b: \operatorname{clo}\mathscr{B} \to \mathscr{V}$ and $\boldsymbol{b}: \operatorname{clo}\mathscr{B} \to \boldsymbol{V}$ thus:

$$b(x) := P(f(x)), \qquad \boldsymbol{b}(x) := \boldsymbol{P}(f(x)) \qquad \text{for all } x \in \operatorname{clo}\mathscr{B}, \tag{7.1}$$

where f is the same as in (3.1). We call b the *external body macroforce* and $\boldsymbol{b}$ the *external body microforce*. By Prop. 3 of Sect. 3 and Assumption 3 we may define the continuous mappings $z: \mathscr{B} \to \mathscr{V}$ and $\boldsymbol{z}: \mathscr{B} \to V$ as

$$z(x) := P(s(x)), \quad \boldsymbol{z}(x) := (\boldsymbol{P}s(x)) \quad \text{for all } x \in \mathscr{B}, \tag{7.2}$$

where s is the same as in (3.7). We call z the *self-macroforce* and $\boldsymbol{z}$ the *self-microforce*. By Theorem 2 of Sect. 3 the mappings $S: \mathscr{B} \to \mathrm{Lin}(\mathscr{V})$ and $\boldsymbol{S}: \mathscr{B} \to \mathrm{Lin}(\mathscr{V}, V)$ defined by

$$S(x) := P(C(x)), \quad \boldsymbol{S}(x) := \boldsymbol{P}(C(x)) \quad \text{for all } x \in \mathscr{B} \tag{7.3}$$

are continuous, if C is the same as in (3.8). We call S the *macrostress tensor* and $\boldsymbol{S}$ the *microstress tensor*. By Assumption 4 we may also define the continuous mappings $p: \partial^*\mathscr{B} \to \mathscr{V}$ and $\boldsymbol{p}: \partial^*\mathscr{B} \to V$ thus

$$p(x) := -P(h(x)), \quad \boldsymbol{p}(x) := -\boldsymbol{P}(h(x)) \quad \text{for all } x \in \partial^*\mathscr{B}. \tag{7.4}$$

We call p the *macrotraction* and $\boldsymbol{p}$ the *microtraction*.

The following Proposition is an immediate consequence of (1)–(3) and Prop. 4 of Sect. 3.

Proposition 1. *If the mappings S and $\boldsymbol{S}$ are of class C^1, then*

$$z + \operatorname{div} S + b = 0 \tag{7.5}$$

and

$$\boldsymbol{z} + \operatorname{div} \boldsymbol{S} + \boldsymbol{b} = 0 \quad \text{in } \mathscr{B}. \tag{7.6}$$

Equation (5) expresses the *balance of macromomentum*, and equation (6) the *balance of micromomentum*. By (4) and the Theorem of Sect. 4 we may write the boundary conditions for (5) and (6).

Proposition 2. *Let $n_{\mathscr{B}}: \partial^*\mathscr{B} \to \mathscr{I}$ be the outer unit normal field of $\mathscr{B}$. If $\partial\mathscr{B}$ is piecewise of class C^1 and if $\lim_{y\to x} S(y)\, n_{\mathscr{B}}(x)$ and $\lim_{y\to x} \boldsymbol{S}(y)\, n_{\mathscr{B}}(x)$ exist for all $x \in \partial^*\mathscr{B}$ then*

$$\lim_{y\to x} S(y)\, n_{\mathscr{B}}(x) = p(x) \tag{7.7}$$

and

$$\lim_{y\to x} \boldsymbol{S}(y)\, n_{\mathscr{B}}(x) = \boldsymbol{p}(x) \quad \text{for all } x \in \partial^*\mathscr{B}. \tag{7.8}$$

8. Noll's axiom of frame-indifference

To complete the set of balance equations a final axiom is required, which is the adaptation to the present, more general circumstances of an axiom of Noll on the working of the totality of actions on a body (see [18] and Sect. I.12 of [2]).

A necessary premiss is the concept of working of a set of actions on a body in a smooth motion. Actually the interest centres on the totality of external actions, be they external body actions or be they external contact actions; for them the definition of working on any subbody is the following:

$$W(\mathscr{C}) := \int_{\mathscr{C}} (b \cdot \dot{x} + \boldsymbol{b} \cdot \dot{\boldsymbol{x}}) + \int_{\partial^*\mathscr{C}} (S n_{\mathscr{C}} \cdot \dot{x} + \boldsymbol{S} n_{\mathscr{C}} \cdot \dot{\boldsymbol{x}}), \tag{8.1}$$

where $\mathscr{C}$ is the part of $\mathscr{B}$ occupied by the subbody. The expression (1) is the trivial generalization of one valid in the classic context.

NOLL's axiom asserts that, *for any choice of* $\mathscr{C}$ *in* $\Omega_{\mathscr{B}}$ *and for any assignment of the actions* b, $\boldsymbol{b}$, S, $\boldsymbol{S}$, *to* $\mathscr{B}$, *the global working* $W(\mathscr{C})$ *is frame-indifferent*, *i.e.*, its value is the same when evaluated in any one of the frames of the same rigid class.

Use of the Divergence Theórem and the balance equations of macromomentum and micromomentum leads us to the following equation for W

$$W(\mathscr{C}) = -\int_{\mathscr{C}} (z \cdot \dot{x} + \boldsymbol{z} \cdot \dot{\boldsymbol{x}} - S \cdot \operatorname{grad} \dot{x} - \boldsymbol{S} \cdot \operatorname{grad} \dot{\boldsymbol{x}}). \tag{8.2}$$

For any complete rigid motion there is a frame in which the body is at rest (*cf.* Sect. 6 above); in that frame the working W on any subbody is null. Since, by NOLL's axiom, the working is frame-indifferent, for a complete rigid motion $W(\mathscr{C}) = 0$ in all frames and for all $\mathscr{C} \in \Omega_{\mathscr{B}}$. By (6.3) and (6.9), for a complete rigid motion equation (2) yields

$$W(\mathscr{C}) = -\dot{x}_0^* \cdot \int_{\mathscr{C}} z - w \cdot \int_{\mathscr{C}} ((x - x_0) \times z + \boldsymbol{A}^T \boldsymbol{z} + \mathrm{e}S - (\operatorname{grad} \boldsymbol{A}^T)\, \boldsymbol{S}). \tag{8.3}$$

The choice of $\dot{x}_0^*$, w in $\mathscr{V}$ and of the region $\mathscr{C}$ in $\Omega_{\mathscr{B}}$ being arbitrary and the integrands being continuous functions of place, the usual process of localization leads to the relations

$$z = 0, \tag{8.4}$$

$$\mathrm{e}S = -\boldsymbol{A}^T \boldsymbol{z} + (\operatorname{grad} \boldsymbol{A}^T)\, \boldsymbol{S}. \tag{8.5}$$

The postulated indifference of (1) to translatory motions of the frame requires the self-force to vanish; the indifference to rotational motions determines the skew part of the macrostress in terms of the self-microforce and the self-microstress. When the microactions vanish or, at least, the operator $\boldsymbol{A}$ vanishes, (5) implies that the macrostress is a symmetric tensor.

9. Constraints

We mentioned in the Introduction that the state of an element in a body with microstructure is best modelled by an element $\boldsymbol{\nu}$ of a connected differentiable manifold $\boldsymbol{M}$ of finite dimension m which is not generally a linear space, which was assumed in the preceding Sections. However, $\boldsymbol{M}$ may always be embedded in a higher-dimensional Euclidean space $\boldsymbol{E}$; the physical body may be imagined as one with a richer microstructure, represented by elements $\boldsymbol{x}$ of $\boldsymbol{E}$, a structure which is constrained, however, and the constraint is expressed locally as follows:

$$\boldsymbol{x} = f(\nu^1, \nu^2, \ldots, \nu^m), \tag{9.1}$$

the variables $\nu^1, \nu^2, \ldots, \nu^m$ being the coordinates of ν in a local chart of an atlas for $\boldsymbol{M}$.

For bodies with the richer microstructure the developments of previous Sections apply together with the conclusions regarding balance laws, the existence of stress and microstress, *etc.* To obtain properties valid for the constrained body, we interpret (1) as a perfect constraint and adopt an extended principle of determinism.

Before we proceed to extract the consequences of that principle we remark that the constraint (1) must be such that if $\hat{\boldsymbol{x}}$ is an allowed value of $\boldsymbol{x}$ corresponding to a choice $\hat{\nu}$ of ν, so is $\hat{\boldsymbol{x}}_Q$, *i.e.* the value assumed by $\hat{\boldsymbol{x}}$ after the rotation defined by Q. $\hat{\boldsymbol{x}}_Q$ corresponds to the choice $\hat{\nu}_Q$ for ν in (1). Thus a group action is defined also over $\boldsymbol{M}$. Its infinitesimal generator will be denoted by $\mathscr{A}_i^\alpha$.

The condition that (1) be acceptable as an internal constraint, *i.e.* the condition of frame-indifference is expressed by the following relation:

$$A_{ki} = B_{k\alpha}\mathscr{A}_i^\alpha, \tag{9.2}$$

in which

$$B_{k\alpha} := \frac{\partial f_k}{\partial \nu^\alpha}.$$

The principle of *extended determinism* (see *e.g.* [2], Sect. IV.7) asserts that S, $\boldsymbol{z}$, $\boldsymbol{S}$ are each the sum of two terms, one active and the other reactive.

Active terms (*viz* $\overset{a}{S}$, $\overset{a}{\boldsymbol{z}}$, $\overset{a}{\boldsymbol{S}}$) are to be specified by constitutive relations; reactive terms (*viz* $\overset{r}{S}$, $\overset{r}{\boldsymbol{z}}$, $\overset{r}{\boldsymbol{S}}$) are bound to remain undetermined but are subject to the global condition that their power density vanish:

$$\overset{r}{S} \cdot \operatorname{grad} \dot{x} + \overset{r}{\boldsymbol{z}} \cdot \dot{\boldsymbol{x}} + \overset{r}{\boldsymbol{S}} \cdot \operatorname{grad} \dot{\boldsymbol{x}} = 0 \tag{9.3}$$

for all kinematic states allowed by the constraint (1).

Because (1) does not restrict $\dot{x}$ at all, (3) ensures that

$$\overset{r}{S} = 0, \tag{9.4}$$

whereas $\overset{r}{\boldsymbol{z}}$, $\overset{r}{\boldsymbol{S}}$ need not vanish but must satisfy the two conditions

$$\begin{aligned} &\overset{r}{\boldsymbol{S}}_{mi} \boldsymbol{B}_{m\alpha} = 0, \\ &(\overset{r}{\boldsymbol{z}}_m - \overset{r}{\boldsymbol{S}}_{mi,i}) \, \boldsymbol{B}_{m\alpha} = 0. \end{aligned} \tag{9.5}$$

Condition (4) implies that the balance equation of macromomentum is 'pure' in the sense that it does not involve reactive terms. Because of (3) also the equation of balance of moment of momentum (8.5) is pure and can be written

$$(\text{skw}\,\overset{a}{S})_{ij} = \tfrac{1}{2}\,e_{ijk}(-A_{mk}\overset{a}{z}_m + A_{mk,s}\overset{a}{S}_{ms}). \tag{9.6}$$

To deduce a pure equation from the condition of balance of micromomentum and the boundary condition for microstresses one only needs to project them locally on the cotangent space by operating on both sides of them with $\boldsymbol{B}_{m\alpha}$ to obtain, respectively,

$$\varrho\beta_\alpha - \zeta_\alpha + \overset{a}{S}_{ms,s}\boldsymbol{B}_{m\alpha} = 0, \tag{9.7}$$

$$\mathscr{S}_{\alpha s}(n_{\mathscr{B}})_s = \sigma_\alpha; \tag{9.8}$$

here

$$\zeta_\alpha := -\boldsymbol{B}_{m\alpha}\overset{a}{z}_m, \qquad \mathscr{S}_{\alpha s} := \overset{a}{S}_{ms}\boldsymbol{B}_{m\alpha}, \tag{9.9}$$

$$\varrho\beta_\alpha := \boldsymbol{B}_{m\alpha}\boldsymbol{b}_m, \qquad \sigma_\alpha := \boldsymbol{B}_{m\alpha}\boldsymbol{p}_m, \tag{9.10}$$

and β comprises actions of microinertia (which are never expressed explicitly in this paper).

Actually with the use of (2) the notation (9) can be introduced also in (6):

$$(\text{skw}\,\overset{a}{S})_{ij} = \tfrac{1}{2}\,e_{ijk}(\mathscr{A}^\alpha_k\zeta_\alpha + \mathscr{A}^\alpha_{k,s}\mathscr{S}_{\alpha s} + \boldsymbol{B}_{m\alpha,s}\mathscr{A}^\alpha_k\overset{a}{S}_{ms}). \tag{9.11}$$

In the situation envisaged in [5] $\mathscr{S}_{\alpha s}$ is not obtained by the projection of a general operator S_{ms} on the cotangent space as in $(9)_2$; rather, as explained repeatedly, it is S_{ms} that derives from $\mathscr{S}_{\alpha s}$ as a consequence of the embedding of $\boldsymbol{M}$ into $\boldsymbol{E}$, though $(9)_2$ remains true.

If one introduces the metric tensor

$$G_{\alpha\beta} := \boldsymbol{B}_{m\alpha}\boldsymbol{B}_{m\beta},$$

its inverse $G^{\alpha\beta}$ with the property

$$G^{\alpha\gamma}G_{\gamma\beta} = \delta^\alpha_\beta$$

and the contravariant base

$$\boldsymbol{B}^\alpha_m := G^{\alpha\beta}\boldsymbol{B}_{m\beta},$$

$\overset{a}{S}_{ms}$ must be imagined restricted as follows

$$\overset{a}{S}_{ms} = \boldsymbol{B}^\alpha_m\mathscr{S}_{\alpha s}.$$

Thus (7) and (11) can be written as follows:

$$\varrho\beta_\alpha - \zeta_\alpha - \mathscr{S}_{\alpha s|s} = 0,$$

$$(\text{skw}\,\overset{a}{S})_{ij} = \tfrac{1}{2}\,e_{ijk}(\mathscr{A}^\alpha_k\zeta_\alpha + \mathscr{A}^\alpha_{k|s}\mathscr{S}_{\alpha s}),$$

where the covariant derivatives

$$\mathscr{S}_{as|s} := \mathscr{S}_{as,s} - B^{\beta}_{m} B_{m\alpha,s} \mathscr{S}_{\beta s},$$

$$\mathscr{A}^{\alpha}_{k|s} := \mathscr{A}^{\alpha}_{k,s} + B^{\alpha}_{m} B_{m\beta,s} \mathscr{A}^{\beta}_{k},$$

have been introduced.

Finally one can write compactly (see (8.3) and (9.4) of [5])

$$\varrho\beta - \zeta + \operatorname{div}_M \mathscr{S} = 0,$$

$$\operatorname{skw} \overset{\mathrm{a}}{S} = \tfrac{1}{2}\,\mathrm{e}(\mathscr{A}^T \zeta + (\operatorname{grad}_M \mathscr{A}^T)\,\mathscr{S}),$$

where the subscript M reminds the reader that the appropriate significance must be attributed to the operators grad and div.

References

1. Noll, W., The foundations of classical mechanics in the light of recent advances in continuum mechanics, pp. 266–281 of "The axiomatic method, with special reference to geometry and physics" (Symposium held at Berkeley, 1957), North-Holland Publ. Co., Amsterdam, 1959.
2. Truesdell, C., *A First Course in Rational Continuum Mechanics*, Vol. 1, Academic Press, New York *etc.*, 1977. Second edition, in press.
3. Beatty, M. F., On the foundation principles of general classical mechanics, Arch. Rational Mech. Anal. **24** (1967) 264–273.
4. Williams, W. O., Axioms for work and energy in general continua I. Smooth velocity fields, Arch. Rational Mech. Anal. **42** (1971), 93–114.
5. Capriz, G., *Continua with Microstructure*, Springer, Berlin *etc.*, 1989.
6. Noll, W., & E. G. Virga, On edge interactions and surface tension, Arch. Rational Mech. Anal., *forthcoming*, 1990.
7. Noll, W., & E. G. Virga, Fit regions and functions of bounded variation, Arch. Rational Mech. Anal. **102** (1988), 1–21.
8. Noll, W., Lectures on the foundations of continuum mechanics and thermodynamics, Arch. Rational. Mech. Anal. **52** (1973), 62–92.
9. Noll, W., Continuum mechanics and geometric integration theory, pp. 17–29 of "Categories in Continuum Physics" (Workshop held at SUNY, Buffalo, 1982), Springer, Berlin *etc.*, 1986.
10. Ziemer, W. P., Cauchy flux and sets of finite perimeter, Arch. Rational Mech. Anal. **84** (1983), 189–201.
11. Šilhavý, M., The existence of the flux vector and the divergence theorem for general Cauchy fluxes, Arch. Rational Mech. Anal. **90** (1985), 195–212.
12. Gurtin, M. E., & W. O. Williams, An axiomatic foundation for continuum thermodynamics, Arch. Rational Mech. Anal. **26** (1967), 83–117.
13. Williams, W. O., On internal interactions and the concept of thermal isolation, Arch. Rational Mech. Anal. **34** (1969), 245–258.
14. Gurtin, M. E., V. J. Mizel & W. O. Williams, A note on Cauchy's stress theorem, J. Math. Anal. Appl. **22** (1968), 398–401.

15. GURTIN, M. E., & L. C. MARTINS, Cauchy's theorem in classical physics, Arch. Rational Mech. Anal. **60** (1976), 305–324.
16. FOSDICK, R. L., & E. G. VIRGA, A variational proof of Cauchy's stress theorem, Arch. Rational Mech. Anal. **105** (1989), 95–103.
17. NOLL, W., Euclidean geometry and Minkowskian chronometry, Amer. Math. Monthly **71** (1964), 129–144.
18. NOLL, W., La mécanique classique, basée sur un axiome d'objectivité, pp. 47–56 of "La méthode axiomatique dans les mécaniques classiques et nouvelles" (Colloque international, Paris, 1959), Gauthier-Villars, Paris, 1963.

Dipartimento di Matematica
Università di Pisa

and

Dipartimento di Matematica
Università di Pavia

(Received June 5, 1989)

Surface Interaction Potentials in Elasticity

PAOLO PODIO-GUIDUGLI & GIORGIO VERGARA CAFFARELLI

Dedicated to Bernard D. Coleman on his sixtieth birthday

Contents

0. Introduction

The problem of modelling body–environment interactions is a rather formidable one: neither is there a wealth of inspiring examples nor is the *corpus* of knowledge in mathematical analysis such as to allow indiscriminate generalization of the few well understood cases.

Primarily, the problem has a *constitutive* nature. In the traditional view, the object of a constitutive theory is the response of the body to deformation processes: modelling body–environment interactions demands *per se* that we assign the constitutive theory the extended task of formalizing our prejudices on the body and the environment under examination, both separately and together.

However, the modelling process is rather inextricably conditioned and guided by the type of initial-value and boundary-value problems that one wishes to formulate, as well as by the mathematical techniques that one wishes to employ, or has to. For example, it is shown in [1] how the choice of admissible interactions is contained by the needs of a local bifurcation analysis by use of formal perturbation methods; on the other hand, paper [2] exemplifies well how that choice may be guided by mathematically reasonable requirements of well-posedness.

In elasticity, the primarily constitutive nature of the modelling process is perhaps best brought about, as done in [3], by confining attention to equilibrium problems with conditions of traction on the entire boundary, in a variational format. The formulation of those equilibrium problems may be "formal", in the sense that a function space setting sufficient to clarify constitutive issues need not allow for, say, existence theorems. In this paper we adopt precisely such a "formal" variational format to arrive at well motivated choices of surface potentials describing the interactions taking place at the common boundary of a body and its environment.

In elastic equilibrium problems, body–environment interactions are accounted for by prescribing a system of loads which are usually of the *dead* type, *i.e.*, they do not depend on the displacement that the body undergoes. More realistic prescriptions of *live*, as opposed to dead, loads were first proposed for study of SEWELL [4], [5], under the heading of "configuration–dependent" loadings (*vid.* also [6], Chapters Q and R).

Although it can be argued with only slight exaggeration that most loadings encountered in applications are of the live type, our subject has not been extensively studied so far. In the context of elasticity linearized about a given equilibrium placement under stress, live loads were considered by CAPRIZ & PODIO-GUIDUGLI [1], PODIO-GUIDUGLI & VERGARA CAFFARELLI [7], PODIO-GUIDUGLI, VERGARA CAFFARELLI & VIRGA [8] and VERGARA CAFFARELLI [9]. In finite elasticity, some results of uniqueness and continuous dependence in the presence of live loads have been obtained by SPECTOR [10], [11]. Remarkably, SEWELL's paper [4], as well as the other papers just cited, have the "formal" character typical of the constitutively oriented approach to the matters alluded to in the preceding paragraph. At variance with this, the results of VALENT [12] concerning local existence, uniqueness and continuous dependence for the case of hydrostatic loading are obtained in a precise function space setting.

From a variational point of view, some body–environment interactions are described by surface potentials, others by volume potentials, others by surface or volume potentials, alternatively.

We here do not deal with genuine volume interaction potentials, for two reasons: first, they are sufficiently well understood from the mechanical point of view and, second, their presence usually does not bring in any mathematical novelty, and sometimes even renders the matters mathematically easier.

The *surface interaction potentials* we consider accomodate many types of live loads,[1] among which the two important examples of *pressure loading* and *membrane loading*; they have the form

$$T\{f\} = \int_{\partial\Omega} \hat{\tau}(\boldsymbol{x}, \boldsymbol{n}, \boldsymbol{f}, \nabla \boldsymbol{f})\, \mathrm{d}\,(\mathrm{Srf}), \tag{0.1}$$

with $\tau(\boldsymbol{x}) := \hat{\tau}(\boldsymbol{x}, \boldsymbol{n}(\boldsymbol{x}), \boldsymbol{f}(\boldsymbol{x}), \nabla \boldsymbol{f}(\boldsymbol{x}))$ interpreted as the energy stored per unit area, when a body undergoes a displacement $\boldsymbol{f}$ from a reference placement Ω, at a point $\boldsymbol{x}$ of the boundary $\partial\Omega$ where $\boldsymbol{n}$ is the outward unit normal field.

Our main concern here is precisely the constitutive problem of selecting the density mapping $\hat{\tau}$ for T; our approach is the same as in [3]:

(i) We introduce the *total potential*

$$E\{f\} = S\{f\} + T\{f\}, \tag{0.2}$$

with

$$S\{f\} = \int_{\Omega} \hat{\sigma}(\boldsymbol{x}, \nabla \boldsymbol{f})\, \mathrm{d}\,(\mathrm{Vol}), \tag{0.3}$$

where the density $\sigma(\boldsymbol{x}) := \hat{\sigma}(\boldsymbol{x}, \nabla \boldsymbol{f}(\boldsymbol{x}))$ of the *body potential* S is interpreted as the elastic energy stored per unit volume at a point of Ω. For $\boldsymbol{S}$ the stress field over $\bar{\Omega}$ and $\boldsymbol{s}$ the surface load field over $\partial\Omega$, we say that the body has a *conservative interaction* with its environment (with respect to the reference placement Ω) if the field pairs $(\boldsymbol{S}, \boldsymbol{s})$ and (σ, τ) satisfy the variational condition

$$\delta S\{f\}\,[\boldsymbol{h}] = \int_{\Omega} \boldsymbol{S} \cdot \nabla \boldsymbol{h}, \qquad \delta T\{f\} = -\int_{\partial\Omega} \boldsymbol{s} \cdot \boldsymbol{h} \tag{0.4}$$

for each $\boldsymbol{f} \in \mathscr{D}$, the collection of all admissible displacements, and for each $\boldsymbol{h} \in \mathscr{H}$, the space of all admissible variations.

(ii) We pose the variational problem

$$\operatorname{extr} \{E\{f\} \mid \boldsymbol{f} \in \mathscr{D}\}, \tag{0.5}$$

and derive necessary conditions to be obeyed by each extremum $\boldsymbol{f}^{(0)}$.

(iii) As a selection criterion for $\hat{\tau}$, we choose to regard S as given and require that $\hat{\tau}$ be such as to obey the necessary conditions for an extremum identically on the domain $\mathscr{D}$ of E.

Our work is divided into two parts. Part I has three sections in which a number of preliminary results of algebra, analysis and mechanics having specific bearing

---

[1] Some interesting examples of live loadings, such as the pressure loading exerted on the inner wall of a deformable gas container [13], [14], escape our analysis here because of their *non-local* nature.

to the developments to come are collected; of these results, some are included so as to explain our notation carefully and make our paper reasonably self-contained; others are new or proved in a new way. Part II begins with a general formulation of elastic equilibria as extremum problems (Section 4); it continues with a variational study of the two important cases of pressure loadings and membrane loadings (Subsections 5.1 and 5.2, respectively)[2], with the purpose of exemplifying the kinds of behavior we shall later incorporate in our general model of interactions among a body, its surface, and its environment, and ends with two other sections, the bulk of this paper, of the contents of which we now give an extensive account.

Section 6 is devoted to obtaining and discussing the equilibrium conditions, *i.e.*, the necessary conditions on the extrema of the total potential E.

We point out that the regions Ω we consider have boundaries which consist either of one smooth surface Σ with no self-intersections, or of two such surfaces Σ_1 and Σ_2, with $\Sigma_1 \cap \Sigma_2 = \emptyset$, and their common boundary Γ, a closed simple curve with tangent $\boldsymbol{t}$; in the second case, the outward normal field has a jump at Γ, and we further distinguish two subcases: (i) $\boldsymbol{n}_1(\boldsymbol{x}) \neq \boldsymbol{n}_2(\boldsymbol{x})$ for all $\boldsymbol{x} \in \Gamma$; (ii) (the cuspidal subcase) $\boldsymbol{n}_1(\boldsymbol{x}) \equiv -\boldsymbol{n}_2(\boldsymbol{x})$ for all $\boldsymbol{x} \in \Gamma$. The necessary conditions for extrema take different forms on the various parts of the body, namely, its interior part Ω and its regular and singular boundary parts $\Sigma_1 \cup \Sigma_2$ and Γ, respectively. Summarizing from Proposition 6.1, we have, as usual, that

$$\operatorname{Div} \hat{\boldsymbol{S}}^{(0)} = \boldsymbol{0}, \qquad \hat{\boldsymbol{S}}^{(0)} := \partial_{\boldsymbol{F}}\hat{\sigma}^{(0)},^{3} \tag{0.6}$$

in the interior part, while on the regular part of the boundary both

$$\hat{\boldsymbol{S}}^{(0)}\boldsymbol{n} = \boldsymbol{s}^{(0)}, \qquad \boldsymbol{s}^{(0)} := -\partial_{\boldsymbol{f}}\hat{\tau}^{(0)} + {}^{s}\operatorname{Div}(\partial_{\boldsymbol{F}}\hat{\tau}^{(0)}) \tag{0.7}$$

and

$$(\partial_{\boldsymbol{F}}\hat{\tau}^{(0)})\,\boldsymbol{n} = \boldsymbol{0} \tag{0.8}$$

prevail; moreover, on the singular part of the boundary, where $\hat{\tau}_\alpha^{(0)}$ and $\nabla_\alpha \boldsymbol{f}^{(0)}$ ($\alpha = 1, 2$) are defined as the limits of $\hat{\tau}^{(0)}|_{\Sigma_\alpha}$ and $\nabla \boldsymbol{f}^{(0)}|_{\Sigma_\alpha}$ for $\boldsymbol{x} \to \Gamma$ from Σ_α, either

$$\partial_{\boldsymbol{F}}\hat{\tau}_1^{(0)}(\boldsymbol{x}, \boldsymbol{n}_1, \boldsymbol{f}^{(0)}, \nabla_1 \boldsymbol{f}^{(0)})\,\boldsymbol{n}_2 + \partial_{\boldsymbol{F}}\hat{\tau}_2^{(0)}(\boldsymbol{x}, \boldsymbol{n}_2, \boldsymbol{f}^{(0)}, \nabla_2 \boldsymbol{f}^{(0)})\,\boldsymbol{n}_1 = \boldsymbol{0} \tag{0.9}$$

or, in the cuspidal subcase,

$$[\partial_{\boldsymbol{F}}\hat{\tau}_1(\boldsymbol{x}, \boldsymbol{n}_1, \boldsymbol{f}^{(0)}, \nabla_1 \boldsymbol{f}^{(0)}) + \partial_{\boldsymbol{F}}\hat{\tau}_2(\boldsymbol{x}, -\boldsymbol{n}_1, \boldsymbol{f}^{(0)}, \nabla_2 \boldsymbol{f}^{(0)})]\,(\boldsymbol{t} \times \boldsymbol{n}_1) = \boldsymbol{0}. \tag{0.10}$$

We take condition $(0.7)_2$ to define the surface load mapping $\hat{\boldsymbol{s}}$ in terms of $\hat{\tau}$ for conservative interactions, just as condition $(0.6)_2$ defines the stress mapping $\hat{\boldsymbol{S}}$ in terms of $\hat{\sigma}$ for all $\boldsymbol{f} \in \mathcal{D}$. The three remaining conditions (0.8)–(0.10) involve $\hat{\tau}$ only through the "membrane stress" $\partial_{\boldsymbol{F}}\hat{\tau}$. In Section 7 we find for what classes of surface potential densities $\hat{\tau}$ one can expect the associated membrane stress

[2] Our present analysis of pressure and membrane loadings has characters of greater generality and detail than the one presented in [3]; in particular, the pressure function may here depend on the displacement, as is the case, *e.g.*, for hydrostatic loading.

[3] We use an affixed $^{(0)}$ as a reminder for the operation of evaluation at $\boldsymbol{f}^{(0)}$.

to satisfy (0.8), alone or else together with (0.9) and (0.10), for all admissible displacements $\boldsymbol{f}$ and all reference placements Ω occupied by the body.

We observe that (0.8) is identically satisfied in $\mathscr{D}$ whenever the membrane stress is tangential, and we show by Proposition 7.1 that this is the case if and only if $\hat{\tau}$ does not depend on normal derivatives of $\boldsymbol{f}$:

$$\hat{\tau}(\boldsymbol{x}, \boldsymbol{n}(\mathrm{x}), \boldsymbol{f}(\mathrm{x}), \nabla \boldsymbol{f}(\mathrm{x})) = \hat{\tau}(\boldsymbol{x}, \boldsymbol{n}(\mathrm{x}), \boldsymbol{f}(\mathrm{x}), {}^s\nabla \boldsymbol{f}(\mathrm{x})), \quad {}^s\nabla \boldsymbol{f} := (\nabla \boldsymbol{f})(\boldsymbol{I} - \boldsymbol{n} \otimes \boldsymbol{n}), \tag{0.11}$$

for all $(\boldsymbol{x}, \boldsymbol{f}) \in \partial\Omega \times \mathscr{D}$ (here ${}^s\nabla \boldsymbol{f}$ is by definition the tangential displacement gradient). We call those surface potentials whose densities satisfy (0.11) *tangential.* Given a surface potential T as in (0.1), we introduce the associated tangential potential T_{sub} with density

$$\hat{\tau}_{\mathrm{sub}}(\boldsymbol{x}, \boldsymbol{n}, \boldsymbol{f}, {}^s\nabla \boldsymbol{f}) := \inf \{\hat{\tau}(\boldsymbol{x}, \boldsymbol{n}, \boldsymbol{f}, {}^s\nabla \boldsymbol{f} + \boldsymbol{y} \otimes \boldsymbol{n}) \mid \boldsymbol{y} \in \mathscr{V}\}, \tag{0.12}$$

and show that under reasonable hypotheses on $\hat{\tau}$ (*e.g.*, when $\hat{\tau}$ is rank-one convex, as in Proposition 7.5) the extremals of the total potential $E = S + T$ are elements of the extremal set of the associated *subpotential* $E_{\mathrm{sub}} = S + T_{\mathrm{sub}}$. This result, together with the fact that both pressure loading and membrane loading admit a tangential potential, motivates us to restrict attention to tangential potentials for the rest of our paper.

We then turn to the question of characterizing, among tangential potentials, those which are *simple*, *i.e.*, correspond to surface load mappings $\hat{s}$ depending at most on tangential derivatives of the first order.[4] Restricting here attention to regular regions Ω, and therefore dropping the explicit dependence on the normal field $\boldsymbol{n}$, we first obtain a representation formula for the density of a simple tangential potential as a rank-one affine function of the tangential displacement gradient:

$$\hat{\tau}(\boldsymbol{x}, \boldsymbol{f}, {}^s\nabla \boldsymbol{f}) = \hat{\gamma}(\boldsymbol{x}, \boldsymbol{f}) + \hat{\boldsymbol{C}}_1(\boldsymbol{x}, \boldsymbol{f}) \cdot {}^s\nabla \boldsymbol{f}^* + \hat{\boldsymbol{C}}_2(\boldsymbol{x}, \boldsymbol{f}) \cdot {}^s\nabla \boldsymbol{f}, \tag{0.13}$$

where ${}^s\nabla \boldsymbol{f}^*$ denotes the cofactor of ${}^s\nabla \boldsymbol{f}$, while $\hat{\gamma}$ and $\hat{\boldsymbol{C}}_1$, $\hat{\boldsymbol{C}}_2$ are arbitrary scalar and tensor-valued mappings, respectively (*cf.* Proposition 7.7). Secondly, we characterize the class of surface loadings alternatively described by surface or volume interaction potentials by showing that, if T is a simple tangential potential, then there is a volume potential

$$F\{\boldsymbol{f}\} = \int_{\Omega} \hat{\varphi}(\boldsymbol{x}, \boldsymbol{f}, \boldsymbol{F}) \, \mathrm{d}\,(\mathrm{Vol}), \tag{0.14}$$

which is a *null Lagrangian* in the sense of [15], [16], such that

$$T\{\boldsymbol{f}\} = F\{\boldsymbol{f}\} \tag{0.15}$$

for all displacements $\boldsymbol{f} \in \mathscr{D}$ (and conversely; *cf.* Proposition 7.9).

---

[4] Simple surface loadings were introduced by SPECTOR [10], [11]; they are easier to handle than nonsimple ones, basically because they lead to a boundary operator $\mathfrak{B}(\boldsymbol{f}) := \hat{\boldsymbol{S}}(\nabla \boldsymbol{f})\,\boldsymbol{n} - \hat{\boldsymbol{s}}(\boldsymbol{f}, {}^s\nabla \boldsymbol{f})$ which, given that the field operator $\mathfrak{F}(\boldsymbol{f}) := -\mathrm{Div}\,\hat{\boldsymbol{S}}(\nabla \boldsymbol{f})$ is of order two, has the "right" order one (*cf.* [3], [7], [8], [9]).

Finally, we return to the extremum conditions (0.9), (0.10), to be obeyed on the singular boundary part Γ of Ω by the pair $(\hat{\tau}_1, \hat{\tau}_2)$, with $\hat{\tau}_\alpha$ the (necessarily homogeneous) surface potential density on the boundary part $\Sigma_\alpha \cup \Gamma$. We interpret (0.9), (0.10) as *joint tangentiality* conditions on the membrane stresses $\partial_F \hat{\tau}_\alpha$ and we seek a class of surface interactions such as to satisfy (0.9), (0.10), with $\nabla_1 \boldsymbol{f}^{(0)} \equiv \nabla_2 \boldsymbol{f}^{(0)}$, identically in $\mathcal{O}$, the collection of all admissible domains.[5] By Proposition 7.13 we then show that, for two mappings $\hat{\tau}_1$, $\hat{\tau}_2$ to compose a jointly tangential surface density pair, they must have the following representation of the rank-one affine type:

$$\hat{\tau}_\alpha(\boldsymbol{n}, \boldsymbol{f}, {}^s\nabla\boldsymbol{f}) - \hat{\gamma}_\alpha(\boldsymbol{n}, \boldsymbol{f}) = \hat{\boldsymbol{c}}(\boldsymbol{f}) \otimes \boldsymbol{n} \cdot {}^s\nabla\boldsymbol{f}^* + \hat{\boldsymbol{C}}(\boldsymbol{f})\, \boldsymbol{N} \cdot {}^s\nabla\boldsymbol{f}, \quad \alpha = 1, 2, \tag{0.16}$$

where $\boldsymbol{N}$ is the skew tensor associated with $\boldsymbol{n}$; thus, not only do they induce a simple loading throughout the boundary, but also they are associated with an energetically equivalent null Lagrangian. In particular, in the important special case of *regular* interactions, *i.e.*, when $\hat{\tau}_1 = \hat{\tau}_2 = \hat{\tau}$, we find (Proposition 7.16) that the associated null Lagrangian F has density

$$\hat{\varphi}(\boldsymbol{f}, \boldsymbol{F}) = \hat{\delta}(\boldsymbol{f}) \det \boldsymbol{F} + \hat{\boldsymbol{D}}(\boldsymbol{f}) \cdot \boldsymbol{F}^*, \tag{0.17}$$

with $\hat{\delta}$ and $\hat{\boldsymbol{D}}$ depending, respectively, on $\hat{\boldsymbol{c}}$ and $\hat{\boldsymbol{C}}$ only,[6] and that (0.15) is paralleled by

$$T\{\boldsymbol{f}\} = F\{\boldsymbol{f}\} + \int_{\partial\Omega} \hat{\gamma}(\boldsymbol{n}, \boldsymbol{f})\, d\,(\mathrm{Srf}). \tag{0.18}$$

Part I. Preliminaries

1. Algebra

In this section we collect, mostly without proof, various results to be of use later; of these results, many are well known, the rest are taken freely from [17], where proofs here omitted may be found.

Our notations are essentially those of [18], with minor variants and additions already introduced in [17] and [3]. As in [18], bold-face type denote vectors, if minuscule and second-order tensors if majuscule; greek letters are reserved for scalar quantities if minuscule, and for regions of space, if majuscule. Our use of parentheses, brackets and braces is completely standard, but we also use brackets to stress linear dependence on the enclosed variable, and braces to enclose the argument of a functional.

---

[5] That is to say, for each admissible choice of a curve Γ and two unit vector fields $\boldsymbol{n}_\alpha$ defined over Γ itself, the latter interpreted as the limits of the outward normal field on the regular part of the boundary on going toward Γ starting from Σ_α.

[6] For pressure loading, a regular interaction, the densities are $\hat{\tau}_p(\boldsymbol{n}, \boldsymbol{f}, {}^s\nabla\boldsymbol{f}) = \hat{\pi}(\boldsymbol{f})\,(\boldsymbol{f} \otimes \boldsymbol{n}) \cdot {}^s\nabla\boldsymbol{f}^*$ for the surface and $\hat{\varphi}_p(\boldsymbol{f}, \nabla\boldsymbol{f}) = \hat{\pi}(\boldsymbol{f}) \det(\nabla\boldsymbol{f})$ for the volume potential (*vid.* Proposition 5.1).

1.1 Collections of Second-Order Tensors

In this paper, $\mathscr{V}$ is an oriented three-dimensional vector space over the real field $\mathbb{R}$, equipped with an inner product denoted by $\cdot$ and a vector product denoted by $\times$.

Lin is the set of all linear transformations of $\mathscr{V}$ into itself (the space of all second-order tensors). In particular, $\boldsymbol{I}$ is the identity transformation; for $\boldsymbol{A} \in \text{Lin}$, $\boldsymbol{A}^{\mathrm{T}}$ is the transpose of $\boldsymbol{A}$; for $\boldsymbol{a}, \boldsymbol{b} \in \mathscr{V}$, $\boldsymbol{a} \otimes \boldsymbol{b} \in \text{Lin}$, the tensor product of $\boldsymbol{a}$ and $\boldsymbol{b}$, is defined by

$$(\boldsymbol{a} \otimes \boldsymbol{b})\, \boldsymbol{v} := (\boldsymbol{b} \cdot \boldsymbol{v})\, \boldsymbol{a} \quad \text{for all } \boldsymbol{v} \in \mathscr{V}. \tag{1.1}$$

The symmetric and skew subcollections of elements of Lin are denoted by Sym and Skw, respectively, so that

$$\text{Lin} = \text{Sym} \oplus \text{Skw} \tag{1.2}$$

and, for all $\boldsymbol{A} \in \text{Lin}$,

$$\begin{gathered} \boldsymbol{A} = \operatorname{sym} \boldsymbol{A} + \operatorname{skw} \boldsymbol{A}, \\ 2 \operatorname{sym} \boldsymbol{A} := (\boldsymbol{A} + \boldsymbol{A}^{\mathrm{T}}) \in \text{Sym}, \quad 2 \operatorname{skw} \boldsymbol{A} := (\boldsymbol{A} - \boldsymbol{A}^{\mathrm{T}}) \in \text{Skw}. \end{gathered} \tag{1.3}$$

Beside Sym and Skw, other subcollections of Lin to be considered here are

$$\text{Lin}^+ := \{\boldsymbol{F} \in \text{Lin} \mid \det \boldsymbol{F} > 0\}, \quad \text{Rot} := \{\boldsymbol{R} \in \text{Lin} \mid \boldsymbol{R}\boldsymbol{R}^{\mathrm{T}} = \boldsymbol{R}^{\mathrm{T}}\boldsymbol{R} = \boldsymbol{I}, \det \boldsymbol{R} = 1\}. \tag{1.4}$$

As is well known, Skw and $\mathscr{V}$ can be set in one-to-one correspondence:

$$\text{Skw} \ni \boldsymbol{W} \leftrightarrow \boldsymbol{w} \in \mathscr{V}, \quad \boldsymbol{W}\boldsymbol{v} = \boldsymbol{w} \times \boldsymbol{v} \text{ for all } \boldsymbol{v} \in \mathscr{V}; \tag{1.5}$$

it follows that

$$|\boldsymbol{W}|^2 = 2\, |\boldsymbol{w}|^2,$$

and, more generally, for $\boldsymbol{w}, \boldsymbol{z} \in \mathscr{V}$ and for $\boldsymbol{W}, \boldsymbol{Z}$ the corresponding elements of Skw,

$$\boldsymbol{W} \cdot \boldsymbol{Z} = 2\boldsymbol{w} \cdot \boldsymbol{z}, \tag{1.6}$$

where we have used the inner product of Lin defined in terms of the trace function by

$$\boldsymbol{A} \cdot \boldsymbol{B} := \operatorname{tr}(\boldsymbol{A}\boldsymbol{B}^{\mathrm{T}}) \tag{1.7}$$

and two bars $|\;|$ denote the norm associated with the inner product of either $\mathscr{V}$ or Lin, so that $\mathscr{S}(1)$, the unit sphere of $\mathscr{V}$, is defined by

$$\mathscr{S}(1) := \{\boldsymbol{v} \in \mathscr{V} \mid \boldsymbol{v} \cdot \boldsymbol{v} = |\boldsymbol{v}|^2 = 1\}. \tag{1.8}$$

For $\boldsymbol{w} \in \mathscr{S}(1)$ and $\boldsymbol{W}$ the element of Skw associated to $\boldsymbol{w}$ by (1.5), we have

$$-\boldsymbol{W}^2 = \boldsymbol{I} - \boldsymbol{P}(\boldsymbol{w}), \tag{1.9}$$

with

$$\boldsymbol{P}(\boldsymbol{w}) := \boldsymbol{w} \otimes \boldsymbol{w} \tag{1.10}$$

the orthogonal projector on the axis of $\boldsymbol{w}$. We record here for future use the following easy result.

1.1. Proposition. *Let* $\boldsymbol{A} \in \mathrm{Lin}$ *and* $\boldsymbol{w} \in \mathscr{S}(1)$ *be such that* $\boldsymbol{A}\boldsymbol{w} = \boldsymbol{0}$. *Then, for* $\boldsymbol{W}$ *the skew tensor associated with* $\boldsymbol{w}$, *we have*

$$\boldsymbol{A} = \boldsymbol{B}\boldsymbol{W}, \quad \text{with } \boldsymbol{B} = -\boldsymbol{A}\boldsymbol{W} \tag{1.11}$$

(*and conversely*).

Proof. It suffices to note that the following chain of implications holds:

$$\boldsymbol{A}\boldsymbol{w} = \boldsymbol{0} \;\Rightarrow\; \boldsymbol{A} = \boldsymbol{A}(\boldsymbol{I} - \boldsymbol{w} \otimes \boldsymbol{w}) \;\Rightarrow\; \boldsymbol{A} = -\boldsymbol{A}\boldsymbol{W}^2. \;\square \tag{1.12}$$

1.2 Cofactor and Determinant

Let $\boldsymbol{k}$ be arbitrarily chosen in $\mathscr{V}$, and let $\boldsymbol{K}$ be the element of Skw associated to $\boldsymbol{k}$ by (1.5). The *cofactor* $\boldsymbol{A}^*$ of $\boldsymbol{A}$ is the unique element of Lin such that $\boldsymbol{A}^*\boldsymbol{k}$ is the vector associated to $\boldsymbol{A}\boldsymbol{K}\boldsymbol{A}^{\mathrm{T}}$ by (1.5). It follows from this definition that, for $\{\boldsymbol{i}, \boldsymbol{j}, \boldsymbol{k}\}$ an orthonormal basis for $\mathscr{V}$,

$$\boldsymbol{A}^*\boldsymbol{k} = \boldsymbol{A}\boldsymbol{i} \times \boldsymbol{A}\boldsymbol{j}. \tag{1.13}$$

A related definition of the *determinant* of $\boldsymbol{A}$ is

$$\det \boldsymbol{A} := \boldsymbol{A}\boldsymbol{i} \times \boldsymbol{A}\boldsymbol{j} \cdot \boldsymbol{A}\boldsymbol{k}. \tag{1.14}$$

We now list some straightforward consequences of the first definition alone (*vid.* also [19] and Section 1.1 of [20]).

1.2. Proposition. *Let* $\alpha \in \mathbb{R}$; $\boldsymbol{a}, \boldsymbol{b} \in \mathscr{V}$; $\boldsymbol{A} \in \mathrm{Lin}$; $\boldsymbol{W} \in \mathrm{Skw}$ *be arbitrarily chosen, Then,*

$$(\alpha \boldsymbol{A})^* = \alpha^2 \boldsymbol{A}^* \; (\textit{in particular}, \; (-\boldsymbol{A})^* = \boldsymbol{A}^*); \tag{1.15}$$

$$\boldsymbol{I}^* = \boldsymbol{I}; \quad (\boldsymbol{a} \otimes \boldsymbol{b})^* = \boldsymbol{0}; \quad (\boldsymbol{I} - \boldsymbol{a} \otimes \boldsymbol{a})^* = \boldsymbol{a} \otimes \boldsymbol{a}; \tag{1.16}$$

$$\boldsymbol{W}^* = \boldsymbol{w} \otimes \boldsymbol{w}. \tag{1.17}$$

Less elementary properties of the cofactor mapping follow by using also the definition of determinant.

1.3. Proposition. *Let* $\boldsymbol{A}, \boldsymbol{B}$ *be arbitrarily chosen in* Lin. *Then*

$$\boldsymbol{A}^*\boldsymbol{A}^{\mathrm{T}} = \boldsymbol{A}^{\mathrm{T}}\boldsymbol{A}^* = (\det \boldsymbol{A})\, \boldsymbol{I}; \tag{1.18}$$

$$\det(\boldsymbol{A} + \boldsymbol{B}) = \det \boldsymbol{A} + \boldsymbol{A}^* \cdot \boldsymbol{B} + \boldsymbol{A} \cdot \boldsymbol{B}^* + \det \boldsymbol{B}. \tag{1.19}$$

Note that it follows from (1.17) and (1.18) that

$$\det \boldsymbol{W} = 0 \quad \text{for all } \boldsymbol{W} \in \mathrm{Skw}. \tag{1.20}$$

It also follows from (1.18) that

$$A^* \cdot A = 3 \det A \quad \text{for all } A \in \text{Lin}, \tag{1.21}$$

and, moreover, that

$$F^* = (\det F)\, F^{-\mathrm{T}} \quad \text{for all } F \in \text{Lin}^+, \tag{1.22}$$

so that, in particular,

$$R^* = R \quad \text{for all } R \in \text{Rot}. \tag{1.23}$$

Other properties of the cofactor result from (1.19) and some well known properties of the determinant, namely,

$$\det A = \det A^{\mathrm{T}} \quad \text{for all } A \in \text{Lin} \tag{1.24}$$

and

$$\det (AB) = (\det A)(\det B) \quad \text{for all } A, B \in \text{Lin}.^7 \tag{1.25}$$

We list these properties in the next proposition.

1.4. Proposition. *Let* A, B *be arbitrarily chosen in* Lin. *Then,*

$$(AB)^* = A^*B^*; \tag{1.26}$$

$$(A^{\mathrm{T}})^* = (A^*)^{\mathrm{T}}. \tag{1.27}$$

With (1.26), $(1.16)_3$ yields a result that will be of use later, when we deal with pressure interactions: for all $A \in \text{Lin}$ and $v \in \mathscr{V}$,

$$(A(I - v \otimes v))^* = A^*v \otimes v. \tag{1.28}$$

We now prove two formulae for the derivatives of the cofactor and the determinant mappings.

1.5. Proposition. *For all* $A, B \in \text{Lin}$,

$$\partial_A \det A = A^*; \tag{1.29}$$

in addition, if $A \in \text{Lin}^+$,

$$(\det A)\, \partial_A A^*[B] = (A^* \cdot B)\, A^* - A^*B^{\mathrm{T}}A^*.^8 \tag{1.30}$$

Proof. A straightforward consequence of definition (1.14) is

$$\det (\alpha A) = \alpha^3 \det A \quad \text{for all } \alpha \in \mathbb{R} \text{ and } A \in \text{Lin}. \tag{1.31}$$

---

[7] These properties of the determinant are best proved by first restricting attention to invertible tensors and then appealing to a continuity argument to establish the desired result for all of Lin.

[8] Notice that, for all $A \in \text{Lin}$, $(\det A)\, \partial_A A^*$ is a *symmetric* mapping from Lin into itself, in the sense that, for all $B, C \in \text{Lin}$,

$$C \cdot (\det A)\, \partial_A A^*[B] = B \cdot (\det A)\, \partial_A A^*\, [C]. \tag{\#}$$

Of course, the symmetry of $(\det A)\, \partial_A A^*$ is a consequence of the fact that, due to (1.29), $\partial_A A^*$ is the second derivative of the determinant mapping.

With this and (1.15) we obtain from (1.19) that

$$\det(A + \alpha B) = \det A + \alpha A^* \cdot B + \alpha^2 A \cdot B^* + \alpha^3 \det B; \qquad (1.32)$$

differentiation of the function

$$\alpha \mapsto \det(A + \alpha B)$$

yields (1.29). Next, we differentiate $(1.18)_1$ and, using (1.29), get

$$(\partial_A A^*[B])\, A^{\mathrm{T}} + A^* B^{\mathrm{T}} = (A^* \cdot B)\, I;\ ^9 \qquad (1.33)$$

multiplying (1.33) on the right by A^* and recalling $(1.18)_2$ we have (1.30). □

We close this section by establishing a formula for the cofactor of the sum of two tensors.

1.6. Proposition. *For all* $A, B \in \mathrm{Lin}$,

$$(A + B)^* = A^* + B^* + \partial_A A^*[B], \quad \textit{with} \quad \partial_A A^*[B] = \partial_B B^*[A]. \qquad (1.34)$$

Proof. Having (1.29), for all $A, B \in \mathrm{Lin}$ we obtain

$$\partial_A(\det(A + B)) = \partial_B(\det(A + B)) = (A + B)^*,$$

so that, on recalling (1.19), we arrive at (1.34). □

2. Analysis

Let $\mathscr{E}$ be a three-dimensional euclidean point space, whose associated translation space we shall identify with the vector space $\mathscr{V}$ introduced in Section 1.1.[10]

Throughout this paper, Ω is a domain, *i.e.*, an open bounded set of $\mathscr{E}$. We shall interpret Ω as the region occupied by a continuous body, and study surface interactions between that body and its environment. Therefore, it is important for us to describe precisely what type of boundary $\partial\Omega$ we shall assign to Ω. Although many of our developments do hold (oftentimes in an obvious fashion) under more general hypotheses, for simplicity and concreteness we shall assume hereafter that one of the following two situations occurs:

(i) $\partial\Omega$ consists of a single surface of class C^3, with no self-intersections;

(ii) $\partial\Omega = \Sigma_1 \cup \Sigma_2 \cup \Gamma$, where both Σ_1 and Σ_2 are C^3-surfaces with no self-intersections, such that $\Sigma_1 \cap \Sigma_2 = \emptyset$ and that their common boundary Γ is a closed simple curve of class C^3; on Γ, the vector fields

$$\Gamma \ni x \mapsto n_\alpha(x) \in \mathscr{V}, \quad \text{with } n_\alpha(x) \in \mathscr{S}(1) \quad \text{for } \alpha = 1, 2, \qquad (2.1)$$

---

[9] Notice that we write $\partial_A A^*[B]$ for the derivative of the cofactor mapping evaluated at A, a linear transformation of Lin into itself, acting on B.

[10] On occasion, having chosen an origin $o \in \mathscr{E}$ once and for all, we shall also tacitly identify a point $x \in \mathscr{E}$ with the corresponding position vector $(x - o) \in \mathscr{V}$.

are well-defined as the limits of the outward unit normal to Σ_α on approaching $\boldsymbol{x} \in \Gamma$ from Σ_α, and it is assumed that

$$\boldsymbol{n}_1(\boldsymbol{x}) \neq \boldsymbol{n}_2(\boldsymbol{x}) \quad \text{for all } \boldsymbol{x} \in \Gamma. \tag{2.2}$$

In case (ii) we shall call $\Sigma_1 \cup \Sigma_2$ and Γ the *regular* and *singular* parts of the boundary of Ω, respectively; in case (i) we shall consistenly say that Ω has a regular boundary, or else that the domain Ω is regular. If at a point $\boldsymbol{x} \in \Gamma$

$$\boldsymbol{n}_1(\boldsymbol{x}) \neq -\boldsymbol{n}_2(\boldsymbol{x}), \tag{2.3}$$

then the unit tangent vector $\boldsymbol{t}(\mathrm{x})$ at $\boldsymbol{x}$ can be chosen so as to coincide with

$$\boldsymbol{t}(\boldsymbol{x}) = \frac{\boldsymbol{n}_1(\boldsymbol{x}) \times \boldsymbol{n}_2(\boldsymbol{x})}{|\boldsymbol{n}_1(\boldsymbol{x}) \times \boldsymbol{n}_2(\boldsymbol{x})|}. \tag{2.4}$$

When condition (2.3) applies throughout Γ (so that there are no cuspidal situations anywhere), we let the tangent field on Γ be defined by (2.4) and choose the corresponding positive path orientation.

2.1. Remark. In the following, when we write $\partial\Omega$ it will be always clear from the context whether we refer exclusively to a regular region or rather to a region with either one of the two types of boundary above; when we write $\Sigma_1 \cup \Sigma_2$, we do so to stress exclusion of the singular part Γ of $\partial\Omega$, but the statement just made will also apply as a rule, perhaps *modulo* some trivial changes, to cases when Ω is regular. □

2.1 Differential Operators on a Surface

When the situation *sub* (i) above occurs, it is well known that any scalar or vector field of class C^2 over Ω can be extended to an equally smooth field defined over an open set containing $\bar{\Omega}$. Likewise, in the situation *sub* (ii), for $\alpha = 1$ or 2 one can regard Σ_α as a part of the complete boundary of a regular domain Ω_α, and extend any field defined over Σ_α to a corresponding field over a neighborhood of Σ_α. Consequently, the definitions of intrinsic differential operators on the regular part of $\partial\Omega$, such as the surface gradient and the surface divergence, can be obtained by restricting to Σ_α the corresponding differential constructs relative to appropriately extended fields defined over an open set containing Σ_α.

For all points of Ω where the outward unit normal $\boldsymbol{n}$ is uniquely defined, let $\boldsymbol{N}$, $\boldsymbol{P}(\boldsymbol{n})$ be the skew tensor and the orthogonal projector associated to $\boldsymbol{n}$ by (1.5) and (1.10), respectively (note that

$$\boldsymbol{N}\boldsymbol{P}(\boldsymbol{n}) = \boldsymbol{P}(\boldsymbol{n})\,\boldsymbol{N} = \boldsymbol{0}). \tag{2.5}$$

The *surface gradient* ${}^s\nabla \boldsymbol{v}$ of a vector field $\boldsymbol{v}$ over $\partial\Omega$ is the tensor field defined over $\partial\Omega$ as

$${}^s\nabla \boldsymbol{v} := (\nabla \boldsymbol{v})\,(\boldsymbol{I} - \boldsymbol{P}(\boldsymbol{n})) = -(\nabla \boldsymbol{v})\,\boldsymbol{N}^2; \tag{2.6}$$

the *surface divergence* is

$$ {}^{\mathrm{s}}\mathrm{Div}\, \boldsymbol{v} := \operatorname{tr}({}^{\mathrm{s}}\nabla \boldsymbol{v}) = (\nabla \boldsymbol{v}) \cdot (\boldsymbol{I} - \boldsymbol{P}(\boldsymbol{n})) = -(\nabla \boldsymbol{v}) \cdot \boldsymbol{N}^2 . \tag{2.7} $$

Moreover, for consistency with definition (2.6), the *normal gradient* of $\boldsymbol{v}$ is the tensor field ${}^{\mathrm{n}}\nabla \boldsymbol{v}$ over $\partial\Omega$ defined by

$$ {}^{\mathrm{n}}\nabla \boldsymbol{v} := \nabla \boldsymbol{v} - {}^{\mathrm{s}}\nabla \boldsymbol{v} = (\nabla \boldsymbol{v})\, \boldsymbol{P}(\boldsymbol{n}) = ({}^{\mathrm{n}}\partial \boldsymbol{v}) \otimes \boldsymbol{n}, \tag{2.8} $$

where

$$ {}^{\mathrm{n}}\partial \boldsymbol{v} := (\nabla \boldsymbol{v})\, \boldsymbol{n} \tag{2.9} $$

is the directional derivative of $\boldsymbol{v}$ with respect to the normal $\boldsymbol{n}$.

2.2. Remark. In anticipation we remark here that, when we consider the gradient and divergence of fields defined over the image $\boldsymbol{f}(\Omega)$ of Ω (or $\boldsymbol{f}(\partial\Omega)$ of $\partial\Omega$) under a deformation $\boldsymbol{f}$ of $\bar{\Omega}$, we shall denote the gradient by grad (or ${}^{\mathrm{s}}$grad) and the divergence by div (or ${}^{\mathrm{s}}$div). □

2.2 Some Properties of Curl

The value at $\boldsymbol{x} \in \Omega$ of the *curl* of a vector field $\boldsymbol{v}$ over Ω is defined to be the vector associated to the skew tensor field $(\nabla \boldsymbol{v} - (\nabla \boldsymbol{v})^{\mathrm{T}})$ evaluated at $\boldsymbol{x}$:

$$ (\mathrm{Curl}\, \boldsymbol{v}(\boldsymbol{x})) \times \boldsymbol{u} := [\nabla \boldsymbol{v}(\boldsymbol{x}) - (\nabla \boldsymbol{v}(\boldsymbol{x}))^{\mathrm{T}}]\, \boldsymbol{u} \quad \text{for all } \boldsymbol{u} \in \mathscr{V}. \tag{2.10} $$

2.3. Proposition. *Let* $\boldsymbol{v}$ *be a smooth vector field over* Ω. *Then,*

(i) *if* $\boldsymbol{V}$ *denotes the skew tensor field associated with* $\boldsymbol{v}$,

$$ \mathrm{Curl}\, \boldsymbol{v} = -\mathrm{Div}\, \boldsymbol{V}; \tag{2.11} $$

(ii) *if* $\boldsymbol{v}(\boldsymbol{x}) \in \mathscr{S}(1)$ *for* $\boldsymbol{x} \in \Omega$,

$$ \boldsymbol{v} \times \mathrm{Curl}\, \boldsymbol{v} = -(\nabla \boldsymbol{v})\, \boldsymbol{v}. \tag{2.12} $$

We omit the proof of this proposition, which amounts to straightforward computations. The proof of the next proposition is less trivial.

2.4. Proposition. *Let* $\boldsymbol{w}$, $\boldsymbol{z}$ *be two smooth vector fields over* Ω, *with* $|\boldsymbol{w}| \equiv 1$, *and let* $\boldsymbol{W}$ *be the skew tensor field associated with* $\boldsymbol{w}$. *Then*

$$ \boldsymbol{W} \cdot \nabla(\boldsymbol{W}\boldsymbol{z}) = -\boldsymbol{W}^2 \cdot \nabla \boldsymbol{z} - (\mathrm{Div}\, \boldsymbol{w})\, (\boldsymbol{w} \cdot \boldsymbol{z}). \tag{2.13} $$

Proof. Firstly, apply the identity

$$ \boldsymbol{A} \cdot \nabla(\boldsymbol{A}\boldsymbol{a}) = (\boldsymbol{A}^{\mathrm{T}}\boldsymbol{A}) \cdot \nabla \boldsymbol{a} + (\mathrm{Div}\,(\boldsymbol{A}^{\mathrm{T}}\boldsymbol{A}) - \boldsymbol{A}^{\mathrm{T}}\, \mathrm{Div}\, \boldsymbol{A}) \cdot \boldsymbol{a}, \tag{2.14} $$

which holds for all sufficiently smooth tensor fields $\boldsymbol{A}$ and vector fields $\boldsymbol{a}$, to the skew tensor field $\boldsymbol{W}$ and the vector field $\boldsymbol{z}$, and obtain

$$ \boldsymbol{W} \cdot \nabla(\boldsymbol{W}\boldsymbol{z}) = -\boldsymbol{W}^2 \cdot \nabla \boldsymbol{z} - (\mathrm{Div}\, \boldsymbol{W}^2 - \boldsymbol{W}\, \mathrm{Div}\, \boldsymbol{W}) \cdot \boldsymbol{z}. \tag{2.15} $$

Secondly, observe that (1.9), (1.10), (2.11) and (2.12) imply that

$$\begin{aligned}\operatorname{Div} \boldsymbol{W}^2 - \boldsymbol{W} \operatorname{Div} \boldsymbol{W} &= \operatorname{Div} (\boldsymbol{w} \otimes \boldsymbol{w}) + \boldsymbol{W} \operatorname{Curl} \boldsymbol{w} \\ &= (\operatorname{Div} \boldsymbol{w})\, \boldsymbol{w} + (\nabla \boldsymbol{w})\, \boldsymbol{w} + \boldsymbol{w} \times \operatorname{Curl} \boldsymbol{w} = (\operatorname{Div} \boldsymbol{w})\, \boldsymbol{w}. \qquad (2.16)\end{aligned}$$

(2.15) and (2.16) together yield the desired result. □

2.3 The Surface Divergence Theorem

Let Σ be a smooth oriented simple surface, with boundary $\partial\Sigma$ a smooth, consistently oriented, closed simple curve: in our present context, typically, $\Sigma \equiv \Sigma_\alpha$ and $\partial\Sigma \equiv \Gamma$. As is well known, for $\boldsymbol{v}$ a vector field over a neighborhood of $\partial\Sigma$, *Stokes' formula* holds, namely,

$$\int_\Sigma (\operatorname{Curl} \boldsymbol{v}) \cdot \boldsymbol{n} = \int_{\partial\Sigma} \boldsymbol{v} \cdot \boldsymbol{t}, \qquad (2.17)$$

where of course $\boldsymbol{n}$ is the normal to Σ, and $\boldsymbol{t}$ is the tangent to $\partial\Sigma$.

2.5. Remark. Only tangential derivatives of $\boldsymbol{v}$ evaluated on Σ occur in the left-hand side of Stokes' formula. Indeed, on the one hand by (1.5) and the definition (2.10),

$$2\,(\operatorname{Curl} \boldsymbol{v}) \cdot \boldsymbol{n} = (\nabla \boldsymbol{v} - (\nabla \boldsymbol{v})^T) \cdot \boldsymbol{N} = 2 \nabla \boldsymbol{v} \cdot \boldsymbol{N}; \qquad (2.18)$$

on the other hand, (2.5) and the definition (2.6) yield

$$\nabla \boldsymbol{v} \cdot \boldsymbol{N} = (\nabla \boldsymbol{v}(\boldsymbol{I} - \boldsymbol{P}(\boldsymbol{n})) \cdot \boldsymbol{N} = {}^s\nabla \boldsymbol{v} \cdot \boldsymbol{N}; \qquad (2.19)$$

in conclusion,

$$(\operatorname{Curl} \boldsymbol{v}) \cdot \boldsymbol{n} = {}^s\nabla \boldsymbol{v} \cdot \boldsymbol{N}. \;\square \qquad (2.20)$$

Our next proposition, the *surface divergence theorem*, has manifold applications in the following, *e.g.*, when we deal with membrane loading.

2.6. Proposition. *Let Σ be a smooth oriented simple surface of normal $\boldsymbol{n}$, having a smooth oriented closed simple curve of tangent $\boldsymbol{t}$ as its boundary $\partial\Sigma$. Furthermore, let $\boldsymbol{u}$ be a vector field of class* C^1, *defined and* tangential *over* Σ:

$$\boldsymbol{u} \cdot \boldsymbol{n} \equiv 0 \quad \text{on } \Sigma. \qquad (2.21)$$

Then

$$\int_\Sigma {}^s\operatorname{Div} \boldsymbol{u} = \int_{\partial\Sigma} \boldsymbol{u} \cdot \boldsymbol{t} \times \boldsymbol{n}. \qquad (2.22)$$

Proof. Let $\boldsymbol{u}$ and $\boldsymbol{n}$ be interpreted as smooth extensions to a neighborhood of Σ of the corresponding vector fields $\boldsymbol{u}$ and $\boldsymbol{n}$ defined over Σ. With a view towards applying Stokes' formula to the vector field

$$\boldsymbol{v} = \boldsymbol{N}\boldsymbol{u}, \qquad (2.23)$$

we observe that (2.20), (2.5), (2.13), (2.21) and (2.7) imply that

$$\begin{aligned} \operatorname{Curl}(\boldsymbol{N}\boldsymbol{u}) \cdot \boldsymbol{n} &= {}^{s}\nabla(\boldsymbol{N}\boldsymbol{u}) \cdot \boldsymbol{N} = \nabla(\boldsymbol{N}\boldsymbol{u}) \cdot \boldsymbol{N} \\ &= -\boldsymbol{N}^2 \cdot \nabla \boldsymbol{u} - (\operatorname{Div} \boldsymbol{n})(\boldsymbol{n} \cdot \boldsymbol{u}) = {}^{s}\operatorname{Div} \boldsymbol{u}. \end{aligned} \tag{2.24}$$

But,

$$\boldsymbol{v} \cdot \boldsymbol{t} = \boldsymbol{N}\boldsymbol{u} \cdot \boldsymbol{t} = \boldsymbol{u} \cdot \boldsymbol{t} \times \boldsymbol{n}, \tag{2.25}$$

so that Stokes' formula directly yields the desired result. □

Let Ω be a domain as described under (i) in the beginning of this section. Firstly, for α and $\boldsymbol{v}$ two sufficiently smooth scalar and vector fields over $\partial\Omega$, we have from (2.7)

$${}^{s}\operatorname{Div}(\alpha \boldsymbol{v}) = \alpha\, {}^{s}\operatorname{Div} \boldsymbol{v} + ((\boldsymbol{I} - \boldsymbol{P}(\boldsymbol{n}))\, \boldsymbol{v}) \otimes \nabla\alpha.$$

Then, choosing $\boldsymbol{v} \equiv \boldsymbol{n}$, we see that

$${}^{s}\operatorname{Div}(\alpha \boldsymbol{n}) = \alpha\, {}^{s}\operatorname{Div} \boldsymbol{n}. \tag{2.26}$$

Writing now

$$\boldsymbol{v} = \boldsymbol{P}(\boldsymbol{n})\, \boldsymbol{v} + (\boldsymbol{I} - \boldsymbol{P}(\boldsymbol{n}))\, \boldsymbol{v},$$

we deduce from (2.22) and (2.26) that

$$\int_{\partial\Omega} {}^{s}\operatorname{Div} \boldsymbol{v} = \int_{\partial\Omega} ({}^{s}\operatorname{Div} \boldsymbol{n})\, \boldsymbol{n} \cdot \boldsymbol{v}. \tag{2.27}$$

Secondly, a tensor field $\boldsymbol{A}$ over $\partial\Omega$ is *tangential* if

$$\boldsymbol{A}\boldsymbol{n} = \boldsymbol{0} \quad \text{on } \partial\Omega; \tag{2.28}$$

using (2.7), we define the *surface divergence* of such a field as

$${}^{s}\operatorname{Div} \boldsymbol{A} \cdot \boldsymbol{v} := {}^{s}\operatorname{Div}(\boldsymbol{A}^{\mathrm{T}}\boldsymbol{v}) - \boldsymbol{A} \cdot {}^{s}\nabla \boldsymbol{v}, \tag{2.29}$$

for each sufficiently smooth vector field $\boldsymbol{v}$ over $\partial\Omega$. It follows from (2.28) that the vector field $\boldsymbol{u} := \boldsymbol{A}^{\mathrm{T}}\boldsymbol{v}$ is tangential on $\partial\Omega$ for each $\boldsymbol{v} \in \mathscr{V}$. Thus, for each sufficiently smooth vector field $\boldsymbol{v}$ over $\partial\Omega$, the surface divergence theorem implies that

$$\int_{\partial\Omega} {}^{s}\operatorname{Div} \boldsymbol{A} \cdot \boldsymbol{v} = -\int_{\partial\Omega} \boldsymbol{A} \cdot {}^{s}\nabla \boldsymbol{v}. \tag{2.30}$$

Finally, let $\partial\Omega$ consist of two regular pieces Σ_1, Σ_2 with a common boundary Γ. Then (2.30) must be replaced by

$$\int_{\Sigma_1 \cup \Sigma_2} {}^{s}\operatorname{Div} \boldsymbol{A} \cdot \boldsymbol{v} = -\int_{\Sigma_1 \cup \Sigma_2} \boldsymbol{A} \cdot {}^{s}\nabla \boldsymbol{v} + \int_{\Gamma} [\boldsymbol{A}_1(\boldsymbol{t} \times \boldsymbol{n}_1) - \boldsymbol{A}_2(\boldsymbol{t} \times \boldsymbol{n}_2)] \cdot \boldsymbol{v}, \tag{2.31}$$

where $\boldsymbol{A}_\alpha$ ($\alpha = 1, 2$) denotes the limit value of the tensor field $\boldsymbol{A}$ on approaching Γ from Σ_α.

2.4 The Weingarten Tensor

For Σ a smooth oriented simple surface of normal $\boldsymbol{n}$, the *Weingarten tensor* field $\boldsymbol{L}$ over Σ is defined to be the negative of the surface gradient of the normal field:

$$\boldsymbol{L} := -{}^{s}\nabla\boldsymbol{n}; \tag{2.32}$$

the trace of $\boldsymbol{L}$ is proportional to the *mean curvature* $\varkappa$ of Σ:

$$\varkappa := \tfrac{1}{2}\,{}^{s}\mathrm{Div}\,\boldsymbol{n}. \tag{2.33}$$

We now prove that the Weingarten tensor field has symmetric values, a classical result in the differential geometry of surfaces obtained here, in a fashion different from usual, as an easy consequence of Stokes' formula and the surface divergence theorem.

2.7. Proposition. *Let Σ be a smooth oriented simple surface of normal $\boldsymbol{n}$, and let Σ be regarded as part of the complete boundary $\partial\Omega$ of a regular domain Ω. Then*

$${}^{s}\nabla\boldsymbol{n}(\boldsymbol{x}) \in \mathrm{Sym} \quad \textit{for all } \boldsymbol{x} \in \Sigma. \tag{2.34}$$

Proof. In place of (2.34), we shall prove the equivalent statement that, for every skew tensor field $\boldsymbol{W}$ of class C^1 over $\partial\Omega$

$$\int_{\partial\Omega} {}^{s}\nabla\boldsymbol{n} \cdot \boldsymbol{W} = 0. \tag{2.35}$$

To this end, we consider two smooth vector fields $\boldsymbol{w}$ and $\boldsymbol{z}$ over Σ and the associated skew tensor fields $\boldsymbol{W}$ and $\boldsymbol{Z}$, and observe that the differential identity

$$\nabla(\boldsymbol{W}\boldsymbol{z}) = \boldsymbol{W}\,\nabla\boldsymbol{z} - \boldsymbol{Z}\,\nabla\boldsymbol{w}$$

yields

$${}^{s}\mathrm{Div}\,(\boldsymbol{W}\boldsymbol{z}) = \boldsymbol{W} \cdot {}^{s}\nabla\boldsymbol{z} - \boldsymbol{Z} \cdot {}^{s}\nabla\boldsymbol{w}. \tag{2.36}$$

Choosing $\boldsymbol{z} = \boldsymbol{n}$ in (2.36) and integrating over $\partial\Omega$ we obtain

$$\int_{\partial\Omega} {}^{s}\nabla\boldsymbol{n} \cdot \boldsymbol{W} = \int_{\partial\Omega} {}^{s}\nabla\boldsymbol{w} \cdot \boldsymbol{N} + \int_{\partial\Omega} {}^{s}\mathrm{Div}\,(\boldsymbol{W}\boldsymbol{n}). \tag{2.37}$$

On the right-hand side of (2.37), the first integral is null by Stokes' formula and (2.20); the second one, as the vector field $\boldsymbol{W}\boldsymbol{n}$ is tangential, by the surface divergence theorem. □

3. Mechanics

Let a regular domain $\Omega \subset \mathscr{E}$ as described in the opening of Section 2 be pointwise identified with a continuous body. Classically, a deformation of the body is the assignment of a *displacement*

$$\boldsymbol{f}: \Omega \to \mathscr{E}, \quad \boldsymbol{y} = \boldsymbol{f}(\boldsymbol{x}), \tag{3.1}$$

from the *reference placement* Ω into the *current placement* $f(\Omega)$, with f smooth (say, of class $C^2(\Omega) \cap C^1(\bar{\Omega})$), injective and locally orientation-preserving in the sense that

$$\det \nabla f(x) > 0 \quad \text{a.e. in } \Omega. \tag{3.2}$$

3.1. Remark. This notion of displacement fits our present needs for a quick, explicit account of the mechanical aspects of elastic equilibrium problems; of course, neither a general existence theory based on such a notion is available for such problems, nor do we propose it as a goal. □

Given a displacement f, we denote by

$$F := \nabla f \quad \text{and} \quad F^* := (\nabla f)^* \tag{3.3}$$

the displacement gradient and its cofactor (*cf.* (1.22)). The ratios of the current to the reference volume and surface measures are expressed in terms of F and F^* by

$$\det F = \frac{\mathrm{d}\,(\mathrm{vol})}{\mathrm{d}\,(\mathrm{Vol})} \quad \text{and} \quad |F^* n| = \frac{\mathrm{d}\,(\mathrm{srf})}{\mathrm{d}\,(\mathrm{Srf})}, \tag{3.4}$$

respectively, $(3.4)_2$ being a consequence of the formula of Nanson for the unit normal m to a point of the boundary $\partial f(\Omega)$[11] of the current placement:

$$m \,\mathrm{d}\,(\mathrm{srf}) = F^* n \,\mathrm{d}\,(\mathrm{Srf}). \tag{3.5}$$

Finally, at a regular point $x \in \partial\Omega$, we shall denote by ${}^sF$ and ${}^nF$ the superficial and normal parts of the displacement gradient F, with ${}^sF$ and ${}^nF$ defined by (2.6) and (2.8), respectively.

For a given displacement f, the *Piola-Kirchhoff* stress field S over Ω and the corresponding *Cauchy* stress field T over $f(\Omega)$ are related by

$$S(x) = T(y)\, F^*(x), \quad \text{with } y = f(x). \tag{3.6}$$

In terms of these stress measures the equilibrium equations are written as

$$\operatorname{Div} S = 0 \quad \text{in } \Omega \quad \text{and} \quad \operatorname{div} T = 0 \quad \text{in } f(\Omega), \tag{3.7}$$

respectively. The traction boundary condition is

$$S n = s \quad \text{on } \partial\Omega \tag{3.8}$$

in terms of the Piola-Kirchhoff stress.

3.2. Remark. In (3.8), we have left unspecified the functional dependence, to be chosen later, of the surface load s on the displacement f; consistently, we refrain from writing the corresponding formula in terms of the Cauchy stress, which would be, rather than generic, vague at this stage. Suffice it to record here the aspect of such a formula in the easy case of dead loading, *i.e.*, when

$$\partial\Omega \ni x \mapsto \hat{s}(x) \in \mathcal{V}, \tag{3.9}$$

[11] Note that, under the present hypotheses, $\partial f(\Omega) = f(\partial\Omega)$.

and we have

$$Tm = |F^*n|^{-1}\, s, \quad \text{with } s = \hat{s}(f^{-1}(y)), \quad \text{on } f(\Sigma_1) \cup f(\Sigma_2), \tag{3.10}$$

where use has been made of (3.5). □

Like s, the Piola-Kirchhoff stress S at a point $x \in \Omega$ will be soon the subject of a constitutive prescription, whose general aspect is

$$\hat{S}: \Omega \times \mathrm{Lin}^+ \to \mathrm{Lin}, \quad S = \hat{S}(x, A), \tag{3.11}$$

with

$$\hat{S}(x, A)\, A^{\mathrm{T}} \in \mathrm{Sym} \quad \text{identically in } \Omega \times \mathrm{Lin}^+. \tag{3.12}$$

Part II. Surface Interactions

4. Variational Equilibrium Problems

A stress field S over $\overline{\Omega}$ and a surface load field s over $\partial\Omega$ are in equilibrium when they obey $(3.7)_1$ and (3.8). Formally, those two equations are equivalent to the single condition

$$\int_\Omega S \cdot \nabla h - \int_{\partial\Omega} s \cdot h = 0 \quad \text{for all test functions } h. \tag{4.1}$$

The question is now to choose constitutive prescriptions for both S and s such as to yield $(3.7)_1$–(3.8) and (4.1) as the strong and weak forms, respectively, of the Euler-Lagrange equation associated to a suitable total energy functional.

We assume that such total energy functional E or, as we also say, the *total potential*, has two parts: a *body potential* S of density σ and a *surface interaction potential* T of density τ:

$$f \mapsto E\{f\} = S\{f\} + T\{f\}, \tag{4.2$_1$}$$

with

$$S = \int_\Omega \sigma \,\mathrm{d}\,(\mathrm{Vol}) \quad \text{and} \quad T = \int_{\partial\Omega} \tau \,\mathrm{d}\,(\mathrm{Srf}). \tag{4.2$_{2,3}$}$$

When the body undergoes a displacement f from the reference placement Ω, σ is interpreted as the elastic energy stored per unit volume at a point of Ω; τ is interpreted as the elastic energy stored per unit area at a point of $\partial\Omega$.

As a domain for the total potential (the collection of all admissible displacements from Ω), we select

$$\mathscr{D} := \{f \in C^2(\Omega \cup \Sigma_1 \cup \Sigma_2) \cap C^0(\overline{\Omega}) \mid f \text{ injective \& locally orientation-preserving; } \nabla f \text{ bounded in } \Omega \text{ \& } \lim_{x \to x_0} \nabla f(x) = \nabla_\alpha f(x_0) \text{ for } x \to x_0 \in \Gamma \text{ from } \Sigma_\alpha\, (\alpha = 1, 2)\}. \tag{4.3}$$

We remark that the classical smoothness requirement on f, namely, $f \in C^2(\Omega) \cap C^1(\overline{\Omega})$, which guarantees the possibility of writing in its strong form the Euler-Lagrange equation associated to E in the interior part, is here adapted because in some applications (*e.g.*, membrane interactions) the surface load $\boldsymbol{s}$ depends on certain second derivatives of $\boldsymbol{f}$ on the regular part of the boundary, specifically on the tangential derivatives of $\nabla \boldsymbol{f}$, while on the singular part of the boundary equilibrium may require that $\boldsymbol{f}$ not be continuously differentiable.

Obviously, the domain $\mathscr{D}$ is neither a flat space nor a convex set. However, not only does $\mathscr{D}$ agree with the classical notion of admissible displacement recalled in the beginning of this section, but also, and crucially for our further developments, it allows for a calculus of variations in a sense made precise by the following proposition.

4.1. Proposition. *For $\mathscr{D}$ defined as in* (4.3), *let* $\mathscr{H} := C^2(\bar{\Omega})$ *be the associated space of admissible variations. Then, for each $\boldsymbol{f} \in \mathscr{D}$ and $\boldsymbol{h} \in \mathscr{H}$, there is a strictly positive number ε_0, depending in general on Ω, $\boldsymbol{f}$ and $\boldsymbol{h}$, such that*

$$(\boldsymbol{f} + \varepsilon \boldsymbol{h}) \in \mathscr{D} \quad \text{for all } \varepsilon \in]-\varepsilon_0, \varepsilon_0[. \tag{4.4}$$

We omit the proof, which amounts to a straightforward adaptation of part (b) in the proof of Theorem 5.5.1 on p. 223 of [20].

We now lay down the following

4.2. Definition. *A continuous body has a* conservative interaction *with its environment (with respect to a reference placement Ω) if the field pairs $(\boldsymbol{S}, \boldsymbol{s})$ and (σ, τ) satisfy the variational condition*

$$\delta S\{\boldsymbol{f}\}[\boldsymbol{h}] = \int_\Omega \boldsymbol{S} \cdot \nabla \boldsymbol{h}, \quad \delta T\{\boldsymbol{f}\}[\boldsymbol{h}] = -\int_{\partial\Omega} \boldsymbol{s} \cdot \boldsymbol{h} \quad \text{for all } \boldsymbol{f} \in \mathscr{D} \text{ and all } \boldsymbol{h} \in \mathscr{H}. \tag{4.5}$$

Whenever, given $(\boldsymbol{S}, \boldsymbol{s})$, densities (σ, τ) can be found for the functionals S and T such that (4.5) prevails, we say that the stress field $\boldsymbol{S}$ and the surface load field $\boldsymbol{s}$ admit, respectively, the body potential S and the surface interaction potential T, defined by $(4.2)_2$ and $(4.2)_3$. Given (σ, τ), we regard (4.5) as a constitutive prescription for $(\boldsymbol{S}, \boldsymbol{s})$. Indeed, with (4.2), (4.5), the differential identity

$$\operatorname{Div}(\boldsymbol{A}^T \boldsymbol{v}) - \operatorname{Div} \boldsymbol{A} \cdot \boldsymbol{v} = \boldsymbol{A} \cdot \nabla \boldsymbol{v}, \tag{4.6}$$

which holds for all smooth tensor fields $\boldsymbol{A}$ and vector fields $\boldsymbol{v}$, and, finally, the divergence theorem, we conclude that

$$\delta E[\boldsymbol{h}] = \int_\Omega \boldsymbol{S} \cdot \nabla \boldsymbol{h} - \int_{\partial\Omega} \boldsymbol{s} \cdot \boldsymbol{h} = -\int_\Omega \operatorname{Div} \boldsymbol{S} \cdot \boldsymbol{h} + \int_{\partial\Omega} (\boldsymbol{S}\boldsymbol{n} - \boldsymbol{s}) \cdot \boldsymbol{h} \tag{4.7}$$

for all $\boldsymbol{h} \in \mathscr{H}$. Therefore, all solutions of the variational problem

$$\operatorname{extr} \{E(\boldsymbol{f}) \mid \boldsymbol{f} \in \mathscr{D}\} \tag{4.8}$$

necessarily satisfy the associated Euler-Lagrange equation, in either one of its forms $(3.7)_1$–(3.8) and (4.1).

As to the density of the body potential, we adhere to common practice in elasticity and choose a smooth mapping

$$\hat{\sigma} : \Omega \times \mathrm{Lin}^+ \to \mathbb{R}^+, \quad \sigma = \hat{\sigma}(\boldsymbol{x}, \boldsymbol{A}) \tag{4.9}$$

which is *spatially symmetric*, *i.e.*, such that

$$\hat{\sigma}(\boldsymbol{x}, \boldsymbol{R}\boldsymbol{A}) = \hat{\sigma}(\boldsymbol{x}, \boldsymbol{A}) \quad \text{for all } \boldsymbol{R} \in \text{Rot and all } (\boldsymbol{x}, \boldsymbol{A}) \in \Omega \times \text{Lin}^+. \tag{4.10}$$

Setting then

$$\boldsymbol{S} = \hat{\boldsymbol{S}}(\boldsymbol{x}, \boldsymbol{A}) = \partial_{\boldsymbol{A}}\hat{\sigma}(\boldsymbol{x}, \boldsymbol{A}) \tag{4.11}$$

we see that $(4.5)_1$ is satisfied identically in $\mathscr{D}$.

4.3. Remark. As is well known, (4.10) yields

$$(\partial_{\boldsymbol{A}}\hat{\sigma}(\boldsymbol{x}, \boldsymbol{A}))\, \boldsymbol{A}^{\mathrm{T}} \in \text{Sym} \quad \text{for all } (\boldsymbol{x}, \boldsymbol{A}) \in \Omega \times \text{Lin}^+. \tag{4.12}$$

Thus, with (4.11), $\hat{\boldsymbol{S}}$ satisfies (3.12). □

We defer our choice of a surface interaction potential density of generality comparable with (4.9), (4.10) until after the discussion of the examples of pressure and membrane loadings. We here recall, however, that a dead interaction (*cf.* (3.9)) admits the potential density

$$\hat{\tau} : \partial\Omega \times \mathscr{V} \to \mathbb{R}, \quad \tau = \hat{\tau}(\boldsymbol{x}, \boldsymbol{v}) = \hat{\boldsymbol{s}}(\boldsymbol{x}) \cdot (\boldsymbol{v} - \boldsymbol{x}); \tag{4.13}$$

with

$$\boldsymbol{s} = \hat{\boldsymbol{s}}(\boldsymbol{x}) = -\partial_{\boldsymbol{v}}\hat{\tau}(\boldsymbol{x}, \boldsymbol{v}), \tag{4.14}$$

$(4.5)_2$ is satisfied identically in $\mathscr{D}$.

5. Examples

In this section we take Ω to be a regular region in the sense of Section 2.

5.1 Pressure Interaction

By definition, a body has a *pressure interaction* with its environment when, given a *pressure function*

$$\hat{\pi} : \mathscr{E} \to \mathbb{R}^+, \quad \pi = \hat{\pi}(\boldsymbol{y}), \tag{5.1}$$

the traction boundary condition in terms of the Cauchy stress is

$$\boldsymbol{T}\boldsymbol{m} = -\pi\boldsymbol{m} \quad \text{on } \partial f(\Omega) \tag{5.2}$$

for all $f \in \mathscr{D}$. In the reference placement, the corresponding boundary condition is

$$\boldsymbol{S}\boldsymbol{n} = -\pi\boldsymbol{F}^*\boldsymbol{n}, \quad \text{with } \pi = \hat{\pi}(f(\boldsymbol{x})), \quad \text{on } \partial\Omega, \tag{5.3}$$

where both (3.5) and (3.6) have been used; we shall call *pressure loading* the vector field

$$\boldsymbol{x} \mapsto \boldsymbol{s}_p(\boldsymbol{x}) := -\hat{\pi}(f(\boldsymbol{x}))\, \boldsymbol{F}^*(\boldsymbol{x})\, \boldsymbol{n}(\boldsymbol{x}) \tag{5.4}$$

over $\partial\Omega$.

5.1. Proposition. *Pressure loading admits the surface interaction potential*

$$T_p\{f\} = \int_{\partial\Omega} \bar{\pi}(f) f \cdot (\nabla f)^* n \,\mathrm{d}\,(\mathrm{Sfr}), \tag{5.5$_1$}$$

with the mapping $\bar{\pi}$ *from* $\mathscr{V}$ *into* $\mathbb{R}$ *defined by*

$$v \mapsto \bar{\pi}(v) := \int_0^1 \hat{\pi}(\gamma v)\,\gamma^2\,\mathrm{d}\gamma \tag{5.5$_2$}$$

Moreover, the volume potential

$$P_p\{f\} = \int_\Omega \hat{\pi}(f) \det(\nabla f)\,\mathrm{d}\,(\mathrm{Vol}) \tag{5.6}$$

is such that

$$T_p\{f\} = P_p\{f\} \quad \textit{for all } f \in \mathscr{D}. \tag{5.7}$$

Proof.[12] We find it expedient to prove (5.7) first, and then to show that

$$\delta T_p\{f\}[h] = \delta P_p\{f\}[h] = \int_{\partial\Omega} s_p \cdot h. \tag{5.8}$$

We begin by observing that the successive use of the divergence theorem, the differential identity (4.6) and the fact known to Euler that the cofactor of the gradient is divergenceless:

$$\mathrm{Div}\, F^* = 0, \tag{5.9}$$

imply that

$$\int_{\partial\Omega} \bar{\pi} f \cdot F^* n = \int_\Omega F^* \cdot \nabla(\bar{\pi} f). \tag{5.10}$$

Moreover,

$$\nabla(\bar{\pi} f) = \bar{\pi} F + f \otimes \nabla\bar{\pi} \quad \text{and} \quad \nabla\bar{\pi} = F^{\mathrm{T}}\,\mathrm{grad}\,\bar{\pi}, \tag{5.11}$$

so that, on recalling (1.18) and its consequence (1.21), we have

$$F^* \cdot \nabla(\bar{\pi} f) = (3\bar{\pi} + f \cdot \mathrm{grad}\,\bar{\pi}) \det F. \tag{5.12}$$

Finally, differentiation of (5.5)$_2$ yields

$$\mathrm{grad}\,\bar{\pi} = (-3\bar{\pi} + \hat{\pi})\,|f|^{-2} f, \tag{5.13}$$

---

[12] In view of (3.4)$_1$ and (3.5), for pressure loading the volume potential may be written as

$$P_p\{f\} = \int_{f(\Omega)} \hat{\pi}(y)\,\mathrm{d}\,(\mathrm{vol}), \tag{\#}$$

and the surface potential as

$$T_p\{f\} = \int_{\partial f(\Omega)} \bar{\pi}(y)\, y \cdot m(y)\,\mathrm{d}\,(\mathrm{srf}), \tag{\#\#}$$

with $\bar{\pi}$ and $\hat{\pi}$ related by (5.5)$_2$. Of course, the appealing greater simplicity of these formulae is counterbalanced by the need of taking variations of integrals with variable domain. Notice also that (5.1) and (5.7) imply that $T_p\{f\} > 0$ throughout $\mathscr{D}$.

so that, in particular,

$$3\bar{\pi} + \boldsymbol{f} \cdot \operatorname{grad} \bar{\pi} = \hat{\pi}. \tag{5.14}$$

With the use of (5.12), and (5.14), (5.10) yields (5.7).

In order to prove $(5.8)_2$, it is sufficient to observe that, due to (4.6) and (5.9),

$$\delta \left(\int_{\Omega} \hat{\pi} \det \boldsymbol{F} \right) [\boldsymbol{h}] = \int_{\Omega} ((\operatorname{grad} \hat{\pi} \cdot \boldsymbol{h}) \det \boldsymbol{F} + \hat{\pi} \operatorname{Div} (\boldsymbol{F}^{*^{\mathrm{T}}} \boldsymbol{h})). \tag{5.15}$$

But, by (1.18) and $(5.11)_2$ with $\hat{\pi}$ in place of $\bar{\pi}$,

$$(\operatorname{grad} \hat{\pi} \cdot \boldsymbol{h}) \det \boldsymbol{F} = \boldsymbol{F}^{*^{\mathrm{T}}} \boldsymbol{h} \cdot \boldsymbol{F}^{\mathrm{T}} \operatorname{grad} \hat{\pi} = \boldsymbol{F}^{*^{\mathrm{T}}} \boldsymbol{h} \cdot \nabla \hat{\pi}. \tag{5.16}$$

Hence, in view of the differential identity

$$\operatorname{Div} (\alpha \boldsymbol{v}) = \alpha \operatorname{Div} \boldsymbol{v} + \boldsymbol{v} \cdot \nabla \alpha \tag{5.17}$$

holding for each smooth scalar field α and vector field $\boldsymbol{v}$, we have

$$\delta \left(\int_{\Omega} \hat{\pi} \det \boldsymbol{F} \right) [\boldsymbol{h}] = \int_{\Omega} \operatorname{Div} (\hat{\pi} \boldsymbol{F}^{*^{\mathrm{T}}} \boldsymbol{h}) = \int_{\partial\Omega} \hat{\pi} \boldsymbol{F}^{*} \boldsymbol{n} \cdot \boldsymbol{h}, \tag{5.18}$$

which, by (5.6) and (5.4), establishes $(5.8)_2$. □

As to the last proposition, some comments are in order, for the purpose also of putting our further developments into perspective.

Firstly, pressure loading $\boldsymbol{s}_p$ depends only on the superficial part ${}^{s}\boldsymbol{F}$ of the displacement gradient $\boldsymbol{F}$: indeed, one immediately sees from (1.28) that

$$\boldsymbol{F}^{*} \boldsymbol{n} = ({}^{s}\boldsymbol{F})^{*} \boldsymbol{n}.\text{[13]} \tag{5.19}$$

This observation suggests that, in generalizing from (5.5), one begin by considering surface interaction potentials of the form[14]

$$T\{\boldsymbol{f}\} = \int_{\partial\Omega} \hat{\tau}(\boldsymbol{f}, {}^{s}\nabla \boldsymbol{f}) \, \mathrm{d}\,(\mathrm{Srf}). \tag{5.20}$$

Secondly, recall that a *null Lagrangian*[15] is, in our present context, a functional on

$$\boldsymbol{f} \mapsto F\{\boldsymbol{f}\} = \int_{\Omega} \hat{\varphi}(\boldsymbol{f}, \nabla \boldsymbol{f}) \, \mathrm{d}\,(\mathrm{Vol}) \tag{5.21}$$

$$\bar{\mathcal{D}} := C^2(\Omega) \cap C^1(\bar{\Omega}) \tag{5.22}$$

---

[13] *Vid.* [3] for a different proof of the fact that $\boldsymbol{F}^{*}\boldsymbol{n}$ depends upon surface derivatives alone. The mechanical relevance of this circumstance was first remarked by GURTIN in [21].

[14] Here and henceforth in this section we silently fail to remark that the functionals we introduce may depend upon $\boldsymbol{x}$.

[15] Null Lagrangians are treated at length in [22] and [23], where appropriate reference to the papers by ERICKSEN [15], EDELEN [16], and others, is also to be found; an interesting paper on the subject, which has just appeared, is [24].

for which a vector field $x \mapsto v_f(x)$ over $\partial\Omega$ can be found, in general depending functionally on the restriction of f to $\partial\Omega$, such that

$$\delta F\{f\}[h] = \int_{\partial\Omega} v_f \cdot h \, \mathrm{d}(\mathrm{Srf}) \quad \text{for all } f \in \bar{\mathscr{D}} \text{ and } h \in \mathscr{H}. \tag{5.23}$$

In view of $(5.8)_2$ and (5.6), we recognize in P_p a null Lagrangian of the form (5.21). Our general results in this direction are the contents of Proposition 7.9 below.

5.2. Remark. Comparing (5.20) and (5.23) in the light of (5.7) one is led to consider live loadings alternatively described by surface and volume potential densities τ and φ such that

$$\int_{\partial\Pi} \hat{\tau}(f, {}^s\nabla f) = \int_{\Pi} \hat{\varphi}(f, \nabla f) \quad \text{for all } f \in \bar{\mathscr{D}} \text{ and parts } \Pi \subset \Omega; \tag{5.24}$$

for such loadings, invariance under change in observer might be assessed by requiring that $\hat{\varphi}$ has the behavior usually postulated for constitutive functionals having the character of volume densities, thus circumventing the nontrivial task of choosing reasonable invariance assumptions for τ or s. This suggestion has been put forward in [3], where it has been also shown that, if $\varphi = \hat{\varphi}(A)$, then (5.24) and the requirement that $\hat{\varphi}$ be spatially symmetric in the sense of (4.10) are compatible (if and) only if

$$\hat{\varphi}(A) = \pi_0 \det A, \quad \text{with } \pi_0 \text{ an arbitrary constant}, \tag{5.25}$$

which is the case for *uniform* pressure loading. □

5.2 Membrane Interaction

We think of a body having a *membrane interaction* with its environment as a body with an elastic membrane glued to it,[16] so that

$$Tm = -2\varepsilon_0 \varkappa m \quad \text{on } \partial f(\Omega), \tag{5.26}$$

where $\varepsilon_0 > 0$ is the material modulus of the membrane and, in view of (2.26) and (3.5),

$$\varkappa = \tfrac{1}{2}\, {}^s\mathrm{div}\, m \quad \text{with} \quad m = \frac{(\nabla f)^* n}{|(\nabla f)^* n|}, \tag{5.27}$$

is the mean curvature of the current boundary of the body. In analogy with (5.4), we call shall *membrane loading* the vector field

$$x \mapsto s_m(x) := -2\varepsilon_0 \varkappa(f(x))\, F^*(x)\, n(x) \tag{5.28}$$

over $\partial\Omega$.

---

[16] Or, alternatively, as a body acted upon by surface tension (*cf.* [3] and [25], p. 153).

5.3. Proposition. *Membrane loading s_m admits the surface interaction potential*

$$T_m\{f\} = \int_{\partial\Omega} \hat{\tau}_m(f, \nabla f)\, \mathrm{d}\,(\mathrm{Srf}), \tag{5.29$_1$}$$

with density

$$\hat{\tau}_m : \mathscr{S}(1) \times \mathrm{Lin}, \quad \tau_m = \hat{\tau}_m(u, A) := \varepsilon_0\, |A^* u|,^{17} \tag{5.29$_2$}$$

and equals the surface divergence of the (referential) membrane stress M:

$$s_m = {}^{s}\mathrm{Div}\, M, \quad M(u, A) := \partial_A \hat{\tau}_m(u, A), \tag{5.30}$$

with the class C^1 *tangential tensor field* M *over* $\partial\Omega$ *appearing in* $(5.30)_1$ *given by*

$$M = \tau_m(I - m \otimes m)\, F^{-\mathrm{T}}, \tag{5.31}$$

as a consequence of definition $(5.30)_2$.

Proof. As preliminary, we recall that the absolute value mapping on $\mathscr{V}$ has derivative

$$\partial_v\, |v| = |v|^{-1}\, v$$

away from the origin of $\mathscr{V}$. This, and the fact pointed out in Footnote 8 that the derivative of the cofactor mapping is a symmetric transformation of Lin into itself, allow us to write

$$\delta T_m\{f\}\,[h] = \varepsilon_0 \int_{\partial\Omega} \delta\, |(\nabla f)^* n|\, [h] = \varepsilon_0 \int_{\partial\Omega} \partial_F F^* \left[\frac{(\nabla f)^* n}{|(\nabla f)^* n|} \otimes n\right] \cdot \nabla h. \tag{5.32}$$

Next, (1.30) and (5.31) imply that

$$\delta T_m\{f\}\,[h] = \int_{\partial\Omega} M \cdot \nabla h. \tag{5.33}$$

Now, in view of (2.7) and because,

$$\nabla h = (\mathrm{grad}\, h)\, \nabla f, \tag{5.34}$$

we have

$$M \cdot \nabla h = \varepsilon_0\, |(\nabla f)^* n|\, {}^{s}\mathrm{div}\, h. \tag{5.35}$$

With (5.35), (2.27) implies that

$$\int_{\partial\Omega} M \cdot \nabla h = \varepsilon_0 \int_{\partial f(\Omega)} {}^{s}\mathrm{div}\, h = \varepsilon_0 \int_{\partial f(\Omega)} ({}^{s}\mathrm{div}\, m)\, m \cdot h; \tag{5.36}$$

---

[17] Notice that, again by (5.19) and just as the pressure potential density, it follows from $(5.29)_2$ that $\hat{\tau}_m(f, \nabla f)$ depends on the displacement gradient through the first tangential derivatives only.

(5.36), with (5.26)–(5.28) and (5.33), is enough to prove that T_m is a surface interaction potential for s_m:

$$\delta T_m\{f\}[h] = -\int_{\partial\Omega} s_m \cdot h. \tag{5.37}$$

We observe that, in view of (3.5), the membrane stress M as given by (5.31) satisfies (2.28), *i.e.*, is tangential. We can then apply (2.30) and obtain

$$\int_{\partial\Omega} M \cdot \nabla h = \int_{\partial\Omega} M \cdot {}^s\nabla h = -\int_{\partial\Omega} {}^s\mathrm{Div}\, M \cdot h,$$

which, together, with (5.37), establishes $(5.30)_1$. □

Define now the *Kirchhoff membrane stress* K to be

$$K := MF^{\mathrm{T}} = \tau_m(I - m \otimes m). \tag{5.38}$$

For all $x \in \Omega$ and $f \in \mathcal{D}$, and for all $t \in \mathcal{V}$ such that $t \cdot m(f(x)) = 0$, we have from (5.38) that

$$Kt = \tau_m t;$$

thus K is a positive-definite multiple of the identity transformation of the tangent plane to $\partial f(\Omega)$ at $f(x)$ into itself. In addition, using among other things the symmetry of the Weingarten tensor (Proposition 2.7) one can show that

$${}^s\mathrm{div}\, K = s_m = {}^s\mathrm{Div}\, M.^{18} \tag{5.39}$$

Thus, as remarked in [3], on the one hand membrane loading cannot account for situations where compressive surface stresses are to be expected (*vid. e.g.* [27] and [28]): more general constitutive prescriptions than $(5.28)_2$ are needed to cover those cases. On the other hand, s_m involves the second derivatives of the displacement field f at the boundary, and therefore points to live surface loads more general than the *simple* loads examined by SPECTOR [10], [11], which by definition depend at most on first tangential derivatives.[19]

6. Properties of Extremals

The examples of dead, pressure and membrane interactions together suggest that we consider general conservative interactions as those described by Definition 4.2, with stored energy per unit volume $\hat{\sigma}$ obeying (4.9), (4.10) and stored energy per unit area $\hat{\tau}$ defined by assigning two class C^2 mappings

$$\hat{\tau}_\alpha : (\Sigma_\alpha \cup \Gamma) \times \mathcal{S}(1) \times \mathcal{V} \times \mathrm{Lin} \to \mathbb{R}, \quad \tau = \hat{\tau}_\alpha(x, u, v, A), \quad \alpha = 1, 2, \tag{6.1}$$

---

[18] For a general treatment of continuous bodies having the form of elastic membranes *vid.* [26].

[19] Notice that, as (5.4) shows, pressure loading s_p does depend only on first tangential derivatives of f, and is therefore simple in the above sense.

and requiring that

$$\hat{\tau}|_{\Sigma_\alpha} := \hat{\tau}_\alpha|_{\Sigma_\alpha}, \quad \alpha = 1, 2.\text{[20]} \tag{6.2}$$

In view of (6.1) and (6.2), for $x \in \Sigma_1 \cup \Sigma_2$ and $f \in \mathscr{D}$, the surface interaction potentials we consider have densities

$$x \mapsto \hat{\tau}(x, n(x), f(x), F(x)). \tag{6.3}$$

It would be desirable to have at our disposal an *a priori* selection criterion for $\hat{\tau}$ playing for (6.1), (6.2) the complementing role played by (4.10) for (4.9). Unfortunately, for live surface interactions the spatial symmetry issue has not received yet a treatment as thorough as it deserves; no generally accepted form such as (4.10) is presently available for invariance requirements under change in observer applicable to surface potentials.[21]

We then address ourselves to deduce other selection criteria for admissible classes of constitutive functionals T from the properties that the extreme points of the variational problem (4.8), with E chosen as just explained, must have: in other words, we shall regard S as given, and require that T be such as to obey the necessary conditions for an extremum identically on the domain $\mathscr{D}$ of E. Our next proposition lists all those conditions.

6.1. Proposition. *Let the total energy functional E be defined by* (4.2), (4.3), (4.9), (4.10) *and* (6.2), *and let* $f^{(0)}$ *be an extreme point of E on its domain* $\mathscr{D}$. *Then,* $f^{(0)}$ *satisfies the following conditions*:

(i) (*equilibrium in the interior*)

$$\operatorname{Div}(\partial_F \hat{\sigma}^{(0)}) = 0 \quad in\ \Omega; \tag{6.4}$$

(ii) (*equilibrium on the regular part of the boundary*)

$$(\partial_F \hat{\sigma}^{(0)})\, n = -\partial_f \hat{\tau}^{(0)} + {}^s\operatorname{Div}(\partial_F \hat{\tau}^{(0)}) \quad on\ \Sigma_1 \cup \Sigma_2; \tag{6.5}$$

---

[20] Thus $\hat{\tau}$ is defined only over the regular part $\Sigma_1 \cup \Sigma_2$ of the boundary of Ω. We remark that, as the normal field and the deformation gradient are in general discontinuous through the singular part Γ of $\partial\Omega$, for all $f \in \mathscr{D}$ the scalar fields

$$x \mapsto \hat{\tau}_\alpha(x, n(x), f(x), \nabla_\alpha f(x)), \quad \alpha = 1, 2,$$

are pointwise different on Γ even if $\hat{\tau}_1|_\Gamma$ and $\hat{\tau}_2|_\Gamma$ and are one and the same function from $\Gamma \times \mathscr{S}(1) \times \mathscr{V} \times \mathrm{Lin}$ into $\mathbb{R}$: it would be difficult to imagine a situation of practical interest where a single-valued $\hat{\tau}$ could be defined over the whole of $\partial\Omega$ (of course, this is not the case if Γ is empty; *vid.* Remark 6.2 below).

[21] As an example of the invariance requirements that one may think of imposing, we recall that in [7] $\hat{\tau}$ was assumed to be *spatially symmetric* in the following reduced sense: for all fixed $x \in \Sigma_1 \cup \Sigma_2$ and $f \in \mathscr{D}$, let $\operatorname{Rot}(x, f)$ be the subgroup of Rot consisting of all rotations about the current normal m at $f(x)$; then $\hat{\tau}$ is such that, for all $(x, u, v, A) \in (\Sigma_1 \cup \Sigma_2) \times \mathscr{S}(1) \times \mathscr{V} \times \mathrm{Lin}$ and for all $f \in \mathscr{D}$,

$$\hat{\tau}(x, u, Rv, RA) = \hat{\tau}(x, u, v, A) \quad \text{for all } R \in \operatorname{Rot}(x, f). \tag{\#}$$

(iii) (*tangency on the regular part of the boundary*)

$$(\partial_F\hat{\tau}^{(0)})\,\boldsymbol{n} = \boldsymbol{0} \quad \text{on } \Sigma_1 \cup \Sigma_2; \tag{6.6}$$

Moreover, if the region Ω is such that

$$\boldsymbol{n}_1(\boldsymbol{x}) \neq -\boldsymbol{n}_2(\boldsymbol{x}) \quad \text{for all } \boldsymbol{x} \in \Gamma, \tag{6.7}$$

then $\boldsymbol{f}^{(0)}$ satisfies the additional condition

(iv) (*equilibrium on the singular part of the boundary*)

$$(\partial_F\hat{\tau}_1^{(0)})\,\boldsymbol{n}_2 + (\partial_F\hat{\tau}_2^{(0)})\,\boldsymbol{n}_1 = \boldsymbol{0} \quad \text{on } \Gamma. \tag{6.8}$$

Proof. For $\boldsymbol{f}^{(0)} \in \mathcal{D}$ an extremum of E, we have

$$\delta E\{\boldsymbol{f}^{(0)}\}[\boldsymbol{h}] = \int_\Omega \partial_F\hat{\sigma}^{(0)} \cdot \nabla\boldsymbol{h} + \int_{\partial\Omega} (\partial_f\hat{\tau}^{(0)} \cdot \boldsymbol{h} + \partial_F\hat{\tau}^{(0)} \cdot \nabla\boldsymbol{h}) = 0 \tag{6.9}$$

for all variations $\boldsymbol{h} \in \mathcal{H}$. Taking firstly $\boldsymbol{h}|_{\partial\Omega} \equiv \boldsymbol{0}$ and applying (4.6) we have (6.4). In addition, in view also of definitions (2.6), (2.8) and (2.9) yielding

$$\nabla\boldsymbol{h} = {}^{s}\nabla\boldsymbol{h} + {}^{n}\partial\boldsymbol{h} \otimes \boldsymbol{n},$$

$(6.9)_2$ reduces to

$$\int_{\partial\Omega} [(\partial_F\hat{\sigma}^{(0)}\boldsymbol{n} + \partial_f\hat{\tau}^{(0)}) \cdot \boldsymbol{h} + \partial_F\hat{\tau}^{(0)} \cdot {}^{s}\nabla\boldsymbol{h} + (\partial_F\hat{\tau}^{(0)})\,\boldsymbol{n} \cdot {}^{n}\partial\boldsymbol{h}] = 0. \tag{6.10}$$

But on the regular part of $\partial\Omega$ the variation fields $\boldsymbol{h}$ and ${}^{n}\partial\boldsymbol{h}$ may be assigned independently. Thus we deduce from (6.10) and (2.28) that $\partial_F\hat{\tau}^{(0)}$ is a tangential field on $\Sigma_1 \cup \Sigma_2$, which is (iii). At the same time, (6.10) and the consequence (2.31) of the surface divergence theorem yield

$$\int_{\Sigma_1\cup\Sigma_2} [(\partial_F\hat{\sigma}^{(0)})\,\boldsymbol{n} + \partial_f\hat{\tau}^{(0)} - {}^{s}\mathrm{Div}\,\partial_F\hat{\tau}] \cdot \boldsymbol{h} + \int_\Gamma [(\partial_F\hat{\tau}_1^{(0)})\,(\boldsymbol{t}\times\boldsymbol{n}_1) - (\partial_F\hat{\tau}_2^{(0)})\,(\boldsymbol{t}\times\boldsymbol{n}_2)] \cdot \boldsymbol{h} = 0. \tag{6.11}$$

Now taking $\boldsymbol{h}|_\Gamma \equiv \boldsymbol{0}$ we have (ii), whereas (6.11) reduces to the line integral. In view of hypothesis (6.7) and (2.4), we have

$$|\boldsymbol{n}_1\times\boldsymbol{n}_2|\,\boldsymbol{t}\times\boldsymbol{n}_1 = \boldsymbol{n}_2 - (\boldsymbol{n}_1\cdot\boldsymbol{n}_2)\,\boldsymbol{n}_1, \; |\boldsymbol{n}_1\times\boldsymbol{n}_2|\,\boldsymbol{t}\times\boldsymbol{n}_2 = (\boldsymbol{n}_1\cdot\boldsymbol{n}_2)\,\boldsymbol{n}_2 - \boldsymbol{n}_1;$$

thus the integrand in the line integral can be written as

$$\{[(\partial_F\hat{\tau}_1^{(0)})\,\boldsymbol{n}_2 + (\partial_F\hat{\tau}_2^{(0)})\,\boldsymbol{n}_1] - (\boldsymbol{n}_1\cdot\boldsymbol{n}_2)\,[(\partial_F\hat{\tau}_1^{(0)})\,\boldsymbol{n}_1 + (\partial_F\hat{\tau}_2^{(0)})\,\boldsymbol{n}_2]\} \cdot \boldsymbol{h} = 0. \tag{6.12}$$

We now observe that, passing to the limit in (6.6), we obtain

$$(\partial_F\hat{\tau}_1^{(0)})\,\boldsymbol{n}_1 = (\partial_F\hat{\tau}_2^{(0)})\,\boldsymbol{n}_2 = \boldsymbol{0} \tag{6.13}$$

identically on Γ. Finally, (6.11), (ii), (6.12), (6.13) and the residual arbitrariness in the choice of $\boldsymbol{h}$ allow us to establish (iv). □

6.2. Remark. As an example of the adjustments alluded to in Remark 2.1, we note that, for Ω a regular region, the normal field is of class C^2, and the dependence of $\hat{\tau}$ on $\boldsymbol{n}$ may then be safely absorbed into the dependence on $\boldsymbol{x}$. Formally, (6.1) and (6.2) together should be read as

$$\hat{\tau}: \partial\Omega \times \mathscr{V} \times \mathrm{Lin} \mapsto \mathbb{R}, \qquad \tau = \hat{\tau}(\boldsymbol{x}, \boldsymbol{v}, \boldsymbol{A}),$$

and (6.3) as

$$\boldsymbol{x} \mapsto \hat{\tau}(\boldsymbol{x}, \boldsymbol{f}(\boldsymbol{x}), \nabla \boldsymbol{f}(\boldsymbol{x}));$$

Proposition 6.1 would then consist of the first three items only, with $\Sigma_1 \cup \Sigma_2$ replaced by $\partial\Omega$. □

6.3. Remark. If $\boldsymbol{n}_1 \equiv -\boldsymbol{n}_2$ on Γ, the integrand of the line integral in (6.11) vanishes, provided that $\nabla_1 \boldsymbol{f}^{(0)} \equiv \nabla_2 \boldsymbol{f}^{(0)}$ and

$$\hat{\tau}_1(\boldsymbol{x}, \boldsymbol{n}, \boldsymbol{u}, \boldsymbol{A}) = -\hat{\tau}_2(\boldsymbol{x}, -\boldsymbol{n}, \boldsymbol{u}, \boldsymbol{A}) \quad \text{for } \boldsymbol{x} \in \Gamma \tag{6.14}$$

(this is the case for a pressure interaction on the whole boundary but not for a membrane interaction); in general, however, (6.11) takes the form

$$[\partial_{\boldsymbol{F}}\hat{\tau}_1, (\boldsymbol{x}, \boldsymbol{n}_1, \boldsymbol{f}^{(0)}, \nabla_1 \boldsymbol{f}^{(0)}) + \partial_{\boldsymbol{F}}\hat{\tau}_2(\boldsymbol{x}, -\boldsymbol{n}_1, \boldsymbol{f}^{(0)}, \nabla_2 \boldsymbol{f}^{(0)})]\,(\boldsymbol{t} \times \boldsymbol{n}_1) = 0 \quad \text{for } \boldsymbol{x} \in \Gamma. \tag{6.15}$$

a condition on minimizers replacing (6.8) in the cuspidal case. □

6.4. Remark. Neither for pressure interactions nor for membrane interactions does the surface potential depend explicitly on the place $\boldsymbol{x}$; when this is the case, we shall say that a surface interaction potential is *homogeneous*. □

In the light of $(3.7)_1$ and (4.11), the equilibrium condition in the interior of Ω comes as no surprise. Likewise, the equilibrium condition on the regular part of $\partial\Omega$ simply suggests that, given a surface potential density $\hat{\tau}$, for $\boldsymbol{x} \in \Sigma_1 \cup \Sigma_2$ and $\boldsymbol{f} \in \mathscr{D}$ we associate with $\hat{\tau}$ the live surface load

$$\boldsymbol{x} \mapsto \hat{\boldsymbol{s}}(\boldsymbol{x}, \boldsymbol{n}(\boldsymbol{x}), \boldsymbol{f}(\boldsymbol{x}), \boldsymbol{F}(\boldsymbol{x})) := (-\partial_{\boldsymbol{f}}\hat{\tau} + {}^{s}\mathrm{Div}\, \partial_{\boldsymbol{F}}\hat{\tau})\,(\boldsymbol{x}, \boldsymbol{n}(\boldsymbol{x}), \boldsymbol{f}(\boldsymbol{x}), \boldsymbol{F}(\boldsymbol{x})). \tag{6.16}$$

Remarkably, all conditions (6.6), (6.8) and (6.15) involve exclusively the "membrane stress" $\partial_{\boldsymbol{F}}\hat{\tau}$.

As to the tangentiality condition (6.6), we observe that both pressure and membrane potentials satisfy it identically. In the next section we shall start from a scrutiny of (6.6), and move on to motivate further consideration of densities $\hat{\tau}$ depending only on $\boldsymbol{x}$, $\boldsymbol{n}$, $\boldsymbol{f}$ and the first tangential derivatives of the displacement field $\boldsymbol{f}$, and then to discuss conditions such as to guarantee that $\hat{\boldsymbol{s}}$, too, shall depend at most on the first tangential derivatives.

As to the equilibrium condition on the singular part Γ of $\partial\Omega$, in either one of its forms (6.8) or (6.15), we repeat that pressure interactions satisfy it whatever the region Ω, and therefore the curve Γ, may be (provided $\nabla_1 \boldsymbol{f}^{(0)} \equiv \nabla_2 \boldsymbol{f}^{(0)}$). By Proposition 7.16 we shall characterize a class of homogeneous surface interaction potentials which satisfy identically both (6.8) and (6.15).

7. Tangential Interactions

We begin by establishing a preparatory result yielding an equivalent version of the tangentiality condition (6.6) for a surface potential density of type (6.2).

7.1. Proposition. *Let* $\hat{\psi}$ *be a smooth mapping from* Lin *into* $\mathbb{R}$. *Then, for each fixed* $a \in \mathscr{V}$,

$$(\partial_A \hat{\psi}(A))\, a \equiv 0 \quad \Leftrightarrow \quad \hat{\psi}(A) \equiv \hat{\psi}(A(I - a \otimes a)). \tag{7.1}$$

Proof. For fixed $a \in \mathscr{V}$. let

$$\hat{\psi}(A) \equiv \hat{\psi}(A(I - a \otimes a)).$$

Then differentiation yields

$$\partial_A \hat{\psi}(A) \equiv (\partial_A \hat{\psi}(A(I - a \otimes a)))\,(I - a \otimes a)$$

and the leftward implication follows. To prove the converse, for $a \in \mathscr{V}$ again fixed we define the real-valued mapping on $\mathrm{Lin} \times \mathscr{V}$

$$\overline{\psi}(Y, y) := \hat{\psi}(Y + y \otimes a). \tag{7.2}$$

Now, differentiating (7.2) with respect to y we have

$$\partial_y \overline{\psi}(Y, y) \equiv (\partial_A \hat{\psi}(A))\, a \quad \text{for } A = Y + y \otimes a.$$

If

$$(\partial_A \hat{\psi}(A))\, a \equiv 0,$$

it follows that $\overline{\psi}$ actually depends on the first variable only. Writing now each $A \in \mathrm{Lin}$ as

$$A = A(I - a \otimes a) + (Aa) \otimes a,$$

and taking

$$Y = A(I - a \otimes a), \qquad y = Aa$$

in (7.2), we have the desired conclusion. □

Now let $\hat{\tau}$ in (6.2) be tangential not only at an extreme point of the total energy functional, as required by (6.6), but rather on the whole of $\mathscr{D}$, as guaranteed by assuming that the membrane stress $\partial_F \hat{\tau}$ be tangential in the sense of (2.28), *i.e.*, that

$$(\partial_A \hat{\tau}(x, u, v, A))\, u = 0 \tag{7.3}$$

for all $(x, u, v, A) \in (\Sigma_1 \cup \Sigma_2) \times \mathscr{S}(1) \times \mathscr{V} \times \mathrm{Lin}$. Putting then, for each fixed triplet $(x, u, v) \in (\Sigma_1 \cup \Sigma_2) \times \mathscr{S}(1) \times \mathscr{V}$,

$$\hat{\tau}(x, u, v, A) = \hat{\psi}(A) \quad \text{for all } A \in \mathrm{Lin}, \tag{7.4}$$

we conclude from (7.1) that (7.3) implies that

$$\hat{\tau}(x, u, v, A) \equiv \hat{\tau}(x, u, v, A(I - u \otimes u)). \tag{7.5}$$

Thus, for the surface interaction potential densities of type (6.2), assuming tangency in the sense of (7.3) is equivalent to restricting attention to the subclass of *tangential* potentials, namely, those having surface densities of the following type:

$$x \mapsto \hat{\tau}(x, n(x), f(x), {}^{s}F(x)). \tag{7.6}$$

7.1 Subpotentials

Beside the hint provided by the necessary property (6.6) of extremals, we shall here offer another reason for considering only tangential potentiels, so putting that choice into better perspective.

We begin by recalling a notion of restricted convexity for functions defined on Lin (*vid.* [22], p. 43ff. and [23], Sect. 4.1.1).

A smooth mapping $\hat{\psi}$ from Lin into $\mathbb{R}$ is said to be *rank-one convex* if, for all $A \in \text{Lin}$ and $a, b \in \mathscr{V}$,

$$\hat{\psi}(A + \alpha a \otimes b) \leqq (1 - \alpha)\, \hat{\psi}(A) + \alpha \hat{\psi}(A + a \otimes b) \quad \text{for all } \alpha \in [0, 1]. \tag{7.7}$$

It is easily seen that $\hat{\psi}$ is rank-one convex if and only if the function on [0, 1] defined by

$$\alpha \mapsto \hat{\psi}(A + \alpha a \otimes b) \tag{7.8}$$

is convex for all choices of A, a and b. If $\hat{\psi}$ is of class C^2, another well known characterization of rank-one convexity is that $\hat{\psi}$ obeys (7.7) if and only if, for each $A \in \text{Lin}$ fixed, the linear mapping $\partial^2_A \hat{\psi}(A)$ from Lin into itself obeys the *Legendre-Hadamard Condition*:

$$a \otimes b \cdot (\partial^2_A \hat{\psi}(A))\, [a \otimes b] \geqq 0 \tag{7.9}$$

for all choices of a and b (*vid.* again [22] or [23], *loc. cit.*).

If both $\hat{\psi}$ and $-\hat{\psi}$ are rank-one convex, $\hat{\psi}$ is said to be *rank-one affine*. For $\hat{\psi}$ a rank-one affine mapping there are two real numbers γ_1, γ_2 and two tensors C_1, C_2 such that the representation formula holds:

$$\hat{\psi}(A) = \gamma_1 + \gamma_2 \det A + C_1 \cdot A^* + C_2 \cdot A. \tag{7.10}$$

The following characterization of rank-one affine mappings of class C^2 will be of use later on.

7.2. Proposition. *A* C^2*-mapping* $\hat{\psi}$ *from* Lin *into* $\mathbb{R}$ *is rank-one affine if and only if, for all* $A \in \text{Lin}$, *the mapping* $\partial^2_A \hat{\psi}(A)$ *obeys the skew-symmetry conditions*

$$\begin{aligned} c \otimes d \cdot (\partial^2_A \hat{\psi}(A))\, [a \otimes b] &= -c \otimes b \cdot (\partial^2_A \hat{\psi}(A))\, [a \otimes d] \\ &= -a \otimes d \cdot (\partial^2_A \hat{\psi}(A))\, [c \otimes b] \end{aligned} \tag{7.11}$$

for all $a, b, c, d \in \mathscr{V}$.

Proof. In view of (7.7) and (7.8), we conclude that $\hat{\psi}$ is rank-one affine if and only if

$$\alpha \mapsto \hat{\psi}(A + \alpha a \otimes b) = (1 - \alpha)\, \hat{\psi}(A) + \alpha \hat{\psi}(A + a \otimes b);$$

differentiating this function twice, we conclude, under the hypotheses, that $\hat{\psi}$ is rank-one affine if and only if

$$a \otimes b \cdot (\partial^2_A \hat{\psi}(A))\, [a \otimes b] = 0 \tag{7.12}$$

for all $a, b \in \mathcal{V}$ (*cf.* (7.9)); in turn, condition (7.12) is known to be equivalent to $(7.11)_1$.[22] Finally, as for all fixed $A \in \text{Lin}$ the linear mapping $\partial_A^{2}\hat{\psi}(A)$ of Lin into itself is symmetric in the sense of Footnote 8, $(7.11)_2$ also follows. □

Consider now the mapping $\overline{\psi}$ defined by (7.2), and put

$$\overline{\psi}_{\text{sub}}(Y) := \inf\{\overline{\psi}(Y, y) \mid y \in \mathcal{V}\}. \qquad (7.13)$$

Using the notation (7.4) again, we see that, for each fixed triplet $(x, u, v) \in (\Sigma_1 \cup \Sigma_2) \times \mathcal{S}(1) \times \mathcal{V}$, the mapping

$$\hat{\tau}_{\text{sub}}(x, u, v, A) := \overline{\psi}_{\text{sub}}(A) \quad \text{for all } A \in \text{Lin}, \qquad (7.14)$$

is well defined. We then have grounds for the following

7.3. Definition. *Given a surface interaction potential T with density* $\hat{\tau}$ *of type* (6.1)–(6.3), *the associated* subpotential *is the tangential functional*

$$T_{\text{sub}}\{f\} := \int_{\partial\Omega} \hat{\tau}_{\text{sub}}(x, n, f, {}^{s}F)\, \mathrm{d}\,(\text{Srf}) \qquad (7.15)$$

with density

$$\hat{\tau}_{\text{sub}}(x, n, r, {}^{s}F) := \inf\{\hat{\tau}(x, n, f, {}^{s}\nabla f + y \otimes n) \mid y \in \mathcal{V}\}. \qquad (7.16)$$

Likewise,

$$E_{\text{sub}} := S + T_{\text{sub}} \qquad (7.17)$$

is the subpotential associated with E.

7.4. Remark. The terminology we use is inspired by the similar but different concept of "subenergy" that ERICKSEN [29] has taken from FLORY [30] in order to model certain types of phase transitions within the framework of thermoelasticity. □

7.5. Proposition. *Let T be a surface interaction potential, with rank-one convex density* $\hat{\tau}$, *and let the associated subpotential* T_{sub} *be Gateaux-differentiable over* $\mathcal{D}$. *Then, if* $f^{(0)}$ *is an extreme point of E jn* $\mathcal{D}$, *it also renders* E_{sub} *stationary.*

Proof. Recall that, on confining attention to conservative interactions by means of Definition 4.2, we already stipulated that E is G-differentiable over $\mathcal{D}$. Assuming now, in addition, that it makes sense to write

$$\lim_{\varepsilon \to 0} \frac{E_{\text{sub}}\{f + \varepsilon h\} - E_{\text{sub}}\{f\}}{\varepsilon} =: \delta E_{\text{sub}}\{f\}\,[h] \qquad (7.18)$$

for all $f \in \mathcal{D}$ and $h \in \mathcal{H}$, we wish to show that

$$\delta E_{\text{sub}}\{f^{(0)}\}\,[h] = 0 \quad \text{for all } h \in \mathcal{H} \qquad (7.19)$$

at each extreme point $f^{(0)}$ of E.

---

[22] Note that $(7.11)_1$ can be written equivalently as

$$((\partial_A^2\hat{\psi}(A))\,[a \otimes b])\, c = -((\partial_A^2\hat{\psi}(A))\,[a \otimes c])\, b \quad \text{for all } a, b, c \in \mathcal{V}. \qquad (\#)$$

First, it follows from the very definitions of E and E_{sub} that

$$E\{f\} \geqq E_{\text{sub}}\{f\} \quad \text{for all } f \in \mathscr{D}. \tag{7.20}$$

A less obvious fact is that, under the hypotheses,

$$E\{\boldsymbol{f}^{(0)}\} = E_{\text{sub}}\{\boldsymbol{f}^{(0)}\} \tag{7.21}$$

at each extreme point $\boldsymbol{f}^{(0)}$ of E. To see this, recall that, at $\boldsymbol{f}^{(0)}$ (6.6) requires that $\hat{\tau}$ is such that

$$(\partial_{\boldsymbol{F}}\hat{\tau}^{(0)})\,\boldsymbol{n} = \boldsymbol{0}. \tag{7.22}$$

Moreover, if $\hat{\tau}$ is rank-one convex, *i.e.*, if the mapping $\hat{\psi}$ defined by (7.4) is rank-one convex for each fixed triplet $(\boldsymbol{x}, \boldsymbol{u}, \boldsymbol{v}) \in (\Sigma_1 \cup \Sigma_1) \times \mathscr{S}(1) \times \mathscr{V}$, then

$$\boldsymbol{b} \mapsto \hat{\psi}({}^{\mathrm{s}}\nabla \boldsymbol{f}^{(0)} + \boldsymbol{b} \otimes \boldsymbol{n})$$

is a convex mapping whose derivative at $\boldsymbol{b} = {}^{\mathrm{n}}\partial \boldsymbol{f}^{(0)}$ equals $(\partial_{\boldsymbol{F}}\hat{\tau}^{(0)})\,\boldsymbol{n}$ and therefore vanishes because of (7.22); but then this mapping has at ${}^{\mathrm{n}}\partial \boldsymbol{f}^{(0)}$ an absolute minimum, and thus, in view also of definition (7.16),

$$\hat{\tau}^{(0)}_{\text{sub}} = \hat{\tau}^{(0)}, \tag{7.23}$$

which implies (7.21). With this and (7.20), we have

$$E_{\text{sub}}\{\boldsymbol{f}^{(0)} + \varepsilon \boldsymbol{h}\} - E_{\text{sub}}\{\boldsymbol{f}^{(0)}\} \leqq E\{\boldsymbol{f}^{(0)} + \varepsilon \boldsymbol{h}\} - E\{\boldsymbol{f}^{(0)}\},$$

so that

$$\lim_{\varepsilon \to 0^+} \frac{E_{\text{sub}}\{\boldsymbol{f}^{(0)} + \varepsilon \boldsymbol{h}\} - E_{\text{sub}}\{\boldsymbol{f}^{(0)}\}}{\varepsilon} \leqq 0 \tag{7.24$_1$}$$

and

$$\lim_{\varepsilon \to 0^-} \frac{E_{\text{sub}}\{\boldsymbol{f}^{(0)} + \varepsilon \boldsymbol{h}\} - E_{\text{sub}}\{\boldsymbol{f}^{(0)}\}}{\varepsilon} \geqq 0. \tag{7.24$_2$}$$

(7.19) then follows from (7.24) and the assumed differentiability of E_{sub}. □

The last proposition shows that, for reasonably general assignments of surface interaction potentials depending in principle on the entire displacement gradient, the extreme points of the resulting total potentials integrate a subset of the extremal set of an associated functional so constructed as to give to normal derivatives the least weight possible in the minimization process.

7.2 Simple Potentials

Here and henceforth we restrict attention to tangential potentials

$$T\{f\} = \int_{\partial\Omega} \hat{\tau}(\boldsymbol{x}, \boldsymbol{n}, \boldsymbol{f}, {}^{\mathrm{s}}\boldsymbol{F})\, \mathrm{d}\,(\mathrm{Srf}). \tag{7.25}$$

As is clear from (6.16), these potentials correspond in general to surface loads depending on first and second tangential derivatives. Our goal in this section is to characterize a class of surface potentials such that the corresponding surface loads are *sinmple, i.e.*, depend at most on first derivatives. Interestingly, the total potentials featuring simple surface interactions may be thought of as defined over a domain larger than $\mathscr{D}$, as second derivatives are not assumed to exist at points of $\Sigma_1 \cup \Sigma_2$.

7.6. Definition. *A tangential potential T with density $\hat{\tau}$ is simple if it gives rise to a simple surface load $\hat{s}$*, i.e., *if*

$$s = -\partial_f \hat{\tau}(x, n, f, {}^sF) + {}^s\mathrm{Div}\, \partial_F \hat{\tau}(x, n, f, {}^sF) = \hat{s}(x, n, f, {}^sF). \qquad (7.26)$$

7.7. Proposition. *A tangential potential T is simple if and only if its density τ admits the following representation formula*:

$$\hat{\tau}(x, u, v, A) = \hat{\gamma}(x, u, v) + (\hat{c}(x, u, v) \otimes u) \cdot A^* + \hat{C}(x, u, v)\, U \cdot A. \qquad (7.27)$$

Proof. Observe first that, by $(7.26)_1$, the part of s containing second derivatives of f vanishes identically on $\Sigma_1 \cup \Sigma_2$ if and only if $\partial_A^2 \hat{\tau}(x, u, v, A)$ has the skew-symmetry property $(7.11)_1$ for all $(x, u, v)) \in (\Sigma_1 \cup \Sigma_2) \times \mathscr{S}(1) \times \mathscr{V}$, *i.e.*, in view of (7.4) and Proposition 7.3, if and only if the mapping

$$A \mapsto \hat{\psi}(A) = \hat{\tau}(x, u, v, A) \qquad (7.28)$$

is rank-one affine. Secondly, by Proposition 7.1, if T is tangential we may replace A with $A(I - u \otimes u)$ in the representation formula (7.10); the result is

$$\hat{\psi}(A) = \gamma_1 + C_1 \cdot (A(I - u \otimes u))^* + C_2 \cdot A(I - u \otimes u)$$

or rather, by (1.28),

$$\hat{\psi}(A) = \gamma_1 + (C_1 u \otimes u) \cdot A^* + C_2(I - u \otimes u) \cdot A;$$

from this, with the notations

$$\gamma_1 =: \gamma_1, \quad C_1 u =: c \quad \text{and} \quad -C_2 U =: C \quad (\textit{cf.}\ (1.11)),$$

(7.27) follows. □

Resuming now the developments at the end of Subsection 5.1 in the present more general context, we turn to characterize those simple surface loadings that are alternatively described by surface and volume interaction potentials. As our attention is restricted, for the rest of this subsection, to bodies comprising regular regions Ω, we shall drop the explicit dependence on the normal n to $\partial\Omega$ in the expression of surface interaction potentials (*cf.* Remark 6.2).

We state without proof a proposition where some results of [15] and [16] are rephrased.

7.8. Proposition. *A functional*

$$F\{f\} = \int_\Omega \hat{\varphi}(x, f, \nabla f)\, \mathrm{d}\,(\mathrm{Vol})$$

on $C^2(\bar{\Omega})$ *is a null Lagrangian if and only if there is a smooth vector-valued mapping*

$$(\boldsymbol{x}, \boldsymbol{u}, A) \mapsto \hat{\boldsymbol{v}}(\boldsymbol{x}, \boldsymbol{u}, A), \tag{7.29}$$

obeying the skew-symmetry condition

$$(\partial_A(\hat{\boldsymbol{v}}(\boldsymbol{x}, \boldsymbol{v}, A) \cdot \boldsymbol{b}))\, \boldsymbol{a} = -(\partial_A(\hat{\boldsymbol{v}}(\boldsymbol{x}, \boldsymbol{v}, A) \cdot \boldsymbol{a}))\, \boldsymbol{b} \tag{7.30}$$

for all $\boldsymbol{a}, \boldsymbol{b} \in \mathscr{V}$ *and for all fixed* $(\boldsymbol{x}, \boldsymbol{u}, A) \in \bar{\Omega} \times \mathscr{V} \times \mathrm{Lin}$, *such that*

$$\hat{\varphi}(\boldsymbol{x}, \boldsymbol{f}(\boldsymbol{x}), \nabla \boldsymbol{f}(\boldsymbol{x})) = \mathrm{Div}\, \hat{\boldsymbol{v}}(\boldsymbol{x}, \boldsymbol{f}(\boldsymbol{x}), \nabla \boldsymbol{f}(\boldsymbol{x})) \tag{7.31}$$

for all $\boldsymbol{x} \in \bar{\Omega}$ *and* $\boldsymbol{f} \in C^2(\bar{\Omega})$.

We are now in a position to state our announced characterization.

7.9. Proposition. *Given a simple tangential potential*

$$T\{\boldsymbol{f}\} = \int_{\partial\Omega} \hat{\tau}(\boldsymbol{x}, \boldsymbol{f}, {}^s\boldsymbol{F})\, \mathrm{d}\,(\mathrm{Srf}), \tag{7.32}$$

there is an associated volume potential

$$F\{\boldsymbol{f}\} = \int_{\Omega} \hat{\varphi}(\boldsymbol{x}, \boldsymbol{f}, \boldsymbol{F})\, \mathrm{d}\,(\mathrm{Vol}) \tag{7.33}$$

such that

$$T\{\boldsymbol{f}\} = F\{\boldsymbol{f}\} \quad \textit{for all } \boldsymbol{f} \in C^2(\bar{\Omega}); \tag{7.34}$$

F is a null Lagrangian. Conversely, if a volume potential F as in (7.33) *is a null Lagrangian, then there is a simple tangential potential T as in* (7.32) *such that* (7.34) *holds.*

Proof. Given $\hat{\tau}$, for $(\boldsymbol{x}, \boldsymbol{u}, A) \in \partial\Omega \times \mathscr{V} \times \mathrm{Lin}$, we consider the vector-valued mapping defined by

$$\hat{\boldsymbol{v}}(\boldsymbol{x}, \boldsymbol{u}, A) := \hat{\tau}(\boldsymbol{x}, \boldsymbol{u}, A)\, \boldsymbol{n}(\boldsymbol{x}) - (\partial_A \hat{\tau}(\boldsymbol{x}, \boldsymbol{u}, A))^{\mathrm{T}} A\boldsymbol{n}(\boldsymbol{x}). \tag{7.35}$$

We observe first that, because T is tangential,

$$(\partial_A \hat{\tau}(\boldsymbol{x}, \boldsymbol{u}, A))\, \boldsymbol{n}(\boldsymbol{x}) \equiv \boldsymbol{0},$$

so that

$$\hat{\boldsymbol{v}}(\boldsymbol{x}, \boldsymbol{u}, A) \cdot \boldsymbol{n}(\boldsymbol{x}) = \hat{\tau}(\boldsymbol{x}, \boldsymbol{u}, A). \tag{7.36}$$

Secondly, differentiating $\hat{\boldsymbol{v}}$ with respect to the third argument, we see that

$$(\partial_A(\hat{\boldsymbol{v}} \cdot \boldsymbol{b}))\, \boldsymbol{a} = -[(\boldsymbol{a} \cdot \boldsymbol{n})\, (\partial_A \hat{\tau})\, \boldsymbol{b} - (\boldsymbol{b} \cdot \boldsymbol{n})\, (\partial_A \hat{\tau})\, \boldsymbol{a}] - ((\partial_A^2 \hat{\tau})\, [\boldsymbol{F}\boldsymbol{n} \otimes \boldsymbol{b}])\, \boldsymbol{a}. \tag{7.37}$$

On the right-hand side of (7.37), the addend within square brackets is obviously skew-symmetric with respect to the interchange of $\boldsymbol{a}$ and $\boldsymbol{b}$, whereas the other addend is likewise skew-symmetric because T is simple and hence, by Proposition 7.7, $\hat{\tau}$ obeys $(7.11)_1$; consequently, $\hat{\boldsymbol{v}}$ satisfies (7.30).[23] As Ω is by assumption

[23] It can be shown that the most general $\hat{\boldsymbol{v}}$ obeying (7.30) and (7.36) has the form $\hat{\boldsymbol{v}}(\boldsymbol{x}, \boldsymbol{u}, A) = \hat{\tau}(\boldsymbol{x}, \boldsymbol{u}, A)\, \boldsymbol{n}(\boldsymbol{x}) - (\partial_A \hat{\tau}(\boldsymbol{x}, \boldsymbol{u}, A))^{\mathrm{T}} A\boldsymbol{n}(\boldsymbol{x}) + (\boldsymbol{c}_1(\boldsymbol{x}, \boldsymbol{u}) + A^{\mathrm{T}} \boldsymbol{c}_2(\boldsymbol{x}, \boldsymbol{u})) \times \boldsymbol{n}(\boldsymbol{x})$, with $\boldsymbol{c}_1(\boldsymbol{x}, \boldsymbol{u})$, $\boldsymbol{c}_2(\boldsymbol{x}, \boldsymbol{u})$ two arbitrary vector-valued mappings.

of class C^3, $\hat{v}$ as in (7.35) can be extended to a class C^2 mapping over $\bar{\Omega}\times\mathscr{V}\times\mathrm{Lin}$; the direct assertion then follows by defining $\hat{\varphi}$ as in (7.31).

A proof of the converse begins by constructing $\hat{\tau}$ from the vector-valued mapping $\hat{v}$ associated with a given null Lagrangian density $\hat{\varphi}$ by Proposition 7.8:

$$\hat{\tau}(\boldsymbol{x}, \boldsymbol{u}, \boldsymbol{A}) := \hat{\boldsymbol{v}}(\boldsymbol{x}, \boldsymbol{u}, \boldsymbol{A}) \cdot \boldsymbol{n}(\boldsymbol{x}). \tag{7.38}$$

It remains to verify that such a $\hat{\tau}$ is both tangential and simple. But, $\hat{\tau}$ is tangential because, by (7.38) and (7.30),

$$(\partial_A\hat{\tau})\,\boldsymbol{n} = (\partial_A(\hat{\boldsymbol{v}}\cdot\boldsymbol{n}))\,\boldsymbol{n} = \boldsymbol{0},$$

so that (7.6) follows from appealing to Proposition 7.1; $\hat{\tau}$ is simple because, by (7.33) and (7.31),

$$\delta F\{\boldsymbol{f}\}[\boldsymbol{h}] = \int_{\partial\Omega} (\partial_{\boldsymbol{F}}\hat{\varphi}(\boldsymbol{x}, \boldsymbol{f}, \boldsymbol{F}))\,\boldsymbol{n}\cdot\boldsymbol{h}\,\mathrm{d}\,(\mathrm{Srf})$$

and thus the corresponding surface loading

$$\hat{\boldsymbol{s}} = -(\partial_{\boldsymbol{F}}\hat{\varphi})\,\boldsymbol{n}$$

depends only on first derivatives. □

7.3 Regular Potentials

Let $\mathcal{O}$ denote the collection of all domains Ω described under (ii) in the beginning of Section 2; moreover, for Ω chosen in $\mathcal{O}$, let the mappings $\hat{\tau}_\alpha$ $(\alpha=1,2)$ define through use of (6.1)–(6.3) a surface interaction potential density $\hat{\tau}$.

On the singular part Γ of $\partial\Omega$ the equilibrium condition (6.8) reads:

$$(\partial_{\boldsymbol{F}}\hat{\tau}_1(\boldsymbol{x},\boldsymbol{n}_1(\boldsymbol{x}),\boldsymbol{f}^{(0)}(\boldsymbol{x}),\nabla_1\boldsymbol{f}^{(0)}(\boldsymbol{x}))\,\boldsymbol{n}_2(\boldsymbol{x}) + (\partial_{\boldsymbol{F}}\hat{\tau}_2(\boldsymbol{x},\boldsymbol{n}_2(\boldsymbol{x}),\boldsymbol{f}^{(0)}(\boldsymbol{x}),\nabla_2\boldsymbol{f}^{(0)}(\boldsymbol{x})),\boldsymbol{n}_1(\boldsymbol{x}) \equiv \boldsymbol{0} \tag{7.39$_1$}$$

at each extremum point $\boldsymbol{f}^{(0)}\in\mathscr{D}$ of the underlying total energy functional, when (6.7) applies; otherwise, if Γ is everywhere cuspidal, *i.e.*, if $\boldsymbol{n}_1 = -\boldsymbol{n}_2$ on Γ, $(7.39)_1$ is replaced by

$$[\partial_{\boldsymbol{F}}\hat{\tau}_1(\boldsymbol{x},\boldsymbol{n}_1(\boldsymbol{x}),\boldsymbol{f}^{(0)}(\boldsymbol{x}),\nabla_1\boldsymbol{f}^{(0)}(\boldsymbol{x})) + \partial_{\boldsymbol{F}}\hat{\tau}_2(\boldsymbol{x},-\boldsymbol{n}_1(\boldsymbol{x}),\boldsymbol{f}^{(0)}(\boldsymbol{x}),\nabla_2\boldsymbol{f}^{(0)}(\boldsymbol{x}))]\,(\boldsymbol{t}(\boldsymbol{x})\times\boldsymbol{n}_1(\boldsymbol{x})) \equiv \boldsymbol{0} \tag{7.39$_2$}$$

(*cf.* Remark 6.3). On the other hand, the companion equilibrium condition (6.6) prevailing on the regular part of $\partial\Omega$ can be split into separate tangentiality conditions for the individual $\hat{\tau}_\alpha$:

$$(\partial_{\boldsymbol{F}}\hat{\tau}_\alpha(\boldsymbol{x},\boldsymbol{n}(\boldsymbol{x}),\boldsymbol{f}^{(0)},\nabla\boldsymbol{f}^{(0)})\,\boldsymbol{n}(\boldsymbol{x}) \equiv \boldsymbol{0} \quad \text{on } \Sigma_\alpha, \quad \alpha = 1, 2. \tag{7.39$_3$}$$

Comparison with $(7.39)_3$ suggests that we regard $(7.39)_{1,2}$ as conditions of joint tangentiality on the pair $(\hat{\tau}_1, \hat{\tau}_2)$, prevailing at extrema. We may ask whether there are surface interactions such as to satisfy all conditions (7.39) identically in $\mathcal{O}$ and $\mathscr{D}$ with $\nabla_1\boldsymbol{f}\equiv\nabla_2\boldsymbol{f}$.

Just as the analogous question which led us to consider potentials with density of type (7.6) obeying condition $(7.39)_3$ identically in $\mathscr{D}$, this question is constitutive in nature.

In order to answer it, we begin by noting, that, as far as condition (7.39) is concerned, the choice of an element $\Omega \in \mathscr{O}$ amounts to choice of a class C^3 closed simple curve Γ and two class C^2 normal fields $\boldsymbol{n}_\alpha = \boldsymbol{n}_\alpha(\boldsymbol{x})$ over Γ. If they have to satisfy (7.39) whatever Γ may be, the mappings $\hat{\tau}_\alpha$ cannot depend explicitly on the place $\boldsymbol{x}$. We then introduce the following definition.

7.10. Definition. *A pair* $(\hat{\tau}_1, \hat{\tau}_2)$ *of class* C^2 *mappings*

$$\hat{\tau}_\alpha : \mathscr{S}(1) \times \mathscr{V} \times \mathrm{Lin} \to \mathbb{R}, \qquad \tau = \hat{\tau}_\alpha(\boldsymbol{u}, \boldsymbol{v}, \boldsymbol{A}), \qquad \alpha = 1, 2,$$

is jointly tangential *if, for all* $(\boldsymbol{v}, \boldsymbol{A}) \in \mathscr{V} \times \mathrm{Lin}$ *fixed, the following two conditions hold*:

$$\partial_{\boldsymbol{A}} \hat{\tau}_1(\boldsymbol{u}_1, \boldsymbol{v}, \boldsymbol{A})\, \boldsymbol{u}_2 + \partial_{\boldsymbol{A}} \hat{\tau}_2(\boldsymbol{u}_2, \boldsymbol{v}, \boldsymbol{A})\, \boldsymbol{u}_1 = \boldsymbol{0} \quad \textit{for all } \boldsymbol{u}_1, \boldsymbol{u}_2 \in \mathscr{S}(1); \qquad (7.40)_1$$

$$\partial_{\boldsymbol{A}} \hat{\tau}_\alpha(\boldsymbol{u}, \boldsymbol{v}, \boldsymbol{A})\, \boldsymbol{u} = \boldsymbol{0}, \alpha = 1, 2, \quad \textit{for all } \boldsymbol{u} \in \mathscr{S}(1). \qquad (7.40)_3$$

7.11. Proposition. *If* $(\hat{\tau}_1, \hat{\tau}_2)$ *is a jointly tangential pair, then for both the associated membrane stresses* $\partial_{\boldsymbol{A}} \hat{\tau}_\alpha$ *one and the same representation formula holds*:

$$\partial_{\boldsymbol{A}} \hat{\tau}_1(\boldsymbol{u}, \boldsymbol{v}, \boldsymbol{A}) = \partial_{\boldsymbol{A}} \hat{\tau}_2(\boldsymbol{u}, \boldsymbol{v}, \boldsymbol{A}) = \hat{\boldsymbol{B}}(\boldsymbol{v}, \boldsymbol{A})\, \boldsymbol{U} \qquad (7.41)$$

for all $(\boldsymbol{u}, \boldsymbol{v}, \boldsymbol{A}) \in \mathscr{S}(1) \times \mathscr{V} \times \mathrm{Lin}$, *with* $\hat{\boldsymbol{B}}$ *a class* C^1 *mapping from* $\mathscr{V} \times \mathrm{Lin}$ *into* Lin, *and* $\boldsymbol{U}$ *the skew tensor associated to* $\boldsymbol{u}$ *by* (1.5).

7.12. Remark. A straightforward, but important, consequence of (7.41) is that

$$\partial_{\boldsymbol{A}} \hat{\tau}_1(\boldsymbol{u}, \boldsymbol{v}, \boldsymbol{A}) + \partial_{\boldsymbol{A}} \hat{\tau}_2(-\boldsymbol{u}, \boldsymbol{v}, \boldsymbol{A}) = \boldsymbol{0} \quad \text{for all } \boldsymbol{u} \in \mathscr{S}(1). \qquad (7.40)_2$$

Thus, if $(\hat{\tau}_1, \hat{\tau}_2)$ is jointly tangential and if, for each Ω in $\mathscr{O}$ and $f \in C^2(\bar{\Omega})$, we define the class C^2 tensor fields

$$\boldsymbol{x} \mapsto \hat{\boldsymbol{B}}(f(\boldsymbol{x}), \nabla f(\boldsymbol{x}))\, \boldsymbol{N}_\alpha(\boldsymbol{x})|_{\Sigma_\alpha \cup \Gamma}, \qquad \alpha = 1, 2, \qquad (7.42)$$

with $\boldsymbol{N}_\alpha(\boldsymbol{x})$ the skew tensor associated to $\boldsymbol{n}_\alpha(\boldsymbol{x})$, then, as $(7.40)_i$ reflects the constitutive significance of $(7.39)_i$ for $i = 1, 2, 3$ in the order, the fields (7.42) satisfy all conditions (7.39) identically in $\mathscr{O}$ and $\mathscr{D}$. □

Proof of Proposition 7.11. Let the variables $(\boldsymbol{v}, \boldsymbol{A})$ be fixed in $\mathscr{V} \times \mathrm{Lin}$ (for convenience, the arguments $\boldsymbol{v}, \boldsymbol{A}$ will not be displayed in the following developments). As $\hat{\tau}_\alpha$ satisfies $(7.40)_3$, we conclude from (1.11) in Proposition 1.1 that, for all $\boldsymbol{u} \in \mathscr{S}(1)$,

$$\partial_{\boldsymbol{A}} \hat{\tau}_\alpha(\boldsymbol{u}) = \hat{\boldsymbol{B}}_\alpha(\boldsymbol{u})\, \boldsymbol{U}, \qquad (7.43)$$

with $\boldsymbol{U}$ the skew tensor associated to $\boldsymbol{u}$ by (1.5) and

$$\hat{\boldsymbol{B}}_\alpha(\boldsymbol{u}) := -\partial_{\boldsymbol{A}} \hat{\tau}_\alpha(\boldsymbol{u})\, \boldsymbol{U}. \qquad (7.44)$$

We then rewrite $(7.40)_1$ as

$$\hat{\boldsymbol{B}}_1(|\boldsymbol{u}_1|^{-1}\,\boldsymbol{u}_1)\,\boldsymbol{U}_1\boldsymbol{u}_2 + \hat{\boldsymbol{B}}_2(|\boldsymbol{u}_2|^{-1}\,\boldsymbol{u}_2)\,\boldsymbol{U}_1\boldsymbol{u}_2 = \boldsymbol{0}, \tag{7.45}$$

this time for all $\boldsymbol{u}_1, \boldsymbol{u}_2 \in \mathscr{V}$. Differentiating (7.45) with respect to $\boldsymbol{u}_1$ and $\boldsymbol{u}_2$ in the direction of $\boldsymbol{h}_1$ and $\boldsymbol{h}_2$, respectively, we have

$$(\partial_{\boldsymbol{u}_1}(\hat{\boldsymbol{B}}_1(|\boldsymbol{u}_1|^{-1}\,\boldsymbol{u}_1)\,\boldsymbol{U}_1)\,[\boldsymbol{h}_1])\,\boldsymbol{h}_2 + (\partial_{\boldsymbol{u}_2}(\hat{\boldsymbol{B}}_2(|\boldsymbol{u}_2|^{-1}\,\boldsymbol{u}_2)\boldsymbol{U}_2)\,[\boldsymbol{h}_2])\,\boldsymbol{h}_1 = \boldsymbol{0}$$

identically in $\boldsymbol{u}_1$, $\boldsymbol{u}_2$ and $\boldsymbol{h}_1$, $\boldsymbol{h}_2$. It follows that there exists a linear mapping $\mathbb{B}$ from $\mathscr{V}$ into Lin, with $\mathbb{B}$ depending at most on $(\boldsymbol{v}, \boldsymbol{A})$, such that

$$\begin{aligned}(\mathbb{B}[\boldsymbol{h}_1])\,\boldsymbol{h}_2 &:= (\partial_{\boldsymbol{u}_1}(\hat{\boldsymbol{B}}_1(|\boldsymbol{u}_1|^{-1}\,\boldsymbol{u}_1)\,\boldsymbol{U}_1)\,[\boldsymbol{h}_1])\,\boldsymbol{h}_2 \\ &= -(\partial_{\boldsymbol{u}_2}(\hat{\boldsymbol{B}}_2(|\boldsymbol{u}_2|^{-1}\,\boldsymbol{u}_2)\,\boldsymbol{U}_2)\,[\boldsymbol{h}_2])\,\boldsymbol{h}_1.\end{aligned} \tag{7.46}$$

Integrating $(7.46)_1$ we obtain

$$\hat{\boldsymbol{B}}_1(|\boldsymbol{u}_1|^{-1}\,\boldsymbol{u}_1)\,\boldsymbol{U}_1 = \mathbb{B}[\boldsymbol{u}_1] + \boldsymbol{C}, \tag{7.47}$$

where $\boldsymbol{C}$ also depends at most on $(\boldsymbol{v}, \boldsymbol{A})$; but, by multiplication of (7.47) on the right by $\boldsymbol{u}_1$, we see that

$$\boldsymbol{0} = (\mathbb{B}[\boldsymbol{u}_1])\,\boldsymbol{u}_1 + \boldsymbol{C}\boldsymbol{u}_1,$$

so that, as $\boldsymbol{u}_1$ is here arbitrary,

$$(\mathbb{B}[\boldsymbol{h}_1])\,\boldsymbol{h}_2 = -(\mathbb{B}([\boldsymbol{h}_2])\,\boldsymbol{h}_1 \quad \text{for all } \boldsymbol{h}_1, \boldsymbol{h}_2 \in \mathscr{V}, \quad \boldsymbol{C} = \boldsymbol{0}. \tag{7.48}$$

(7.43), (7.47) and (7.48) yield

$$\partial_{\boldsymbol{A}}\hat{\tau}_1(\boldsymbol{u}) = \partial_{\boldsymbol{A}}\hat{\tau}_2(\boldsymbol{u}) = \mathbb{B}[\boldsymbol{u}]$$

for all $\boldsymbol{u} \in \mathscr{S}(1)$; moreover, due to $(7.48)_1$, the condition

$$\mathbb{B}[\boldsymbol{u}] = \boldsymbol{B}\boldsymbol{U} \quad \text{for all } \boldsymbol{u} \in \mathscr{V}$$

uniquely determines the second-order tensor $\boldsymbol{B}$ in terms of $\mathbb{B}$.[24] □

Proposition 7.11 indicates that the mappings $\hat{\tau}_1, \hat{\tau}_2$ cannot differ much from each other if $(\hat{\tau}_1, \hat{\tau}_2)$ has to be a jointly tangential pair. Actually, an even more stringent result than (7.41) is valid.

7.13. Proposition. *If $(\hat{\tau}_1, \hat{\tau}_2)$ is a jointly tangential pair, then there are two class C^2 mappings*

$$\hat{\gamma}_\alpha : \mathscr{S}(1) \times \mathscr{V} \to \mathbb{R}, \quad \gamma_\alpha = \hat{\gamma}_\alpha(\boldsymbol{u}, \boldsymbol{v}) \tag{7.49}$$

such that the following representation formulae hold:

$$\hat{\tau}_\alpha(\boldsymbol{u}, \boldsymbol{v}, \boldsymbol{A}) - \hat{\gamma}_\alpha(\boldsymbol{u}, \boldsymbol{v}) = (\hat{\boldsymbol{c}}(\boldsymbol{v}) \otimes \boldsymbol{u}) \cdot \boldsymbol{A}^* + \hat{\boldsymbol{C}}(\boldsymbol{v})\,\boldsymbol{U} \cdot \boldsymbol{A}, \quad \alpha = 1, 2. \tag{7.50}$$

---

[24] To detail the algebra, we note that, in view of $(7.46)_1$, $\mathbb{B}$ can be repesented as $\mathbb{B} = \boldsymbol{b} \otimes \boldsymbol{W}$, with $\boldsymbol{b} \in \mathscr{V}$ and $\boldsymbol{W} \in$ Skw, so that in turn $\boldsymbol{B}$ has the representation $\boldsymbol{B} = \boldsymbol{b} \otimes \boldsymbol{w}$, where $\boldsymbol{w}$ is of course the vector associated with $\boldsymbol{W}$.

Proof. Recall that, in view of (7.41), the membrane stress $\partial_A \hat{\tau}_\alpha(\boldsymbol{u}, \boldsymbol{v}, \boldsymbol{A})$ depends linearly on $\boldsymbol{u}$:

$$\partial_A \hat{\tau}_\alpha(\boldsymbol{u}, \boldsymbol{v}, \boldsymbol{A}) = \hat{\boldsymbol{B}}(\boldsymbol{v}, \boldsymbol{A})\, \boldsymbol{U} \tag{7.51}$$

for $\alpha = 1$ or 2.

Our first step is to prove that $\hat{\tau}_\alpha$ is rank-one affine, *i.e.*, that, for all $(\boldsymbol{u}, \boldsymbol{v}, \boldsymbol{A}) \in \mathscr{S}(1) \times \mathscr{V} \times \mathrm{Lin}$,

$$\boldsymbol{c} \otimes \boldsymbol{d} \cdot (\partial_A^2 \hat{\tau}_\alpha(\boldsymbol{u}, \boldsymbol{v}, \boldsymbol{A}))\, [\boldsymbol{c} \otimes \boldsymbol{d}] = 0 \tag{7.52}$$

for all $\boldsymbol{c}, \boldsymbol{d} \in \mathscr{V}$ (*cf.* (7.12)). We begin by noticing a consequence of (7.51):

$$(\partial_A^2 \hat{\tau}_\alpha(\boldsymbol{u}, \boldsymbol{v}, \boldsymbol{A}))\, [\boldsymbol{H}] = \partial_A \hat{\boldsymbol{B}}(\boldsymbol{v}, \boldsymbol{A})\, [\boldsymbol{H}]\, \boldsymbol{U} \quad \text{for all } \boldsymbol{H} \in \mathrm{Lin}, \tag{7.53}$$

and by writing

$$\begin{aligned} \boldsymbol{c} \otimes \boldsymbol{d} \cdot (\partial_A^2 \hat{\tau}_\alpha(\boldsymbol{u}, \boldsymbol{v}, \boldsymbol{A}))\, [\boldsymbol{c} \otimes \boldsymbol{d}] &= \boldsymbol{c} \cdot (\partial_A \boldsymbol{B}(\boldsymbol{v}, \boldsymbol{A})\, [\boldsymbol{c} \otimes \boldsymbol{d}]\, \boldsymbol{U})\, \boldsymbol{d} \\ &= -\boldsymbol{c} \otimes \boldsymbol{u} \cdot (\partial_A \hat{\boldsymbol{B}}(\boldsymbol{v}, \boldsymbol{A})\, [\boldsymbol{c} \otimes \boldsymbol{d}]\, \boldsymbol{D}) = -\boldsymbol{c} \otimes \boldsymbol{u} \cdot \partial_A^2 \hat{\tau}_\alpha(\boldsymbol{d}, \boldsymbol{v}, \boldsymbol{A})\, [\boldsymbol{c} \otimes \boldsymbol{d}], \end{aligned} \tag{7.54}$$

where repeated use has been made of (7.53) and $\boldsymbol{D}$ is the skew tensor uniquely associated with $\boldsymbol{d}$. We notice next that, as $\partial_A^2 \hat{\tau}_\alpha(\boldsymbol{u}, \boldsymbol{v}, \boldsymbol{A})$ is symmetric,

$$\partial_A^2 \hat{\tau}_\alpha(\boldsymbol{u}, \boldsymbol{v}, \boldsymbol{A})\, [\boldsymbol{H}] = \partial_A(\partial_A \hat{\tau}_\alpha(\boldsymbol{u}, \boldsymbol{v}, \boldsymbol{A}) \cdot \boldsymbol{H}) \quad \text{for all } \boldsymbol{H} \in \mathrm{Lin}, \tag{7.55}$$

so that in particular, for $\boldsymbol{H} = \boldsymbol{c} \otimes \boldsymbol{d}$,

$$\partial_A^2 \hat{\tau}_\alpha(\boldsymbol{u}, \boldsymbol{v}, \boldsymbol{A})\, [\boldsymbol{c} \otimes \boldsymbol{d}] = \partial_A(\boldsymbol{c} \cdot \partial_A \hat{\tau}_\alpha(\boldsymbol{u}, \boldsymbol{v}, \boldsymbol{A})\, \boldsymbol{d}). \tag{7.56}$$

Combining then (7.54) and (7.56), and recalling that $\hat{\tau}_\alpha$ is tangential:

$$\partial_A \hat{\tau}_\alpha(\boldsymbol{u}, \boldsymbol{v}, \boldsymbol{A})\, \boldsymbol{u} = \boldsymbol{0} \quad \text{for all } (\boldsymbol{u}, \boldsymbol{v}, \boldsymbol{A}) \in \mathscr{S}(1) \times \mathscr{V} \times \mathrm{Lin}, \tag{7.57}$$

we establish (7.52), as

$$-\boldsymbol{c} \otimes \boldsymbol{u} \cdot \partial_A^2 \hat{\tau}_\alpha(\boldsymbol{d}, \boldsymbol{v}, \boldsymbol{A})\, [\boldsymbol{c} \otimes \boldsymbol{d}] = -\boldsymbol{c} \otimes \boldsymbol{u} \cdot \partial_A(\boldsymbol{c} \cdot \partial_A \hat{\tau}_\alpha(\boldsymbol{u}, \boldsymbol{v}, \boldsymbol{A})\, \boldsymbol{d}) = 0.$$

Our second step consists in adapting in the obvious way the argument in the proof of Proposition 7.7 to conclude that $\hat{\tau}_\alpha$, being rank-one affine and tangential (and being, moreover, homogeneous with associated membrane stress linear in $\boldsymbol{u}$ as (7.51) requires), must have the form

$$\hat{\tau}_\alpha(\boldsymbol{u}, \boldsymbol{v}, \boldsymbol{A}) = \hat{\gamma}_\alpha(\boldsymbol{u}, \boldsymbol{v}) + (\hat{\boldsymbol{c}}_\alpha(\boldsymbol{v}) \otimes \boldsymbol{u}) \cdot \boldsymbol{A}^* + \hat{\boldsymbol{C}}_\alpha(\boldsymbol{v})\, \boldsymbol{U} \cdot \boldsymbol{A}, \tag{7.58}$$

with

$$\hat{\boldsymbol{C}}_\alpha(\boldsymbol{v})\, \boldsymbol{U} + \partial_A \boldsymbol{A}^*(\boldsymbol{A})\, [\hat{\boldsymbol{c}}_\alpha(\boldsymbol{v}) \otimes \boldsymbol{u}] = \hat{\boldsymbol{B}}(\boldsymbol{v}, \boldsymbol{A})\, \boldsymbol{U}, \quad \alpha = 1, 2. \tag{7.59}$$

Finally, from (7.59) we deduce that, for each fixed $\boldsymbol{v} \in \mathscr{V}$,

$$[\hat{\boldsymbol{C}}_1(\boldsymbol{v}) - \hat{\boldsymbol{C}}_2(\boldsymbol{v})]\, \boldsymbol{U} \cdot \boldsymbol{A} + 2(\hat{\boldsymbol{c}}_1(\boldsymbol{v}) - \hat{\boldsymbol{c}}_2(\boldsymbol{v})) \otimes \boldsymbol{u} \cdot \boldsymbol{A}^* = 0, \tag{7.60}$$

identically in $\boldsymbol{u}$ and $\boldsymbol{A}$.[25] Exploiting the arbitrariness in the choice of $\boldsymbol{A}$, we see

---

[25] To arrive at (7.60) we have made use of the fact, following from the symmetry of $\partial_A \boldsymbol{A}^*$, (1.18), (1.21) and (1.30), that

$$\partial_A \boldsymbol{A}^*[\boldsymbol{A}] = 2\boldsymbol{A}^*.$$

with the use of (1.15) that (7.60) is equivalent to

$$[\hat{C}_1(v) - \hat{C}_2(v)]\, U = 0 \quad \text{and} \quad (\hat{c}_1(v) - \hat{c}_2(v)) \otimes u = 0$$

identically in u, or rather, to

$$\hat{C}_1(v) = \hat{C}_2(v) \quad \text{and} \quad \hat{c}_1(v) = \hat{c}_2(v) \quad \text{for all } v \in \mathscr{V}. \tag{7.61}$$

(7.58) and (7.61) together then yield the desired conclusion. □

In the light of the above proposition we see that, given a jointly tangential pair $(\hat{\tau}_1, \hat{\tau}_2)$ and a domain $\Omega \in \mathcal{O}$, one can construct an associated homogeneous and tangential surface interaction potential T with density

$$\hat{\tau}: \mathscr{S}(1) \times \mathscr{V} \times \mathrm{Lin} \to \mathbb{R}, \quad \tau = \hat{\tau}(u, v, A) = \hat{\gamma}(u, v) + (\hat{c}(v) \otimes u) \cdot A^* + \hat{C}(v)\, U \cdot A,$$

by requiring that, for all $f \in C^2(\bar{\Omega})$,

$$\hat{\gamma}(n(x), f(x))|_{\Sigma_\alpha} := \hat{\gamma}_\alpha(n(x), f(x))|_{\Sigma_\alpha}, \quad \alpha = 1, 2. \tag{7.62}$$

As such a T is simple, it is obvious from Proposition 7.9 that one can also construct a null Lagrangian

$$F\{f\} = \int_\Omega \hat{\varphi}(f, F)\, \mathrm{d}\,(\mathrm{Vol}) \tag{7.63}$$

such that

$$T\{f\} = F\{f\} + \int_{\partial\Omega} \hat{\gamma}(n, f)\, \mathrm{d}\,(\mathrm{Srf}) \quad \text{for all } f \in C^2(\bar{\Omega}). \tag{7.64}$$

Conversely, given a null Lagrangian density $\hat{\varphi}$ as in (7.63) and two mappings

$$\hat{\gamma}_\alpha: \mathscr{S}(1) \times \mathscr{V} \to \mathbb{R}, \quad \gamma_\alpha = \hat{\gamma}_\alpha(u, v), \quad \alpha = 1, 2, \tag{7.65}$$

it is not difficult to construct a jointly tangential pair of mappings $(\hat{\tau}_1, \hat{\tau}_2)$ such that the homogeneous tangential potential T defined by (7.64) be regular with respect to $(\hat{\tau}_1, \hat{\tau}_2)$.

In this way, surface interactions obeying all conditions (7.39) identically in $\mathcal{O}$ and $\mathscr{D}$ are characterized as those whose surface densities are specified by a jointly tangential pair of mappings, and an answer to our original question is thereby provided. Without great loss of generality, as Proposition 7.13 makes clear, we finally turn to consider interactions specified by one density mapping, rather than two. In this connection, we introduce a definition which parallels strictly Definition 7.10.

7.14. Definition. *A class* C^2 *mapping*

$$\hat{\tau}: \mathscr{S}(1) \times \mathscr{V} \times \mathrm{Lin} \to \mathbb{R}, \quad \tau = \hat{\tau}(u, v, A),$$

is regular *if, for all* $(v, A) \in \mathscr{V} \times \mathrm{Lin}$ *fixed, the following condition holds:*

$$\partial_A \hat{\tau}(u_1, v, A)\, u_2 + \partial_A \hat{\tau}(u_2, v, A)\, u_1 = 0 \quad \textit{for all } u_1, u_2 \in \mathscr{S}(1). \tag{7.66}$$

7.15. Remark. Clearly, for $\hat{\tau}_1 \equiv \hat{\tau}_2$ $(7.40)_1$ reduces to (7.66); moreover, $(7.40)_3$ reduces to

$$(\partial_A \hat{\tau}(u, v, A))\, u = 0 \quad \text{for all } u \in \mathscr{S}(1), \tag{7.67}$$

and taking $u_1 = u_2 = u$ in (7.66) we see that a regular $\hat{\tau}$ satisfies (7.67); finally, again as a consequence of the linearity of $\partial_A \hat{\tau}$ (*cf.* Remark 7.12), we see that a regular $\hat{\tau}$ also satisfies

$$\partial_A \hat{\tau}(u, v, A) + \partial_A \hat{\tau}(-u, v, A) = 0 \quad \text{for all } u \in \mathscr{S}(1), \tag{7.68}$$

which is the form presently taken by $(7.40)_2$. □

In view of Proposition 7.13, a regular $\hat{\tau}$ admits the representation

$$\hat{\tau}(u, v, A) = \hat{\gamma}(u, v) + (\hat{c}(v) \otimes u) \cdot A^* + \hat{C}(v)\, U \cdot A. \tag{7.69}$$

For each $\Omega \in \mathcal{O}$ fixed, we say that the homogeneous tangential potential

$$T\{f\} = \int_{\partial\Omega} \hat{\tau}(n, f, {}^sF)\, \mathrm{d}\,(\mathrm{Srf}), \tag{7.70}$$

with density $\hat{\tau}$ given by (7.69), is a *regular* surface interaction potential. Regular interactions can be regarded as a generalization of a pressure interaction on the entire boundary, whose density is (*cf.* $(5.5)_1$)

$$\hat{\tau}_p(u, v, A) = \hat{\pi}(v)\, (v \otimes u) \cdot A^*; \tag{7.71}$$

they are characterized by our last proposition.

7.16 Proposition. *A tangential potential T as in* (7.70) *is regular if and only if there is a null Lagrangian F as in* (7.63) *and a mapping*

$$\hat{\gamma} : \mathscr{S}(1) \times \mathscr{V} \to \mathbb{R}, \quad \gamma = \hat{\gamma}(u, v),$$

such that (7.64) *holds.*

Proof. At this stage, it suffices to show how to construct the density $\hat{\varphi}$ of F from the density $\hat{\tau}$ of T (and conversely). To this end, given $\hat{\tau}$ as in (7.69), for each $f \in \mathscr{D}$ we write

$$\hat{\tau}(n, f, F) - \hat{\gamma}(n, f) = \hat{v}(f, F) \cdot n, \tag*{$(7.72)_1$}$$

with

$$\hat{v}(f, F) \cdot w := F^{*\mathrm{T}} \hat{c}(f) \cdot w - F^{\mathrm{T}} \hat{C}(f) \cdot W \tag*{$(7.72)_2$}$$

for all $w \in \mathscr{V}$ and W the skew tensor associated with w. By virtue of divergence theorem, (7.72) yields

$$\hat{\varphi}(f, F) = \mathrm{Div}\, \hat{v}(f, F). \tag{7.73}$$

Moreover, it is not difficult to give $\hat{\varphi}$ the canonical form of a null Lagrangian density [15], [16]: performing the differentiation indicated in (7.73) we obtain, with the use of (4.6) and (5.9),

$$\hat{\varphi}(f, F) = \hat{\delta}(f) \det F + \hat{D}(f) \cdot F^*, \tag*{$(7.74)_1$}$$

with

$$\hat{\delta}(f) := \mathrm{div}\, \hat{c}(f), \quad 3\hat{D}(f) \cdot a \otimes b := -b \cdot \mathrm{curl}\, (\hat{C}(f)\, b) \quad \text{for all } a, b \in \mathscr{V}. \;\square \tag*{$(7.74)_2$}$$

References

1. G. Capriz & P. Podio-Guidugli, Duality and stability questions for the linearized traction problem with live loads in elasticity. Pp. 70–78 of *Stability in the Mechanics of Continua*, F. H. Schroeder, Ed., Springer-Verlag, 1982.
2. P. Podio-Guidugli, G. Vergara Caffarelli & E. G. Virga, The role of ellipticity and normality assumptions in formulating live-boundary conditions in elasticity. Quart. Appl. Math. **44**, pp. 659–664, 1987.
3. P. Podio-Guidugli, A variational approach to live loadings in finite elasticity. J. Elasticity **19**, pp. 25–36, 1988.
4. M. J. Sewell, On configuration-dependent loading. Arch. Rational Mech. Anal. **23**, pp. 327–351, 1967.
5. M. J. Sewell, On the calculation of potential functions defined on curved boundaries. Proc. Royal Soc. London Ser. A **286**, pp. 402–411, 1966.
6. R. J. Knops & E. W. Wilkes, *Theory of Elastic Stability*. Handbuch der Physik VIa/3, C. Truesdell, Ed., Springer-Verlag, 1973.
7. P. Podio-Guidugli & G. Vergara Caffarelli, On a class of live traction problems in elasticity. Pp. 291–304 of *Trends and Applications of Pure Mathematics to Mechanics*, P. G. Ciarlet & M. Roseau, Eds., Springer-Verlag, 1984.
8. P. Podio-Guidugli, G. Vergara Caffarelli & E. G. Virga, Una formula di Green per il problema di carichi vivi dell'elastostatica linearizzata. Pp. 1–10 of *Atti VII Congresso AIMETA*, Vol. II, 1984.
9. G. Vergara Caffarelli, Green's formulas for linearized problems with live loads. In *Metastability and Incompletely Posed Problems*, I.M.A. Vol. Math. Appl. **3** (S. Antman and coeditors), Springer-Verlag, 1987.
10. S. J. Spector, On uniqueness in finite elasticity with general loading. J. Elasticity **10**, pp. 145–161, 1980.
11. S. J. Spector, On uniqueness for the traction problem in finite elasticity. J. Elasticity **12**, pp. 367–383, 1982.
12. T. Valent, *Boundary Value Problems of Finite Elasticity*. Springer-Verlag, 1988.
13. P. Podio-Guidugli & E. G. Virga, Sull'equilibrio di un involucro elastico pieno di gas. Atti Acc. Lincei Rend. Fis. (8) LXXXI. Nota I, pp. 287–292; Nota II, pp. 293–298, 1987.
14. D. Fisher, Configuration dependent pressure potentials. J. Elasticity **19**, pp. 77–84, 1988.
15. J. L. Ericksen, Nilpotent energies in liquid crystal theory. Arch. Rational Mech. Anal. **10**, pp. 189–196, 1962.
16. D. G. B. Edelen, The null set of the Euler-Lagrange operator. Arch. Rational Mech. Anal. **11**, pp. 117–121, 1962.
17. L. C. Martins & P. Podio-Guidugli, On the local measures of mean rotation in continuum mechanics. Forthcoming.
18. M. E. Gurtin, *An Introduction to Continuum Mechanics*. Academic Press, 1981.
19. D. De Tommasi & S. Marzano, On the local analysis of continuum deformations. To appear in Meccanica.
20. P. G. Ciarlet, *Mathematical Elasticity. Vol. I: Three-Dimensional Elasticity*. North-Holland, 1988.
21. M. E. Gurtin, On the nonlinear theory of elasticity. Pp. 237–253 of *Contemporary Developments in Continuum Mechanics and Partial Differential Equations*, G. M. de la Penha & L. A. J. Medeiros, Eds., North-Holland, 1978.
22. B. Dacorogna, *Weak Continuity and Weak Lower Semicontinuity of Non-Linear Functionals*. Lecture Notes in Mathematics, Vol. 922, Springer-Verlag, Berlin, 1982.
23. B. Dacorogna, *Direct Methods in the Calculus of Variations*. In press.

24. P. L. OLVER & J. SIVALOGANATHAN, The structure of null Lagrangians. Nonlinearity, **1**, pp. 389–398, 1988.
25. C. TRUESDELL, *A First Course in Rational Continuum Mechanics*. Academic Press, 1977. (Second edition in press.)
26. M. E. GURTIN & A. I. MURDOCH, A continuum theory of elastic material surfaces. Arch. Rational Mech. Anal. **57**, pp. 291–323, 1975.
27. F. ANDREUSSI & M. E. GURTIN, On the wrinkling of a free surface. J. Appl. Phys. **48**, pp. 3798–3799, 1977.
28. M. E. GURTIN & A. I. MURDOCH, Surface stress in solids. Int. J. Solids Structures, pp. 431–440, 1978.
29. J. L. ERICKSEN, Some simpler cases of the Gibbs problem for thermoelastic solids. J. Thermal Stresses **4**, pp. 13–30, 1981.
30. P. J. FLORY, Thermodynamic relations for high elastic materials. Trans. Faraday Soc. **57**, pp. 829–838, 1961.

Dipartimento di Ingegneria Civile Edile
Università di Roma "Tor Vergata"
Roma, Italy

and

Dipartimento di Matematica
Università di Trento
Povo, Italy

(Received December 5, 1988)

Some Kinematical Results Concerning Steady Flows and Extensional Flows

WAN-LEE YIN

Dedicated to Professor Bernard D. Coleman on his Sixtieth Birthday

1. Introduction

Vector fields used as local reference directions (*i.e.*, base vectors) for the representation of tensorial physical quantities need not be directly associated with any curvilinear coordinate system. Such vector fields form an "anholonomic" system and their use in differential geometry has been expounded by mathematicians *. In classical fluid mechanics the significance of the streamline direction or the vorticity direction and their use as reference directions for tensorial physical quantities were long recognized. Besides, in certain classes of hydrodynamic flows there are material surfaces and one or more families of material lines whose tangent or normal vectors may be naturally included in a convected anholonomic reference system. However, because the constitutive equations of classical fluid mechanics are not history-dependent, there is no compelling need to trace the evolution of kinematical variables of a material element and, therefore, the use of convected anholonomic reference systems is not essential. On the other hand, the response functionals of rheological fluids with memory depend on the history of kinematical variables. The component forms of such relationships may be considerably simplified when an appropriate intrinsic reference system is used.

Recent works by MARRIS and me demonstrate the usefulness of intrinsic vector fields in obtaining the solutions of overdetermined systems of differential equations for specific classes of flows or in proving that solutions of such systems do not exist. In some problems, an anholonomic reference system is chosen from the start and the analysis proceeds to show that the anholonomic system reduces to a holonomic system as a result of the over-determinacy of the system of equations. The solutions of the problem, if there are any, are then obtained in terms of the curvilinear coordinates associated with the holonomic reference system.

In the present work, we demonstrate that intrinsic vector fields may be used along a different line of approach to obtain more general results concerning steady

---

* See Reference [1] for an account of the work of CARTAN.

flows and a certain special class of generally unsteady flows called extensional flows. In a sense, the analysis of the following two sections illustrates a more constructive usage of anholonomic systems in the kinematics of flow.

In Section 2, we solve the general problem concerning the determination of the deformation gradient tensor from the velocity field of a steady flow. In Section 2 we obtain a necessary and sufficient condition (in terms of the first two Rivlin-Ericksen tensors) for a class of motions in which three families of material axes remain mutually orthogonal. Such motions, which are here called extensional flows, include a special class of motions with constant stretch history first studied by COLEMAN, *i.e.*, steady extensions.

2. Determination of the deformation gradient history of a steady flow

In this section, we solve the general problem of determining the deformation gradient history of a material element in a steady flow from the given velocity field of that flow. This problem is of interest to experimental determination or verification of constitutive equations of rheological fluids, because the response functionals of such fluids contain the deformation gradient history as a principal independent variable. While the steady velocity field and the stress field may be experimentally measured, at least in principle, the formulation of their relationship requires the determination of the deformation gradient history of each fluid element from the measured velocity field.

Let $\boldsymbol{v}$ denote the steady velocity field and let

$$\boldsymbol{L} = \nabla \boldsymbol{v} \tag{1}$$

be the velocity gradient. Then the deformation gradient history $\boldsymbol{F}$ of a material element is governed by the differential equation [2]:

$$\dot{\boldsymbol{F}} = \boldsymbol{L}\boldsymbol{F}. \tag{2}$$

For unsteady flows this differential equation is extremely difficult to solve, except in some trivial special cases, because the velocity gradient is given in a spatial description whereas the deformation gradient history and its material derivative must be found in a material description. In steady flows, however, the material time derivative is related to the directional derivative along the streamline. One has

$$d/dt = V\boldsymbol{s} \cdot \nabla,$$

where V is the magnitude of velocity and $\boldsymbol{s}$ is the unit tangent vector of the streamline. We consider any particular point in the flow, and denote that point by the symbol O. Let the velocity vector at O be given by

$$\boldsymbol{v}_{\mathrm{O}} = V_{\mathrm{O}} \boldsymbol{s}_{\mathrm{O}}.$$

Furthermore, we choose two smooth curves C_1 and C_2 such that these curves and the streamline through O intersect orthogonally at O. Let $\boldsymbol{e}_{\mathrm{O}}$ and $\boldsymbol{f}_{\mathrm{O}}$ denote, respectively, the unit tangent vectors of C_1 and C_2 at O. We next consider the two stream surfaces S_1 and S_2, containing, respectively, the curves C_1 and C_2. At any point on the streamline passing through O, let $\boldsymbol{e}$ and $\boldsymbol{f}$ be the unit tangent vectors of S_1 and S_2, respectively, that are orthogonal to the streamline direction $\boldsymbol{s}$. It should be emphasized that although $\boldsymbol{e}_O \cdot \boldsymbol{f}_O = 0$, in general $\boldsymbol{e}$ is *not* perpendicular to $\boldsymbol{f}$. We then have the following expression for the history of deformation gradient relative to the configuration of the material element at the current position O:

$$\boldsymbol{F} = \lambda_1 \boldsymbol{s} \otimes \boldsymbol{s}_O + (\lambda_2 \boldsymbol{e} + \mu_2 \boldsymbol{s}) \otimes \boldsymbol{e}_O + (\lambda_3 \boldsymbol{f} + \mu_3 \boldsymbol{s}) \otimes \boldsymbol{f}_O, \tag{3}$$

where $\lambda_1(t)$, $\lambda_2(t)$, $\lambda_3(t)$ and $\mu_3(t)$ are unknown functions to be determined. Here, along the streamline, the arc length parameter s starting from the point O is related to the time variable t according to the following equation:

$$t = \int_0^s \{1/V(s')\}\, ds'. \tag{4}$$

Furthermore, since $\boldsymbol{F}$ reduces to the identity tensor at $t = 0$, we have

$$\lambda_1(0) = \lambda_2(0) = \lambda_3(0) = 1, \qquad \mu_2(0) = \mu_3(0) = 0.$$

Since the response functional depends on the past history of the deformation gradient, the five scalar functions $\lambda_1, \ldots, \mu_3$ need only be evaluated for negative values of t or s.

It is easy to determine the function λ_1. We consider two particles A and B which momentarily occupy the two points on the streamline with arc lengths 0 and s. In an infinitesimal time interval dt, the two particles move through distances $ds_0 = V_0\, dt$ and $ds = V\, dt$, respectively. Hence,

$$\lambda_1 = ds/ds_0 = V/V_0,$$

where V denotes the flow speed at B.

Next, we have

$$\boldsymbol{L}\boldsymbol{e} = \boldsymbol{e} \cdot \nabla(V\boldsymbol{s}) = V\boldsymbol{e} \cdot \nabla \boldsymbol{s} + (\boldsymbol{e} \cdot \nabla V)\, \boldsymbol{s}, \tag{5}$$

$$\boldsymbol{L}\boldsymbol{s} = \boldsymbol{s} \cdot \nabla(V\boldsymbol{s}) = V\boldsymbol{s} \cdot \nabla \boldsymbol{s} + (\boldsymbol{s} \cdot \nabla V)\, \boldsymbol{s}. \tag{6}$$

Furthermore, from equation (2) we obtain

$$\boldsymbol{F}\boldsymbol{e}_0 = \lambda_2 \boldsymbol{e} + \mu_2 \boldsymbol{s}.$$

Differentiating with respect to time, and making use of equations (2), (5) and (6), from the last equation we show that

$$\dot{\lambda}_2 \boldsymbol{e} + \lambda_2 V\boldsymbol{s} \cdot \nabla \boldsymbol{e} + \dot{\mu}_2 \boldsymbol{s} + \mu_2 V\boldsymbol{s} \cdot \nabla \boldsymbol{s} = \dot{\boldsymbol{F}}\boldsymbol{e}_0 = \boldsymbol{L}\boldsymbol{F}\boldsymbol{e}_0 = \boldsymbol{L}(\lambda_2 \boldsymbol{e} + \mu_2 \boldsymbol{s}).$$

The components of the last equation in the directions $\boldsymbol{e}$ and $\boldsymbol{s}$ give, respectively,

$$(d/ds) \ln \lambda_2 = -\boldsymbol{s}(\boldsymbol{e} \cdot \nabla \boldsymbol{e}),$$

$$(d/ds)\,(\mu_2/V) = (\lambda_2/V^2)\, \boldsymbol{s} \times \boldsymbol{e} \cdot (\mathrm{curl}\, \boldsymbol{v}).$$

Hence,

$$\lambda_2(s) = \exp\left\{-\int_0^s s\cdot(e\cdot\nabla e)\,ds'\right\}, \tag{7}$$

$$\mu_2(s) = V(s)\int_0^s \lambda_2(s')\,\{1/V(s')\}^2\,\{s\times e\cdot(\text{curl}\,v)\}\,ds', \tag{8}$$

where

$$\omega_2 = s\times e\cdot(\text{curl}\,v)$$

is the component of the vorticity vector along the local normal direction of the stream surface S_1. Results similar to equations (7) and (8) may be obtained for the functions λ_3 and μ_3.

Since λ_2 is proportional to the spacing between neighboring streamlines on the stream surface S_1, it can be obtained from the geometry of streamlines without actually evaluating the integral in equations (7). The integrand in Eq. (8) depends on the values of V, λ_2 and ω_2 along the streamline from $s'=0$ to $s'=s$. The functions μ_2 and μ_3 may be evaluated by numerical integration. Although the accuracy of the results decreases with increasing arc length s, this may not seriously affect the response functional if the latter has the property of fading memory.

For a steady flow, the preceding analysis yields an analytical expression of the deformation gradient history of a fluid element in terms of the geometrical quantities associated with the streamline passing through the element and the speed and the vorticity along that streamline. For the purpose of approximate numerical computation, however, the deformation gradient may be evaluated alternatively, in a more straightforward manner, from the geometrical pattern of the streamlines and the isochronal surfaces. Starting from an arbitrary referential isochronal surface corresponding to $t=0$, an isochronal surface corresponding to another time instant t may be determined by marking the point on each streamline with the arc length parameter s which satisfies equation (4). We then consider a stream tube bounded by two pairs of stream surfaces. An upstream volume element of this stream tube cut by a pair of isochronals with the time lapse Δt is deformed into a downstream volume element cut by another pair of isochronals with the same time lapse Δt. Direct comparison of the shape and size of the two volume elements yields the relative deformation gradient. Alternatively, both volume elements may be compared to a unit cube and the deformation gradients with respect to the latter, $\boldsymbol{F}_1$ and $\boldsymbol{F}_2$, determine the relative deformation gradient of the second volume element with respect to the first, $\boldsymbol{F}_2(\boldsymbol{F}_1)^{-1}$.

For steady extrusion of metals in axisymmetric dies, the streamline pattern and the isochronal surfaces may be obtained experimentally using a technique described by BERGHAUS & PEACOCK [3]. The deformation gradient and the associated finite strain measures may be directly calculated by comparing a volume element in the remote upstream region and a corresponding volume element in a near region, without going through the elaborate computational procedure described in their paper (which, incidently, introduces an "accumulated shear" that is not really a component of strain).

Flows for which the vorticity vector curl v is everywhere parallel to the velocity vector have been studied by BELTRAMI and more recently by TRUESDELL, BJØRGUM,

GODAL, MARRIS, and other authors [4]. These motions are called screw motions or Beltrami motions. For such motions, and for all irrotational motions, the integrands in the expressions of μ_2 and μ_3 vanish identically. Consequently, in steady Beltrami motions and steady irrotational motions the deformation gradient history reduces to the simpler form

$$F = \lambda_1 s \otimes s_0 + \lambda_2 e \otimes e_0 + \lambda_3 f \otimes f_0 .$$

As was mentioned above, the vector fields e and f are generally *not* orthogonal, even though e_0 and f_0 are orthogonal by definition.

3. Extensional flows

In their kinematical study of motions with constant stretch history, COLEMAN & NOLL introduced a class of motions called steady extensions [5]. In such motions the material elements are continually stretched (or compressed) along three mutually perpendicular material directions as they move and rotate in space. The rate of stretching of each material direction is required to be constant in time.

One may generalize the concept and consider the larger class of motions in which each material element is similarly stretched or compressed along three mutually orthogonal material axes, although the rate of stretching of the these material axes need no longer be constant in time. Since there is no shear deformation associated with the three material axes, such motions will be called extensional motions or extensional flows. They are generally unsteady motions.

For a material element in an extensional flow, the deformation gradient at time s referred to a local reference configuration at time t has the following form:

$$F_t(s) = \sum_{i=1}^{3} \lambda_i(s)\, e_i(s) \otimes e_i(t), \tag{9}$$

where e_1, e_2 and e_3 are orthonormal material axes and

$$\lambda_1(t) = \lambda_2(t) = \lambda_3(t) = 1 .$$

The relative Cauchy-Green tensor $C_t(s)$ is given by

$$C_t(s) = F_t(s)^T F_t(s) = \sum_{i=1}^{3} \lambda_i(s)^2\, e_i(t) \otimes e_i(t). \tag{10}$$

In the characterization of motions with constant stretch history, the Rivlin-Ericksen tensors play an important role [6, 7]. These kinematical tensors are defined as follows:

$$A_n(t) = \frac{d^n}{ds^n} C_t(s)\big|_{s=t} \quad (n = 1, 2, \ldots). \tag{11}$$

For example, the fourth and all subsequent Rivlin-Ericksen tensors vanish identically in viscometric flows [6]. The Rivlin-Ericksen tensors satisfy the recurrence relation

$$A_{n+1} = \dot{A}_n + L^T A_n + A_n L, \tag{12}$$

where L stands for the velocity gradient tensor.

In the following, we show that the class of extensional motions defined above may be characterized by an important relation between the first and second Rivlin-Ericksen tensors. Specifically, we shall prove the following theorem:*

Theorem. *A motion is an extensional motion if and only if the first and second Rivlin-Ericksen tensors have common eigenvectors for all time. This is the case if and only if all Rivlin-Ericksen tensors have common eigenvectors for all time.*

The "only if" part of the first statement of the theorem is easily seen from equations (10) and (11). The "if" part of the second statement is obvious. The converse statements of the theorem may be proved as follows.

Let $\alpha(t)$, $\beta(t)$ and $\gamma(t)$ denote the eigenvalues of the first Rivlin-Ericksen tensor $A_1(t)$. Let $e_1(t)$, $e_2(t)$ and $e_3(t)$ be the corresponding eigenvectors (some of the eigenvectors may fail to be unique if the eigenvalues are not all distinct). Let $\{i, j, k\}$ be a *constant* orthonormal basis. We define the tensors

$$Q(t) = e_1(t) \otimes i + e_2(t) \otimes j + e_3(t) \otimes k, \tag{13}$$

$$\Lambda_1(t) = \alpha(t)\, i \otimes i + \beta(t) j \otimes j + \gamma(t)\, k \otimes k. \tag{14}$$

Then

$$\dot{\Lambda}_1(t) = \dot{\alpha}(t)\, i \otimes i + \dot{\beta}(t) j \otimes j + \dot{\gamma}(t)\, k \otimes k, \tag{15}$$

$$\begin{aligned} A_1(t) &= \alpha e_1 \otimes e_1 + \beta e_2 \otimes e_2 + \gamma e_3 \otimes e_3 \\ &= Q(t)\, \Lambda_1(t)\, Q(t)^T, \end{aligned} \tag{16}$$

where $Q(t)^T$ denotes the transpose of $Q(t)$. Let $L(t)$ be the stretching tensor at time t and $W(t)$ be the spin tensor. Then

$$L = W + \tfrac{1}{2} A_1 .$$

Differentiating Eq. (16) and substituting the result into

$$A_2 = \dot{A}_1 + L^T A_1 + A_1 L,$$

we get

$$\begin{aligned} A_2 &= \dot{Q} Q^T A_1 + A_1 Q \dot{Q}^T + Q \dot{\Lambda}_1 Q^T + (-W + A_1/2)\, A_1 + A_1(W + A_1/2) \\ &= -(\Omega + W)\, A_1 + A_1(\Omega + W) + Q(\dot{\Lambda}_1 + \Lambda_1^2)\, Q^T. \end{aligned}$$

Hence

$$A_2 - Q(\dot{\Lambda}_1 + \Lambda_1^2)\, Q^T = -(\Omega + W)\, A_1 + A_1(\Omega + W), \tag{17}$$

where

$$\Omega \equiv -\dot{Q} Q^T. \tag{18}$$

* Without reference to extensional motions and geometrical implications, MARRIS [8] discovered the theorem under the condition that the first Rivlin-Ericksen tensor has three distinct eigenvalues. I am indebted to Professor MARRIS for showing me his results. However, the general proof produced here is essentially different from MARRIS' original proof (for the case of distinct eigenvalues).

If A_1 and A_2 have the same eigenvectors (e_1, e_2 and e_3), then the left-hand side of equation (17) has a diagonal matrix when referred to the rotating basis $\{e_1, e_2, e_3\}$. However, referred to the same basis, the matrix of the right-hand side has off-diagonal elements only, because the matrix of A_1 has only diagonal elements and because $\boldsymbol{\Omega} + \boldsymbol{W}$ is skew-symmetric. It follows that both sides of equations (17) must vanish. Consequently,

$$A_2 = Q(\dot{\Lambda}_1 + \Lambda_1^2)\, Q^T, \tag{19}$$

$$(\Omega + W)\, A_1 = A_1(\Omega + W). \tag{20}$$

Equation (19) implies that if A_1 has repeated eigenvalues, so does A_2, and the multiplicity of any eigenvalue of A_2 is equal to the multiplicity of the corresponding eigenvalue of A_1. This conclusion follows whenever the two kinematical tensors are coaxial, *i.e.*, when their eigenvectors coincide.

We now proceed to prove the "only if" part of the second statement of the theorem by induction. It has already been shown that if in some time interval the two tensors A_1 and A_2 are coaxial, then any repeated eigenvalue of the first tensor has the same multiplicity as the corresponding eigenvalue of the second tensor. We write down the induction hypothesis

$$A_n = Q\Lambda_n Q^T \tag{21}$$

for any integer equal to or greater than 2, where the matrix of Λ_n with respect to the basis $\{\boldsymbol{i}, \boldsymbol{j}, \boldsymbol{k}\}$ is diagonal and the multiplicity of a repeated eigenvalue of A_n equals the multiplicity of the corresponding eigenvalue of A_1. Then,

$$\begin{aligned} A_{n+1} &= \dot{A}_n + L^T A_n + A_n L \\ &= \dot{Q}Q^T A_n + A_n Q\dot{Q}^T + Q\dot{\Lambda}_n Q^T \\ &\quad + (-W + A_1/2)\, A_n + A_n(A_1/2 + W) \\ &= -(\Omega + W)\, A_n + A_n(\Omega + W) + Q(\dot{\Lambda}_n + \Lambda_1\Lambda_n/2 + \Lambda_n\Lambda_1/2)\, Q^T. \end{aligned} \tag{22}$$

We now make use of a lemma given by WANG [7]: Let $[S]$ be a 3×3 diagonal matrix and $[W]$ be a 3×3 skew-symmetric matrix,

$$[S] = \begin{bmatrix} \alpha & 0 & 0 \\ 0 & \beta & 0 \\ 0 & 0 & \gamma \end{bmatrix}, \quad [W] = \begin{bmatrix} 0 & x & y \\ -x & 0 & z \\ -y & -z & 0 \end{bmatrix}.$$

Then

i) if $\alpha \neq \beta \neq \gamma$, $[SW] = [WS]$ if and only if $x = y = z = 0$.

ii) if $\alpha = \beta \neq \gamma$, $[SW] = [WS]$ if and only if $y = z = 0$.

iii) if $\alpha = \beta = \gamma$, $[SW] = [WS]$ for all x, y, z.

From equation (20), and from the induction hypothesis that any repeated eigenvalues of A_1 and A_n have the same multiplicity, it follows that for all three cases considered in Wang's lemma, we have

$$(\Omega + W)\, A_n = A_n(\Omega + W).$$

Consequently, equation (22) reduces to

$$A_{n+1} = Q(\dot{A}_n + A_1 A_n/2 + A_n A_1/2)\, Q^T. \tag{23}$$

Hence the tensor A_{n+1} also has a diagonal matrix when referred to the basis $\{Qi, Qj, Qk\} = \{e_1, e_2, e_3\}$, and to a repeated eigenvalue of A_1 there is a corresponding repeated eigenvalue of A_{n+1} with the same multiplicity. This completes the induction proof of the "only if" part of the second statement of the theorem.

Substituting equations (21) and (23) into the following series expansion

$$C_t(s) = I + (s - t)\, A_1(t) + \{(s + t)^2/2!\}\, A_2(t) + \ldots,$$

we obtain,

$$\begin{aligned} C_t(s) &= Q(t)\, \{I + (s - t)\, A_1(t) + \{(s - t)^2/2!\}\, A_2(t) + \ldots\}\, Q(t)^T \\ &= Q(t) \exp\left\{\int_t^s A_1(\xi)\, d\xi\right\} Q(t)^T. \end{aligned}$$

Thus the eigenvectors of $C_t(s)$ are independent of s; only its eigenvalues depend on s. The material element is continually stretched along three material axes $e_1(s)$, $e_2(s)$ and $e_3(s)$ which remain orthogonal as they translate and rotate in space. The proof of the theorem is now complete.

References

1. Schouten, J. A.: *Ricci Calculus*, 2nd Ed., Springer (1954).
2. Truesdell, C., & Noll, W.: *The Non-linear Field Theories of Mechanics*. Encyclopedia of Physics III/3 (S. Flügge, Ed.), Springer (1965).
3. Berghaus, D. G., & Peacock, H. B., "Deformation and strain analysis for high-extrusion ratios and elevated temperatures," *Experimental Mech.*, Vol. **25**, pp. 301–308 (1985).
4. Truesdell, C., *The Kinematics of Vorticity*, Indiana Univ. Press (1953).
5. Coleman, B. D., & Noll, W., "Steady extensions of incompressible simple fluids," *Phys. Fluids*, Vol. **5**, pp. 840–843 (1962).
6. Noll, W., "Motions with constant stretch history," *Archive Rational Mech. Anal.*, Vol. **11**, pp. 97–105 (1962).
7. Wang, C.-C., "A representation theorem for the constitutive equation for a simple material in motions with constant stretch history," *Archive Rational Mech. Anal.*, Vol. **20**, pp. 329–340 (1966).
8. Marris, A. W., Private communication; to be published.

School of Civil Engineering
Georgia Institute of Technology
Atlanta

(Received June 30, 1989)

Material Symmetry and Crystals

ROGER L. FOSDICK & BRIAN HERTOG

Dedicated to Bernard Coleman on the occasion of his sixtieth birthday.

1. Introduction

When a crystal undergoes a change of symmetry during a phase transition, it suffers a change in the point group that characterizes the crystal structure. This kind of phase transition also is associated with the existence of variant forms whose particular arrangements are closely related to the symmetry change itself [1, 21]. Thus, the standard practice of considering a *fixed* point group as the material symmetry group clearly is inadequate to describe such phenomena. This observation was given an interesting perspective ten years ago by ERICKSEN [2] when he suggested that a more local theory of material symmetry was needed. Others [19, 20], already had begun to question the classical development of constitutive theory concerning material symmetry on the grounds that the symmetry was not allowed to change with either the temperature or the deformation gradient. The ideas of ERICKSEN in [2], and later in [17, 18], have motivated some additional development of the local theory for crystals by PITTERI [3] and others, wherein point groups serve as local material symmetry groups that are embedded in a basic global invariance group; the latter group identifies the crystal lattice structure. We, also, shall pursue this direction here.

Sections 2 and 3 of this paper are preliminary and basic in that they, respectively, introduce and discuss the concept of a local theory of material symmetry, and provide the elements of a common lattice description of crystals. As a minimal requirement on a local theory, it is reasonable to expect that in the pure mechanics of elastic bodies for which the stored energy response function W_0 depends on the deformation gradient, for example, the invariance considerations of material symmetry would include the possibility that the degree of symmetry depends on the level of the deformation gradient, itself. That is, given a deformation gradient $\boldsymbol{F}_0$, the usual invariance condition of material symmetry $W_0(\boldsymbol{F}) = W_0(\boldsymbol{F}\boldsymbol{H})$ may be expected to hold for a *fixed* group of unimodular transformations $\boldsymbol{H}$ only when the (admissible) deformation gradients $\boldsymbol{F}$ lie in an appropriate open set $\mathscr{D}_{\boldsymbol{F}_0}$ containing $\boldsymbol{F}_0$. In this case, the particular subgroup of unimodular transformations, denoted by $\mathfrak{S}(\mathscr{D}_{F_0})$, is called the local material symmetry group on $\mathscr{D}_{F_0}$. Moreover, it also is reasonable to expect that the open set $\mathscr{D}_{F_0}$ is sufficiently rich that material symmetry transformations H applied to any of its members will, again, generate members of $\mathscr{D}_{F_0}$, *i.e.* $\mathscr{D}_{F_0}H = \mathscr{D}_{F_0}$. In addition, in order to accommodate the requirement of observer invariance, the open set $\mathscr{D}_{F_0}$ also is expected to satisfy the condition that observer transformations are local in the sense that $Q\mathscr{D}_{F_0} = \mathscr{D}_{F_0}$ for all orthogonal Q; of course, then, $W_0(F) = W_0(QF)$ for all such Q and for all F in $\mathscr{D}_{F_0}$.

The invariance condition of material symmetry may be interpreted as follows: After having deformed a body so that the deformation gradient at a given particle is F_0, it makes no difference to the value of the stored energy response function at that particle whether it is then further deformed with an arbitrary *relative* deformation gradient whose composition with F_0 is in $\mathscr{D}_{F_0}$, or whether it is first placed in an intermediate configuration whose deformation gradient is equal to $F_0HF_0^{-1}$, with $H \in \mathfrak{S}(\mathscr{D}_{F_0})$, and then distorted by the *same* relative deformation gradient as noted above. This interpretation suggests that for a *solid* body it is reasonable to expect that $F_0\mathfrak{S}(\mathscr{D}_{F_0})F_0^{-1} \subset \mathfrak{O}$, where $\mathfrak{O}$ is the orthogonal group. That is, the intermediate configurations that qualify should correspond to a subgroup of *rigid* transformations.

In Section 4 of this paper, we show (*cf.* Theorem 1) that for any deformation gradient F_0 of a crystal whose structure is described by one of the fourteen Bravais lattices, there is an open set $\mathscr{D}_{F_0}$, containing F_0, that satisfies $Q\mathscr{D}_{F_0}H = \mathscr{D}_{F_0}$ for all orthogonal Q and all H in the local material symmetry group on $\mathscr{D}_{F_0}$, *i.e.*, all H in $\mathfrak{S}(\mathscr{D}_{F_0})$. In addition, it is shown that $\mathfrak{S}(\mathscr{D}_{F_0})$ must be conjugate through F_0 to a subgroup of the orthogonal group in the sense that $F_0\mathfrak{S}(\mathscr{D}_{F_0})F_0^{-1} \subset \mathfrak{O}$. It then follows (*cf.* Remark 3) that $F_0\mathfrak{S}(\mathscr{D}_{F_0})F_0^{-1}$ will correspond to one of the crystallographic point groups, and will be of cubic type and/or hexagonal type depending on whether it is a subgroup of a cubic or a hexagonal point group. All triclinic, monoclinic, rhombic, trigonal, tetragonal, and cubic point groups are of cubic type while all triclinic, monoclinic, rhombic, trigonal and hexagonal point groups are of hexagonal type; clearly, these types are not distinct and have considerable overlap. Thus, for the theory of crystals as described by Bravais lattices, the prejudices, as mentioned above and applied within the continuum theory of material symmetry and objectivity for elastic solids, are provable properties. While our considerations and arguments employ group theory, PITTERI [3] has obtained similar properties from a different point of view. His main theorem is summarized briefly in our Remark 5.

Now, suppose that F_0 is a deformation gradient of a crystal, and that $\mathscr{D}_{F_0}$ is the open set of Theorem 1, and $\mathfrak{S}(\mathscr{D}_{F_0})$ is the local material symmetry group on $\mathscr{D}_{F_0}$. Further, suppose that H is a symmetry transformation in the conjugate group to the basic symmetry group which describes the Bravais lattice structure of the crystal. Then, if a deformation gradient F is such that FH has a local material symmetry group contained in $\mathfrak{S}(\mathscr{D}_{F_0})$, the deformation gradient F is said to lie in the symmetry class of the deformation gradient F_0. In this case, the local constitutive and invariance properties of the crystal at F_0 on $\mathscr{D}_{F_0}$ and at F on the corresponding open set of Theorem 1 are equivalent even though F may not be contained in $\mathscr{D}_{F_0}$. For such situations, we show in Theorem 2 that there is an open set of deformation gradients $\mathscr{D}^*_{F_0}$ that has the same properties as $\mathscr{D}_{F_0}$ and which serves to cover the same range of values of the stored energy response function as is covered over the complete set of deformation gradients that are in the symmetry class of F_0. The set $\mathscr{D}^*_{F_0}$ is called a fundamental domain for the symmetry class of F_0.

At the end of Section 4, we show (*cf.*, Theorem 3 and its supporting lemma) that for any crystal there are always deformation gradients F_c and F_h whose associated local material symmetry groups are of maximal cubic and maximal hexagonal type, respectively, and which are such that an arbitrary deformation gradient F is in the symmetry class of either F_c or F_h, or both. Thus, for a given deformation gradient F, there is a symmetry transformation H in the group conjugate to the basic symmetry group which describes the Bravais lattice structure of the crystal such that FH is in the union of the fundamental domains for the symmetry classes of F_c and F_h. In other words, we show that the union of the two fundamental domains for the symmetry classes of F_c and F_h will serve as a master fundamental domain for the crystal itself.

Finally, in Section 5, we consider briefly the main elements of a local theory of material symmetry for a more general stored energy response function; one which would apply to the theory of elastic dielectrics. We state our results in Theorems 4, 5, and 6, which are mild generalizations of the earlier three theorems of this work.

2. The local theory of material symmetry

The purpose of this section is to introduce the concept of a local theory of material symmetry from a continuum mechanics point of view. To do this, first let $\mathscr{B}_0 \subset \mathbb{R}^3$ be an undistorted, regular reference configuration of a homogeneous body whose particles occupy the points $X \in \mathscr{B}_0$. We assume that at $X \in \mathscr{B}_0$ the stored energy density of the body, $W(X)$, is given by a response function $W_0\colon \mathrm{Inv} \to \mathbb{R}$, measured per unit reference volume, such that

$$W(X) = W_0(\nabla\chi(X)), \tag{2.1}$$

where the mapping $X \to \chi(X)$ of $\mathscr{B}_0$ into $\mathbb{R}^3$ denotes the deformation of $\mathscr{B}_0$, and $\nabla\chi(X)$ is the deformation gradient at $X \in \mathscr{B}_0$. Inv denotes the set of all invertible linear transformations of $\mathbb{R}^3$ into itself.

The notion of material symmetry of a body is concerned with the concept that the response of the body is invariant to certain changes in reference configuration. The invariance may be described mathematically by a material symmetry group $\mathfrak{S}$ that consists in all unimodular transformations H that satisfy

$$W_0(F) = W_0(FH) \quad \forall\, F \in \mathrm{Inv},$$

where

$$F \equiv \nabla\chi(X), \quad X \in \mathscr{B}_0.$$

Each $H \in \mathfrak{S}$ may by interpreted as being a gradient of a mapping from $\mathscr{B}_0$ to one of its peers (see TRUESDELL [4]). That is, if $y\colon \mathscr{B}_0 \to \mathscr{B}_y$ and if $H \equiv \nabla y(X) \in X$, then $\mathscr{B}_0$ and $\mathscr{B}_y$ are peers and $H \in \mathfrak{S}$ whenever

$$W_0(F) = W_y(F) \quad \forall F \in \mathrm{Inv},$$

where $W_y: \mathrm{Inv} \to \mathbb{R}$ is the response function defined with respect to $\mathscr{B}_y$.

The local theory of material symmetry differs from the usual global theory in that it is concerned with the invariance properties of (2.1) on subsets of Inv. Given any $F_0 \in \mathrm{Inv}$, an open set $\mathscr{D}_{F_0} \subsetneqq \mathrm{Inv}$ is envisioned — a set which contains F_0 — on which the local material symmetry properties of (2.1) are described by the group $\mathfrak{S}(\mathscr{D}_{F_0})$ of all unimodular transformations H for which

$$W_0(F) = W_0(FH) \quad \forall F \in \mathscr{D}_{F_0}, \tag{2.2$_1$}$$

and

$$\mathscr{D}_{F_0}H = \mathscr{D}_{F_0}, \tag{2.2$_2$}$$

where

$$\mathscr{D}_{F_0}H \equiv \{A \in \mathrm{Inv} \mid A = FH \quad \text{if } F \in \mathscr{D}_{F_0}\}.$$

At this level of generality, a set of the form $\mathscr{D}_{F_0}$ is simply assumed to exist. Note that $(2.2)_2$ embodies the local nature of $\mathfrak{S}(\mathscr{D}_{F_0})$, a group which we shall call the *local material symmetry group on* $\mathscr{D}_{F_0}$.

Besides the invariance properties of material symmetry, we also require (2.1) to be objective, or, equivalently, observer invariant under rigid body rotations. Thus, for each $Q \in \mathfrak{O}$, where $\mathfrak{O}$ denotes the orthogonal group of linear transformations, we require†

$$W_0(F) = W_0(QF) \quad \forall F \in \mathrm{Inv}. \tag{2.3$_1$}$$

In addition, in order to eliminate preferred observers in the local theory, it is natural to require $\mathscr{D}_{F_0}$ to satisfy

$$Q\mathscr{D}_{F_0} = \mathscr{D}_{F_0} \quad \forall Q \in \mathfrak{O}, \tag{2.3$_2$}$$

where

$$Q\mathscr{D}_{F_0} \equiv \{A \in \mathrm{Inv} \mid A = QF \quad \text{if } F \in \mathscr{D}_{F_0}\}.$$

Together, (2.2) and (2.3) lead to the following local invariance statement for (2.1) on $\mathscr{D}_{F_0} \subsetneqq \mathrm{Inv}$:

$$W_0(F) = W_0(QFH) \quad \forall F \in \mathscr{D}_{F_0}, \tag{2.4$_1$}$$

and

$$Q\mathscr{D}_{F_0}H = \mathscr{D}_{F_0}, \tag{2.4$_2$}$$

whenever $Q \in \mathfrak{O}$ and $H \in \mathfrak{S}(\mathscr{D}_{F_0})$.

Let us rewrite, temporarily, the invariance conditions of material symmetry by introducing the relative deformation gradient

$$F^* \equiv FF_0^{-1}, \quad F \in \mathscr{D}_{F_0}.$$

† Strictly speaking, rigid body rotations are restricted to lie in the *proper* subgroup of $\mathfrak{O}$. Here, however, we have included in $(2.3)_1$ the useful and harmless assumption that $W_0(F) = W_0(-F)\ \forall F \in \mathrm{Inv}$.

Then $(2.2)_1$ may be written as

$$W_0(F^*F_0) = W_0(F^*F_0H),$$

or

$$W_0(F^*F_0) = W_0(F^*KF_0), \tag{2.5}$$

where we have defined the unimodular transformation K by

$$K \equiv F_0HF_0^{-1}, \quad H \in \mathfrak{S}(\mathcal{D}_{F_0}).$$

In these terms, according to (2.5) the invariance condition of material symmetry states that after deforming a particle of $\mathcal{B}_0$ to a configuration whose deformation gradient is F_0, it makes no difference in the value of the stored energy density of the particle whether it is then further deformed simply with the *relative* deformation gradient F^* or whether it is next placed in an intermediate configuration, defined by the relative deformation gradient K and then further deformed relative to this new configuration by the same relative deformation gradient F^*. The relative deformation gradient K characterizes a material symmetry transformation that involves no relative change of shape if it is orthogonal. This motivates the following

Definition 1. *$\mathcal{B}_0$ is locally solid at $F_0 \in \mathrm{Inv}$ if there is an open set $\mathcal{D}_{F_0} \subset \mathrm{Inv}$ containing F_0 such that $F_0\mathfrak{S}(\mathcal{D}_{F_0})F_0^{-1} \subset \mathfrak{O}$.*

The only difference between this definition of a solid and the more conventional one (see, for example, TRUESDELL [4]) is that the present definition is local and and applies to subsets of Inv.

In the remainder of this paper, we will first summarize some elements concerning the symmetry of crystals, and then we will show that for each possible deformation gradient F_0 of a crystal there is an open set $\mathcal{D}_{F_0} \subset \mathrm{Inv}$, containing F_0, which is such that $Q\mathcal{D}_{F_0}H = \mathcal{D}_{F_0}$ for all $Q \in \mathfrak{O}$ and all $H \in \mathfrak{S}(\mathcal{D}_{F_0})$, where $\mathfrak{S}(\mathcal{D}_{F_0})$ is the local material symmetry group on $\mathcal{D}_{F_0}$. Moreover, we will show that $\mathfrak{S}(\mathcal{D}_{F_0})$ is conjugate through F_0 to a subgroup of $\mathfrak{O}$, *i.e.*, $F_0\mathfrak{S}(\mathcal{D}_{F_0})F_0^{-1} \subset \mathfrak{O}$. At least for crystals, then, the local theory of material symmetry, as discussed above, has a meaningful interpretation, and, according to the definition above, at any of its possible deformation gradients a crystal is locally solid.

3. Crystals

We shall now account for the fundamental periodicity of crystalline structure and determine the material symmetry properties of crystals. We begin by describing the crystal structure in terms of 1-lattices. PITTERI [3] defines a 1-lattice† as a countable set of points $Y \in \mathbb{R}^3$ each of which is a linear combination of three linearly independent *lattice vectors* E_a, $a = 1, 2, 3$. More precisely, a 1-lattice

† A 1-lattice is also a Bravais lattice of which there are 14 types (see BRADLEY & CRACKNELL [5]).

consists in those points in $\mathbb{R}^3$ that can be represented in the form

$$Y = n^a E_a + Y_0, \tag{3.1}$$

where $Y_0 \in \mathbb{R}^3$ is a fixed reference point and n^a, $a = 1, 2, 3$, are integers. Thus the set

$$\mathscr{L}(E_a) \equiv \{Y \in \mathbb{R}^3 \mid (3.1) \text{ holds}\} \tag{3.2}$$

is a 1-lattice and each $Y \in \mathscr{L}(E_a)$ represents an atom or perhaps a set of atoms in the crystal.

Lattice vectors are not unique. In fact, it is easy to show that $\mathscr{L}(E_a) = \mathscr{L}(E'_a)$ whenever

$$E'_a = h_a^b E_b, \tag{3.3$_1$}$$

where h_a^b, $a, b = 1, 2, 3$, are the integer entries of a 3×3 matrix $\|h_a^b\|$ that satisfies

$$\det \|h_a^b\| = \pm 1. \tag{3.3$_2$}$$

The set of all such matrices $\|h_a^b\|$ forms a discrete, non-compact group $\mathfrak{G}$ whose group operation is matrix multiplication:

$$\mathfrak{G} \equiv \{\|h_a^b\| \mid (3.3)_2 \text{ holds and } h_a^b,\ a, b = 1, 2, 3 \text{ are integers}\}. \tag{3.4}$$

For each $\|h_a^b\| \in \mathfrak{G}$, then, we have

$$\mathscr{L}(E_a) = \mathscr{L}(h_a^b E_b), \tag{3.5}$$

and it is in this sense that $\mathfrak{G}$ is considered to be the basic symmetry group for elastic crystals [8, 15, 16, 17]. Further, for each $\|h_a^b\| \in \mathfrak{G}$ we can identify an $H \in \mathfrak{U}$, where $\mathfrak{U}$ denotes the unimodular group of linear transformations, such that

$$HE_a = h_a^b E_b, \quad a = 1, 2, 3. \tag{3.6}$$

The set of all H that corresponds through the conjugate relation (3.6) to the group $\mathfrak{G}$ forms the basic conjugate symmetry group:

$$\mathfrak{G}(E_a) \equiv \{H \in \mathfrak{U} \mid H = h_a^b E_b \otimes E^a,\ \|h_a^b\| \in \mathfrak{G}\}, \tag{3.7}$$

where the reciprocal lattice vectors E^a, $a = 1, 2, 3$, satisfy

$$E_a \cdot E^b = \delta_a^b.$$

Hence we may infer from (3.5) and (3.6) that

$$\mathscr{L}(E_a) = \mathscr{L}(HE_a) \quad \forall\ H \in \mathfrak{G}(E_a). \tag{3.8}$$

Of special interest is that subset of $\mathfrak{G}(E_a)$ which is contained in the orthogonal group $\mathfrak{O}$. Thus, consider

$$\mathfrak{P}(E_a) \equiv \mathfrak{G}(E_a) \cap \mathfrak{O}. \tag{3.9}$$

Since $\mathfrak{P}(E_a)$ consists in those orthogonal transformations in $\mathfrak{G}(E_a)$ that leave the 1-lattice $\mathscr{L}(E_a)$ invariant, it follows that $\mathfrak{P}(E_a)$ must be one of the crystallographic point groups. In Table 1 we list the 32 crystallographic point groups along with

Table 1. *The 32 crystallographic point groups and their generators. The indices i and j (≠ i) may be set equal to* 1, 2, *or* 3 *depending on the choice of crystallographic axes. An asterisk is used to identify the holohedral point groups.*

Triclinic		*Trigonal*	
1	**1**	3	R_i^3
$\bar{1}$*	$-$**1**	$\bar{3}$	$-R_i^3$
		3*m*	$R_i^3, -R_j^2$
Monoclinic		32	R_i^3, R_j^2
m	$-R_i^2$	$\bar{3}m$*	$-R_i^3, -R_j^2$
2	R_i^2	*Cubic*	
$\frac{2}{m}$*	$R_i^2, -R_i^2$	23	R_i^2, R_4^3
Rhombic		$\frac{2}{m}\bar{3}$	$R_i^2, -R_i^2, -R_4^3$
2*m*	$R_i^2, -R_j^2$	$\bar{4}3m$	$-R_i^4, R_4^3, -R_j^2$
22	R_i^2, R_j^2	432	R_i^4, R_4^3, R_j^2
*mmm**	$-R_1^2, -R_2^2, -R_3^2$	$\frac{4}{m}\bar{3}$*	$R_i^4, -R_i^2, -R_4^3$
Tetragonal		*Hexagonal*	
$\bar{4}$	$-R_i^4$	$\bar{6}$	$-R_i^6$
4	R_i^4	$\bar{6}2$	$-R_i^6, R_j^2$
$\frac{4}{m}$	$R_i^4, -R_i^4$	6	R_i^6
$\bar{4}2$	$-R_i^4, R_j^2$	$\frac{6}{m}$	$R_i^6, -R_i^2$
4*m*	$R_i^4, -R_j^2$	6*m*	$R_i^6, -R_j^2$
42	R_i^4, R_j^2	62	R_i^6, R_j^2
$\frac{4}{m}m$*	$R_i^4, -R_i^2, R_j^2$	$\frac{6}{m}2$*	$R_i^6, -R_i^2, R_j^2$

a set of generators, *i.e.*, a set of elements whose products and powers exhaust the group. In this table, R_α^n denotes a rotation of $2\pi/n$ radians about the unit vector i_α of a right-handed orthonormal basis for $\alpha = 1, 2$, or 3, and about $i_4 \equiv (i_1 + i_2 + i_3)/\sqrt{3}$ for $\alpha = 4$. In addition, the international symbols in the left hand column† corresponding to each point group have the following meaning: (i) a single integer n denotes an n-fold axis of rotational symmetry while strings of two or three integers denote additional orthogonal axes of such type of symmetry; (ii) $\bar{n}$ denotes an n-fold axis of rotational symmetry with a polar inversion

† In some cases, here , our notation is slightly more compact than the conventiona one (see Bradley & Cracknell [5]).

along the axis; (iii) $m \equiv \bar{2}$ denotes a reflection or mirror plane; (iv) $\frac{n}{m}$ denotes an n-fold axis of rotational symmetry together with a mirror plane which is orthogonal to the axis.

Remark 1. A 1-lattice description of the crystal structure is not always adequate. More generally, for some integer $n \geqq 0$, the crystal structure may be described in terms of an $(n+1)$-lattice, *i.e.*, as a collection of $n+1$ identical 1-lattices $\mathscr{L}_0, \mathscr{L}_1, \ldots, \mathscr{L}_n$ that differ only by uniform translations (see PITTERI [6]). The $(n+1)$-lattice may be described in terms of a given 1-lattice, $\mathscr{L}_0$ say, and the *shifts*, $\boldsymbol{P}_i$, $i = 1, 2, \ldots, n$, where each $\boldsymbol{P}_i$ is the vector difference between a fixed point in $\mathscr{L}_0$ and a fixed point in $\mathscr{L}_i$. In particular, if $\mathscr{L}_0 = \mathscr{L}(\boldsymbol{E}_a)$ for some set of lattice vectors $\boldsymbol{E}_a$, $a = 1, 2, 3$, then an $(n+1)$-lattice consists in $\mathscr{L}(\boldsymbol{E}_a)$ and the 1-lattices $\mathscr{L}_i$, $i = 1, 2, \ldots, n$, defined by

$$\mathscr{L}_i \equiv \{\boldsymbol{Y} + \boldsymbol{P}_i \in \mathbb{R}^3 \mid \boldsymbol{Y} \in \mathscr{L}(\boldsymbol{E}_a)\}. \tag{3.10}$$

Note from (3.10) that the shift $\boldsymbol{P}_i$ is only unique up to transformations of the form

$$\boldsymbol{P}_i' = \boldsymbol{P}_i + \boldsymbol{Y}, \qquad \boldsymbol{Y} \in \mathscr{L}(\boldsymbol{E}_a).$$

Thus, although $\mathscr{L}(\boldsymbol{E}_a)$ satisfies the invariance condition (3.8), the invariance group of the $(n+1)$-lattice may only be a subset of $\mathfrak{G}(\boldsymbol{E}_a)$ since the $(n+1)$-lattice is invariant only with respect to those $\boldsymbol{H} \in \mathfrak{G}(\boldsymbol{E}_a)$ that satisfy

$$\boldsymbol{H}\boldsymbol{P}_i = \boldsymbol{P}_i + \boldsymbol{Y}, \qquad \boldsymbol{Y} \in \mathscr{L}(\boldsymbol{E}_a) \tag{3.11}$$

for $i = 1, 2, \ldots, n$. This also means that the point group corresponding to an $(n+1)$-lattice contains only those elements of $\mathfrak{P}(\boldsymbol{E}_a)$ that satisfy (3.11). □

Next we shall use the 1-lattice description of crystals and describe briefly how a stored energy function of the form (2.1) and a material symmetry group for crystals are determined. To do this, first let $\mathscr{B}_0 \subset \mathbb{R}^3$ be an undistorted regular reference configuration of a homogeneous body and assign a fixed set of lattice vectors $\bar{\boldsymbol{E}}_a$, $a = 1, 2, 3$, to $\mathscr{B}_0$†. In effect, this ascribes a lattice or crystal structure to $\mathscr{B}_0$. Now, if the body undergoes a deformation $\chi: \mathscr{B}_0 \to \mathbb{R}^3$, it is common, though not without reservation, to impose the *Cauchy-Born hypothesis* which requires the lattice vectors to deform as material vectors, *i.e.*,

$$\bar{\boldsymbol{E}}_a \to \boldsymbol{e}_a \equiv \boldsymbol{F}\bar{\boldsymbol{E}}_a, \tag{3.12}$$

at least for deformation gradients $\boldsymbol{F}$ that belong to some open set in Inv containing the identity.†† Further, motivated by molecular theories of elasticity (see, for example, ERICKSEN [7]), it is common to assume that the stored energy density at $\boldsymbol{X} \in \mathscr{B}_0$ depends on the deformation of the crystal through an explicit dependence

† We can impose selection rules such as those introduced by PARRY [8] to determine a unique set of fixed lattice vectors.

†† According to ZANZOTTO [22], there is strong evidence that the Cauchy-Born rule fails to hold for certain crystals.

on its deformed lattice vectors e_a. Thus we assume the existence of a function $\varphi: \mathbb{R}^3 \times \mathbb{R}^3 \times \mathbb{R}^3 \to \mathbb{R}$ such that†

$$W = \varphi(e_a) = \varphi(F\bar{E}_a). \tag{3.13}$$

The deformed lattice vectors e_a, $a = 1, 2, 3$, are not unique and, in fact, they must satisfy a condition of the form (3.5), *i.e.*, $\mathscr{L}(e_a) = \mathscr{L}(h_a^b e_b)$ for all $\|h_a^b\| \in \mathfrak{G}$. Thus the choice of lattice vectors should not affect W and consequently we shall assume that

$$\varphi(e_a) = \varphi(h_a^b e_b) \quad \forall \, \|h_a^b\| \in \mathfrak{G}.$$

Moreover, since

$$h_a^b e_b = F h_a^b \bar{E}_b = F H \bar{E}_b,$$

where $H = h_a^b \bar{E}_b \otimes \bar{E}^a \in \mathfrak{G}(\bar{E}_a)$, we see that

$$\varphi(e_a) = \varphi(FH\bar{E}_a), \quad \forall \, H \in \mathfrak{G}(\bar{E}_a). \tag{3.14}$$

In addition to this invariance condition we shall require (3.13) to be objective, and so

$$\varphi(e_a) = \varphi(Qe_a) = \varphi(QF\bar{E}_a) \quad \forall \, Q \in \mathfrak{O}. \tag{3.15}$$

It follows from (3.13), (3.14) and (3.15) that, since the lattice vectors $\bar{E}_a$ are fixed, we may write

$$W = W(F) \tag{3.16}$$

for some function $W\colon \mathrm{Inv} \to \mathbb{R}$, where

$$W(F) = W(QFH) \tag{3.17$_1$}$$

for all $F \in \mathrm{Inv}$ for which the Cauchy-Born hypothesis holds, all $Q \in \mathfrak{O}$, and all $H \in \mathfrak{G}(\bar{E}_a)$. If we let $\mathscr{D}$ denote the open set of deformation gradients in Inv which is available to the function W, it is clear that this set will have to satisfy the properties that $\mathbf{1} \in \mathscr{D}$, $\mathscr{D} \subseteq \mathrm{Inv}$, and

$$\mathscr{D} = Q\mathscr{D}H \quad \forall \, Q \in \mathfrak{O} \quad \text{and} \quad \forall \, H \in \mathfrak{G}(\bar{E}_a). \tag{3.17$_2$}$$

We shall call $\mathfrak{G}(\bar{E}_a)$ the *global*†† *material symmetry group on* $\mathscr{D}$ and establish the following

Definition 2. *$\mathscr{B}_0$ is crystalline on $\mathscr{D} \subset$ Inv, and $\mathscr{D}$ is called the* **crystal domain** *for $\mathscr{B}_0$, if the (local) material symmetry group on $\mathscr{D}$, $\mathfrak{S}(\mathscr{D})$, corresponds to $\mathfrak{G}(\bar{E}_a)$ for some set of lattice vectors $\bar{E}_a$, $a = 1, 2, 3$.*

Remark 2. Suppose that in addition to a set of lattice vectors $\bar{E}_a$, $a = 1, 2, 3$, we also assign shifts $\bar{P}_i$, $i = 1, 2, \ldots, n$, to $\mathscr{B}_0$. Further, suppose in analogy to

† Here, to simplify, we suppress the dependence on X.

†† We use the term *global* here rather than local in order to emphasize that $\mathscr{D}$ is the entire domain of the stored energy function (3.16).

the Cauchy-Born hypothesis that these shifts transform as material vectors, *i.e.*, $\bar{P}_i \to p_i \equiv F\bar{P}_i$. Then, if these shifts are also arguments of (3.13), a stored energy function of the form (3.16) would still emerge, *i.e.*, $W = \varphi(e_a, p_i) = \varphi(F\bar{E}_a, F\bar{P}_i) \to W(F)$. We know from Remark 1, however, that the invariance group of an $(n+1)$-lattice may only be a subset of $\mathfrak{G}(\bar{E}_a)$, $\mathfrak{G}^*(\bar{E}_a)$ say, and so $(3.17)_1$ should hold only for those $H \in \mathfrak{G}^*(\bar{E}_a)$. In this case, Definition 2 must be revised by replacing $\mathfrak{G}(\bar{E}_a)$ with $\mathfrak{G}^*(\hat{E}_a) \subsetneqq \mathfrak{G}(\hat{E}_a)$. □

4. The local theory of material symmetry and crystals

In the first part of this section we shall prove that whenever $\mathscr{B}_0$ is crystalline on $\mathscr{D} \subset \text{Inv}$ it must also be locally solid at every $F \in \mathscr{D}$. That is, if the stored energy function (3.16) is defined on a crystal domain $\mathscr{D}$, then, given any $F_0 \in \mathscr{D}$, we show that there is a corresponding open set $\mathscr{D}_{F_0} \subset \mathscr{D}$, containing F_0, such that $Q\mathscr{D}_{F_0}H = \mathscr{D}_{F_0}$ for all $Q \in \mathfrak{O}$ and all $H \in \mathfrak{S}(\mathscr{D}_{F_0})$, where the local material symmetry group $\mathfrak{S}(\mathscr{D}_{F_0})$ is conjugate through F_0 to a subgroup of $\mathfrak{O}$, *i.e.*, $F_0\mathfrak{S}(\mathscr{D}_{F_0})F_0^{-1} \subset \mathfrak{O}$. Before proving this result in Theorem 1 and its corollary, we shall first establish some preliminary results in Lemmas 1 and 2. Further, having shown that for each $F_0 \in \mathscr{D}$ such an open set $\mathscr{D}_{F_0} \subset \mathscr{D}$ exists, we then introduce the useful concept of a symmetry class of F_0 as the set of all deformation gradients $F \in \text{Inv}$ that satisfy the requirement that for some $H \in \mathfrak{S}(\mathscr{D})$ the local material symmetry group associated with FH is contained in $\mathfrak{S}(\mathscr{D}_{F_0})$. While the set of deformation gradients in the symmetry class of F_0 may not be contained in the open set $\mathscr{D}_{F_0}$, the local material symmetry group that is associated with each of these deformation gradients F through the composition FH for some $H \in \mathfrak{S}(\mathscr{D})$ is, in fact, included in $\mathfrak{S}(\mathscr{D}_{F_0})$. This observation suggests the possibility of Theorem 2 which shows that for each $F_0 \in \mathscr{D}$ there is an open domain $\mathscr{D}^*_{F_0} \subset \mathscr{D}$ *having exactly the same properties as* $\mathscr{D}_{F_0}$, which is "sufficiently large" to cover the complete range of values taken on by the stored energy function over the symmetry class of F_0. This domain is called a fundamental domain for the symmetry class of F_0. Finally, in Theorem 3, we prove that there is a set, $\mathscr{D}^* \subset \mathscr{D}$, with the property that for each $F \in \mathscr{D}$ there is an $H \in \mathfrak{S}(\mathscr{D})$ such that

$$FH \in \bar{\mathscr{D}}^*,$$

where $\bar{\mathscr{D}}^*$ is the closure of $\mathscr{D}^*$. The significance of this last result is revealed by the invariance condition $(3.17)_1$, *i.e.*, the entire range of values exhibited by the stored energy function (3.16) on $\mathscr{D}$ is exhausted by $\bar{\mathscr{D}}^*$.

Now, to serve as a reminder and also as motivation for the work leading to Theorem 1, we emphasize, from Section 2, that the desired open set $\mathscr{D}_{F_0} \subset \mathscr{D}$ must satisfy

$$Q\mathscr{D}_{F_0}H = \mathscr{D}_{F_0} \tag{4.1}$$

whenever $Q \in \mathfrak{O}$ and $H \in \mathfrak{S}(\mathscr{D}_{F_0})$. Essentially, once we have shown that $\mathscr{D}_{F_0}$ satisfies (4.1) and that $F_0\mathfrak{S}(\mathscr{D}_{F_0})F_0^{-1} \subset \mathfrak{O}$, then we will have proved Theorem 1 and its corollary.

To begin, suppose $\mathscr{K}$ and $\mathscr{L}$ are subsets of Inv, and let

$$\mathscr{G}(\mathscr{K}, \mathscr{L}) \equiv \{T \in \text{Inv} \mid LT = K \text{ whenever } K \in \mathscr{K} \text{ and } L \in \mathscr{L}\}. \quad (4.2)$$

We then have

Lemma 1. *If* $\mathscr{K}$ *and* $\mathscr{L}$ *are compact subsets of* Inv, *then* $\mathscr{G}(\mathscr{K}, \mathscr{L})$ *is compact.*

Proof.† Consider the homeomorphism

$$f: (T, L) \to (LT, L)$$

of Inv×Inv onto Inv×Inv whose inverse image is given by

$$f^{-1}: (K, L) \to (L^{-1}K, L).$$

Since $\mathscr{K} \times \mathscr{L}$ is compact it follows that the inverse image $f^{-1}(\mathscr{K} \times \mathscr{L})$ must also be compact. Thus, to complete the proof, we note that

$$\text{proj}_1 f^{-1}(\mathscr{K} \times \mathscr{L}) = \mathscr{G}(\mathscr{K}, \mathscr{L}),$$

and, hence, $\mathscr{G}(\mathscr{K}, \mathscr{L})$ must be compact. □

Next we need

Lemma 2. *Suppose that the set* $\mathscr{S} \subset \text{Inv}$ *is discrete and, further, that* $\mathscr{K}$ *and* $\mathscr{L}$ *are compact subsets of* Inv. *Then the set*

$$\mathscr{S} \cap \mathscr{G}(\mathscr{K}, \mathscr{L})$$

is finite, i.e., *it contains only a finite number of elements.*

Proof. By Lemma 1, the set $\mathscr{G}(\mathscr{K}, \mathscr{L})$ is compact, and so for every open cover of $\mathscr{G}(\mathscr{K}, \mathscr{L})$ there exists a finite subcover or, equivalently, each collection of open sets that contain $\mathscr{G}(\mathscr{K}, \mathscr{L})$ has a finite subcollection of open sets whose union also contains $\mathscr{G}(\mathscr{K}, \mathscr{L})$. That is, suppose we choose as a cover of $\mathscr{G}(\mathscr{K}, \mathscr{L})$ a collection of open sets such that no member of this collection contains more than one element of the discrete set $\mathscr{S}$. Then, since any finite subcover formed out of such a collection of open sets must contain $\mathscr{G}(\mathscr{K}, \mathscr{L})$ and since each open set in this finite subcover can contain at most one element of $\mathscr{S}$, it follows that $\mathscr{S} \cap \mathscr{G}(\mathscr{K}, \mathscr{L})$ contains only a finite number of elements. □

The preceding two lemmas are the basis for establishing the proof of our Theorem 1, below. Before continuing, however, suppose $\mathscr{B}_0$ is crystalline on $\mathscr{D} \subset \text{Inv}$ and define

$$\mathfrak{G}_F \equiv \{H \in \mathfrak{S}(\mathscr{D}) \mid FHF^{-1} \in \mathfrak{D}\} \quad (4.3)$$

for any $F \in \text{Inv}$.

---

† This proof is an adaptation of one given on page 257 of Bourbaki [9].

Remark 3. The set $\mathfrak{G}_F$ is conjugate to a crystallographic point group and, hence, it is finite. To see this, recall from Definition 2 that $\mathfrak{S}(\mathscr{D}) = \mathfrak{G}(\bar{E}_a)$ for some set of lattice vectors $\bar{E}_a$, $a = 1, 2, 3$.† Thus, it follows from (3.7) that for each $H \in \mathfrak{G}(\bar{E}_a)$ there is a $\|h_a^b\| \in \mathfrak{G}$ such that

$$H\bar{E}_a = h_a^b \bar{E}_b, \quad a = 1, 2, 3.$$

In turn, this implies

$$FH\bar{E}_a = h_a^b F\bar{E}_b = h_a^b e_b, \tag{4.4}$$

where (3.12) has been used, or equivalently,

$$FHF^{-1}F\bar{E}_a = FHF^{-1}e_a = h_a^b e_b.$$

Thus we see that

$$\mathfrak{G}(e_a) = F\mathfrak{G}(\bar{E}_a)\, F^{-1}, \tag{4.5}$$

or

$$\mathfrak{G}(e_a) = \{A \in \text{Inv} \mid A = FHF^{-1} \text{ and } H \in \mathfrak{G}(\bar{E}_a)\},$$

where $\mathfrak{G}(e_a)$ is the group obtained from (3.7) when E_a is replaced by e_a. Finally, observe that the crystallographic point group

$$\mathfrak{P}(e_a) = \mathfrak{G}(e_a) \cap \mathfrak{O}$$

is conjugate to $\mathfrak{G}_F$, *i.e.*, $\mathfrak{P}(e_a) = F\mathfrak{G}_F F^{-1}$. □

We now have

Theorem 1. *Let* F_0 *be a fixed element of* $\mathscr{D}$, *and let* $\mathscr{D} \subset \text{Inv}$ *be a crystal domain for* $\mathscr{B}_0$. *Then, there is an open neighborhood of* F_0, $\mathscr{N}_{F_0} \subset \mathscr{D}$, *and a related open set*

$$\mathscr{D}_{F_0} \equiv \{A \in \text{Inv} \mid A = QFH \text{ where } Q \in \mathfrak{O},\ H \in \mathfrak{G}_{F_0}, \text{ and } F \in \mathscr{N}_{F_0}\}, \tag{4.6$_1$}$$

containing F_0, *such that*††

$$\mathscr{D}_{F_0} H \cap \mathscr{D}_{F_0} = \emptyset \quad \forall\, H \in \mathfrak{S}(\mathscr{D}) \setminus \mathfrak{G}_{F_0}. \tag{4.6$_2$}$$

Proof.††† Let $\mathscr{N}'_{F_0} \subset \text{Inv}$ be an open neighborhood of F_0 whose closure, $\bar{\mathscr{N}}'_{F_0}$, is compact. Then, since $\mathfrak{O}$ is compact, it follows that the set

$$\mathfrak{O}\bar{\mathscr{N}}'_{F_0} \equiv \bigcup_{Q \in \mathfrak{O}} Q\bar{\mathscr{N}}'_{F_0}$$

is also compact. To see this consider the homeomorphism

$$f: (T, L) \to (LT, L)$$

† More generally, as noted in Remark 2, $\mathfrak{S}(\mathscr{D})$ may be a subgroup of $\mathfrak{G}(\bar{E}_a)$.

†† Note that $\mathscr{D}_{F_0}$ also satisfies the condition $Q\mathscr{D}_{F_0}H = \mathscr{D}_{F_0}\ \forall\, Q \in \mathfrak{O}$ and $\forall\, H \in \mathfrak{G}_{F_0}$.

††† This proof is based partially on Proposition 8, page 256, of BOURBAKI [9].

of Inv×Inv onto Inv×Inv (see Lemma 1). The set $\bar{\mathcal{N}}'_{F_0}\times\mathfrak{O}$ is compact in Inv×Inv, which implies that $f(\bar{\mathcal{N}}'_{F_0}\times\mathfrak{O})$ is compact in Inv×Inv. Then, since

$$\mathrm{proj}_1 f(\bar{\mathcal{N}}'_{F_0}\times\mathfrak{O}) = \mathfrak{O}\bar{\mathcal{N}}'_{F_0},$$

we may conclude that $\mathfrak{O}\bar{\mathcal{N}}'_{F_0}$ is compact. Also note that the set

$$\mathscr{D}'_{F_0} \equiv \bigcup_{H\in\mathfrak{G}_{F_0}} \mathfrak{O}\mathcal{N}'_{F_0}H \tag{4.7$_1$}$$

is such that its closure,

$$\bar{\mathscr{D}}'_{F_0} = \bigcup_{H\in\mathfrak{G}_{F_0}} \mathfrak{O}\bar{\mathcal{N}}'_{F_0}H, \tag{4.7$_2$}$$

is compact since $\bar{\mathscr{D}}'_{F_0}$ is the union of a finite collection of compact sets.

Next, recall (4.2) and consider the set

$$\mathscr{S}' \equiv \mathfrak{S}(\mathscr{D}) \cap \mathscr{T}(\bar{\mathscr{D}}'_{F_0}, \bar{\mathscr{D}}'_{F_0}).$$

Since $\mathfrak{S}(\mathscr{D})$ is discrete in Inv we know from Lemma 2 that $\mathscr{S}'$ is finite. Moreover, (4.7) implies that $\mathfrak{G}_{F_0} \subseteqq \mathscr{S}'$.

Now suppose that $\mathfrak{G}_{F_0} \equiv \mathscr{S}'$. Then, there is no $H\in\mathfrak{S}(\mathscr{D})\setminus\mathfrak{G}_{F_0}$ such that

$$FH\in\bar{\mathscr{D}}'_{F_0}$$

for any $F\in\bar{\mathscr{D}}'_{F_0}$. Alternatively, this means that

$$\bar{\mathscr{D}}'_{F_0}H\cap\bar{\mathscr{D}}'_{F_0}=\emptyset \tag{4.8}$$

for all $H\in\mathfrak{S}(\mathscr{D})\setminus\mathfrak{G}_{F_0}$. Thus, whenever $\mathfrak{G}_{F_0}\equiv\mathscr{S}'$, we simply define

$$\mathcal{N}_{F_0}\equiv\mathcal{N}'_{F_0}\cap\mathscr{D},$$

and

$$\mathscr{D}_{F_0}\equiv\mathscr{D}'_{F_0}\cap\mathscr{D}. \tag{4.9}$$

Note that, since $\mathscr{D}_{F_0}\subset\bar{\mathscr{D}}'_{F_0}$, it follows that

$$\mathscr{D}_{F_0}H\cap\mathscr{D}_{F_0}=\emptyset$$

for all $H\in\mathfrak{S}(\mathscr{D})\setminus\mathfrak{G}_{F_0}$. Moreover, (4.7), (4.9) and the group properties of $\mathfrak{O}$ and $\mathfrak{G}_{F_0}$ imply that

$$Q\mathscr{D}_{F_0}H=\mathscr{D}_{F_0}$$

for all $Q\in\mathfrak{O}$ and for all $H\in\mathfrak{G}_{F_0}$. Thus for the case $\mathfrak{G}_{F_0}\equiv\mathscr{S}'$ the proof is complete.

Alternatively, suppose that $\mathfrak{G}_{F_0}\neq\mathscr{S}'$. Then, (4.8) is satisfied by all $H\in\mathfrak{S}(\mathscr{D})\setminus\mathscr{S}'$, but not by those $H\in\mathscr{S}'\setminus\mathfrak{G}_{F_0}$ as required. However, if there is an open neighborhood $\mathcal{N}''_{F_0}$ of F_0, and a related set

$$\mathscr{D}''_{F_0}\equiv\{A\in\mathrm{Inv}\mid A=QFH \text{ where } Q\in\mathfrak{O},\ H\in\mathfrak{G}_{F_0}, \text{ and } F\in\mathcal{N}''_{F_0}\},$$

such that

$$\mathscr{D}''_{F_0}H \cap \mathscr{D}''_{F_0} = \emptyset \qquad \forall\, H \in \mathscr{S}' \setminus \mathfrak{G}_{F_0},$$

then we can define

$$\mathscr{N}_{F_0} \equiv \mathscr{N}'_{F_0} \cap \mathscr{N}''_{F_0} \cap \mathscr{D} \tag{4.10$_1$}$$

and

$$\mathscr{D}_{F_0} \equiv \mathscr{D}'_{F_0} \cap \mathscr{D}''_{F_0} \cap \mathscr{D} \tag{4.10$_2$}$$

to complete this proof. Thus, in the remainder of this proof we show that such a neighborhood $\mathscr{N}''_{F_0}$ and a related set $\mathscr{D}''_{F_0}$ do, in fact, exist.

To begin with, since $\mathscr{S}'$ is finite, we may write

$$\mathscr{S}' \setminus \mathfrak{G}_{F_0} = \{H_1, H_2, \ldots, H_n\},$$

where $n < \infty$. Next, consider the set

$$\mathscr{A}_{F_0} = \mathfrak{O}F_0\mathfrak{G}_{F_0} \equiv \{A \in \mathrm{Inv} \mid A = QF_0H \text{ where } Q \in \mathfrak{O}, \text{ and } H \in \mathfrak{G}_{F_0}\},$$

which is compact since $\mathfrak{O}$ is compact and $\mathfrak{G}_{F_0}$ is finite. Clearly,

$$\mathscr{A}_{F_0}H_i \cap \mathscr{A}_{F_0} = \emptyset \tag{4.11}$$

for every $i = 1, 2, \ldots, n$, since $H_i \notin \mathfrak{G}_{F_0}$ and, hence,

$$QF_0HH_i \neq Q'F_0H'$$

whenever Q and $Q' \in \mathfrak{O}$, and H and $H' \in \mathfrak{G}_{F_0}$. For each index i we now follow the procedure described in steps 1 through 5 below and illustrated in Figure 1.

Step 1. Both $\mathscr{A}_{F_0}$ and $\mathscr{A}_{F_0}H_i$ are compact and, by (4.11), disjoint. Hence, there is no sequence of points contained in one of these sets that converges to a point in the other set, *i.e.*, any convergent sequence of points in a compact set can only converge to another point in that set (see, for example, MUNKRES [10]). This means that as Q runs over $\mathfrak{O}$ and H over $\mathfrak{G}_{F_0}$ there are open neighborhoods, $\mathscr{N}_{QF_0HH_i}$ of each $QF_0HH_i \in \mathscr{A}_{F_0}H_i$ and $\mathscr{N}_{QF_0H}$ of each $QF_0H \in \mathscr{A}_{F_0}$, such that

$$\mathscr{N}_{QF_0HH_i} \cap \mathscr{N}_{Q'F_0H'} = \emptyset \qquad \forall\, Q, Q' \in \mathfrak{O}, \text{ and } \forall\, H, H' \in \mathfrak{G}_{F_0}. \tag{4.12}$$

Step 2. Let

$$\mathscr{N}_{F_0H} \equiv \bigcap_{Q \in \mathfrak{O}} Q^T \mathscr{N}_{QF_0H} \tag{4.13}$$

for each $H \in \mathfrak{G}_{F_0}$ Clearly, $F_0H \in \mathscr{N}_{F_0H}$ since $F_0H \in Q^T\mathscr{N}_{QF_0H}$ for every $Q \in \mathfrak{O}$. Note that $\mathscr{N}_{F_0H}$ will be closed and will consist in the single point F_0H only if there is a sequence of points $\{Q_iF_0H\}$, $Q_i \in \mathfrak{O}$, such that the diameter of the set $\mathscr{N}_{Q_iF_0H}$ converges to zero in the limit as $i \to \infty$. However, since $\mathfrak{O}F_0H$ is compact, it follows that any such sequence $\{Q_iF_0H\}$ can only converge to some point in $\mathfrak{O}F_0H$, Q^*F_0H say. But the set $\mathscr{N}_{Q^*F_0H}$ is open regardless of the value of $Q^* \in \mathfrak{O}$ (see *Step 1*) and hence the diameter of $\mathscr{N}_{Q^*F_0H}$ cannot vanish and $\mathscr{N}_{F_0H}$ must be open.

Step 3. The set $\mathcal{N}_{F_0H}H^{-1}$, $H \in \mathfrak{G}_{F_0}$, contains F_0 and is open. Thus, the set

$$\mathcal{N}_{F_0}^{(0)} \equiv \bigcap_{H \in \mathfrak{G}_{F_0}} \mathcal{N}_{F_0H}H^{-1} \tag{4.14}$$

contains F_0 and is open because the intersection is over a finite set. From (4.13) and (4.14), we then may conclude that

$$Q\mathcal{N}_{F_0}^{(0)}H \subseteqq \mathcal{N}_{QF_0H} \quad \forall\, Q \in \mathfrak{O}, \text{ and } \forall\, H \in \mathfrak{G}_{F_0}. \tag{4.15}$$

Step 4. As in (4.13) and (4.14), we define an open set

$$\mathcal{N}_{F_0HH_i} \equiv \bigcap_{Q \in \mathfrak{O}} Q^T \mathcal{N}_{QF_0HH_i}, \tag{4.16}$$

for each $H \in \mathfrak{G}_{F_0}$, and the open set

$$\mathcal{N}_{F_0}^{(i)} \equiv \bigcap_{H \in \mathfrak{G}_{F_0}} \mathcal{N}_{F_0HH_i}(HH_i)^{-1} \tag{4.17}$$

which contains F_0. Together, (4.16) and (4.17) yield

$$Q\mathcal{N}_{F_0}^{(i)}HH_i \subseteqq \mathcal{N}_{QF_0HH_i} \quad \forall\, Q \in \mathfrak{O}, \text{ and } \forall\, H \in \mathfrak{G}_{F_0}. \tag{4.18}$$

Step 5. Define

$$\mathcal{N}_{F_0}^{\prime(i)} \equiv \mathcal{N}_{F_0}^{(0)} \cap \mathcal{N}_{F_0}^{(i)}. \tag{4.19}$$

From (4.12), (4.15) and (4.18), we then have

$$Q\mathcal{N}_{F_0}^{\prime(i)}HH_i \cap Q^*\mathcal{N}_{F_0}^{\prime(i)}H^* = \emptyset \quad \forall\, Q, Q^* \in \mathfrak{O}, \text{ and } \forall\, H, H^* \in \mathfrak{G}_{F_0}. \tag{4.20}$$

After repeating the above procedure for each $i = 1, 2, \ldots, n$, we define

$$\mathcal{N}_{F_0}^{\prime\prime} \equiv \bigcap_{i=1}^{n} \mathcal{N}_{F_0}^{\prime(i)}. \tag{4.21}$$

Then, for each index i, since

$$\mathcal{N}_{F_0}^{\prime\prime} \subseteqq \mathcal{N}_{F_0}^{\prime(i)},$$

it follows from (4.20) that

$$Q\mathcal{N}_{F_0}^{\prime\prime}HH_i \cap Q'\mathcal{N}_{F_0}^{\prime\prime}H' = \emptyset \quad \forall\, Q, Q' \in \mathfrak{O}, \text{ and } \forall\, H, H' \in \mathfrak{G}_{F_0}.$$

In turn, this implies that the set

$$\mathcal{D}_{F_0}^{\prime\prime} \equiv \bigcup_{H \in \mathfrak{G}_{F_0}} \mathfrak{O}\mathcal{N}_{F_0}^{\prime\prime}H \tag{4.22}$$

has the property

$$\mathcal{D}_{F_0}^{\prime\prime}H \cap \mathcal{D}_{F_0}^{\prime\prime} = \emptyset \quad \forall\, H \in \mathcal{S}' \setminus \mathfrak{G}_{F_0}.$$

Therefore, to complete this proof, it suffices to substitute (4.21) and (4.22) into (4.10). □

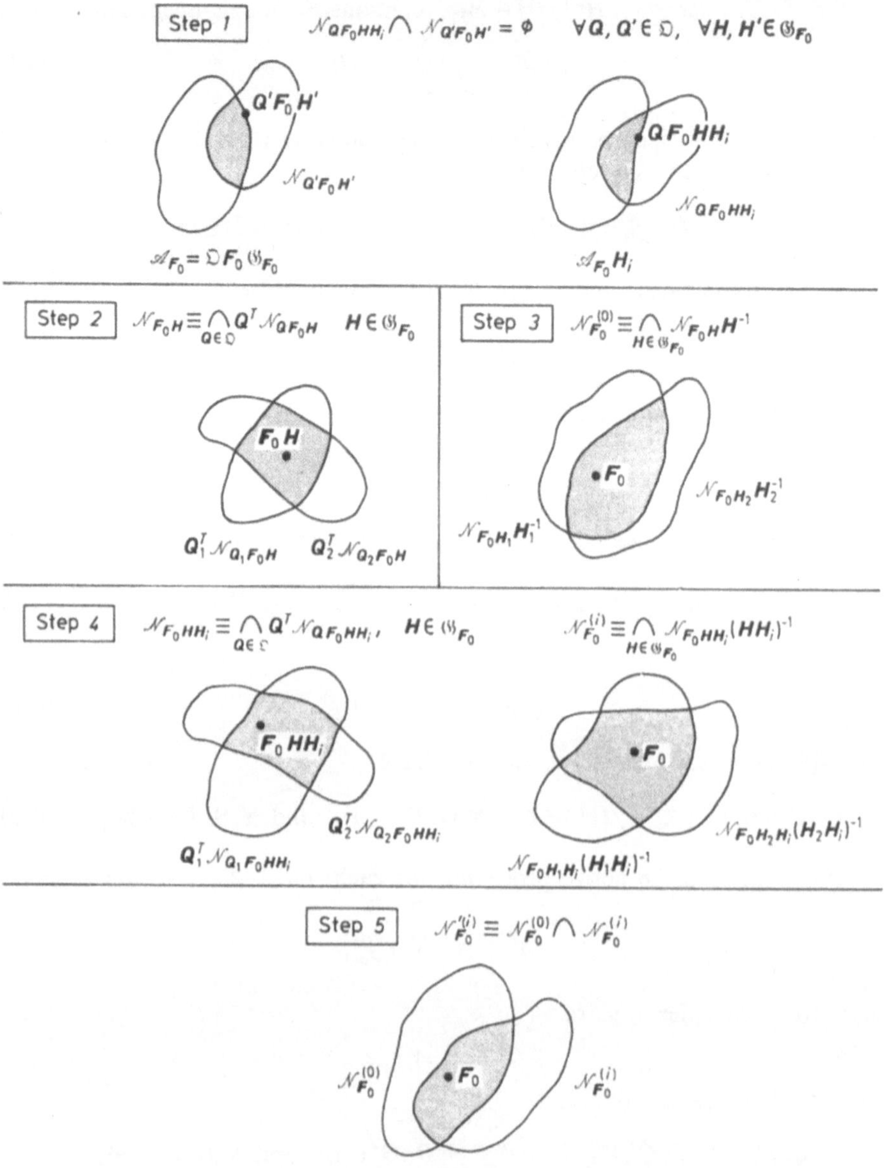

Fig. 1. Illustration of the procedure described by steps 1 through 5 in the proof of Theorem 1.

An immediate consequence of Theorem 1, Definition 1, and the observation that for $\boldsymbol{F}_0 \in \mathscr{D}$ the local material symmetry group on $\mathscr{D}_{\boldsymbol{F}_0}$, as introduced in (2.2), is given by

$$\mathfrak{S}(\mathscr{D}_{\boldsymbol{F}_0}) = \mathfrak{G}_{\boldsymbol{F}_0}, \tag{4.23}$$

is the following

Corollary. *$\mathscr{B}_0$ is locally solid at every $F_0 \in \mathscr{D}$.*

Remark 4. There are some additional interesting results that follow from Theorem 1. First, for any $F_0 \in \mathscr{D}$, the set $\mathscr{D}_{F_0} \subset \mathscr{D}$ given in $(4.6)_1$ has the property

$$\mathfrak{G}_F \subseteq \mathfrak{G}_{F_0} \quad \forall\, F \in \mathscr{D}_{F_0}. \tag{4.24}$$

To see this, suppose (4.24) is not true for some $F \in \mathscr{D}_{F_0}$. Then, by (4.3), there is in $\mathfrak{G}_F$ an $H' \in \mathfrak{S}(\mathscr{D}) \setminus \mathfrak{G}_{F_0}$ and a $Q' \in \mathfrak{O}$ such that

$$Q'FH' = F.$$

However, since $Q'F \in \mathscr{D}_{F_0}$, this implies

$$\mathscr{D}_{F_0}H' \cap \mathscr{D}_{F_0} \neq \emptyset,$$

which contradicts Theorem 1 and hence (4.24) must hold. In physical terms, the group $\mathfrak{G}_F$ associated with $F \in \mathscr{D}$ may be interpreted as being such that $F\mathfrak{G}_F F^{-1} \subset \mathfrak{O}$ is the point group that describes the lattice structure of $\mathscr{B}_0$. Thus, (4.24) says that $\mathfrak{G}_{F_0}$ corresponds to the "maximal" local symmetry group allowable relative to those that can be associated with other points in $\mathscr{D}_{F_0}$. It also follows from Theorem 1 that the local material symmetry group can vary within a crystal domain $\mathscr{D}$. In particular, we known from Theorem 1 and (4.23) that there is an open set $\mathscr{D}_{F_0} \subset \mathscr{D}$ containing a given point $F_0 \subset \mathscr{D}$ such that $\mathfrak{S}(\mathscr{D}_{F_0}) = \mathfrak{G}_{F_0}$. Consequently, given any two points F_1 and $F_2 \in \mathscr{D}$ such that

$$\mathfrak{G}_{F_1} \neq \mathfrak{G}_{F_2},$$

Theorem 1 implies the existence of an open set $\mathscr{D}_{F_1}$ and $\mathscr{D}_{F_2}$ such that

$$\mathfrak{G}(\mathscr{D}_{F_1}) \neq \mathfrak{G}(\mathscr{D}_{F_2}). \quad \square$$

Remark 5. PITTERI [3] has obtained a result comparable to Theorem 1. In his work, however, the stored energy function is defined on the set of 3×3 positive definite, symmetric matrices $\mathscr{M}$, i.e., $W = \overline{W}(C_{ab})$ where $C_{ab} = e_a \cdot e_b$ and e_a, $a = 1, 2, 3$, are the deformed lattice vectors. The appropriate material symmetry group in this case is the matrix group $\mathfrak{G}$ defined in (3.4) and the corresponding invariance statement is

$$\overline{W}(C_{ab}) = \overline{W}((h_a^r)^T\, C_{rs} h_b^s)$$

for all $C_{ab} \in \mathscr{M}$ and all $\|h_b^a\| \in \mathfrak{G}$. Explicitly, for any $C_{ab} \in \mathscr{M}$, PITTERI shows that there is a neighborhood $\mathscr{N} \subset \mathscr{M}$ of C_{ab} such that every $\|h_b^a\| \in \mathfrak{G}$ satisfies either $(h_a^r)^T C_{rs} h_b^s = C_{ab}$ or $(h_a^r)^T C'_{rs} h_b^s \notin \mathscr{N}$ for all $C'_{rs} \in \mathscr{N}$. $\square$

To prepare for Theorem 2, we have

Definition 4. $F \in \mathrm{Inv}$ *is said to lie in the symmetry class of* $F_0 \in \mathrm{Inv}$ *if there is an* $H \in \mathfrak{S}(\mathscr{D})$ *such that*

$$H^{-1}\mathfrak{G}_F H = \mathfrak{G}_{FH} \subseteq \mathfrak{G}_{F_0}.$$

The symmetry class of a fixed $F_0 \in \mathrm{Inv}$ identifies all of the deformation gradients $F \in \mathrm{Inv}$ that are such that under some symmetry transformation of F to FH, $H \in \mathfrak{S}(\mathscr{D})$, the new *local* symmetry group that is associated with FH, *i.e.*, $\mathfrak{G}_{FH}$, is contained in $\mathfrak{G}_{F_0}$. Theorem 1 shows that for any $F_0 \in \mathscr{D}$ there is an open set $\mathscr{D}_{F_0} \subset \mathscr{D}$ that satisfies $(4.6)_2$ and is such that the local material symmetry group on $\mathscr{D}_{F_0}$ is given by $\mathfrak{S}(\mathscr{D}_{F_0}) = \mathfrak{G}_{F_0}$. In Theorem 2 below we show that there is a "largest" set, $\mathscr{D}^*_{F_0} \subset \mathscr{D}$, with these properties, it being largest in the sense that it contains a set of deformation gradients sufficient to cover the full range of values taken on by the stored energy function over the complete symmetry class of F_0. We shall state this result more precisely in terms of the following

Definition 5. *Given any* $F_0 \in \mathrm{Inv}$, *then the open set* $\mathscr{D}^*_{F_0} \subset \mathscr{D}$ *is said to be a "largest" set or, more precisely, a fundamental domain for the symmetry class of* F_0, *if it contains* F_0 *and satisfies the three conditions*

(i) $Q\mathscr{D}^*_{F_0}H = \mathscr{D}^*_{F_0} \quad \forall\, Q \in \mathfrak{O}$, and $\forall\, H \in \mathfrak{G}_{F_0}$,

(ii) $\mathscr{D}^*_{F_0}H \cap \mathscr{D}^*_{F_0} = \emptyset \quad \forall\, H \in \mathfrak{S}(\mathscr{D}) \setminus \mathfrak{G}_{F_0}$,

and

(iii) *for each* $F \in \mathscr{D}$ *lying in the symmetry class of* F_0 *there is an* $H \in \mathfrak{S}(\mathscr{D})$ *such that*

$$FH \in \overline{\mathscr{D}}^*_{F_0},$$

where $\overline{\mathscr{D}}^*_{F_0}$ *is the closure of* $\mathscr{D}^*_{F_0}$.

Remark 6. Suppose that $\mathscr{D}^*_{F_0} \subset \mathscr{D}$ is a fundamental domain for the symmetry class of $F_0 \in \mathscr{D}$ and that

$$\mathfrak{S}(\mathscr{D}^*_{F_0}) = \mathfrak{G}_{F_0} \not\subset \mathfrak{O}.$$

Then, it may prove useful to redefine the stored energy function (3.16) by replacing the lattice vectors $\overline{E}_a$ with $\overline{E}'_a = F_0\overline{E}_a$, $a = 1, 2, 3$. Then $\mathscr{D}$ is transformed into a new crystal domain, $\mathscr{D}'$, say. In particular, $F_0 \in \mathscr{D}$ is transformed into $\mathbf{1} \in \mathscr{D}'$ and $\mathscr{D}^*_{F_0} \subset \mathscr{D}$ into $\mathscr{D}'_1 \subset \mathscr{D}'$. The end result of this procedure is that the point group

$$\mathfrak{S}(\mathscr{D}'_1) = \mathfrak{G}'_1 = \mathfrak{S}(\mathscr{D}') \cap \mathfrak{O}$$

replaces $\mathfrak{S}(\mathscr{D}^*_{F_0}) \not\subset \mathfrak{O}$. □

The importance of fundamental domains is related to the global invariance condition

$$W(F) = W(FH) \quad \forall\, F \in \mathscr{D}, \quad \forall\, H \in \mathfrak{S}(\mathscr{D}). \tag{4.25}$$

In particular, (4.25) and condition (iii) of Definition 6 imply that a fundamental domain for the symmetry class of any point $F_0 \in \mathrm{Inv}$ is sufficient to cover the entire range of values exhibited by the stored energy function (3.16) over that subset of $\mathscr{D}$ which consists in *all* points in the symmetry class of F_0. Alternatively, if $F \in \mathscr{D}$ is in the symmetry class of $F_0 \in \mathrm{Inv}$ and if $\mathscr{D}^*_{F_0} \subset \mathscr{D}$ is a fundamental domain for the symmetry class of F_0, then there exists a point $K \in \bar{\mathscr{D}}^*_{F_0}$ such that

$$W(F) = W(K).$$

This now leads to

Theorem 2. *Let* $\mathscr{D} \subset \mathrm{Inv}$ *be a crystal domain for* $\mathscr{B}_0$ *and let* $F_0 \in \mathrm{Inv}$. *Then there is a fundamental domain for the symmetry class of* F_0.

Proof. Let $F \in \mathscr{D}$ be any point in the symmetry class of F_0. Then, consider the procedure described in Steps 1 through 4 below and illustrated in Figure 2.

Step 1. Since F is in the symmetry class of F_0, it follows that there is an $H \in \mathfrak{S}(\mathscr{D})$ and a point

$$K = FH \in \mathscr{D}$$

such that

$$\mathfrak{G}_K \subseteqq \mathfrak{G}_{F_0}.$$

Further, as noted in Remark 3, $\mathfrak{G}_{F_0}$ and $\mathfrak{G}_K$ are finite and hence we may write

$$\mathfrak{G}_{F_0} \setminus \mathfrak{G}_K = \{H_1, H_2, \ldots, H_n\},$$

where $n < \infty$. Now let

$$K_j \equiv KH_j$$

for $j = 1, 2, \ldots, n$.

Step 2. For convenience, define $K_0 \equiv K$. Then, from Theorem 1 we know that for each $j = 0, 1, 2, \ldots, n$ there is an open neighborhood of K_j, $\mathscr{N}_{K_j} \subset \mathscr{D}$, and a related set

$$\mathscr{D}_{K_j} \equiv \{A \in \mathscr{D} \mid A = QMH \text{ where } Q \in \mathfrak{O},\ H \in \mathfrak{G}_{K_j}, \text{ and } M \in \mathscr{N}_{K_j}\}$$

such that

$$\mathscr{D}_{K_j} H \cap \mathscr{D}_{K_j} = \emptyset \qquad \forall\, H \in \mathfrak{S}(\mathscr{D}) \setminus \mathfrak{G}_{K_j}. \tag{4.26}$$

Now, define the open set

$$\mathscr{N}'_K \equiv \bigcap_{j=0}^{n} \mathscr{N}_{K_j} H_j^{-1},$$

where $H_0 \equiv 1$.

Step 3. Let

$$\mathscr{N}'_{K_j} \equiv \mathscr{N}'_K H_j \tag{4.27}$$

for $j = 1, 2, \ldots, n$. Then, since $\mathscr{N}'_{K_j} \subseteqq \mathscr{N}_{K_j}$ for each $j = 0, 1, 2, \ldots, n$, it follows that $\mathscr{D}'_{K_j} \subseteqq \mathscr{D}_{K_j}$, where

$$\mathscr{D}'_{K_j} \equiv \{A \in \mathscr{D} \mid A = QMH \text{ where } Q \in \mathfrak{O},\ H \in \mathfrak{G}_{K_j}, \text{ and } M \in \mathscr{N}'_{K_j}\}. \tag{4.28}$$

Further, from (4.27) and (4.28), we conclude that

$$\mathscr{D}'_{K_i} H_i^{-1} H_j = \mathscr{D}'_{K_j}, \qquad i, j = 0, 1, 2, \ldots, n. \tag{4.29}$$

Hence, for every pair of indices i and j, (4.26) and (4.29) yield

$$\mathcal{D}'_{K_i} H_i^{-1} H_j H \cap \mathcal{D}'_{K_j} = \emptyset \quad \forall\, H \in \mathfrak{S}(\mathcal{D}) \setminus \mathfrak{G}_{K_j},$$

or, equivalently,

$$\mathcal{D}'_{K_i} H \cap \mathcal{D}'_{K_j} = \emptyset \quad \forall\, H \in \mathfrak{S}(\mathcal{D}) \setminus H_i^{-1} \mathfrak{G}_K H_j, \tag{4.30}$$

where we have used the fact that $\mathfrak{G}_{K_j} = H_j^{-1} \mathfrak{G}_K H_j$.

Step 4. It is well known from group theory (see, for example, HALL [11]) that the group $\mathfrak{G}_{F_0}$ can be decomposed into the right cosets of its subgroup $\mathfrak{G}_K$, *i.e.*,

$$\mathfrak{G}_{F_0} = \mathfrak{G}_K + \mathfrak{G}_K H_1 + \ldots + \mathfrak{G}_K H_n$$

or, since $H_i^{-1} \mathfrak{G}_{F_0} = \mathfrak{G}_{F_0}$ for $i = 0, 1, 2, \ldots, n$,

$$\mathfrak{G}_{F_0} = H_i^{-1} \mathfrak{G}_K + H_i^{-1} \mathfrak{G}_K H_1 + \ldots + H_i^{-1} \mathfrak{G}_K H_n. \tag{4.31}$$

Thus it follows from (4.28), (4.30) and (4.31) that the open set

$$\mathcal{D}_1 \equiv \bigcup_{j=0}^{n} \mathcal{D}'_{K_j}$$

satisfies

$$Q \mathcal{D}_1 H = \mathcal{D}_1 \quad \forall\, Q \in \mathfrak{O}, \text{ and } \forall\, H \in \mathfrak{G}_{F_0} \tag{4.32$_1$}$$

and

$$\mathcal{D}_1 H \cap \mathcal{D}_1 = \emptyset \quad \forall\, H \in \mathfrak{S}(\mathcal{D}) \setminus \mathfrak{G}_{F_0}. \tag{4.32$_2$}$$

Having completed the procedure described in steps 1 through 4 above, we suppose that for each $F \in \mathcal{D}$ in the symmetry class of F_0 there is an $H \in \mathfrak{S}(\mathcal{D})$ such that

$$FH \in \overline{\mathcal{D}}_1.$$

Then, (4.32) implies that a fundamental domain for the symmetry class of F_0 is given by

$$\mathcal{D}^*_{F_0} \equiv \mathcal{D}_1.$$

If, however, there is a point $F \in \mathcal{D}$ in the symmetry class of F_0 such that

$$FH \notin \overline{\mathcal{D}}_1 \quad \forall\, H \in \mathfrak{S}(\mathcal{D}),$$

then we repeat steps 1 through 3 and proceed to step 5 below.

Step 5. In order to eliminate an overlapping of $\mathcal{N}'_{K_j}$, and hence $\mathcal{D}'_{K_j}$, with $\overline{\mathcal{D}}_1 H$ for $H \in \mathfrak{S}(\mathcal{D}) \setminus \mathfrak{G}_{F_0}$, we first set

$$\mathcal{N}''_{K_j} \equiv \{A \in \mathcal{N}'_{K_j} \mid A \notin \overline{\mathcal{D}}_1 H \;\forall\, H \in \mathfrak{S}(\mathcal{D}) \setminus \mathfrak{G}_{F_0}\} \tag{4.33}$$

and

$$\mathcal{D}''_{K_j} \equiv \{A \in \mathcal{D} \mid A = QMH \text{ where } Q \in \mathfrak{O},\, H \in \mathfrak{G}_{K_j}, \text{ and } M \in \mathcal{N}''_{K_j}\} \tag{4.34}$$

for $j = 0, 1, 2, \ldots, n$. Then, since $\mathcal{D}''_{K_j} \subsetneq \mathcal{D}'_{K_j}$ for each index j, it follows from (4.30), (4.31) and (4.33) that the open set

$$\mathcal{D}_2 \equiv \bigcup_{j=0}^{n} \mathcal{D}''_{K_j} \tag{4.35}$$

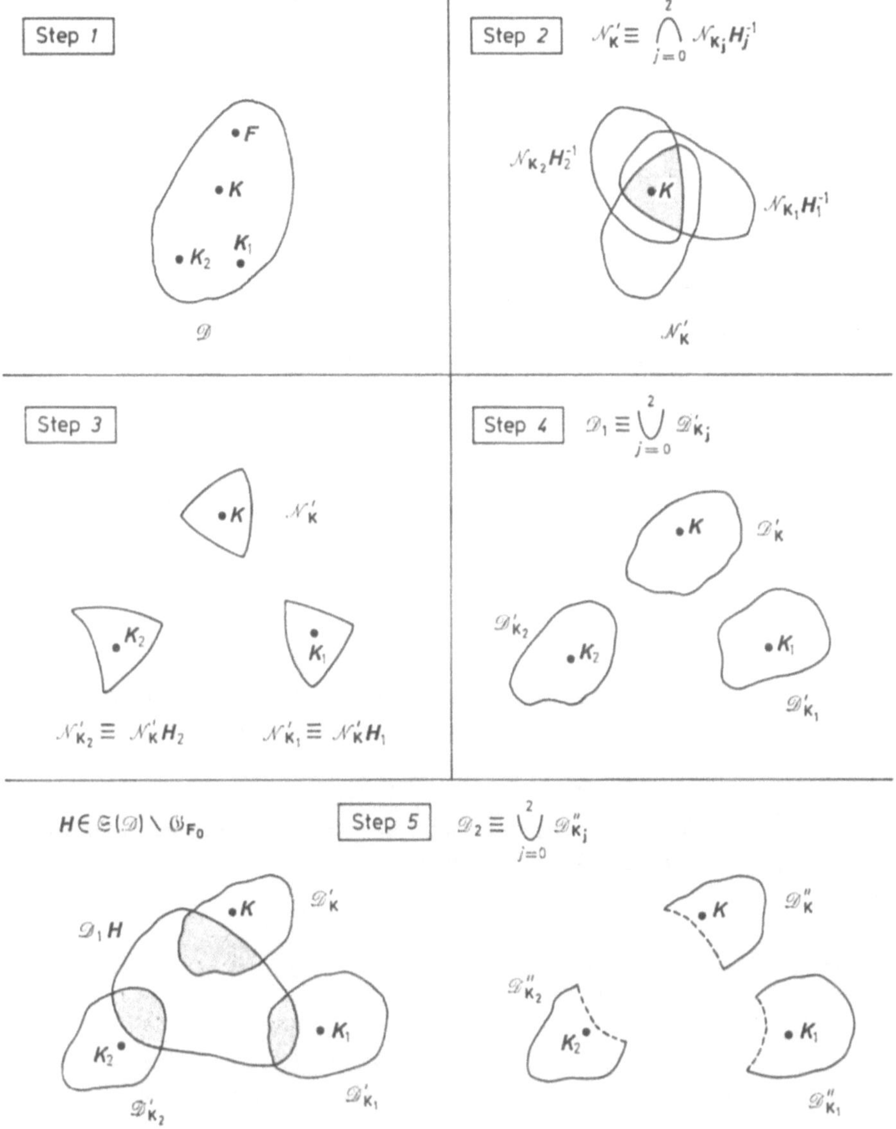

Fig. 2. Illustration of steps 1 through 5 in Theorem 2 for the case when $\mathfrak{G}_{F_0} \setminus \mathfrak{G}_K = \{H_1, H_2\}$.

satisfies

$$Q\mathscr{D}_2 H = \mathscr{D}_2 \quad \forall\, Q \in \mathfrak{O}, \quad \forall\, H \in \mathfrak{G}_{F_0}, \tag{4.36$_1$}$$

and

$$\mathscr{D}_2 H \cap \mathscr{D}_2 = \emptyset \quad \forall\, H \in \mathfrak{S}(\mathscr{D}) \setminus \mathfrak{G}_{F_0}. \tag{4.36$_2$}$$

Now suppose that for each $F \in \mathscr{D}$ in the symmetry class of F_0 there is an $H \in \mathfrak{S}(\mathscr{D})$ such that

$$FH \in \overline{\mathscr{D}}_1 \cup \overline{\mathscr{D}}_2.$$

Then, since $\mathscr{D}_1$ and $\mathscr{D}_2$ do not overlap, *i.e.*, by (4.33), (4.34) and (4.35)

$$\overline{\mathscr{D}}_1 H \cap \mathscr{D}_2 = \emptyset \quad \forall\, H \in \mathfrak{S}(\mathscr{D}) \setminus \mathfrak{G}_{F_0},$$

we see from (4.32) and (4.36) that the set

$$\mathscr{D}^*_{F_0} \equiv \mathscr{D}_1 \cup \mathscr{D}_2$$

is a fundamental domain for the symmetry class of F_0. Alternatively, if there is a point $F \in \mathscr{D}$ in the symmetry class of F_0 such that

$$FH \notin \overline{\mathscr{D}}_1 \cup \overline{\mathscr{D}}_2 \quad \forall\, H \in \mathfrak{S}(\mathscr{D}),$$

then we return to the procedure outlined in steps 1 through 3 and in step 5, above.

Continuing in this manner, we arrive at the existence of sets $\mathscr{D}_1, \mathscr{D}_2, \ldots,$ and $\mathscr{D}_i$, $i \geqq 1$, such that

$$\mathscr{D}^*_{F_0} \equiv \mathscr{D}_1 \cup \mathscr{D}_2 \ldots \cup \mathscr{D}_i$$

is a fundamental domain for the symmetry class of F_0. To see this, it suffices to note that if there is a point $F \in \mathscr{D}$ in the symmetry class of F_0 such that

$$FH \notin \overline{\mathscr{D}}_1 \cup \overline{\mathscr{D}}_2 \ldots \cup \overline{\mathscr{D}}_j \quad \forall\, H \in \mathfrak{S}(\mathscr{D})$$

for some $j \geqq 1$, then the procedure described in Steps 1 through 3 and Step 5 will produce an open set $\mathscr{D}_{j+1}$ such that

$$FH \in \overline{\mathscr{D}}_1 \cup \overline{\mathscr{D}}_2 \ldots \cup \overline{\mathscr{D}}_j \cup \overline{\mathscr{D}}_{j+1}$$

for some $H \in \mathfrak{S}(\mathscr{D})$. This completes the proof. □

Remark 7. Theorem 2 does not determine a unique fundamental domain $\mathscr{D}^*_{F_0} \subset \mathscr{D}$ for the symmetry class of a given point $F_0 \in \text{Inv}$. This is due to the presence of points $F \in \mathscr{D}$ such that either $\mathfrak{G}_F = \{-\mathbf{1}, \mathbf{1}\}$ or $\mathfrak{G}_F = \{\mathbf{1}\}$ and, hence,

$$\mathfrak{G}_{FH} = \mathfrak{G}_F \subset \mathfrak{G}_{F_0} \quad \forall\, H \in \mathfrak{S}(\mathscr{D}).$$

Thus, the choice of the point $K = FH$ in step 1 of the proof of Theorem 2 is essentially arbitrary and the set $\mathscr{D}^*_{F_0}$ is clearly not unique. We can, however, introduce a measure of distance, *i.e.*, a norm, in order to eliminate this non-uniqueness by requiring $H \in \mathfrak{S}(\mathscr{D})$ to be so chosen that K is as close to F_0 as possible. □

Before proceeding to Theorem 3, recall from Remark 3 that for any $F \in \text{Inv}$ the group $\mathfrak{G}_F$ is such that $F\mathfrak{G}_F F^{-1} \subset \mathfrak{O}$ is a crystallographic point group. Further, each of these point groups is a subgroup of a cubic point group and/or a hexagonal point group (see Table 1). Hence, $F\mathfrak{G}_F F^{-1} \subset \mathfrak{O}$ is a crystallographic point group of the *cubic* type and/or the *hexagonal* type depending on whether it is a subgroup of a cubic or a hexagonal point group. In particular, the cubic type consists in all triclinic, monoclinic, rhombic, trigonal, tetragonal and cubic point groups. Alternatively, the hexagonal type consists in all triclinic, monoclinic, rhombic, trigonal and hexagonal point groups. Note that these two types of point groups are not disjoint and have considerable overlap.

As a preliminary to Theorem 3, we now give

Lemma 3. *Let* $\mathscr{D} \subset \text{Inv}$ *be a crystal domain for* $\mathscr{B}_0$. *Then there are points* F_c *and* $F_h \in \text{Inv}$ *called, respectively, points of maximal cubic and maximal hexagonal symmetry with the property that for each* $F \in \mathscr{D}$ *there is an* $H \in \mathfrak{S}(\mathscr{D})$ *such that*

$$H^{-1}\mathfrak{G}_F H \subseteqq \mathfrak{G}_{F_c} \tag{4.37$_1$}$$

and/or

$$H^{-1}\mathfrak{G}_F H \subseteqq \mathfrak{G}_{F_h}. \tag{4.37$_2$}$$

That is, each point $F \in \mathscr{D}$ *lies in the symmetry class of* F_c *and/or* F_h. *Here* $F_c\mathfrak{G}_{F_c}F_c^{-1} \subset \mathfrak{O}$ *and* $F_h\mathfrak{G}_{F_h}F_h^{-1} \subset \mathfrak{O}$ *are, respectively, crystallographic point groups of the cubic and hexagonal types.*

Proof. First, recall from Remark 2 that

$$\mathfrak{S}(\mathscr{D}) \subseteqq \mathfrak{G}(\bar{E}_a)$$

for some fixed set of lattice vectors $\bar{E}_a$, $a = 1, 2, 3$. Further, by (4.5) we have

$$\mathfrak{G}(e_a) \equiv F\mathfrak{G}(\bar{E}_a)\, F^{-1}$$

for any $F \in \text{Inv}$, where $e_a = F\bar{E}_a$. Thus,

$$F\mathfrak{S}(\mathscr{D})\, F^{-1} \subseteqq \mathfrak{G}(e_a).$$

Now, let $F_c \in \text{Inv}$ be such that for $e'_a = F_c\bar{E}_a$, $a = 1, 2, 3$,

$$\mathfrak{P}_c \equiv \mathfrak{G}(e'_a) \cap \mathfrak{O}$$

is the holohedral cubic point group $\dfrac{4}{m}\bar{3}$ (see Table 1). Then, every crystallographic point group in $\mathfrak{S}(\mathscr{D})$ of the cubic type is a subgroup of $\mathfrak{P}_c$ and, hence, of $\mathfrak{G}(e'_a)$.† Thus, any subgroup of $\mathfrak{G}(e'_a)$ that is conjugate to a point group $\mathfrak{P}' \subseteqq \mathfrak{P}_c$ must be expressible in the form $H^{-1}\mathfrak{P}'H$, for some $H \in \mathfrak{G}(e'_a)$. Moreover, since $F_c\mathfrak{S}(\mathscr{D})\, F_c^{-1} \subseteqq \mathfrak{G}(e'_a)$, any subgroup of $F_c\mathfrak{S}(\mathscr{D})\, F_c^{-1}$ that is conjugate to a point

† While $Q^{-1}\mathfrak{P}Q$ is a cubic point group for any $Q \in \mathfrak{O}$ whenever $\mathfrak{P} \subseteqq \mathfrak{P}_c$, it is generally not true that $Q^{-1}\mathfrak{P}Q \subseteqq \mathfrak{S}(\mathscr{D})$.

group $\mathfrak{P}'' \subsetneqq \mathfrak{P}(e_a') \subsetneqq \mathfrak{P}_c$, where $\mathfrak{P}(e_a') = \boldsymbol{F}_c \mathfrak{S}(\mathscr{D}) \boldsymbol{F}_c^{-1} \cap \mathfrak{O}$, must be expressible in the form $\boldsymbol{H}^{-1} \mathfrak{P}'' \boldsymbol{H}$ for some $\boldsymbol{H} \in \boldsymbol{F}_c \mathfrak{S}(\mathscr{D}) \boldsymbol{F}_c^{-1}$. Therefore, we may conclude that any subgroup of $\mathfrak{S}(\mathscr{D})$ that is conjugate to a point group of the cubic type must be expressible in the form $\boldsymbol{H}^{-1} \mathfrak{G}_c \boldsymbol{H}$ for some $\boldsymbol{H} \in \mathfrak{S}(\mathscr{D})$ and some group $\mathfrak{G}_c \subsetneqq \mathfrak{G}_{\boldsymbol{F}_c} = \boldsymbol{F}^{-1} \mathfrak{P}(e_a') \boldsymbol{F}$.

Next, let $\boldsymbol{F}_h \in \mathrm{Inv}$ be such that for $e_a'' = \boldsymbol{F}_h \overline{\boldsymbol{E}}_a$, $a = 1, 2, 3$,

$$\mathfrak{P}_h \equiv \mathfrak{G}(e_a'') \cap \mathfrak{O}$$

is the holohedral hexagonal point group $\frac{6}{m} 2$. Then, by an argument similar to that used above, it follows that any subgroup of $\mathfrak{S}(\mathscr{D})$ that is conjugate to a point group of the hexagonal type must be expressible in the form $\boldsymbol{H}^{-1} \mathfrak{G}_h \boldsymbol{H}$ for some $\boldsymbol{H} \in \mathfrak{S}(\mathscr{D})$ and some group $\mathfrak{G}_h \subsetneqq \mathfrak{G}_{\boldsymbol{F}_h}$.

Finally, since for any $\boldsymbol{F} \in \mathscr{D}$ the group $\mathfrak{G}_{\boldsymbol{F}}$ is conjugate to a crystallographic point group, there exists an $\boldsymbol{H} \in \mathfrak{S}(\mathscr{D})$ for each $\boldsymbol{F} \in \mathscr{D}$ such that

$$\boldsymbol{H}^{-1} \mathfrak{G}_{\boldsymbol{F}} \boldsymbol{H} \subsetneqq \mathfrak{G}_{\boldsymbol{F}_c}$$

and/or

$$\boldsymbol{H}^{-1} \mathfrak{G}_{\boldsymbol{F}} \boldsymbol{H} \subsetneqq \mathfrak{G}_{\boldsymbol{F}_h}.$$

Thus, $\boldsymbol{F}_c$ and $\boldsymbol{F}_h$ are, respectively, points in Inv of maximal cubic and hexagonal symmetry. This, completes the proof. □

For the final result of this section we combine Theorem 2 and Lemma 3 to obtain

Theorem 3. *Let $\mathscr{D} \subset \mathrm{Inv}$ be a crystal domain for $\mathscr{B}_0$. Then, there is an open set, $\mathscr{D}^* \subset \mathscr{D}$, with the following property: for each $\boldsymbol{F} \in \mathscr{D}$ there is an $\boldsymbol{H} \in \mathfrak{S}(\mathscr{D})$ such that*

$$\boldsymbol{F}\boldsymbol{H} \in \overline{\mathscr{D}}^*.$$

In particular, whenever $\boldsymbol{F}_c$ and $\boldsymbol{F}_h$ are, respectively, points of maximal cubic and hexagonal symmetry in Inv, *we may let*

$$\mathscr{D}^* = \mathscr{D}^*_{\boldsymbol{F}_c} \cup \mathscr{D}^*_{\boldsymbol{F}_h}$$

where $\mathscr{D}^*_{\boldsymbol{F}_c}$ and $\mathscr{D}^*_{\boldsymbol{F}_h}$ are, respectively, fundamental domains for the symmetry classes of $\boldsymbol{F}_c$ and $\boldsymbol{F}_h$.

The proof of this theorem follows immediately from Theorem 2 and Lemma 3.

5. A generalization

In this section we shall consider briefly the main elements of a local theory of material symmetry for a constitutive assumption slightly more general than that covered by (2.1). Here, the stored energy density at $\boldsymbol{X} \in \mathscr{B}_0$ is supposed to be given by a response function $W_0 : \mathscr{D} \to \mathbb{R}$ such that

$$W(X) = W_0(\nabla\chi(X), g(X), s(X)),$$

where, now, the open set $\mathscr{D} \subsetneqq \mathrm{Inv} \times \mathbb{R}^3 \times \mathbb{R}$. Depending on the particular application, the constitutive variables g and s may represent a variety of different physical quantities. For example, the representation (5.1) would apply to the theory of elastic dielectrics if g and s were interpreted, respectively, as the referential electric polarization (or, electric displacement) and the temperature (or, entropy).†

Now to describe the local invariance properties which represent objectivity and local material symmetry for (5.1), we first specify a fixed point $\Sigma_0 \equiv (F_0, g_0, s_0)$ in $\mathscr{D}$ and suppose there is an open set $\mathscr{D}_{\Sigma_0} \subset \mathscr{D}$ containing Σ_0 such that

$$W_0(\Sigma_0) = W_0(\mathbf{L}_Q \mathbf{R}_H \Sigma) \quad \forall\, \Sigma \equiv (F, g, s) \in \mathscr{D}_{\Sigma_0},$$

where††, for any $\Sigma = (F, g, s)$ in $\mathscr{D}$,

$$\mathbf{L}_Q \Sigma \equiv (QF, g, s) \quad \forall\, Q \in \mathfrak{O},$$

$$\mathbf{R}_H \Sigma \equiv (FH, H^{-1}g, s) \quad \forall\, H \in \mathfrak{S}(\mathscr{D}_{\Sigma_0}).$$

Here $\mathfrak{S}(\mathscr{D}_{\Sigma_0})$, which represents the local material symmetry group on $\mathscr{D}_{\Sigma_0}$, is a subgroup of the unimodular group, and the open set $\mathscr{D}_{\Sigma_0}$ is supposed to satisfy the locality requirement that

$$\mathbf{L}_Q \mathbf{R}_H \mathscr{D}_{\Sigma_0} = \mathscr{D}_{\Sigma_0}$$

for all $Q \in \mathfrak{O}$ and $H \in \mathfrak{S}(\mathscr{D}_{\Sigma_0})$.

For the remainder of this section we consider $\mathscr{D}$ to be a crystal domain for $\mathscr{B}_0$, so that, in analogy to Definition 2,

$$\mathfrak{S}(\mathscr{D}) = \mathfrak{G}(\bar{E}_a)$$

for some set of lattice vectors $\bar{E}_a$, $a = 1, 2, 3$. Before proceeding, however, we first introduce some additional notation. In particular, if $\mathscr{V} \subset \mathbb{R}^3$ and if H is any linear transformation of $\mathbb{R}^3$ to itself, then let

$$H\mathscr{V} \equiv \{a \in \mathbb{R}^3 \mid a = Hb \text{ where } b \in \mathscr{V}\}.$$

Further, for $F \in \mathrm{Inv}$ and $g \in \mathbb{R}^3$, we suppose $\mathfrak{G}_F$ is as defined in (4.3), *i.e.*,

$$\mathfrak{G}_F \equiv \{H \in \mathfrak{S}(\mathscr{D}) \mid FHF^{-1} \in \mathfrak{O}\},$$

and we introduce

$$\mathfrak{G}_F(g) \equiv \{H \in \mathfrak{G}_F \mid Hg = g\}.$$

Remark 8. Clearly, the set $\mathfrak{G}_F(g)$ is a finite group. In fact, $F\mathfrak{G}_F(g)F^{-1} \subset \mathfrak{O}$ is a crystallographic point group since $F\mathfrak{G}_F F^{-1} \subset \mathfrak{O}$ is (see Remark 3). To show that $\mathfrak{G}_F(g)$ is a group requires a standard argument. □

---

† See, for example HERTOG [1], GRINDLAY [12], or TOUPIN [13].

†† Constitutive variables with these transformation properties can be found in HERTOG [1] and HUTTER [14].

In support of Theorem 4, which follows, we have

Lemma 4. *Suppose* $F \in \mathrm{Inv}$ *and* $g \in \mathbb{R}^3$. *Then there is an open neighborhood of* g, $\mathscr{V}_g \subset \mathbb{R}^3$, *such that*

(i) $H\mathscr{V}_g = \mathscr{V}_g \quad \forall\, H \in \mathfrak{G}_F(g)$,

and

(ii) $H\mathscr{V}_g \cap \mathscr{V}_g = \emptyset \quad \forall\, H \in \mathfrak{G}_F \setminus \mathfrak{G}_F(g)$.

Proof. Consider the case $\mathfrak{G}_F(g) = \mathfrak{G}_F$, *i.e.*, the case

$$Hg = g \quad \forall\, H \in \mathfrak{G}_F.$$

Let $\mathscr{V}'_g \subset \mathbb{R}^3$ be an open neighborhood of g and set

$$\mathscr{V}_g = \bigcup_{H \in \mathfrak{G}_F} H\mathscr{V}'_g. \tag{5.2}$$

Clearly $\mathscr{V}_g$ is open and contains g. Moreover, the group properties of $\mathfrak{G}_F$ imply that

$$H\mathscr{V}_g = \mathscr{V}_g \quad \forall\, H \in \mathfrak{G}_F.$$

Therefore, the set $\mathscr{V}_g$ defined in (5.2) satisfies the conditions of the lemma for the case $\mathfrak{G}_F(g) \equiv \mathfrak{G}_F$.

Next, consider the case $\mathfrak{G}_F(g) \subset \mathfrak{G}_F$. To begin with, since $\mathfrak{G}_F$ is finite, we may write

$$\mathfrak{G}_F \setminus \mathfrak{G}_F(g) \equiv \{H_1, H_2, \ldots, H_n\}$$

for some $n < \infty$. Now, for each $i = 1, 2, \ldots, n$, let

$$g_i \equiv H_i g,$$

and note that $g_i \neq g$ since $H_i g \neq g$ (see Figure 3). Then, choose open neighborhoods $\mathscr{V}'_g$ and $\mathscr{V}'_{g_i}$ of, respectively, g and g_i such that

$$\mathscr{V}'_g \cap \mathscr{V}'_{g_i} = \emptyset. \tag{5.3}$$

Further, for each index i, let

$$\mathscr{V}^{(i)}_g \equiv \mathscr{V}'_g \cap H_i^{-1}\mathscr{V}'_{g_i},$$

which is an open set containing g.

Note that

$$H_i\mathscr{V}^{(i)}_g \subseteq \mathscr{V}'_{g_i},$$

and, hence, by (5.3), that

$$H_i\mathscr{V}^{(i)}_g \cap \mathscr{V}^{(i)}_g = \emptyset \tag{5.4}$$

for each index i. Thus, by introducing the open set

$$\mathscr{V}''_g \equiv \bigcap_{i=1}^{n} \mathscr{V}^{(i)}_g,$$

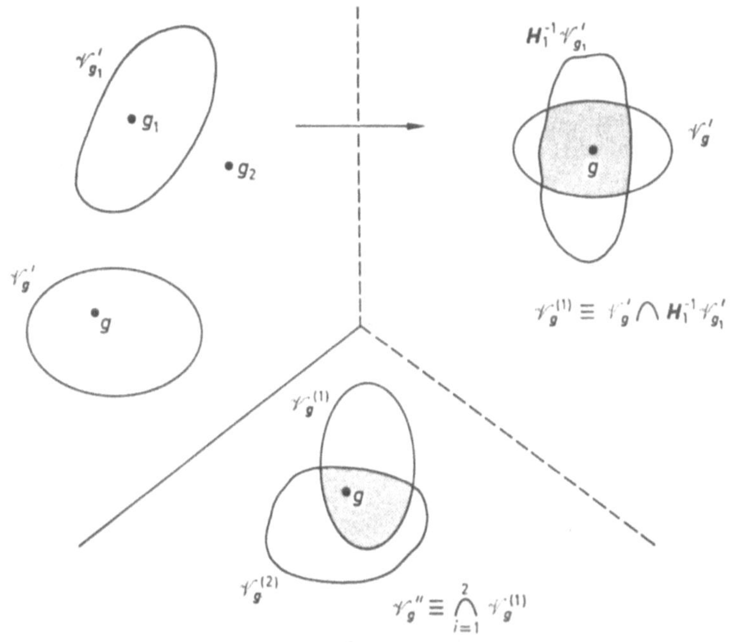

Fig. 3. Illustration of the procedure used for determining $\mathscr{V}_g''$ in Lemma 4 when $\mathfrak{G}_F \setminus \mathfrak{G}_F(g) = \{H_1, H_2\}$.

we see from (5.4) that

$$H_i \mathscr{V}_g'' \cap \mathscr{V}_g'' = \emptyset, \qquad i = 1, 2, \ldots, n.$$

Finally, to complete this proof, define

$$\mathscr{V}_g \equiv \bigcap_{H \in \mathfrak{G}_F(g)} H \mathscr{V}_g'',$$

which is an open neighborhood of g. □

Remark 9. Suppose that $\mathscr{V}_g \subset \mathbb{R}^3$ is an open neighborhood of $g \in \mathbb{R}^3$ that satisfies the conditions of Lemma 4 for a fixed $F \in \text{Inv}$. Then, $\mathscr{V}_g \subset \mathbb{R}^3$ has the property

$$\mathfrak{G}_F(g') \subseteqq \mathfrak{G}_F(g) \qquad \forall\, g' \in \mathscr{V}_g. \tag{5.5}$$

To see this, observe that if (5.5) is not valid, then in $\mathfrak{G}_F(g')$ there is some $H' \in \mathfrak{G}_F \setminus \mathfrak{G}_F(g)$ such that

$$H'g' = g'.$$

In this case,

$$H' \mathscr{V}_g \cap \mathscr{V}_g \neq \emptyset$$

which contradicts Lemma 4 and so (5.5) must hold. □

Analogous to Theorem 1, we now have

Theorem 4. *Suppose that* $\mathcal{D} \subsetneqq \mathrm{Inv} \times \mathbb{R}^3 \times \mathbb{R}$ *is a crystal domain for* $\mathcal{B}_0$ *and that* $\Sigma_0 \equiv (F_0, g_0, s_0) \in \mathcal{D}$. *Then there is an open neighborhood of* Σ_0, $\mathcal{N}_{\Sigma_0} \subset \mathcal{D}$, *and a related open set*

$$\mathcal{D}_{\Sigma_0} \equiv \{(A, a, \alpha) \in \mathcal{D} \mid (A, a, \alpha) = (QFH, H^{-1}g, s) \quad \text{where } Q \in \mathfrak{O},\ H \in \mathfrak{G}_{F_0}(g_0), \text{ and } (F, g, s) \in \mathcal{N}_{\Sigma_0}\},$$

containing Σ_0 *such that*[†]

$$\mathbf{R}_H \mathcal{D}_{\Sigma_0} \cap \mathcal{D}_{\Sigma_0} = \emptyset \qquad \forall\, H \in \mathfrak{S}(\mathcal{D}) \setminus \mathfrak{G}_{F_0}(g_0). \tag{5.6}$$

Proof. First, by employing the method used in the proof of Theorem 1, it is straightforward to show that there is a neighborhood $\mathcal{N}_{F_0} \subset \mathrm{Inv}$ of F_0 and a related set

$$\mathcal{D}_{F_0} \equiv \{A \in \mathrm{Inv} \mid A = QFH \text{ where } Q \in \mathfrak{O},\ H \in \mathfrak{G}_{F_0}, \text{ and } F \in \mathcal{N}_{F_0}\}$$

such that

$$\mathcal{D}_{F_0} H \cap \mathcal{D}_{F_0} = \emptyset \qquad \forall\, H \in \mathfrak{S}(\mathcal{D}) \setminus \mathfrak{G}_{F_0}. \tag{5.7}$$

Next, let $\mathcal{V}_{g_0} \subset \mathbb{R}^3$ be an open neighborhood of g_0 that satisfies the conditions of Lemma 4. and let $\mathcal{I} \subset \mathbb{R}$ be an open interval containing s_0. Now, consider the sets

$$\mathcal{N}_{\Sigma_0} \equiv \{\mathcal{N}_{F_0} \times \mathcal{V}_{g_0} \times \mathcal{I}\} \cap \mathcal{D}$$

and

$$\mathcal{D}_{\Sigma_0} \equiv \{\mathcal{D}_{F_0} \times \mathcal{V}_{g_0} \times \mathcal{I}\} \cap \mathcal{D}.$$

Since (5.7) holds and since Lemma 4 requires[††]

$$H^{-1}\mathcal{V}_{g_0} \cap \mathcal{V}_{g_0} = \emptyset \qquad \forall\, H \in \mathfrak{G}_{F_0} \setminus \mathfrak{G}_{F_0}(g_0),$$

it is clear that the sets $\mathcal{N}_{\Sigma_0}$ and $\mathcal{D}_{\Sigma_0}$ satisfy the conditions of this theorem, to complete its proof. □

Remark 10. Observations similar to those in Remark 4 also apply to Theorem 4. In particular, for a given $\Sigma_0 \equiv (F_0, g_0, s_0) \in \mathcal{D}$, suppose that the open set $\mathcal{D}_{\Sigma_0} \subset \mathcal{D}$ satisfies the conditions of Theorem 4. Then, it follows from Remarks 4 and 9 that

$$\mathfrak{G}_F(g) \subsetneqq \mathfrak{G}_{F_0}(g_0) \qquad \forall\, H \in (F, g, s) \in \mathcal{D}_{\Sigma_0}. \quad \square$$

Before proceeding to the analogs of Theorems 2 and 3, it is first convenient to introduce a terminology similar to that given in Definitions 4 and 5. In particular, for $\Sigma \equiv (F, g, s)$ and $\Sigma_0 \equiv (F_0, g_0, s_0) \in \mathrm{Inv} \times \mathbb{R}^3 \times \mathbb{R}$, we shall say that Σ *lies in the symmetry class of* Σ_0 if there is an $H \in \mathfrak{S}(\mathcal{D})$ such that

$$H^{-1}\mathfrak{G}_F(g)\, H = \mathfrak{G}_{FH}(g) \subsetneqq \mathfrak{G}_{F_0}(g_0).$$

† Note that $\mathfrak{S}(\mathcal{D}_{\Sigma_0}) = \mathfrak{G}_{F_0}(g_0)$ since, in addition to (5.6), $\mathcal{D}_{\Sigma_0}$ satisfies the condition $\mathbf{L}_Q \mathbf{R}_H \mathcal{D}_{\Sigma_0} = \mathcal{D}_{\Sigma_0}$ $\forall\, Q \in \mathfrak{O}$ and $\forall\, H \in \mathfrak{G}_{F_0}(g_0)$.

†† The group properties of $\mathfrak{G}_{F_0}$ and $\mathfrak{G}_{F_0}(g_0)$ allow H to be replaced by H^{-1} in Lemma 4.

Further, for any $\Sigma_0 \equiv (F_0, g_0, s_0) \in \mathscr{D}$, the open set $\mathscr{D}^*_{\Sigma_0} \subset \mathscr{D}$ is said to be a *fundamental domain for the symmetry class of* Σ_0 if it contains Σ_0 and satisfies

(i) $\mathbf{L}_Q \mathbf{R}_H \mathscr{D}^*_{\Sigma_0} = \mathscr{D}^*_{\Sigma_0} \quad \forall\, Q \in \mathfrak{O}$, *and* $\forall\, H \in \mathfrak{G}_{F_0}(g_0)$,

(ii) $\mathbf{R}_H \mathscr{D}^*_{\Sigma_0} \cap \mathscr{D}^*_{\Sigma_0} = \emptyset \quad \forall\, H \in \mathfrak{S}(\mathscr{D}) \setminus \mathfrak{G}_{F_0}(g_0)$,

and

(iii) *for each* $\Sigma \equiv (F, g, s) \in \mathscr{D}$ *lying in the symmetry class of* Σ_0 *there is an* $H \in \mathfrak{S}(\mathscr{D})$ *such that* $\mathbf{R}_H \Sigma \in \bar{\mathscr{D}}^*_{\Sigma_0}$.

We now have

Theorem 5. *If* $\mathscr{D} \in \text{Inv} \times \mathbb{R}^3 \times \mathbb{R}$ *is a crystal domain for* $\mathscr{B}_0$ *and if* $\Sigma_0 \equiv (F_0, g_0, s_0) \in \text{Inv} \times \mathbb{R}^3 \times \mathbb{R}$, *then there is a fundamental domain for the symmetry class of* Σ_0.

Proof. The proof of Theorem 2 readily may be adapted to prove this Theorem by making identifications of the type $F_0 \to \Sigma_0$ and $\mathfrak{G}_{F_0} \to \mathfrak{G}_{F_0}(g_0)$. Hence we omit the details. □

Finally, analogous to Theorem 3, we have

Theorem 6. *There is an open set* $\mathscr{D}^* \subset \mathscr{D}$ *with the property that for each* $\Sigma \equiv (F, g, s) \in \mathscr{D}$ *there is an* $H \in \mathfrak{S}(\mathscr{D})$ *such that*

$$\mathbf{R}_H \Sigma \in \bar{\mathscr{D}}^*.$$

Proof. Let F_c and F_h be, respectively, points of maximal cubic and maximal hexagonal symmetry in Inv.† Then, since $\mathfrak{G}_F(g) \subseteq \mathfrak{G}_F$ for all $F \in \text{Inv}$ and all $g \in \mathbb{R}^3$, it follows that there are points $\Sigma_c \equiv (F_c, g_c, s_c)$ and $\Sigma_h = (F_h, g_h, s_h)$ in $\text{Inv} \times \mathbb{R}^3 \times \mathbb{R}$ such that for each $\Sigma \equiv (F, g, s) \in \mathscr{D}$ there is an $H \in \mathfrak{S}(\mathscr{D})$ satisfying the following relations:

$$H^{-1} \mathfrak{G}_F(g)\, H \subseteq \mathfrak{G}_{F_c}(g_c)$$

and/or

$$H^{-1} \mathfrak{G}_F(g)\, H \subseteq \mathfrak{G}_{F_h}(g_h).$$

In particular, then, these also hold if $g_c = g_h = 0$. Now, to complete this proof, we define

$$\mathscr{D}^* \equiv \mathscr{D}^*_{\Sigma_c} \cup \mathscr{D}^*_{\Sigma_h},$$

where $\mathscr{D}^*_{\Sigma_c}$ and $\mathscr{D}^*_{\Sigma_h}$ are, respectively, fundamental domains for the symmetry classes of Σ_c and Σ_h. □

Acknowledgement. We gratefully acknowledge the National Science Foundation's support of this work.

† Lemma 3 may be readily adapted to the present case by replacing $\mathscr{D}$ in Lemma 3 with the present interpretation of $\mathscr{D} \cap \text{Inv}$.

References

1. Hertog, B., The Elastic Dielectric: Minimizing States in Ferroelectrics. Ph.D. Thesis, University of Minnesota, 1987.
2. Ericksen, J. L., On the Symmetry and Stability of Thermoelastic solids. J. Applied Mech. **45**, 740–744 (1978).
3. Pitteri, M., Reconciliation of Local and Global Symmetries of Crystals. J. Elasticity **14**, 175–190 (1984).
4. Truesdell, C., *A First Course in Rational Continuum Mechanics, Vol. 1*. New York: Academic Press 1977.
5. Bradley, C. J., & A. P. Cracknell, *The Mathematical Theory of Symmetry in Solids*. Oxford: Clarendon Press 1972.
6. Pitteri, M., On $\nu + 1$ Lattices. J. Elasticity **15**, 3–25 (1985).
7. Ericksen, J. L., Changes in Symmetry in Elastic Crystals. Proc. IUTAM Symposium on Finite Elasticity, Lehigh University, 167–177 (1981).
8. Parry, G. P., On the Elasticity of Monatomic Crystals. Math. Proc. Camb. Phil. Soc. **80**, 189–211 (1976).
9. Bourbaki, N., *Elements of Mathematics. General Topology, Part I*. Reading, Massachusetts: Addison-Wesley 1966.
10. Munkres, J. R., *Topology: A First Course*. Englewood Cliffs, New Jersey: Prentice-Hall 1974.
11. Hall, M., *The Theory of Groups*. New York: Macmillan 1964.
12. Grindlay, J., *An Introduction to the Phenomenological Theory of Ferroelectricity*. New York: Pergamon 1970.
13. Toupin, R. A., The Elastic Dielectric. J. Rational Mech. Anal. **5**, 849–916 (1956).
14. Hutter, K., On Thermodynamics and Thermostatics of Viscous Thermoelastic Solids in the Electromagnetic Fields. A Lagrangian Formulation. Arch. Rational Mech. Anal. **58**, 339–368 (1975).
15. Ericksen, J. L., Nonlinear Elasticity of Diatomic Crystals. Int. J. Solids Structures **6**, 951–957 (1970).
16. Ericksen, J. L., Special Topics in Nonlinear Elastostatics. *Advances in Applied Mechanics*, ed. C.-S. Yih, vol. 17. Academic Press, New York 1977.
17. Ericksen, J. L., On the Symmetry of Deformable Crystals. Arch. Rational Mech. Anal. **72**, 1–13 (1979).
18. Ericksen, J. L., Some Phase Transitions in Crystals. Arch. Rational Mech. Anal. **73**, 99–124 (1980).
19. Gurtin, M. E., & Williams, W. O., Phases of Elastic Material. ZAMP **18**, 132–135 (1967).
20. Wang, C.-C., Generalized Simple Bodies. Arch. Rational Mech. Anal. **32**, 1–28 (1969).
21. James, R. D., Displacive Phase Transformations in Solids. J. Mech. Phys. Solids **34**, 359–394 (1986).
22. Zanzotto, G., Twinning in Minerals and Metals: Remarks on the Comparison of a Thermoelastic Theory with some Experimental Results. To appear. Atti. Acc. Lincei Rend. Fis. (8) **82** (1988).

Department of Aerospace Engineering
and Mechanics
University of Minnesota
Minneapolis

(Received September 25, 1988)

A Mathematical Theory of the Guillotine

PIERO VILLAGGIO

To Bernard Coleman for his 60th birthday

1. Prologue

In the year 1989, in a not yet well identified country of the world, a violent revolution exploded, unleashed by the excessive privileges of the upper class. The interim government, instituted by the democratic rebels, decided to reintroduce capital punishment by decapitation, in analogy to what was done in France two centuries before, and consequently the guillotine made its reappearance in history.

A man called *** was appointed superintendent of the executions and was charged with deciding the political and technical questions generated by the cruel decision of the government. Mr. *** was absolutely unable to carry out his task. He was an aristocrat masquerading as a populist, interested only in his own theoretical studies, and, above all, pathologically revolted by human blood.

Unfortunately he was not courageous enough to resign from his position for fear of raising suspicions about his past. He thus decided to exploit his knowledge of mechanics so as to render the guillotine more efficient and, perhaps, as painless as possible. In terms customary in his time, he solved an "optimization problem" in solid mechanics.

2. The Problem of Mr. ***

The traditional shape of the blade of the guillotine is trapezoidal (Figure 1) with the skew side sharpened in order to make penetration easier. The typical cross section of the blade, hatched in the figure, has a sharp lower edge where the first contact with the neck of the condemned occurs. But, while the lengths of the sides and thickness of the blade are prescribed, there is a certain freedom in modelling the profile which bounds the lower part of the cross section. Though in machines of the past it was an interval of a straight line, it is not necessarily true that this is the most favorable form of penetration. Granted that the thickness must be constant, the question arises whether a sharp or a blunt edge is more convenient.

Converted into more abstract terms, the problem faced by Mr. *** was that of optimizing the profile of the cutting edge of a rigid blade in order that, the load being equal, the penetration into a soft body would be highest, avoiding breaks.

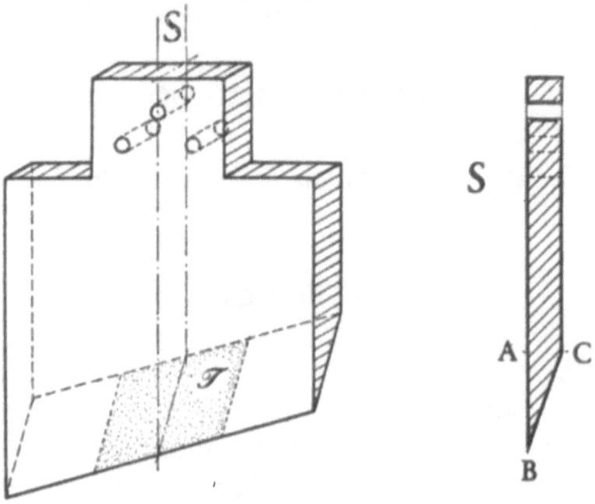

Figure 1

In practice the first contact does not occur along the entire lower side of the blade, but on a subinterval $\mathcal{T}$. If, however, the diameter of the body that must be cut is sufficiently large, this interval of first contact should be large enough that a substantial part of the edge enters into action at the same instant. On the other hand, if the blade is sufficiently thin, and the height of the sharpened part of the edge is also small compared with the thickness, the transverse penetration of the blade will be equal to its thickness just after the first instant of contact. These two considerations make plausible the assumption of regarding the state of deformation induced by the cross section of the blade into the flesh of a fat neck as that generated by the complete indentation of a rigid punch of variable profile on an elastic half-plane in a state of plane strain. The scheme of the position of the punch with respect to the half-plane after the first penetration is that indicated in Figure 2, where the length of AC, the thickness of the blade, and that of AB,

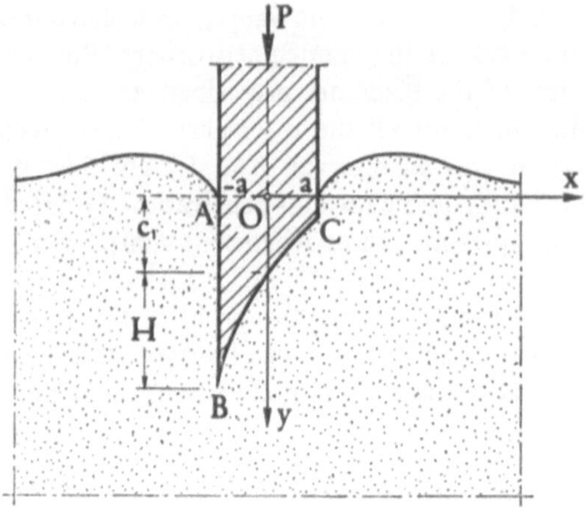

Figure 2

the height of the sharpened edge, are prescribed, while the shape of BC must be determined so as to improve the elastic penetration of the punch under constant load P.

Of course such a model requires other tacit simplifications which must be kept in mind: the assumption that all the cross sections of the blade have the same initial penetration of magntiude H implies that the obliqueness of the lower side has been neglected; in addition, it is assumed that the contact is so smooth that the possible tangential stress due to friction can be disregarded; finally, the process is sufficiently slow to ignore dynamic effects. It must be also recalled that the hypothesis that H is small with respect to the thickness $2a$ is necessary to infer the strains small enough to be regarded as infinitesimal and consequently to determine them by the methods of plane linear elasticity. For, in particular, a flat punch ($H =$ zero), the first solution of the corresponding elastic problem is due to SADOWSKI [1928]; and the same result was found through other techniques by MUSKHELISHVILI [1941] and by L. FÖPPL [1941].

The problem of modelling the shape of the edge is, however, different. In this case of force P acting on the punch, its thickness and height are prescribed, and the profile BC must be found which yields the largest penetration, that is the maximum displacement in the direction of the y-axis of a characteristic point of the blade taken to be the point with initial coordinates (0, 0). It may appear that the best shape is such that the profile BC has a very sharp vertex at B, but this solution is not physically acceptable for, if the vertex is too pointed, the excess of localized stress may cause the blade to break. The conclusion is that the solution must be a compromise between the need to sharpen the profile at B to favor penetration and the need to maintain a certain slope to avoid rupture.

The problem formulated in these terms is no longer a classical problem but one of shape optimization. Although these problems are very common in technical mechanics, their solution is difficult. That found here is only partial as it is assumed that the curve BC can be expressed by a polynomial, keeping in mind the obvious fact that, if the properties of regularity imposed on minimizing functions are removed, the minimum may be much lower or even not exist at all.

3. The Rectilinear Profile

Before adressing the case in which the profile is analytically represented by a polynomial it is expedient to consider a simpler example, like that in which BC is rectilinear and, therefore, the punch is triangular. The shape of the profile being prescribed, the question of optimizing it does not arise, since BC is simply the segment connecting these two points. Again, the condition that the angle with vertex at B is not too small imposes a limit on the magnitude of H.

With the x and y axes chosen as is shown in Figure 2, the Cartesian equation of BC is

$$y = f(x) = \frac{H}{2}\left(1 - \frac{x}{a}\right) + G, \quad -a \leqq x \leqq a, \tag{3.1}$$

where G is a constant for the moment not determined. This curve also represents the normal displacement $v(x, 0)$ imparted to the boundary of the half-plane $y \geqq 0$ in the interval $-a \leqq x \leqq a$, the remaining part being free from stresses. Under the hypothesis that the half-plane is linearly elastic with Young's modulus E and Poisson's ratio σ, the problem was solved by FLORIN [1936]. The very simple solution is found by assuming the normal pressure $p(x)$, exerted upon the interval where penetration occurs, to have the form

$$p(x) = \frac{c_0 - c_1 \dfrac{x}{a}}{\sqrt{1 - \left(\dfrac{x}{a}\right)^2}}, \tag{3.2}$$

where c_0, c_1 are two indeterminate constants. This pressure is singular at the ends of the interval where the boundary conditions change type. The constants c_0, c_1 are determined by imposing the condition that the displacement $v(x, 0)$, or, better, its derivative $\dfrac{\partial v}{\partial x}(x, 0)$ is $f'(x)$:

$$\frac{\partial v}{\partial x}(x, 0) = -\frac{2(1-\sigma^2)}{\pi E} \int_{-a}^{a} \frac{p(u)\, du}{x - u} = -\frac{H}{2a}. \tag{3.3}$$

In addition to this relation, it is known that the resultant of the pressure $p(x)$ is equivalent to the vertical load for unit of length P, that is

$$\int_{-a}^{a} p(u)\, du = P. \tag{3.4}$$

When the expression (3.2) for $p(u)$ is put into the above equations, with the abbreviation $\theta = \dfrac{\pi E}{2(1-\sigma^2)}$ they become

$$-\frac{1}{\theta} \int_{-a}^{a} \frac{c_0 - c_1 \dfrac{u}{a}}{\sqrt{1 - \left(\dfrac{u}{a}\right)^2}\,(x - u)}\, du = -\frac{H}{2a}, \tag{3.5}$$

$$\int_{-a}^{a} \frac{c_0 - c_1 \dfrac{u}{a}}{\sqrt{1 - \left(\dfrac{u}{a}\right)^2}}\, du = P, \tag{3.6}$$

and these integrals can be evaluated in finite form by the change of variables $x = -a \cos \alpha$, $u = -a \cos \beta$ (*cf.* GRÖBNER & HOFREITER [1966, II 332]). After introducing the values of the integrals it is possible to derive c_0 and c_1:

$$c_0 = \frac{P}{\pi a}, \qquad c_1 = \frac{\theta H}{2\pi a}. \tag{3.7}$$

This is FLORIN's result, but the method used here is due to L. FÖPPL (*cf.* SZABÓ [1964, § 11]).

Once the local pressure is known, the partial resultant of this pressure starting from the end point $x = -a$, is given by

$$\mathscr{P}(x) = \int_{-a}^{x} p(u)\,du = \int_{-a}^{x} \frac{c_0 - c_1 \dfrac{u}{a}}{\sqrt{1 - \left(\dfrac{u}{a}\right)^2}}\,du$$

$$= \frac{P}{\pi}\left(\arcsin\frac{x}{a} + \frac{\pi}{2}\right) + \frac{\theta H}{2\pi}\sqrt{1 - \left(\frac{x}{a}\right)^2}, \tag{3.8}$$

which, of course, is P for $x = a$.

Still from the formula giving the pressure it is possible to obtain the displacement $v(x, 0)$ in the load region $-a \leqq x \leqq a$:

$$v(x, 0) = -\frac{1}{\theta}\int_{-a}^{a} p(u) \ln (x - u)\,du + \text{constant}; \tag{3.9}$$

since this displacement is defined to within a constant, it is customary to designate two points of the boundary $(\pm c, 0)$ with $c > a$ as fixed (*cf.* GIRKMANN [1963, § 32]). Thus $v(x, 0)$ assumes the form

$$v(x, 0) = -\frac{1}{\theta}\int_{-a}^{a} p(u) \ln \frac{|x - u|}{c}\,du. \tag{3.10}$$

The displacement of a characteristic point of the base, for example, the origin (0, 0) can be taken as a measure of the penetration of the blade under the load P. Then the formula (3.10) becomes

$$v_0 = v(0, 0) = -\frac{1}{\theta}\int_{-a}^{a} \frac{c_0 - c_1 \dfrac{u}{a}}{\sqrt{1 - \left(\dfrac{u}{a}\right)^2}}\left(\ln\frac{|u|}{a} - \ln\frac{c}{a}\right) du, \tag{3.11}$$

that is, by again using the results of GRÖBNER & HOFREITER [1966, II 324, 60]

$$v_0 = \frac{\pi a}{\theta} c_0 \ln\left(2\frac{c}{a}\right) = \frac{P}{\theta}\ln\left(2\frac{c}{a}\right). \tag{3.12}$$

Because v_0 is determined, also the constant G is known since $f(0)$ must be equal to v_0, and its value is $G = \dfrac{P}{\theta}\ln\left(2\dfrac{c}{a}\right) - \dfrac{H}{2}$.

At this point the problem would be solved if not for the fact that, if H is too large and the edge too sharp, the blade may break from excess pressure. In practice, since the pressure is infinite at B (as well at C), this condition must be rendered more precise: what is important to bound is the product

$$K = \lim_{x \to -a} p(x)\sqrt{1 + \frac{x}{a}}, \tag{3.13}$$

which is called the "stress concentration factor" and is a measure of how the pressure tends to infinity in the neighborhood of the vertex B. This quantity must obey a constraint of the type $K \leqq K_0$, this latter being a constant of the material composing the blade. Then, recalling (3.2), it is found that

$$K = \frac{1}{\sqrt{2}}(c_0 + c_1) = \frac{1}{\sqrt{2}}\left(\frac{P}{\pi a} + \frac{\theta H}{2\pi a}\right) \leqq K_0. \tag{3.14}$$

The concluding result is thus the following: at constant load P, the penetration v_0 is independent of H, but this cannot be too large in order to not violate (3.14) and the condition $f(a) = G \geqq 0$, ensuring complete penetration.

4. The Parabolic Profile

An alternative solution of the problem is that of adopting for BC, instead of a rectilinear profile, a parabolic one, described by the equation

$$y = f(x) = \frac{H}{4}\left(1 - \frac{x}{a}\right)^2 + G, \quad -a \leqq x \leqq a, \tag{4.1}$$

where G is a constant.

Also in this case an integral equation like (3.3) must be solved, but with the difference that $f'(x)$ is now $-\frac{H}{2}\left(1 - \frac{x}{a}\right)$. By taking the expression for the pressure to be

$$p(x) = \frac{c_0 - c_1 \frac{x}{a} + c_2\left(\frac{x}{a}\right)^2}{\sqrt{1 - \left(\frac{x}{a}\right)^2}}, \tag{4.2}$$

with c_0, c_1, c_2, being new constants, and placing it into (3.3) after modifying the term $f'(x)$, we obtain the new equation

$$-\frac{1}{\theta}\int_{-a}^{a} \frac{c_0 - c_1 \frac{u}{a} + c_2\left(\frac{u}{a}\right)^2}{\sqrt{1 - \left(\frac{u}{a}\right)^2}\,(x - u)}\,du = -\frac{H}{2a}\left(1 - \frac{x}{a}\right). \tag{4.3}$$

The condition that the resultant pressure is still P is

$$\int_{-a}^{a} \frac{c_0 - c_1 \frac{u}{a} + c_2\left(\frac{u}{a}\right)^2}{\sqrt{1 - \left(\frac{u}{a}\right)^2}}\,du = P. \tag{4.4}$$

By again using the tables of GRÖBNER & HOFREITER the two foregoing integrals can be evaluated explicitly so as to arrive at the two equations

$$c_1 - c_2\frac{x}{a} = \frac{\theta H}{2\pi a}\left(1 - \frac{x}{a}\right), \qquad c_0 + \frac{c_2}{2} = \frac{P}{\pi a},$$

and hence

$$c_0 = \frac{P}{\pi a} - \frac{\theta H}{4\pi a}, \qquad c_1 = \frac{\theta H}{2\pi a}, \qquad c_2 = \frac{\theta H}{2\pi a}. \tag{4.5}$$

The constants c_0, c_1, c_2 being known, penetration is now given by the expression

$$v_0 = v(0,0) = -\frac{1}{\theta}\int_{-a}^{a} \frac{c_0 - c_1\frac{u}{a} + c_2\left(\frac{u}{a}\right)^2}{\sqrt{1-\left(\frac{u}{a}\right)^2}}\left(\ln\frac{|u|}{a} - \ln\frac{c}{a}\right)du,$$

the calculation of which furnishes

$$v_0 = \frac{\pi a}{\theta} c_0 \ln\left(2\frac{c}{a}\right) - \frac{\pi a}{\theta} c_2\left(\tfrac{1}{4} - \tfrac{1}{2}\ln\left(2\frac{c}{a}\right)\right), \tag{4.4}$$

or, after introducing the values of c_0 and c_2,

$$v_0 = \frac{P}{\theta}\ln\left(2\frac{c}{a}\right) + \frac{H}{8}, \tag{4.5}$$

and a comparison with the formula (3.12) shows that penetration is increased by the amount $\frac{H}{8}$ for a parabolic profile. From (4.5) the derivation of the constant $G = \frac{P}{\theta}\ln\left(2\frac{c}{a}\right) - \frac{H}{8}$ is also immediate, and G must be non-negative in order to guarantee complete penetration.

The factor K, which is still defined by (3.13), becomes

$$K = \frac{1}{\sqrt{2}}(c_0 + c_1 + c_2) = \frac{1}{\sqrt{2}}\left(\frac{P}{\pi a} + 3\frac{\theta H}{4\pi a}\right), \tag{4.6}$$

and, to prevent possible ruptures, K must not be larger than K_0.

Since K is now higher than in the case of a rectilinear profile, the conclusion is that a parabolic profile improves the elastic penetration but worsens the specific pressure at the vertex.

5. The Optimal Profile

As the two simple cases of rectilinear and parabolic profile are, in some sense, in conflict with each other, the first more effective for local pressure concentration at the vertex, the second for elastic penetration, it is natural to ask whether it is possible to design a mixed profile, effecting a compromise between the two counterposed requirements.

For this purpose it is convenient to start with a profile having the equation

$$y = f(x) = a_1\left(1 - \frac{x}{a}\right) + a_2\left(1 - \frac{x}{a}\right)^2 + G, \tag{5.1}$$

where, this time, a_1 and a_2 are constants to be determined and G is the constant introduced before. However, the coefficient of the equation (5.1) must obey the condition that the height remains H, that is

$$f(-a) - f(a) = 2a_1 + 4a_2 = H; \tag{5.2}$$

and that the graph of BC nowhere crosses the x-axis within the interval $-a \leqq x \leqq a$. A simple, though more restrictive, condition ensuring the latter constraint is that

$$f(a) \geqq 0, \quad f'(x) \leqq 0,$$

or, in terms of the coefficients a_1, a_2, G,

$$G \geqq 0, \quad a_1 \geqq 0, \quad a_1 + 2a_2 \geqq 0. \tag{5.3}$$

The pressure $p(x)$ can still be assumed to have the form (4.2), determining it in terms of the three constants c_0, c_1, c_2, which must satisfy an equation like (4.3) but with the right-hand side modified: $f'(x) = -\dfrac{a_1}{a} - 2\dfrac{a_2}{a}\left(1 - \dfrac{x}{a}\right)$. The condition on the resultant remains (4.4). Calculation of the integrals yields the relations

$$c_1 - c_2\frac{x}{a} = \frac{\theta}{\pi a}\left[a_1 + 2a_2\left(1 - \frac{x}{a}\right)\right], \quad c_0 + \frac{c_2}{2} = \frac{P}{\pi a},$$

and hence

$$c_0 = \frac{P}{\pi a} - \frac{\theta a_2}{\pi a}, \quad c_1 = \frac{\theta}{\pi a}(a_1 + 2a_2), \quad c_2 = +\frac{\theta}{\pi a}2a_2.$$

The elastic penetration is still given by (4.4), but, with the new values for the constants it becomes

$$v_0 = \frac{P}{\theta}\ln\left(2\frac{c}{a}\right) + \frac{a_2}{2}, \tag{5.4}$$

and the goemetrical condition $v_0 = f(0)$, gives the relation determining G:

$$v_0 = a_1 + a_2 + G.$$

Lastly, (4.6) gives the factor of pressure concentration, while the condition of safety against breaking assumes the form

$$K = \frac{1}{\sqrt{2}}(c_0 + c_1 + c_2) = \frac{1}{\sqrt{2}}\left(\frac{P}{\pi a} + \frac{\theta}{\pi a}a_1 + 3\frac{\theta}{\pi a}a_2\right) \leqq K_0. \tag{5.5}$$

The optimum profile will then be that which maximizes v_0, regarded as a function of a_1, a_2, G, with the condition (5.2) and the constraints (5.3), (5.5). Since

(5.2) permits the elimination of a_2, the other constraints reduce to:

$$a_1 \leqq \tfrac{4}{3}\frac{P}{\theta}\ln\left(2\frac{c}{a}\right)-\frac{H}{6},\qquad a_1 \geqq 0,\qquad a_1 \geqq \tfrac{3}{2}H+2\frac{P}{\theta}-2\sqrt{2}\frac{\pi a}{\theta}K_0, \tag{5.6}$$

and v_0 can be written

$$v_0=\frac{P}{\theta}\ln\left(2\frac{c}{a}\right)+\frac{H}{8}-\frac{a_1}{4}. \tag{5.7}$$

The discussion of the problem is simple. It is clear that the constraints are compatible only if

$$\tfrac{4}{3}\frac{P}{\theta}\ln\left(2\frac{c}{a}\right)-\frac{H}{6}\geqq \max\left\{0,\tfrac{3}{2}H+2\frac{P}{\theta}-2\sqrt{2}\frac{\pi a}{\theta}K_0\right\}. \tag{5.8}$$

Provided that this inequality is satisfied, two cases are possible: if the maximum of the right-hand side is 0, then the solution is $a_1=0$ and $a_2=\frac{H}{4}$, which corresponds to the simple parabola; if, alternatively, the second term on the right-hand side of (5.8) is the maximum, then the solution is $a_1=\tfrac{3}{2}H+2\frac{P}{\theta}-2\sqrt{2}\frac{\pi a}{\theta}K_0$, $a_2=\sqrt{2}\frac{\pi a}{\theta}K_0-\frac{P}{\theta}-\frac{H}{2}$, and the profile is mixed. In either case it is convenient to thin the profile at the vertex.

6. Epilogue

The reprisals of 1989 never took place because every citizen could be proven to be a revolutionary. Mr. *** received solemn awards and his theory was applied to grind knives for slicing bread and salami. Only the pigs protested.

References

[1928] Sadowski, M. A.: „Zweidimensionale Probleme der Elastizitätstheorie.“ Zeit. Angew. Math. Mech. V. 8, pp. 107–121.

[1941] Muskhelishvili, N. I.: "The fundamental boundary problems of the theory of elasticity for the half-plane." Soobshceniya A. N. Gruz. S.S.E. V. III, n. 10, pp. 873–880.

[1941] Föppl, L.: „Elastische Beanspruchung des Erdbodens unter Fundamenten.“ Forschg. Ing. Wes. V. 12, p. 31.

[1948] Florin, V. A.: *Analysis of Foundation of Hydraulic Structures*. Stroiizdat: Moscow.

[1963] Girkmann, K.: *Flächentragwerke*. 6th Ed. Wien: Springer.

[1964] Szabó, I.: *Höhere Technische Mechanik*. Berlin/Göttingen/Heidelberg: Springer.

[1965] Gröbner, W., & Hofreither, N.: *Integraltafel*. Wien/New York: Springer.

Istituto di Scienza delle Costruzioni
Università di Pisa

(Received June 27, 1989)

Global Properties of Buckled States of Plates that can Suffer Thickness Changes

STUART S. ANTMAN

This paper is dedicated to Bernard Coleman on the occasion of his sixtieth birthday

Abstract

This paper treats the axisymmetric buckling of nonlinearly elastic Cosserat plates, which can suffer thickness changes, as well as flexure, midplane extension, and shear. The governing equations are accordingly quite complicated. Nevertheless, it is shown that all solutions, bifurcating or not, have a simple, detailed nodal structure that distinguishes branches globally.

1. Introduction

In this paper we study the large buckled states of a circular plate subject to a pressure applied to its edge. The material of the body is described by a very general system of nonlinear constitutive equations for elastic materials. The novelty of the analysis is that it supplies a detailed global description of buckled states of plates that can suffer not only flexure, midplane extension, and shear, but also thickness changes. Accounting for thickness changes increases the order of the governing system of equations. Despite the resulting complexity, we are able to determine detailed nodal properties of solutions that globally distinguish both bifurcating and nonbifurcating branches. Indeed, the global qualitative theory hinges on a simple, but not obvious, symmetry property of the constitutive equations. The heart of this paper accordingly lies in Theorem 3.9, which asserts that the requisite symmetry property is physically natural. (The development leading to this theorem contains some refinements of shell theory.) To avoid obscuring the essential ideas by grappling with the technical difficulties associated with polar singularities, which are adequately treated elsewhere, we restrict our attention to annular plates. We comment on intact plates in Section 7.

Notation. Vectors (which are elements of Euclidean 3-space) are denoted by lower-case bold-face symbols and second-order tensors (which are linear transformations of Euclidean 3-space into itself) are denoted by upper-case bold-face symbols. The dot product of (vectors) $\boldsymbol{u}$ and $\boldsymbol{v}$ is denoted by $\boldsymbol{u} \cdot \boldsymbol{v}$. The value of tensor $\boldsymbol{A}$ at vector $\boldsymbol{v}$ is denoted by $\boldsymbol{A} \cdot \boldsymbol{v}$ (in place of the more usual $\boldsymbol{Av}$) and the product of $\boldsymbol{A}$ and $\boldsymbol{B}$ is denoted by $\boldsymbol{A} \cdot \boldsymbol{B}$ (in place of the more usual $\boldsymbol{AB}$). The transpose of $\boldsymbol{A}$ is denoted by $\boldsymbol{A}^*$. We write $\boldsymbol{u} \cdot \boldsymbol{A} = \boldsymbol{A}^* \cdot \boldsymbol{u}$. The inner product of $\boldsymbol{A}$ and $\boldsymbol{B}$ (which equals the trace of $\boldsymbol{A} \cdot \boldsymbol{B}$) is denoted $\boldsymbol{A} : \boldsymbol{B}$. The dyadic product of vectors $\boldsymbol{a}$ and $\boldsymbol{b}$ is denoted by $\boldsymbol{ab}$ (in place of the more usual $\boldsymbol{a} \otimes \boldsymbol{b}$). It is defined by $(\boldsymbol{ab}) \cdot \boldsymbol{u} = (\boldsymbol{b} \cdot \boldsymbol{u})\, \boldsymbol{a}$ for all $\boldsymbol{u}$. Thus $(\boldsymbol{ab}) : (\boldsymbol{uv}) = (\boldsymbol{a} \cdot \boldsymbol{u})\,(\boldsymbol{b} \cdot \boldsymbol{v})$. Twice repeated lower-case Latin indices are summed from 1 to 3.

The (Gâteaux) differential of $\boldsymbol{u} \mapsto \boldsymbol{f}(\boldsymbol{u})$ at $\boldsymbol{v}$ in the direction $\boldsymbol{h}$ is $\frac{d}{dt}\boldsymbol{f}(\boldsymbol{v} + t\boldsymbol{h})\big|_{t=0}$. When it is linear in $\boldsymbol{h}$ we denote this differential by $\frac{\partial \boldsymbol{f}}{\partial \boldsymbol{u}}(\boldsymbol{v}) \cdot \boldsymbol{h}$. We often denote the function $\boldsymbol{u} \mapsto \boldsymbol{f}(\boldsymbol{u})$ by $\boldsymbol{f}(\cdot)$. Obvious analogs of these notations will also be used.

Background. The COSSERATS (1909) introduced statical theories of rods and shells that could suffer flexure, extension, and shear. In their formulations, the configuration of a shell is a surface each point of which is endowed with a rigid orthonormal triad that can rotate independently of the surface. A variety of refined linear and nonlinear shell theories, which are related to that of the COSSERATS, have been subsequently developed by REISSNER, MINDLIN, and many others in an extensive literature. If the rigid triad of the Cosserat theory is replaced by a unit vector, then the governing equations have the virtue that they can be posed in a form in which the only stress resultants that appear are the classical contact force and couple. (*Cf.* NAGHDI (1972) for extensive references and commentary on the remarks of this paragraph.) ERICKSEN & TRUESDELL (1958) extended the Cosserat theory of shells by replacing the rigid triad with a triple of independent extensible vectors, called directors. These models thus account for thickness changes. They developed the full theory of strain for these general Cosserat theories and formulated the equilibrium equations for forces and moments, but left open the question of additional equilibrium equations to match the additional kinematical versatility the general theory allows and the question of appropriate acceleration terms for dynamical theories. These issues were soon resolved by GREEN, NAGHDI, & WAINWRIGHT (1965) and COHEN & DESILVA (1966), *inter alia.* The theory we analyze below, which has but one director, is equivalent to that of GREEN *et al.* ANTMAN (1971) gave a general existence theory for axisymmetric equilibrium states of such Cosserat shells by direct methods of the calculus of variations and introduced convenient strain measures.

Perhaps the most important problem in which thickness changes are significant is that of necking. ANTMAN (1973) and ANTMAN & CARBONE (1977) studied the necking of nonlinearly elastic bars and NEGRÓN-MARRERO (1990) studied the axisymmetric necking of a circular plate as global bifurcation problems for Cosserat theories. Their constitutive equations satisfy a one-dimensional version of the strong ellipticity condition. Thus the instabilities they found are consequences of the nonlinearity of response rather than of the loss of ellipticity. ERICKSEN (1975, 1977) on the other hand proposed that necking be studied as a phase change consequent on the nonconvexity of the stored energy. His ideas were extensively developed by CARR, GURTIN, & SLEMROD (1984), COLEMAN (1983, 1985), COLEMAN & NEWMAN (1988), among others. COLEMAN (1983) and OWEN (1987) have shown how the nonconvexity of energy characterizing phase-change theories can be regarded as coming from a singular limit of the equations for a Cosserat theory with a strongly elliptic energy.

Thickness changes are also very important for bodies subject to large compression. In particular, the trivial state of a buckling problem for a rod or shell depends upon the compressive load. The nature of this dependence is significantly influenced by the susceptibility of the the body to thickness changes and in turn significantly influences the spectrum of the linearization of the problem about the trivial solution. (*Cf.* ANTMAN & PIERCE (1990) and NEGRÓN-MARRERO & ANTMAN (1990).)

ANTMAN (1978), GAUSS & ANTMAN (1984), and NEGRÓN-MARRERO & ANTMAN (1990) have carried out global studies of buckling of nonlinearly elastic plates not susceptible to thickness changes. Below it is shown how these works can be readily generalized.

2. Formulation of the Governing Equations

Geometry of deformation. Let $\{\boldsymbol{i}, \boldsymbol{j}, \boldsymbol{k}\}$ be a fixed right-handed orthonormal basis for Euclidean 3-space and let (s, ϕ) be polar coordinates for the $\{\boldsymbol{i}, \boldsymbol{j}\}$-plane. Set

$$(2.1) \qquad \boldsymbol{e}_1(\phi) = \cos\phi\,\boldsymbol{i} + \sin\phi\,\boldsymbol{j}, \qquad \boldsymbol{e}_2(\phi) = -\sin\phi\,\boldsymbol{i} + \cos\phi\,\boldsymbol{j}, \qquad \boldsymbol{e}_3 = \boldsymbol{k}.$$

The *axisymmetric configuration of an annular Cosserat plate* that can suffer flexure, shear, and both in-plane and transverse extension is determined by a pair of vector-valued functions $\boldsymbol{r}$, $\boldsymbol{d}$ of the form

$$(2.2) \qquad [a, 1]\times\mathbb{R} \ni (s, \phi) \mapsto \boldsymbol{r}(s, \phi) \equiv \varrho(s)\,\boldsymbol{e}_1(\phi) + \zeta(s)\,\boldsymbol{k}, \qquad \boldsymbol{d}(s, \phi) \equiv \delta(s)\,\boldsymbol{a}_3(s, \phi)$$

where a is a given number in (0, 1) and where $\boldsymbol{a}_3$ is a unit vector normal to $\boldsymbol{e}_2$. We take the orthonormal triad $\boldsymbol{a}_1 \equiv \boldsymbol{e}_2 \times \boldsymbol{a}_3$, $\boldsymbol{a}_2$, and $\boldsymbol{a}_3$ to have the form

$$(2.3) \qquad \begin{aligned} \boldsymbol{a}_1(s, \phi) &\equiv \cos\theta(s)\,\boldsymbol{e}_1(\phi) + \sin\theta(s)\,\boldsymbol{k}, \\ \boldsymbol{a}_2(\phi) = \boldsymbol{e}_2(\phi), \qquad \boldsymbol{a}_3(s, \phi) &\equiv -\sin\theta(s)\,\boldsymbol{e}_1(\phi) + \cos\theta(s)\,\boldsymbol{k}. \end{aligned}$$

Thus such a configuration is determined by the four real-valued functions ϱ, ζ, δ, and θ. To interpret these variables, we assume that in its natural (unstressed) configuration a circular plate, regarded as a thin three-dimensional body, has a midplane occupying the annulus of the $\{\boldsymbol{i}, \boldsymbol{j}\}$-plane defined by the inequalities $a \leqq s \leqq 1$. The coordinates (s, ϕ) identify a typical material point on this plane. $\boldsymbol{r}(s, \phi)$ is interpreted as the deformed position of such a material point. $\boldsymbol{d}(s, \phi)$ is interpreted as characterizing the deformed configuration of the material fiber whose reference configuration is on the normal to the midplane through the material point with coordinates (s, ϕ). In particular, $\boldsymbol{a}_3$ or θ determines the orientation and δ determines the stretch of this fiber. $\boldsymbol{d}$ is called a *director*.

We set

$$(2.4) \qquad \frac{\partial \boldsymbol{r}}{\partial s}(s, \phi) \equiv \nu(s)\,\boldsymbol{a}_1(s, \phi) + \eta(s)\,\boldsymbol{a}_3(s, \phi).$$

The set of *strain* variables for our theory is

$$(2.5) \qquad \boldsymbol{w} \equiv (\tau, \nu, \eta, \sigma, \mu, \delta, \omega)$$

where

$$(2.6) \qquad \tau(s) \equiv \varrho(s)/s, \qquad \sigma(s) \equiv \sin\theta(s)/s, \qquad \mu \equiv \theta', \qquad \omega \equiv \delta'.$$

Here and below the prime denotes differentiation with respect to s; τ measures azimuthal stretch, ν stretch in the $\boldsymbol{a}_1$-direction, η shear, σ flexure about rays, μ flexure of material ways, and δ transverse stretch. Note that $\boldsymbol{w}$ determines a configuration to within a translation along the $\boldsymbol{k}$-axis and a rotation about this axis.

The *reference configuration* is characterized by $\varrho(s) = s$, $\theta(s) = 0$, $\delta(s) = 1$. In this configuration $\boldsymbol{w}$ reduces to $(1, 1, 0, 0, 0, 1, 0)$. We shall assume that this configuration is stress-free.

In the course of our study we must impose certain restrictions on our constitutive relations. To show that these restrictions are reasonable, we motivate them by interpreting the Cosserat theory as describing a family of constrained deformations of a three-dimensional body (or equivalently, as describing an approximation of the deformation of such a body). Such an approach assigns a concrete interpretation to $\boldsymbol{r}$ and $\boldsymbol{d}$ and also to the stress resultants we introduce below. We accordingly assume that the reference configuration of the body occupies a region

$$\mathscr{B} \equiv \{s\boldsymbol{e}_1(\phi) + z\boldsymbol{k} \colon s \in [a, 1], \phi \in [0, 2\pi), z \in [-h(s), h(s)]\} \tag{2.7}$$

where the half-thickness h is a given positive-valued continuously differentiable function. The material points of $\mathscr{B}$ for which $z = \pm h(s)$ form the *faces* of $\mathscr{B}$ and those for which $s = a, 1$ form the *edges* of $\mathscr{B}$.

We assume that the deformed position $\boldsymbol{p}(\boldsymbol{z})$ of the material point $\boldsymbol{z} \equiv s\boldsymbol{e}_1(\phi) + z\boldsymbol{k}$ of $\mathscr{B}$ is constrained to have the form

$$\boldsymbol{p}(\boldsymbol{z}) = \boldsymbol{r}(s, \phi) + \beta(\delta, s, z)\, \boldsymbol{a}_3(s, \phi) \tag{2.8}$$

where $\boldsymbol{r}$ and $\boldsymbol{d}$ have the form (2.2) and where β is a given continuously differentiable function on $\{(\delta, s, z) \colon \delta \in (0, \infty), s \in [a, 1], z \in [-h(s), h(s)]\}$ with

$$\beta(\delta, s, \cdot) \text{ odd}, \quad \frac{\partial \beta}{\partial z}(\delta, s, z) > 0 \quad \forall\, (\delta, s, z) \text{ with } z \in (-h(s), h(s)),$$

$$\beta(1, s, z) = z. \tag{2.9}$$

It is traditional to take $\beta(\delta, s, z) = z\,\delta$.

To give a more accurate two-dimensional description of boundary conditions on the faces of the three-dimensional plate it is often advantageous to avoid this restriction. *Cf.* ANTMAN (1972, Sec. 5). On the other hand, choosing $\beta(\delta, s, z)$ to have the form $\gamma(z)\,\delta$ does not represent a genuine generalization because its value in the reference configuration is just $\gamma(z)$, which represents a true thickness variable, while z is relegated to an arbitrary parameter. A simple change of variables reduces this case to the traditional case.

When $\boldsymbol{p}$ has the constrained form (2.8), we find that

$$\partial \boldsymbol{p}/\partial \boldsymbol{z} = [(\nu - \beta\mu)\, \boldsymbol{a}_1 + (\eta + \beta_s + \beta_\delta \omega)\, \boldsymbol{a}_3]\, \boldsymbol{e}_1 + (\tau - \beta\sigma)\, \boldsymbol{e}_2\boldsymbol{e}_2 + \beta_z \boldsymbol{a}_3 \boldsymbol{k}. \tag{2.10}$$

Here and below we denote partial derivatives by subscripts. (In particular, β_s denotes the partial derivative of β with respect to its second argument, and not a total partial derivative with respect to s.) The right Cauchy-Green deformation tensor $C \equiv (\partial \boldsymbol{p}/\partial \boldsymbol{z})^* \cdot (\partial \boldsymbol{p}/\partial \boldsymbol{z})$ is then given by

$$C = C_{11}\boldsymbol{e}_1\boldsymbol{e}_1 + C_{22}\boldsymbol{e}_2\boldsymbol{e}_2 + C_{33}\boldsymbol{k}\boldsymbol{k} + C_{13}(\boldsymbol{e}_1\boldsymbol{k} + \boldsymbol{k}\boldsymbol{e}_1). \tag{2.11}$$

where

$$C_{11} \equiv [\nu - \beta\mu]^2 + [\eta + \beta_s + \beta_\delta\omega]^2, \quad C_{22} \equiv [\tau - \beta\sigma]^2,$$
$$C_{33} \equiv \beta_z^2, \quad C_{13} \equiv [\eta + \beta_s + \beta_\delta\omega]\,\beta_z.$$

(We define $C_{ij} \equiv \boldsymbol{e}_i \cdot \boldsymbol{C} \cdot \boldsymbol{e}_j$.) Note that $C_{21} = 0 = C_{23}$.

When (2.8) holds, it follows from (2.10) that the requirement that $\boldsymbol{p}$ preserve orientation, *i.e.*, that its Jacobian be positive everywhere, is equivalent to the inequalities

$$\tau > \beta(s, h(s), \delta)\,|\sigma|, \quad \nu > \beta(s, h(s), \delta)\,|\mu|, \quad \delta > 0. \tag{2.12a,b,c}$$

The set of $\boldsymbol{w}$'s satisfying (2.12) is denoted by $\mathscr{W}(s)$. For fixed δ this set is convex.

Stress resultants and equilibrium equations. It is a straightforward matter to obtain the standard differential equations for the equilibrium of forces and moments by summing their resultants on arbitrary subsets on the annulus. The resulting equations, when supplemented with constitutive equations, are insufficient to determine the configuration of a Cosserat plate. Specifically, we need an additional equation to correspond to the additional variable δ. We could get this equation by postulating it, or equivalently by invoking a principle of virtual work for shells, or, for hyperelastic shells, by using a variational principle, which is also tantamount to the first two alternatives. But it is more illuminating to derive all of our equilibrium equations directly from the three-dimensional equations of equilibrium, which we pose in the form of the three-dimensional Principle of Virtual Work (*cf.* Antman & Osborn (1979)).

Let

$$\boldsymbol{T}(\boldsymbol{z}) = \boldsymbol{T}(\boldsymbol{z}) \cdot \boldsymbol{e}_k(\phi)\,\boldsymbol{e}_k(\phi) \tag{2.13}$$

denote the first Piola-Kirchhoff stress tensor at material point $\boldsymbol{z}$. It is related to the symmetric second Piola-Kirchhoff stress tensor $\boldsymbol{S}$ by

$$\boldsymbol{T}(\boldsymbol{z}) = \frac{\partial \boldsymbol{p}}{\partial \boldsymbol{z}}(\boldsymbol{z}) \cdot \boldsymbol{S}(\boldsymbol{z}). \tag{2.14}$$

In consonance with our assumption of axisymmetry we require that

$$\boldsymbol{e}_2 \cdot \boldsymbol{T} \cdot \boldsymbol{e}_1 = 0, \quad \boldsymbol{e}_2 \times \boldsymbol{T} \cdot \boldsymbol{e}_2 = \boldsymbol{0}, \quad \boldsymbol{e}_2 \cdot \boldsymbol{T} \cdot \boldsymbol{k} = 0. \tag{2.15}$$

To postpone prescribing boundary conditions, we employ a restricted form of the principle of virtual work in which the test functions (virtual displacements) vanish at the edges of $\mathscr{B}$. If there are no body forces and if there are no surface tractions applied to the faces of $\mathscr{B}$, then the weak form of the equilibrium equations of continuum mechanics is equivalent to the principle of virtual work:

$$\int_a^1 \int_0^{2\pi} \int_{-h}^{h} s\boldsymbol{T}(\boldsymbol{z}) : \frac{\partial \dot{\boldsymbol{p}}}{\partial \boldsymbol{z}}(\boldsymbol{z})\, dz\, d\phi\, ds = 0 \tag{2.16}$$

for all smooth $\dot{\boldsymbol{p}}$ that vanish at the edges $s = a, 1$. We could take $\dot{\boldsymbol{p}}$ to be the variation of (2.8) (*i.e.*, to lie in the tangent space at $\boldsymbol{p}$ to the manifold of position fields defined by (2.8)):

$$\dot{\boldsymbol{p}} = \dot{\boldsymbol{r}} - \beta\dot{\theta}\boldsymbol{a}_1 + \beta_\delta\,\dot{\delta}\boldsymbol{a}_3 \tag{2.17}$$

where $\dot{\boldsymbol{r}} \cdot \boldsymbol{e}_2 = 0$, and where $\dot{\boldsymbol{r}}$, $\dot{\theta}$, $\dot{\delta}$ are arbitrary smooth functions of s that vanish at a and 1.

Let us first take $\dot{\boldsymbol{p}}$ to have the form (2.17) with $\dot{\theta} = 0 = \dot{\delta}$, *i.e.*, we take $\dot{\boldsymbol{p}}$ to be independent of z and to be perpendicular to $\boldsymbol{e}_2$. Then (2.16) reduces to

$$\int_a^1 [s(N\boldsymbol{a}_1 + H\boldsymbol{a}_3) \cdot \dot{\boldsymbol{r}}_s + T\boldsymbol{e}_1 \cdot \dot{\boldsymbol{r}}]\, ds = 0 \tag{2.18}$$

for all such $\dot{\boldsymbol{r}}$, where

$$N\boldsymbol{a}_1 + H\boldsymbol{a}_3 \equiv \int_{-h}^{h} \boldsymbol{T} \cdot \boldsymbol{e}_1\, dz, \qquad T \equiv \boldsymbol{e}_2 \cdot \int_{-h}^{h} \boldsymbol{T} \cdot \boldsymbol{e}_2\, dz. \tag{2.19}$$

The form of (2.19) is in accord with (2.15). $N(s)\,\boldsymbol{a}_1(s, \phi) + H(s)\,\boldsymbol{a}_3(s, \phi)$ is the resultant contact force per unit reference length of the circle $\phi \mapsto s\boldsymbol{e}_1(\phi)$ exerted at (s, ϕ, z) by the material outside the cylinder $(\phi, z) \mapsto s\boldsymbol{e}_1(\phi) + z\boldsymbol{k}$ on the material inside this cylinder. Similarly, $T(s)\,\boldsymbol{e}_2(\phi)$ is the resultant contact force per unit reference length of the ray $s \mapsto s\boldsymbol{e}_1(\phi)$ acting across the plane $(s, z) \mapsto s\boldsymbol{e}_1(\phi) + z\boldsymbol{k}$. Equation (2.18) is the weak form of

$$\frac{\partial}{\partial s}[s(N\boldsymbol{a}_1 + H\boldsymbol{a}_3)] - T\boldsymbol{e}_1 = \boldsymbol{0}. \tag{2.20}$$

We could get another consequence of (2.16) by choosing $\dot{\boldsymbol{p}}$ to have the form (2.17) with $\dot{\boldsymbol{r}} = \boldsymbol{0}$, $\dot{\delta} = 0$, but it is more illuminating to make an analogous choice of $\dot{\boldsymbol{p}}$ corresponding to an unconstrained $\boldsymbol{p}$:

$$\dot{\boldsymbol{p}} = \boldsymbol{e}_2 \times (\boldsymbol{p} - \boldsymbol{r})\,\dot{\theta}. \tag{2.21}$$

Substituting (2.21) into (2.16) we find the expression

$$\boldsymbol{T} : \{[\boldsymbol{e}_2 \times (\boldsymbol{p} - \boldsymbol{r})_s]\,\boldsymbol{e}_1 + [\boldsymbol{e}_2 \times (\boldsymbol{p} - \boldsymbol{r})_z]\,\boldsymbol{k}\} \tag{2.22a}$$

in the integrand. By substituting the identity $\partial\boldsymbol{p}/\partial\boldsymbol{z} = \boldsymbol{p}_s\boldsymbol{e}_1 + s^{-1}\,\boldsymbol{p}_\phi\boldsymbol{e}_2 + \boldsymbol{p}_z\boldsymbol{k}$ into (2.14), exploiting the symmetry of $\boldsymbol{S}$, and using (2.15) we can ultimately reduce (2.22a) to

$$(\boldsymbol{r}_s \times \boldsymbol{e}_2) \cdot \boldsymbol{T} \cdot \boldsymbol{e}_1. \tag{2.22b}$$

Thus (2.16) with (2.21) is the weak version of

$$(sM)' - \Sigma \cos\theta + s(H\nu - N\eta) = 0 \tag{2.23}$$

where

$$M \equiv -\boldsymbol{e}_2 \cdot \int_{-h}^{h} (\boldsymbol{p} - \boldsymbol{r}) \times \boldsymbol{T} \cdot \boldsymbol{e}_1\, dz, \qquad \Sigma \equiv \boldsymbol{e}_1 \cdot \int_{-h}^{h} (\boldsymbol{p} - \boldsymbol{r}) \times \boldsymbol{T} \cdot \boldsymbol{e}_2\, dz. \tag{2.24}$$

$M(s)\,\boldsymbol{e}_2(\phi)$ is the bending couple, *i.e.*, the resultant contact torque about $\boldsymbol{r}(s, \phi)$ per unit reference length of the circle $\phi \mapsto s\boldsymbol{e}_1(\phi)$ exerted at (s, ϕ, z) by the material outside the cylinder $(\phi, z) \mapsto s e_1(\phi) + z k$ on the material inside this cylinder. Similarly, $\Sigma(s)\, a_1(s, \phi)$ is the resultant contact couple per unit reference length of the ray $s \mapsto s e_1(\phi)$.

We note that (2.23) can also be derived from the weak form of the symmetry condition for S (*cf.* ANTMAN & OSBORN (1979, Sec. 5)). By using (2.17) with $\dot{r} = 0, \dot{\delta} = 0$, we would obtain an equation similar to (2.23), but not identical to it.

Motivated by the choice of (2.17) with $\dot{r} = 0, \dot{\theta} = 0$ we substitute $\dot{p} = \gamma(s, z)\, \dot{\delta} a_3$ into (2.16) to obtain the weak form of

(2.25) $$(s\Omega)' - s\Delta = 0,$$

where

$$\Omega \equiv a_3 \cdot \int_{-h}^{h} \gamma T \cdot e_1 \, dz, \quad \Delta \equiv a_3 \cdot \int_{-h}^{h} (T \cdot e_1 \gamma_s + T \cdot k \gamma_z)\, dz + \delta^{-1} (\tilde{\Sigma}\sigma + \tilde{M}\mu),$$

(2.26a,b)

(2.26c,d) $$\delta^{-1} \tilde{\Sigma} \equiv -e_2 \cdot \int_{-h}^{h} \gamma T \cdot e_2 \, dz, \quad \delta^{-1} \tilde{M} \equiv -a_1 \cdot \int_{-h}^{h} \gamma T \cdot e_1 \, dz.$$

If $\gamma(s, z) = z$, which corresponds to the choice of $\beta(\delta, s, z) = z\,\delta$, then $\tilde{\Sigma} = \Sigma$, $\tilde{M} = M$, and the expression for Δ simplifies appreciably. Note that (2.20), (2.23) and (2.25) are exact consequences of the laws of continuum mechanics (under the assumption of axisymmetry). Equations (2.20) and (2.23) are standard.

Nonlinear elasticity. A Cosserat plate, constrained to undergo axisymmetric deformations, is *nonlinearly elastic* if there are functions $\hat{T}, \hat{N}, \hat{H}, \hat{\Sigma}, \hat{M}, \hat{\Delta}, \hat{\Omega}$ defined on $\{(w, s)\colon w \in \mathscr{W}(s), s \in [a, 1]\}$ such that

(2.27) $$T(s) = \hat{T}(w(s), s), \quad etc.$$

We assume that these constitutive functions are continuously differentiable. In the next section we describe constitutive restrictions that are essential for our analysis.

Boundary conditions. We assume that the edge $s = 1$ of the plate is constrained to be parallel to k so that

(2.28) $$\theta(1) = 0.$$

A normal pressure of intensity λ units of force per unit reference length is applied to the edge $s = 1$ so that

(2.29) $$N(1) = -\lambda, \quad H(1) = 0.$$

We finally assume that either the thickness of the plate at this edge is fixed so that there is a prescribed number δ_1 such that

(2.30a) $$\delta(1) = \delta_1$$

or else there is no restraint preventing such a change so that

(2.30b) $$\Omega(1) = 0.$$

The inner edge is subjected to similar boundary conditions:

$$\varrho(a) = a, \quad \theta(a) = 0, \quad H(a) = 0, \quad \Omega(a) = 0. \tag{2.31}$$

We can readily handle variants of these boundary conditions. We must solve (2.4), (2.20), (2.23), (2.25), (2.27), (2.28)–(2.31), which constitute our *boundary value problem* BVP.

3. Constitutive Restrictions

To carry out our analysis, we require that the constitutive functions (2.27) have certain monotonicity, growth, and symmetry. The monotonicity conditions can be chosen so that substitution of (2.27) into the equilibrium equations yields a quasilinear system of ordinary differential equations having a suitable kind of ellipticity, which supports standard techniques of analysis for boundary value problems. The same conditions have natural physical interpretations. These monotonicity conditions also suggest natural compatible growth conditions. It is easy to impose symmetry conditions that ensure we can carry out our global qualitative analysis. But it is not evident whether such conditions, designed merely to meet mathematical exigencies, are physically natural or artificial. To resolve this issue we again resort to our three-dimensional interpretation of the Cosserat theory. In this process we could also deduce the monotonicity conditions from three-dimensional considerations.

The three-dimensional constitutive equations for nonlinearly elastic materials have the frame-indifferent form

$$\boldsymbol{T}(\boldsymbol{z}) = \frac{\partial \boldsymbol{p}}{\partial \boldsymbol{z}}(\boldsymbol{z}) \cdot \hat{\boldsymbol{S}}(\boldsymbol{C}(\boldsymbol{z}), \boldsymbol{z}) \tag{3.1}$$

(*cf.* (2.14)). When $\boldsymbol{p}$ has the constrained form (2.8), we find that $\boldsymbol{C}$ is given by (2.11).

We obtain constitutive equations for our constrained three-dimensional theory by substituting (2.8) and (2.11) into (3.1) and then substituting the resulting expression into (2.19), (2.24), (2.26) with $\gamma(s, z)$ chosen to equal $\beta_\delta(\delta(s), s, z)$. If we set $S^{ij} \equiv \boldsymbol{e}_i \cdot \hat{\boldsymbol{S}} \cdot \boldsymbol{e}_j$ and assume in consonance with (2.15) that these components are independent of ϕ, then the versions of (2.27) so obtained, called the *induced constitutive equations*, have the form

$$\hat{N} = \int_{-h}^{h} (\nu - \beta\mu) S^{11}\, dz, \quad \hat{M} = -\int_{-h}^{h} \beta(\nu - \beta\mu) S^{11}\, dz,$$

$$\hat{H} = \int_{-h}^{h} [(\eta + \beta_s + \beta_\delta\omega) S^{11} + \beta_z S^{31}]\, dz, \quad \hat{\Omega} = \int_{-h}^{h} \beta_\delta[(\eta + \beta_s + \beta\omega) S^{11} + \beta_z S^{31}]\, dz,$$

$$\hat{T} = \int_{-h}^{h} (\tau - \beta\sigma) S^{22}\, dz, \quad \hat{\Sigma} = -\int_{-h}^{h} \beta(\tau - \beta\sigma) S^{22}\, dz, \tag{3.2}$$

$$\hat{\Delta} = \int_{-h}^{h} [(\eta + \beta_s + \beta\omega) S^{13} + \beta_{\delta z} S^{33}]\, dz - \int_{-h}^{h} \beta_\delta[\mu(\nu - \beta\mu) S^{11} + \sigma(\tau - \beta\sigma) S^{22}]\, dz,$$

where the S^{ij} depend on (2.11) and (s, z).

The following is a special case of a result proved by ANTMAN (1976, Sec. 4d):

3.3. Proposition. *If* (3.1) *satisfies the strict form of the strong ellipticity condition, then the induced constitutive functions* $\hat{T}$, etc., *satisfy the following monotonicity conditions: The matrices*

$$\text{(3.4a,b,c)} \qquad \frac{\partial(\hat{N}, \hat{H}, \hat{M}, \hat{\Omega})}{\partial(\nu, \eta, \mu, \omega)}, \quad \frac{\partial(\hat{T}, \hat{\Sigma})}{\partial(\tau, \sigma)}, \quad \frac{\partial\hat{\Delta}}{\partial\delta} \quad \textit{are positive-definite}.$$

We adopt (3.4) as a fundamental constitutive restriction. It ensures that an increase in the bending strain μ be accompanied by an increase in the bending couple M, *etc.* Rather than supplement (3.4) with a compatible system of growth conditions (requiring that infinite resultants accompany extreme strains, *cf.* NEGRÓN-MARRERO & ANTMAN (1990), *e.g.*), we are content to prescribe a major consequence of a suitable set of such conditions:

3.5. Hypothesis. *For any given numbers* (N, M, H, Ω) *the equations*

$$\text{(3.5)} \qquad \hat{N}(w, s) = N, \quad \hat{H}(w, s) = H, \quad \hat{M}(w, s) = M, \quad \hat{\Omega}(w, s) = \Omega$$

have a unique solution for (ν, η, μ, ω) *as a function of* $(\tau, N, H, \sigma, M, \delta, \Omega, s)$.

We denote this solution by

$$\text{(3.7a)} \qquad \nu = \nu^{\#}(\tau, N, H, \sigma, M, \delta, \Omega, s), \ldots,$$

where the ellipsis stands for analogous expressions for $\eta^{\#}$, $\mu^{\#}$, $\omega^{\#}$. Let us set

$$\text{(3.7b)} \qquad T^{\#}(\tau, N, H, \sigma, M, \delta, \Omega, s) \equiv \hat{T}(\tau, \nu^{\#}, \eta^{\#}, \sigma, \mu^{\#}, \delta, \omega^{\#}, s), \ldots$$

where the ellipsis stands for analogous definitions of $\Sigma^{\#}$, $\Delta^{\#}$. The arguments of $\nu^{\#}, \eta^{\#}, \mu^{\#}, \omega^{\#}$ are those shown in (3.7a). The functions $\nu^{\#}$, *etc.*, of (3.7) are necessarily continuously differentiable. The constitutive equations (3.7) are equivalent to (2.27).

Now we turn to the symmetry conditions, which constitute the raison d'être for our development of the Cosserat theory from the three-dimensional theory. We assume that (3.1) satisfies the symmetry conditions:

$$\text{(3.8a)} \qquad S^{21} = 0 = S^{23} \quad \text{if } C_{21} = 0 = C_{23},$$

$$\text{(3.8b)} \qquad S^{11}, S^{22}, S^{33} \text{ are even functions of } C_{13} \text{ if } C_{21} = 0 \text{ or } C_{23} = 0,$$

$$\text{(3.8c)} \qquad S^{13} \text{ is an odd function of } C_{13} \text{ if } C_{21} = 0 \text{ or } C_{23} = 0.$$

A cylindrically orthotropic material satisfies these conditions. Note that (2.8), (2.11), and (3.8) ensure (2.15).

The central result of our analysis is

3.9. Theorem. *Let* (3.8) *hold. Then under the transformation*

$$\text{(3.10)} \qquad (\eta, \sigma, \mu) \mapsto -(\eta, \sigma, \mu)$$

the induced constitutive functions $\hat{H}$, $\hat{\Sigma}$, $\hat{M}$ change sign while the other induced constitutive functions remain unchanged.

Proof. We replace the arguments η, σ, μ of $\hat{M}$ in (3.2) with their negatives. We then make the change of variable $z \mapsto -z$ in the resulting integral. Since $\beta(\delta, s, \cdot)$ is odd, $\beta(\delta, s, z)$ is thereby replaced by $-\beta(\delta, s, z)$, *etc.* The arguments now occupying the C_{11}, C_{22}, C_{33}-slots in S^{11} are unchanged, while the argument now occupying the C_{31}-slot is the negative of that originally occupying this slot. We use (3.8b) to restore the original sign. We thereby find that $\hat{M}$ changes sign under (3.10). The other symmetries are proved likewise. □

The induced constitutive equations enjoy many other symmetries, which can be determined by similar methods. It is easy to impose mild constitutive restrictions on $\hat{S}$ that ensure that the induced constitutive functions satisfy Hypothesis 3.5.

The symmetry condition of Theorem 3.8 is much weaker than the analogous conditions for plates for which δ is constrained to equal 1 (*cf.* ANTMAN (1978)). In such theories the shear response is uncoupled from the flexural response. Equation (4.1a) below can be exploited in a way that renders Theorem 3.9 as useful as the stronger results for the more restricted theory.

Throughout the rest of this paper we assume that (2.27) satisfies (3.4), Hypothesis 3.5, and the conclusion of Theorem 3.9.

4. Alternative Formulation of the Governing Equations

We recast BVP in a form that makes the role of symmetry explicit and that promotes the ensuing analysis. We integrate (2.20) subject to the boundary condition (2.29) and use (3.7) to obtain

$$-N(s) = \Lambda^{\#}(s) \cos \theta(s), \tag{4.1a}$$

$$H(s) = \Lambda^{\#}(s) \sin \theta(s) \tag{4.1b}$$

where

$$s\Lambda^{\#}(s) \equiv \lambda + \int_s^1 T(\tau(t), \ldots)\, dt. \tag{4.1c}$$

We next deduce from (2.23), (2.25), (3.7), and (4.1) that

$$\theta' = \mu^{\#}(\tau, N, \Lambda^{\#} \sin \theta, s^{-1} \sin \theta, M, \delta, \Omega, s), \tag{4.2}$$

$$(sM)' = \Sigma^{\#} \cos \theta - s\Lambda^{\#}\nu^{\#} \sin \theta + sN\eta^{\#}, \tag{4.3}$$

$$\delta' = \delta^{\#}(\tau, N, \Lambda^{\#} \sin \theta, s^{-1} \sin \theta, M, \delta, \Omega, s), \tag{4.4}$$

$$(s\Omega)' = s\,\Delta^{\#} \tag{4.5}$$

where the arguments of $\Sigma^{\#}$, $\Delta^{\#}$, $\nu^{\#}$, $\eta^{\#}$ in (4.3) and (4.5) are the same as those of $\mu^{\#}$ in (4.2).

Hypothesis 3.5 and the consequences of Theorem 3.9 imply that under the transformation

$$(4.6) \qquad (H, \sigma, M) \mapsto -(H, \sigma, M)$$

the constitutive functions $\eta^{\#}, \Sigma^{\#}, \mu^{\#}$ of (3.7) change sign while the other constitutive functions of (3.7) remain unchanged. It then follows from the mean value theorem that $\eta^{\#}, \Sigma^{\#}, \mu^{\#}$ can be represented as

$$(4.7\text{a}) \qquad \eta^{\#} = \overline{\eta_H} H + \overline{\eta_\sigma}\sigma + \overline{\eta_M} M, \qquad etc.$$

where

$$(4.7\text{b}) \qquad \overline{\eta_H}(\tau, N, H, \sigma, M, \delta, \Omega, s) \equiv \int_0^1 \eta_H(\tau, N, tH, t\sigma, tM, \delta, \Omega, s)\, dt, \qquad etc.$$

Then equations (4.2), (4.3) assume the forms

$$(4.9) \qquad \theta' = (\overline{\mu_H} \Lambda^{\#} + \overline{\mu_\sigma} s^{-1}) \sin\theta + \overline{\mu_M} M,$$

$$(sM)' = \{[\overline{\Sigma_H}\Lambda^{\#} + \overline{\Sigma_\sigma} s^{-1} - s\Lambda^{\#}(\overline{\eta_H}\Lambda^{\#} + \overline{\eta_\sigma} s^{-1})] \cos\theta - s\Lambda^{\#}\nu^{\#}\} \sin\theta$$

$$(4.10) \qquad + [\overline{\Sigma_M} - s\Lambda^{\#}\overline{\eta_M}]\, M \cos\theta .$$

BVP is thus equivalent to the system (4.1), (4.4), (4.5), (4.9), (4.10), (2.28)–(2.31) for

$$(4.11) \qquad \boldsymbol{u} \equiv (\varrho, N, H, \theta, M, \delta, \Omega).$$

A pair $(\boldsymbol{u}, \lambda)$ satisfying BVP is called a *solution pair*. A connected family of solution pairs is called a *branch of solutions*.

5. Unbuckled States

If we regard the N, H, δ, Ω appearing in (4.9), (4.10) as given well behaved functions, then the second-order system (4.9), (4.10) admits the solution $\theta = 0 = M$. Any solution of BVP for which $\theta = 0 = M$ is said to describe an *unbuckled state.*

Unbuckled states are typically not unique. For λ positive there can be such states with a necking unstability (*cf.* ANTMAN (1973), ANTMAN & CARBONE (1977), and NEGRÓN-MARRERO (1990)). The nonuniqueness can be manifested in far more prosaic circumstances, as we now show. It is easy to see that unbuckled states are governed by the system

$$(5.1\text{a}) \qquad \hat{N} = -\lambda - \int_s^1 \hat{T}\, dt.$$

$$(5.1\text{b}) \qquad (s\hat{\Omega})' - s\hat{\Delta} = 0$$

where the arguments of $\hat{N}$, $\hat{\Omega}$, and $\hat{\Delta}$ are $(\varrho(s)/s, \varrho'(s), 0, 0, 0, \delta(s), \delta'(s), s)$ and those of $\hat{T}$ are the same with s replaced by t. Let us now make several simplifying assumptions. We suppose that these constitutive functions are independent of s, as would happen if we should construct our constitutive equations as in Section 3 by using three-dimensional constitutive equations for a homogeneous material and by taking the plate thickness h constant. Next we suppose that $\hat{\Omega} = 0$ when $\eta = \sigma = \mu = \omega = 0$. This property is a consequence of (3.8) for induced constitutive equations. Finally we assume that $\hat{T}(\tau, \nu, 0, 0, 0, \delta, 0) = \hat{N}(\nu, \tau, 0, 0, 0, \delta, 0)$. A transversely isotropic plate would have this property.

Suppose that boundary condition (2.30b) holds. Then we can seek unbuckled states with $\varrho' = \nu_0$ (const), $\delta = \delta_0$ (const). These numbers satisfy

(5.2) $$\hat{N}(\nu_0, \nu_0, 0, 0, 0, \delta_0, 0) = -\lambda, \qquad \hat{\Delta}(\nu_0, \nu_0, 0, 0, 0, \delta_0, 0) = 0.$$

It is easy to impose natural growth conditions ensuring that (5.2) has a solution (by an argument using degree theory), but our monotonicity conditions are insufficient to ensure that such a solution is unique. (The positive-definiteness of the matrix of partial derivatives of $\hat{T}, \hat{N}, \hat{\Delta}$ with respect to τ, ν, δ would suffice to ensure the uniqueness. This positive-definiteness is not a consequence of the strong ellipticity condition, although it follows from the condition of COLEMAN & NOLL (1963)).

Were we to impose conditions less restrictive than those that led to (5.2), then the study of the mere existence of unbuckled states could lead to challenging exercises in analysis. *E.g.*, an inkling of the complexity that can occur for plates not satisfying the transverse isotropy condition can be found in the work of ANTMAN & NEGRÓN-MARRERO (1987) and of NEGRÓN-MARRERO & ANTMAN (1990) on uniform Cosserat plates with inextensible directors. The unbuckled state need not depend continuously on the load parameter λ. The corresponding analysis for nonuniform plates for our problem would be far more formidable. The requisite tools for treating these singular problems can be modelled on those used by NEGRÓN-MARRERO (1985).

6. Nodal Properties of Buckled Plates

Let C^0 denote the space of continuous functions on $[a, 0]$. We use analogous expressions for other spaces. We denote the Cartesian product of m copies of C^0 by $[C^0]^m$.

Let us suppose that all the arguments of the overlined functions in (4.9), (4.10) are continuous functions of s. (These arguments are the same as those of $\mu^\#$ in (4.2).) Thus (4.9) implies that $\theta \in C^1$ if and only if $(\theta, M) \in [C^0]^2$. Moreover, θ has a simple zero at s^* if and only if $\theta(s^*) = 0, M(s^*) \neq 0$, and θ has a double zero at s^* if and only if $\theta(s^*) = 0, M(s^*) = 0$. (By a *double zero* we mean a zero of order 2 or more.) Let

$$\mathscr{X}_k \equiv \{(\theta, M) \in [C^0]^2 : \theta(a) = 0 = \theta(1);$$

(6.1) $$\theta \text{ has exactly } k + 1 \text{ zeros on } [a, 1], \text{ each of which is simple.}\}$$

$\mathscr{Z}_k$ is an open set in $C^0 \times C^0$. If (θ, M) belongs to its boundary ∂Z_k, then θ has a double zero on $[a, 1]$.

Let $\mathscr{S}$ be a connected family of solution pairs containing a solution pair $(\tilde{\boldsymbol{u}}, \tilde{\lambda})$ with $(\tilde{\theta}, \tilde{M})$ in $\mathscr{Z}_k$. Since $\mathscr{Z}_k$ is open in $[C^0)]^2$, there is a neighborhood $\mathscr{N}$ of $(\tilde{\boldsymbol{u}}, \tilde{\lambda})$ in the topology of $[C^0]^7 \times \mathbb{R}$ such that if $(\boldsymbol{u}, \lambda) \in \mathscr{N} \cap \mathscr{S}$, then $(\theta, M) \in \mathscr{Z}_k$. Let $\tilde{\mathscr{S}}$ be the largest connected component of $\mathscr{S}$ containing $(\tilde{\boldsymbol{u}}, \tilde{\lambda})$ and having the further property that there is no θ corresponding to one its solution pairs having a double zero. From the preceding paragraph we know that (θ, M) in $\mathscr{Z}_k$ for each solution pair of $\tilde{\mathscr{S}}$.

Now if θ has a double zero, then the corresponding zero values for (θ, M) can be taken as initial data for (4.9), (4.10). The resulting initial value problem has a unique solution, which must be the zero solution (by virtue of the special form of (4.9), (4.10) consequent on the constitutive symmetry under the transformations (3.10)). This conclusion holds no matter what continuous values the hidden variables in (4.9), (4.10) assume. Hence we have

6.2. Theorem. *If a connected set $\mathscr{S}$ of solution pairs contains no unbuckled state and if it has one solution pair with $(\tilde{\theta}, \tilde{M})$ in $\mathscr{Z}_k$, then each of its solution pairs (θ, M) is in $\mathscr{Z}_k$. Thus the number of simple zeros of θ is constant on such a set of solution pairs and characterizes it globally.*

To prove the existence of branches of buckled states we first need to show that we can write BVP as a fixed-point problem involving a compact operator on $[C^0]^7$. We can easily do this by integrating (4.4), (4.5), (4.9), (4.10) and using (3.4) to show that the arbitrary constants of integration can be adjusted to ensure that all the boundary conditions are satisfied. (For details of a much more difficult version of this process, *cf.* Negrón-Marrero & Antman (1990) and the works cited therein.) We can now invoke the theory of Rabinowitz (1971) and Crandall & Rabinowitz (1971) to obtain

6.3. Theorem. *Let $\bar{\lambda}$ be an eigenvalue of odd algebraic multiplicity of the linearization of BVP about a branch of unbuckled states and let $\overline{\mathbf{u}}(\bar{\lambda})$ be the corresponding unbuckled state. From $(\overline{\mathbf{u}}(\bar{\lambda}), \bar{\lambda})$ bifurcates a branch of buckled states that is either unbounded in $[C^0]^7 \times \mathbb{R}$ or else returns to another such pair $(\overline{\mathbf{u}}(\bar{\varkappa}), \bar{\varkappa})$. If the eigenvalue is simple, then the bifurcating branch inherits the nodal properties of the eigenfunction.*

7. Comments

For intact plates, the polar singularity causes severe difficulties in converting BVP to a problem suitably involving compact operators. Methods for treating such questions can be based on those used by Antman (1978), Gauss & Antman (1984), and Negrón-Marrero & Antman (1990). There seems to be an attractive alternative to their procedures. We could attempt to construct the bifurcation diagram for the intact plate as a limit of those for annular plates as the radius *a* of the hole goes to zero. Such procedures have been successfully applied to handle singularities for other problems. They rely on the derivation of suitable *a priori* estimates for the disposition of bifurcating branches. The requisite mathematical theory is given by ALEXANDER (1981).

The results we have presented here are easily adapted to handle the problem for the planar buckling of a Cosserat column. On the other hand, a preliminary study indicates that if a comparable treatment of spatial buckling were possible, it would require a much deeper analysis. The same remarks apply to the axisymmetric buckling of a spherical shell under hydrostatic pressure.

ANTMAN & PIERCE (1990) have shown how one can determine detailed qualitative properties of nonbifurcating solution branches for the equations describing the planar buckling of columns that cannot suffer thickness changes. There seems to be no obstacle preventing their techniques, which combine global bifurcation theory with singularity theory, from being applied to our problem.

Acknowledgment. The work reported here was supported in part by NSF Grant DMS-85-03317 and AFOSR-URI Grant 87-0073.

References

J. C. ALEXANDER (1981), A primer on connectivity, in *Proc. Conf. on Fixed Point Theory*, E. FADELL & G. FOURNIER, eds., Lect. Notes in Math. # **886**, Springer-Verlag, New York-Heidelberg-Berlin, 455–483.

S. S. ANTMAN (1971), Existence and nonuniqueness of axisymmetric equilibrium states of nonlinearly elastic shells, *Arch. Rational Mech. Anal.* **40**, 329–371.

S. S. ANTMAN (1972), The Theory of Rods, in *Handbuch der Physik, Vol. VIa/2*, C. TRUESDELL, ed., Springer-Verlag, 641–703.

S. S. ANTMAN (1973), Nonuniqueness of equilibrium states for bars in tension, *J. Math. Anal. Appl.* **44**, 333–349.

S. S. ANTMAN (1976), Ordinary differential equations of one-dimensional nonlinear elasticity I: Foundations of the theories of nonlinearly elastic rods and shells, *Arch. Rational Mech. Anal.* **61**, 307–351.

S. S. ANTMAN (1978), Buckled states of nonlinearly elastic plates, *Arch. Rational Mech. Anal.* **67**, 111–149.

S. S. ANTMAN & E. R. CARBONE (1977), Shear and necking instabilities in nonlinear elasticity, *J. Elasticity* **7**, 125–151.

S. S. ANTMAN & P. V. NEGRÓN-MARRERO (1987), The remarkable nature of radially symmetric equilibrium states of aeolotropic nonlinearly elastic bodies, *J. Elasticity* **18**, 131–164.

S. S. ANTMAN & J. E. OSBORN (1979), The principle of virtual work and integral laws of motion, *Arch. Rational Mech. Anal.* **69**, 231–262.

S. S. ANTMAN & J. F. PIERCE (1990), The intricate global structure of buckled states of compressible columns, SIAM *J. Appl. Math.* to appear.

J. CARR, M. E. GURTIN, & M. SLEMROD (1984), Structured phase transitions on a finite interval, *Arch. Rational Mech. Anal.* **86**, 317–351.

H. COHEN & C. N. DESILVA (1966), Theory of directed surfaces, *J. Math. Phys.* **7**, 960–966.

B. D. COLEMAN (1983), Necking and drawing in polymer fibers under tension, *Arch. Rational Mech. Anal.* **83**, 115–137.

B. D. Coleman (1985), On the cold drawing of polymers, *Comp. & Maths. with Appls.* **11**, 35–65.
B. D. Coleman & D. C. Newman (1988), On the rheology of cold drawing. I. Elastic materials, *J. Polymer Sci: Part B: Polymer Phys.* **26**, 1801–1822.
B. D. Coleman & W. Noll (1963), The thermodynamics of elastic materials with heat conduction and viscosity, *Arch. Rational Mech. Anal.* **13**, 167–178.
E. & F. Cosserat (1909), *Théorie des Corps Déformables*, Hermann.
M. G. Crandall, & P. H. Rabinowitz (1971), Bifurcation from simple eigenvalues, *J. Funct. Anal.* **8**, 321–340.
J. L. Ericksen (1975), Equilibrium of bars, *J. Elasticity* **5**, 191–202.
J. L. Ericksen (1977), Special Topics in Nonlinear Elastostatics, in *Advances in Applied Mechanics, Vol.* 17, C.-S. Yih, ed., Academic Press.
J. L. Ericksen & C. Truesdell (1958), Exact theory of stress and strain in rods and shells, *Arch. Rational Mech. Anal.* **1**, 295–323.
R. C. Gauss & S. S. Antman (1984), Large thermal buckling of nonuniform beams and plates, *Int. J. Solids Structures* **20**, 979–1000.
A. E. Green P. M. Naghdi, & W. L. Wainwright (1965), A general theory of a Cosserat surface, *Arch. Rational Mech. Anal.* **20**, 287–308.
P. M. Naghdi (1972) The Theory of Shells, in *Handbuch der Physik, Vol. VIa/2*, C. Truesdell, ed., Springer-Verlag, 425–640.
P. V. Negrón-Marrero (1985), Large buckling of circular plates with singularities due to anisotropy, Univ. of Maryland, Dissertation.
P. V. Negrón-Marrero (1990), Necked states of nonlinearly elastic plates, *Proc. Roy. Soc. Edinburgh*, to appear.
P. V. Negrón-Marrero & S. S. Antman (1990), Singular global bifurcation problems for buckling of anisotropic plates, *Proc. Roy. Soc. London*, to appear.
N. C. Owen (1987), Existence and stability of necking deformations for nonlinearly elastic rods, *Arch. Rational Mech. Anal.* **98**, 357–383.
P. H. Rabinowitz (1971), Some global results for nonlinear eigenvalue problems, *J. Funct. Anal.* **7**, 487–513.

Department of Mathematics and
Institute for Physical Science and Technology
University of Maryland
College Park

(Received June 28, 1989)

Regular Precessions in a Central Newtonian Field of Forces

GIUSEPPE GRIOLI

For Bernard Coleman on his Sixtieth Birthday

I will consider the problem of the motion of a gyroscope about its mass center in a central Newtonian field of forces. This study may be of interest in Celestial Mechanics, particularly in the dynamics of the solar system.

Specifically, I will consider the motion of a rigid body subjected to the gravitational forces due to a mass placed at a point of a plane containing the trajectory of the body's center of mass.

The effect of the gravitational forces is evaluated according to the usual approximation in Celestial Mechanics, in which the extent of the body is taken into account.

The case in which the motion of the mass center is a Keplerian motion is particularly interesting but very difficult in general. After some general preliminary statements, I will suppose that the trajectory of the mass center is a circumference described at a constant speed, with the aim of characterizing some significant solutions. It is well known that this case is often close to reality in problems of the solar system.

By using a special, kinematical, and general property of rigid motions, which I demonstrated earlier [1], I will determine a class of regular precessions which are dynamically possible, that is a class of motions in which the angular velocity is the sum of two vectors: one constant in space, one fixed in the body.

1. General premisses

Let S be a gyroscope, G its mass center and $G\boldsymbol{j}_i$ $(i = 1, 2, 3)$ a rectangular coordinate trihedral fixed in the body, with origin at G and coincident with a system of central axes of inertia. Further, I suppose that $\boldsymbol{j}_3$ is parallel to the gyroscopic axis.

The condition that the body is a gyroscope means only that its central ellipsoid of inertia is spherical.

Let σ be the matrix of inertia, A the moment of inertia of the body with respect to the equatorial axes, while C denotes the moment of inertia with respect to the gyroscopic axis.

Let L be the trajectory of G in a plane, π, fixed in an inertial space, and let O be the point of π where there is a mass m from which derives the Newtonian field of forces.

Denoting by $\boldsymbol{e}$ the unit vector along the perpendicular to π, let us put

$$\boldsymbol{H} = \text{vers}\, OG, \qquad OG = r(t)\,\boldsymbol{H} \qquad [r(t) > 0]. \tag{1}$$

If r is supposed very large in comparison with the dimensions of S and if the effect of the gravitational forces is taken according to the usual approximation in Celestial Mechanics, the instantaneous moment vector, $\boldsymbol{M}$, of the external forces with respect to G is [2], [3],

$$\boldsymbol{M} = a\boldsymbol{H} \times \sigma\boldsymbol{H}, \qquad a = 3mhr^{-3} > 0, \tag{2}$$

where h denotes the constant of gravitation.

2. Basic equations

I will study the motion of the body about its mass center, that is with respect to a coordinate trihedral whose origin is at G and whose axes have invariable directions in the inertial space [to which the plane π belongs].

Denoting by $\boldsymbol{\omega} = p_i \boldsymbol{j}_i$ the angular velocity of S, according to the basic equations of dynamics one has

$$\frac{d\sigma\boldsymbol{\omega}}{dt} = a\boldsymbol{H} \times \sigma\boldsymbol{H}. \tag{3}$$

Together with (3), the following equations, which express the invariability of the unit vector $\boldsymbol{e}$ and the motion of $\boldsymbol{H}$, are to be considered:

$$\frac{d\boldsymbol{e}}{dt} = 0, \qquad \frac{d\boldsymbol{H}}{dt} = \nu\boldsymbol{e} \times \boldsymbol{H}, \tag{4}$$

where $\nu(t)$ is a scalar coefficient depending on the motion of $\boldsymbol{H}$ in the plane π.

Obviously, putting $\boldsymbol{j}_3 = \boldsymbol{k}$, one has

$$\sigma\boldsymbol{\omega} = A\boldsymbol{\omega} + (C - A)\, p_3 \boldsymbol{k}, \qquad \sigma\boldsymbol{H} = A\boldsymbol{H} + (C - A)\, H_3 \boldsymbol{k}. \tag{5}$$

Putting

$$\xi = (C - A)\, A^{-1} \tag{6}$$

and denoting by a dot the derivative with respect to time, we find that the equations (3), (4) are equivalent to the following set of eight scalar equations:

$$\dot{p}_1 + \xi p_2 p_3 = a\xi H_2 H_3, \quad \dot{p}_2 - \xi p_1 p_3 = -a\xi H_1 H_3, \tag{7}$$

$$\dot{e}_1 + p_2 e_3 - p_3 e_2 = 0, \quad \dot{e}_2 + p_3 e_1 - p_1 e_3 = 0, \quad \dot{e}_3 + p_1 e_2 - p_2 e_1 = 0, \tag{8}$$

$$\begin{aligned} &\dot{H}_1 + (p_2 - \nu e_2)\, H_3 - (p_3 - \nu e_3)\, H_2 = 0, \\ &\dot{H}_2 + (p_3 - \nu e_3)\, H_1 - (p_1 - \nu e_1)\, H_3 = 0, \\ &\dot{H}_3 + (p_1 - \nu e_1)\, H_2 - (p_2 - \nu e_2)\, H_1 = 0. \end{aligned} \tag{9}$$

and to the condition $p_3 = \text{const}$.

The solutions of differential equations (7), (8), (9) which interest us are those which satisfy the conditions

$$\boldsymbol{e} \cdot \boldsymbol{e} = 1, \quad \boldsymbol{H} \cdot \boldsymbol{H} = 1, \quad \boldsymbol{e} \cdot \boldsymbol{H} = 0. \tag{10}$$

3. Regular precessions

The system of differential equations (7), (8), (9) is rather complicated but it is possible to find some solutions which are significant from the geometrical and kinematical points of view and of interest in the field of celestial mechanics, such as the rotations and the regular precessional rotations. In the following considei this last kind of motions.

It may be observed that if the system of equations (7), (8), (9) admits a solution wherein $H_3 = 0$, the corresponding motion is a regular precession. In fact, in this case one has a Poinsot motion, the vector $\sigma\omega$ is invariable in the inertial space, and from (5) it follows that the angular velocity is the sum of a vector fixed in space and another fixed in the body.

The problem of finding the dynamically possible precessional motions of a rigid body may be easier if one applies a certain general kinematical property which they satisfy. Namely, as a particular case of a more general property of rigid motions, the following theorem subsists [1]: the condition

$$z = p_1 \dot{p}_2 - p_2 \dot{p}_1 + p_3(p_1^2 + p_2^2) - y(p_1^2 + p_2^2)^{\frac{3}{2}} = 0, \tag{11}$$

where y is a constant, is necessary and sufficient for a rigid motion with a fixed point to be a precessional motion [not necessarily a regular precession] having the axis $\boldsymbol{j}_3$ as the axis of the figure.

That means that in a dynamical problem the precessional motions are only those solutions that satisfy the condition $z = 0$, if they exist. If we put $y = \operatorname{ctg} \theta$, the unit vector $\boldsymbol{c}$ which characterizes the direction of the axis of precession, fixed in the space, is given by

$$c = \frac{\sin\theta}{\varrho}\,\omega + \left(\cos\theta - \frac{\sin\theta}{\varrho}\,p_3\right)k, \tag{12}$$

where

$$\varrho = \sqrt{p_1^2 + p_2^2} > 0.$$

With the definitions

$$p_1 = \varrho\cos\alpha(t), \qquad p_2 = \varrho\sin\alpha(t), \tag{13}$$

(11) is equivalent to the equation

$$\dot\alpha = \varrho y - p_3. \tag{14}$$

If we suppose that $H_3 = 0$ or that $H_1 = H_2 = 0$, keeping in mind (13), from (7), (13) one has

$$\begin{aligned} &\dot\varrho\cos\alpha - \varrho(\dot\alpha - \xi p_3)\sin\alpha = 0,\\ &\dot\varrho\sin\alpha + \varrho(\dot\alpha - \xi p_3)\cos\alpha = 0, \end{aligned} \tag{15}$$

from which it follows [see (14)]

$$\dot\varrho = 0, \quad \xi p_3 - \dot\alpha = (\xi + 1)\,p_3 - \varrho y = 0. \tag{16}$$

That means that in each motion during which $H_3 = 0$, ϱ and $\dot\alpha$ are constant, as previously observed.

The condition that e is a vector constant in space certainly is satisfied if one assumes $e = c$. In this hypothesis, if $\dot\nu = 0$, $\dot a = 0$, ϱ and $\dot\alpha$ are constant also if $H_3 \neq 0$. In order to show that, let us observe that if $e = c$ and $H_3 \neq 0$, from equations (9) and (10), according to (12), (13), (14), it follows that

$$H_3\dot H_3 + \frac{\eta}{a\xi}\varrho\dot\varrho = 0, \quad \eta = 1 - \frac{\nu}{\varrho}\sin\theta, \tag{17}$$

$$a\xi H_3^2 y + \varrho[(\xi + 1)\,p_3 - \varrho y] = 0. \tag{18}$$

Equation (18) shows that $y \neq 0$. If, in fact, $y = 0$ from (16) it follows that $p_3 = \dot\alpha = 0$ or $\varrho = 0$, conditions which are compatible only with rotary motions but not with real precessional motions.

Differentiating equation (18) with respect to time and comparing the result with (17), we find

$$\left[2\varrho - \nu\sin\theta - p_3\,\frac{1+\xi}{2y}\right]\frac{\dot\varrho}{a\xi} = 0 \tag{19}$$

from which follows $\varrho = \text{const.}$

One concludes that if ν and a are constant, the only dynamically possible precessional motions, whose axis of precession is parallel to e, are regular precessions.

In the following I will suppose that $\boldsymbol{v}$ and $\boldsymbol{a}$ are constant, that is that the trajectory of the mass center of the body is a circumference traversed with constant velocity. These conditions are often close to reality in problems of the solar system.

Keeping in mind (7), (9), (10), (12), (14), (17), (18), we may show that only two kinds of regular precessions with axes of precession parallel to $\boldsymbol{e}$ are dynamically possible. They are characterized by the following equations:

First kind of regular precessions.

$$\boldsymbol{\omega} = \nu \left[\boldsymbol{e} - \frac{\xi}{1+\xi} \cos\theta \boldsymbol{k}\right], \qquad \cos\theta \neq 0,$$

$$\boldsymbol{e} = \sin\theta\,(\cos\alpha \boldsymbol{j}_1 + \sin\alpha \boldsymbol{j}_2) + \cos\theta \boldsymbol{k}, \tag{20}$$

$$\boldsymbol{H} = -\sin\alpha \boldsymbol{j}_1 + \cos\alpha \boldsymbol{j}_2, \qquad \dot{\alpha} = \frac{\nu\cos\theta}{1+\xi}\xi.$$

The above precessional motions for $\theta = 0$ become the uniform rotations

$$\boldsymbol{k} = \boldsymbol{e}, \qquad \boldsymbol{\omega} = \frac{\nu}{1+\xi}\boldsymbol{e},$$

$$\boldsymbol{H} = -\sin\alpha \boldsymbol{j}_1 + \cos\alpha \boldsymbol{j}_2, \qquad \dot{\alpha} = \frac{\nu\xi}{1+\xi}. \tag{20'}$$

Second kind of regular precessions.

$$\boldsymbol{\omega} = \nu \boldsymbol{e} - \frac{\nu^2 + a}{(1+\xi)\,\nu}\xi \cos\theta \boldsymbol{k}, \qquad \cos\theta \neq 0,$$

$$\boldsymbol{e} = \sin\theta(\cos \boldsymbol{j}_1 + \sin\alpha\alpha \boldsymbol{j}_2) + \cos\theta \boldsymbol{k}, \tag{21}$$

$$\boldsymbol{H} = -\cos\theta\,(\cos\alpha \boldsymbol{j}_1 + \sin\alpha \boldsymbol{j}_2) + \sin\theta \boldsymbol{k}, \qquad \dot{\alpha} = \frac{\nu^2 + a}{(1+\xi)\,\nu}\xi\cos\theta.$$

The above precessional motions for $\theta = 0$ become the uniform rotations

$$\boldsymbol{k} = \boldsymbol{e}, \qquad \boldsymbol{\omega} = \frac{\nu^2 - a\xi}{(1+\xi)\,\nu}\boldsymbol{e},$$

$$\boldsymbol{H} = -\,(\cos\alpha \boldsymbol{j}_1 + \sin\alpha \boldsymbol{j}_2), \qquad \dot{\alpha} = \frac{\nu^2 + a}{(1+\xi)\,\nu}\xi. \tag{21'}$$

Remark. Supposing $\xi \neq 0$, $(A \neq C)$, one sees that in the rotary motions (20'), (21') always is $\boldsymbol{\omega} \neq \nu\boldsymbol{e}$. That means that the body in none of those motions shows the same face permanently to the center of attraction. Rotary motions with this property exist, also in the case of asymmetrical bodies [4], but they cannot be obtained as limits for $\theta = 0$ from the regular precessions characterized above.

Further, the first kind of regular precessions and equations (7), for $H_3 = 0$, do not depend on $a(t)$. Therefore, the motions (20), (20′) are dynamically possible also if the trajectory of the mass center of the body is not circular, that is, $a(t)$ is not a constant.

References

1. GRIOLI, G., "Qualche teorema di cinematica dei moti rigidi". Rend. Acc. Naz. dei Lincei, S. VIII, vol. XXXIV, f. 6 (1963).
2. LEIMANIS, E., "The general Problem of the Motion of Coupled Rigid Bodies about a Fixed Point". Springer Tracts in Natural Philosophy, vol. 7.
3. GRIOLI, G., "Particular Solutions in Stereodynamics" Corso CIME 1971, Bressanone, 2-12, June.
4. BENTSIK, E., Su possibili moti semplici di un satellite artificiale soggetto a forze newtoniane". Rend. Sem. Mat. Univ. di Padova, vol. XXXIX (1967).

Istituto di Matematica e
Meccanica Razionale
Università di Padova

(Received May 23, 1989)

A Mechanical Theory for Crystallization of a Rigid Solid in a Liquid Melt; Melting-Freezing Waves

MORTON E. GURTIN

Dedicated to Bernard Coleman on his sixtieth birthday

1. Introduction

Crystals of solid helium in their melt exhibit a phenomenon generally not found in other materials: oscillations of the solid-liquid interface in which atoms of the solid move only when they melt and enter the liquid. Such *melting-freezing waves* were predicted by ANDREEV & PARSHIN [AP] in 1978 and exhibited experimentally[1] by KESHISHEV, PARSHIN, & BABKIN [KPB] in 1979.

Motivated by this classic paper of ANDREEV & PARSHIN, I develop a continuum theory of crystallization using, as a basis, a framework developed in [G1][2]. I restrict attention to a purely mechanical[3] theory, and, to avoid geometric complications that accompany evolving surfaces, confine the derivation to a two-dimensional theory in which the interface evolves as a plane curve.

One of the chief differences between theories involving phase transitions and the more classical theories of continuum mechanics is the *creation and deletion of material points* as the phase interface moves relative to the underlying material. To describe this process I introduce an *interactive energy-balance* in which the

---

[1] KESHISHEV, PARSHIN, & BABKIN observed frequencies of 3–15 kHz; since then frequencies up to 10^7 kHz have been reported (*cf.* CASTAING, BALIBAR, & LAROCHE [CBL] and the references cited by MARIS & ANDREEV [MA]).

[2] See also [G2, G3, AG, GS]. GURTIN & PODIO GUIDUGLI [GPG] develop a (possibly oversimplified) theory of melting-freezing waves in which the melt is considered only as a source of atoms for the crystallization process and the inertia — which in the current theory arises from the motion of the melt — is presumed to be an effective inertia endowed to the interface. The results of this simplified model show fair qualitative agreement with the current theory, a major deficiency being a dispersion relation of the form $\omega^2 \sim \lambda^2$ rather than $\omega^2 \sim \lambda^3$ (*cf.* (1.8)). The chief advantage of the model of [GPG] is a hyperbolic evolution equation for the interface; for an isotropic crystal this equation has the simple form $\varrho V^\circ + \beta V = \psi K + F$ (ϱ, β, ψ = constants > 0, F = constant).

[3] As noted by MARIS & ANDREEV [MA], for superconductors such as solid helium solidification is "essentially a mechanical process, rather than a thermal process as it is for ordinary materials".

energy and power transferred from the bulk material to the phase interaction is related to the *interactive dissipation*, which is the energy associated with the exchange of atoms between phases. This energy balance and balance of mass and linear momentum are the basic balance laws of the theory.

I suppose that the crystal is rigid, incapable of deformation, and model the melt as an incompressible,[4] inviscid fluid. The interface is characterized by constitutive equations which allow the interfacial energy, the interfacial force, and the interactive dissipation to depend on the orientation and normal velocity of the interface; and a mechanical version of the second law is used to deduce suitable constitutive restrictions.

The theory leads to a free-boundary problem for the evolution of the interface; this problem consists in solving

$$\varrho \boldsymbol{v}^{\cdot} = -\operatorname{grad} p, \qquad \operatorname{div} \boldsymbol{v} = 0 \tag{1.1}$$[5]

in the melt and

$$\begin{gathered}[\psi(\theta) + \psi''(\theta)]\, K = \beta(\theta)\, V + F - \tfrac{1}{2}\varrho_c v^2 + (1-\zeta)\, p, \\ \boldsymbol{v} \cdot \boldsymbol{m} = (1-\zeta)\, V \end{gathered} \tag{1.2}$$

on the interface, subject to suitable initial conditions and far-field conditions. Here K and V are the curvature and normal velocity of the interface, $\psi(\theta)$ is the interfacial energy[6] as a function of the angle θ to the interface-normal $\boldsymbol{m}$, $\beta(\theta)$ is the kinetic coefficient of the interface, $\boldsymbol{v}$ and p are the velocity and pressure of the melt. Further, $\zeta = \varrho_c/\varrho$ with ϱ_c and ϱ the constant crystal and melt densities, while $F = \Psi_c - \zeta\Psi$ with Ψ_c and Ψ the constant crystal and melt energies.

I show, as a consequence of the *crystallization equations* (1.1) and (1.2), that if the far-field does not supply melt, then

$$\begin{gathered}(d/dt)\operatorname{area}(\mathscr{C}) = 0, \\ (d/dt)\left\{\int_{\partial\mathscr{C}} \psi(\theta)\, ds + \int_{\mathscr{M}} \tfrac{1}{2}\varrho v^2\, da\right\} = -\int_{\partial\mathscr{C}} \beta(\theta)\, V^2\, ds \leqq 0, \end{gathered} \tag{1.3}$$

where $\mathscr{C} = \mathscr{C}(t)$ and $\mathscr{M} = \mathscr{M}(t)$ denote the regions occupied by the crystal and the melt. These relations furnish Lyapunov functions for the crystallization problem.

When the melt is irrotational, $\boldsymbol{v}(\boldsymbol{x}, t)$ is proportional to the gradient of a potential $u(\boldsymbol{x}, t)$; if, in addition, the melt velocity is sufficiently small that the term

[4] ANDREEV & PARSHIN [AP] note that the phase velocity of melting-freezing waves is generally well below the sound velocity.

[5] grad, div, and Δ are the *spatial gradient*, *spatial divergence*, and *spatial laplacian*. For $\Phi(\boldsymbol{x}, t)$ defined in the melt: Φ_t is the *spatial time-derivative*, $\Phi^{\cdot}$ the *material time-derivative*; thus $\Phi^{\cdot} = \Phi_t + \boldsymbol{v} \cdot \operatorname{grad} \Phi$ for Φ scalar-valued, *etc.* (*cf.*, *e.g.*, [G4], § 8).

[6] I use the term *energy* (for $\psi(\theta)$, Ψ, and Ψ_c) in a generic sense; which thermodynamic potentials (free energy, internal energy, *etc.*) these quantities actually represent depends on which thermodynamic theory this *purely mechanical theory* is meant to "approximate"; the current theory is, of course, independent of such considerations.

involving v^2 in (1.2) is negligible, then the crystallization equations reduce to

$$\Delta u = 0 \tag{1.4}$$

in the melt and

$$\begin{aligned} [\psi(\theta) + \psi''(\theta)]\, K &= \beta(\theta)\, V - u_t\,, \\ \partial u/\partial m &= \alpha V \end{aligned} \tag{1.5}$$

on the interface, with $\alpha = (\varrho_c - \varrho)^2/\varrho$.

The system (1.4) and (1.5) with $\beta(\theta) = 0$, *linearized about a flat interface at equilibrium*, reduces to a system proposed by ANDREEV & PARSHIN [AP]:

$$\begin{aligned} &u_{xx} + u_{yy} = 0 \qquad (-\infty < x < \infty, y > 0, t > 0)\,, \\ &(\psi + \psi'')_0\, h_{xx} = -u_t, \qquad u_y = \alpha h_t \qquad (-\infty < x < \infty, y = 0, t > 0)\,, \end{aligned} \tag{1.6}$$

where $y = h(x, t)$ defines the interface, the subscript zero indicates evaluation at equilibrium, and the subscripts x and y denote partial differentiation with respect to the corresponding variable. As noted by ANDREEV & PARSHIN [AP], this system has an oscillatory solution of the form

$$h(x, t) = e^{i\lambda x}\, e^{-i\omega t} \tag{1.7}$$

with

$$\omega^2 = \frac{\varrho(\psi + \psi'')_0\, \lambda^3}{(\varrho_c - \varrho)^2}\,, \tag{1.8}$$

yielding a proportionality of ω^2 to λ^3 found experimentally by KESHISHEV, PARSHIN, & BABKIN [KPB].

Global growth relations, similar to (1.3), are established for the linearized equations (1.6) and used to establish uniqueness for the associated initial-value problem.

I discuss the form the basic equations take when the theory is three-dimensional, and, within that context, solve the problem of spherically symmetric crystallization in an infinite melt whose far-field pressure P is finite. The behavior of the crystal is governed by the sign of the constant

$$C = \Psi_c + P - \zeta(\Psi + P):$$

(i) for $C \geqq 0$, crystals melt in finite time; (ii) for $C < 0$ and the melt initially at rest, crystals melt in finite time if initially small, but grow unboundedly if initially large.

2. Crystals

We consider an infinite *crystal lattice* modelled as a two-dimensional continuum, in fact as $\mathbb{R}^2$. A **crystal** $\mathscr{C}$ is then a compact subset of the lattice with boundary, $\partial\mathscr{C}$, a smooth, simple closed curve (Figure 1). $\partial\mathscr{C}$ represents the **interface** between the crystal and its melt; we write $\boldsymbol{m}(\boldsymbol{x})$ for the outward unit normal to $\partial\mathscr{C}$ and define a unit tangent $\boldsymbol{l}(\boldsymbol{x})$ (the direction of increasing arc length) so that $\{\boldsymbol{l}(\boldsymbol{x}), \boldsymbol{m}(\boldsymbol{x})\}$ is a positively oriented basis of $\mathbb{R}^2$. We write f_s for the derivative, sometimes partial, of f with respect to arc length on $\partial\mathscr{C}$. We then have the Frenet formulas

$$\boldsymbol{m}_s = -K\boldsymbol{l}, \quad \boldsymbol{l}_s = K\boldsymbol{m}, \tag{2.1}$$

with $K(\boldsymbol{x})$ the **curvature** of $\partial\mathscr{C}$. We define the **angle** $\theta(\boldsymbol{x})$, as a smooth function of $\boldsymbol{x}$, through

$$\boldsymbol{m} = (\cos\theta, \sin\theta), \quad \boldsymbol{l} = (\sin\theta, -\cos\theta); \tag{2.2}$$

then

$$K = \theta_s. \tag{2.3}$$

Since our goal is to model crystallization, we consider crystals $\mathscr{C}(t)$ that evolve with time t, under the assumption that $\partial\mathscr{C}(t)$ is a smooth evolving curve (in the sense of [AG]).

We write $V(\boldsymbol{x}, t)$ for the **normal velocity** of $\partial\mathscr{C}(t)$ in the direction $\boldsymbol{m}(\boldsymbol{x}, t)$. Let $\boldsymbol{V}(\boldsymbol{x}, t) = V(\boldsymbol{x}, t)\,\boldsymbol{m}(\boldsymbol{x}, t)$. Fix t and $\boldsymbol{x} \in \partial\mathscr{C}(t)$ and (for β sufficiently close to t) let $\boldsymbol{y}(\beta)$ denote the curve that passes through $\boldsymbol{x}$ at time t and has

$$d\boldsymbol{y}(\beta)/d\beta = \boldsymbol{V}(\boldsymbol{y}(\beta), \beta). \tag{2.4}$$

Then the **normal time-derivative** $\Phi^\circ(\boldsymbol{x}, t)$ (following $\partial\mathscr{C}(t)$) of a scalar or vector function $\Phi(\boldsymbol{x}, t)$ is defined by

$$\Phi^\circ(\boldsymbol{x}, t) = (d/d\beta)\,\Phi(\boldsymbol{y}(\beta), \beta)\big|_{\beta=t}. \tag{2.5}$$

The identities

$$\begin{aligned} \theta^\circ &= \boldsymbol{l}^\circ \cdot \boldsymbol{m} = -\boldsymbol{m}^\circ \cdot \boldsymbol{l} = V_s, \\ (V\boldsymbol{m})_s &= \boldsymbol{m}\theta^\circ - KV\boldsymbol{l}, \\ (V\boldsymbol{m})^\circ &= V^\circ\boldsymbol{m} - \theta^\circ V\boldsymbol{l}. \end{aligned} \tag{2.6}$$

are standard.[7]

By an **interfacial chunk** we mean a smoothly evolving curve $\mathscr{A}(t)$ with $\mathscr{A}(t) \subset \partial\mathscr{C}(t)$ at each time t. We write $(\boldsymbol{v}_{\partial\mathscr{A}})_1(t)$ and $(\boldsymbol{v}_{\partial\mathscr{A}})_2(t)$ for the **endpoint velocities** of $\mathscr{A}(t)$:

$$(\boldsymbol{v}_{\partial\mathscr{A}})_1(t) = d\boldsymbol{x}_1(t)/dt, \quad (\boldsymbol{v}_{\partial\mathscr{A}})_2(t) = d\boldsymbol{x}_2(t)/dt, \tag{2.7}$$

with $\boldsymbol{x}_1(t)$ and $\boldsymbol{x}_2(t)$ the initial and terminal points of $\mathscr{A}(t)$. Then

$$\boldsymbol{m}(\boldsymbol{x}_i(t), t)) \cdot (\boldsymbol{v}_{\partial\mathscr{A}})_i(t) = V(\boldsymbol{x}_i(t), t)) \quad (i = 1, 2). \tag{2.8}$$

Given a smooth function $\Phi(\boldsymbol{x}, t)$ and a smooth vector function $\boldsymbol{\Phi}(\boldsymbol{x}, t)$, we write

$$\begin{aligned} \int_{\partial\mathscr{A}(t)} \Phi &= \Phi(\boldsymbol{x}_2(t), t) - \Phi(\boldsymbol{x}_1(t), t), \\ \int_{\partial\mathscr{A}(t)} \boldsymbol{\Phi} \cdot \boldsymbol{v}_{\partial\mathscr{A}} &= \boldsymbol{\Phi}(\boldsymbol{x}_2(t), t)) \cdot (\boldsymbol{v}_{\partial\mathscr{A}})_2(t) - \boldsymbol{\Phi}(\boldsymbol{x}_1(t), t)) \cdot (\boldsymbol{v}_{\partial\mathscr{A}})_1(t), \end{aligned} \tag{2.9}$$

---

[7] *Cf., e.g.,* [AG], eqts. (2.4), (2.18).

so that

$$\int_{\partial\mathscr{A}(t)} \Phi = \int_{\mathscr{A}(t)} \Phi_s \, ds. \tag{2.10}$$

Less trivial is the **transport identity**[8]

$$(d/dt) \int_{\mathscr{A}(t)} \Phi \, ds = \int_{\mathscr{A}(t)} (\Phi^\circ - \Phi K V) \, ds + \int_{\partial\mathscr{A}(t)} \Phi \boldsymbol{l} \cdot \boldsymbol{v}_{\partial\mathscr{A}}. \tag{2.11}$$

In what follows, we will generally omit the argument t when writing such integrals.

We write

$$\mathscr{M}(t) = \text{closure of } \mathbb{R}^2 \setminus \mathscr{C}(t)$$

for the region occupied by the **melt**. By a **control volume** we mean a *bounded* region $D \subset \mathbb{R}^2$ with D fixed in time;

$$D_{\mathscr{C}}(t) = \mathscr{C}(t) \cap D, \quad D_{\mathscr{M}}(t) = \mathscr{M}(t) \cap D, \quad \mathscr{d}(t) = \partial\mathscr{C}(t) \cap D \tag{2.12}$$

are then the portions of D contained in the crystal, in the melt, and on the interface; we will generally denote by $\boldsymbol{n}$ the outward unit normal to ∂D (Figure 1).

Let $\varphi(\boldsymbol{x}, t)$ and $\Phi(\boldsymbol{x}, t)$ be smooth functions defined for $\boldsymbol{x} \in \mathscr{C}(t)$ and $\boldsymbol{x} \in \mathscr{M}(t)$, respectively. We will repeatedly use the identities

$$\begin{aligned} (d/dt) \int_{D_{\mathscr{C}}} \varphi \, da &= \int_{D_{\mathscr{C}}} \varphi_t \, da + \int_{\mathscr{d}} \varphi V \, ds, \\ (d/dt) \int_{D_{\mathscr{M}}} \Phi \, da &= \int_{D_{\mathscr{M}}} \Phi_t \, da - \int_{\mathscr{d}} \Phi V \, ds, \end{aligned} \tag{2.13}$$

as well as the special case

$$(d/dt) \, \text{area} \, (D_{\mathscr{C}}) = -(d/dt) \, \text{area} \, (D_{\mathscr{M}}) = \int_{\mathscr{d}} V \, ds. \tag{2.14}$$

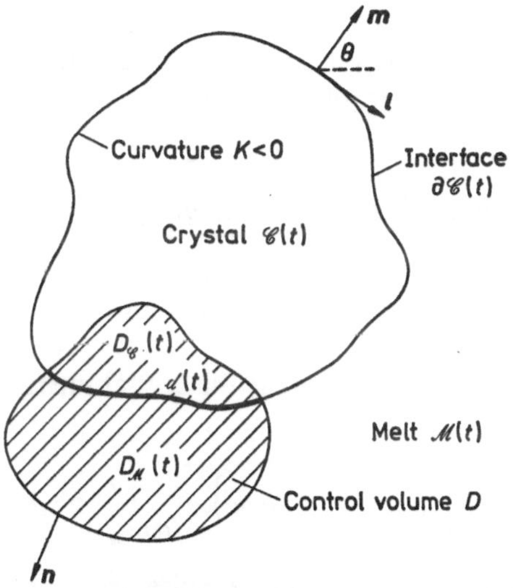

Fig. 1. Crystal, melt, and control volume.

[8] *Cf., e.g.,* [AG], eqts. (2.20), (2.34).

3. Capillary force. Balance laws. Dissipation inequality

3.1. Basic quantities

The crystals we consider are rigid, incapable of deformation, with corresponding melt an incompressible, inviscid fluid. Let $\mathscr{C}(t)$ be an evolving crystal. The mechanics of $\mathscr{C}(t)$ is described by five functions,

$\mathsf{C}(\boldsymbol{x}, t)$ **interfacial force,** defined for $\boldsymbol{x} \in \partial\mathscr{C}(t)$,

$\psi(\boldsymbol{x}, t)$ **interfacial energy,** defined for $\boldsymbol{x} \in \partial\mathscr{C}(t)$,

$\boldsymbol{T}_c(\boldsymbol{x}, t)$ **crystal stress,** defined for $\boldsymbol{x} \in \mathscr{C}(t)$,

$p(\boldsymbol{x}, t)$ **melt pressure,** defined for $\boldsymbol{x} \in \mathscr{M}(t)$,

$\boldsymbol{v}(\boldsymbol{x}, t)$ **melt velocity,** defined for $\boldsymbol{x} \in \mathscr{M}(t)$,

and four constants,

ϱ_c **crystal density,**

ϱ **melt density,**

Ψ_c **crystal energy,**

Ψ **melt energy.**

$\mathsf{C}(\boldsymbol{x}, t)$ represents the force *within the interface* exerted across $\boldsymbol{x}$ at time t; if $\mathscr{s}^-$ and $\mathscr{s}^+$, respectively, denote left and right neighborhoods of $\boldsymbol{x}$ in $\partial\mathscr{C}(t)$, then $\mathsf{C}(\boldsymbol{x}, t)$ is the force exerted *on* $\mathscr{s}^-$ *by* $\mathscr{s}^+$. We decompose this force into normal and tangential components:

$$\mathsf{C} = \sigma \boldsymbol{l} + \xi \boldsymbol{m}; \tag{3.1}$$

$\sigma(\boldsymbol{x}, t)$ is then the **surface tension,** $\xi(\boldsymbol{x}, t)$ the **surface shear.**

Because the crystal is **rigid**, the stress $\boldsymbol{T}_c$ is indeterminate. Similarly, the **incompressibility** of the melt leads to the constraint

$$\operatorname{div} \boldsymbol{v} = 0 \tag{3.2}$$

and to the indeterminacy of the melt pressure p.

3.2. *Balance of mass*

We neglect interfacial mass. Then, as the crystal interior is immobile, **balance of mass** requires that

$$(d/dt)\left\{\int_{D_{\mathscr{C}}} \varrho_c \, da + \int_{D_{\mathscr{M}}} \varrho \, da\right\} + \int_{(\partial D)_{\mathscr{M}}} \varrho \boldsymbol{v} \cdot \boldsymbol{n} \, ds = 0 \tag{3.3}$$

for every control volume D. Since the densities are constant, if we apply (2.14) and then shrink D to the interface, we are led to the **interfacial mass balance**

$$\varrho(\boldsymbol{v} \cdot \boldsymbol{m} - V) = -\varrho_c V, \tag{3.4}$$

or alternatively,

$$\boldsymbol{v} \cdot \boldsymbol{m} = (1 - \zeta)\, V, \tag{3.5}$$

where

$$\zeta = \varrho_c/\varrho \tag{3.6}$$

is the **mass ratio.** We assume throughout that the crystal and melt densities are unequal:

$$\zeta \neq 1\,. \tag{3.7}$$

3.3. Balance of momentum

Balance of momentum is the assertion that, given any control volume D,

$$(d/dt) \int\limits_{D_{\mathcal{M}}} \varrho \boldsymbol{v}\, da + \int\limits_{(\partial D)_{\mathcal{M}}} \varrho \boldsymbol{v}(\boldsymbol{v} \cdot \boldsymbol{n})\, ds = \int\limits_{\partial \mathcal{A}} \mathsf{C} + \int\limits_{(\partial D)_{\mathcal{C}}} \boldsymbol{T}_c \boldsymbol{n}\, ds - \int\limits_{(\partial D)_{\mathcal{M}}} p \boldsymbol{n}\, ds\,. \tag{3.8}$$

Since D is arbitrary, this leads (in the standard manner) to the local balance laws

$$\operatorname{div} \boldsymbol{T}_c = \boldsymbol{0}, \qquad \varrho \boldsymbol{v}^{\cdot} = -\operatorname{grad} p \tag{3.9}$$

in the crystal and in the melt, respectively, and, by use of (2.13), to

$$\mathsf{C}_s = \boldsymbol{T}_c \boldsymbol{m} + p \boldsymbol{m} + \varrho \boldsymbol{v}(\boldsymbol{v} \cdot \boldsymbol{m} - V)$$

on the interface. By (3.4), this last relation reduces to the **interfacial balance law**

$$\mathsf{C}_s = \boldsymbol{T}_c \boldsymbol{m} + p \boldsymbol{m} - \varrho_c V \boldsymbol{v}, \tag{3.10}$$

or, more succinctly,

$$\mathsf{C}_s + \boldsymbol{b} = \boldsymbol{0}, \tag{3.11}$$

with

$$\boldsymbol{b} = -\boldsymbol{T}_c \boldsymbol{m} - p \boldsymbol{m} + \varrho_c V \boldsymbol{v} \tag{3.12}$$

the total force (including inertia) exerted by the crystal and the melt *on* the interface.

Next, (2.1), (2.6), and (3.1) yield

$$(\mathsf{C} \cdot V \boldsymbol{m})_s = \xi \theta^\circ - \sigma K V + V \mathsf{C}_s \cdot \boldsymbol{m}; \tag{3.13}$$

thus, by (2.8), (2.11), (3.1), and (3.11), we have the **power identity**[9]

$$\int\limits_{\partial \mathscr{A}} \mathsf{C} \cdot \boldsymbol{v}_{\partial \mathscr{A}} + \int\limits_{\mathscr{A}} \boldsymbol{b} \cdot (V \boldsymbol{m})\, ds = \int\limits_{\mathscr{A}} (\xi \theta^\circ - \sigma K V)\, ds + \int\limits_{\partial \mathscr{A}} \sigma \boldsymbol{l} \cdot \boldsymbol{v}_{\partial \mathscr{A}} \tag{3.14}$$

for any interfacial chunk $\mathscr{A}(t)$. This identity renders plausible the assumption, made in the next section, that the interfacial force expends power on an interfacial chunk over the velocities of its endpoints. The term $-\sigma K V$ represents power expended in creating new interface; $\xi \theta^\circ$ represents power expended in changing the

[9] *Cf.* [G1], eqt. (3.4).

orientation of the interface; the term involving $\sigma \boldsymbol{l} \cdot \boldsymbol{v}_{\partial \mathscr{s}}$ compensates for the tangential motion of the endpoints of $\partial \mathscr{s}$.

The scalar field

$$p_c = -\boldsymbol{m} \cdot \boldsymbol{T}_c \boldsymbol{m} \tag{3.15}$$

describes the normal interaction between the interface and the crystal. Using (3.5), (3.12), and (3.15), we can write the power identity in the form

$$\int_{\partial \mathscr{s}} \mathsf{C} \cdot \boldsymbol{v}_{\partial \mathscr{s}} + \int_{\mathscr{s}} \{p_c - p + (1 - \zeta) \varrho_c V^2\} V \, ds = \int_{\mathscr{s}} (\xi \theta^\circ - \sigma K V) \, ds + \int_{\partial \mathscr{s}} \sigma \boldsymbol{l} \cdot \boldsymbol{v}_{\partial \mathscr{s}} . \tag{3.16}$$

The term involving $(1 - \zeta) \varrho_c V^2$ represents a supply of kinetic energy induced by the difference in crystal and melt densities.

3.4. Dissipation inequality

The version of the *second law* that we shall use is the assertion that, for any control volume D, the rate of energy increase plus the energy outflow cannot be greater than the power supplied. The terms

$$\int_{D_{\mathscr{C}}} \Psi_c \, da + \int_{D_{\mathscr{M}}} (\Psi + \tfrac{1}{2} \varrho v^2) \, da + \int_{d} \psi \, ds \tag{3.17}$$

sum to the total energy of D, while

$$\int_{(\partial D)_{\mathscr{M}}} (\Psi + \tfrac{1}{2} \varrho v^2) (\boldsymbol{v} \cdot \boldsymbol{n}) \, ds \tag{3.18}$$

represents the loss in energy due to the flow of melt across $(\partial D)_{\mathscr{M}}$. A basic postulate of the theory is that the interfacial force expend power over the velocity[10] of the endpoints of d; the total power expended on D is therefore given by

$$\int_{\partial d} \mathsf{C} \cdot \boldsymbol{v}_{\partial d} - \int_{(\partial D)_{\mathscr{M}}} p \boldsymbol{v} \cdot \boldsymbol{n} \, ds . \tag{3.19}$$

In view of this discussion, we assume that the **dissipation inequality**[11]

$$(d/dt) \left\{ \int_{D_{\mathscr{C}}} \Psi_c \, da + \int_{D_{\mathscr{M}}} (\Psi + \tfrac{1}{2} \varrho v^2) \, da + \int_{d} \psi \, ds \right\} + \int_{(\partial D)_{\mathscr{M}}} (\Psi + \tfrac{1}{2} \varrho v^2) (\boldsymbol{v} \cdot \boldsymbol{n}) \, ds$$
$$\leqq \int_{\partial d} \mathsf{C} \cdot \boldsymbol{v}_{\partial d} - \int_{(\partial D)_{\mathscr{M}}} p \boldsymbol{v} \cdot \boldsymbol{n} \, ds \tag{3.20}$$

hold for every control volume D.

---

[10] The formulation in [AG] (eqt. (3.7); *cf.* [G1], eqt. (1.3)) is based on the assumption that the capillary force expend power over the *normal* interfacial velocity $V\boldsymbol{m}$, while the dissipation inequality in [AG] contains a term to account for an outflow of energy across ∂d, a term which I do not include here. These two views are consistent, since both theories yield the identity $\sigma = \psi$. I believe the current formulation to be conceptually preferable.

[11] *Cf.* [G1], § 3; [G2], § 5; [AG], eqt. (3.7).

This relation is satisfied trivially when D lies solely in the crystal, and, by $(3.9)_2$, also when D lies solely in the melt. On the other hand, if we apply (3.20) to a control volume D which contains the interface, appeal to (2.11), (2.13), (3.4), (3.6), and (3.16), and shrink D to the interface, we see that

$$\int_{\mathscr{d}} \{\psi^\circ - \xi\theta^\circ + (\sigma - \psi) KV + V\{\Psi_c + p_c - \zeta(\Psi + p) - \tfrac{1}{2}\varrho_c v^2 + \varrho_c(1 - \zeta) V^2\} ds + \int_{\partial \mathscr{d}} \{(\psi - \sigma) \boldsymbol{l} \cdot \boldsymbol{v}_{\partial\mathscr{d}}\} \leqq 0. \tag{3.21}$$

This inequality must hold for every control volume D. Given a time t_0 and a subcurve $\mathscr{d}_0$ of $\mathscr{s}(t_0)$, we can always find a control volume D such that $\mathscr{d}(t_0) = \mathscr{d}_0$, but the tangential velocities of the endpoints of $\mathscr{d}$ are arbitrary at t_0. Thus, as a consequence of (3.21), we have the well known **identification of surface tension with energy:**

$$\sigma = \psi. \tag{3.22}$$

Finally, it is clear from (3.22) that (3.21) can hold for all control volumes only if

$$\psi^\circ - \xi\theta^\circ + V\{\Psi_c + p_c - \zeta(\Psi + p) - \tfrac{1}{2}\varrho_c v^2 + \varrho_c(1 - \zeta) V^2\} \leqq 0. \tag{3.23}$$

3.5. Interactive energy-balance

One of the chief differences between theories involving phase transitions and the more classical theories of continuum mechanics is the *creation and deletion of material points* as the phase interface moves relative to the underlying material. This is essentially a *bulk* interaction between phases, which we now isolate by restricting attention to the bulk material arbitrarily close to the interface. We represent the action of the interface on the bulk material immediately adjacent to it by two functions of $\boldsymbol{x} \in \partial\mathscr{C}(t)$ and t:

$\delta(\boldsymbol{x}, t)$ **interactive dissipation,**

$\boldsymbol{g}(\boldsymbol{x}, t)$ **interactive force.**

$\delta(\boldsymbol{x}, t)$ represents the net *outflow* of energy from the bulk material at the interface, per unit length; this is energy associated with the kinetics of attachment in the exchange of atoms between the crystal and the melt. $\boldsymbol{g}(\boldsymbol{x}, t)$ is the net force, per unit length, exerted by the interface on the bulk material; this force is assumed to expend power over the normal velocity $V\boldsymbol{m}$.

The interactive fields enter our theory through the **interactive energy-balance**

$$(d/dt)\left\{\int_{D_{\mathscr{C}}} \Psi_c \, da + \int_{D_{\mathscr{M}}} (\Psi + \tfrac{1}{2}\varrho v^2) \, da\right\} + \int_{(\partial D)_{\mathscr{M}}} (\Psi + \tfrac{1}{2}\varrho v^2)(\boldsymbol{v} \cdot \boldsymbol{n}) \, ds + \int_{\mathscr{d}} \delta \, ds = - \int_{(\partial D)_{\mathscr{M}}} p\boldsymbol{v} \cdot \boldsymbol{n} \, ds + \int_{\mathscr{d}} \boldsymbol{g} \cdot (V\boldsymbol{m}) \, ds, \tag{3.24}$$

which we assume to hold in every control volume D (*cf.* Figure 2). Using steps analogous to those used to derive (3.21), we find that (3.24) is equivalent to

$$\int_{d} \{(\Psi + \tfrac{1}{2}\varrho v^2)(\boldsymbol{v}\cdot\boldsymbol{m} - V) + \Psi_c V + p\boldsymbol{v}\cdot\boldsymbol{m} - \boldsymbol{g}\cdot(V\boldsymbol{m})\}\,ds = -\int_{d} \delta\,ds. \tag{3.25}$$

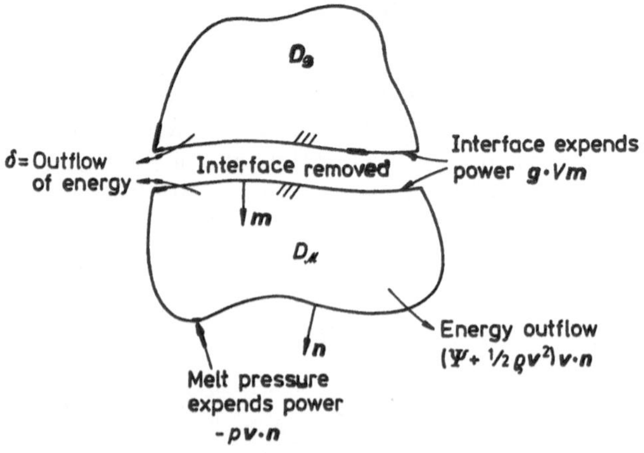

Fig. 2. Free-body diagram used to write the interactive energy balance.

We now determine the specific form of the force $\boldsymbol{g}$ using the invariance of (3.25) under **Galilean changes in observer.** Under such a change the velocities transform according to[12]

$$\boldsymbol{v} \to \boldsymbol{v} + \boldsymbol{a}, \quad V\boldsymbol{m} \to V\boldsymbol{m} + \boldsymbol{a}, \tag{3.26}$$

with $\boldsymbol{a}$ the (constant) relative velocity of the observers. Under (3.26) the term $\Psi_c V$ is invariant, since V there represents a velocity *relative to the crystal.* However, since the crystal is now not at rest, but instead has velocity $\boldsymbol{a}$, the term

$$\int_{d} \{(\tfrac{1}{2}\varrho_c \boldsymbol{a}^2)\,V + \boldsymbol{T}_c\boldsymbol{m}\cdot\boldsymbol{a}\}\,ds,$$

should be added to the left side of (3.25). Granted this, and assuming that δ is invariant under (3.26), we find, since $\boldsymbol{a}$ and d are arbitrary, that[13]

$$\boldsymbol{g} = -\boldsymbol{b}.$$

[12] General transformations of observer during crystallization are studied by GURTIN & STRUTHERS [GS].

[13] In view of (3.11), we might have taken $\boldsymbol{g} = -\boldsymbol{b}$ from the outset. Because of the presence of inertial terms and because of the nonstandard nature of the power expended by an interface, I find it more compelling to derive $\boldsymbol{g} = -\boldsymbol{b}$ as a consequence of a standard requirement of invariance.

(The transformed energy-balance contains terms linear and quadratic in $\boldsymbol{a}$; the linear terms, set equal to zero, yield $\boldsymbol{g} = -\boldsymbol{b}$; the quadratic terms yield balance of mass (3.4).) Thus, appealing to (3.5), (3.15), and (3.22), we are led to the (local) **interactive energy-balance:**

$$V\{\Psi_c + p_c - \zeta(\Psi + p) - \tfrac{1}{2}\varrho_c \boldsymbol{v}^2 + \varrho_c(1-\zeta)\,V^2\} = -\delta\,. \tag{3.27}$$

The interactive dissipation δ is a basic physical quantity of the theory, to be specified by a constitutive equation, and (3.27) is a fundamental balance law to be satisfied in all crystallization processes.

An important feature of (3.27) is that it allows us to rewrite (3.23) as the **dissipation inequality**

$$\psi^\circ - \xi\theta^\circ \leqq \delta\,. \tag{3.28}$$

4. Constitutive equations for the interface

As **constitutive equations** we allow the interfacial energy, the interfacial force, and the interactive dissipation to depend on the orientation of the interface through the angle θ and on the kinetics of the interface through the normal velocity V:

$$\psi = \psi(\theta, V), \quad \mathsf{C} = \mathsf{C}(\theta, V), \quad \delta = \delta(\theta, V)\,. \tag{4.1}$$

We assume that the relations $(4.1)_{1,2}$ are consistent with (3.22), and, in view of (3.27), we suppose that

$$\delta(\theta, 0) = 0\,. \tag{4.2}$$

Given an evolving crystal $\mathscr{C}(t)$, the constitutive equations (4.1) may be used to compute a corresponding **constitutive process.** A basic hypothesis[14] of our theory is the **dissipation axiom:**

constitutive processes must be consistent
with the local dissipation inequality (3.28),

[14] I rationalize this postulate as follows: Generalize the theory to include body forces $\boldsymbol{f}_c$ and $\boldsymbol{f}_m$ in the crystal and melt, respectively, so that (3.9) are replaced by $\operatorname{div} \boldsymbol{T}_c + \boldsymbol{f}_c = \boldsymbol{0}$, $\varrho \boldsymbol{v}^\cdot = -\operatorname{grad} p + \boldsymbol{f}_m$ (*). Assume that a constitutive process is given. Construct a velocity field $\boldsymbol{v}$ in the melt that is consistent with balance of mass (3.5) and incompressibility (3.2); use (3.10) and (3.27) (supplemented by (3.7)) to compute p and $\boldsymbol{T}_c\boldsymbol{m}$ on the interface; extend p and $\boldsymbol{T}_c$ into the melt and into the crystal; compute $\boldsymbol{f}_c$ and $\boldsymbol{f}_m$ through (*). The corresponding set of fields is consistent with the balance aws of mass and momentum, as well as with the interactive energy-balance, and all that remains is the dissipation inequality (3.28).

an axiom which yields the following important constitutive restrictions:[15]

(i) *the energy* ψ *and interfacial force* C *are independent of the normal velocity* V, *and* $\psi(\theta)$ *generates* $\mathsf{C}(\theta)$ *through*[16]

$$\mathsf{C}(\theta) = \psi(\theta)\, \boldsymbol{l}(\theta) + \psi'(\theta)\, \boldsymbol{m}(\theta); \tag{4.3}$$

(ii) *the interactive dissipation is given by a relation of the form*

$$\delta(\theta, V) = \beta(\theta, V)\, V^2, \quad \beta(\theta, V) \geqq 0. \tag{4.4}$$

(Conditions (i) and (ii) are also sufficient that (3.28) hold in all constitutive processes.)

We assume that the **kinetic coefficient** $\beta(\theta, V)$ depends only on θ, and that

$$\psi(\theta), \quad \beta(\theta) > 0. \tag{4.5}$$

Remarks.

1° By (4.3), $\sigma = \sigma(\theta)$ and $\xi = \xi(\theta)$ with

$$\sigma(\theta) = \psi(\theta), \quad \xi(\theta) = \psi'(\theta). \tag{4.6}$$

2° In view of (3.27) and (4.3), the left side of (3.23) is equal to $-\delta$, which identifies $\delta = \beta(\theta)\, V^2$ as the *sole rate of energy dissipation*; indeed, tracing backwards the argument leading to (3.28), we find that

$$-\int_{\mathscr{A}} \beta(\theta)\, V^2\, ds$$

represents the left side of (3.20) minus the right.

3° (2.1), (2.6), and (4.3) yield the identity

$$\mathsf{C} \cdot (V\boldsymbol{m})_s = \psi^\circ - \psi K V. \tag{4.7}$$

4° (3.27) and (4.4) furnish us with a formula for the "crystal pressure":

$$p_c = -\beta(\theta)\, V + \zeta p - F + \tfrac{1}{2}\varrho_c \mathfrak{v}^2 - \varrho_c(1 - \zeta)\, V^2, \tag{4.8}$$

with F the *constant*

$$F = \Psi_c - \zeta\Psi. \tag{4.9}$$

5° It might seem reasonable to allow the constitutive equations (4.1) to depend on the tangential velocity $\mathfrak{v} = \boldsymbol{v} \cdot \boldsymbol{l}$ of the melt at the interface. This generalization results in no essential change: the dissipation inequality rules out the dependence of ψ and C on $\mathfrak{v}$, so that (4.3) remains valid, and (4.5) is replaced by

$$\delta(\theta, V, \mathfrak{v}) = \beta(\theta, V, \mathfrak{v})\, V^2, \quad \beta(\theta, V, \mathfrak{v}) \geqq 0. \tag{4.10}$$

---

[15] These restrictions are obvious modifications of results given in [G1, AG] (Compatibility Theorem), and I refer to [G1, AG] for their proof. The use of the second law to find restrictions on constitutive equations traces back to COLEMAN & NOLL [CN], who used this proceedure for single-phase thermoelastic materials.

[16] $f'(\theta) = df(\theta)/d\theta$.

5. Partial differential equations. Free-boundary conditions

The thermodynamic restrictions (4.3) have important consequences when applied to the interfacial balance law (3.10). Indeed, (2.1), (2.3), and (4.3) imply that

$$\mathsf{C}_s = [\psi(\theta) + \psi''(\theta)]\, K\boldsymbol{m}, \tag{5.1}$$

while (3.5), (3.15), (3.10), and (4.8) yield

$$\begin{aligned} \boldsymbol{m}\cdot\mathsf{C}_s &= p - p_c - \varrho_c(1-\zeta)\,V^2 = \beta(\theta)\,V + F - \tfrac{1}{2}\varrho_c \boldsymbol{v}^2 - (\zeta - 1)\,p, \\ \boldsymbol{l}\cdot\mathsf{C}_s &= \boldsymbol{l}\cdot T_c\boldsymbol{m} - \varrho_c V\boldsymbol{v}\cdot\boldsymbol{l}. \end{aligned} \tag{5.2}$$

The interfacial balance law therefore has the *normal component*

$$[\psi(\theta) + \psi''(\theta)]\, K = \beta(\theta)\,V + F - \tfrac{1}{2}\varrho_c\boldsymbol{v}^2 + (1-\zeta)\,p, \tag{5.3}$$

a crucial relation as it and (3.5) comprise the free-boundary conditions of the theory.

The tangential component of the interfacial balance law,

$$\boldsymbol{l}\cdot T_c\boldsymbol{m} = \varrho_c V\boldsymbol{v}\cdot\boldsymbol{l}, \tag{5.4}$$

gives the tangential component of the crystal traction at the interface;[17] since the crystal stress is indeterminate, (5.4) is of little importance.

Summarizing, the basic system of equations consists of (3.2), (3.5), $(3.9)_2$, and (5.3):

$$\varrho\boldsymbol{v}^{\cdot} = -\operatorname{grad} p, \quad \operatorname{div}\boldsymbol{v} = 0, \quad \text{(melt)} \tag{5.5}$$

$$\left.\begin{aligned} [\psi(\theta) + \psi''(\theta)]\, K &= \beta(\theta)\,V + F - \tfrac{1}{2}\varrho_c\boldsymbol{v}^2 + (1-\zeta)\,p, \\ \boldsymbol{v}\cdot\boldsymbol{m} &= (1-\zeta)\,V. \end{aligned}\right\} \quad \text{(interface)} \tag{5.6}$$

We will refer to (5.5) and (5.6) as the **crystallization equations**; the melt equations (5.5) are to hold in $\mathcal{M}(t)$ for $t > 0$, the interface conditions (5.6) on $\partial\mathcal{C}(t)$ for $t > 0$.

There should be cases of interest in which the melt velocity (and its gradient) are small. Granted this, it would appear reasonable to neglect the term $\frac{1}{2}\varrho_c\boldsymbol{v}^2$ in (5.6) and to replace the material time-derivative $\boldsymbol{v}^{\cdot}$ in (5.5) with the spatial time-derivative $\boldsymbol{v}_t$; we will refer to this proceedure as the **weak-inertia approximation.**

The crystallization equations can be simplified when the flow is **irrotational**: we write $\boldsymbol{v}(\boldsymbol{x}, t)$ as the gradient of a *potential* $\varphi(\boldsymbol{x}, t)$,

$$\boldsymbol{v} = \operatorname{grad}\varphi, \tag{5.7}$$

and replace (5.5) by the Bernoulli equation[18]

$$p = -\varrho\varphi_t - \tfrac{1}{2}\varrho\boldsymbol{v}^2 \tag{5.8}$$

---

[17] In addition, the interfacial shear induces a couple per unit length along the interface which must be balanced by torques in the crystal (*cf.* [G1], Remark 3.2).

[18] *Cf., e.g.*, [G4], p. 120.

and the requirement that φ be harmonic. It is actually more convenient to use the potential

$$u(x, t) = \varrho(1 - \zeta)\,\varphi(x, t) - Ft; \tag{5.9}$$

then

$$\varrho(\zeta - 1)\,v = -\operatorname{grad} u \tag{5.10}$$

and the crystallization equations take the form:

$$\Delta u = 0, \quad \text{(melt)} \tag{5.11}$$

$$\left.\begin{aligned} [\psi(\theta) + \psi''(\theta)]\,K &= \beta(\theta)\,V - u_t - \tfrac{1}{2}\alpha^{-1}(\operatorname{grad} u)^2, \\ \partial u/\partial m &= \alpha V, \end{aligned}\right\} \quad \text{(interface)} \tag{5.12}$$

where $\partial u/\partial m = m \cdot \operatorname{grad} u$, while

$$\alpha = \varrho(1 - \zeta)^2. \tag{5.13}$$

Under the *weak-inertia approximation* the term $(\operatorname{grad} u)^2$ in $(5.12)_1$ is dropped, so that $(5.12)_1$ becomes

$$[\psi(\theta) + \psi''(\theta)]\,K = \beta(\theta)\,V - u_t, \tag{5.14}$$

but (5.11) and $(5.12)_2$ remain unchanged. Consistent with this approximation, we neglect the term involving v^2 in (5.8), so that, by (5.9),

$$(\zeta - 1)\,p = u_t + F. \tag{5.15}$$

Remark. For an isotropic crystal both ψ and β are constants; for ψ constant and $\beta = 0$, and under the weak inertia approximation, we have the equations

$$\Delta u = 0, \quad \text{(melt)} \tag{5.16}$$

$$u_t = -\psi K, \qquad \partial u/\partial m = \alpha V. \quad \text{(interface)} \tag{5.17}$$

It is interesting to compare the simplified equations (5.16) and (5.17) to those of the (quasi-static) theory of MULLINS & SEKERKA for thermally-driven solidification when one of the two phases does not conduct heat.[19] There the temperature u of the conducting phase satisfies (5.16) and $(5.17)_2$, but the condition $(5.17)_1$ is replaced by $u = \psi K$.

6. The crystallization problem. Global balance relations

Initial conditions appropriate to the crystallization equations would appear to be a prescription of the region occupied by the crystal (and hence also that occupied by the melt) together with a prescription of the velocity v:

$$\mathscr{C}(0) \textit{ and } v(x, 0) \textit{ prescribed.} \tag{6.1}$$

---

[19] *Cf., e.g.*, [G2], eqt. (11.9). (The curvature in [G2] has sign opposite to that here).

Since the melt is infinite, restrictions should be imposed on its behavior at infinity. We here consider the **far-field conditions:**[20]

$$v(x, t) = o(|x|^{-1}), \quad p(x, t) = O(1) \quad \text{as } |x| \to \infty, \tag{6.2}$$

for each t. Note that, by (6.2),

$$\int_{\mathcal{S}_r} v \cdot n \, ds \to 0 \quad \text{as } r \to \infty, \tag{6.3}$$

where $\mathcal{S}_r$ is a circle of radius r and n is the outward unit normal on $\mathcal{S}_r$. Thus the *net flow of melt vanishes at infinity.*

We will use the term **crystallization problem** to designate the free-boundary problem defined by the crystallization equations (5.5) and (5.6), the initial conditions (6.1), and the far-field conditions (6.2). A solution of this problem will be termed **regular** if

$$(d/dt) \int_{\mathcal{M}_r(t)} v^2 \, da \to (d/dt) \int_{\mathcal{M}(t)} v^2 \, da \quad \text{as } r \to \infty, \tag{6.4}$$

where $\mathcal{M}_r(t)$ is the portion of $\mathcal{M}(t)$ interior to the circle $\mathcal{S}_r$.

The next result yields Lyapunov functions for the crystallization problem.

Global balance relations. *Regular solutions of the crystallization problem satisfy*

$$\begin{gathered} (d/dt) \operatorname{area}(\mathcal{C}) = 0, \\ (d/dt) \left\{ \int_{\partial \mathcal{C}} \psi(\theta) \, ds + \int_{\mathcal{M}} \tfrac{1}{2} \varrho v^2 \, da \right\} = - \int_{\partial \mathcal{C}} \beta(\theta) V^2 \, ds \leqq 0 . \end{gathered} \tag{6.5}$$

Proof. By the divergence theorem, (2.14), (3.5), $(5.5)_2$, and (6.3), for r sufficiently large,

$$0 = \int_{\mathcal{S}_r} v \cdot n \, ds - \int_{\partial \mathcal{C}} v \cdot m \, ds = (\zeta - 1) \int_{\partial \mathcal{C}} V \, ds = (\zeta - 1) (d/dt) \operatorname{area}(\mathcal{C}), \tag{6.6}$$

which, by virtue of (3.7), implies $(6.5)_1$.

The verification of $(6.5)_2$ is more complicated. First of all, (2.11), (4.7), and an integration by parts yield

$$(d/dt) \int_{\partial \mathcal{C}} \psi \, ds = - \int_{\partial \mathcal{C}} V \mathrm{C}_s \cdot m \, ds . \tag{6.7}$$

Less trivial is the identity

$$\begin{gathered} (d/dt) \int_{\mathcal{M}} k \, da = \int_{\partial \mathcal{C}} [(1 - \zeta) p - \tfrac{1}{2} \varrho_c v^2] V \, ds, \\ k = \tfrac{1}{2} \varrho v^2, \end{gathered} \tag{6.8}$$

---

[20] Another possibility will be considered in Section 11.

which we now prove. By $(2.13)_2$,

$$(d/dt) \int_{\mathcal{M}_r} \mathscr{k}\, da = \int_{\mathcal{M}_r} \mathscr{k}_t\, da - \int_{\partial\mathscr{C}} V\mathscr{k}\, ds. \tag{6.9}$$

Also, we may use (6.2) to conclude that, as $r \to \infty$,

$$\int_{\mathscr{S}_r} \mathscr{k}\boldsymbol{v} \cdot \boldsymbol{n}\, ds = o(1), \qquad \int_{\mathscr{S}_r} p\boldsymbol{v} \cdot \boldsymbol{n}\, ds = o(1). \tag{6.10}$$

Therefore, by (3.4), (3.5), (5.5), and (6.4), as $r \to \infty$,

$$\begin{aligned} -\int_{\partial\mathscr{C}} \tfrac{1}{2}\varrho_c v^2 V\, ds &= \int_{\partial\mathscr{C}} \mathscr{k}(\boldsymbol{v}\cdot\boldsymbol{m} - V)\, ds \\ &= -\int_{\mathcal{M}_r} [\mathscr{k}_t + \operatorname{div}(\mathscr{k}\boldsymbol{v})]\, da + (d/dt)\int_{\mathcal{M}_r} \mathscr{k}\, da + o(1) \\ &= -\int_{\mathcal{M}_r} \mathscr{k}^{\cdot}\, da + (d/dt)\int_{\mathcal{M}_r} \mathscr{k}\, da + o(1) \\ &= \int_{\mathcal{M}_r} \boldsymbol{v}\cdot\operatorname{grad} p\, da + (d/dt)\int_{\mathcal{M}} \mathscr{k}\, da + o(1) \\ &= -\int_{\partial\mathscr{C}} p\boldsymbol{v}\cdot\boldsymbol{m}\, ds + (d/dt)\int_{\mathcal{M}} \mathscr{k}\, da + o(1) \\ &= -\int_{\partial\mathscr{C}} (1-\zeta)\, pV\, ds + (d/dt)\int_{\mathcal{M}} \mathscr{k}\, da + o(1), \end{aligned} \tag{6.11}$$

and, letting $r \to \infty$, we obtain (6.8).

In view of (5.1), $\mathsf{C}_s \cdot \boldsymbol{m}$ is equal to the right side of $(5.6)_1$. Thus we may conclude from (6.7) and (6.8) that the left side of $(6.5)_2$ is equal to

$$-\int_{\partial\mathscr{C}} V[\beta(\theta)\, V + F]\, ds, \tag{6.12}$$

which yields $(6.5)_2$, since (6.6) implies that the integral of FV over $\partial\mathscr{C}$ vanishes. □

Remark. The global balance relation $(6.5)_2$ actually follows from the global dissipation inequality (3.20) in conjunction with $(6.5)_1$ and Remark 2° of Section 4. Indeed, let D in (3.20) be the interior of the circle $\mathscr{S}_r$ (with r large enough that D contains the crystal); note that, by $(6.5)_1$, (d/dt) area $(D_{\mathscr{C}}) = (d/dt)$ area $(D_{\mathcal{M}}) = 0$; using (6.3), (6.4), and (6.10), let $r \to \infty$ in (3.20) (strengthened by Remark 2° of Section 4).

7. Small motions about a flat interface

7.1. Basic equations

A flat interface furnishes an *equilibrium* of the crystallization equations (5.5), (5.6) provided

$$F + (1-\zeta)\, p = 0, \tag{7.1}$$

and this relation, (4.8), and (4.9) yield

$$\Psi_c + p_c = \zeta(\Psi + p), \qquad p_c = p. \tag{7.2}$$

We now consider solutions which are close to an equilibrium of this form. We choose x and y as coordinates in $\mathbb{R}^2$ and, without loss in generality, stipulate that the x-axis represent the equilibrium position of the interface, oriented so that arc length increases with x. We assume that:

(i) the interface is represented as a graph $y = h(x, t)$ (Figure 3) with *interfacial height* h "small";
(ii) the melt velocity $\boldsymbol{v}$ is irrotational and "small";

and we *formally linearize* the crystallization equations with respect to h and $\boldsymbol{v}$.

We begin the system (5.11), (5.12). Considering V, K, and θ as functions of (x, t), we obtain

$$V = h_t \sin\theta, \qquad K = h_{xx} \sin^3\theta, \tag{7.3}$$

where subscripts denote partial differentiation with respect to the corresponding variable. Noting that the interface has angle $\theta = \frac{\pi}{2}$ at equilibrium, we define

$$A = (\psi + \psi'')_0, \qquad B = \beta_0, \tag{7.4}$$

where the subscript zero denotes evaluation at $\theta = \frac{\pi}{2}$. Then the interface condition $(5.12)_1$, linearized formally with respect to h and $\boldsymbol{v}$, is given by

$$Ah_{xx} = Bh_t - u_t. \tag{7.5}$$

Consistent with the approximations leading to (7.5) is the assumption that the melt equation (5.11) hold in the halfspace $y > 0$, and that (7.5) and the interface condition $(5.12)_2$ be satisfied on the x-axis with $\partial u/\partial m$ replaced by u_y. Thus the crystallization equations, *linearized about equilibrium*, take the form:[21]

$$\begin{aligned} u_{xx} + u_{yy} &= 0 \qquad (-\infty < x < \infty, y > 0, t > 0), \\ Ah_{xx} = Bh_t - u_t, \qquad u_y &= \alpha h_t \qquad (-\infty < x < \infty, y = 0, t > 0). \end{aligned} \tag{7.6}$$

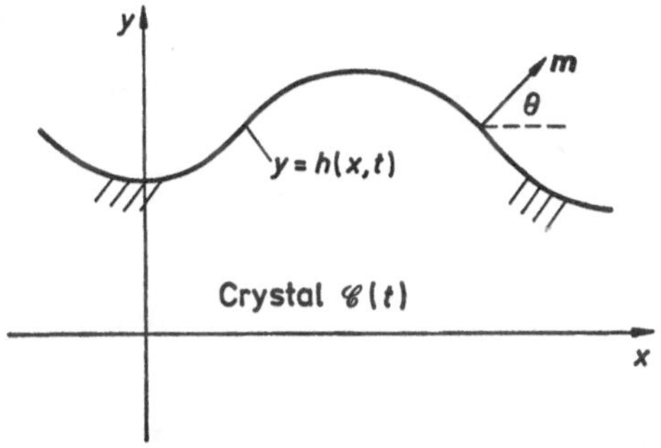

Fig. 3. Sign conventions when the interface is a graph $y = h(x, t)$.

[21] These equations, for $F = 0$, are due to ANDREEV & PARSHIN [AP]. F manifests itself when boundary conditions (or conditions at infinity) are expressed in terms of the pressure $p = -(1 - \zeta)^{-1}[u_t + F]$ (*cf.* Section 11).

7.2. *Melting-freezing waves*

ANDREEV & PARSHIN [AP] note that (7.6) has the solution

$$\begin{aligned} h(x, t) &= C_1 e^{i\lambda x}\, e^{-(i\omega+\gamma)t}, \\ u(x, y, t) &= C_2 e^{-\lambda y}\, e^{i\lambda x}\, e^{-(i\omega+\gamma)t} \end{aligned} \tag{7.7}$$

with

$$\omega^2 = \frac{A\lambda^3}{\alpha} - \frac{B^2\lambda^2}{4\alpha^2}, \qquad \gamma = \frac{B\lambda}{2\alpha}, \tag{7.8}$$

a solution which represents *damped oscillations of the interface*. For $B = 0$ (and hence approximately for B small), ω^2 is proportional to λ^3,

$$\omega^2 = \frac{\varrho(\psi + \psi'')_0\, \lambda^3}{(\varrho_c - \varrho)^2}, \tag{7.9}$$

a proportionality found in the experiments of KESHISHEV, PARSHIN, & BABKIN [KPB].

Remark. The relation (7.8) corresponds to oscillations on a flat interface at angle θ only if

$$\psi(\theta) + \psi''(\theta) > 0. \tag{7.10}$$

This inequality is essentially a condition of static stability for the interface: it follows from the requirement that straight line-segments locally minimize interfacial energy. There is no compelling physical reason to suppose that (7.10) is satisfied; in fact, materials' scientists often consider energies which violate (7.10) for particular ranges of θ (*cf.* GJOSTEIN [G], CAHN & HOFFMAN [CH]).

7.3. *The linearized crystallization problem. Uniqueness*

Initial conditions for the linearized crystallization equations are

$$\begin{aligned} &\operatorname{grad} u(x, y, 0) \quad (-\infty < x < \infty, y > 0) \quad \textit{and} \\ &h(x, 0) \quad (-\infty < x < \infty) \quad \textit{prescribed}, \end{aligned} \tag{7.11}$$

while possible **far-field conditions** are the following counterparts of (6.2) (*cf.* (5.10), (5.15)):

$$\operatorname{grad} u(\mathbf{x}, t) = o(|\mathbf{x}|^{-1}), \quad u_t(\mathbf{x}, t) = O(1) \quad \text{as } |\mathbf{x}| \to \infty. \tag{7.12}$$

Further, since the interface is not a closed curve, but begins and ends at infinity, an **interfacial far-field condition** is needed; one possibility is:

$$h_x(\pm\infty, t) \quad \textit{or} \quad h(\pm\infty, t) \quad \textit{prescribed}. \tag{7.13}$$

We will refer to the problem defined by (7.6) and (7.11)–(7.13) as the **linearized crystallization problem.**[22]

The next result, an analog of the global balance relation $(6.5)_2$, yields uniqueness for the linearized problem. To state this result precisely, let

$$\mathcal{M}_\infty = \{(x, y):\ -\infty < x < \infty,\ y > 0\},$$
$$\mathscr{s}_\infty = \{(x, y):\ -\infty < x < \infty,\ y = 0\}; \tag{7.14}$$

let $\mathcal{M}_r$ denote the portion of $\mathcal{M}_\infty$ interior to a circle of radius r; let $\boldsymbol{n}$ denote the outward unit normal on $\partial\mathcal{M}_r$; let $\mathcal{S}_r$ denote the *circular portion* of $\partial\mathcal{M}_r$; let $\mathscr{s}_r$ denote the *flat portion* of $\partial\mathcal{M}_r$. We will refer to a solution of the linearized problem as **regular** if

$$(d/dt) \int_{\mathcal{M}_r} |\operatorname{grad} u|^2\, da \to (d/dt) \int_{\mathcal{M}_\infty} |\operatorname{grad} u|^2\, da,$$
$$(d/dt) \int_{\mathscr{s}_r} (h_x)^2\, ds \to (d/dt) \int_{\mathscr{s}_\infty} (h_x)^2\, ds \tag{7.15}$$

as $r \to \infty$.

Global balance relation. *Consider a regular solution of the linearized crystallization problem with*

$$(h_t h_x)(\pm\infty, t) = 0 \tag{7.16}$$

for all $t > 0$. *Then*

$$(d/dt)\left\{\int_{\mathcal{M}_\infty} |\operatorname{grad} u|^2\, da + \alpha A \int_{\mathscr{s}_\infty} (h_x)^2\, ds\right\} = -2\alpha B \int_{\mathscr{s}_\infty} (h_t)^2\, ds \leqq 0. \tag{7.17}$$

Proof. Since $(\partial u/\partial n) = -u_y$ on $\mathscr{s}_r$, if we apply the divergence theorem to the integral of $u_t(\partial u/\partial n)$ over $\partial\mathcal{M}_r$, we conclude, with the aid of $(7.6)_1$ and (7.12), that

$$\int_{\mathscr{s}_r} u_t u_y\, ds = -\tfrac{1}{2}(d/dt) \int_{\mathcal{M}_r} |\operatorname{grad} u|^2\, da + o(1)$$

as $r \to \infty$. On the other hand, $(7.6)_2$ and (7.16) imply that

$$\int_{\mathscr{s}_r} u_t u_y\, ds = \alpha \int_{\mathscr{s}_r} u_t h_t\, ds = \alpha \int_{\mathscr{s}_r} h_t(Bh_t - Ah_{xx})\, ds,$$
$$= \alpha B \int_{\mathscr{s}_r} (h_t)^2\, ds + \tfrac{1}{2}\alpha A (d/dt) \int_{\mathscr{s}_r} (h_x)^2\, ds + o(1).$$

---

[22] An existence theorem for this problem has been established by Rogers [R], who shows that: (i) for sufficiently smooth data the solution goes asymptotically to a steady state; (ii) solutions propagate with infinite speed. Rogers analyzes the crystallization equations in the alternative form $h_{tt} = (\alpha\pi)^{-1}\, \mathscr{L}[Ah_{xxx} - Bh_{xt}]$, where $\mathscr{L}$ is the integral operator consisting of spatial convolution on $(-\infty, \infty)$ with respect to x^{-1}. This integral equation was derived independently by MacCamy (private communication).

If we equate the last two relations and let $r \to \infty$ using (7.15), we are led to (7.17). □

The global balance relation yields a certain degree of *stability* for the linear crystallization problem, since for α, A, $B > 0$ it implies decay with time of spatial L^2 norms of $|\text{grad}\, u|$ and h_x, as well as the boundedness of the integral of $(h_t)^2$ over $\mathcal{M}_\infty \times (0, t)$.

An immediate corollary of (7.17) is the following result, in which the assumptions $\alpha > 0$, $A > 0$, $B > 0$ should be kept in mind. (*Cf.* (3.7), (4.5), (5.13), (7.4), and (7.10); actually, $B \geqq 0$ rather than $B > 0$ is used.)

Uniqueness theorem. *The linearized crystallization problem* (*with given data*) *has at most one regular solution* (*modulo an additive constant for* u).

Proof. Let h and u denote the interfacial height and potential corresponding to the difference between two solutions. Since the problem is linear, h and u furnish a regular solution of the crystallization problem with *null data*, so that, in particular, (7.16) is satisfied and $\text{grad}\, u(x, y, 0) \equiv \mathbf{0}$, $h(x, 0) \equiv 0$. We may therefore use (7.17) to conclude that $u = u(t)$, $h = h(t)$, so that, by (7.6), h and u are constant. But $h(0) = 0$; thus $h \equiv 0$. □

8. The crystallization problem in a bounded container

If the crystallization process takes place in a bounded **container** Ω, then Ω is the union of the crystal $\mathscr{C}(t)$ and the melt $\mathscr{M}(t)$, and the **interface** between $\mathscr{C}(t)$ and $\mathscr{M}(t)$ is generally not $\partial\mathscr{C}(t)$, but rather

$$\mathscr{I}(t) = \partial\mathscr{C}(t) \cap \partial\mathscr{M}(t) \tag{8.1}$$

(Figure 4). In this case we need boundary conditions on

$$(\partial\Omega)_{\mathscr{M}}(t) = \partial\Omega(t) \cap \partial\mathscr{M}(t), \tag{8.2}$$

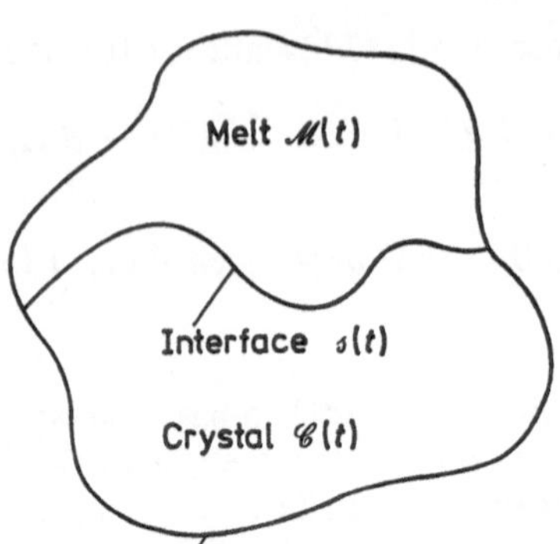

Fig. 4. Crystallization in a bounded container Ω.

the portion of the container that bounds the melt, as well as a contact condition on $\partial\mathscr{A}(t)$, the portion of the interface that meets the container. Let n denote the outward unit normal to $\partial\Omega$. We shall restrict attention to the **boundary condition**

$$v \cdot n = 0 \quad \text{on} \quad (\partial\Omega)_{\mathscr{M}} \tag{8.3}$$

in conjunction with the **contact condition**[23]

$$\mathsf{C}(\theta) \text{ parallel to } n \text{ at points of } \partial\mathscr{A}. \tag{8.4}$$

The free-boundary problem defined by the crystallization equations (5.5) and (5.6)[24], the boundary condition (8.3), and the contact condition (8.4) will be referred to as the **crystallization problem in a bounded container.**

Global balance relations. *Solutions of the crystallization problem in a bounded container satisfy*

$$\begin{gathered}(d/dt)\,\text{area}\,(\mathscr{C}) = 0,\\ (d/dt)\left\{\int_{\mathscr{A}} \psi(\theta)\,ds + \int_{\mathscr{M}} \tfrac{1}{2}\varrho v^2\,da\right\} = -\int_{\mathscr{A}} \beta(\theta)\,V^2\,ds \leqq 0.\end{gathered} \tag{8.5}$$

Proof. The first of (8.5) follows using (8.3) and the argument shown in (6.6) with $\partial\mathscr{C}$ replaced by $\mathscr{A}$ and $\mathscr{S}_r$ by $(\partial\Omega)_{\mathscr{M}}$.

The verification of $(8.5)_2$ is based on the identities:

$$\begin{gathered}(d/dt)\int_{\mathscr{A}} \psi\,ds = -\int_{\mathscr{A}} V\,\mathsf{C}_s \cdot m\,ds,\\ (d/dt)\int_{\mathscr{M}} k\,da = \int_{\mathscr{A}} [(1-\zeta)\,p - \tfrac{1}{2}\varrho_c v^2]\,V\,ds,\end{gathered} \tag{8.6}$$

with $k = \frac{1}{2}\varrho v^2$, and follows as in the paragraph containing (6.12).

Thus we have only to establish (8.6). By (2.7) and (8.4),

$$\int_{\partial\mathscr{A}} \mathsf{C} \cdot v_{\partial\mathscr{A}} = 0,$$

and this relation, (2.8), (2.11), (3.22), and (4.7), yield $(8.6)_1$. The result $(8.6)_2$ follows from (3.4), (3.5), (5.5), and (8.3) using the steps outlined in (6.9) and (6.11) with the $o(1)$ symbol not present, and with $\partial\mathscr{C}$ replaced by $\mathscr{A}$, $\mathscr{M}_r$, by $\mathscr{M}$. □

9. Viscous melt

We could also allow the melt to be an incompressible, viscous fluid. In this case the melt *stress* is given by the constitutive relation

$$T = -p\mathbf{1} + \mu\,(\text{grad}\,v + \text{grad}\,v^T), \tag{9.1}$$

[23] For convenience, I neglect contact energy between the container and the melt and between the container and the crystal. In the experiments of Keshishev, Parshin, & Babkin [KPB] the solid helium poorly wets the container, indicating the importance of including contact energy.

[24] (5.7) are here required to hold on $\mathscr{A}(t)$.

with $\mu > 0$ the (constant) melt *viscosity*, and the melt equations take the form

$$\varrho v^{\cdot} = -\operatorname{grad} p + 2\mu \Delta v, \qquad \operatorname{div} v = 0. \tag{9.2}$$

In accord with the presence of viscosity is the assumption that *the melt adhere to the crystal*, so that the tangential component of v vanishes at the interface; by (3.5), this is equivalent to the requirement that

$$v = (1 - \zeta) V m. \tag{9.3}$$

Balance of momentum now leads to an obvious counterpart of (3.10):

$$\mathsf{C}_s = T_c m - T m - \varrho_c V v. \tag{9.4}$$

Regarding the dissipation inequality (3.20) and the interactive energy-balance (3.24), we replace $-pv \cdot n$ by $v \cdot Tn$, but the end results are still (3.27) and (3.28). Here we use the fact that, because of $(9.2)_2$ and (9.3),

$$m \cdot (\operatorname{grad} v)\, m = 0, \tag{9.5}$$

so that, by (9.1),

$$m \cdot T m = -p. \tag{9.6}$$

We continue to postulate constitutive equations of the form (4.1); the resulting thermodynamic restrictions (4.3) and (4.4) remain valid.

Summarizing, the basic equations for an **incompressible, viscous melt** are:

$$\left.\begin{aligned} \varrho[v_t + (\operatorname{grad} v)\, v] &= -\operatorname{grad} p + 2\mu \Delta v, \\ \operatorname{div} v &= 0, \end{aligned}\right\} \quad \text{(melt)} \tag{9.7}$$

$$\left.\begin{aligned} [\psi(\theta) + \psi''(\theta)]\, K &= \beta(\theta)\, V + F - \tfrac{1}{2}\varrho_c v^2 + (1 - \zeta)\, p, \\ v &= (1 - \zeta) V m, \end{aligned}\right\} \quad \text{(interface)} \tag{9.8}$$

with the terms $(\operatorname{grad} v)\, v$ and $\frac{1}{2}\varrho_c v^2$ dropped in the weak inertia approximation.

For a bounded container we replace the boundary condition (8.3) by

$$v = 0 \quad \text{on} \quad (\partial\Omega)_{\mathscr{M}}. \tag{9.9}$$

Granted these changes, *the global balance relations* (6.5) *and* (8.5) *remain valid* provided we account for the additional (dissipative) term

$$-\tfrac{1}{2}\mu \int_{\mathscr{M}} |\operatorname{grad} v + \operatorname{grad} v^T|^2 \, da \tag{9.10}$$

on the right sides of $(6.5)_2$ and $(8.5)_2$.

10. Three-dimensional theory

When the crystallization process takes place in $\mathbb{R}^3$, the interface evolves as a *surface*, rather than as a curve, but apart from notation the theory is identical. Following the notation and terminology of [G1], we write ∇_{surf} for the *surface gradient*, $\boldsymbol{L} = -\nabla_{\text{surf}}\, \boldsymbol{m}$ for the *curvature tensor*, and $\boldsymbol{H} = \operatorname{tr} \boldsymbol{L}$ for twice the *mean curvature*. Then the only essential changes regarding the equations presented in Sections 5 and 9 are the replacement of $\psi(\theta)$ by $\psi(\boldsymbol{m})$, $\beta(\theta)$ by $\beta(\boldsymbol{m})$, and

$$[\psi(\theta) + \psi''(\theta)]\, K \quad \text{by} \quad \psi(\boldsymbol{m})\, H + \psi_{\boldsymbol{mm}}(\boldsymbol{m}) \cdot L, \tag{10.1}$$

where $\psi_{\boldsymbol{mm}}(\boldsymbol{m})$ is the second gradient of $\psi(\boldsymbol{m})$ on the surface of the unit ball.

In the three-dimensional theory the interfacial force is replaced by an **interfacial stress**[25] C, with $\mathsf{C}(\boldsymbol{x}, t)$ a linear transformation from the tangent space to $\partial\mathscr{C}(t)$ at $\boldsymbol{x}$ into $\mathbb{R}^3$. For crystallization in a bounded container, as discussed in Section 8, the contact condition takes the form

$$\mathsf{C}(\boldsymbol{x}, t)\, \boldsymbol{t}(\boldsymbol{x}) \quad \text{parallel to } \boldsymbol{n}(\boldsymbol{x}) \quad \text{at points } \boldsymbol{x} \in \partial\mathscr{A}(t), \tag{10.2}$$

where $\boldsymbol{t}(\boldsymbol{x})$, a vector tangent to $\mathscr{A}(t)$ at $\boldsymbol{x}$, is the outward unit normal to the boundary curve $\partial\mathscr{A}(t)$. The global balance relations (6.5) and (8.5) then hold with the obvious replacements (area $\to$ vol, $ds \to da$, $da \to dv$) privided we require, for (6.5), that

$$\boldsymbol{v}(\boldsymbol{x}, t) = o(|\boldsymbol{x}|^{-2}), \quad p(\boldsymbol{x}, t) = O(1) \quad \text{as } |\boldsymbol{x}| \to \infty, \tag{10.3}$$

estimates which imply (6.3).

11. Radial solutions for an isotropic crystal

Consider an isotropic crystal, in the shape of a sphere, undergoing spherically symmetric crystallization. For convenience, we use the weak-inertia approximation, so that the underlying equations are (5.11), $(5.12)_2$, and (5.14), as modified by (10.1). *We do not require that the far-field conditions* (10.3) *hold*, as the only spherically symmetric solution consistent with (10.3) has the melt and interface stationary. (This is as expected: because of the incompressibility of the melt, a net flow of melt at infinity is required for a nontrivial spherically symmetric solution.)

The general radial solution $u(r, t)$ of (5.11), exterior to a sphere, is

$$u(r, t) = A(t) + B(t)/r, \tag{11.1}$$

and, if the *melt pressure is constant at infinity*, with value P, then, in view of (4.9) and (5.15),

$$\begin{gathered} A(t) = -Ct, \\ C = \Psi_c + P - \zeta(\Psi + P). \end{gathered} \tag{11.2}$$

By (5.3), $C = 0$ for a flat interface at equilibrium; in fact, the sign of C is related to the statical stability of the crystal and melt: the **melt is stable or unstable relative to the crystal** according as $C \geqq 0$ or $C < 0$.[26]

---

[25] *Cf.* [G1]. The thermodynamic restrictions concerning $\mathsf{C}(\boldsymbol{m})$ are as given in [G1], eqt. $(4.5)_{2,3}$ and are strictly analogous to (4.3).

[26] I take this as a *formal definition* justified by the growth theorem. The energies Ψ_c and Ψ are measured per unit volume; the definition may be more transparent in terms of energies per unit mass; that is, in terms of $\varrho_c^{-1} C = \varrho_c^{-1}(\Psi_c + P) - \varrho^{-1}(\Psi + P)$.

The function $B(t)$ determines the melt velocity: by (5.10), the melt velocity is radial with radial component v given by

$$\varrho(\zeta - 1)\, v(r, t) = B(t)/r^2 . \tag{11.3}$$

We assume, as is natural, that the crystal density is larger than the melt density. Then

$$\zeta > 1, \tag{11.4}$$

and the velocity field is **(initially) directed outward, directed inward,** or **null** according as $B(0) > 0$, $B(0) < 0$, or $B(0) = 0$.

Since the crystal is isotropic, ψ and β are constant. Further, if $r = R(t)$ designates the interface, then the mean curvature is $-R(t)^{-1}$, while the normal velocity is $dR(t)/dt$. Thus, defining constants

$$\varphi = 2\psi\alpha > 0, \quad \gamma = C\alpha, \tag{11.5}$$

and writing

$$\tau = t/\alpha, \quad R^{\cdot} = dR/d\tau, \quad B^{\cdot} = dB/d\tau, \tag{11.6}$$

the free-boundary conditions $(5.12)_2$ and (5.14), with the change indicated by (10.1), reduce to a pair of ordinary differential equations:

$$\begin{aligned} B^{\cdot} &= \varphi + \gamma R - \beta B R^{-1}, \\ R^{\cdot} &= -BR^{-2} . \end{aligned} \tag{11.7}$$

The phase portrait for this system yields the following

Growth theorem. *Let the melt be stable relative to the crystal. Then (irrespective of the initial conditions) the crystal melts in finite time.*

Let the melt be unstable relative to the crystal.

(i) *If the melt velocity is initially null, then crystals of radius*

$$R(0) > R_{\text{crit}} = \varphi/|\gamma| = 2\psi/[\zeta(\Psi + P) - (\Psi_c + P)]$$

grow unboundedly as $t \to \infty$, crystals of radius $R(0) < R_{\text{crit}}$ melt completely in finite time.

(ii) *For any initial radius, but for a melt velocity of sufficiently large magnitude: if the melt velocity is directed outward, then the crystal melts in finite time; if the melt velocity is directed inward, then the crystal grows unboundedly as $t \to \infty$.*

Assume that the melt is *unstable* relative to the crystal, and suppose that the initial data are such that the crystal grows unboundedly as $t \to \infty$. The *large-time approximation* of (11.7) is then obtained formally by setting $\varphi = 0$ in (11.7):

$$\begin{aligned} B^{\cdot} &= \gamma R - \beta B R^{-1}, \\ R^{\cdot} &= -BR^{-2} . \end{aligned} \tag{11.8}$$

This equation has a simple solution which, when expressed in terms of t rather than τ (*cf.* (11.6)), has the form

$$R(t) = V_0 t, \quad B(t) = -\alpha V_0^3 t^2, \tag{11.9}$$

with

$$V_0 = \frac{-\beta + (\beta^2 - 8\gamma)^{\frac{1}{2}}}{4\alpha}. \tag{11.10}$$

Thus, at least formally, the normal velocity of the interface has the limiting value V_0 as $t \to \infty$. This limiting velocity has a more transparent form when $\beta = 0$; namely,

$$V_0 = \frac{\{\frac{1}{2}\varrho[\zeta(\Psi + P) - (\Psi_c + P)]\}^{\frac{1}{2}}}{\varrho_c - \varrho}. \tag{11.11}$$

Acknowledgment. The research presented here was supported by the Army Research Office and the National Science Foundation. I am greatly indebted to valuable discussions with PAOLO PODIO GUIDUGLI and ALLAN STRUTHERS.

References

[AG] ANGENENT, S., & M. E. GURTIN, Multiphase thermomechanics with interfacial structure. 2. Evolution of an isothermal interface. Arch. Rational Mech. Anal. **108**, 323–391 (1989).

[AP] ANDREEV, A. F., & A. Y. PARSHIN, Equilibrium shape and oscillations of the surface of quantum crystals, Sov. Phys. JETP **48**, 763–766 (1978).

[CBL] CASTAING, B., BALIBAR, S., & C. LAROCHE, Mobilité à 1 MHz du font de fusion de ^{4}He, J. Physique **41**, 897–903 (1980).

[CH] CAHN, J. W., & D. W. HOFFMANN, A vector thermodynamics for anisotropic surfaces. 2. Curved and faceted surfaces, Act. Metall. **22**, 1205–1214 (1974).

[CN] COLEMAN, B. D., & W. NOLL, The thermodynamics of elastic materials with heat conduction and viscosity, Arch. Rational Mech. Anal. **13**, 167–178 (1963).

[G] GJOSTEIN, N. A., Adsorption and surface energy. 2. Thermal faceting from minimization of surface energy, Act. Metall. **11**, 969–978 (1963).

[G1] GURTIN, M. E., Multiphase thermomechanics with interfacial structure. 1. Heat conduction and the capillary balance law, Arch. Rational Mech. Anal. **104**, 195–221 (1988).

[G2] GURTIN, M. E., On the two-phase Stefan problem with interfacial energy and entropy, Arch. Rational Mech. Anal. **96**, 199–241 (1986).

[G3] GURTIN, M. E., Toward a nonequilibrium thermodynamics of two phase materials, Arch. Rational Mech. Anal. **100**, 275–312 (1988).

[G4] GURTIN, M. E.. *An Introduction to Continuum Mechanics*, Academic Press (1981).

[GPG] GURTIN, M. E., & P. PODIO GUIDUGLI, A hyperbolic theory for the evolution of plane curves, Forthcoming.

[GS] GURTIN, M. E., & A. STRUTHERS, Multiphase thermomechanics with interfacial structure. 3. Evolving phase boundaries in the presence of bulk deformation, Arch. Rational Mech. Anal. Forthcoming.

[KPB] KESHISHEV, K. O., PARSHIN, A. Y., & A. V. BABKIN, Experimental detection of crystallization waves in He^4, JETP Lett. **30**, 56–59 (1979).

[MA] MARIS, J. H., & A. F. ANDREEV, The surface of crystalline helium 4, Phys. Today, 25–30 (Feb. 1987).

[R] ROGERS, R. C., Notes on a linear model of mechanical phase transitions, Forthcoming.

Department of Mathematics
Carnegie Mellon University
Pittsburgh, Pennsylvania

(Received September 13, 1989)

On Edge Interactions and Surface Tension

WALTER NOLL & EPIFANIO G. VIRGA

Dedicated to Bernard D. Coleman on the occasion of his sixtieth birthday

Contents

Introduction

The idea of a "line distribution of force over the edges of a body" seems to have occurred first to TOUPIN in 1962 (see p. 403 of [To1]). However, his treatment suffered from the same defect as most treatments of distributions of forces prior to 1960, namely these distributions were only implicit in formulas for resultant forces. The systems of forces giving rise to these resultants were not brought into the open. (See the remark by C. TRUESDELL in "General References" on p. 156 of [Tr].) The present paper is a first attempt to bring into the open the systems of forces that give rise to distributions of forces over edges.

The problem of how to give a precise definition of an edge interaction and TOUPIN's early work on this problem were brought to our attention by MAURIZIO VIANELLO in a discussion in Pisa in June 1987. During this discussion, the three of us came up with the germ of an idea from which the concept of regions in edge contact as described by Definition 4 of Section 3 and the corresponding Assumption I of Section 5 were developed later.

A precise concept of a "system of forces" was first proposed by one of us in 1957 [N1] and has been extensively developed since. GURTIN & WILLIAMS realized in 1967 [GW] that an analogous concept describes "systems of heat fluxes". Both kinds of systems were later subsumed under the abstract concept of an "interaction" (see Section 8 in [N2]). A definition of "interaction" that is suitable for the present context is presented in Section 4. In Sections 5 and 6 we show how "edge interactions" can be identified as special contact interactions. In Section 7 we derive the consequences of balance laws when edge interactions are present and when certain *ad hoc* assumptions are made. In Section 8 we discuss the boundary conditions that apply when edge interactions are present. In Section 9 we show how some of TOUPIN's results fit into the scheme of the present paper. Finally, we show how surface tension can be viewed as the manifestation of the presence of certain edge interactions.

Mathematicians often use terms such as "region with piecewise smooth boundary" without any further explanation (see *e.g.* axiom (S2) in [N1]), and they often give inequivalent and even obviously unintended precise definitions when challenged to make this concept precise (see Appendix II of [F]). In the context of the present paper it is crucial to have a very precise definition that does not exclude "cusped" edges. Sections 1, 2, and 3 deal with this and related issues. Our definition of "regular region" seems to be equivalent to the one given by KELLOGG in [K], p. 112. We also need to consider a somewhat more restricted concept, for which we use the term "biregular region".

What we call "edge interactions" should not be confused with external actions concentrated along curves; our theory is not related to theories of "stress-concentration" in any sense. An edge interaction between two internal parts of a given body, in the sense of the present paper, can occur only when a side of one of the parts is in "everywhere-cusped" contact with a side of the other part along a common edge of the two parts (see Definition 2 of Section 3). The additivity properties of interactions then show that if the edge is a "cusped edge" of one of the two parts, the interaction must be zero. It is for this reason that one must pay careful attention to "cusped edges", and that one can only consider what we call "biregular" contact (see Definition 4 of Section 3).

There are at least two issues which we could not settle. The first concerns the Assumptions I, II, III, and IV in Sections 5, 6, and 7. It would be desirable to derive them from other assumptions that have more transparent and natural physical interpretations. For classical surface interactions, this issue has been studied at length only recently (see [S] and [GWZ] and the literature cited there). For edge interactions, the issue is likely to be much more difficult. The second issue concerns a class of regions appropriate to represent parts of a given body (see conditions (i)–(iv) of Section 4). We do not specify explicitly such a class here, and we describe our interactions only for special situations. Even for the case of classical surface interaction, we found only recently a class (we call its members "fit regions") that we consider completely appropriate (see [NV]). The problem of finding a class that is appropriate when edge-interactions are present is likely to be extremely difficult and to require concepts from geometric measure theory that have not yet been invented.

The present paper does not consider at all the type of constitutive assumptions that should govern edge interactions. For a long time there has been evidence that non-simple materials cannot be accommodated by classical surface interactions alone and that edge interactions must necessarily occur in materials of second grade. Indeed, it seems to us that it was VIANELLO's interest in the theory of materials of second grade that motivated him to urge re-examination of TOUPIN's earlier work. We hope that the present paper will open a way to deal with these matters.

Notation and Terminology

Generally we use the notation and terminology of [FDSI]. For example, $\mathbb{P}$ denotes the set of all positive reals including zero and $\mathbb{P}^\times := \mathbb{P} \setminus \{0\}$ denotes the set of all strictly positive reals. The range of a mapping f is denoted by $\operatorname{Rng} f$. If A is a subset of the domain of a mapping f, then $f|_A$ denotes the restriction of f to A. If B is a set that includes the range of a mapping f, then $f|^B$ is obtained from f by changing its codomain to B. If B is a subset of a given set A, then $1_{B\subset A}$ denotes the *inclusion-mapping* from B to A, *i.e.*, the mapping whose domain is B, whose codomain is A, and whose values are given by $1_{B\subset A}(x) = x$ for all $x \in B$.

If $\mathscr{V}$ and $\mathscr{W}$ are linear spaces, then $\operatorname{Lin}(\mathscr{V}, \mathscr{W})$ denotes the set of all linear mappings from $\mathscr{V}$ to $\mathscr{W}$. The *dual space* $\mathscr{W}^*$ of $\mathscr{W}$ is defined by $\mathscr{W}^* := \operatorname{Lin}(\mathscr{W}, \mathbb{R})$.

When we use the terms "inner-product space" and "Euclidean space", it is understood that they are *genuine* in the sense of the definitions in Chapter 4 of [FDSI]. Let $\mathscr{V}$ be an inner-product space. The unit ball of $\mathscr{V}$ is denoted by $\operatorname{Ubl}\mathscr{V} := \{v \in \mathscr{V} \mid |v| < 1\}$ and the unit-sphere of $\mathscr{V}$ by $\operatorname{Usph}\mathscr{V} := \{v \in \mathscr{V} \mid |v| = 1\}$. If $\mathscr{S}$ is a subset of $\mathscr{V}$, then $\mathscr{S}^\perp := \{v \in \mathscr{V} \mid v \cdot u = 0 \text{ for all } u \in \mathscr{S}\}$.

We often abbreviate $\mathscr{U} := \operatorname{Usph}\mathscr{V}$ and use the notation

$$(\mathscr{U}^2)_\perp := \{(u, w) \in \mathscr{U}^2 \mid u \cdot w = 0\}$$

for the set of all perpendicular pairs of unit vectors. It is easily seen that $(\mathscr{U}^2)_\perp$ is a 3-dimensional analytic manifold imbedded in the 6-dimensional inner-product space $\mathscr{V}^2$.

The gradient of a mapping φ at a point x in the domain of φ is denoted by $\nabla_x\varphi$ if meaningful. Similarly, $\operatorname{div}_x h$ denotes the divergence of a vector field h at a given point x of the domain of h.

Let $\mathscr{D}$ be a subset of a given Euclidean space $\mathscr{E}$. The interior, closure, and boundary of $\mathscr{D}$ are denoted by $\operatorname{Int}\mathscr{D}$, $\operatorname{Clo}\mathscr{D}$, and $\operatorname{Bdy}\mathscr{D}$, respectively. We say that $\mathscr{D}$ is *regularly open* if $\mathscr{D} = \operatorname{IntClo}\mathscr{D}$.

If n and m are integers, then $n \,..\, m$ denotes the set of all integers k such that $n \leqq k \leqq m$. We abbreviate $n^] := 1 \,..\, n$, $n^[:= 0 \,..\, (n-1)$.

To distinguish between integrals with respect to volume, surface-area, and curve-length, we write dv, da, or dl, respectively, after the integrand.

1. Regular regions, surfaces, and curves

We assume that a 3-dimensional Euclidean space $\mathscr{E}$ with translation space $\mathscr{V} := \mathscr{E} - \mathscr{E}$ is given.

Let a bounded connected C^2-manifold $\mathscr{M}$ imbedded in $\mathscr{E}$ be given (see Chapter 3 of [FDSII]). If $\dim \mathscr{M} = 3$ then $\mathscr{M}$ is just a **region**, *i.e.* a connected open set.

If $\dim \mathcal{M} = 2$, we call $\mathcal{M}$ a **surface**, and if $\dim \mathcal{M} = 1$, we call $\mathcal{M}$ a **curve**. If $\dim \mathcal{M} = 0$, then $\mathcal{M}$ is just a singleton.

We use the notation

$$\mathbf{E}_{\mathcal{M}} : \mathcal{M} \to \operatorname{Lin} \mathcal{V} \tag{1.1}$$

for the mapping whose value at $x \in \mathcal{M}$ is the symmetric idempotent whose range is the tangent space $\operatorname{Tan}_{\mathcal{M}}(x)$ to $\mathcal{M}$ at x, so that

$$\mathbf{E}_{\mathcal{M}}(x)^2 = \mathbf{E}_{\mathcal{M}}(x), \qquad \mathbf{E}_{\mathcal{M}}(x)^T = \mathbf{E}_{\mathcal{M}}(x) \tag{1.2}$$

and

$$\operatorname{Rng} \mathbf{E}_{\mathcal{M}}(x) = \operatorname{Tan}_{\mathcal{M}}(x) \quad \text{for all } x \in \mathcal{M}. \tag{1.3}$$

The **border** of $\mathcal{M}$ is defined by

$$\operatorname{Bo} \mathcal{M} := \operatorname{Clo} \mathcal{M} \setminus \mathcal{M}. \tag{1.4}$$

The border of $\mathcal{M}$ coincides with the boundary $\operatorname{Bdy} \mathcal{M}$ of $\mathcal{M}$ if $\dim \mathcal{M} = 3$, *i.e.* if $\mathcal{M}$ is a region. If $\dim \mathcal{M} < 3$, then $\operatorname{Bdy} \mathcal{M} = \operatorname{Clo} \mathcal{M}$.

Definition 1. *We say that the bounded connected C^2-manifold $\mathcal{M}$ is* **regular** *if* Clo $\mathcal{M}$ *has a finite partition $\mathfrak{P}$ with the following properties.*

(i) *Every piece of $\mathfrak{P}$ is a connected C^2-manifold.*

(ii) $\mathcal{M} \in \mathfrak{P}$.

(iii) *For every piece $\mathcal{P} \in \mathfrak{P}$, $\mathbf{E}_{\mathcal{P}}$ has a continuous extension to* Clo $\mathcal{P}$. *Using poetic license, we denote this extension also by*

$$\mathbf{E}_{\mathcal{P}} : \operatorname{Clo} \mathcal{P} \to \operatorname{Lin} \mathcal{V}. \tag{1.5}$$

(iv) *For every $\mathcal{P} \in \mathfrak{P}$ and every $\mathcal{C} \in \mathfrak{P}$ with $\mathcal{C} \subset \operatorname{Bo} \mathcal{P}$ and $\dim \mathcal{P} = \dim \mathcal{C} + 2$, there are exactly two pieces $\mathcal{S} \in \mathfrak{P}$ such that*

$$\mathcal{S} \subset \operatorname{Bo} \mathcal{P}, \qquad \mathcal{C} \subset \operatorname{Bo} \mathcal{S}. \tag{1.6}$$

We call a finite partition of Clo $\mathcal{M}$ *with the properties* (i)–(iv) a **regular partition for** $\mathcal{M}$. *For every $k \in 0..3$, we denote by $\mathfrak{P}_k$ the set of all manifolds in $\mathfrak{P}$ that have dimension k.*

Let a regular manifold $\mathcal{M}$ and a regular partition $\mathfrak{P}$ for $\mathcal{M}$ be given. The only piece of $\mathfrak{P}$ of dimension $m := \dim \mathcal{M}$ is $\mathcal{M}$, *i.e.*

$$\mathfrak{P}_m = \{\mathcal{M}\} \quad \text{when } m := \dim \mathcal{M}. \tag{1.7}$$

We have

$$\operatorname{Bo} \mathcal{M} = \bigcup_{k \in m[} \bigcup \mathfrak{P}_k. \tag{1.8}$$

Every piece $\mathcal{P}$ of $\mathfrak{P}$ is again a regular manifold and $\{\mathcal{S} \in \mathfrak{P} \mid \mathcal{S} \subset \operatorname{Clo} \mathcal{P}\}$ is a regular partition of $\mathcal{P}$. We define the tangent space to $\mathcal{M}$ at a point $x \in \operatorname{Bo} \mathcal{M}$ by $\operatorname{Tan}_{\mathcal{M}}(x) := \operatorname{Rng} \mathbf{E}_{\mathcal{M}}(x)$, using the extended $\mathbf{E}_{\mathcal{M}}$ of (1.5). Then (1.3) becomes valid for all $x \in \text{Clo}\,\mathscr{M}$. If $\mathscr{M}$ is a **regular region** (*i.e.* if $\dim \mathscr{M} = 3$), we call the surfaces in $\mathfrak{P}_2$ **sides** of $\mathscr{M}$. If $\mathscr{M}$ is a regular region or a **regular surface** (*i.e.* if $\dim \mathscr{M} = 2$), we call the curves in $\mathfrak{P}_1$ **edges** and the singletons in $\mathfrak{P}_0$ **vertices** of $\mathscr{M}$. The only member of one of the latter is called a **vertex-point**. The condition (iv) of Definition 1 may then be restated as follows: Each edge $\mathscr{C}$ of a regular region is included in the border of exactly two sides; we call them the **sides adjacent** to $\mathscr{C}$. Each side of a regular region is a regular surface. Each vertex-point x of a regular surface belongs to the border of exactly two edges; we call them the **edges adjacent** to x. The concepts of side, edge, vertex, and vertex-point depend, of course, not only on $\mathscr{M}$ but also on the given regular partition $\mathfrak{P}$ of $\mathscr{M}$. Usually no ambiguity will arise if the appropriate partition $\mathfrak{P}$ is not explicitly described.

Remark 1. A regular manifold of dimension 1 or greater always admits infinitely many regular partitions. Indeed given such a partition, new ones can always be obtained by subdivision. Of course, there always are regular partitions with a minimum number of pieces, but there may be more than one such partition.

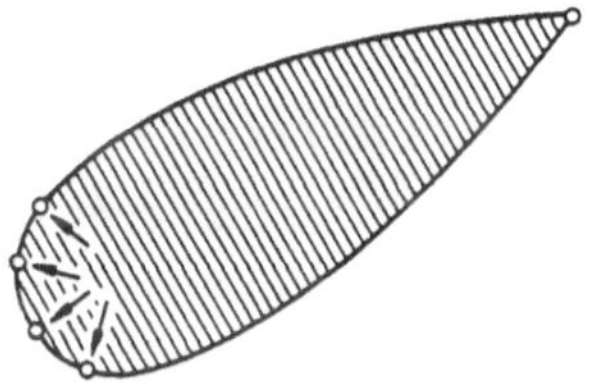

Figure 1

Figure 1 shows a regular surface with this property. A regular partition for the surface must have at least 5 pieces (the surface itself, two edges, and two vertices). There are infinitely many regular partitions with 5 pieces because one of the vertex-points can be put into infinitely many positions. □

Definition 2. Let $\mathscr{M}$ *be a regular manifold. Then the* **reduced border** Rbo $\mathscr{M}$ of $\mathscr{M}$ *is defined to be the union of all regular manifolds of dimension* $\dim \mathscr{M} - 1$ *that are included in* Bo $\mathscr{M}$.

Let $\mathfrak{P}$ be a regular partition of a given regular manifold $\mathscr{M}$ with $m := \dim \mathscr{M}$. Clearly, we then have

$$\bigcup \mathfrak{P}_{m-1} \subset \text{Rbo}\,\mathscr{M}. \tag{1.9}$$

The example given in Remark 1 shows that it may not be possible to find a partition $\mathfrak{P}$ such that the inclusion (1.9) reduces to equality. However, one can easily show that for every $x \in \text{Rbo}\,\mathscr{M}$ one can find a regular partition $\mathfrak{P}$ for $\mathscr{M}$ such that $x \in \mathscr{P}$ for some $\mathscr{P} \in \mathfrak{P}_{m-1}$.

Remark 2. Let $\mathscr{R}$ be a regular region. Then $\mathscr{R}$ is a **fit region** in the sense described in [NV]. Also, the reduced border of $\mathscr{R}$ is a subset of the reduced boundary of $\mathscr{R}$ as defined in [NV]: $\text{Rbo}\,\mathscr{R} \subset \text{Rby}\,\mathscr{R}$. Moreover, the set-difference $\text{Rby}\,\mathscr{R} \setminus \text{Rbo}\,\mathscr{R}$ is a set of area-measure zero. □

Proposition 1. *Let a regular manifold $\mathcal{M}$ with $\dim \mathcal{M} \geqq 1$ and $x \in \operatorname{Rbo} \mathcal{M}$ be given. Then there is exactly one $\boldsymbol{u} \in \operatorname{Usph} \mathcal{V}$ such that*
(i) $\boldsymbol{u} \in (\operatorname{Tan}_{\mathcal{K}}(x))^{\perp} \cap \operatorname{Tan}_{\mathcal{M}}(x)$ *for every regular manifold $\mathcal{K}$ of dimension $\dim \mathcal{M} - 1$ that contains x and is included in $\operatorname{Bo} \mathcal{M}$.*
(ii) *There is a $\delta \in \mathrm{P}^{\times}$ such that*

$$(x + (\operatorname{Tan}_{\mathcal{M}}(x))^{\perp} + \mathrm{P}\boldsymbol{u}) \cap \mathcal{M} \cap (x + \delta \operatorname{Ubl} \mathcal{V}) = \emptyset. \tag{1.10}$$

Rather than giving a formal proof of Proposition 1, we illustrate its geometric meaning in Figure 2 for the case when $\dim \mathcal{M} = 2$. We note that $\mathcal{H} := x + (\operatorname{Tan}_{\mathcal{M}}(x))^{\perp} + \mathrm{P}\boldsymbol{u}$ is, in this case, the half-plane that touches $\mathcal{M}$ at x and is perpendicular to both the tangent-plane to $\mathcal{M}$ at x and to the edge $\mathcal{K}$ that contains x and is included in $\operatorname{Bo} \mathcal{M}$.

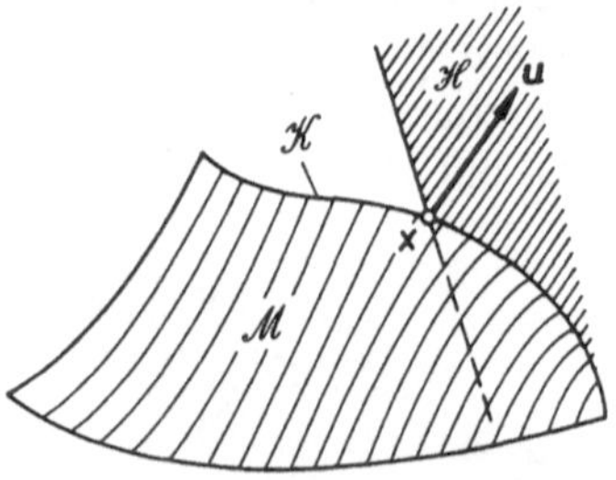

Figure 2

Definition 3. *Let x be a point on the reduced border of a regular manifold $\mathcal{M}$. We call the unit vector $\boldsymbol{u}$ described in Proposition 1 the* **outer unit normal** *to $\mathcal{M}$ at x. The function that associates this outer unit normal with $x \in \operatorname{Rbo} \mathcal{M}$ will be denoted by*

$$\boldsymbol{v}_{\mathcal{M}} : \operatorname{Rbo} \mathcal{M} \to \operatorname{Usph} \mathcal{V}. \tag{1.11}$$

We note that $\boldsymbol{v}_{\mathcal{M}}$ is continuous and that

$$\boldsymbol{E}_{\mathcal{M}}(x)\, \boldsymbol{v}_{\mathcal{M}}(x) = \boldsymbol{v}_{\mathcal{M}}(x) \quad \text{for all } x \in \operatorname{Rbo} \mathcal{M}. \tag{1.12}$$

If $\mathfrak{P}$ is a regular partition for $\mathcal{M}$ we have

$$\boldsymbol{E}_{\mathcal{S}}(x)\, \boldsymbol{v}_{\mathcal{M}}(x) = \boldsymbol{0} \quad \text{for all } x \in \mathcal{S} \tag{1.13}$$

and all $\mathcal{S} \in \mathfrak{P}_{m-1}$ when $m := \dim \mathcal{M}$.

Let a regular region $\mathcal{R}$ and a regular partition $\mathfrak{P}$ for $\mathcal{R}$ be given. Also, let $\mathcal{S}$ be a side of $\mathcal{R}$ belonging to $\mathfrak{P}$, *i.e.* $\mathcal{S} \in \mathfrak{P}_2$. It follows from condition (iii) of Definition 1, applied to $\mathcal{S}$, that $\boldsymbol{v}_{\mathcal{R}|\mathcal{S}} : \mathcal{S} \to \operatorname{Usph} \mathcal{V}$ has a continuous extension to $\operatorname{Clo} \mathcal{S}$. We denote this extension by

$$\boldsymbol{v}_{\mathcal{R};\mathcal{S}} : \operatorname{Clo} \mathcal{S} \to \operatorname{Usph} \mathcal{V}. \tag{1.14}$$

It follows from Definition 3 and condition (i) of Proposition 1 that

$$\boldsymbol{v}_{\mathcal{S}}(x) \cdot \boldsymbol{v}_{\mathcal{R};\mathcal{S}}(x) = 0 \quad \text{for all } x \in \operatorname{Rbo} \mathcal{S}. \tag{1.15}$$

Let $\mathcal{C}$ be an edge of $\mathcal{R}$ belonging to $\mathfrak{P}$, *i.e.* $\mathcal{C} \in \mathfrak{P}_1$, and let $\mathcal{S}$ and $\mathcal{S}'$ be the two sides of $\mathcal{R}$ adjacent to $\mathcal{C}$. Given $x \in \mathcal{C}$, we have $\boldsymbol{v}_{\mathcal{R};\mathcal{S}}(x) = \boldsymbol{v}_{\mathcal{R};\mathcal{S}'}(x)$ only in the exceptional case when x belongs to the reduced boundary of $\mathcal{R}$.

2. Functions on manifolds, surface divergence

Let $\mathscr{E}$ and $\mathscr{V}$ and a manifold $\mathscr{M}$ as in Section 1 be given. Also, let a finite-dimensional linear space $\mathscr{W}$ be given. If a function $\eta : \mathscr{M} \to \mathscr{W}$ is differentiable one can define, for each $x \in \mathscr{M}$, the gradient $\nabla_x \eta \in \operatorname{Lin}(\operatorname{Tan}_{\mathscr{M}}(x), \mathscr{W})$, of η at x (see, *e.g.*, (33.4) of [FDSII]).

Definition 1. *Let* $\eta : \mathscr{M} \to \mathscr{W}$ *be differentiable. We then define*

$$\nabla \eta : \quad \mathscr{M} \to \operatorname{Lin}(\mathscr{V}, \mathscr{W})$$

by

$$\nabla \eta(x) := \nabla_x \eta := \nabla_x \eta \, \boldsymbol{E}_{\mathscr{M}}(x)|^{\operatorname{Tan}_{\mathscr{M}}(x)} \quad \text{for all } x \in \mathscr{M}. \tag{2.1}$$

We say that η *is* **of class** C^1 *if* $\nabla \eta$ *is continuous. If* $\mathscr{M}$ *is a surface*, we call $\nabla \eta$ the **surface-gradient** of η.

We say *that a function* $\eta : \operatorname{Clo} \mathscr{M} \to \mathscr{W}$ *is* **of class** C^1 *if* $\eta|_{\mathscr{M}}$ *is of class* C^1 *and if* $\nabla(\eta|_{\mathscr{M}}) : \mathscr{M} \to \operatorname{Lin}(\mathscr{V}, \mathscr{W})$ *has a continuous extension to* $\operatorname{Clo} \mathscr{M}$, *which will then be denoted by* $\nabla \eta$.

We note that, if $\eta : \mathscr{M} \to \mathscr{W}$ is differentiable, we have

$$\operatorname{Tan}_{\mathscr{M}}(x)^{\perp} \subset \operatorname{Null} \nabla_x \eta \quad \text{for all } x \in \mathscr{M}. \tag{2.2}$$

Definition 2. *Let* $\mathscr{S}$ *be a surface of class* C^2 *and let* $\boldsymbol{h} : \mathscr{S} \to \mathscr{V}$ *be differentiable. The* **surface-divergence**

$$\operatorname{div}_{\mathscr{S}} \boldsymbol{h} : \quad \mathscr{S} \to \mathbb{R} \tag{2.3}$$

is defined to be the value-wise trace of $\nabla \boldsymbol{h}$, i.e.,

$$(\operatorname{div}_{\mathscr{S}} \boldsymbol{h})(x) := \operatorname{tr}(\nabla \boldsymbol{h}(x)) \quad \text{for all } x \in \mathscr{S}. \tag{2.4}$$

If $\boldsymbol{H} : \mathscr{S} \to \operatorname{Lin}(\mathscr{V}, \mathscr{W})$ *is differentiable then its* **surface-divergence**

$$\operatorname{div}_{\mathscr{S}} \boldsymbol{H} : \quad \mathscr{S} \to \mathscr{W} \tag{2.5}$$

is characterized by

$$\omega(\operatorname{div}_{\mathscr{S}} \boldsymbol{H}(x)) = \operatorname{div}_{\mathscr{S}}(\boldsymbol{H}^T \omega)(x) \quad \text{for all } x \in \mathscr{S}, \omega \in \mathscr{W}^*. \tag{2.6}$$

(Compare this with the definitions of divergence in Section 67 of [FDSI].)

Definition 3. *Let* $\mathscr{S}$ *be a surface of class* C^2. *We say that a function* $\boldsymbol{h} : \mathscr{S} \to \mathscr{V}$ *is* **tangential** *if* $\boldsymbol{h}(x) \in \operatorname{Tan}_{\mathscr{S}}(x)$ *for all* $x \in \mathscr{S}$. *We say that a function* $\boldsymbol{H} : \mathscr{S} \to \operatorname{Lin}(\mathscr{V}, \mathscr{W})$ *is* **tangential** *if*

$$\operatorname{Tan}_{\mathscr{S}}(x)^{\perp} \subset \operatorname{Null} \boldsymbol{H}(x) \quad \text{for all } x \in \mathscr{S}. \tag{2.7}$$

From use of (22.9) of [FDSI], it is evident that $\boldsymbol{H} : \mathscr{S} \to \operatorname{Lin}(\mathscr{V}, \mathscr{W})$ is tangential if and only if $\boldsymbol{H}^T \omega : \mathscr{S} \to \mathscr{V}$ is tangential for every $\omega \in \mathscr{W}^*$. If $\eta : \mathscr{S} \to \mathscr{W}$ is differentiable, then $\nabla \eta : \mathscr{S} \to \operatorname{Lin}(\mathscr{V}, \mathscr{W})$ is tangential by (2.2).

The following result is an easy consequence of the classical Stokes Theorem (a proof will be given in [FDSII]).

Surface-divergence Theorem. *Let* $\mathscr{S}$ *be a regular surface* (in the sense of Definition 1 of Section 1). *For every function* $\boldsymbol{h}: \operatorname{Clo}\mathscr{S} \to \mathscr{V}$ *that is of class* C^1 (in the sense of Definition 1) *and tangential* (in the sense of Definition 3) *we have*

$$\int_{\mathscr{S}} \operatorname{div}_{\mathscr{S}} \boldsymbol{h}\, da = \int_{\operatorname{Rbo}\mathscr{S}} (\boldsymbol{h} \cdot \boldsymbol{v}_{\mathscr{S}})\, dl, \tag{2.8}$$

where $\boldsymbol{v}_{\mathscr{S}}$ *is the outer unit normal function of* $\mathscr{S}$ (as defined by Definition 3 of Section 1).

Using (2.6), one immediately obtains the following

Corollary. *Let* $\mathscr{S}$ *be a regular surface and* $\mathscr{W}$ *a finite-dimensional linear space. For every tangential function* $\boldsymbol{H}: \operatorname{Clo}\mathscr{S} \to \operatorname{Lin}(\mathscr{V}, \mathscr{W})$ *of class* C^1 *we have*

$$\int_{\mathscr{S}} \operatorname{div}_{\mathscr{S}} \boldsymbol{H}\, da = \int_{\operatorname{Rbo}\mathscr{S}} (\boldsymbol{H}\boldsymbol{v}_{\mathscr{S}})\, dl. \tag{2.9}$$

3. Contact of surfaces and of regions

Definition 1. *Let* $\mathscr{M}$ and $\mathscr{M}'$ *be regular manifolds with* $\dim \mathscr{M} = \dim \mathscr{M}' \geqq 1$. *The* **contact** *of* $\mathscr{M}$ *and* $\mathscr{M}'$ *is defined by*

$$\operatorname{Ctc}(\mathscr{M}, \mathscr{M}') := \operatorname{Bo}\mathscr{M} \cap \operatorname{Bo}\mathscr{M}' \tag{3.1}$$

and the **reduced contact** *by*

$$\operatorname{Rtc}(\mathscr{M}, \mathscr{M}') := \operatorname{Rbo}\mathscr{M} \cap \operatorname{Rbo}\mathscr{M}'. \tag{3.2}$$

If $\mathscr{C}$ *is a regular manifold included in* $\operatorname{Ctc}(\mathscr{M}, \mathscr{M}')$, *we say that* $\mathscr{M}$ *and* $\mathscr{M}'$ *are* **in contact along** $\mathscr{C}$; *if* $\mathscr{C} \subset \operatorname{Rtc}(\mathscr{M}, \mathscr{M}')$ *we say that* $\mathscr{M}$ *and* $\mathscr{M}'$ *are* **in smooth contact** *along* $\mathscr{C}$.

Definition 2. *Let* $\mathscr{S}$ *and* $\mathscr{S}'$ *be regular surfaces. We say that* $\mathscr{S}$ *and* $\mathscr{S}'$ *are* **in cusped contact** *at a given* $x \in \operatorname{Rtc}(\mathscr{S}, \mathscr{S}')$ *if*

$$\boldsymbol{v}_{\mathscr{S}}(x) = \boldsymbol{v}_{\mathscr{S}'}(x). \tag{3.3}$$

We say that $\mathscr{S}$ *and* $\mathscr{S}'$ *are* **in everywhere-cusped contact** *along a given curve* $\mathscr{C} \subset \operatorname{Rtc}(\mathscr{S}, \mathscr{S}')$ *if* (3.3) *holds for all* $x \in \mathscr{C}$. *We say that* $\mathscr{S}$ *and* $\mathscr{S}'$ *are in* **nowhere-cusped contact** *along* $\mathscr{C}$ *if*

$$\boldsymbol{v}_{\mathscr{S}}(x) \neq \boldsymbol{v}_{\mathscr{S}'}(x) \tag{3.4}$$

for all $x \in \mathscr{C}$.

Every regular surface is in everywhere-cusped contact with itself along each of its edges. Of course, if $\mathscr{S}$ and $\mathscr{S}'$ are regular surfaces in contact along a curve $\mathscr{C}$, (3.3) may hold for some $x \in \mathscr{C}$ while (3.4) holds for other $x \in \mathscr{C}$. Then $\mathscr{S}$ and $\mathscr{S}'$ are neither in everywhere-cusped nor in nowhere-cusped contact along $\mathscr{C}$.

Definition 3. *We say that a subset* $\mathscr{R}$ *of* $\mathscr{E}$ *is a* **biregular region** *if it is a regular region and if there is a regular partition* $\mathfrak{P}$ *for* $\mathscr{R}$ *such that for every edge* $\mathscr{C}$ *in* $\mathfrak{P}$, *the two sides in* $\mathfrak{P}$ *adjacent to* $\mathscr{C}$ *are either in everywhere-cusped contact or in nowhere-cusped contact along* $\mathscr{C}$. *We call such a partition a* **biregular partition** *for* $\mathscr{R}$. *We say that an edge* $\mathscr{C}$ *in* $\mathfrak{P}$ *is a* **cusped edge** *if the two sides in* $\mathfrak{P}$ *adjacent to* $\mathscr{C}$ *are in everywhere-cusped contact along* $\mathscr{C}$.

Remark 1. Given a regular partition $\mathfrak{P}$ for a biregular region $\mathscr{R}$, one can always find a refinement of $\mathfrak{P}$ that is biregular partition for $\mathscr{R}$. □

Definition 4. *Let* $\mathscr{R}$ *and* $\mathscr{R}'$ *be disjoint biregular regions. We say that* $\mathscr{R}$ *and* $\mathscr{R}'$ *are in* **biregular contact** *if one can find biregular partitions* $\mathfrak{P}$ *and* $\mathfrak{P}'$ *for* $\mathscr{R}$ *and* $\mathscr{R}'$, *respectively, such that*

$$\operatorname{Ctc}(\mathscr{R}, \mathscr{R}') = \bigcup (\mathfrak{P} \cap \mathfrak{P}') \tag{3.5}$$

and if for every $\mathscr{C} \in \mathfrak{P}_1 \cap \mathfrak{P}_1'$, *every* $\mathscr{S} \in \mathfrak{P}_2$, *and every* $\mathscr{S}' \in \mathfrak{P}_2'$ *such that* $\mathscr{C} \subset \operatorname{Rtc}(\mathscr{S}, \mathscr{S}')$, *the surfaces* $\mathscr{S}$ *and* $\mathscr{S}'$ *are either in everywhere-cusped or in nowhere-cusped contact along* $\mathscr{C}$.

We say that $\mathscr{R}$ *and* $\mathscr{R}'$ *are in* **simple surface contact** *along a given regular surface* $\mathscr{S}$ *if they are in biregular contact and if the partitions* $\mathfrak{P}$ *and* $\mathfrak{P}'$ *above can be chosen such that*

$$\mathfrak{P}_2 \cap \mathfrak{P}_2' = \{\mathscr{S}\} \quad \textit{and} \quad \operatorname{Ctc}(\mathscr{R}, \mathscr{R}') = \operatorname{Clo} \mathscr{S}. \tag{3.6}$$

We say that $\mathscr{R}$ *and* $\mathscr{R}'$ *are in* **simple edge contact** *along a given regular curve* $\mathscr{C}$ *if they are in biregular contact and if the partitions* $\mathfrak{P}$ *and* $\mathfrak{P}'$ *above can be chosen such that*

$$\mathfrak{P}_1 \cap \mathfrak{P}_1' = \{\mathscr{C}\} \quad \textit{and} \quad \operatorname{Ctc}(\mathscr{R}, \mathscr{R}') = \operatorname{Clo} \mathscr{C}. \tag{3.7}$$

Partitions $\mathfrak{P}$ *and* $\mathfrak{P}'$ *that satisfy the conditions* (3.5), (3.6), *or* (3.7), *as appropriate, will be called* **appropriate partitions.**
If $\mathscr{R}$ and $\mathscr{R}'$ are in biregular contact and if $\mathfrak{P}$ and $\mathfrak{P}'$ are appropriate partitions, we have

$$\bigcup (\mathfrak{P}_2 \cap \mathfrak{P}_2') \subset \operatorname{Rtc}(\mathscr{R}, \mathscr{R}'). \tag{3.8}$$

Remark 2. One can easily define a concept of **regular contact** of disjoint regular regions by a definition analogous to (and simpler than) Definition 4. At the present time however, we do not know how to deal with contact interactions when the contact is only regular and not biregular. □

Definition 5. *Let* $\mathscr{R}$ *be a biregular region. We say that a subset* $\mathscr{P}$ *of* $\mathscr{R}$ *is a* **section** *of* $\mathscr{R}$ *if*
(i) $\mathscr{P}$ *is a biregular region,*
(ii) $\operatorname{Int}(\mathscr{R} \setminus \mathscr{P})$ *is a biregular region,*
(iii) $\mathscr{P}$ *is in biregular contact with* $\operatorname{Int}(\mathscr{R} \setminus \mathscr{P})$.

Let $\mathscr{R}$ be a biregular region and let $\mathscr{P}$ be a biregular region included in $\mathscr{R}$. If $\operatorname{Clo} \mathscr{P} \subset \mathscr{R}$ then $\mathscr{P}$ is necessarily a section of $\mathscr{R}$, but if $\operatorname{Clo} \mathscr{P} \not\subset \mathscr{R}$, then $\mathscr{P}$ need not be a section of $\mathscr{R}$.

4. Contact interactions

We assume now that a biregular bounded region $\mathscr{B}$ in $\mathscr{E}$ and a collection Ω of regularly open subsets of $\mathscr{B}$ with the following properties are given:
(i) All sections of $\mathscr{B}$ (see Definition 5 of Section 3) belong to Ω.
(ii) The intersection of any two members of Ω belongs to Ω, *i.e.*

$$\mathscr{P} \cap \mathscr{Q} \in \Omega \quad \text{for all } \mathscr{P}, \mathscr{Q} \in \Omega.$$

(iii) The **join** of any two members of Ω belongs to Ω, *i.e.*

$$\mathscr{P} \vee \mathscr{Q} := \operatorname{Int} \operatorname{Clo}(\mathscr{P} \cup \mathscr{Q}) \in \Omega \quad \text{for all } \mathscr{P}, \mathscr{Q} \in \Omega. \tag{4.1}$$

(iv) The **exterior in** $\mathscr{B}$ of any member of Ω belongs to Ω, *i.e.*

$$\mathscr{P}^b := \operatorname{Int}(\mathscr{B} \setminus \mathscr{P}) \in \Omega \quad \text{for all } \mathscr{P} \in \Omega. \tag{4.2}$$

We call the members of Ω the **parts** of $\mathscr{B}$. We denote the set of all disjoint pairs of parts of $\mathscr{B}$ by

$$(\Omega^2)_{\text{dis}} := \{(\mathscr{P}, \mathscr{Q}) \in \Omega^2 \mid \mathscr{P} \cap \mathscr{Q} = \emptyset\}. \tag{4.3}$$

The set of all sections of $\mathscr{B}$ will be denoted by Ω_{sec}, so that $\Omega_{\text{sec}} \subset \Omega$ by condition (i). The elements of

$$\Omega_{\text{int}} := \{\mathscr{P} \in \Omega \mid \operatorname{Clo} \mathscr{P} \subset \mathscr{B}\} \tag{4.4}$$

will be called **interior parts** of $\mathscr{B}$. If $\mathscr{P} \in \Omega_{\text{int}}$ is a biregular region, then $\mathscr{P} \in \Omega_{\text{sec}}$ and $\operatorname{Bdy} \mathscr{P} = \operatorname{Ctc}(\mathscr{P}, \mathscr{P}^b)$.

Given any part $\mathscr{R} \in \Omega$, we put

$$\Omega_{\mathscr{R}} := \{\mathscr{Q} \in \Omega \mid \mathscr{Q} \subset \mathscr{R}\}$$

and call the members of $\Omega_{\mathscr{R}}$ the **parts of** $\mathscr{R}$. Of course, we have $\Omega_{\mathscr{B}} = \Omega$. The conditions (ii) and (iii) remain satisfied if Ω is replaced by $\Omega_{\mathscr{R}}$ or Ω_{int} and we use the notation (4.3) also when Ω is replaced by $\Omega_{\mathscr{R}}$ or by Ω_{int}.

Now let a finite-dimensional linear space $\mathscr{W}$ be given. Given $\mathscr{R} \in \Omega$ and $F: \Omega_{\mathscr{R}} \to \mathscr{W}$, we say that F is **additive** if

$$F(\mathscr{P} \vee \mathscr{Q}) = F(\mathscr{P}) + F(\mathscr{Q}) \quad \text{for all } (\mathscr{P}, \mathscr{Q}) \in (\Omega_{\mathscr{R}}{}^2)_{\text{dis}}. \tag{4.5}$$

We say that

$$I: (\Omega^2)_{\text{dis}} \to \mathscr{W}$$

is an **interaction** if, for every $\mathscr{R} \in \Omega$, both $I(\cdot, \mathscr{R}^b): \Omega_{\mathscr{R}} \to \mathscr{W}$ and $I(\mathscr{R}^b, \cdot): \Omega_{\mathscr{R}} \to \mathscr{W}$ are additive. We say that I is a **contact-interaction** if for all $(\mathscr{P}, \mathscr{Q}) \in (\Omega^2)_{\text{dis}}$

$$\operatorname{Ctc}(\mathscr{P}, \mathscr{Q}) = \emptyset \;\Rightarrow\; I(\mathscr{P}, \mathscr{Q}) = 0. \tag{4.6}$$

Definition 1. *We say that a given contact-interaction is* **quasi-balanced** *if there is a continuous function* $f: \operatorname{Clo} \mathscr{B} \to \mathscr{W}$ *such that*

$$I(\mathscr{P}, \mathscr{P}^b) = \int_{\mathscr{P}} f\, dv \quad \text{for all } \mathscr{P} \in \Omega_{\text{int}}. \tag{4.7}$$

Proposition 1. *If I is a quasi-balanced contact-interaction, then I is* **skew** *in the sense that*

$$I(\mathscr{P}, \mathscr{Q}) = -I(\mathscr{Q}, \mathscr{P}) \quad \text{for all } (\mathscr{P}, \mathscr{Q}) \in (\Omega_{\text{int}}{}^2)_{\text{dis}}. \tag{4.8}$$

The proof follows immediately from Theorem A, p. 65, of [N2] because $\mathscr{P} \mapsto I(\mathscr{P}, \mathscr{P}^b) : \Omega_{\text{int}} \to \mathscr{W}$ is additive when (4.7) holds.

Remark 1. If the region $\mathscr{B}$ is identified with a continuous body that occupies $\mathscr{B}$ in a "placement", then a contact-interaction I may be interpreted as a system of internal forces or of internal heat-transfers as explained in Sections 9 and 10 of [N2]. In the former case we have $\mathscr{W} := \mathscr{V}$ and in the latter $\mathscr{W} := \mathbb{R}$.

The assumption that I is quasi-balanced is related, in the former interpretation, to the law of balance of forces, and in the latter interpretation to the law of balance of energy. □

Intermezzo: Wedge Interactions

This short section aims to motivate Assumption I of the following section.

Let $\mathscr{D}$ be an open disc in a plane. We use the term **wedge** for an open sector of $\mathscr{D}$. We consider a collection $\Omega_{\mathscr{D}}$ generated by wedges in the following sense: The members of $\Omega_{\mathscr{D}}$ are unions of finite pairwise disjoint collections of wedges no two of which have a side in common. In particular, the wedges belong to $\Omega_{\mathscr{D}}$. Figure 1 illustrates a member of $\Omega_{\mathscr{D}}$.

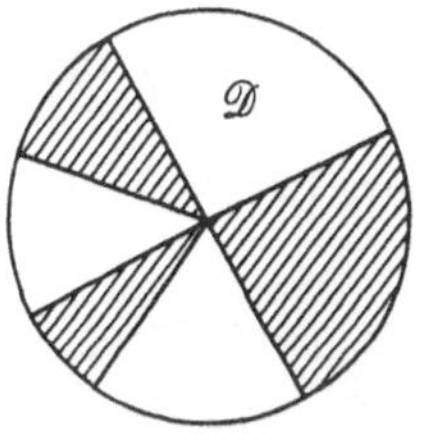

Figure 1

It is easily seen that $\Omega_{\mathscr{D}}$ is a **material universe** as described in the Appendix of [N2] if inclusion is taken as the relation "part of" and **meet, joint,** and **exterior** are defined, respectively, thus:

$$\mathscr{P} \wedge \mathscr{Q} := \mathscr{P} \cap \mathscr{Q},$$

$$\mathscr{P} \vee \mathscr{Q} := \text{Int Clo}\,(\mathscr{P} \cup \mathscr{Q}),$$

$$\mathscr{P}^e := \text{Int}\,(\mathscr{D} \setminus \mathscr{P}).$$

$\mathscr{D}$ is the **maximum** of $\Omega_{\mathscr{D}}$ and the empty set is the **minimum** of $\Omega_{\mathscr{D}}$. The collection Ω defined in Section 4 is a material universe far richer than $\Omega_{\mathscr{D}}$ as defined here.

We denote by $(\Omega_{\text{int}}{}^2)_{\text{dis}}$ the collection of all pairs of disjoint elements of $\Omega_{\mathscr{D}}$. Let $\mathscr{W}$ be a linear space as in Section 4.

Problem. *Find all interactions* $I: (\Omega_{\mathscr{D}}{}^2)_{\text{dis}} \to \mathscr{W}$ *such that* $I(\mathscr{P}, \mathscr{Q}) = 0$ *whenever* $\mathscr{P}$ *and* $\mathscr{Q}$ *are wedges that have no side in common.*

Solution. Let $\mathscr{U}$ be the set of all plane unit vectors. We denote by

$$(\mathscr{U}^2)_\perp := \{(\boldsymbol{u}, \boldsymbol{w}) \in \mathscr{U}^2 \mid \boldsymbol{u} \cdot \boldsymbol{w} = 0\}$$

the collection of all disjoint pairs of unit vectors. Given an interaction I as in the problem there is exactly one function

$$\gamma : (\mathscr{U}^2)_\perp \to \mathscr{W} \tag{I.1}$$

such that

$$I(\mathscr{P}, \mathscr{Q}) = \gamma(\boldsymbol{u}, \boldsymbol{w}) \tag{I.2}$$

whenever $\mathscr{P}$ and $\mathscr{Q}$ are wedges in contact along exactly one side and the unit vectors $\boldsymbol{w}$ and $\boldsymbol{u}$ are defined as follows: $\boldsymbol{w}$ is directed away from the side that $\mathscr{P}$ and $\mathscr{Q}$ have in common and $\boldsymbol{u}$ is orthogonal to $\boldsymbol{w}$ and away from $\mathscr{P}$ (see Figure 2).

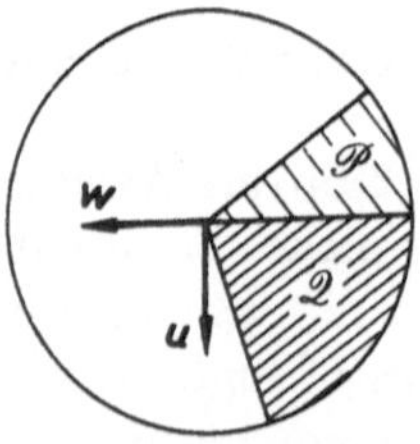

Figure 2

On the other hand, given a function γ as in (I.1), an interaction I satisfying (I.2) is uniquely determined by the biadditivity of I on the whole $(\Omega_{\mathscr{D}}{}^2)_{\text{dis}}$. □

Given any wedge $\mathscr{P}$, the pairs $(\boldsymbol{u}_1, \boldsymbol{w}_1)$, $(\boldsymbol{u}_2, \boldsymbol{w}_2) \in (\mathscr{U}^2)_\perp$ are determined as shown in Figure 3.

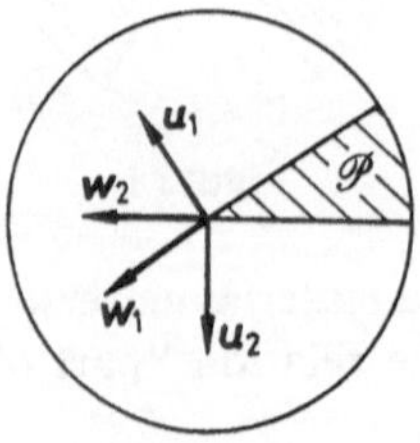

Figure 3

It follows from (I.2) and the additivity of the interaction I that

$$I(\mathscr{P}, \mathscr{P}^e) = \gamma(\boldsymbol{u}_1, \boldsymbol{w}_1) + \gamma(\boldsymbol{u}_2, \boldsymbol{w}_2). \tag{I.3}$$

Thus, by Theorem A of [N2], I is **skew** in the sense that

$$I(\mathscr{P}, \mathscr{Q}) + I(\mathscr{Q}, \mathscr{P}) = 0 \quad \text{for all } \mathscr{P}, \mathscr{Q} \in (\Omega_{\mathscr{D}}{}^2)_{\text{dis}}$$

if and only if

$$\gamma(\boldsymbol{u}, \boldsymbol{w}) + \gamma(-\boldsymbol{u}, \boldsymbol{w}) = 0 \quad \text{for all } (\boldsymbol{u}, \boldsymbol{w}) \in (\mathscr{U}^2)_\perp .$$

Let $\mathscr{D}' \subset \mathscr{D}$ be a given open sector of $\mathscr{D}$. We now consider the collection $\Omega_{\mathscr{D}'}$ of all members of $\Omega_{\mathscr{D}}$ that are included in $\mathscr{D}'$. $\Omega_{\mathscr{D}'}$ is a material universe itself and $(\Omega_{\mathscr{D}'}{}^2)_{\text{dis}}$ is defined in the same way as $(\Omega_{\mathscr{D}}{}^2)_{\text{dis}}$. A wedge of $\mathscr{D}'$ is **internal** if it has no side in common with $\mathscr{D}'$, it is **peripheral** otherwise.

Let $I: (\Omega^2_{\mathscr{D}'})_{\text{dis}} \to \mathscr{W}$ be an interaction satisfying (I.2) for all pairs of *internal* wedges. Clearly, $I(\mathscr{P}, \mathscr{Q})$ cannot be computed by mere use of biadditivity if $\mathscr{P}$ or $\mathscr{Q}$ or both are peripheral. We assume that

$$I(\mathscr{P}, \mathscr{Q}) = \gamma(\boldsymbol{u}, \boldsymbol{w})$$

whenever *both* $\mathscr{P}$ and $\mathscr{Q}$ are peripheral wedges and the unit vectors $\boldsymbol{w}$ and $\boldsymbol{u}$ are defined as follows: $\boldsymbol{w}$ is directed away from the side that $\mathscr{P}$ and $\mathscr{D}'$ have in common and $\boldsymbol{u}$ is orthogonal to $\boldsymbol{w}$ and is directed away from $\mathscr{P}$ (see Figure 4).

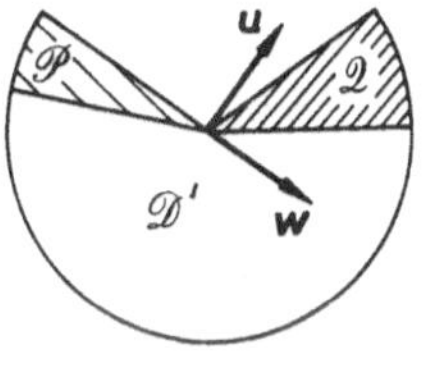

Figure 4

Thus it is easily seen that $I: (\Omega^2_{\mathscr{D}'})_{\text{dis}} \to \mathscr{W}$ is again uniquely determined by γ.

5. Edge interactions

We assume that a biregular bounded region $\mathscr{B}$, a collection of parts of $\mathscr{B}$, and a contact-interaction $I: (\Omega^2)_{\text{dis}} \to \mathscr{W}$ are given as described in Section 4. In this section we will discuss the consequences of two special assumptions about the nature of I.

Let a section $\mathscr{P}$ of $\mathscr{B}$ be given. It easily follows from Definition 5 of Section 3 that, if $\mathfrak{P}$ is an appropriate biregular partition for $\mathscr{P}$, every piece of $\mathfrak{P}$ either is included entirely in $\mathscr{B}$ or is included entirely in Bdy $\mathscr{B}$. In the former case, we say that the piece is **internal,** in the latter, **peripheral.** Thus a given side, edge, or vertex of $\mathscr{P}$, if it belongs to $\mathfrak{P}$, is either internal or peripheral. All pieces of $\mathfrak{P}$ are internal if and only if $\mathscr{P}$ is an interior part of $\mathscr{B}$.

Definition 1. *Let $\mathscr{P}$ and $\mathscr{Q}$ be sections of $\mathscr{B}$ that are in biregular contact and let $\mathfrak{P}$ and $\mathfrak{Q}$ be appropriate biregular partitions for $\mathscr{P}$ and $\mathscr{Q}$, respectively. Also, let an edge $\mathscr{C} \in \mathfrak{P}_1 \cap \mathfrak{Q}_1$ be given, so that $\mathscr{P}$ and $\mathscr{Q}$ are in contact along $\mathscr{C}$.*

We say that $\mathscr{P}$ and $\mathscr{Q}$ are in **inessential contact** *along $\mathscr{C}$ either if $\mathscr{C}$ is a cusped edge of $\mathscr{P}$ and both sides of $\mathscr{P}$ adjacent to $\mathscr{C}$ are internal, or if $\mathscr{C}$ is a cusped edge of $\mathscr{Q}$ and both sides of $\mathscr{Q}$ adjacent to $\mathscr{C}$ are internal; otherwise we say that $\mathscr{P}$ and $\mathscr{Q}$ are in* **essential contact** *along $\mathscr{C}$.*

We define the set $\mathfrak{K}(\mathscr{C})$, consisting of sides of $\mathscr{P}$, as follows:

(i) *If* $\mathscr{P}$ *and* $\mathscr{Q}$ *are in inessential contact along* $\mathscr{C}$, *then* $\mathfrak{R}(\mathscr{C}) := \emptyset$.

(ii) *If* $\mathscr{P}$ *and* $\mathscr{Q}$ *are in essential contact along* $\mathscr{C}$, *then* $\mathfrak{R}(\mathscr{C}) := \{\mathscr{S} \in \mathfrak{P}_2 \mid \mathscr{S}$ *is adjacent to* $\mathscr{C}$ *and either* (a) *or* (b) *below holds*$\}$, *where* (a) *and* (b) *are the following conditions*:

(a) $\mathscr{S}$ *is an internal side of* $\mathscr{P}$ *that is in cusped contact with an internal side of* $\mathscr{Q}$.

(b) $\mathscr{S}$ *is a peripheral side of* $\mathscr{P}$ *that is in contact with a peripheral side of* $\mathscr{Q}$.

If we wish to emphasize the dependence of $\mathfrak{R}(\mathscr{C})$ on $\mathscr{P}$ and $\mathscr{Q}$, we replace $\mathfrak{R}$ by $\mathfrak{R}_{\mathscr{P},\mathscr{Q}}$. It is clear that $\mathfrak{R}(\mathscr{C})$ is a set of cardinality 0, 1, or 2. We illustrate the various situations that determine $\mathfrak{R}(\mathscr{C})$ in Figures 1–6 below. Each of these shows a cross-section orthogonal to $\mathscr{C}$.

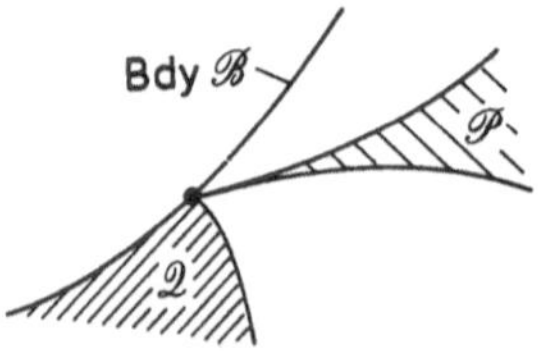

Figure 1

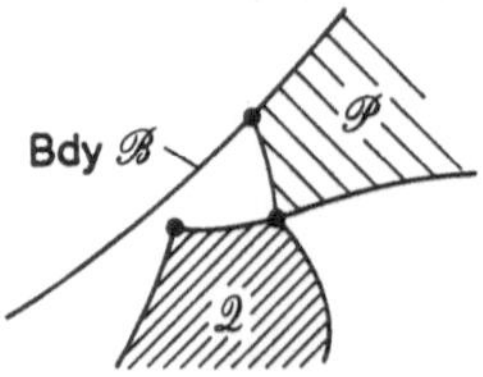

Figure 2

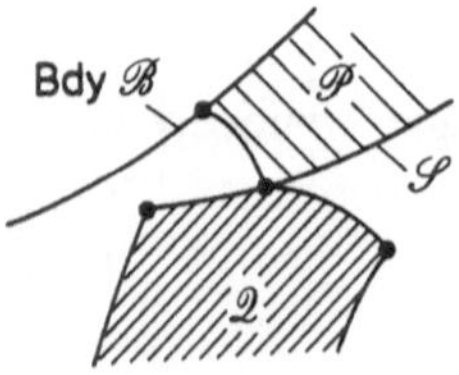

Figure 3

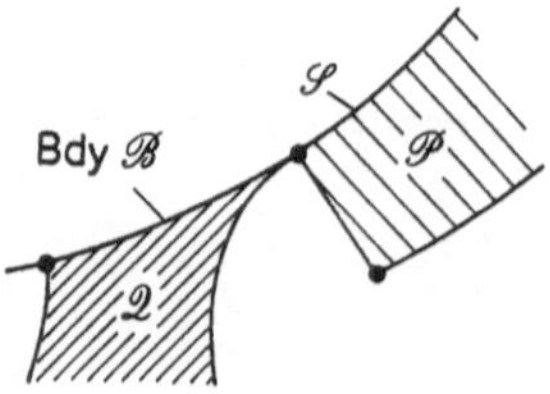

Figure 4

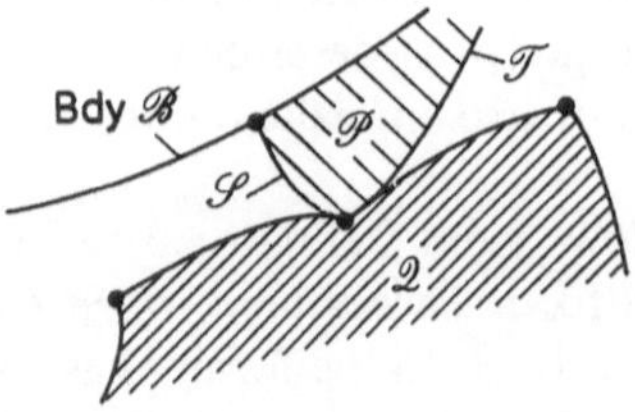

Figure 5

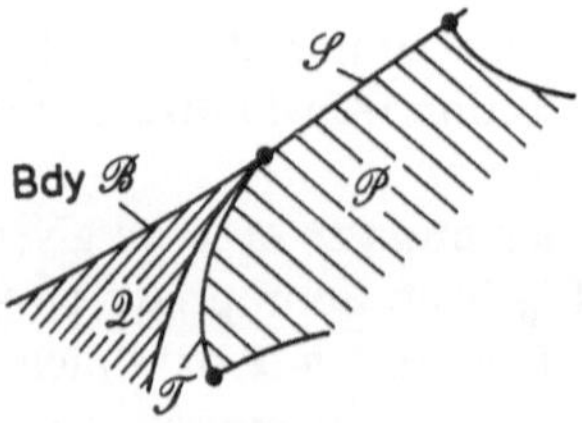

Figure 6

In Figure 1, we have $\mathfrak{R}(\mathscr{C}) = \emptyset$ because $\mathscr{P}$ and $\mathscr{Q}$ are in inessential contact along $\mathscr{C}$. The contact is essential in Figures 2–6. Specifically, we have $\mathfrak{R}(\mathscr{C}) = \emptyset$ in Figure 2, $\mathfrak{R}(\mathscr{C}) = \{\mathscr{S}\}$ in Figures 3 and 4, and $\mathfrak{R}(\mathscr{C}) = \{\mathscr{S}, \mathscr{T}\}$ in Figures 5 and 6. In Figures 3 and 5 we have $\mathscr{S} \in \mathfrak{R}(\mathscr{C})$ because condition (a) is satisfied. In Figures 4 and 6 we have $\mathscr{S} \in \mathfrak{R}(\mathscr{C})$ because condition (b) is satisfied.

If $\mathscr{P}$ and $\mathscr{Q}$ are in simple edge contact along the curve $\mathscr{C}$, then every cross-section of $\mathscr{C}$ in a small neighborhood of $\mathscr{C}$ looks like a wedge contact as described in the Intermezzo. The following assumption states, roughly, that the edge interaction along $\mathscr{C}$ is obtained by integrating wedge interactions on the cross-sections along $\mathscr{C}$.

Assumption I. *There is a function*

$$\gamma : \mathrm{Clo}\,\mathscr{B}\times(\mathscr{U}^2)_\perp \to \mathscr{W} \tag{5.1}$$

of class C^1 *such that*

$$I(\mathscr{P}, \mathscr{Q}) = \int_{\mathscr{C}} \gamma(x, \boldsymbol{v}_{\mathscr{P};\mathscr{S}}(x), \boldsymbol{v}_{\mathscr{S}}(x))\, dl_x \tag{5.2}$$

whenever $\mathscr{P}$ *and* $\mathscr{Q}$ *are disjoint sections of* $\mathscr{B}$ *in simple edge contact along the curve* $\mathscr{C}$ *and* $\mathfrak{K}(\mathscr{C}) = \{\mathscr{S}\}$ (see Definition 1 above and Definition 4 of Section 3).

Proposition 1. *We have*

$$I(\mathscr{P}, \mathscr{Q}) = \int_{\mathscr{C}} \sum_{\mathscr{S}\in\mathfrak{K}(\mathscr{C})} \gamma(x, \boldsymbol{v}_{\mathscr{P};\mathscr{S}}(x), \boldsymbol{v}_{\mathscr{S}}(x))\, dl_x \tag{5.3}$$

whenever $\mathscr{P}$ *and* $\mathscr{Q}$ *are disjoint sections of* $\mathscr{B}$ *in simple edge contact along the curve* $\mathscr{C}$.

Proof. First, we assume that $\mathscr{P}$ and $\mathscr{Q}$ are in essential contact along $\mathscr{C}$. We then have the following three cases.

(α) $\mathfrak{K}(\mathscr{C})$ *is empty.* It is easily seen that we can then construct a section $\mathscr{Q}'$ of $\mathscr{B}$, disjoint from both $\mathscr{P}$ and $\mathscr{Q}$, such that $\mathscr{P}$ is in simple edge contact along $\mathscr{C}$ with both $\mathscr{Q}'$ and $\mathscr{Q} \vee \mathscr{Q}'$ and such that

$$\mathfrak{K}_{\mathscr{P},\mathscr{Q}'}(\mathscr{C}) = \mathfrak{K}_{\mathscr{P},\mathscr{Q}\vee\mathscr{Q}'}(\mathscr{C}) = \{\mathscr{S}\}$$

for some surface $\mathscr{S} \subset \mathrm{Bo}\,\mathscr{P}$ (see Figure 7). It follows from Assumption I that

$$I(\mathscr{P}, \mathscr{Q}') = I(\mathscr{P}, \mathscr{Q} \vee \mathscr{Q}').$$

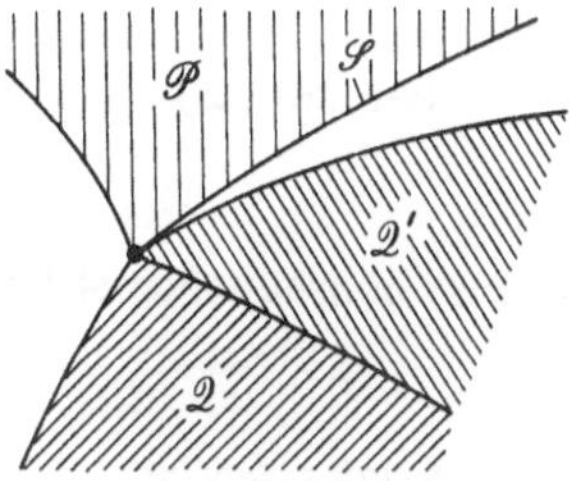

Figure 7

Using the additivity of $I(\mathscr{P}, \cdot)$, we conclude that $I(\mathscr{P}, \mathscr{Q}) = 0$, which is the same as (5.3) when $\mathfrak{K}(\mathscr{C}) = \emptyset$.

(β) $\mathfrak{K}(\mathscr{C})$ *is a singleton.* Then (5.3) simply reduces to (5.2).

(γ) $\mathfrak{K}(\mathscr{C})$ *is a doubleton*, say $\mathfrak{K}(\mathscr{C}) := \{\mathscr{T}_1, \mathscr{T}_2\}$. It is easily seen that we can then construct disjoint sections $\mathscr{Q}_1$ and $\mathscr{Q}_2$, both in simple edge contact with $\mathscr{P}$ along $\mathscr{C}$, such that $\mathscr{Q} = \mathscr{Q}_1 \vee \mathscr{Q}_2$ and $\mathfrak{K}_{\mathscr{P},\mathscr{Q}_1}(\mathscr{C}) = \{\mathscr{T}_1\}$, $\mathfrak{K}_{\mathscr{P},\mathscr{Q}_2}(\mathscr{C}) = \{\mathscr{S}_2\}$ (see Figure 8).

Using the additivity of $I(\mathscr{P}, \cdot)$, we get

$$I(\mathscr{P}, \mathscr{Q}) = I(\mathscr{P}, \mathscr{Q}_1) + I(\mathscr{P}, \mathscr{Q}_2).$$

Now $I(\mathscr{P}, \mathscr{Q}_1)$ and $I(\mathscr{P}, \mathscr{Q}_2)$ can be evaluated using Assumption I. The result is (5.3) with $\mathfrak{K}(\mathscr{C}) = \{\mathscr{T}_1, \mathscr{T}_2\}$.

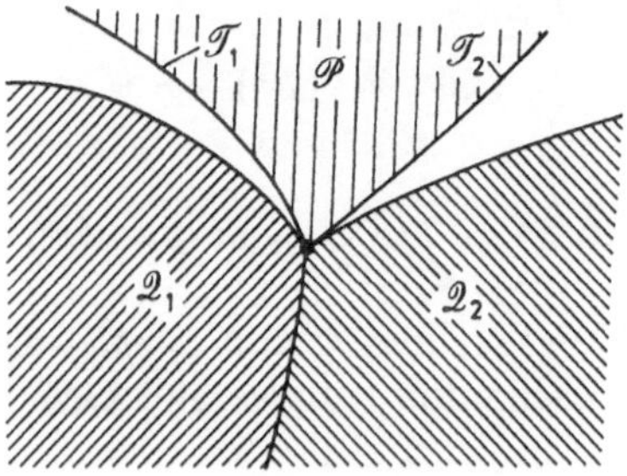

Figure 8

We now assume that $\mathscr{P}$ and $\mathscr{Q}$ are in inessential contact along $\mathscr{C}$. In particular, assume that $\mathscr{C}$ is a cusped edge of $\mathscr{P}$ and both sides of $\mathscr{P}$ adjacent to $\mathscr{C}$ are internal, but that $\mathscr{C}$ is *not* a cusped edge of $\mathscr{Q}$. We can then construct a section $\mathscr{P}'$ of $\mathscr{B}$, disjoint from both $\mathscr{P}$ and $\mathscr{Q}$, such that both $\mathscr{P}'$ and $\mathscr{P} \vee \mathscr{P}'$ are in essential simple edge contact with $\mathscr{Q}$ along $\mathscr{C}$ and $\mathfrak{R}_{\mathscr{P}\vee\mathscr{P}',\mathscr{Q}}(\mathscr{C}) = \mathfrak{R}_{\mathscr{P}',\mathscr{Q}}(\mathscr{C})$ (see Figure 9).

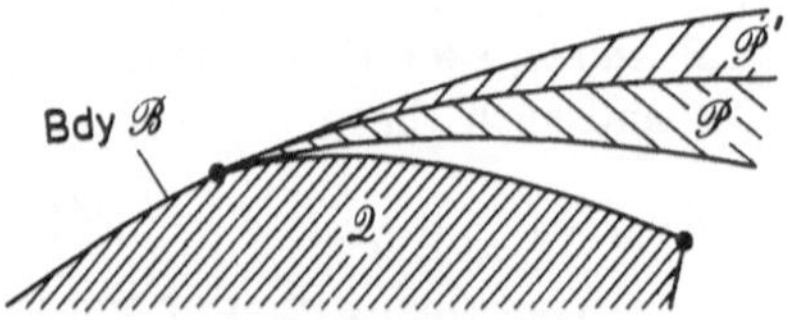

Figure 9

Using the additivity of $I(\cdot, \mathscr{Q})$, we have $I(\mathscr{P} \vee \mathscr{P}', \mathscr{Q}) = I(\mathscr{P}, \mathscr{Q}) + I(\mathscr{P}', \mathscr{Q})$. Since (5.3) can be used to determine $I(\mathscr{P} \vee \mathscr{P}', \mathscr{Q})$ and $I(\mathscr{P}', \mathscr{Q})$ and since they have the same value, it follows that $I(\mathscr{P}, \mathscr{Q}) = 0$. All other cases can be treated in a similar way. □

The following result is an easy consequence of Proposition 1 and the additivity of $I(\mathscr{P}, \cdot)$.

Proposition 2. *Let $\mathscr{P}$ and $\mathscr{Q}$ be disjoint sections of $\mathscr{B}$ in biregular contact and let* $\mathfrak{P}$ and $\mathfrak{Q}$ *be appropriate biregular partitions for $\mathscr{P}$ and $\mathscr{Q}$, respectively, so that*

$$\operatorname{Ctc}(\mathscr{P}, \mathscr{Q}) = \bigcup (\mathfrak{P} \cap \mathfrak{Q}) \tag{5.4}$$

(see Definition 4 of Section 3). *If $\mathfrak{P}$ and $\mathfrak{Q}$ have no surfaces in common,* i.e. *if* $\mathfrak{P}_2 \cap \mathfrak{Q}_2 = \emptyset$, *we have*

$$I(\mathscr{P}, \mathscr{Q}) = \sum_{\mathscr{C}\in\mathfrak{P}_1\cap\mathfrak{Q}_1} \int_{\mathscr{C}} \sum_{\mathscr{S}\in\mathfrak{R}(\mathscr{C})} \gamma(x, \boldsymbol{v}_{\mathscr{P};\mathscr{S}}(x), \boldsymbol{v}_{\mathscr{S}}(x))\, dl_x. \tag{5.5}$$

Assumption II. *For every biregular part $\mathscr{P}$ of $\mathscr{B}$ there is a continuous function*

$$\tau_{\mathscr{P}}: \ \operatorname{Rbo}\mathscr{P} \setminus \operatorname{Bdy}\mathscr{B} \ \to \ \mathscr{W} \tag{5.6}$$

such that

$$I(\mathscr{P}, \mathscr{Q}) = \int_{\mathscr{S}} \tau_{\mathscr{P}}\, da \tag{5.7}$$

whenever (a) $\mathcal{Q}$ *is a section of* $\mathcal{B}$ *disjoint from* $\mathcal{P}$ *that is in simple surface contact with* $\mathcal{P}$ *along the regular surface* $\mathcal{S}$. (b) $\mathcal{Q}$ *has exactly one side other than* $\mathcal{S}$, *say* $\mathcal{T}$, *and* $\mathcal{T}$ *is an internal side, and* (c) *all edges of* $\mathcal{Q}$ *are cusped.*

Figure 10 illustrates a situation for which (5.7) is valid.

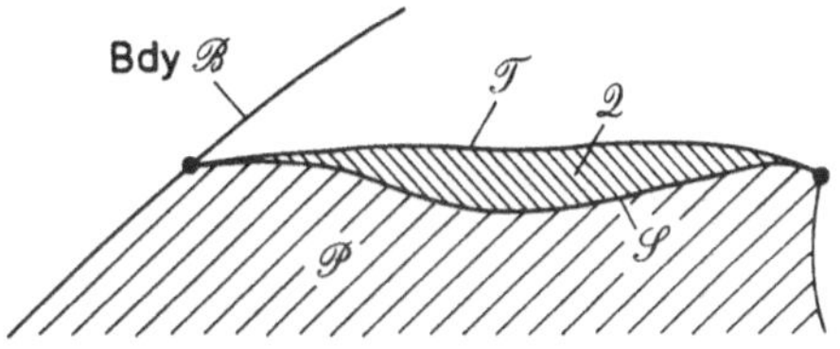

Figure 10

Proposition 3. *We have*

$$\gamma(x, \boldsymbol{u}, -\boldsymbol{w}) = -\gamma(x, \boldsymbol{u}, \boldsymbol{w}) \quad \text{for all } x \in \operatorname{Clo} \mathcal{B}, \quad (\boldsymbol{u}, \boldsymbol{w}) \in (\mathcal{U}^2)_\perp. \tag{5.8}$$

Proof. Let $x \in \mathcal{B}$ and $(\boldsymbol{u}, \boldsymbol{w}) \in (\mathcal{U}^2)_\perp$ be given. We consider a line-segment $\mathcal{C}$ included in $\mathcal{B}$ whose midpoint is x and which is perpendicular to both $\boldsymbol{u}$ and $\boldsymbol{w}$. We can easily construct five interior sections $\mathcal{P}$, $\mathcal{Q}_1$, $\mathcal{Q}_1'$, $\mathcal{Q}_2$, $\mathcal{Q}_2'$ with the following properties (see Figure 11): (i) The sections are pairwise disjoint and in biregular contact. (ii) The sections $\mathcal{Q}_1$ and $\mathcal{Q}_2$ are in simple surface contact with $\mathcal{P}$ along the two sides $\mathcal{S}_1$, $\mathcal{S}_2$ of $\mathcal{P}$ that are adjacent to $\mathcal{C}$ in some biregular partition for $\mathcal{P}$. (iii) The sections $\mathcal{Q}_1$ and $\mathcal{Q}_2$ satisfy the conditions (b) and (c) of Assumption II. (iv) The sections $\mathcal{Q}_1'$ and $\mathcal{Q}_2'$ are in simple edge contact with $\mathcal{P}$ along $\mathcal{C}$ such that $\mathfrak{R}_{\mathcal{P},\mathcal{Q}_1'}(\mathcal{C}) = \{\mathcal{S}_1\}$ and $\mathfrak{R}_{\mathcal{P},\mathcal{Q}_2'}(\mathcal{C}) = \{\mathcal{S}_2\}$. (v) $\mathcal{S} := \mathcal{S}_1 \cup \mathcal{S}_2 \cup \mathcal{C}$ is a regular surface and $\mathcal{Q} := \mathcal{Q}_1 \vee \mathcal{Q}_1' \vee \mathcal{Q}_2 \vee \mathcal{Q}_2'$ is a section that is in simple surface contact with $\mathcal{P}$ along $\mathcal{S}$ such that the conditions (b) and (c) of Assumption II are satisfied. (vi) $\mathcal{S}$ is a subset of a plane and $\boldsymbol{v}_{\mathcal{P}}|_{\mathcal{S}}$ is a constant with value $\boldsymbol{u}$. The functions $\boldsymbol{v}_{\mathcal{S}_1}|_{\mathcal{C}}$ and $\boldsymbol{v}_{\mathcal{S}_2}|_{\mathcal{C}}$ are constants with values $\boldsymbol{w}$ and $-\boldsymbol{w}$, respectively.

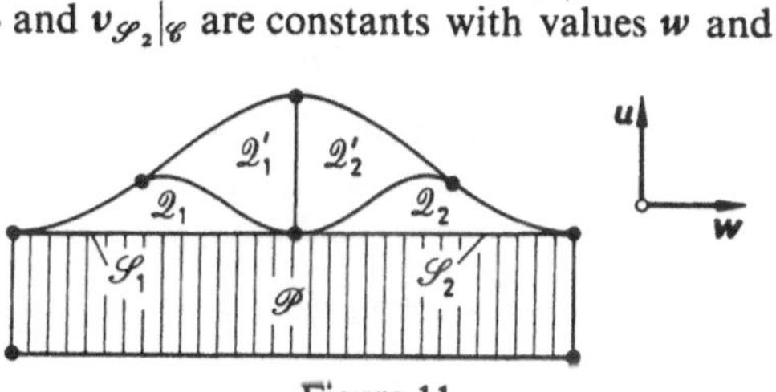

Figure 11

It follows from (i) and (v) and Assumption II that

$$I(\mathcal{P}, \mathcal{Q}_1 \vee \mathcal{Q}_1' \vee \mathcal{Q}_2 \vee \mathcal{Q}_2') = I(\mathcal{P}, \mathcal{Q}) = \int_{\mathcal{S}} \boldsymbol{\tau}_{\mathcal{P}}\, da. \tag{5.9}$$

It follows from (i), (ii), (iii), and Assumption II that

$$I(\mathcal{P}, \mathcal{Q}_1) = \int_{\mathcal{S}_1} \boldsymbol{\tau}_{\mathcal{P}}\, da, \quad I(\mathcal{P}, \mathcal{Q}_2) = \int_{\mathcal{S}_2} \boldsymbol{\tau}_{\mathcal{P}}\, da \tag{5.10}$$

and from (i), (iv), (vi), and Assumption I that

$$\begin{aligned} I(\mathcal{P}, \mathcal{Q}_1') &= \int_{\mathcal{C}} \gamma(y, \boldsymbol{v}_{\mathcal{P};\mathcal{S}_1}(y), \boldsymbol{v}_{\mathcal{S}_1}(y))\, dl_y = \int_{\mathcal{C}} \tau(y, \boldsymbol{u}, \boldsymbol{w})\, dl_y, \\ I(\mathcal{P}, \mathcal{Q}_2') &= \int_{\mathcal{C}} \gamma(y, \boldsymbol{v}_{\mathcal{P};\mathcal{S}_2}(y), \boldsymbol{v}_{\mathcal{S}_2}(y))\, dl_y = \int_{\mathcal{C}} \gamma(y, \boldsymbol{u}, -\boldsymbol{w})\, dl_y. \end{aligned} \tag{5.11}$$

Using the additivity of $I(\mathcal{P}, \cdot)$ and the additivity of surface integrals, we conclude from (5.9)–(5.11) that

$$0 = \int_{\mathcal{C}} (\gamma(y, \boldsymbol{u}, \boldsymbol{w}) + \gamma(y, \boldsymbol{u}, -\boldsymbol{w}))\, dl_y.$$

Since the line-segment $\mathcal{C}$ centered at x can be made arbitrarily short, it follows that (5.8) holds. If $x \in \operatorname{Bdy} \mathcal{B}$ rather than $x \in \mathcal{B}$, then (5.8) follows from the assumed continuity of γ. □

Definition 1. *Let $\mathcal{S}$ be a regular surface. A continuous function* $\boldsymbol{n} : \operatorname{Clo} \mathcal{S} \to \mathcal{U}$ *is called an* **orientation** *of $\mathcal{S}$ if $\boldsymbol{n}(x) \in (\operatorname{Tan}_{\mathcal{S}}(x))^{\perp}$ for all $x \in \operatorname{Clo} \mathcal{S}$. We say that $\mathcal{S}$ is* **orientable** *if admits an orientation $\boldsymbol{n}$.*

By an **oriented regular surface** *we mean a pair $(\mathcal{S}, \boldsymbol{n})$ of a regular surface $\mathcal{S}$ and an orientation $\boldsymbol{n}$ of $\mathcal{S}$.*

An orientable regular surface $\mathcal{S}$ admits exactly two orientations, and if $\boldsymbol{n}$ is one then $-\boldsymbol{n}$ is the other. Let $(\mathcal{S}, \boldsymbol{n})$ be an oriented surface. Then

$$\boldsymbol{E}_{\mathcal{S}} = \mathbf{1}_{\mathcal{V}} - \boldsymbol{n} \otimes \boldsymbol{n} \tag{5.12}$$

and $\boldsymbol{n}$ is of class C^1. If $\boldsymbol{n}$ is constant then $\mathcal{S}$ must be subset of a plane.

Proposition 4. *For every oriented regular surface $(\mathcal{S}, \boldsymbol{n})$ with $\mathcal{S} \subset \mathcal{B}$ there is a continuous function*

$$\sigma_{(\mathcal{S},\boldsymbol{n})} : \operatorname{Clo} \mathcal{S} \to \mathcal{W}$$

such that

$$\sigma_{(\mathcal{S},\boldsymbol{n})} = \tau_{\mathcal{P}}|_{\mathcal{S}} \tag{5.13}$$

whenever $\mathcal{P}$ is a biregular part of $\mathcal{B}$ with $\mathcal{S} \subset \operatorname{Rbo} \mathcal{P}$ and $\boldsymbol{n} = \boldsymbol{v}_{\mathcal{P}|\mathcal{S}}$.

Moreover, if $(\mathcal{S}, \boldsymbol{n})$ and $(\mathcal{S}', \boldsymbol{n}')$ are oriented regular surfaces such that $\mathcal{S}' \subset \mathcal{S}$ and $\boldsymbol{n}' = \boldsymbol{n}_{|\mathcal{S}'}$, then

$$\sigma_{(\mathcal{S}',\boldsymbol{n}')} = \sigma_{(\mathcal{S},\boldsymbol{n})}|_{\mathcal{S}'}. \tag{5.14}$$

The proof of Proposition 4, which will not be given here, is only slightly more complicated than the proof of the corresponding proposition for classical surface interactions (see *e.g.* Theorem 1 on p. 271 of [N1]).

Proposition 5. *Let $\mathcal{P}$ and $\mathcal{Q}$ be sections of $\mathcal{B}$ in biregular contact and let $\mathfrak{P}$ and $\mathfrak{Q}$ be appropriate biregular partitions of $\mathcal{P}$ and $\mathcal{Q}$, respectively. Then*

$$I(\mathcal{P}, \mathcal{Q}) = \sum_{\mathcal{S} \in \mathfrak{P}_2 \cap \mathfrak{Q}_2} \int_{\mathcal{S}} \sigma_{(\mathcal{S}, \boldsymbol{v}_{\mathcal{P}|\mathcal{S}})}\, da + \sum_{\mathcal{C} \in \mathfrak{P}_1 \cap \mathfrak{Q}_1} \int_{\mathcal{C}} \sum_{\mathcal{S} \in \mathfrak{K}(\mathcal{C})} \gamma(x, \boldsymbol{v}_{\mathcal{P};\mathcal{S}}(x), \boldsymbol{v}_{\mathcal{P}}(x))\, dl_x. \tag{5.15}$$

Proof. Let $\mathscr{S} \in \mathfrak{P}_2 \cap \mathfrak{Q}_2$ be given. It is easily seen that one can construct a section $\mathscr{R}_{\mathscr{S}}$ of $\mathscr{Q}$ with the following properties (see Figure 12): (i) $\mathscr{R}_{\mathscr{S}}$ is in simple surface contact with $\mathscr{P}$ along $\mathscr{S}$; (ii) $\mathscr{R}_{\mathscr{S}}$ has exactly one side other than $\mathscr{S}$, say $\mathscr{S}'$, and $\mathscr{S}'$ is included in $\mathscr{Q}$; (iii) all edges of $\mathscr{R}_{\mathscr{S}}$ are cusped. Moreover, the construction can be done in such a way that the collection $\{\mathscr{R}_{\mathscr{S}} \mid \mathscr{S} \in \mathfrak{P}_2 \cap \mathfrak{Q}_2\}$ is pairwise disjoint and $\operatorname{Ctc}(\mathscr{R}_{\mathscr{S}}, \mathscr{R}_{\mathscr{S}'}) \subset \operatorname{Bdy} \mathscr{P}$ for any two distinct $\mathscr{S}, \mathscr{S}' \in \mathfrak{P}_2 \cap \mathfrak{Q}_2$.

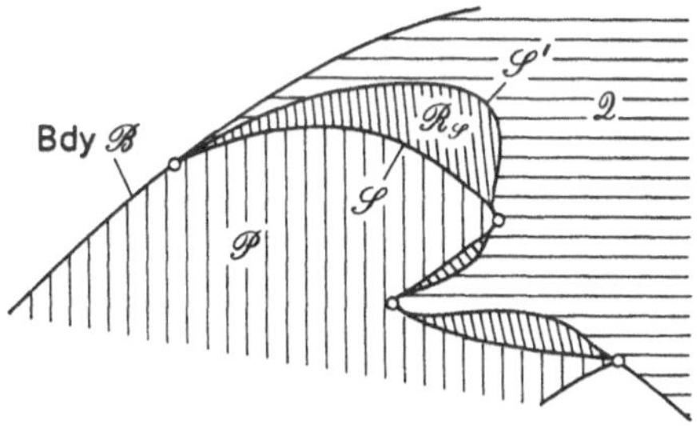

Figure 12

It is easily seen that

$$\mathscr{Q}' := \operatorname{Int}(\mathscr{Q} \setminus \bigcup \{\mathscr{R}_{\mathscr{S}} \mid \mathscr{S} \in \mathfrak{P}_2 \cap \mathfrak{Q}_2\}) \tag{5.16}$$

is a section of $\mathscr{Q}$ that has an appropriate biregular partition $\mathfrak{Q}'$ such that $\mathfrak{P}_2 \cap \mathfrak{Q}'_2 = \emptyset$ and $\mathfrak{P}_1 \cap \mathfrak{Q}'_1 = \mathfrak{P}_1 \cap \mathfrak{Q}_1$. The section $\mathscr{Q}'$ is in biregular contact with $\mathscr{P}$ and Proposition 2 can be used to evaluate $I(\mathscr{P}, \mathscr{Q}')$. In view of condition (iii) above, we also have $\mathfrak{R}_{\mathscr{P},\mathscr{Q}}(\mathscr{C}) = \mathfrak{R}_{\mathscr{P},\mathscr{Q}'}(\mathscr{C})$ for all $\mathscr{C} \in \mathfrak{Q}_1 \cap \mathfrak{P}_1$, so that (5.5) gives

$$I(\mathscr{P}, \mathscr{Q}') = \sum_{\mathscr{C} \in \mathfrak{P}_1 \cap \mathfrak{Q}_1} \int_{\mathscr{C}} \sum_{\mathscr{S} \in \mathfrak{R}(\mathscr{C})} \gamma(x, \boldsymbol{v}_{\mathscr{P};\mathscr{S}}(x), \boldsymbol{v}_{\mathscr{S}}(x)) \, dl_x \tag{5.17}$$

when $\mathfrak{R} := \mathfrak{R}_{\mathscr{P},\mathscr{Q}}$. Using (5.16) and the additivity of $I(\mathscr{P}, \cdot)$ one easily proves using induction that

$$I(\mathscr{P}, \mathscr{Q}) = I(\mathscr{P}, \mathscr{Q}') + \sum_{\mathscr{S} \in \mathfrak{P}_2 \cap \mathfrak{Q}_2} I(\mathscr{P}, \mathscr{R}_{\mathscr{S}}). \tag{5.18}$$

Now, in view of the conditions (i), (ii), and (iii) above. Assumption II can be used to evaluate $I(\mathscr{P}, \mathscr{R}_{\mathscr{S}})$. Using also Proposition 4, we get

$$I(\mathscr{P}, \mathscr{R}_{\mathscr{S}}) = \int_{\mathscr{S}} \sigma_{(\mathscr{S}, \boldsymbol{v}_{\mathscr{P}}|_{\mathscr{S}})} \, da. \tag{5.19}$$

Combining (5.17), (5.18), and (5.19), we get the desired result (5.15). □

Proposition 6 shows that the values of the interaction I for sections in biregular contact are uniquely determined by the function γ and the functions $\sigma_{(\mathscr{S},\boldsymbol{n})}$. The problem of how to express the values of I in more general situations is open.

If $\gamma = 0$ we say that the interaction I is a **pure surface interaction.** If $\sigma_{(\mathscr{S},\boldsymbol{n})} = 0$ for all oriented regular surfaces $(\mathscr{S}, \boldsymbol{n})$ we say that I is a **pure edge interaction.** Every interaction satisfying Assumptions I and II is then the value-wise sum of a pure surface interaction and a pure edge interaction.

Remark 1. Suppose that a function

$$\gamma : \mathrm{Clo}\, \mathscr{B} \times (\mathscr{U}^2)_\perp \to \mathscr{W}$$

of class C^1 satisfying $\gamma(x, \boldsymbol{u}, -\boldsymbol{w}) = -\gamma(x, \boldsymbol{u}, \boldsymbol{w})$ for all $(x, \boldsymbol{u}, \boldsymbol{w}) \in \mathrm{Clo}\, \mathscr{B} \times (\mathscr{U}^2)_\perp$ has been prescribed. Also suppose that, for each oriented regular surface $(\mathscr{S}, \boldsymbol{n})$ with $\mathscr{S} \subset \mathscr{B}$, a continuous function $\sigma_{(\mathscr{S},\boldsymbol{n})} : \mathrm{Clo}\, \mathscr{S} \to \mathscr{W}$ has been *prescribed* in such a way that (5.14) holds. We can then *define* $I(\mathscr{P}, \mathscr{Q})$ by (5.15) for all disjoint pairs $(\mathscr{P}, \mathscr{Q})$ of sections of $\mathscr{B}$ in biregular contact. One can then prove that I is *conditionally biadditive* in the sense that

$$I(\mathscr{P}, \mathscr{Q}_1 \vee \mathscr{Q}_2) = I(\mathscr{P}, \mathscr{Q}_1) + I(\mathscr{P}, \mathscr{Q}_2)$$

and

$$I(\mathscr{Q}_1 \vee \mathscr{Q}_2, \mathscr{P}) = I(\mathscr{Q}_1, \mathscr{P}) + I(\mathscr{Q}_2, \mathscr{P})$$

hold provided that: (i) $\mathscr{P}$, $\mathscr{Q}_1$ and $\mathscr{Q}_2$ are pairwise disjoint sections of $\mathscr{B}$; (ii) $\mathscr{Q}_1 \vee \mathscr{Q}_2$ is a section of $\mathscr{B}$; (iii) $\mathscr{P}$ is in biregular contact with $\mathscr{Q}_1$, $\mathscr{Q}_2$, and $\mathscr{Q}_1 \vee \mathscr{Q}_2$. □

6. Resultant actions

We assume that a region $\mathscr{B}$ and a contact interaction I as described in Section 5 are given. In particular, we assume that the Assumptions I and II are satisfied.

The following result describes the *resultant* action on a given section of $\mathscr{B}$, *i.e.* the action $I(\mathscr{P}, \mathscr{P}^b)$ on a section $\mathscr{P}$ of $\mathscr{B}$ by the exterior $\mathscr{P}^b$ of $\mathscr{P}$ in $\mathscr{B}$, provided that the condition (6.1) below is valid. We will see in Section 7 that (6.1) is automatically satisfied if I is quasi-balanced in the sense of Definition 1 of Section 4.

Proposition 1. *Assume that the function* γ *of* Assumption I *satisfies*

$$\gamma(x, -\boldsymbol{u}, \boldsymbol{w}) = -\gamma(x, \boldsymbol{u}, \boldsymbol{w}) \quad \textit{for all} \quad x \in \mathrm{Clo}\, \mathscr{B}, \quad (\boldsymbol{u}, \boldsymbol{w}) \in (\mathscr{U}^2)_\perp. \tag{6.1}$$

Let $\mathscr{P}$ *be a section of* $\mathscr{B}$ *and let* $\mathfrak{P}$ *be an appropriate biregular partition for* $\mathscr{P}$. *Then*

$$I(\mathscr{P}, \mathscr{P}^b) = \sum_{\mathscr{S} \in \mathfrak{P}_2^{\mathrm{int}}} \left(\int_{\mathscr{S}} \sigma_{(\mathscr{S}, \boldsymbol{v}_{\mathscr{P}}|_{\mathscr{S}})}\, da + \int_{\mathrm{Rbo}\, \mathscr{S}} \gamma(x, \boldsymbol{v}_{\mathscr{P};\mathscr{S}}(x), \boldsymbol{v}_{\mathscr{S}}(x))\, dl_x \right)$$
$$+ \sum_{\mathscr{S} \in \mathfrak{P}_2^{\mathrm{per}}} \int_{\mathrm{Rbo}\mathscr{S} \cap \mathrm{Bdy}\mathscr{P}^b} \gamma(x, \boldsymbol{v}_{\mathscr{P};\mathscr{S}}(x), \boldsymbol{v}_{\mathscr{S}}(x))\, dl_x \tag{6.2}$$

where $\mathfrak{P}_2^{\mathrm{int}}$ *is the set of all internal sides of* $\mathscr{P}$ *and* $\mathfrak{P}_2^{\mathrm{per}}$ *is the set of all peripheral sides of* $\mathscr{P}$.

Proof. It is easily seen that we may choose an appropriate biregular partition $\mathfrak{Q}$ for $\mathscr{P}^b$ such that

$$\mathfrak{P}_2 \cap \mathfrak{Q}_2 = \mathfrak{P}_2^{\mathrm{int}}. \tag{6.3}$$

i.e. such that the internal sides of $\mathscr{P}$ are exactly the sides that $\mathscr{P}$ and $\mathscr{P}^b$ have in common.

Let $\mathscr{C} \in \mathfrak{P}_1 \cap \mathfrak{Q}_1$, *i.e.* an edge that $\mathscr{P}$ and $\mathscr{P}_b$ have in common, be given, and let $\mathscr{S}_1, \mathscr{S}_2 \in \mathfrak{P}_2$ be the two sides of $\mathscr{P}$ adjacent to $\mathscr{C}$. There are two cases: (i) $\mathscr{C}$ is *not* an internal cusped edge of $\mathscr{P}$. Then $\mathscr{C}$ cannot be an internal cusped edge of $\mathscr{P}^b$ either and $\mathscr{P}$ and $\mathscr{P}^b$ are in essential contact along $\mathscr{C}$ (see Definition 1 of Section 5). In this case, the set $\mathfrak{R}(\mathscr{C})$ is easily seen to be given by $\mathfrak{R}(\mathscr{C}) = \{\mathscr{S}_1, \mathscr{S}_2\}$ so that

$$\int_{\mathscr{C}} \gamma(x, \boldsymbol{v}_{\mathscr{P};\mathscr{S}_1}(x), \boldsymbol{v}_{\mathscr{S}_1}(x))\, dl_x + \int_{\mathscr{C}} \gamma(x, \boldsymbol{v}_{\mathscr{P};\mathscr{S}_2}(x), \gamma_{\mathscr{S}_2}(x))\, dl_x$$

$$= \int_{\mathscr{C}} \sum_{\mathscr{S} \in \mathfrak{R}(\mathscr{C})} \gamma(x, \boldsymbol{v}_{\mathscr{P};\mathscr{S}}(x), \boldsymbol{v}_{\mathscr{S}}(x))\, dl_x. \tag{6.4}$$

(ii) $\mathscr{C}$ is an internal cusped edge of $\mathscr{P}$. Then $\mathscr{P}$ and $\mathscr{P}^b$ are in inessential contact along $\mathscr{C}$ and hence $\mathfrak{R}(\mathscr{C}) = \emptyset$. Hence the right side of (6.4) is zero. We also have $\boldsymbol{v}_{\mathscr{P};\mathscr{S}_1}(x) = -\boldsymbol{v}_{\mathscr{P};\mathscr{S}_2}(x)$ and $\boldsymbol{v}_{\mathscr{S}_1}(x) = \boldsymbol{v}_{\mathscr{S}_2}(x)$ for all $x \in \mathscr{C}$. Hence, by the assumption (6.1), the left side of (6.4) is also zero.

We conclude that (6.4) is valid no matter what $\mathscr{C} \in \mathfrak{P}_1 \cap \mathfrak{Q}_1$ is. If $\mathscr{C}$ is an internal edge of $\mathscr{P}$, then $\mathscr{S}_1, \mathscr{S}_2 \in \mathfrak{P}_2^{\text{int}}$. If $\mathscr{C}$ is a peripheral edge of $\mathscr{P}$, then one of $\mathscr{S}_1$ and $\mathscr{S}_2$, say $\mathscr{S}_1$, is internal while the other, $\mathscr{S}_2$, is peripheral and $\mathscr{C} \subset \text{Rbo}\, \mathscr{S}_2 \cap \text{Bdy}\, \mathscr{P}^b$.

Since the edges of every $\mathscr{S} \in \mathfrak{P}_2^{\text{int}}$ and the edges included in Bdy $\mathscr{P}^b$ of every $\mathscr{S} \in P_2^{\text{per}}$ all belong to $\mathfrak{P}_1 \cap \mathfrak{Q}_1$, it follows from (6.3) and (6.4) that (5.15) reduces to (6.2) when $\mathscr{Q} := \mathscr{P}^b$. □

Remark 1. If a given edge $\mathscr{C} \in \mathfrak{P}_1$ is included in the reduced boundary Rby $\mathscr{P}$ of $\mathscr{P}$, then the contributions from $\mathscr{C}$ to the line integrals in (6.2) corresponding to the two sides adjacent to $\mathscr{C}$ cancel by Proposition 3 of Section 5. Hence the right side of (6.2) does not change when $\mathfrak{P}$ is replaced by a refinement, as it shouldn't.

□

We now add an assumption which is consistent with (5.8), but does not follow from it.

Assumption III. *There is a function*

$$G\colon\ \text{Clo}\, \mathscr{B} \times \mathscr{U} \to \text{Lin}\, (\mathscr{V}, \mathscr{W}) \tag{6.5}$$

of class C^1 *such that*

$$\gamma(x, \boldsymbol{u}, \boldsymbol{w}) = \boldsymbol{G}(x, \boldsymbol{u})\, \boldsymbol{w} \quad \textit{for all } x \in \text{Clo}\, \mathscr{B}, \quad (\boldsymbol{u}, \boldsymbol{w}) \in (\mathscr{U}^2)_\perp. \tag{6.6}$$

The function $\boldsymbol{G}$ is not uniquely determined by γ. Indeed, if $\boldsymbol{G}'$ is a C^1-function of the type (6.5) that satisfies (6.6), then

$$\boldsymbol{G}(x, \boldsymbol{u}) := \boldsymbol{G}'(x, \boldsymbol{u})\, (\mathbf{1}_{\mathscr{V}} - (\boldsymbol{u} \otimes \boldsymbol{u})) \quad \textit{for all } x \in \text{Clo}\, \mathscr{B}, \quad \boldsymbol{u} \in \mathscr{U} \tag{6.7}$$

defines another one. The function $\boldsymbol{G}$ defined by (6.7) satisfies, in addition to (6.6), the following *normalization condition*:

$$\boldsymbol{u} \in \text{Null}\, \boldsymbol{G}(x, \boldsymbol{u}) \quad \text{for all} \quad x \in \text{Clo}\, \mathscr{B}, \quad \boldsymbol{u} \in \mathscr{U}, \tag{6.8}$$

It is easily seen that γ determines uniquely a C^1-function G that satisfies both (6.6) and (6.8), and from now on we will assume that G denotes this function.

Remark 2. One can prove that Assumption III is implied by the following

Assumption III′. *There is a* $k \in P^{\times}$ *such that for every biregular oriented surface* $(\mathcal{S}, \boldsymbol{n})$ *with* $\mathcal{S} \subset \mathcal{B}$ *we have*

$$\left| \int_{\mathrm{Rbo}\,\mathcal{S}} \gamma(x, \boldsymbol{n}(x), \boldsymbol{v}_{\mathcal{S}}(x)\, dl_x \right| \leqq k \;\mathrm{area}\,(\mathcal{S}). \tag{6.9}$$

Unfortunately, we do not know of any convincing physical reasons that one might use to justify Assumption III′ or Assumption III. □

Now let an oriented regular surface $(\mathcal{S}, \boldsymbol{n})$ with $\mathcal{S} \subset \mathrm{Clo}\,\mathcal{B}$ be given. Then

$$G(1_{\mathcal{S}\subset\mathcal{E}}, \boldsymbol{n}) = (x \mapsto G(x, \boldsymbol{n}(x)) : \;\; \mathrm{Clo}\,\mathcal{S} \to \mathrm{Lin}\,(\mathcal{V}, \mathcal{W}) \tag{6.10}$$

is a mapping of Class C^1 in the sense of Definition 1 of Section 2. We use the abbreviation

$$\varrho_{(\mathcal{S},\boldsymbol{n})} := \mathrm{div}_{\mathcal{S}}\,(G(1_{\mathcal{S}\subset\mathcal{E}}, \boldsymbol{n})) : \;\; \mathcal{S} \to \mathcal{W} \tag{6.11}$$

for the surface-divergence of $G(1_{\mathcal{S}\subset\mathcal{E}}, \boldsymbol{n})$ as defined in Definition 2 of Section 2. Using (5.12) and the fact that

$$\mathrm{Tan}_{\mathcal{S}}(x) = \{\boldsymbol{n}(x)\}^{\perp} \quad \text{for all } x \in \mathrm{Clo}\,\mathcal{S}, \tag{6.12}$$

one can use the chain rule and the definitions of Section 2 to show that $\varrho_{(\mathcal{S},\boldsymbol{n})}$ is determined by the following explicit formula:

$$\begin{aligned} \omega\varrho_{(\mathcal{S},\boldsymbol{n})}(x) = {} & \omega\,\mathrm{div}_x\,(G(\cdot, \boldsymbol{n}(x))) - \boldsymbol{n}(x) \cdot \nabla_x(\omega G(\cdot, \boldsymbol{n}(x)))\, \boldsymbol{n}(x) \\ & + \mathrm{tr}\,(\nabla_{\boldsymbol{n}(x)}(\omega G(x, \cdot))\, \nabla_x \boldsymbol{n}|^{\{\boldsymbol{n}(x)\}^{\perp}}) \quad \text{for all } x \in \mathrm{Clo}\,\mathcal{S}, \quad \omega \in \mathcal{W}^*. \end{aligned} \tag{6.13}$$

In the case when $\mathcal{W} = \mathbb{R}$ and $\mathrm{Lin}\,(\mathcal{V}, \mathbb{R}) = \mathcal{V}^* \cong \mathcal{V}$, (6.13) reduces to

$$\begin{aligned} \varrho_{(\mathcal{S},\boldsymbol{n})}(x) = {} & \mathrm{div}_x\,(G(\cdot, \boldsymbol{n}(x))) - \boldsymbol{n}(x) \cdot \nabla_x G(\cdot, \boldsymbol{n}(x))\, \boldsymbol{n}(x) \\ & + \mathrm{tr}(\nabla_{\boldsymbol{n}(x)}\, G(x, \cdot)\, \nabla_x \boldsymbol{n}|^{\{\boldsymbol{n}(x)\}^{\perp}}) \quad \text{for all } x \in \mathrm{Clo}\,\mathcal{S}. \end{aligned} \tag{6.14}$$

The normalization condition (6.8) ensures that $G(1_{\mathcal{S}\subset\mathcal{E}}, \boldsymbol{n})$ is a tangential mapping in the sense of Definition 3 of Section 2. Hence we can apply the Corollary to surface-divergence Theorem stated in Section 2 to obtain the following

Proposition 2. *Let* $\mathcal{P}$ *be a section of* $\mathcal{B}$ *and let* $\mathfrak{P}$ *be an appropriate biregular partition for* $\mathcal{P}$. *For every internal side* $\mathcal{S} \in \mathfrak{P}_2^{\mathrm{int}}$ *of* $\mathcal{P}$ *we then have*

$$\begin{aligned} & \int_{\mathcal{S}} \sigma_{(\mathcal{S},\boldsymbol{v}_{\mathcal{P}}|_{\mathcal{S}})}\, da + \int_{\mathrm{Rbo}\,\mathcal{S}} \gamma(x, \boldsymbol{v}_{\mathcal{P};\mathcal{S}}(x), \boldsymbol{v}_{\mathcal{S}}(x))\, dl_x \\ & \quad = \int_{\mathcal{S}} (\sigma_{(\mathcal{S},\boldsymbol{v}_{\mathcal{P}}|_{\mathcal{S}})} + \varrho_{(\mathcal{S},\boldsymbol{v}_{\mathcal{P}}|_{\mathcal{S}})})\, da, \end{aligned} \tag{6.15}$$

where ϱ *is characterized by* (6.13).

Since $\mathscr{S} \in \mathfrak{P}^2_{\text{int}}$ implies $\mathscr{S} \subset \mathscr{B}$, one needs functions of the form $\varrho_{(\mathscr{S},\boldsymbol{n})}$ only for surfaces $\mathscr{S}$ included in $\mathscr{B}$ when applying Proposition 2. Nevertheless, it is important to note, for use in Section 8, that $\varrho_{(\mathscr{S},\boldsymbol{n})}$, as given by (6.11) or (6.13), is meaningful even when $\mathscr{S}$ is a regular surface included in the *boundary* of $\mathscr{B}$.

7. Quasi-balanced interactions

We assume again that a region $\mathscr{B}$ and a contact interaction I as described in Section 5 are given. We assume that Assumption III of Section 6 as well as the Assumptions I and II of Section 5 are satisfied. Finally, we assume that the interaction I is quasi-balanced in the sense of Definition 1 of Section 4. The continuous function f: Clo $\mathscr{B} \to \mathscr{W}$ for which (4.7) holds is clearly uniquely determined by I.

For each oriented biregular surface $(\mathscr{S}, \boldsymbol{n})$ with $\mathscr{S} \subset \mathscr{B}$ we now define

$$\varphi_{(\mathscr{S},\boldsymbol{n})} := \sigma_{(\mathscr{S},\boldsymbol{n})} + \varrho_{(\mathscr{S},\boldsymbol{n})} : \ \mathscr{S} \to \mathscr{W}, \tag{7.1}$$

where $\sigma_{(\mathscr{S},\boldsymbol{n})}$ is characterized by Proposition 4 of Section 5 and $\varrho_{(\mathscr{S},\boldsymbol{n})}$ is given by (6.11) or (6.13).

It follows from Proposition 4 of Section 5 and from (6.13) that the functions $\varphi_{(\mathscr{S},\boldsymbol{n})}$ defined by (7.1) satisfy the following two conditions:

(A) For every oriented biregular surface $(\mathscr{S}, \boldsymbol{n})$ with $\mathscr{S} \subset \mathscr{B}$, $\varphi_{(\mathscr{S},\boldsymbol{n})}$ is continuous.

(B) If $(\mathscr{S}, \boldsymbol{n})$ and $(\mathscr{S}', \boldsymbol{n}')$ are oriented biregular surfaces such that $\mathscr{S}' \subset \mathscr{S} \subset \mathscr{B}$ and $\boldsymbol{n}' = \boldsymbol{n}|_{\mathscr{S}'}$, then

$$\varphi_{(\mathscr{S}',\boldsymbol{n}')} = \varphi_{(\mathscr{S},\boldsymbol{n})}|_{\mathscr{S}'}. \tag{7.2}$$

It follows from Proposition 1 of Section 4, applied to the case when $\mathscr{P}$ and $\mathscr{Q}$ are suitably chosen biregular interior parts of $\mathscr{B}$ in simple edge contact, that the function γ of Assumption I of Section 5 satisfies $\gamma(x, -\boldsymbol{u}, \boldsymbol{w}) = -\gamma(x, \boldsymbol{u}, \boldsymbol{w})$ for all $x \in \mathscr{B}$ and all $(\boldsymbol{u}, \boldsymbol{w}) \in (\mathscr{U}^2)_\perp$. In view of the assumed continuity of γ on its entire domain Clo $\mathscr{B} \times (\mathscr{U}^2)_\perp$, we conclude that

$$\gamma(\cdot, -\boldsymbol{u}, \boldsymbol{w}) = -\gamma(\cdot, \boldsymbol{u}, \boldsymbol{w}) \quad \text{for all } (\boldsymbol{u}, \boldsymbol{w}) \in (\mathscr{U}^2)_\perp \tag{7.3}$$

i.e. that the assumption (6.1) of Proposition 1 of Section 6 is automatically satisfied. Using this fact and the fact that an interior section $\mathscr{P}$ does not have any peripheral sides, we conclude from (4.7) and Propositions 1 and 2 of Section 6 that the functions $\varphi_{(\mathscr{S},\boldsymbol{n})}$ also satisfy the following third condition:

(C) There is a continuous function f: Clo $\mathscr{B} \to \mathscr{W}$ such that, for every biregular part $\mathscr{P} \in \Omega_{\text{int}}$ and every biregular partition $\mathfrak{P}$ for $\mathscr{P}$, we have

$$\sum_{\mathscr{S} \in \mathfrak{P}_2} \int_{\mathscr{S}} \varphi_{(\mathscr{S},\boldsymbol{n})} \, da = \int_{\mathscr{P}} f \, dv. \tag{7.4}$$

Now let $x \in \mathscr{B}$ and $\boldsymbol{u} \in \mathscr{U}$ be given. Then the plane section $(x + \{\boldsymbol{u}\}^{\perp}) \cap \mathscr{B}$ includes biregular plane surfaces $\mathscr{T}$ that contain x. For each such surface $\mathscr{T}$, the constant $\boldsymbol{u}$ is an orientation of $\mathscr{T}$ and hence $\varphi_{(\mathscr{T},\boldsymbol{u})}$, $\sigma_{(\mathscr{T},\boldsymbol{u})}$, and $\varrho_{(\mathscr{T},\boldsymbol{u})}$ are all meaningful. It follows from condition (B) that

$$\delta(x, \boldsymbol{u}) := \varphi_{(\mathscr{T},\boldsymbol{u})}(x) \tag{7.5}$$

depends only on x and $\boldsymbol{u}$ and not on what particular plane surface $\mathscr{T}$ has been selected. Similar statements apply to

$$\alpha(x, \boldsymbol{u}) := \sigma_{(\mathscr{T},\boldsymbol{u})}(x) \tag{7.6}$$

and

$$\beta(x, \boldsymbol{u}) := \varrho_{(\mathscr{T},\boldsymbol{u})}(x) \tag{7.7}$$

because of Proposition 4 of Section 5 and (6.13).

We now add one more assumption.

Assumption IV. *For each* $\boldsymbol{u} \in \mathscr{U}$, *the function* $\alpha(\cdot, \boldsymbol{u}) : \mathscr{B} \to \mathscr{W}$ *given by* (7.6) *is continuous.*

It follows from (6.13) and (7.7) that

$$\omega\beta(x, \boldsymbol{u}) = \omega \operatorname{div}_x (\boldsymbol{G}(\cdot, \boldsymbol{u})) - \boldsymbol{u} \cdot \nabla_x(\omega \boldsymbol{G}(\cdot, \boldsymbol{u}))\, \boldsymbol{u} \tag{7.8}$$

for all $x \in \mathscr{B}$, $\boldsymbol{u} \in \mathscr{U}$, and $\omega \in \mathscr{W}^*$. Since the function $\boldsymbol{G}$ of Assumption III was assumed to be of class C^1, it follows from (7.8) that the function $\beta(\cdot, \boldsymbol{u}) : \mathscr{B} \to \mathscr{W}$ is continuous. Hence, since $\delta = \alpha + \beta$ by (7.1) and (7.5)–(7.7), we conclude that the functions $\varphi_{(\mathscr{S},\boldsymbol{u})}$ also satisfy the following fourth condition:

(D) For every $\boldsymbol{u} \in \mathscr{U}$, the function $\delta(\cdot, \boldsymbol{u}) : \mathscr{B} \to \mathscr{W}$ given by (7.5) in terms of $\varphi_{(\mathscr{T},\boldsymbol{u})}$ is continuous.

The main result of [N3] is that the conditions (A)–(D) imply the existence of a continuous function

$$\boldsymbol{F} : \mathscr{B} \to \operatorname{Lin}(\mathscr{V}, \mathscr{W}) \tag{7.9}$$

such that

$$\boldsymbol{F}(x)\, \boldsymbol{n}(x) = \varphi_{(\mathscr{S},\boldsymbol{n})}(x) = \sigma_{(\mathscr{S},\boldsymbol{n})}(x) + \varrho_{(\mathscr{S},\boldsymbol{n})}(x) \quad \text{for all } x \in \mathscr{S} \tag{7.10}$$

for all oriented biregular surfaces $(\mathscr{S}, \boldsymbol{n})$ with $\mathscr{S} \subset \mathscr{B}$.

Remark 1. If (D) is replaced by a stronger condition that need not be satisfied by the functions $\varphi_{(\mathscr{S},\boldsymbol{n})}$ here (namely that $\varphi_{(\mathscr{S},\boldsymbol{n})}$ is uniformly bounded with respect to $\mathscr{S}$), the conclusion above was already obtained by one of us in 1957 (see [N1]). If one assumes that the values of $\tau_{\mathscr{P}}(x)$ of the functions $\tau_{\mathscr{P}}$ of Assumption II of Section 5 depend on x and $\mathscr{P}$ only through x, $\boldsymbol{v}_{\mathscr{P}}(x)$, and the tangential gradient of $\boldsymbol{v}_{\mathscr{P}}$ at x and if one also assumes that this dependence is of class C^1, then the conclusion can also be obtained from the recent result of Fosdick & Virga [FV]. □

It follows from (7.3) that, under the present hypotheses, the function G of Assumption III of Section 6 must satisfy

$$G(\cdot, -u) = -G(\cdot, u) \quad \text{for all } u \in \mathscr{U}. \tag{7.11}$$

Putting the results above together with Proposition 5 of Section 5, we obtain the following conclusion

Theorem I. *Let I be a quasi-balanced contact interaction for which Assumptions I-IV are satisfied. Then there is a function*

$$G: \text{ Clo } \mathscr{B} \times \text{Usph } \mathscr{V} \to \text{Lin}(\mathscr{V}, \mathscr{W}) \tag{7.12}$$

of class C^1 *satisfying* (7.11) *and a continuous function*

$$F: \mathscr{B} \to \text{Lin}(\mathscr{V}, \mathscr{W}) \tag{7.13}$$

such that, for any two disjoint sections $\mathscr{P}$ and $\mathscr{Q}$ of $\mathscr{B}$ in biregular contact, we have

$$I(\mathscr{P}, \mathscr{Q}) = \sum_{\mathscr{S} \in \mathfrak{P}_2 \cap \mathfrak{Q}_2} \int_{\mathscr{S}} (F v_{\mathscr{P}} - \text{div}_{\mathscr{S}} G(1_{\mathscr{S} \subset \mathscr{E}}, v_{\mathscr{P}}|_{\mathscr{S}}))\, da$$
$$+ \sum_{\mathscr{C} \in \mathfrak{P}_1 \cap \mathfrak{Q}_1} \int_{\mathscr{C}} \left(\sum_{\mathscr{S} \in \mathfrak{K}(\mathscr{C})} G(x, v_{\mathscr{P};\mathscr{S}}(x))\, v_{\mathscr{S}}(x) \right) dl_x, \tag{7.14}$$

whenever $\mathfrak{P}$ and $\mathfrak{Q}$ *are appropriate biregular partitions of $\mathscr{P}$ and $\mathscr{Q}$, and when* $\mathfrak{K}(\mathscr{C})$ *for each* $\mathscr{C} \in \mathfrak{P}_1 \cap \mathfrak{Q}_1$, *is defined according to Definition 1 of Section 5.*

Remark 2. If an interaction I satisfies the hypotheses of Theorem I, it cannot be a pure edge interaction as defined at the end of Section 5. To see this, assume that I is a pure edge interaction and that $x \in \mathscr{B}$ is given. Then, by (7.1) and (7.10),

$$0 = \sigma_{(\mathscr{S},n)}(x) = F(x)\, n(x) - \varrho_{(\mathscr{S},n)}(x) \tag{7.15}$$

for all oriented surfaces $(\mathscr{S}, n)$ with $x \in \mathscr{S}$ and $\mathscr{S} \subset \mathscr{B}$. If (7.15) holds, then $\varrho_{(\mathscr{S},n)}(x)$ can depend on $(\mathscr{S}, n)$ only through the value $n(x)$. But (6.13) shows that this can be the case only when $G(x, \cdot)$ has a zero gradient, *i.e.* only when $G(x, u)$ does not depend on $u \in \mathscr{U}$. In view of (7.11) it follows that $G(x, \cdot) = 0$. Since $x \in \mathscr{B}$ was arbitrary, G must be zero. □

The following result follows from Propositions 1 and 2 of Section 6 and from (7.10) and (6.6).

Theorem II. *Let I be a quasi-balanced interaction for which Assumptions I–IV are satisfied and let G and F be the functions described in Theorem I. For every section $\mathscr{P}$ of $\mathscr{B}$ and every appropriate biregular partition $\mathfrak{P}$ for $\mathscr{P}$ we then have*

$$I(\mathscr{P}, \mathscr{P}^b) = \sum_{\mathscr{S} \in \mathfrak{P}_2^{\text{int}}} \int_{\mathscr{S}} F v_{\mathscr{P}}\, dv + \sum_{\mathscr{S} \in \mathfrak{P}_2^{\text{per}}} \int_{\text{Rbo}\mathscr{S} \cap \text{Bdy}\mathscr{P}^b} G(x, v_{\mathscr{P};\mathscr{S}}(x))\, v_{\mathscr{S}}(x)\, dl_x. \tag{7.16}$$

In particular, if $\mathscr{P}$ is an interior biregular part of $\mathscr{P}$ we have

$$I(\mathscr{P}, \mathscr{P}^b) = \int_{\text{Rbo}\mathscr{P}} F v_{\mathscr{P}}\, dv = \int_{\mathscr{P}} f\, dv. \tag{7.17}$$

If the function $F : \mathscr{B} \to \operatorname{Lin}(\mathscr{V}, \mathscr{W})$ is of class C^1, it follows from the divergence theorem and the fact that (7.17) holds for all biregular $\mathscr{P} \in \Omega_{\text{int}}$ that

$$f = \operatorname{div} F. \tag{7.18}$$

8. Boundary conditions

We continue to assume that an interaction I satisfying the hypotheses of Theorems I and II of the preceding section is given. We have seen in the preceding section that the field equation (7.18) is the same as it would be if there are no edge interactions, *i.e.* if $G = 0$. The presence of edge interaction becomes manifest only when suitable boundary conditions are imposed. Physically, such boundary conditions describe the action of the environment on the boundary of the body $\mathscr{B}$ under consideration. We consider only the case when the action of the environment is a pure surface action. *i.e.* when there are no edge interactions between $\mathscr{B}$ and the environment. Thus, we strengthen our assumption that I be quasi-balanced as follows:

Assumption V. *There are continuous functions* $f : \operatorname{Clo} \mathscr{B} \to \mathscr{W}$ *and* $h : \operatorname{Rbo} \mathscr{B} \to \mathscr{W}$ *such that*

$$I(\mathscr{P}, \mathscr{P}^b) = \int_{\mathscr{P}} f\, dv + \int_{\operatorname{Clo}\mathscr{P} \cap \operatorname{Rby}\mathscr{B}} h\, da \tag{8.1}$$

for all parts $\mathscr{P} \in \Omega$.

The function h describes the surface action of the environment on $\mathscr{B}$. Proposition 1 of Section 4 can now be strengthened as follows:

Proposition 1. *If Assumption V is valid, then I is skew in the sense that*

$$I(\mathscr{P}, \mathscr{Q}) = -I(\mathscr{Q}, \mathscr{P}) \quad \textit{for all } \mathscr{P}, \mathscr{Q} \in (\Omega^2)_{\text{dis}}. \tag{8.2}$$

The proof follows again from Theorem A, p. 76, of [N2] because $(\mathscr{P} \mapsto I(\mathscr{P}, \mathscr{P}^b)) : \Omega \to \mathscr{W}$ is additive when (8.1) holds.

Theorem III. *Let I be an interaction for which Assumptions I–V are valid and let G be the function determined by I as described in Theorem I. If* $\mathfrak{P}$ *is a biregular partition for* $\mathscr{B}$ *and* $\mathscr{C} \in \mathfrak{P}_1$ *is an edge of* $\mathscr{B}$, *we have*

$$G(x, \boldsymbol{v}_{\mathscr{B};\mathscr{S}_1}(x))\, \boldsymbol{v}_{\mathscr{S}_1}(x) + G(x, \boldsymbol{v}_{\mathscr{B};\mathscr{S}_2}(x))\, \boldsymbol{v}_{\mathscr{S}_2}(x) = 0 \quad \textit{for all } x \in \mathscr{C}, \tag{8.3}$$

when $\mathscr{S}_1, \mathscr{S}_2 \in \mathfrak{P}_2$ *are the two sides adjacent to the edge* $\mathscr{C}$.

Proof. If $\mathscr{C}$ is a cusped edge of $\mathscr{B}$, we have $\boldsymbol{v}_{\mathscr{S}_1} = \boldsymbol{v}_{\mathscr{S}_2}$ and $\boldsymbol{v}_{\mathscr{B};\mathscr{S}_1} = -\boldsymbol{v}_{\mathscr{B};\mathscr{S}_2}$ and (8.3) is valid because of (7.11). Assume, then, that $\mathscr{C}$ is not a cusped edge of $\mathscr{B}$ and let $x \in \mathscr{C}$ be given. For every curve $\mathscr{C}'$ with $x \in \mathscr{C}'$ and $\mathscr{C}' \subset \mathscr{C}$, we can construct disjoint sections $\mathscr{P}_1$ and $\mathscr{P}_2$ of $\mathscr{B}$ with the following properties: (i) $\mathscr{P}_1$ and $\mathscr{P}_2$ are in simple edge contact along $\mathscr{C}'$, (ii) one of the sides of $\mathscr{P}_1$ adjacent to $\mathscr{C}'$, say $\mathscr{S}'_1$, is included in $\mathscr{S}_1$, (iii) one of the sides of $\mathscr{P}_2$ adjacent to $\mathscr{C}'$, say $\mathscr{S}'_2$, is included in $\mathscr{S}_2$, and (iv) the internal sides of $\mathscr{P}_1$ and $\mathscr{P}_2$ that are adjacent to $\mathscr{C}'$ are not in cusped contact. Figure 1 illustrates this situation by showing a cross-section orthogonal to $\mathscr{C}$. It follows from Definition 1 of Section 5 that

$$\mathfrak{K}_{\mathscr{P}_1,\mathscr{P}_2}(\mathscr{C}') = \{\mathscr{S}'_1\}, \quad \mathfrak{K}_{\mathscr{P}_2,\mathscr{P}_1}(\mathscr{C}') = \{\mathscr{S}'_2\}.$$

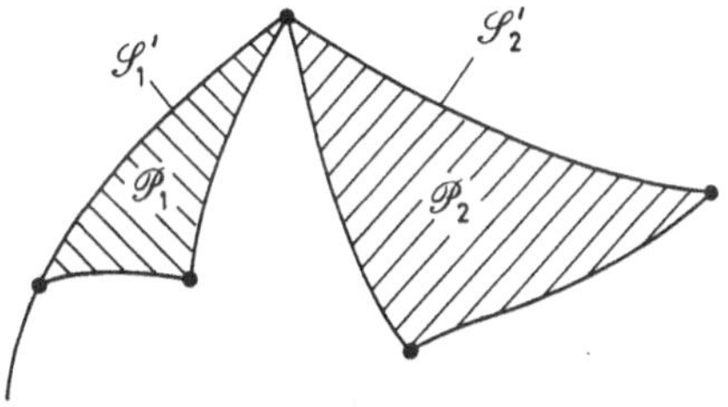

Figure 1

Since $I(\mathscr{P}_1, \mathscr{P}_2) = -I(\mathscr{P}_2, \mathscr{P}_1)$ by Proposition 1 above, it follows from Assumption I of Section 5 that

$$\int_{\mathscr{C}'} \gamma(z, \mathbf{v}_{\mathscr{B};\mathscr{S}_1}(z), \mathbf{v}_{\mathscr{S}_1}(z))\, dl_z = -\int_{\mathscr{C}'} \gamma(z, \mathbf{v}_{\mathscr{B};\mathscr{S}_2}(z), \mathbf{v}_{\mathscr{S}_2}(z))\, dl_z \tag{8.4}$$

because $\mathbf{v}_{\mathscr{B};\mathscr{S}_1}|_{\mathscr{C}'} = \mathbf{v}_{\mathscr{P}_1;\mathscr{S}'_1}|_{\mathscr{C}'}$, $\mathbf{v}_{\mathscr{B};\mathscr{S}_2}|_{\mathscr{C}'} = \mathbf{v}_{\mathscr{P}_2;\mathscr{S}'_2}|_{\mathscr{C}'}$, $\mathbf{v}_{\mathscr{S}'_1}|_{\mathscr{C}'} = \mathbf{v}_{\mathscr{S}_1}|_{\mathscr{C}'}$ and $\mathbf{v}_{\mathscr{S}'_2}|_{\mathscr{C}'} = \mathbf{v}_{\mathscr{S}_2}|_{\mathscr{C}'}$. Since the curve $\mathscr{C}'$ satisfying $x \in \mathscr{C}'$ and $\mathscr{C}' \subset \mathscr{C}$ was arbitrary, it follows from (8.4) and the continuity of γ that

$$\gamma(x, \mathbf{v}_{\mathscr{B};\mathscr{S}_1}(x), \mathbf{v}_{\mathscr{S}_1}(x)) = -\gamma(x, \mathbf{v}_{\mathscr{B};\mathscr{S}_2}(x), \mathbf{v}_{\mathscr{S}_2}(x)).$$

In view of Assumption III, (6.6), the desired result (8.3) follows. □

If $\mathscr{C}$ is not a genuine edge, *i.e.* if $\mathbf{v}_{\mathscr{S}_1}(x) = -\mathbf{v}_{\mathscr{S}_2}(x)$ for all $x \in \mathscr{C}$, then $\mathbf{v}_{\mathscr{B};\mathscr{S}_1}(x) = \mathbf{v}_{\mathscr{B};\mathscr{S}_2}(x) = \mathbf{v}_{\mathscr{B}}(x)$ for all $x \in \mathscr{C}$ and hence (8.3) is trivially satisfied.

Theorem IV. *Let I be an interaction for which Assumptions I–V are satisfied and which therefore determines the functions* $\mathbf{F}$ *and* $\mathbf{G}$ *as described in Theorem I. For all regular surfaces $\mathscr{S}$ included in the boundary of the body $\mathscr{B}$ we then have*

$$\lim_{y \to x} (\mathbf{F}(y)\, \mathbf{v}_{\mathscr{B}}(x)) = \varrho_{(\mathscr{S}, \mathbf{v}_{\mathscr{B}}|_{\mathscr{S}})}(x) - h(x) \quad \text{for all } x \in \mathscr{S}, \tag{8.5}$$

where $\varrho_{(\mathscr{S}, \mathbf{v}_{\mathscr{B}}|_{\mathscr{S}})}$ is given by (6.11) *or* (6.13).

Proof. Let a regular surface $\mathscr{S}$ included in Bdy $\mathscr{B}$ and $x \in \mathscr{S}$ be given. It is not hard to construct a section $\mathscr{P}$ of $\mathscr{B}$ with an appropriate biregular partition $\mathfrak{P} := \{\mathscr{P}, \mathscr{T}, \mathscr{T}', \mathscr{C}\}$ for $\mathscr{P}$ such that (i) $\mathscr{T}$ is an external side of $\mathscr{P}$ with $x \in \mathscr{T}$ and Clo $\mathscr{T} \subset \mathscr{S}$. (ii) $\mathscr{T}'$ is an internal side of $\mathscr{P}$, and (iii) $\mathscr{C} = \mathrm{Bo}\,\mathscr{T} = \mathrm{Bo}\,\mathscr{T}'$ is the only edge of $\mathscr{P}$ and $\mathscr{C}$ is a cusped edge (see Figure 2). Furthermore, $\mathscr{P}$ can be constructed in such a way that $\mathscr{P}$ is included in an arbitrarily small neighborhood of x and the values of $\mathbf{v}_{\mathscr{P}}|_{\mathscr{T}'}$ differ arbitrarily little from $-\mathbf{v}_{\mathscr{B}}(x)$. Since $\mathfrak{P}_2^{\mathrm{int}} = \{\mathscr{T}'\}$, $\mathfrak{P}_2^{\mathrm{per}} = \{\mathscr{T}\}$ and Rbo $\mathscr{T} \subset$ Bdy $\mathscr{P}^b$, we see that the formula (7.16) of Theorem II gives

$$I(\mathscr{P}, \mathscr{P}^b) = \int_{\mathscr{T}'} \mathbf{F} \mathbf{v}_{\mathscr{P}}\, da + \int_{\mathrm{Rbo}\mathscr{T}} \mathbf{G}(z, \mathbf{v}_{\mathscr{P};\mathscr{T}}(z))\, \mathbf{v}_{\mathscr{S}}(z)\, dl_z. \tag{8.6}$$

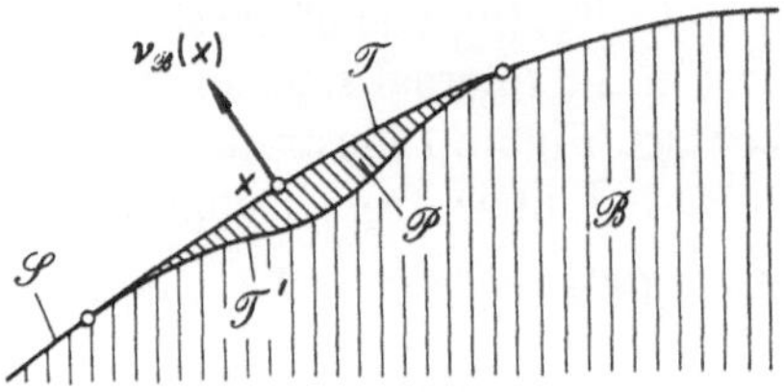

Figure 2

Using the corollary to the surface-divergence theorem stated in Section 2, the notation (6.11), and the fact that $\boldsymbol{v}_{\mathscr{P};\mathscr{T}} = \boldsymbol{v}_{\mathscr{B}}|_{\mathrm{Clo}\,\mathscr{T}}$, it follows from (8.6) that

$$I(\mathscr{P}, \mathscr{P}^b) = \int_{\mathscr{T}'} \boldsymbol{F}\, \boldsymbol{v}_{\mathscr{P}}\, da + \int_{\mathscr{T}} \varrho_{(\mathscr{T}, \boldsymbol{v}_{\mathscr{B}}|_{\mathscr{T}})}\, da. \tag{8.7}$$

On the other hand, since $\mathrm{Clo}\,\mathscr{P} \cap \mathrm{Rby}\,\mathscr{B} = \mathrm{Clo}\,\mathscr{T}$, it follows from (8.1) that

$$I(\mathscr{P}, \mathscr{P}^b) = \int_{\mathscr{P}} f\, dv + \int_{\mathscr{T}} h\, da$$

and hence, by (8.7), that

$$\int_{\mathscr{T}'} \boldsymbol{F} \boldsymbol{v}_{\mathscr{P}}\, da = \int_{\mathscr{T}} (h - \varrho_{(\mathscr{T}, \boldsymbol{v}_{\mathscr{B}}|_{\mathscr{T}})})\, da + \int_{\mathscr{P}} f\, dv. \tag{8.8}$$

Since $\boldsymbol{v}_{\mathscr{P}}|_{\mathscr{T}'}$ differs arbitrarily little from $-\boldsymbol{v}_{\mathscr{B}}(x)$, we obtain the desired result (8.5) by dividing (8.8) by the area of $\mathscr{T}$ and then taking the limit as $\mathscr{P}$ shrinks to $\{x\}$. □

9. Examples, surface tension

As a special example, we now consider the case when the function $\boldsymbol{G}$ of Assumption III, Section 6 can be expressed in the form (6.7) when $\boldsymbol{G}'(x, \boldsymbol{u})$ depends *linearly* on $\boldsymbol{u}$. More precisely, we assume that there is a function

$$\boldsymbol{A}\colon\ \mathrm{Clo}\,\mathscr{B} \to \mathrm{Lin}\,(\mathscr{V}, \mathrm{Lin}\,(\mathscr{V}, \mathscr{W})) \tag{9.1}$$

of class C^1 such that $\boldsymbol{G}\colon\ \mathrm{Clo}\,\mathscr{B} \times \mathscr{U} \to \mathrm{Lin}\,(\mathscr{V}, \mathscr{W})$ is given by

$$\boldsymbol{G}(x, \boldsymbol{u}) = (\boldsymbol{A}(x)\,\boldsymbol{u})\,(\mathbf{1}_{\mathscr{V}} - \boldsymbol{u} \otimes \boldsymbol{u}) \quad \text{for all } x \in \mathrm{Clo}\,\mathscr{B}, \quad \boldsymbol{u} \in \mathscr{U}. \tag{9.2}$$

Let an oriented regular surface $(\mathscr{S}, \boldsymbol{n})$ with $\mathscr{S} \subset \mathrm{Clo}\,\mathscr{B}$ be given. If we substitute (9.2) into (6.11), we obtain

$$\varrho_{(\mathscr{S}, \boldsymbol{n})} = \mathrm{div}_{\mathscr{S}}\,((\boldsymbol{A}\,|_{\mathscr{S}}\,\boldsymbol{n})\,(\mathbf{1}_{\mathscr{V}} - \boldsymbol{n} \otimes \boldsymbol{n})). \tag{9.3}$$

In view of (2.6), (9.3) is equivalent to

$$\omega \varrho_{(\mathscr{S}, \boldsymbol{n})} = \mathrm{div}_{\mathscr{S}}\,((\mathbf{1}_{\mathscr{V}} - \boldsymbol{n} \otimes \boldsymbol{n})\,(\omega \boldsymbol{A}\,|_{\mathscr{S}}\,\boldsymbol{n})) \quad \text{for all } \omega \in \mathscr{W}^*, \tag{9.4}$$

where $\omega \boldsymbol{A} \in \mathrm{Lin}\,(\mathscr{V}, \mathscr{V}^*) \cong \mathrm{Lin}\,\mathscr{V}$ is characterized by

$$((\omega \boldsymbol{A})\,\boldsymbol{u}) \cdot \boldsymbol{v} = \omega((\boldsymbol{A}\boldsymbol{u})\,\boldsymbol{v}) \quad \text{for all } \boldsymbol{u}, \boldsymbol{v} \in \mathscr{V}, \omega \in \mathscr{W}^*. \tag{9.5}$$

Now let $\omega \in \mathscr{W}^*$ be given and put

$$\boldsymbol{B} := \omega \boldsymbol{A}\,|_{\mathscr{S}}\colon\ \mathscr{S} \to \mathrm{Lin}\,\mathscr{V} \tag{9.6}$$

so that (9.4) gives

$$\omega\varrho_{(\mathscr{S},n)} = \operatorname{div}_{\mathscr{S}}(\boldsymbol{B}\boldsymbol{n} - (\boldsymbol{n}\cdot\boldsymbol{B}\boldsymbol{n})\,\boldsymbol{n}). \tag{9.7}$$

In order to evaluate the right side of (9.7) we observe that the surface divergence satisfies rules that are analogous to the rules stated in Section 67 of [FDSI] for the ordinary divergence. In these rules, ordinary gradients must be replaced by surface-gradients defined according to Definition 1 of Section 2. The analogue of Proposition 2 of Section 67 gives

$$\operatorname{div}_{\mathscr{S}}(\boldsymbol{B}\boldsymbol{n}) = \boldsymbol{n}\cdot\operatorname{div}_{\mathscr{S}}(\boldsymbol{B}^T) + \operatorname{tr}(\boldsymbol{B}^T\nabla\boldsymbol{n}). \tag{9.8}$$

The analogue of Proposition 1 of Section 67 gives

$$\operatorname{div}_{\mathscr{S}}((\boldsymbol{n}\cdot\boldsymbol{B}\boldsymbol{n})\,\boldsymbol{n}) = (\boldsymbol{n}\cdot\boldsymbol{B}\boldsymbol{n})\operatorname{div}_{\mathscr{S}}\boldsymbol{n} + \nabla(\boldsymbol{n}\cdot\boldsymbol{B}\boldsymbol{n})\cdot\boldsymbol{n}. \tag{9.9}$$

Now, if $\eta:\mathscr{S}\to\mathbb{R}$ is of class C^1, then $\nabla\eta$ is tangential (see Definition 3 of Section 2) by (2.2), which means that $(\nabla\eta)\cdot\boldsymbol{n} = 0$. Hence the second term on the right side of (9.9) is zero. Therefore, using (9.9), (9.8), and (9.6), we see that (9.7) yields

$$\omega\varrho_{(\mathscr{S},n)} = \boldsymbol{n}\cdot\operatorname{div}_{\mathscr{S}}(\boldsymbol{B}^T) + \operatorname{tr}(\boldsymbol{B}^T\nabla\boldsymbol{n}) - (\boldsymbol{n}\cdot\boldsymbol{B}\boldsymbol{n})\operatorname{div}_{\mathscr{S}}\boldsymbol{n}. \tag{9.10}$$

Given $x\in\mathscr{S}$, the lineon $\overline{\boldsymbol{L}} = -\nabla_x\boldsymbol{n}$ has the following significance (see Chapter 3 of [FDSII]): $\overline{\boldsymbol{L}}$ is symmetric, one of its spectral values is zero and the other two are the principal curvatures of $\mathscr{S}$ at x. The mean curvature of $\mathscr{S}$ at x is given by

$$\mathrm{H}_{\mathscr{S}}(x) := \tfrac{1}{2}\operatorname{tr}\overline{\boldsymbol{L}} = -\tfrac{1}{2}\operatorname{tr}\nabla_x\boldsymbol{n} = -\tfrac{1}{2}(\operatorname{div}_{\mathscr{S}}\boldsymbol{n})(x) \tag{9.11}$$

($\overline{\boldsymbol{L}}$ is closely related to what are often called the "Weingarten map" and the "Second Fundamental Form"). Using (9.11), (9.6), and the fact that $\nabla\boldsymbol{n}$ has values in $\operatorname{Sym}\mathscr{V}$, we see that (9.10) becomes

$$\omega\varrho_{(\mathscr{S},n)} = \boldsymbol{n}\cdot\operatorname{div}_{\mathscr{S}}((\omega\boldsymbol{A})^T|_{\mathscr{S}}) + \operatorname{tr}(\nabla\boldsymbol{n}(\omega\boldsymbol{A}|_{\mathscr{S}})) + 2\mathrm{H}_{\mathscr{S}}(\boldsymbol{n}\cdot(\omega\boldsymbol{A}|_{\mathscr{S}})\,\boldsymbol{n}), \tag{9.12}$$

valid for all $\omega\in\mathscr{W}^*$. The use of (9.12) leads to a more explicit form for the boundary condition (8.5).

If we substitute (9.2) into (8.3) and observe that $\boldsymbol{v}_{\mathscr{S}_1}(x)\cdot\boldsymbol{v}_{\mathscr{B};\mathscr{S}_1}(x) = 0 = \boldsymbol{v}_{\mathscr{S}_2}(x)\cdot\boldsymbol{v}_{\mathscr{B};\mathscr{S}_2}(x)$ for all $x\in\mathscr{C}$ we see that the boundary-edge condition (8.3) becomes

$$(\boldsymbol{A}(x)\,\boldsymbol{v}_{\mathscr{B};\mathscr{S}_1}(x))\,\boldsymbol{v}_{\mathscr{S}_1}(x) + (\boldsymbol{A}(x)\,\boldsymbol{v}_{\mathscr{B};\mathscr{S}_2}(x))\,\boldsymbol{v}_{\mathscr{S}_2}(x)) = 0 \quad \text{for all } x\in\mathscr{C} \tag{9.13}$$

when $\boldsymbol{G}$ has the form (9.2).

Remark 1. If we substitute (9.12) into the boundary condition (8.5), we obtain a result that is consistent with one of boundary conditions that Toupin obtained in his treatment of non-simple elastic materials (see equations (10.9) or (10.14) on p. 102 or [To2] or equation (7.9) on p. 402 of [To1]). The boundary-edge condition (9.13) is also consistent with a condition found by Toupin (see (7.11) on p. 402 of [To1]). □

We now specialize further. We assume that the space $\mathscr{W}$ coincides with $\mathscr{V}$ and that the function $\boldsymbol{A}$ of (9.1) is defined by

$$\boldsymbol{A}(x)\,\boldsymbol{u} := (\boldsymbol{k}(x)\cdot\boldsymbol{u})\,\mathbf{1}_{\mathscr{V}} \quad \text{for all } x\in\operatorname{Clo}\mathscr{B},\ \boldsymbol{u}\in\mathscr{V} \tag{9.14}$$

in terms of a given function

$$\boldsymbol{k} : \operatorname{Clo} \mathscr{B} \to \mathscr{V} \tag{9.15}$$

of class C^1. Let $\boldsymbol{w} \in \mathscr{V}$ be given, so that $\boldsymbol{w}\cdot \in \mathscr{V}^*$ (see the explanation of the identification $\mathscr{V} \cong \mathscr{V}^*$ in Section 41, p. 134, of [FDSI]). We then have

$$\boldsymbol{w} \cdot \boldsymbol{A} = \boldsymbol{w} \otimes \boldsymbol{k}, \quad (\boldsymbol{w} \cdot \boldsymbol{A})^T = \boldsymbol{k} \otimes \boldsymbol{w}. \tag{9.16}$$

Substitution of (9.16) and $\omega := \boldsymbol{w}\cdot$ into (9.10) gives

$$\boldsymbol{w}\cdot \varrho_{(\mathscr{S},\boldsymbol{n})} = \boldsymbol{n} \cdot \operatorname{div}_{\mathscr{S}} (\boldsymbol{k}|_{\mathscr{S}} \otimes \boldsymbol{w}) + \operatorname{tr} (\nabla \boldsymbol{n}(\boldsymbol{w} \otimes \boldsymbol{k}|_{\mathscr{S}})) + 2\mathrm{H}_{\mathscr{S}}(\boldsymbol{k}|_{\mathscr{S}} \cdot \boldsymbol{n}) (\boldsymbol{n} \cdot \boldsymbol{w}).$$

Using analogues of the rules (67.10) and (66.9) of [FDSI] we obtain

$$\boldsymbol{w}\cdot \varrho_{(\mathscr{S},\boldsymbol{n})} = \nabla(\boldsymbol{k}|_{\mathscr{S}} \cdot \boldsymbol{n}) \cdot \boldsymbol{w} + 2\mathrm{H}_{\mathscr{S}}(\boldsymbol{k}|_{\mathscr{S}} \cdot \boldsymbol{n}) (\boldsymbol{n} \cdot \boldsymbol{w}).$$

Since this equation is valid for all $\boldsymbol{w} \in \mathscr{V}$, we conclude that

$$\varrho_{(\mathscr{S},\boldsymbol{n})} = \nabla \sigma_{\mathscr{S}} + 2\mathrm{H}_{\mathscr{S}}\, \sigma_{\mathscr{S}} \boldsymbol{n} \quad \text{where } \sigma_{\mathscr{S}} := \boldsymbol{k}|_{\mathscr{S}} \cdot \boldsymbol{n}. \tag{9.17}$$

If (9.14) and hence (9.17) is valid and if the function $\boldsymbol{F} : \mathscr{B} \to \operatorname{Lin} \mathscr{V}$ has a continuous extension (also denoted by $\boldsymbol{F}$) to Clo $\mathscr{B}$, then the boundary condition (8.5) becomes

$$\boldsymbol{F}\boldsymbol{v}_{\mathscr{B}}|_{\mathscr{S}} = \nabla \sigma_{\mathscr{S}} + 2\mathrm{H}_{\mathscr{S}} \sigma_{\mathscr{S}} \boldsymbol{v}_{\mathscr{B}|\mathscr{S}} - \boldsymbol{h}|_{\mathscr{S}}. \tag{9.18}$$

We now assume, in addition, that $\boldsymbol{F}$ has the form $\boldsymbol{F} := -p\mathbf{1}_{\mathscr{V}}$ for a given

$$p : \operatorname{Clo} \mathscr{B} \to \mathbb{R} \tag{9.19}$$

and that $\boldsymbol{h} = -p_0 \boldsymbol{v}_{\mathscr{B}}$ for a given

$$p_0 : \operatorname{Rbo} \mathscr{B} \to \mathbb{R}. \tag{9.20}$$

In this case, (19.8) reduces to

$$(p_0 - p)\, \boldsymbol{v}_{\mathscr{B}}|_{\mathscr{S}} = \nabla \sigma_{\mathscr{S}} + 2\mathrm{H}_{\mathscr{S}} \sigma_{\mathscr{S}} \boldsymbol{v}_{\mathscr{B}}|_{\mathscr{S}}. \tag{9.21}$$

Since $\nabla \sigma_{\mathscr{S}}$ is tangential by (2.2), it follows from (9.21) that $\nabla \sigma_{\mathscr{S}} = 0$ and

$$p_0 - p = 2\mathrm{H}_{\mathscr{S}} \sigma_{\mathscr{S}}, \quad \sigma_{\mathscr{S}} = \text{constant}. \tag{9.22}$$

Remark 2. (9.22) is the classical formula for surface tension in fluids. It has been derived here *without* assuming that the interface between different fluids behaves like a material skin. The classical derivations of (9.22) are based on the assumption that Bdy $\mathscr{B}$ is a "material surface" in some sense. A rigorous modern treatment of such material surfaces is given in [GM], which also contains the derivation of a formula similar to (9.18). □

Acknowledgement. The research of E. G. VIRGA was partially supported by a grant from the Applied Mathematics Program of the U.S. National Science Foundation to The Johns Hopkins University.

Ringraziamento. Ringraziamo la Sig.ra PHYLLIS MIZEL per aver suggerito il titolo di *Intermezzo* per il paragrafo tra il quarto ed il quinto.

References

[FDSI] NOLL, W., "Finite-Dimensional Spaces: Algebra, Geometry, and Analysis, Vol. I", Martinus Nijhoff Publishers, 1987.

[FDSII] NOLL, W., "Finite-Dimensional Spaces: Algebra, Geometry, and Analysis, Vol. II", to be published. A preliminary version of Chapt. 3 is available in the form of lecture notes. These notes were the basis of Sect. 3 of [GM].

[F] FRAENKEL, L. E., "On Regularity of the Boundary in the Theory of Sobolev Spaces", *Proc. London Math. Soc.* (3) **39**, 385–427 (1979).

[FV] FOSDICK, R. L., & E. G. VIRGA, "A Variational Proof of the Stress Theorem of Cauchy", *Arch. Rational Mech. Anal.* **105**, 95–103 (1989).

[GM] GURTIN, M. E., & A. I. MURDOCH, "A Continuum Theory of Elastic Material Surfaces", *Arch. Rational Mech. Anal.* **57**, 291–323 (1974).

[GW] GURTIN, M. E., & W. O. WILLIAMS, "An Axiomatic Foundation for Continuum Thermodynamics", *Arch. Rational Mech. Anal.* **26**, 83–117 (1967).

[GWZ] GURTIN, M. E., W. O. WILLIAMS & W. P. ZIEMER, "Geometric Measure Theory and the Axioms of Continuum Thermodynamics", *Arch. Rational Mech. Anal.* **92**, 1–22 (1986).

[K] KELLOGG, O. D., "Foundations of Potential Theory", Springer, Berlin, 1929.

[N1] NOLL, W., "The Foundations of Classical Mechanics in the Light of Recent Advances in Continuum Mechanics", *Proceedings of the Berkeley Symposium on the Axiomatic Method*, 226–281, Amsterdam, 1959.

[N2] NOLL, W., "Lectures on the Foundations of Continuum Mechanics and Thermodynamics", *Arch. Rational Mech. Anal.* **52**, 62–92 (1973).

[N3] NOLL, W., "On Contactors for Surface Interactions", to appear.

[NV] NOLL, W., & E. G. VIRGA, "Fit Regions and Functions of Bounded Variation", *Arch. Rational Mech. Anal.* **102**, 1–21 (1988).

[S] ŠILHAVÝ, M., "The Existence of the Flux Vector and the Divergence Theorem for General Cauchy Fluxes", *Arch. Rational Mech. Anal.* **90**, 195–212 (1985).

[To1] TOUPIN, R. A., "Elastic Materials with Couple-stresses", *Arch. Rational Mech. Anal.* **11**, 385–414 (1962).

[To2] TOUPIN, R. A., "Theories of Elasticity with Couple-stress", *Arch. Rational Mech. Anal.* **17**, 85–112 (1964).

[Tr] TRUESDELL, C., "A First Course in Rational Continuum Mechanics", Vol. I, Academic Press, 1977. Second edition, corrected, revised, and augmented, in press.

Department of Mathematics
Carnegie Mellon University
Pittsburgh, Pennsylvania

&

Dipartimento di Matematica
Università di Pavia
Pavia, Italy

(Received August 1, 1989)

On a Statistical-Kinetic Model for Generalized Continua

MARIO PITTERI

Dedicated to Bernard Coleman on the occasion of his 60th birthday

1. Introduction

In recent times special attention has been devoted to modelling general continua with structure, both for theoretical reasons and for interest in the applications. Some contributions to this analysis are contained in the proceedings of a meeting held in Canada. Elsewhere [1987] I analyzed a paper of GODDARD [1986] which is contained in those proceedings.

As is not unusual in continuum mechanics, considerations based on kinematics or mechanics of systems of material points can be used to motivate assumptions for continua. In the case of continua with structure, besides the early contributions of GREEN & RIVLIN [1964a] and ERINGEN [1966], [1967], models of increasing generality have been presented by CAPRIZ & PODIO-GUIDUGLI [1976]–[1983] and CAPRIZ [1984]–[1989]. Along a different line of thought, MURDOCH [1983]–[1987] has averaged molecular quantities over space and time to obtain gross fields and equations of balance of a continuum.

A still different line of thought has been adopted by physical chemists and fluid dynamicists, who use a statistical-kinetic approach to obtain gross fields and the equations of balance they satisfy. In this paper I shall adopt this view, the starting point for which can be considered to be a paper of IRVING & KIRKWOOD [1950]. NOLL [1955] provides a more sound and simple mathematical setting for the IRVING & KIRKWOOD theory. PITTERI [1986]–[1988] extends NOLL's theory to include forces that are not necessarily conservative, and applies this more general theory to analyze the motions of mixtures and the behavior of gross fields under change of observer.

In the aforementioned statistical-kinetic models the microscopic individuals, whatever they may be, atoms, molecules, or larger aggregates, are treated as unconstrained material points. There have been attempts to construct statistical-kinetic models of more general continua, by attributing to each microscopic individual more degrees of freedom than a material point has. DAHLER & SCRIVEN [1963], for instance, are mostly interested in the continuum description, in particular in a reasonable equation of balance of rotational momentum for structured continua, to be added to the equations of conservation of mass and of balance of linear momentum. The statistical model, which DAHLER & SCRIVEN [1963] apply to a diatomic fluid, is used to justify a certain form for the equations of balance, and is unnecessarily general as well as not very explicit. For instance, DAHLER & SCRIVEN do not specify whether the 2 atoms in a diatomic molecule should be regarded as constrained or free, and the phase space variables describing the molecule depend heavily on this choice. If the atoms are free, then, according to PITTERI [1986], the stress tensor is symmetric if pairwise interactions of the most general kind compatible with Newtonian mechanics of mass-points and with the physical equivalence of all inertial frames are presumed: according to BRESSAN [1962], the internal force F_{ij} that the point P_j exerts upon P_i has necessarily the constitutive equation

$$F_{ij} = n_{ij}\phi_{ij}(r_{ij}, \|v_j - v_i\|, (v_j - v_i)\cdot n_{ij}), \qquad \phi_{ij}: \mathbb{R}^+\times\mathbb{R}^+\times\mathbb{R} \to \mathbb{R} \tag{1.1}$$

for some choice of functions ϕ_{ij} such that $\phi_{ij} \equiv \phi_{ji}$. Here $n_{ij}[r_{ij}]$ is the unit vector [is the norm] of P_iP_j, and v_i $[v_j]$ is the velocity of P_i [of P_j].

The remark above on the specification of constraints applies in part to a paper of CURTISS, BYRD & HASSAGER [1976], which is in certain aspects more satisfactory than DAHLER & SCRIVEN's. Indeed in the former paper the treatment of constraints is explicit and correct but, for instance, for a diatomic molecule, the authors apply the same theory both to rigid dumbbells and to elastic dumbbells, in which the rigid rod connecting the 2 atoms is replaced by a spring. In this case, in principle, each atom should be considered separately, and the existence and stability of molecules should come out of the dynamics of the whole system, perhaps under certain assumptions on the initial probability distribution function. A rational and realistic treatment of such a dynamical problem seems pretty hard to accomplish.

CURTISS, BYRD & HASSAGER [1976], like DAHLER & SCRIVEN [1963], do not consider the equation of balance of energy. Unlike DAHLER & SCRIVEN, CURTISS, BYRD & HASSAGER do not discuss balance of angular momentum in general. From their analysis of the stress tensor under certain assumptions about microscopic forces, they seem to expect that tensor to be generally symmetric.

ALBLAS [1976], too, provides a statistical-kinetic model for structured continua, and he considers the equation of balance of energy. On the other hand he does not seem to regard a molecule as a constrained system, certainly less so than do DAHLER & SCRIVEN [1963], who at least integrate over Lagrangean co-ordinates and momenta, regarded as the independent variables. ALBLAS instead obtains gross fields by integrating over positions and velocities of all the atoms, so excluding constraints. Also, ALBLAS mentions that multipole stresses should occur also between molecules without internal degrees of freedom, an opinion with which I disagree.

In this paper we analyze the statistical-kinetic evolution of a system composed of a finite number of subsystems, each of which is composed of a finite number of mass-points subjected to ideal constraints. The mass-points are assumed to interact pairwise through a force of the general form (1.1), and to be possibly subjected to external forces. We write the equations of motion in semi-Hamiltonian form; hence, by the general results in PITTERI [1986], we can write an equation of evolution for the probability density function W, an equation which generalizes the standard Liouville equation to systems of mass-points which are not Hamiltonian.

Among the various fields that one can construct, particulary important are those related to the microstructure. Unlike CURTISS, BYRD & HASSAGER [1976], I refrain from writing an equation of balance for the microstructural conjugate momenta, since these are, roughly speaking, the components of a co-vector field on a manifold. To obtain gross quantities we should integrate that field over the manifold and, in general, this integration is not meaningful because the outcome depends on the choice of co-ordinates. In addition, it is not completely obvious which terms and equations the presence of the microstructure should add to the usual equations of balance of mass, linear momentum, rotational momentum and energy. For instance, MURDOCH [1987] proposes that a tensor, rather than an axial vector, should enter an equation of balance of generalized rotational momentum, the skew part of which corresponds, roughly, to the usual balance of rotational momentum.

For the aforementioned reasons I decided to regard the equation of balance of energy as the basic equation of balance, and to use an argument of invariance under a change of observer to obtain the other relevant equations of balance. More explicitly: the validity of the equation of balance of energy with respect to any two observers φ^* and φ, which is a theorem here, and the calculated relation of the various gross fields, mass density, gross velocity, stress tensor, and so forth, with respect to φ^* and φ, imply the validity of equations of balance for *mass density*, *linear momentum*, *microinertia*, and *rotational momentum*. The analysis of the transformation rules of gross fields under a change of observer closely follows the one in PITTERI [1987]. Therefore a different approach is necessary to justify, within the present theory, a tensorial equation of balance of generalized rotational momentum like the one proposed in MURDOCH [1987].

We recall that, within continuum mechanics, a scalar equation from which the vector and tensor equations of balance of linear and rotational momentum, respectively, follow, is the Piola Power Theorem, which is a version of the virtual work theorem, and is presented in TRUESDELL & TOUPIN [1960, pp. 596–597], for instance. According to a common prejudice, presented in MARSDEN & HUGHES [1983], for instance, GREEN & RIVLIN [1964b] is the first paper in which the equation of balance of linear momentum is deduced from an equation of balance of energy and certain requirements of frame-indifference. Although this idea is appealing as such, in GREEN & RIVLIN [1964b] there are so many assumptions that the main argument becomes trivial. Within a different framework, in NOLL [1963] it is proved that the assumption of frame-indifference of the power of all forces, and the frame-indifference of the forces themselves, imply that the equations of balance of linear and of rotational momentum hold. In NOLL [1963] the existence and the frame-indifference of the stress tensor are not postulated at the outset, as is done in GREEN & RIVLIN [1964b] instead. Rather, both existence and frame-indifference are outcomes of the theory. We also recall that in CAPRIZ, PODIO-GUIDUGLI & WILLIAMS [1982] an energy balance and certain assumptions of frame-indifference are used to deduce equations of balance for microinertia and for generalized rotational momentum. In this paper balance of energy and frame-indifference are used in a way different from that of GREEN & RIVLIN [1964b], NOLL [1963] or CAPRIZ, PODIO-GUIDUGLI & WILLIAMS [1982]. For instance, here the equation of balance of energy with respect to any observer is not postulated but deduced, as are the transformation rules of gross fields under a change of observer. Nevertheless the results in this paper have been inspired, in a weak sense, by the results in NOLL [1963] and CAPRIZ, PODIO-GUIDUGLI & WILLIAMS [1982], as well as those in CAPRIZ & PODIO-GUIDUGLI [1983].

To construct gross fields and derive the equations of balance we follow NOLL [1955] rather than the fluid dynamicists. Therefore we employ straightforward mathematics, and avoid, for instance, the unnecessary and cumbersome use of δ-functions and infinite series expansions of them. Also, unlike ALBLAS [1976], we avoid assuming the molecules identical.

In the aforementioned statistical-kinetic analyses two ways of averaging emerge as possible when we aim at constructing gross fields. This fact is best seen through an example, which for brevity we only sketch. Let $\boldsymbol{x}_{r\alpha}$ denote the position vector of the typical atom $P_{r\alpha}$ of the rth molecule of our system, and let $m_{r\alpha}$ denote its mass. We denote by $\boldsymbol{x}_{r0}$ the position of the center of mass of the rth molecule, to be regarded as an independent variable, and assume that the relative position $\xi_{r\alpha} = \boldsymbol{x}_{r\alpha} - \boldsymbol{x}_{r0}$ is a given function of a finite number of Lagrangean co-ordinates $q_r^1, q_r^2, \ldots, q_r^{Nr}$. A possible way of defining the mass density field is the following:

$$\varrho(\boldsymbol{x}, t) = \sum_r m_r \langle W \mid \boldsymbol{x}_{r0} = \boldsymbol{x} \rangle, \qquad m_r := \sum_\alpha m_{r\alpha}, \tag{1.2}$$

where the bracket indicates that the probability density W is integrated over the Lagrangean co-ordinates, conjugate momenta, positions and velocities of the centers of mass of all the molecules, with the exception of $\boldsymbol{x}_{r0}$ which is set equal to $\boldsymbol{x}$. In words, $\varrho(\boldsymbol{x}, t)$ is the expected mass of molecules whose center of mass is at $\boldsymbol{x}$. The second possibility is to set

$$\varrho(\boldsymbol{x}, t) = \sum_{r,\alpha} m_{r\alpha} \langle W \mid \boldsymbol{x}_{r\alpha} = \boldsymbol{x} \rangle . \tag{1.3}$$

Now the probability density W is integrated over all the independent variables except $\boldsymbol{x}_{r0}$, which is set equal to $\boldsymbol{x} - \xi_{r\alpha}$. In words, $\varrho(\boldsymbol{x}, t)$ is the expected mass of atoms whose position is $\boldsymbol{x}$. The two definitions are not equivalent in general and, for instance, the choice exemplified by (1.2) is made in DAHLER & SCRIVEN [1963] and ALBLAS [1976], whereas (1.3) and its analogues for the other gross fields are adopted in CURTISS, BYRD & HASSAGER [1976]. Here we choose (1.2) because it is simpler. By adopting (1.3) we should get equations of balance of the same general form as the ones we obtain here, but with extra contributions from the forces that atoms in a molecule exert on other atoms of the same molecule. I did not find in the paper of CURTISS, BYRD & HASSAGER sufficient physical evidence that these extra contributions are really necessary. If that were the case, their choice of averaging would be preferable.

2. Preliminaries

Let us consider a system of n material points that can be divided into a finite number, say m, of subsystems, not necessarily all consisting of the same number of points. Each subsystem is subjected to ideal holonomic constraints that are "internal" to the subsystem in a sense that will be made clear below. We agree to use greek minuscules, $\alpha, \beta, \ldots$ to label material points belonging to the same subsystem, whereas different subsystems will be labelled by the last latin minuscules, $r, s, \ldots$. Latin minuscules like i, j, k, l, will label Lagrangean co-ordinates and conjugate momenta. Thus, for instance, two indices will be used to label a material point $P_{r\alpha}$, the first to select the subsystem, the second to select the point within it.

Following PITTERI [1986], to keep the notation as simple as possible we shall not write the limits in a summation. These limits are understandable from the type of index we use, in accord with the conventions set down above. We denote by $\boldsymbol{x}_{r\alpha}$ the position of the point $P_{r\alpha}$, $m_{r\alpha}$ its mass, and we let

$$\boldsymbol{x}_{r\alpha} = \boldsymbol{x}_{r\alpha}(\boldsymbol{x}_{r0}, q_r^1, q_r^2, \ldots, q_r^{Nr}) = \boldsymbol{x}_{r0} + \boldsymbol{\xi}_{r\alpha}(q_r^1, q_r^2, \ldots, q_r^{Nr}) \tag{2.1}$$

be a parametric representation of the constraints, which, for simplicity, we presume sufficient to cover with a single chart. We assume $\boldsymbol{x}_{r0}$ to be the position of the center of mass of the rth subsystem. This is equivalent to assuming the following equalities:

$$\sum_\alpha m_{r\alpha}\boldsymbol{\xi}_{r\alpha} = \mathbf{0}, \quad \text{or} \quad \sum_\alpha m_{r\alpha}\boldsymbol{x}_{r\alpha} = m_r\boldsymbol{x}_{r0} \quad \text{if } m_r := \sum_\alpha m_{r\alpha}. \tag{2.2}$$

By differentiating (2.1) we obtain the following expression for the velocity $\boldsymbol{v}_{r\alpha}$ of $P_{r\alpha}$:

$$\begin{gathered}\boldsymbol{v}_{r\alpha} = \boldsymbol{v}_{r\alpha}(\boldsymbol{x}_{r0}, \boldsymbol{v}_{r0}, q_r^1, q_r^2, \ldots, q_r^{Nr}, \dot{q}_r^1, \dot{q}_r^2, \ldots, \dot{q}_r^{Nr}) = \boldsymbol{v}_{r0} + \dot{\boldsymbol{\xi}}_{r\alpha},\\ \dot{\boldsymbol{\xi}}_{r\alpha} = \frac{\partial \boldsymbol{\xi}_{r\alpha}}{\partial q_r^i}\,\dot{q}_r^i,\end{gathered} \tag{2.3}$$

where $\boldsymbol{v}_{r0}$ is the velocity of the center of mass of the rth subsystem. Here and below the summation convention over repeated indices like i, j, k, is assumed. Then (2.2) implies that

$$\sum_\alpha m_{r\alpha}\frac{\partial \boldsymbol{\xi}_{r\alpha}}{\partial q_r^i} = \mathbf{0} \quad \text{for any } r \text{ and } i. \tag{2.4}$$

Following ideas of CAPRIZ & PODIO-GUIDUGLI [1983] and CAPRIZ [1984]–[1989], we think of the constraints on a subsystem as being realized by means of devices acting on the points of that subsystem alone. The requirement that the constraints be so "internal" to a subsystem expresses the fact that rotating the whole subsystem about its center of mass is compatible with the constraints. Formally, if $\boldsymbol{\xi}_{r\alpha}$ is an allowed position with respect to the center of mass, then so is $\boldsymbol{Q}\boldsymbol{\xi}_{r\alpha}$, for any proper orthogonal tensor $\boldsymbol{Q}$. By taking 1-parameter families of $\boldsymbol{Q}$ and differentiating with respect to the parameter, we see that

(†) for any choice of $\boldsymbol{\xi}_{r\alpha}$ as in (2.1), the vectors $\boldsymbol{A}\boldsymbol{\xi}_{r\alpha}$ are tangent to the constraints, for any skew tensor $\boldsymbol{A}$.

We assume the motion of the material points to be governed by the equations

$$m_{r\alpha}\boldsymbol{a}_{r\alpha} = \boldsymbol{f}_{r\alpha} + \boldsymbol{\phi}_{r\alpha}, \tag{2.5}$$

where $\boldsymbol{f}_{r\alpha}$ is the total active force, and $\boldsymbol{\phi}_{r\alpha}$ is the total reaction of the constraints on $P_{r\alpha}$. Next, we assume the constraints to be ideal, that is, to satisfy the equality

$$\sum_{r,\alpha} \boldsymbol{\phi}_{r\alpha} \cdot \delta \boldsymbol{x}_{r\alpha} = 0 \tag{2.6}$$

for any choice of virtual displacements $\delta \boldsymbol{x}_{r\alpha}$. By (2.1) this equality is equivalent to the following:

$$\sum_{\alpha} \boldsymbol{\phi}_{r\alpha} = 0 \quad \text{for any } r, \quad \text{and} \quad \sum_{\alpha} \boldsymbol{\phi}_{r\alpha} \cdot \frac{\partial \boldsymbol{\xi}_{r\alpha}}{\partial q_r^i} = 0 \quad \text{for any } r \text{ and } i. \tag{2.7}$$

As is well known, (2.7) imply the following equations of motion:

$$\boldsymbol{R}_r := \sum_{\alpha} \boldsymbol{f}_{r\alpha} = \sum_{\alpha} m_{r\alpha} \boldsymbol{a}_{r\alpha} = m_r \boldsymbol{a}_{r0}, \quad \text{where } \boldsymbol{a}_{r0} = \dot{\boldsymbol{v}}_{r0}, \tag{2.8}$$

$$Q_{ri} := \sum_{\alpha} \boldsymbol{f}_{r\alpha} \cdot \frac{\partial \boldsymbol{\xi}_{r\alpha}}{\partial q_r^i} = \sum_{\alpha} m_{r\alpha} \boldsymbol{a}_{r\alpha} \cdot \frac{\partial \boldsymbol{\xi}_{r\alpha}}{\partial q_r^i}; \tag{2.9}$$

here the total force $\boldsymbol{R}_r$ on the rth group, and the ith Lagrangean component of force Q_{ri} on the rth group, are all functions of $\boldsymbol{x}_{s0}, \boldsymbol{v}_{s0}, q_s^1, q_s^2, \ldots, q_s^{Ns}, \dot{q}_s^1, \dot{q}_s^2, \ldots, \dot{q}_s^{Ns}$ for $s = 1, \ldots, m$. Moreover, (2.8) and (2.9) are the Lagrangean equations for the total kinetic energy

$$\begin{aligned} T &= \tfrac{1}{2} \sum_{r,\alpha} m_{r\alpha} (\boldsymbol{v}_{r\alpha})^2 = \tfrac{1}{2} \sum_{r,\alpha} m_{r\alpha} (\boldsymbol{v}_{r0}^2 + \dot{\boldsymbol{\xi}}_{r\alpha}^2 + 2\boldsymbol{v}_{r0} \cdot \dot{\boldsymbol{\xi}}_{r\alpha}) \\ &= \tfrac{1}{2} \sum_{r} m_r (\boldsymbol{v}_{r0})^2 + \sum_{r} \bar{\mathscr{T}}_r (q_r^1, q_r^2, \ldots, q_r^{Nr}, \dot{q}_r^1, \dot{q}_r^2, \ldots, \dot{q}_r^{Nr}), \end{aligned} \tag{2.10}$$

where

$$\bar{\mathscr{T}}_r = \tfrac{1}{2} (a_r)_{ij} \dot{q}_r^i \dot{q}_r^j \quad \text{and} \quad (a_r)_{ij} := \sum_{\alpha} m_{r\alpha} \frac{\partial \boldsymbol{\xi}_{r\alpha}}{\partial q_r^i} \cdot \frac{\partial \boldsymbol{\xi}_{r\alpha}}{\partial q_r^j}. \tag{2.11}$$

Let us assume that

$$\begin{aligned} \boldsymbol{f}_{r\alpha} &= \boldsymbol{g}_{r\alpha}(\boldsymbol{x}_{r\alpha}, \boldsymbol{v}_{r\alpha}) + \hat{\boldsymbol{f}}_{r\alpha}(\{\boldsymbol{x}_{s\alpha}\}, \{\boldsymbol{v}_{s\alpha}\}, t) - \sum_{s,\beta} \frac{\partial V_{r\alpha s\beta}(\| \boldsymbol{x}_{r\alpha} - \boldsymbol{x}_{s\beta} \|)}{\partial \boldsymbol{x}_{r\alpha}} \\ &= \boldsymbol{g}_{r\alpha}(\boldsymbol{x}_{r\alpha}, \boldsymbol{v}_{r\alpha}) + \hat{\boldsymbol{f}}_{r\alpha}(\{\boldsymbol{x}_{s\alpha}\}, \{\boldsymbol{v}_{s\alpha}\}, t) - \sum_{s,\beta} \frac{V'_{r\alpha s\beta}}{\| \boldsymbol{x}_{r\alpha} - \boldsymbol{x}_{s\beta} \|} (\boldsymbol{x}_{r\alpha} - \boldsymbol{x}_{s\beta}). \end{aligned} \tag{2.12}$$

Here $\{\boldsymbol{x}_{s\alpha}\}$ indicates the set of position vectors of all the points, the analogue being true for $\{\boldsymbol{v}_{s\alpha}\}$, and, for $P_{r\alpha}$, $\boldsymbol{g}_{r\alpha}$ is the total external force; $\hat{\boldsymbol{f}}_{r\alpha}$ is the total non-conservative internal force, and $V_{r\alpha s\beta}$ is the potential energy of the conservative part of the interaction with $P_{s\beta}$. Also, by $V'_{r\alpha s\beta}$ we denote the derivative of $V_{r\alpha s\beta}$ with respect to the positive real variable on which it depends. We understand that $V_{r\alpha s\beta}$ vanishes identically when $r = s$ and $\alpha = \beta$, and we use the symbol

$$\sum_{r,\alpha,s,\beta}{}' \tag{2.13}$$

to indicate that, in the summation, s assumes all the values different from r. As is well known, the Lagrangean equations based on T and Q_{ri} can be put into normal form; hence they are equivalent to a first-order system, also in normal form. The most convenient transformation of the Lagrangean equations into a first-order normal system turns out to be the one that leads to semi-Hamiltonian equations. To obtain them, let us introduce the conjugate momenta

$$\boldsymbol{p}_{r0} := \frac{\partial T}{\partial \boldsymbol{v}_{r0}} = m_r \boldsymbol{v}_{r0} \quad \text{and} \quad p_{ri} := \frac{\partial T}{\partial \dot{q}_r^i} = (a_r)_{ij}\, \dot{q}_r^j. \tag{2.14}$$

Since (a_r) is positive definite, (2.14) are globally invertible for $\boldsymbol{v}_{r0}$ and $\dot{q}_r^j$:

$$\boldsymbol{v}_{r0} = (m_r)^{-1}\, \boldsymbol{p}_{r0} \quad \text{and} \quad \dot{q}_r^j = (a_r^{-1})^{jk}\, p_{rk}. \tag{2.15}$$

Introducing the Hamiltonian function

$$\mathscr{H} = \mathscr{T}_{r0} + \mathscr{T}_r + V, \tag{2.16}$$

where

$$V((\boldsymbol{x}_{t0}, q_t^1, q_t^2, \ldots, q_t^{N_t})_{t=1,\ldots,m}) := \tfrac{1}{2} \sum_{r,\alpha,s,\beta} V_{r\alpha s\beta}, \tag{2.17}$$

$$\mathscr{T}_{r0} := \tfrac{1}{2} \sum_r \frac{(\boldsymbol{p}_{r0})^2}{m_r}, \quad \text{and} \quad \mathscr{T}_r := \tfrac{1}{2} (a_r^{-1})^{ij}\, p_{ri} p_{rj}, \tag{2.18}$$

we can write the equations of motion of the system of material points in the form

$$\dot{\boldsymbol{x}}_{r0} = \frac{\partial \mathscr{H}}{\partial \boldsymbol{p}_{r0}} = \frac{\boldsymbol{p}_{r0}}{m_r}, \quad \dot{q}_r^i = \frac{\partial \mathscr{H}}{\partial p_{ri}} = (a_r^{-1})^{ik}\, p_{rk}, \tag{2.19}$$

$$\dot{\boldsymbol{p}}_{r0} = \boldsymbol{R}_r = \hat{\boldsymbol{R}}_r - \frac{\partial \mathscr{H}}{\partial \boldsymbol{x}_{r0}}, \quad \dot{p}_{ri} = \hat{Q}_{ri} - \frac{\partial \mathscr{H}}{\partial q_r^i}, \tag{2.20}$$

where

$$\hat{\boldsymbol{R}}_r = \sum_\alpha (\hat{\boldsymbol{f}}_{r\alpha} + \boldsymbol{g}_{r\alpha}), \qquad \hat{Q}_{ri} = (\hat{Q}_I)_{ri} + (\hat{Q}_E)_{ri}, \tag{2.21}$$

$$(\hat{Q}_I)_{ri} = \sum_\alpha \hat{\boldsymbol{f}}_{r\alpha} \cdot \frac{\partial \boldsymbol{\xi}_{r\alpha}}{\partial q_r^i} \quad \text{and} \quad (\hat{Q}_E)_{ri} = \sum_\alpha \boldsymbol{g}_{r\alpha} \cdot \frac{\partial \boldsymbol{\xi}_{r\alpha}}{\partial q_r^i}. \tag{2.22}$$

This system is of the general form considered in PITTERI [1986]. Hence, according to the results of that paper, the probability density function

$$W = W(t, (\boldsymbol{x}_{r0}, \boldsymbol{p}_{r0}, q_r^1, q_r^2, \ldots, q_r^{N_r}, p_{r1}, p_{r2}, \ldots, p_{rN_r})_{r=1,\ldots,m}) \tag{2.23}$$

satisfies the Liouville equation

$$0 = \frac{\partial W}{\partial t} + \sum_r \left[\frac{\partial}{\partial \boldsymbol{x}_{r0}} \cdot \left(W \frac{\partial \mathscr{H}}{\partial \boldsymbol{p}_{r0}}\right) + \frac{\partial}{\partial \boldsymbol{p}_{r0}} \cdot \left(W\left(\hat{\boldsymbol{R}}_r - \frac{\partial \mathscr{H}}{\partial \boldsymbol{x}_{r0}}\right)\right)\right]$$
$$+ \sum_r \left[\frac{\partial}{\partial q_r^i}\left(W \frac{\partial \mathscr{H}}{\partial p_{ri}}\right) + \frac{\partial}{\partial p_{ri}}\left(W\left(\hat{Q}_{ri} - \frac{\partial \mathscr{H}}{\partial q_r^i}\right)\right)\right]. \tag{2.24}$$

As in PITTERI [1986]–[1988], we can construct gross fields and use (2.24) to write the equation of evolution that they satisfy. For the constrained system we are considering here there are, though, two difficulties. First, gross fields are constructed by integrating over a manifold, and so, roughly, only scalar functions can be multiplied by W and then integrated with respect to the q's and p's. Therefore we may wish to write equations of balance for a micromomentum density, but the quantity obtained by integrating the co-vector $(p_{r1}, p_{r2}, \dots, p_{rN_r})$ is not independent of the choice of local co-ordinates, in general, and hence may fail to mean anything. Secondly, we do not have a good feeling for which equations of balance connected with the microstructure have to be added to the standard ones of balance of mass, linear momentum, rotational momentum and energy. According to the remarks in the introduction, we deduce the extra balance equations for the microstructure from the balance of energy.

3. Balance of energy

Let us introduce the following gross fields:

$$\varrho(\boldsymbol{x}, t) = \sum_r m_r \langle W \mid \boldsymbol{x}_{r0} = \boldsymbol{x} \rangle, \tag{3.1}$$

$$\varrho \boldsymbol{u}(\boldsymbol{x}, t) = \sum_{r,\alpha} m_{r\alpha} \langle \boldsymbol{v}_{r\alpha} W \mid \boldsymbol{x}_{r0} = \boldsymbol{x} \rangle = \sum_r m_r \langle \boldsymbol{v}_{r0} W \mid \boldsymbol{x}_{r0} = \boldsymbol{x} \rangle, \tag{3.2}$$

$$\varepsilon(\boldsymbol{x}, t) = \varepsilon_K(\boldsymbol{x}, t) + \varepsilon_V(\boldsymbol{x}, t), \qquad \boldsymbol{\eta}_{r0} = \boldsymbol{v}_{r0} - \boldsymbol{u}, \tag{3.3}$$

$$\varepsilon_K(\boldsymbol{x}, t) = \sum_r \langle (\mathcal{T}_{r0} + \mathcal{T}_r) W \mid \boldsymbol{x}_{r0} = \boldsymbol{x} \rangle, \tag{3.4}$$

$$\varepsilon_V(\boldsymbol{x}, t) = \tfrac{1}{2} \sum_{r,\alpha,s,\beta} \langle V_{r\alpha s\beta} W \mid \boldsymbol{x}_{r0} = \boldsymbol{x} \rangle, \tag{3.5}$$

$$\boldsymbol{T} = \boldsymbol{T}_K + \boldsymbol{T}_V, \qquad \boldsymbol{q} = \boldsymbol{q}_K + \boldsymbol{q}_V + \boldsymbol{q}_T, \tag{3.6}$$

$$\boldsymbol{T}_K = -\sum_r m_r \langle \boldsymbol{\eta}_{r0} \otimes \boldsymbol{\eta}_{r0} W \mid \boldsymbol{x}_{r0} = \boldsymbol{x} \rangle, \tag{3.7}$$

$$\boldsymbol{T}_V = \sum_{r,\alpha,s,\beta}{}' \int_{\mathbb{R}^3} d\boldsymbol{z} \int_0^\infty d\theta \left\langle \frac{V'_{r\alpha s\beta}}{\| \boldsymbol{x}_{r\alpha} - \boldsymbol{x}_{s\beta} \|} (\boldsymbol{x}_{r\alpha} - \boldsymbol{x}_{s\beta}) W \,\middle|\, \boldsymbol{x}_{r0} = \boldsymbol{x} + \theta \boldsymbol{z}, \; \boldsymbol{x}_{s0} = \boldsymbol{x} - (1-\theta)\boldsymbol{z} \right\rangle \otimes \boldsymbol{z}, \tag{3.8}$$

$$\boldsymbol{q}_K = \tfrac{1}{2} \sum_r m_r \langle (\boldsymbol{\eta}_{r0})^2 \boldsymbol{\eta}_{r0} W \mid \boldsymbol{x}_{r0} = \boldsymbol{x} \rangle + \sum_r \langle \mathcal{T}_r \boldsymbol{\eta}_{r0} W \mid \boldsymbol{x}_{r0} = \boldsymbol{x} \rangle, \tag{3.9}$$

$$\boldsymbol{q}_V = -\tfrac{1}{2} \sum_{r,\alpha,s,\beta}{}' \int_{\mathbb{R}^3} d\boldsymbol{z}\, \boldsymbol{z} \int_0^\infty d\theta \left\langle (\boldsymbol{\eta}_{r0} + \boldsymbol{\eta}_{s0}) \cdot (\boldsymbol{x}_{r\alpha} - \boldsymbol{x}_{s\beta}) \frac{V'_{r\alpha s\beta}}{\| \boldsymbol{x}_{r\alpha} - \boldsymbol{x}_{s\beta} \|} W \,\middle|\, \boldsymbol{x}_{r0} = \boldsymbol{x} + \theta \boldsymbol{z}, \; \boldsymbol{x}_{s0} = \boldsymbol{x} - (1-\theta)\boldsymbol{z} \right\rangle, \tag{3.10}$$

$$\boldsymbol{q}_T = \tfrac{1}{2} \sum_{r,\alpha,s,\beta} \langle V_{r\alpha s\beta}\, \boldsymbol{\eta}_{r0} W \mid \boldsymbol{x}_{r0} = \boldsymbol{x} \rangle, \tag{3.11}$$

$$b = \sum_r \langle \hat{\boldsymbol{R}}_r W \mid \boldsymbol{x}_{r0} = \boldsymbol{x} \rangle, \tag{3.12}$$

$$r = \sum_r \langle \hat{\boldsymbol{R}}_r \cdot \boldsymbol{\eta}_{r0} W \mid \boldsymbol{x}_{r0} = \boldsymbol{x} \rangle, \tag{3.13}$$

$$\mathfrak{P}_I = (\mathfrak{P}_I)_1 + (\mathfrak{P}_I)_2, \quad (\mathfrak{P}_I)_2 = \sum_r \left\langle \frac{\partial \mathcal{T}_r}{\partial p_{ri}} (\hat{Q}_I)_{ri} \, W \,\middle|\, \boldsymbol{x}_{r0} = \boldsymbol{x} \right\rangle, \tag{3.14}$$

$$(\mathfrak{P}_I)_1 = -\tfrac{1}{2} \sum_{r,\alpha,s,\beta}{}' \left\langle \left(\frac{\partial \mathcal{T}_r}{\partial p_{ri}} \frac{\partial V_{r\alpha s\beta}}{\partial q_r^i} - \frac{\partial \mathcal{T}_s}{\partial p_{si}} \frac{\partial V_{r\alpha s\beta}}{\partial q_s^i} \right) W \,\middle|\, \boldsymbol{x}_{r0} = \boldsymbol{x} \right\rangle, \tag{3.15}$$

$$\mathfrak{P}_E = \sum_r \left\langle \frac{\partial \mathcal{T}_r}{\partial p_{ri}} (\hat{Q}_E)_{ri} \, W \,\middle|\, \boldsymbol{x}_{r0} = \boldsymbol{x} \right\rangle. \tag{3.16}$$

Then the following equation of balance of total gross energy ε holds:

$$\frac{\partial \varepsilon}{\partial t} + \operatorname{div} (\boldsymbol{q} - \boldsymbol{T}^{\mathsf{T}} \boldsymbol{u} + \varepsilon \boldsymbol{u}) = \boldsymbol{b} \cdot \boldsymbol{u} + r + \mathfrak{P}_I + \mathfrak{P}_E, \tag{3.17}$$

where $\boldsymbol{T}^{\mathsf{T}}$ is the transpose of $\boldsymbol{T}$.

Here we give only a sketched proof, without fully justifying the procedure. This is essentially the same as in PITTERI [1986]–[1988], to which we refer the reader who is interested in the details. First, by differentiating under the integral sign and applying (2.24), we see that

$$\begin{aligned} \frac{\partial \varepsilon_K}{\partial t} = &-\tfrac{1}{2} \sum_{r,s} (m_r)^{-1} \left\langle (\boldsymbol{p}_{r0})^2 \left(\frac{\partial}{\partial \boldsymbol{x}_{s0}} \cdot \left(\frac{\boldsymbol{p}_{s0}}{m_s} W \right) + \frac{\partial}{\partial \boldsymbol{p}_{s0}} \cdot (W \boldsymbol{R}_s) \right) \middle|\, \boldsymbol{x}_{r0} = \boldsymbol{x} \right\rangle \\ &- \sum_{r,s} \left\langle \mathcal{T}_r \frac{\boldsymbol{p}_{s0}}{m_s} \cdot \frac{\partial W}{\partial \boldsymbol{x}_{s0}} \,\middle|\, \boldsymbol{x}_{r0} = \boldsymbol{x} \right\rangle \\ &- \sum_{r,s} \left\langle \mathcal{T}_r \left(\frac{\partial}{\partial q_s^i} \left(W \frac{\partial \mathcal{H}}{\partial p_{si}} \right) + \frac{\partial}{\partial p_{si}} \left(W \left(\hat{Q}_{si} - \frac{\partial \mathcal{H}}{\partial q_s^i} \right) \right) \right) \middle|\, \boldsymbol{x}_{r0} = \boldsymbol{x} \right\rangle. \end{aligned} \tag{3.18}$$

Integration by parts yields

$$\begin{aligned} \frac{\partial \varepsilon_K}{\partial t} = &-\operatorname{div} \left(\tfrac{1}{2} \sum_r (m_r)^{-2} \langle (\boldsymbol{p}_{r0})^2 \boldsymbol{p}_{r0} W \mid \boldsymbol{x}_{r0} = \boldsymbol{x} \rangle \right) + \sum_r (m_r)^{-1} \langle \boldsymbol{p}_{r0} \cdot \boldsymbol{R}_r W \mid \boldsymbol{x}_{r0} = \boldsymbol{x} \rangle \\ &- \operatorname{div} \left(\sum_r (m_r)^{-1} \langle \mathcal{T}_r \boldsymbol{p}_{r0} W \mid \boldsymbol{x}_{r0} = \boldsymbol{x} \rangle \right) + \sum_r \left\langle \frac{\partial \mathcal{T}_r}{\partial p_{ri}} \hat{Q}_{ri} W \,\middle|\, \boldsymbol{x}_{r0} = \boldsymbol{x} \right\rangle \\ &+ \sum_r \left\langle \left(\frac{\partial \mathcal{T}_r}{\partial q_r^i} \frac{\partial \mathcal{H}}{\partial p_{ri}} - \frac{\partial \mathcal{T}_r}{\partial p_{ri}} \frac{\partial \mathcal{H}}{\partial q_r^i} W \right) \middle|\, \boldsymbol{x}_{r0} = \boldsymbol{x} \right\rangle \end{aligned} \tag{3.19}$$

where, according to (2.20),

$$\boldsymbol{R}_r = \hat{\boldsymbol{R}}_r - \sum_{\alpha,s,\beta} \frac{\partial V_{r\alpha s\beta}}{\partial \boldsymbol{x}_{r0}}. \tag{3.20}$$

But

$$\sum_r \left\langle \left(\frac{\partial \mathcal{T}_r}{\partial q_r^i} \frac{\partial \mathcal{H}}{\partial p_{ri}} - \frac{\partial \mathcal{T}_r}{\partial p_{ri}} \frac{\partial \mathcal{H}}{\partial q_r^i} W \right) \middle| \boldsymbol{x}_{r0} = \boldsymbol{x} \right\rangle = - \sum_{r,\alpha,s,\beta} \left\langle \frac{\partial \mathcal{T}_r}{\partial p_{ri}} \frac{\partial V_{r\alpha s\beta}}{\partial q_r^i} W \middle| \boldsymbol{x}_{r0} = \boldsymbol{x} \right\rangle. \tag{3.21}$$

Furthermore,

$$\frac{\partial \varepsilon_V}{\partial t} = -\tfrac{1}{2} \sum_{r,\alpha,t} \left\langle \sum_{s,\beta} V_{r\alpha s\beta} \left(\frac{\partial}{\partial \boldsymbol{x}_{t0}} \cdot \left(\frac{\boldsymbol{p}_{t0}}{m_t} W \right) + \frac{\partial}{\partial q_t^i} \left(W \frac{\partial \mathcal{H}}{\partial p_{ti}} \right) \right) \middle| \boldsymbol{x}_{r0} = \boldsymbol{x} \right\rangle. \tag{3.22}$$

Integration by parts yields

$$\begin{aligned} \frac{\partial \varepsilon_V}{\partial t} &= -\tfrac{1}{2} \operatorname{div} \sum_{r,\alpha,s,\beta} \left\langle V_{r\alpha s\beta} \frac{\boldsymbol{p}_{r0}}{m_r} W \middle| \boldsymbol{x}_{r0} = \boldsymbol{x} \right\rangle \\ &+ \tfrac{1}{2} \sum_{r,\alpha,s,\beta} \left\langle \left(\frac{\boldsymbol{p}_{r0}}{m_r} \cdot \frac{\partial V_{r\alpha s\beta}}{\partial \boldsymbol{x}_{r0}} + \frac{\boldsymbol{p}_{s0}}{m_s} \cdot \frac{\partial V_{r\alpha s\beta}}{\partial \boldsymbol{x}_{s0}} \right) W \middle| \boldsymbol{x}_{r0} = \boldsymbol{x} \right\rangle \\ &+ \tfrac{1}{2} \sum_{r,\alpha,s,\beta} \left\langle \left(\frac{\partial V_{r\alpha s\beta}}{\partial q_r^i} \frac{\partial \mathcal{T}_r}{\partial p_{ri}} + \frac{\partial V_{r\alpha s\beta}}{\partial q_s^i} \frac{\partial \mathcal{T}_s}{\partial p_{si}} \right) W \middle| \boldsymbol{x}_{r0} = \boldsymbol{x} \right\rangle. \end{aligned} \tag{3.23}$$

Applying PITTERI's [1987] version of NOLL's [1955] lemma to express the divergence of $\boldsymbol{q}_V - (\boldsymbol{T}_V)^{\mathsf{T}} \boldsymbol{u}$, we obtain (3.17).

4. Change of observer

Following TRUESDELL [1977], we consider two observers φ and φ^*, and assume φ to be inertial. Denoting by $(\boldsymbol{x}^*, t^*)$ and $(\boldsymbol{x}, t)$ the positions and times that φ^* and φ associate with the same arbitrary event-point, respectively, we have

$$\boldsymbol{x} = \boldsymbol{x}_0(t) + \boldsymbol{Q}(t)\,(\boldsymbol{x}^* - \boldsymbol{x}_0^*) \quad \text{and} \quad t = t^* + a, \tag{4.1}$$

where $\boldsymbol{Q}(t)$ is a proper orthogonal tensor for any t, and a is a real number. Then, in a common notation, we deduce the standard formulae for velocities and accelerations:

$$\boldsymbol{v}^* = \boldsymbol{Q}^{\mathsf{T}}(\boldsymbol{v} - \boldsymbol{v}^{\tau}), \quad \boldsymbol{v}^{\tau} = \dot{\boldsymbol{x}}_0 + \boldsymbol{A}\boldsymbol{Q}(\boldsymbol{x}^* - \boldsymbol{x}_0^*), \quad \boldsymbol{A}(t) := \dot{\boldsymbol{Q}}(t)\,\boldsymbol{Q}^{\mathsf{T}}(t), \tag{4.2}$$

and

$$\boldsymbol{a}^* = \boldsymbol{Q}^{\mathsf{T}}(\boldsymbol{a} - \ddot{\boldsymbol{x}}_0 - (\boldsymbol{A}^2 + \dot{\boldsymbol{A}})\,\boldsymbol{Q}(\boldsymbol{x}^* - \boldsymbol{x}_0^*) - 2\boldsymbol{A}\boldsymbol{Q}\boldsymbol{v}^*), \tag{4.3}$$

where $\boldsymbol{v}^{\tau}$ is the dragging velocity, and $\boldsymbol{A}$ is the angular velocity of φ^* with respect to φ.

With respect to φ^*, we consider the same system of points $\{P_{r\alpha}\}$ as above, constrained in such a way that, at any time t, the configurations compatible with the constraints in φ^* are those and only those that, at the same time, are given by (2.1) with respect to φ. This seems to be a reasonable way of stating that a system of material points is subjected to the "same constraints" with respect to both φ^* and φ.

A parametric representation of the aforementioned time-dependent holonomic constraints with respect to φ^* is

$$\boldsymbol{x}_{r\alpha} = \boldsymbol{x}_0(t) + Q(t)\,(\boldsymbol{x}^*_{r\alpha} - \boldsymbol{x}^*_0) \tag{4.4}$$

when $\boldsymbol{x}_{r\alpha}$ is given by (2.1). By simple algebra we can show that

$$\boldsymbol{x}^*_{r\alpha} = \boldsymbol{x}^*_{r\alpha}(\boldsymbol{x}^*_{r0}, q^1_r, q^2_r, \ldots, q^{Nr}_r, t) = \boldsymbol{x}^*_{r0} + \xi^*_{r\alpha}(q^1_r, q^2_r, \ldots, q^{Nr}_r, t), \tag{4.5}$$

where $\boldsymbol{x}^*_{r0}$ is the position of the center of mass of the rth subsystem with respect to φ^*,

$$\boldsymbol{x}_{r0} = \boldsymbol{x}_0(t) + Q(t)\,(\boldsymbol{x}^*_{r0} - \boldsymbol{x}^*_0), \tag{4.6}$$

and

$$\xi^*_{r\alpha}(q^1_r, q^2_r, \ldots, q^{Nr}_r, t) = \boldsymbol{x}^*_{r\alpha} - \boldsymbol{x}^*_{r0} = Q^{\mathsf{T}}(t)\,\xi_{r\alpha}(q^1_r, q^2_r, \ldots, q^{Nr}_r). \tag{4.7}$$

It is also quite clear that, for virtual displacements,

$$\delta\boldsymbol{x}^*_{r\alpha} = Q^{\mathsf{T}}(t)\,\delta\boldsymbol{x}_{r\alpha}; \tag{4.8}$$

hence the constraint with respect to φ^* is ideal if and only if the set of the reactions $\boldsymbol{\Phi}^*_{r\alpha}$ it can produce coincides with the set given by

$$\boldsymbol{\Phi}^*_{r\alpha} = Q^{\mathsf{T}}(t)\boldsymbol{\Phi}_{r\alpha} \tag{4.9}$$

when $\boldsymbol{\Phi}_{r\alpha}$ ranges over the reactions allowed by the ideal constraint with respect to φ. We assume (4.9) to hold henceforth.

In (4.5) we have implicitly adopted the position $\boldsymbol{x}^*_{r0}$ of the center of mass of the rth subsystem as a new Lagrangean co-ordinate, by analogy with the choice made in PITTERI [1987]. Following the aforementioned ideas of CAPRIZ & PODIO-GUIDUGLI [1983] and CAPRIZ [1984]–[1989], we do similarly for the q's. Explicitly, to any proper orthogonal tensor Q we associate the diffeomorphism

$$q^{*i}_r = \mathfrak{q}^i_r(q^1_r, q^2_r, \ldots, q^{Nr}_r, Q^{\mathsf{T}}) \tag{4.10}$$

such that

$$\xi_{r\alpha}(\mathfrak{q}^i_r(q^1_r, q^2_r, \ldots, q^{Nr}_r, Q^{\mathsf{T}})) = Q^{\mathsf{T}}\xi_{r\alpha}(q^1_r, q^2_r, \ldots, q^{Nr}_r). \tag{4.11}$$

Then, to any smooth function $Q = Q(t)$ taking values on the manifold of proper orthogonal tensors, we can associate a smooth change of Lagrangean co-ordinates:

$$q^{*i}_r = \chi^{*i}_r(q^1_r, q^2_r, \ldots, q^{Nr}_r, t) := \mathfrak{q}^i_r(q^1_r, q^2_r, \ldots, q^{Nr}_r, Q^{\mathsf{T}}(t)), \tag{4.12}$$

which, together with its inverse

$$q^i_r = \chi^i_r(q^{*1}_r, q^{*2}_r, \ldots, q^{*Nr}_r, t), \tag{4.13}$$

will be used in the formulas below. As a consequence, we can represent the constraints with respect to φ^* by the parametric equations

$$\boldsymbol{x}^*_{r\alpha} = \boldsymbol{x}^*_{r\alpha}(\boldsymbol{x}^*_{r0}, q^{*1}_r, q^{*2}_r, \ldots, q^{*Nr}_r) = \boldsymbol{x}^*_{r0} + \xi_{r\alpha}(q^{*1}_r, q^{*2}_r, \ldots, q^{*Nr}_r). \tag{4.14}$$

From (4.10)–(4.14) we easily deduce the following identities:

$$\frac{\partial \xi_{r\alpha}}{\partial q_r^i}(\chi_r^{*j})\frac{\partial \chi_r^{*i}}{\partial q_r^k} = Q^{\mathrm{T}}(t)\frac{\partial \xi_{r\alpha}}{\partial q_r^k}(\{q_r^m\}), \tag{4.15}$$

$$Q^{\mathrm{T}}(t)\frac{\partial \xi_{r\alpha}}{\partial q_r^{*i}}(\{q_r^m\})\frac{\partial \chi_r^i}{\partial q_r^{*k}}(\chi_r^{*j})\frac{\partial \chi_r^{*k}}{\partial t} = \dot{Q}^{\mathrm{T}}(t)\,\xi_{r\alpha}(\{q_r^m\}); \tag{4.16}$$

hence

$$\frac{\partial \xi_{r\alpha}}{\partial q_r^i}(\chi_r^{*j})\frac{\partial \chi_r^{*i}}{\partial t} = \dot{Q}^{\mathrm{T}}(t)\,\xi_{r\alpha}(\{q_r^m\}) \quad \text{and} \quad \dot{q}_r^{*i} = \frac{\partial \chi_r^{*i}}{\partial q_r^k}\dot{q}_r^k + \frac{\partial \chi_r^{*i}}{\partial t}. \tag{4.17}$$

Here and below we understand that, whenever in an equality both functions of $\boldsymbol{x}^*$, t^* and of $\boldsymbol{x}$, t [both t, $\boldsymbol{x}_{r\alpha}$, $\boldsymbol{v}_{r\alpha}$, and t^*, $\boldsymbol{x}^*_{r\alpha}$, $\boldsymbol{v}^*_{r\alpha}$] appear, then $\boldsymbol{x}^*$ and t^* are expressed as functions of $\boldsymbol{x}$ and t corresponding through (4.1) [then they are related by the obvious analogues of (4.1) and (4.2)].

According to PITTERI [1987], when referred to φ^*, the motion of the points $P_{r\alpha}$ is governed by the "Newtonian" equations

$$m_{r\alpha}\boldsymbol{a}^*_{r\alpha} = \boldsymbol{f}^*_{r\alpha}(\{\boldsymbol{x}^*_{s\alpha}\}, \{\boldsymbol{v}^*_{s\alpha}\}, t^*) + \boldsymbol{\phi}^*_{r\alpha} \tag{4.18}$$

where

$$\boldsymbol{f}^*_{r\alpha} = Q^{\mathrm{T}}(\boldsymbol{f}_{r\alpha}(\{\boldsymbol{x}_{s\alpha}\}, \{\boldsymbol{v}_{s\alpha}\}, t) - m_{r\alpha}\boldsymbol{\gamma}^*_{r\alpha}(\boldsymbol{x}^*_{r\alpha}, \boldsymbol{v}^*_{r\alpha}, t)) \tag{4.19}$$

and, by the analogue of (4.2) for $\boldsymbol{x}^*_{r\alpha}$

$$\begin{aligned}\boldsymbol{\gamma}^*_{r\alpha} &= \ddot{\boldsymbol{x}}_0 + (A^2 + \dot{A})\,Q(\boldsymbol{x}^*_{r\alpha} - \boldsymbol{x}^*_0) + 2AQ\boldsymbol{v}^*_{r\alpha} \\ &= \ddot{\boldsymbol{x}}_0 + (A - \dot{A}^2)\,(\boldsymbol{x}_{r\alpha} - \boldsymbol{x}_0) + 2A(\boldsymbol{v}_{r\alpha} - \dot{\boldsymbol{x}}_0).\end{aligned} \tag{4.20}$$

Therefore the differential system of equations governing the motion of $\{P_{r\alpha}\}$ with respect to φ^* has the same general form as the one with respect to φ, which is given by (2.8), (2.9). Using (4.14), we write the kinetic energy T^* as follows:

$$T^* = \tfrac{1}{2}\sum_r m_r(\boldsymbol{v}^*_{r0})^2 + \bar{\mathcal{T}}^*_r, \quad \text{where} \quad (a^*_r)_{ij} = \sum_\alpha m_{r\alpha}\frac{\partial \xi_{r\alpha}}{\partial q_r^{*j}}\cdot\frac{\partial \xi_{r\alpha}}{\partial q_r^{*j}}, \tag{4.21}$$

$$\bar{\mathcal{T}}^*_r = \tfrac{1}{2}\sum_\alpha m_{r\alpha}(\dot{\xi}^*_{r\alpha})^2 = \tfrac{1}{2}(a^*_r)_{ij}\,\dot{q}_r^{*i}\dot{q}_r^{*j}, \tag{4.22}$$

and, in terms of the co-ordinates q_r^i,

$$\bar{\mathcal{T}}^*_r = \bar{\mathcal{T}}_r + b_{ri}\dot{q}_r^i + d_r, \tag{4.23}$$

$$b_{ri} := -\sum_\alpha m_{r\alpha}\frac{\partial \xi_{r\alpha}}{\partial q_r^i}\cdot A(t)\,\xi_{r\alpha} = A\cdot\mathcal{K}_{ri}, \tag{4.24}$$

$$\mathcal{K}_{ri} := \sum_\alpha m_{r\alpha}\xi_{r\alpha}\otimes\frac{\partial \xi_{r\alpha}}{\partial q_r^i}, \quad \mathcal{I}_r := \sum_\alpha m_{r\alpha}\xi_{r\alpha}\otimes\xi_{r\alpha}, \tag{4.25}$$

$$d_r = \tfrac{1}{2}\sum_\alpha m_{r\alpha}(A(t)\,\xi_{r\alpha})^2 = -\tfrac{1}{2}A^2\cdot\mathcal{I}_r. \tag{4.26}$$

In (4.21) v_{r0}^* is the velocity of the center of mass of the rth subsystem with respect to φ^*, and is related to x_{r0} and v_{r0} by

$$v_{r0}^* = Q^\mathsf{T}(v_{r0} - \dot{x}_0(t)) - A(t)\,(x_{r0} - x_0)). \tag{4.27}$$

The conjugate momenta with respect to the q^*'s are

$$p_{r0}^* = \frac{\partial T^*}{\partial v_{r0}^*} = m_r v_{r0}^* \quad \text{and} \quad p_{ri}^* = \frac{\partial T^*}{\partial \dot{q}_r^{*i}} = (a_r^*)_{ij}\, \dot{q}_r^{*j}. \tag{4.28}$$

Equivalently

$$v_{r0}^* = (m_r)^{-1}\, p_{r0}^* \quad \text{and} \quad \dot{q}_r^{*j} = (a_r^{*-1})^{jk}\, p_{rk}^*. \tag{4.29}$$

The equations of motion can be put in semi-Hamiltonian form again:

$$\dot{x}_{r0}^* = \frac{\partial \mathscr{H}^*}{\partial p_{r0}^*} = \frac{p_{r0}^*}{m_r}, \qquad \dot{q}_r^{*i} = \frac{\partial \mathscr{H}^*}{\partial p_{ri}^*} = \frac{\partial \mathscr{T}_r^*}{\partial p_{ri}^*} = (a_r^{*-1})^{ij}\, p_{rj}^*, \tag{4.30}$$

$$\dot{p}_{r0}^* = R_r^* = \hat{R}_r^* - \frac{\partial \mathscr{H}^*}{\partial x_{r0}^*}, \qquad \dot{p}_{ri}^* = \hat{Q}_{ri}^* - \frac{\partial \mathscr{H}^*}{\partial q_r^{*i}}, \tag{4.31}$$

where, by (4.15)–(4.17), on the understanding that functions of the q and of the q^* are related by either (4.12) or (4.13),

$$\hat{Q}_{ri}^* = (\hat{Q}_I^*)_{ri} + (\hat{Q}_E^*)_{ri}, \quad (\hat{Q}_I^*)_{ri} = \sum_\alpha Q^\mathsf{T} \hat{f}_{r\alpha} \cdot \frac{\partial \xi_{r\alpha}}{\partial q_r^{*i}} = \frac{\partial \chi_r^k}{\partial q_r^{*i}} (\hat{Q}_I)_{rk}, \tag{4.32}$$

$$(\hat{Q}_E^*)_{ri} = \sum_\alpha Q^\mathsf{T}(g_{r\alpha} - m_{r\alpha}\gamma_{r\alpha}^*) \cdot \frac{\partial \xi_{r\alpha}}{\partial q_r^{*i}} = \frac{\partial \chi_r^k}{\partial q_r^{*i}} \left((\hat{Q}_E)_{rk} - \sum_\alpha m_{r\alpha}\gamma_{r\alpha}^* \cdot \frac{\partial \xi_{r\alpha}}{\partial q_r^k} \right), \tag{4.33}$$

$$\hat{R}_r^* = \sum_\alpha Q^\mathsf{T}(\hat{f}_{r\alpha} + g_{r\alpha} - m_{r\alpha}\gamma_{r\alpha}^*) = Q^\mathsf{T}\left(\hat{R}_r - \sum_\alpha m_{r\alpha}\gamma_{r\alpha}^*\right), \tag{4.34}$$

$$\mathscr{H}^* = \mathscr{H}^*(t^*, (x_{r0}^*, p_{r0}^*, \{q_r^{*i}\}, \{p_{ri}^*\})_{r=1,\ldots,m}) = \tfrac{1}{2} \sum_r \left(\frac{(p_{r0}^*)^2}{m_r} + \mathscr{T}_r^* \right) + V, \tag{4.35}$$

and

$$\mathscr{T}_r^* = \tfrac{1}{2}\,(a_r^{*-1})^{ij}\, p_{ri}^* p_{rj}^* = \tfrac{1}{2}\,(a_r^{-1})^{ij}\, p_{ri} p_{rj} + (a_r^{-1})^{ij}\, p_{ri} b_{rj} + d_r. \tag{4.36}$$

Notice that, by (2.4), (4.12), (4.5)–(4.7), and the analogue of (4.2) for $v_{r\alpha}^*$ and $x_{r\alpha}^*$

$$\sum_\alpha m_{r\alpha}\gamma_{r\alpha}^* = m_r(\ddot{x}_0 + (A^2 + \dot{A})\, Q(x_{r0}^* - x_0^*) + 2AQv_{r0}^*), \tag{4.37}$$

and

$$\sum_\alpha m_{r\alpha}\gamma_{r\alpha}^* \cdot \frac{\partial \xi_{r\alpha}}{\partial q_r^i} = -\,\mathscr{K}_{ri} \cdot (\dot{A} + A^2) - 2A \cdot \sum_\alpha m_{r\alpha} \frac{\partial \xi_{r\alpha}}{\partial q_r^j} \otimes \frac{\partial \xi_{r\alpha}}{\partial q_r^i}\, (a_r^{-1})^{jk}\, p_{rk}. \tag{4.38}$$

We shall denote by

$$W^* = W^*(t^*, (x_{r0}^*, p_{r0}^*, \{q_r^{*i}\}, \{p_{ri}^*\})_{r=1,\ldots,m}) \tag{4.39}$$

the probability density in φ^* that produces the same distribution of initial states as W in φ. That is,

$$W^*(t^*, (\boldsymbol{x}^*_{r0}, \boldsymbol{p}^*_{r0}, \{q^{*i}_r\}, \{p^*_{ri}\})_{r=1,\dots,m}) = W(t, (\boldsymbol{x}_{r0}, \boldsymbol{p}_{r0}, \{q^i_r\}, \{p_{ri}\})_{r=1,\dots,m}) \quad (4.40)$$

whenever t, $\boldsymbol{x}_{r0}$, $\boldsymbol{p}_{r0}$, q^i_r, p_{ri} and t^*, $\boldsymbol{x}^*_{r0}$, $\boldsymbol{p}^*_{r0}$, q^{*i}_r, p^*_{ri} are related by $(4.1)_2$, (4.6), (4.12)–(4.13), (4.27), (4.28), (2.14) and (2.15), since this transformation has jacobian 1. Indeed, by (4.21),

$$(a^*_r)_{ij} = \frac{\partial \chi^k_r}{\partial q^{*i}_r} (a_r)_{kl} \frac{\partial \chi^l_r}{\partial q^{*j}_r} \quad (4.41)$$

which, together with (4.17), (4.28) and (2.14), implies that

$$\begin{aligned} p^*_{ri} &= \frac{\partial \chi^k_r}{\partial q^{*i}_r} (a_r)_{kj} \dot{q}^j_r + \frac{\partial \chi^k_r}{\partial q^{*i}_r} (a_r)_{kl} \frac{\partial \chi^l_r}{\partial q^{*j}_r} \frac{\partial \chi^{*j}_r}{\partial t} \\ &= \frac{\partial \chi^k_r}{\partial q^{*i}_r} p_{rk} - \frac{\partial \chi^k_r}{\partial q^{*i}_r} (a_r)_{kj} \frac{\partial \chi^j_r}{\partial t}. \end{aligned} \quad (4.42)$$

Then, in obvious matrix notation, the jacobian of the transformation of the q and the p into the q^* and the p^* satisfies the equations

$$\begin{aligned} \frac{\partial(q^*, p^*)}{\partial(q, p)} &= \det \begin{pmatrix} \dfrac{\partial \boldsymbol{q}^*}{\partial \boldsymbol{q}} & \dfrac{\partial \boldsymbol{q}^*}{\partial \boldsymbol{p}} \\ \dfrac{\partial \boldsymbol{p}^*}{\partial \boldsymbol{q}} & \dfrac{\partial \boldsymbol{p}^*}{\partial \boldsymbol{p}} \end{pmatrix} = \det \begin{pmatrix} \dfrac{\partial \boldsymbol{\chi}^*}{\partial \boldsymbol{q}} & \boldsymbol{0} \\ \dfrac{\partial \boldsymbol{p}^*}{\partial \boldsymbol{q}} & \left(\dfrac{\partial \boldsymbol{\chi}}{\partial \boldsymbol{q}^*}\right)^{\mathsf{T}} \end{pmatrix} \\ &= \det \frac{\partial \boldsymbol{\chi}^*}{\partial \boldsymbol{q}} \det \left(\frac{\partial \boldsymbol{\chi}}{\partial \boldsymbol{q}^*}\right)^{\mathsf{T}} = 1. \end{aligned} \quad (4.43)$$

It is now clear that the jacobian of the full transformation of the independent variables with respect to φ into the analogues with respect to φ^* is 1.

Since the equations of motion of the system of material points with respect to φ^* are semi-Hamiltonian, W^* satisfies the analogue of (2.24) for t^*, $\boldsymbol{x}^*_{r0}$, $\boldsymbol{p}^*_{r0}$, q^{*i}_r, p^*_{ri}, $\mathcal{H}^*$, $\hat{\boldsymbol{R}}^*_r$ and $\hat{Q}^*_{ri}$. Hence, according to PITTERI [1986], [1987], we can construct gross quantities and equations of balance with respect to φ^*. In particular, we can write the equation of balance of energy:

$$\frac{\partial \varepsilon^*}{\partial t^*} + \operatorname{div}^* (\boldsymbol{q}^* - (\boldsymbol{T}^*)^{\mathsf{T}} \boldsymbol{u}^* + \varepsilon^* \boldsymbol{u}^*) = \boldsymbol{b}^* \cdot \boldsymbol{u}^* + r^* + \mathfrak{P}^*_I + \mathfrak{P}^*_E. \quad (4.44)$$

Here the starred fields are obtained by the analogues of (3.1)–(3.15) when all the variables and functions are starred, and integration is performed with respect to all the $\boldsymbol{x}^*_{s0}$, $s \neq r$, $\boldsymbol{p}^*_{s0}$, q^{*i}_s and p^*_{si}. As an example, the analogues of (3.1)–(3.5) are

$$\varrho^*(\boldsymbol{x}, t) = \sum_r m_r \langle W^* \mid \boldsymbol{x}^*_{r0} = \boldsymbol{x}^* \rangle, \quad (4.45)$$

$$\varrho^* \boldsymbol{u}^*(\boldsymbol{x}^*, t^*) = \sum_{r,\alpha} m_{r\alpha} \langle \boldsymbol{v}^*_{r\alpha} W^* \mid \boldsymbol{x}^*_{r0} = \boldsymbol{x}^* \rangle = \sum_r m_r \langle \boldsymbol{v}^*_{r0} W^* \mid \boldsymbol{x}^*_{r0} = \boldsymbol{x}^* \rangle, \quad (4.46)$$

$$\varepsilon^*(\boldsymbol{x}^*, t^*) = \varepsilon^*_K(\boldsymbol{x}^*, t^*) + \varepsilon^*_V(\boldsymbol{x}^*, t^*), \qquad \boldsymbol{\eta}^*_{r0} = \boldsymbol{v}^*_{r0} - \boldsymbol{u}^*, \tag{4.47}$$

$$\varepsilon^*_K(\boldsymbol{x}^*, t^*) = \tfrac{1}{2} \sum_r (m_r)^{-1} \langle (\boldsymbol{p}^*_{r0})^2 \, W^* \mid \boldsymbol{x}^*_{r0} = \boldsymbol{x}^* \rangle + \sum_r \langle \mathcal{T}^*_r W^* \mid \boldsymbol{x}^*_{r0} = \boldsymbol{x}^* \rangle, \tag{4.48}$$

$$\varepsilon^*_V(\boldsymbol{x}^*, t^*) = \tfrac{1}{2} \sum_{r,\alpha,s,\beta} \langle V_{r\alpha s\beta} W^* \mid \boldsymbol{x}^*_{r0} = \boldsymbol{x}^* \rangle. \tag{4.49}$$

5. Comparison of gross fields

We express now the gross fields for the observer φ^*, to be constructed as we mentioned in the preceding section, in terms of their analogues with respect to φ, these having been defined in Section 3. For some of these fields we follow closely the procedure of Pitteri [1987]. First of all, since $\boldsymbol{x}_{r0} = \boldsymbol{x}$ if and only if $\boldsymbol{x}^*_{r0} = \boldsymbol{x}^*$, by (4.1) and (4.40)

$$\varrho^*(\boldsymbol{x}, t) = \sum_r m_r \langle W^* \mid \boldsymbol{x}^*_{r0} = \boldsymbol{x}^* \rangle = \sum_r m_r \langle W \mid \boldsymbol{x}_{r0} = \boldsymbol{x} \rangle = \varrho. \tag{5.1}$$

By (4.40) and (2.2) and the analogue of (4.2) for $\boldsymbol{x}_{r\alpha}$

$$\varrho^* \boldsymbol{u}^* = Q^{\mathsf{T}} \sum_{r,\alpha} m_{r\alpha} \langle (\boldsymbol{v}_{r\alpha} - \dot{\boldsymbol{x}}_0 - A(\boldsymbol{x}_{r\alpha} - \boldsymbol{x}_0)) \, W \mid \boldsymbol{x}_{r0} = \boldsymbol{x} \rangle \tag{5.2}$$

$$= \varrho Q^{\mathsf{T}} (\boldsymbol{u} - \dot{\boldsymbol{x}}_0 - A(\boldsymbol{x} - \boldsymbol{x}_0)).$$

Since $\varrho^* = \varrho$,

$$\boldsymbol{u}^* = Q^{\mathsf{T}} (\boldsymbol{u} - \dot{\boldsymbol{x}}_0 - A(\boldsymbol{x} - \boldsymbol{x}_0)); \tag{5.3}$$

hence the gross velocities with respect to φ^* and φ satisfy the analogue of (4.2), in accord with standard assumptions and with the results in Pitteri [1987]. Since $\| \boldsymbol{x}^*_{r\alpha} - \boldsymbol{x}^*_{s\beta} \| = \| \boldsymbol{x}_{r\alpha} - \boldsymbol{x}_{s\beta} \|$, we easily deduce from (4.46) that $\varepsilon^*_V = \varepsilon_V$. On the other hand, by (4.2) and (4.45), denoting by $\boldsymbol{v}^\tau_{r0}$ the analogue of $\boldsymbol{v}^\tau$ in (4.2) when $\boldsymbol{x}^*_{r0}$ replaces $\boldsymbol{x}^*$, we see that

$$\varepsilon^*_K = \tfrac{1}{2} \sum_r m_r \langle ((\boldsymbol{v}_{r0})^2 + (\boldsymbol{v}^\tau_{r0})^2 - 2\boldsymbol{v}_{r0} \cdot \boldsymbol{v}^\tau_{r0}) \, W \mid \boldsymbol{x}_{r0} = \boldsymbol{x} \rangle \tag{5.4}$$

$$+ \sum_r \langle (\tfrac{1}{2} (a_r^{-1})^{ij} p_{ri} p_{rj} + (a_r^{-1})^{ij} p_{ri} b_{rj} + d_r) \, W \mid \boldsymbol{x}_{r0} = \boldsymbol{x} \rangle$$

$$= \varepsilon_K + \tfrac{1}{2} \varrho (\boldsymbol{v}^\tau)^2 - \varrho \boldsymbol{u} \cdot \boldsymbol{v}^\tau + \boldsymbol{A} \cdot \boldsymbol{K} - \tfrac{1}{2} \boldsymbol{A}^2 \cdot \boldsymbol{I},$$

where

$$\boldsymbol{K} = \sum_r \langle (a_r^{-1})^{ij} p_{ri} \mathcal{K}_{rj} \, W \mid \boldsymbol{x}_{r0} = \boldsymbol{x} \rangle, \tag{5.5}$$

and

$$\boldsymbol{I} = \sum_r \langle \mathcal{I}_r \, W \mid \boldsymbol{x}_{r0} = \boldsymbol{x} \rangle. \tag{5.6}$$

As in PITTERI [1987], the stress tensor is objective. Indeed, by (4.1) and (4.40),

$$\int_{\mathbb{R}^3} dz \int_0^\infty d\theta \left\langle \frac{V'_{r\alpha s\beta}}{\| x^*_{r\alpha} - x^*_{s\beta} \|} (x^*_{r\alpha} - x^*_{s\beta}) W^* \right| x^*_{r0} = x^* + \theta z, x^*_{s0} = x^* - (1-\theta) z \Big\rangle \otimes z$$

$$= Q^{\mathsf{T}} \int_{\mathbb{R}^3} dz \int_0^\infty d\theta \left\langle \frac{V'_{r\alpha s\beta}}{\| x_{r\alpha} - x_{s\beta} \|} (x_{r\alpha} - x_{s\beta}) W \right| x_{r0} = x + \theta\, Qz,$$

$$x_{s0} = x - (1-\theta)\, Qz \Big\rangle \otimes z. \quad (5.7)$$

Since T^*_V is given by the analogue of (3.8) when all the quantities, except z and θ, are starred, we can replace the variable z of integration in the right-hand side of (5.7) by Qz to show that

$$T^*_V = Q^{\mathsf{T}} T_V Q. \quad (5.8)$$

To analyze T^*_K, let us notice that, by $(4.47)_2$ and (5.3),

$$\eta^*_{r0} = Q^{\mathsf{T}}(\eta_{r0} - A(x_{r0} - x)). \quad (5.9)$$

Then

$$T^*_K = -\sum_r m_r \langle (Q^{\mathsf{T}}\eta_{r0} \otimes Q^{\mathsf{T}}\eta_{r0} - Q^{\mathsf{T}}\eta_{r0} \otimes Q^{\mathsf{T}}A(x_{r0} - x) - Q^{\mathsf{T}}A(x_{r0} - x)$$

$$\otimes Q^{\mathsf{T}}\eta_{r0} - Q^{\mathsf{T}}A(x_{r0} - x) \otimes Q^{\mathsf{T}}A(x_{r0} - x))\, W \mid x_{r0} = x \rangle \quad (5.10)$$

$$= Q^{\mathsf{T}} T_K Q.$$

Next, we show that, in constrast with what happens in PITTERI [1987], the heat-flux vector is not generally objective although internal forces may all be conservative. First, (5.9) and the analogue of (3.11) for q^*_T easily imply that

$$q^*_T = Q^{\mathsf{T}} q_T. \quad (5.11)$$

On the other hand, by (5.9), (4.36), and (4.24)–(4.26)

$$q^*_K = \tfrac{1}{2} \sum_r m_r \langle (\eta^*_{r0})^2\, \eta^*_{r0} W^* \mid x^*_{r0} = x^* \rangle + \sum_r \langle \mathcal{T}^*_r \eta^*_{r0} W^* \mid x^*_{r0} = x^* \rangle$$

$$= Q^{\mathsf{T}} q_K + Q^{\mathsf{T}}(A : \Xi_K - A^2 : \Xi_I), \quad (5.12)$$

where

$$\Xi_K = \tfrac{1}{2} \sum_r \langle (a_r^{-1})^{ij}\, p_{ri} (\mathcal{K}_{rj} - \mathcal{K}^{\mathsf{T}}_{rj}) \otimes \eta_{r0} W \mid x_{r0} = x \rangle, \quad (5.13)$$

$$\Xi_I = \tfrac{1}{2} \sum_r \langle \mathcal{I}_r \otimes \eta_{r0} W \mid x_{r0} = x \rangle, \quad (5.14)$$

and, for instance, in terms of Cartesian components in $\mathbb{R}^3$, $(A : \Xi_K)_i := (A)_{jk} (\Xi_K)^{jk}{}_i$. Moreover, for any choice of vectors u and v, let us denote by $u \wedge v$ the skew tensor $u \otimes v - v \otimes u$. By (5.9) and its analogue for η^*_{s0} we have

$$q^*_V = -\tfrac{1}{2} \sum_{r,\alpha,s,\beta}{}' \int_{\mathbb{R}^3} dz\, z \int_0^\infty d\theta \Big\langle (\eta^*_{r0} + \eta^*_{s0}) \cdot (x^*_{r\alpha} - x^*_{s\beta}) \times \frac{V'_{r\alpha s\beta}}{\| x^*_{r\alpha} - x^*_{s\beta} \|} W^* \Big| x^*_{r0} = x^* + \theta z,\, x^*_{s0} = x^* - (1-\theta)\, z \Big\rangle \tag{5.15}$$
$$= Q^{\mathsf{T}}(q_V - A : \Psi),$$

in which

$$\Psi = \tfrac{1}{4} \sum_{r,\alpha,s,\beta}{}' \int_{\mathbb{R}^3} dz \int_0^\infty d\theta \Big\langle (x_{r0} + x_{s0} - 2x) \wedge (x_{r\alpha} - x_{s\beta}) \frac{V'_{r\alpha s\beta}}{\| x_{r\alpha} - x_{s\beta} \|} W \Big| x_{r0} = x + \theta z,\, x_{s0} = x - (1-\theta)\, z \Big\rangle \otimes z. \tag{5.16}$$

By (4.18), (4.19) and (4.37)

$$b^* = \sum_r \langle R^*_r W^* \mid x^*_{r0} = x^* \rangle = \sum_r \Big\langle Q^{\mathsf{T}} \Big(\hat{R}_r - \sum_\alpha m_{r\alpha} \gamma^*_{r\alpha} \Big) W \Big| x_{r0} = x \Big\rangle$$
$$= Q^{\mathsf{T}} \Big(b - \sum_r m_r \langle (\ddot{x}_0 + (\dot{A} - A)^2 (x_{r0} - x_0) + 2A(v_{r0} - \dot{x}_0(t)))\, W \mid x_{r0} = x \rangle \Big)$$
$$= Q^{\mathsf{T}}(b - \varrho(\ddot{x}_0 + (\dot{A} - A^2)(x - x_0) + 2A(u - \dot{x}_0))). \tag{5.17}$$

Therefore, as in Pitteri [1987], the body force b satisfies the field analogue of (4.11)–(4.12), as is usually assumed in continuum mechanics, for instance in Truesdell [1977].

The field r is objective, since, by (4.27), (4.34), (4.37) and (5.9), and by the equality

$$\sum_r \langle m_r \eta_{r0} W \mid x_{r0} = x \rangle = 0 \tag{5.18}$$

we deduce that

$$r^* = \sum_r \langle \hat{R}^*_r \cdot \eta^*_{r0} W^* \mid x^*_{r0} = x^* \rangle$$
$$= r - \sum_r m_r \langle \eta_{r0} \cdot (\ddot{x}_0 + (A^2 + \dot{A})(x_{r0} - x_0) + 2AQv^*_{r0})\, W \mid x_{r0} = x \rangle$$
$$= r - \sum_r m_r \langle \eta_{r0} \cdot 2A v_{r0} W \mid x_{r0} = x \rangle \tag{5.19}$$
$$= r - \sum_r m_r \langle \eta_{r0} \cdot 2A \eta_{r0}\, W \mid x_{r0} = x \rangle = r$$

since A is skewsymmetric.

To analyze the remaining terms in the equation of balance of energy, let us set

$$\mathfrak{P}_I^* = (\mathfrak{P}_I^*)_1 + (\mathfrak{P}_I^*)_2, \qquad (\mathfrak{P}_I^*)_2 = \sum_r \left\langle \frac{\partial \mathcal{T}_r^*}{\partial p_{ri}^*} (\hat{Q}_I^*)_{ri} W^* \,\middle|\, \boldsymbol{x}_{r0}^* = \boldsymbol{x}^* \right\rangle, \tag{5.20}$$

$$(\mathfrak{P}_I^*)_1 = -\tfrac{1}{2} \sum_{r,\alpha,s,\beta}{}' \left\langle \left(\frac{\partial \mathcal{T}_r^*}{\partial p_{ri}^*} \frac{\partial V_{r\alpha s\beta}}{\partial q_r^{*i}} - \frac{\partial \mathcal{T}_s^*}{\partial p_{si}^*} \frac{\partial V_{r\alpha s\beta}}{\partial q_s^{*i}} \right) W^* \,\middle|\, \boldsymbol{x}_{r0}^* = \boldsymbol{x}^* \right\rangle, \tag{5.21}$$

and

$$\mathfrak{P}_E^* = \sum_r \left\langle \frac{\partial \mathcal{T}_r^*}{\partial p_{ri}^*} (\hat{Q}_E^*)_{ri}\, W^* \,\middle|\, \boldsymbol{x}_{r0}^* = \boldsymbol{x}^* \right\rangle, \tag{5.22}$$

which are the analogous for φ^* of (3.14)–(3.16). By (4.30), $(4.17)_2$ and (2.18)

$$\frac{\partial \mathcal{T}_r^*}{\partial p_{ri}^*} = \frac{\partial \chi_r^{*i}}{\partial q_r^k} \frac{\partial \mathcal{T}_r}{\partial p_{rk}} + \frac{\partial \chi_r^{*i}}{\partial t} \tag{5.23}$$

and, easily,

$$\frac{\partial V_{r\alpha s\beta}}{\partial q_r^{*i}} = \frac{\partial \xi_{r\alpha}}{\partial q_r^i} (\chi_r^{*j}) \cdot (\boldsymbol{x}_{r\alpha}^* - \boldsymbol{x}_{s\beta}^*) \frac{V'_{r\alpha s\beta}}{\| \boldsymbol{x}_{r\alpha}^* - \boldsymbol{x}_{s\beta}^* \|}. \tag{5.24}$$

Then, by (4.15) and $(4.17)_1$,

$$\frac{\partial \mathcal{T}_r^*}{\partial p_{ri}^*} \frac{\partial V_{r\alpha s\beta}}{\partial q_r^{*i}} = \frac{\partial \mathcal{T}_r}{\partial p_{ri}} \frac{\partial V_{r\alpha s\beta}}{\partial q_r^i} + \dot{Q}^{\mathsf{T}} \xi_{r\alpha}(\{q^m\}) \cdot Q^{\mathsf{T}} (\boldsymbol{x}_{r\alpha} - \boldsymbol{x}_{s\beta}) \frac{V'_{r\alpha s\beta}}{\| \boldsymbol{x}_{r\alpha} - \boldsymbol{x}_{s\beta} \|}, \tag{5.25}$$

the analogue being true when s replaces r; hence

$$\begin{aligned} (\mathfrak{P}_I^*)_1 &= (\mathfrak{P}_I)_1 + \tfrac{1}{2} \sum_{r,\alpha,s,\beta}{}' \left\langle A(\xi_{r\alpha} + \xi_{s\beta}) \cdot (\boldsymbol{x}_{r\alpha} - \boldsymbol{x}_{s\beta}) \frac{V'_{r\alpha s\beta}}{\| \boldsymbol{x}_{r\alpha} - \boldsymbol{x}_{s\beta} \|} W \,\middle|\, \boldsymbol{x}_{r0} = \boldsymbol{x} \right\rangle \\ &= (\mathfrak{P}_I)_1 + A \cdot \operatorname{div} \boldsymbol{\Phi}, \end{aligned} \tag{5.26}$$

where

$$\boldsymbol{\Phi} = \tfrac{1}{4} \sum_{r,\alpha,s,\beta}{}' \int_{\mathbb{R}^3} dz \int_0^\infty d\theta \left\langle (\xi_{r\alpha} + \xi_{s\beta}) \wedge (\boldsymbol{x}_{r\alpha} - \boldsymbol{x}_{s\beta}) \frac{V'_{r\alpha s\beta}}{\| \boldsymbol{x}_{r\alpha} - \boldsymbol{x}_{s\beta} \|} W \,\middle|\, \boldsymbol{x}_{r0} = \boldsymbol{x} + \theta \boldsymbol{z}, \boldsymbol{x}_{s0} = \boldsymbol{x} - (1 - \theta)\, \boldsymbol{z} \right\rangle \otimes \boldsymbol{z}. \tag{5.27}$$

Moreover, by (4.32), (5.23) and (4.16),

$$\begin{aligned} (\mathfrak{P}_I^*)_2 &= (\mathfrak{P}_I)_2 + \sum_r \left\langle \frac{\partial \chi_r^{*i}}{\partial t} \frac{\partial \chi_r^k}{\partial q_r^{*i}} (\hat{Q}_I)_{rk}\, W \,\middle|\, \boldsymbol{x}_{r0} = \boldsymbol{x} \right\rangle \\ &= (\mathfrak{P}_I)_2 + \sum_r \left\langle \hat{\boldsymbol{f}}_{r\alpha} \cdot \frac{\partial \xi_{r\alpha}}{\partial q_r^k} \frac{\partial \chi_r^k}{\partial q_r^{*i}} \frac{\partial \chi_r^{*i}}{\partial t} W \,\middle|\, \boldsymbol{x}_{r0} = \boldsymbol{x} \right\rangle \\ &= (\mathfrak{P}_I)_2 - \sum_r \langle \hat{\boldsymbol{f}}_{r\alpha} \cdot A \xi_{r\alpha} W \mid \boldsymbol{x}_{r0} = \boldsymbol{x} \rangle. \end{aligned} \tag{5.28}$$

By (5.22), (4.33), (5.23), (3.15), (4.16)

$$\mathfrak{P}_E^* = \sum_r \left\langle \left(\frac{\partial \chi_r^{*i}}{\partial q_r^k} \frac{\partial \mathscr{T}_r}{\partial p_{rk}} + \frac{\partial \chi_r^{*i}}{\partial t} \right) \frac{\partial \chi_r^j}{\partial q_r^{*i}} \left((\hat{Q}_E)_{rj} - \sum_\alpha m_{r\alpha} \gamma_{r\alpha}^* \cdot \frac{\partial \xi_{r\alpha}}{\partial q_r^j} \right) W \,\middle|\, \boldsymbol{x}_{r0} = \boldsymbol{x} \right\rangle$$

$$= \mathfrak{P}_E + \sum_{r,\alpha} \left\langle \left(\dot{Q}^\mathsf{T} \xi_{r\alpha} \cdot Q^\mathsf{T} g_{r\alpha} - m_{r\alpha} \gamma_{r\alpha}^* \cdot \frac{\partial \xi_{r\alpha}}{\partial q_r^k} \frac{\partial \mathscr{T}_r}{\partial p_{rk}} \right.\right. \tag{5.29}$$

$$\left.\left. - m_{r\alpha} \gamma_{r\alpha}^* \cdot Q \dot{Q}^\mathsf{T}(t)\, \xi_{r\alpha} \right) W \,\middle|\, \boldsymbol{x}_{r0} = \boldsymbol{x} \right\rangle.$$

Moreover, by (4.20), (2.2), (5.6) and (5.5)

$$\sum_{r,\alpha} \langle m_{r\alpha} \gamma_{r\alpha}^* \cdot Q \dot{Q}^\mathsf{T} \xi_{r\alpha} W \mid \boldsymbol{x}_{r0} = \boldsymbol{x} \rangle$$

$$= A \cdot \sum_{r,\alpha} \langle m_{r\alpha} \xi_{r\alpha} \otimes \gamma_{r\alpha}^* W \mid \boldsymbol{x}_{r0} = \boldsymbol{x} \rangle$$

$$= A \cdot \sum_{r,\alpha} \langle m_{r\alpha} \xi_{r\alpha} \otimes ((\dot{A} - A^2)(\boldsymbol{x}_{r\alpha} - \boldsymbol{x}_0) + 2A \boldsymbol{v}_{r\alpha}) W \mid \boldsymbol{x}_{r0} = \boldsymbol{x} \rangle$$

$$= A \cdot \sum_{r,\alpha} \langle m_{r\alpha} \xi_{r\alpha} \otimes ((\dot{A} - A^2)\, \xi_{r\alpha} + 2A \dot{\xi}_{r\alpha}) W \mid \boldsymbol{x}_{r0} = \boldsymbol{x} \rangle$$

$$= A \cdot I \dot{A}^\mathsf{T} - A \cdot I A^2 + 2A^2 \cdot K. \tag{5.30}$$

Also, by (4.20), (2.4), (2.3), (5.6), (5.5) and the skewsymmetry of A

$$\sum_{r,\alpha} \left\langle m_{r\alpha} \gamma_{r\alpha}^* \cdot \frac{\partial \xi_{r\alpha}}{\partial q_r^k} \frac{\partial \mathscr{T}_r}{\partial p_{rk}} W \,\middle|\, \boldsymbol{x}_{r0} = \boldsymbol{x} \right\rangle$$

$$= \sum_{r,\alpha} \left\langle m_{r\alpha} \frac{\partial \mathscr{T}_r}{\partial p_{rk}} \frac{\partial \xi_{r\alpha}}{\partial q_r^k} \cdot ((\dot{A} - A^2)\, \xi_{r\alpha} + 2A \dot{\xi}_{r\alpha}) W \,\middle|\, \boldsymbol{x}_{r0} = \boldsymbol{x} \right\rangle \tag{5.31}$$

$$= -\dot{A} \cdot K - A^2 \cdot K.$$

We can now write (4.41) in terms of fields related to φ, taking into account that, for any function $f = f(\boldsymbol{x}, t)$

$$\frac{\partial}{\partial t^*} f(\boldsymbol{x}(\boldsymbol{x}^*, t^*), t(t^*)) = \left(\frac{\partial f}{\partial t} + \frac{\partial f}{\partial x^i} \frac{\partial x^i}{\partial t} \right)_{\boldsymbol{x} = \boldsymbol{x}(\boldsymbol{x}^*, t^*), t = t(t^*)} \tag{5.32}$$

whenever the functions $\boldsymbol{x} = \boldsymbol{x}(\boldsymbol{x}^*, t^*)$ and $t = t(t^*)$ are given by (4.1). With the understanding that the derivative $\partial \boldsymbol{v}^\tau / \partial t$ is taken at constant $\boldsymbol{x}^*$ and not $\boldsymbol{x}$, we have

$$\frac{\partial \varepsilon}{\partial t} + \operatorname{grad} \varepsilon \cdot \boldsymbol{v}^\tau + \tfrac{1}{2} (\boldsymbol{v}^\tau)^2 \left(\frac{\partial \varrho}{\partial t} + \operatorname{grad} \varrho \cdot \boldsymbol{v}^\tau \right) + \varrho \boldsymbol{v}^\tau \cdot \frac{\partial \boldsymbol{v}^\tau}{\partial t} - \varrho \boldsymbol{u} \cdot \frac{\partial \boldsymbol{v}^\tau}{\partial t}$$

$$- \boldsymbol{v}^\tau \cdot \left(\frac{\partial (\varrho \boldsymbol{u})}{\partial t} + \operatorname{grad} (\varrho \boldsymbol{u})\, \boldsymbol{v}^\tau \right) + \dot{A} \cdot K + A \cdot \left(\frac{\partial K}{\partial t} + (\operatorname{grad} K)\, \boldsymbol{v}^\tau \right) - A \dot{A} \cdot I$$

$$- \tfrac{1}{2} A^2 \cdot \left(\frac{\partial I}{\partial t} + (\operatorname{grad} I)\, \boldsymbol{v}^\tau \right) + \operatorname{div} \{ q + A : \Xi_K - \tfrac{1}{2} A^2 : \Xi_I - A : \Psi$$

$$- T^\mathsf{T}(\boldsymbol{u} - \boldsymbol{v}^\tau) + (\varepsilon + \tfrac{1}{2} \varrho (\boldsymbol{v}^\tau)^2 - \varrho \boldsymbol{u} \cdot \boldsymbol{v}^\tau + A \cdot K - \tfrac{1}{2} A^2 \cdot I)(\boldsymbol{u} - \boldsymbol{v}^\tau) \} =$$

$$= (b - \varrho(\ddot{x}_0 + (\dot{A} - A^2)(x - x_0) + 2A(u - \dot{x}_0))) \cdot (u - v^\tau) + r + \mathfrak{P}_I$$

$$+ A \cdot \operatorname{div} \Phi + \tfrac{1}{2} A \cdot \sum_r \langle \xi_{r\alpha} \wedge \hat{f}_{r\alpha} W \mid x_{r0} = x \rangle + \mathfrak{P}_E$$

$$+ \tfrac{1}{2} A \cdot \sum_r \langle \xi_{r\alpha} \wedge g_{r\alpha} W \mid x_{r0} = x \rangle + \dot{A} \cdot K + A^2 \cdot K - A \cdot I\dot{A}^\mathrm{T}$$

$$+ A \cdot IA^2 - 2A^2 \cdot K. \tag{5.33}$$

The terms $\dot{A} \cdot K$ and $-A\dot{A} \cdot I$ on the left-hand side of this equation are equal to the terms $\dot{A} \cdot K$ and $-A \cdot I\dot{A}^\mathrm{T}$ on the right-hand side, respectively, so they all cancel out. Moreover, as is easy to deduce from (4.2),

$$\frac{\partial v^\tau}{\partial t} = \ddot{x}_0 + (A^2 + \dot{A})(x - x_0). \tag{5.34}$$

These results, the validity of the equation of balance of energy (3.17), and the vanishing of certain terms, as discussed below, imply that, both on the left-hand and the right-hand sides we are left with terms that multiply either v^τ or $(v^\tau)^2$ or A or A^2. Since v^τ and A can be chosen independently, it is not difficult to see that each one of the 4 aforementioned contributions on the right-hand side must equal the corresponding one on the left-hand side. Therefore, writing the divergence in (5.33) explicitly, and denoting by K' the skew part of K, we have

$$\tfrac{1}{2}(v^\tau)^2 \left(\frac{\partial \varrho}{\partial t} + \operatorname{grad} \varrho \cdot u + \varrho \operatorname{div} u\right) = 0, \tag{5.35}$$

$$-v^\tau \cdot \left(\frac{\partial(\varrho u)}{\partial t} + \operatorname{grad}(\varrho u)\, u + \varrho u \operatorname{div} u - \operatorname{div} T\right) = -v^\tau \cdot b, \tag{5.36}$$

$$A \cdot \left(\frac{\partial K'}{\partial t} + (\operatorname{grad} K')\, u + \operatorname{div} \Xi_K + K' \operatorname{div} u - \operatorname{div} \Psi + T\right)$$

$$= A \cdot \operatorname{div} \Phi + \tfrac{1}{2} A \cdot \sum_r \langle \xi_{r\alpha} \wedge (\hat{f}_{r\alpha} + g_{r\alpha})\, W \mid x_{r0} = x \rangle, \tag{5.37}$$

and

$$-\tfrac{1}{2} A^2 \cdot \left(\frac{\partial I}{\partial t} + (\operatorname{grad} I)\, u + I \operatorname{div} u + \operatorname{div} \Xi_I\right) = -A^2 \cdot K. \tag{5.38}$$

The remaining terms in (5.33) cancel out. Indeed, first of all, $A \cdot IA^2$ identically vanishes since I is symmetric whereas A is skew. Then (5.33) reduces to

$$-\varrho(u - v^\tau) \cdot \frac{\partial v^\tau}{\partial t} - (u - v^\tau) \cdot \varrho\,(\operatorname{grad} v^\tau)(u - v^\tau) \tag{5.39}$$

$$= -(u - v^\tau) \cdot \varrho(\ddot{x}_0 + (A^2 + \dot{A})(x - x_0) + 2A(u - v^\tau)).$$

Since A, which coincides with $\operatorname{grad} v^\tau$, is skewsymmetric, equality (5.39) is an identity by (5.34) and the fact that the quadratic terms in $u - v^\tau$ are both identically zero.

The equations implied by (5.35)–(5.38), that is,

$$\frac{\partial \varrho}{\partial t} + \operatorname{div}(\varrho \boldsymbol{u}) = 0, \tag{5.40}$$

$$\frac{\partial(\varrho \boldsymbol{u})}{\partial t} + \operatorname{div}(\varrho \boldsymbol{u} \otimes \boldsymbol{u}) - \operatorname{div} \boldsymbol{T} = \boldsymbol{b}, \tag{5.41}$$

$$\frac{\partial \boldsymbol{K}'}{\partial t} + \operatorname{div}(\boldsymbol{K}' \otimes \boldsymbol{u}) + \operatorname{div} \boldsymbol{\Xi}_K - \operatorname{div}(\boldsymbol{\Psi} + \boldsymbol{\Phi}) + \tfrac{1}{2}(\boldsymbol{T} - \boldsymbol{T}^{\mathsf{T}})$$

$$= \tfrac{1}{2} \sum_r \langle \xi_{r\alpha} \wedge \hat{f}_{r\alpha} W \mid \boldsymbol{x}_{r0} = \boldsymbol{x} \rangle + \tfrac{1}{2} \sum_r \langle (\xi_{r\alpha} \wedge g_{r\alpha}) W \mid \boldsymbol{x}_{r0} = \boldsymbol{x} \rangle, \tag{5.42}$$

and

$$\frac{\partial \boldsymbol{I}}{\partial t} + \operatorname{div}(\boldsymbol{I} \otimes \boldsymbol{u}) + \operatorname{div} \boldsymbol{\Xi}_I = \boldsymbol{K} + \boldsymbol{K}^{\mathsf{T}} \tag{5.43}$$

can be interpreted as the equations of balance of mass, linear momentum, rotational momentum and microinertia. The 3rd-order tensor $\boldsymbol{\Psi} + \boldsymbol{\Phi}$ can be interpreted as the couple-stress tensor, and the last term in (5.42) as the body couple.

Acknowledgements. I acknowledge with gratitude useful discussions with G. F. Capriz and P. Podio-Guidugli on structured continua. This work was performed as part of the CNR research project on Mathematical Models for Modern Continua.

References

J. B. Alblas [1976], "A note on the physical foundation of the theory of multipole-stresses", *Archives of Mechanics* **28**, 279–298.

A. Bressan [1962], "Metodo di assiomatizzazione in senso stretto della meccanica classica. Applicazione di esso ad alcuni problemi di assiomatizzazione non ancora completamente risolti", *Rendiconti Seminario Matematico Università di Padova* **32**, 55–212.

G. F. Capriz [1984], "Continui con microstruttura", *Bollettino U.M.I.* (6) 3 **A**, 181–195.

G. F. Capriz [1985], "Continua with latent microstructure", *Archive for Rational Mechanics and Analysis* **90**, 43–56.

G. F. Capriz [1988], *Continui con microstruttura*, ETS Editrice, Pisa.

G. F. Capriz [1989], *Continua with microstructure*, Springer-Verlag, New York, *etc.*

G. F. Capriz & P. Podio-Guidugli [1976], "Discrete and continuous bodies with affine structure", *Annali di Matematica Pura e Appl.*, (IV) **111**, 195–211.

G. F. Capriz & P. Podio-Guidugli [1981], "Materials with finite-dimensional structure", *Mechanics of Structured Media*, A. P. S. Selvadurai, Ed., Proc. of the Int. Symposium on the Mechanical behavior of Structured Media, Ottawa, Elsevier Publishing Co., Amsterdam.

G. F. Capriz & P. Podio-Guidugli [1983], "Structured continua from a lagrangian point of view", *Annali di Matematica Pura e Appl.* (IV) **135**, 1–25.

G. F. Capriz, P. Podio-Guidugli & W. Williams [1982], "On balance equations for materials with affine structure", *Meccanica* **17**, 80–84.

C. F. Curtiss, R. B. Byrd & O. Hassager [1976], "Kinetic theory and rheology of macromolecular solutions", *Advances in Chemical Physics* **35**, 31–117.

J. S. DAHLER & L. E. SCRIVEN [1963], "Theory of structured continua. I", *Proceedings of the Royal Society of London* A **275**, 504–527.

A. C. ERINGEN [1966], "Mechanics of micromorphic materials", pp. 131–138 of the *Proceedings of the XI International Congress of Applied Mechanics*, Springer-Verlag, Berlin.

A. C. ERINGEN [1967], "Theory of micropolar continua", pp. 23–40 of the *Proceedings of the IX Midwestern Mechanics Conference*, Wiley, New York.

J. D. GODDARD [1986], "Microstructural origins of continuum stress fields — A brief history and some unresolved issues", *Recent developments in structured continua* (D. DE KEE & P. N. KALONI Ed.), Pitman Research Notes in Mathematics Series, Longman Scientific & Technical, Harlow.

A. E. GREEN & R. S. RIVLIN [1964a], "Multipolar continuum mechanics", *Archive for Rational Mechanics and Analysis* **17**, 113–147.

A. M. GREEN & R. S. RIVLIN [1964b], "On Cauchy's equations of motion", *Journal of Applied Mathematics and Physics* (*ZAMP*) **15**, 290–292.

J. H. IRVING & J. G. KIRKWOOD [1960], "The statistical mechanical theory of transport processes, IV: the equations of hydrodynamics", *Journal of Chemical Physics* **18**, 817–829.

J. E. MARSDEN & T. J. R. HUGHES [1983], *Mathematical foundations of elasticity*, Prentice-Hall, Englewood Cliffs, NJ.

A. I. MURDOCH [1985], "A corpuscular approach to continuum mechanics: basic considerations", *Archive for Rational Mechanics and Analysis* **88**, 291–321.

A. I. MURDOCH [1987], "On the relationship between balance relations for generalised continua and molecular behaviour", *International Journal of Engineering Science* **25**, 883–914.

W. NOLL [1955], "Die Herleitung der Grundgleichungen der Thermomechanik der Kontinua aus der Statistischen Mechanik", *Journal of Rational Mechanics and Analysis* **4**, 627–646.

W. NOLL [1963], "La mécanique classique, basée sur un axiome d'objectivité", pp. 47–56 of *La Méthode Axiomatique dans les Mécaniques Classiques et Modernes* (Colloque International à Paris, 1959), Paris, Gauthier-Villars, reprinted in W. NOLL, *The foundations of Mechanics and Thermodynamics*, Springer-Verlag, New York, etc., 1974.

M. PITTERI [1986], "Continuum equations of balance in classical statistical mechanics", *Archive for Rational Mechanics and Analysis* **94**, 291–305, and Corrigendum, *Archive for Rational Mechanics and Analysis* **100** (1988), 315–316.

M. PITTERI [1987], "On the objectivity of certain gross fields constructed within classical statistical mechanics", *Proceedings of the ISIMM Symposium on Kinetic Theory and Extended Thermodynamics* (I. MÜLLER & T. RUGGERI Eds)., Pitagora, Bologna.

M. PITTERI [1988], "On the equations of balance for mixtures constructed by means of classical statistical mechanics", *Meccanica* **23**, 3–10.

C. TRUESDELL [1977], *A first course in rational continuum mechanics*, Vol. 1, Academic Press, New York, etc.

C. TRUESDELL & R. A. TOUPIN [1960], "The classical field theories", *Handbuch der Physik* III/1, S. FLÜGGE ed., Springer-Verlag, Berlin, Heidelberg and New York.

Dipartimento di Metodi e Modelli Matematici
per le Scienze Applicate
Università di Padova

(Received November 9, 1989)

Memory Effects and Homogenization

Luc Tartar

Dedicated to Bernard D. Coleman *on his sixtieth birthday*

In the Fall of 1987 I was putting in order some of my results on memory effects induced by homogenization which I used for a contribution to a book in honor of E. De Giorgi on the occasion of his sixtieth birthday, for I had discussed some of these questions with him a long time earlier, during one of my visits to Pisa [7].

I had considered a linear case in which a memory effect had to be given by a convolution operator as a consequence of invariance under translation in time, and I had even determined which convolution kernels could appear in my example but when I had tried a simple nonlinear problem I could not identify the class of nonlinear memory effects that would be induced by homogenization.

Ten years before, I had advocated the point of view that characterizing special classes of partial differential equations stable by homogenization would help to discover natural classes of constitutive laws in continuum mechanics, but my efforts were not very convincing then [6]. Having found a simple linear setting where linear memory effects could appear by homogenization, my hope was that solving nonlinear cases could point out natural classes of materials with fading memory, a subject whose foundations had been laid by Bernard Coleman in collaboration with Walter Noll and Victor Mizel [1, 2, 3, 4].

I thought that it could interest Bernard Coleman, and indeed he made some interesting comments. However there were some questions which had puzzled me for many years and which I could still not solve. I will try here to go a little further in the analysis of these questions.

1. A linear problem

The model problem that I had chosen led to an equation of the form

$$\frac{\partial u^\varepsilon(x, t)}{\partial t} + a^\varepsilon(x)\, u^\varepsilon(x, t) = f(x, t) \text{ in } \Omega \times (0, T) \text{ with } u^\varepsilon(x, 0) = 0 \text{ in } \Omega, \tag{1.1}$$

in which Ω is an open set in R^N and the sequence of measurable functions a^ε satisfies the bounds

$$0 \leqq \alpha \leqq a^\varepsilon(x) \leqq \beta \quad \text{a.e. } x \in \Omega$$

and converges only weakly to a function a^0. One expected then that a subsequence of u^ε would converge weakly to u^0, and the question was to find whether u^0 would satisfy an equation of the same kind as (1.1). The answer was in general negative, and the characterization of the limiting equation involved more than the weak limit a^0. The result that I proved in [7] was

Theorem 1.2. *After extraction of a subsequence, there is a kernel K defined on $\Omega \times [0, \infty)$ such that for every function f (bounded and measurable, for instance) the sequence u^ε converges weakly to u^0, a solution of the equation*

$$\frac{\partial u^0(x,t)}{\partial t} + a^0(x)\, u^0(x,t) - \int_0^t K(x, t-s)\, u^0(x,s)\, ds = f(x,t) \quad \text{in } \Omega \times (0,T) \text{ with } u^0(x,0) = 0 \text{ in } \Omega. \tag{1.3}$$

The kernel K is measurable in $x \in \Omega$ and analytic in t and admits a representation

$$K(x,t) = \int e^{-\lambda t}\, d\mu_x(\lambda) \quad \text{a.e. } x \in \Omega \tag{1.4}$$

where the nonnegative measure $d\mu$ is measurable in x, has its support in $[\alpha, \beta]$, and satisfies the bound

$$\int \frac{d\mu_x(\lambda)}{(\beta - \lambda)(\lambda - \alpha)} \leqq 1 \quad \text{a.e. } x \in \Omega. \tag{1.5}$$

Remark 1.6. Knowledge of the YOUNG measure associated to a subsequence of a^ε is necessary for computing $d\mu$, and this involves a nonlinear transformation whose properties are easily deduced by using the integral representation of Pick functions. One should notice that the spatial domain Ω need not be an open set of R^N and may be any measure space endowed with a measure having no atoms.

This result shows that *a memory effect can be induced by oscillations in spatial properties.* At a microscopic level the equation (1.1) defines a semi-group, and only its value at the present time is necessary for deducing the future, but at a macroscopic level one has performed some averaging procedures, here modeled by taking weak limits, and the effective equation (1.3) no longer defines a semi-group. The memory effect found here fades exponentially according to (1.4) if $\alpha > 0$.

Constitutive relations involving a memory effect appear then to be a natural framework if one wishes to understand the behavior of fine mixtures of materials satisfying constitutive relations without memory. In [7] there is another example in which transport with fluctuating velocity induces some kind of a diffusion effect with memory.

2. The time-dependent linear case, I

A few years before and partly as a step towards the nonlinear case, I had asked one of my students, LUISA MASCARENHAS, to study the time-dependent case and she had done so as part of her thesis (Lisbon, July 1983) [5]. As her proofs involve a lot of technical computations, we shall now approach the time-dependent case in a different way. We consider an equation

$$\frac{\partial u^\varepsilon(x,t)}{\partial t} + a^\varepsilon(x,t)\,u^\varepsilon(x,t) = f(x,t) \quad \text{in } \Omega\times(0,T) \text{ with } u^\varepsilon(x,0)=0 \text{ in } \Omega \tag{2.1}$$

and assume that the sequence of measurable functions a^ε satisfies the bounds

$$0 \leqq \alpha \leqq a^\varepsilon(x,t) \leqq \beta \quad \text{a.e. in } \Omega\times(0,T) \tag{2.2}$$

and is equicontinuous in t, i.e. there is a function ω such that $\omega(\sigma)\to 0$ as $\sigma\to 0$ and

$$|a^\varepsilon(x,t) - a^\varepsilon(x,s)| \leqq \omega(|t-s|) \text{ a.e. } (x,t,s)\in\Omega\times(0,T)\times(0,T). \tag{2.3}$$

Let us remark that by changing u^ε into $e^{kt}v^\varepsilon$ one can replace a^ε by $a^\varepsilon + k$ (and f by fe^{-kt}), and so one can apply these ideas to bounded sequences which are equicontinuous in t. Notice also that (2.3) is implied by the stronger requirement

$$\left|\frac{\partial a^\varepsilon(x,t)}{\partial t}\right| \leqq M \quad \text{a.e. } (x,t,s)\in\Omega\times(0,T)\times(0,T).$$

Let us first define the functions A^ε and B^ε by the formulas

$$A^\varepsilon(x,s,t) = \exp\left[-\int_s^t a^\varepsilon(x,\sigma)\,d\sigma\right] \quad \text{in } \Omega\times(0,T)\times(0,T) \tag{2.5}$$

and

$$B^\varepsilon(x,s,t) = a^\varepsilon(x,t)\exp\left[-\int_s^t a^\varepsilon(x,\sigma)\,d\sigma\right] \quad \text{in } \Omega\times(0,T)\times(0,T). \tag{2.6}$$

The bounds (2.2) imply that the functions A^ε are bounded with bounded derivatives in s and t and that the functions B^ε are bounded with bounded derivatives in s, while the bounds (2.3) imply that the functions B^ε are equicontinuous in t. Using equicontinuity properties in time variables, let us then extract a subsequence such that

$$\begin{aligned} a^\varepsilon(\cdot,t) &\to a^0(\cdot,t) && \text{in } L^\infty(\Omega) \text{ weak* for every } t\in(0,T),\\ A^\varepsilon(\cdot,s,t) &\to A^0(\cdot,s,t) && \text{in } L^\infty(\Omega) \text{ weak* for every } s,t\in(0,T),\\ B^\varepsilon(\cdot,s,t) &\to B^0(\cdot,s,t) && \text{in } L^\infty(\Omega) \text{ weak* for every } s,t\in(0,T). \end{aligned} \tag{2.7}$$

Let us finally define the function C by the formula

$$C(x,s,t) = B^0(x,s,t) - a^0(x,t)\,A^0(x,s,t) \quad \text{in } \Omega\times(0,T)\times(0,T). \tag{2.8}$$

We can now state

Theorem 2.9. *Under hypotheses (2.2)–(2.3) and after extraction of a subsequence in order to dispose of (2.7), there is a kernel K defined on* $\Omega\times(0,T)\times(0,T)$ *such that for every function f, bounded and measurable, the sequence* u^ε *of solutions of (2.1) converges weakly to the solution* u^0 *of*

$$\frac{\partial u^0(x,t)}{\partial t}+a^0(x,t)\,u^0(x,t)-\int_0^t K(x,s,t)\,u^0(x,s)\,ds=f(x,t) \tag{2.10}$$

$$\text{in } \Omega\times(0,T) \quad \text{with } u^0(x,0)=0 \text{ in } \Omega.$$

The kernel K is given by

$$K(x,s,t)=\frac{\partial D(x,s,t)}{\partial s}-D(x,s,t)\,a^0(x,s) \quad \text{in } \Omega\times(0,T)\times(0,T) \tag{2.11}$$

where the kernel D defined on $\Omega\times(0,T)\times(0,T)$ *is the solution of the Volterra equation*

$$C(x,s,t)=D(x,s,t)-\int_s^t C(x,s,\sigma)\,D(x,\sigma,t)\,d\sigma \quad \text{in } \Omega\times(0,T)\times(0,T) \tag{2.12}$$

with C defined by (2.7)–(2.8).

Proof. From the explicit expression for u^ε

$$u^\varepsilon(x,t)=\int_0^t A^\varepsilon(x,s,t)\,f(x,s)\,ds \quad \text{in } \Omega\times(0,T)$$

we see that u^ε converges weakly to u^0, given as follows:

$$u^0(x,t)=\int_0^t A^0(x,s,t)\,f(x,s)\,ds \quad \text{in } \Omega\times(0,T)$$

while $a^\varepsilon u^\varepsilon$ converges weakly to w^0, given as follows:

$$w^0(x,t)=\int_0^t B^0(x,s,t)\,f(x,s)\,ds \quad \text{in } \Omega\times(0,T).$$

Taking the limit in (2.1), we see then that u^0 satisfies the equation

$$\frac{\partial u^0(x,t)}{\partial t}+a^0(x,t)\,u^0(x,t)+\int_0^t C(x,s,t)\,f(x,s)\,ds=f(x,t) \quad \text{in } \Omega\times(0,T) \tag{2.13}$$

in which the function C is given by (2.8). Notice that C is bounded and measurable, has a bounded derivative in s, is equicontinuous in t and satisfies

$$C(x,t,t)=0 \quad \text{in } \Omega\times(0,T)$$

because $A^0(x,t,t)=1$ and $B^0(x,t,t)=a^0(x,t)$ in $\Omega\times(0,T)$. We wish to show that (2.13) can be written in the form (2.10) with a kernel K, extending the convolution case $K(x,t-s)$, found when a^ε was assumed independent of t.

In order to do this we notice that the family of Volterra equations indexed by x, namely

$$f(x, t) - \int_0^t C(x, s, t) f(x, s)\, ds = g(x, t) \quad \text{in } \Omega(\times 0, T), \tag{2.14}$$

is solved by the explicit formula

$$f(x, t) = g(x, t) + \int_0^t D(x, s, t)\, g(x, s)\, ds \quad \text{in } \Omega\times(0, T) \tag{2.15}$$

in which the kernel D is the solution of (2.12). D inherits the properties of being bounded and equicontinuous in t and satisfies

$$D(x, t, t) = 0 \quad \text{in } \Omega\times(0, T) \tag{2.16}$$

and has a bounded derivative in s, which actually satisfies the equation

$$\frac{\partial C(x, s, t)}{\partial s} = \frac{\partial D(x, s, t)}{\partial s} - \int_s^t \frac{\partial C(x, s, \sigma)}{\partial s} D(x, \sigma, t)\, d\sigma \quad \text{in } \Omega\times(0, T)\times(0, T).$$

We notice then that (2.13) has the form (2.14) with the function g defined by

$$g(x, t) = \frac{\partial u^0(x, t)}{\partial t} + a^0(x, t)\, u^0(x, t) \quad \text{in } \Omega\times(0, T)$$

so that (2.15) implies

$$\begin{aligned} f(x, t) = {} & \frac{\partial u^0(x, t)}{\partial t} + a^0(x, t)\, u^0(x, t) \\ & + \int_0^t D(x, s, t) \left[\frac{\partial u^0(x, s)}{\partial s} + a^0(x, s)\, u^0(x, s)\right] ds \quad \text{in } \Omega\times(0, T) \end{aligned}$$

which after integration by parts and use of (2.16) becomes

$$\frac{\partial u^0(x, t)}{\partial t} + a^0(x, t)\, u^0(x, t) + \int_0^t \left[D(x, s, t)\, a^0(x, s) - \frac{\partial D(x, s, t)}{\partial s}\right] u^0(x, s)\, ds$$
$$\text{in } \Omega\times(0, T)$$

and so formula (2.11) for the kernel K energes.

Remark 2.17. To characterize the possible kernels K and generalize the conditions (1.4)–(1.5) is an open problem.

3. The time-dependent linear case, II

As it seems difficult to extend the preceding analysis to a nonlinear setting, we consider now another approach. Defining b^ε in $\Omega\times(0, T)$ by

$$b^\varepsilon(x, t) = a^\varepsilon(x, t) - a^0(x, t) \quad \text{in } \Omega\times(0, T) \tag{3.1}$$

we consider an equation with a parameter γ:

$$\frac{\partial U^\varepsilon(x,t;\gamma)}{\partial t} + [a^0(x,t) + \gamma b^\varepsilon(x,t)]\, U^\varepsilon(x,t;\gamma) = f(x,t) \quad \text{in } \Omega\times(0,T) \text{ with } U^\varepsilon(x,0;\gamma)=0 \quad \text{in } \Omega \tag{3.2}$$

so that our preceding problem corresponds to the case $\gamma = 1$, and our analysis will be based on analyticity properties in the parameter γ.

First let us extract a subsequence such that for each $k \geqq 1$ and for every $s_1, \dots, s_k \in (0,T)$

$$b^\varepsilon(\cdot, s_1) \dots b^\varepsilon(\cdot, s_k) \to M_k(\cdot, s_1, \dots, s_k) \quad \text{in } L^\infty(\Omega) \text{ weak*} \tag{3.3}$$

which is indeed possible due to the equicontinuity property (2.3). With the functions M_k defined by (3.3), let us then define the functions N_k and P_{k+1} by

$$N_k(x,s,t) = \int \cdot \int_{\Delta'} M_k(x,s,s_2,\dots,s_k)\, ds_2 \dots ds_k \quad \text{in } \Omega\times(0,T)\times(0,T) \tag{3.4}$$

and

$$P_{k+1}(x,s,t) = \int \cdot \int_{\Delta'} M_{k+1}(x,s,s_2,\dots,s_k,t)\, ds_2 \dots ds_k \quad \text{in } \Omega\times(0,T)\times(0,T) \tag{3.5}$$

for $k \geqq 2$, where $0 \leqq s \leqq t$ and the domain of integration Δ' is $s \leqq s_2 \leqq \dots \dots \leqq s_k \leqq t$; we also define $N_1 = P_1 = 0$ and $P_2 = M_2$.

We can now state the

Theorem 3.6. *The kernel K described in Theorem 2.9 can be obtained by taking the value $\gamma = 1$ in the power-series expansion*

$$K(x,s,t;\gamma) = \sum_{k=2}^{\infty} \gamma^k K_k(x,s,t) \quad \text{in } \Omega\times(0,T)\times(0,T) \tag{3.7}$$

where K_2 is defined by

$$K_2(x,s,t) = M_2(x,s,t) \exp\left[-\int_s^t a^0(x,\sigma)\, d\sigma\right] \tag{3.8}$$

and K_k is obtained by induction for $k \geqq 3$:

$$K_k(x,s,t) = (-1)^k P_k(x,s,t) \exp\left[-\int_s^t a^0(x,\sigma)\, d\sigma\right] - \sum_{j=2}^{k-1} (-1)^{k-j} \int_s^t K_j(x,z,t)\, N_{k-j}(x,s,z) \exp\left[-\int_s^z a^0(x,\tau)\, d\tau\right] dz \tag{3.9}$$

with N_j and P_j defined by (3.4)–(3.5).

Proof. U^ε admits the expansion

$$U^\varepsilon(x,t;\gamma) = \sum_{k=0}^{\infty} \gamma^k U_k^\varepsilon(x,t) \quad \text{in } \Omega\times(0,T)$$

in which the functions U_k^ε are defined by induction:

$$\frac{\partial U_k^\varepsilon(x,t)}{\partial t} + a^0(x,t)\,U_k^\varepsilon(x,t) + b^\varepsilon(x,t)\,U_{k-1}^\varepsilon(x,t) = 0 \quad \text{in } \Omega\times(0,T) \text{ with } U_k^\varepsilon(x,0)=0 \quad \text{in } \Omega \text{ for } k\geqq 1 \tag{3.10}$$

and U_0^ε is independent of ε and is a solution of

$$\frac{\partial U_0(x,t)}{\partial t} + a^0(x,t)\,U_0(x,t) = f(x,t) \quad \text{in } \Omega\times(0,T) \text{ with } U_0(x,0)=0 \text{ in } \Omega. \tag{3.11}$$

Because each function U_k^ε in the sequence is bounded, one can extract a subsequence such that for every k the sequence U_k^ε converges weakly to a function U_k^0. One notices then that U_0^0 coincides with U_0 and that $U_1^0 = 0$ because b^ε converges weakly to 0. The first correction to equation (3.11) is then of order γ^2.

From (3.10) we obtain the functions U_k^ε by induction:

$$U_k^\varepsilon(x,t) = -\int_0^t \exp\left[-\int_s^t a^0(x,\sigma)\,d\sigma\right] b^\varepsilon(x,s)\,U_{k-1}^\varepsilon(x,s)\,ds \quad \text{in } \Omega\times(0,T) \text{ for } k\geqq 1,$$

implying that

$$U_k^\varepsilon(x,t) = (-1)^k \int\cdot\int_\Delta b^\varepsilon(x,s_1)\dots b^\varepsilon(x,s_k) \times \exp\left[-\int_{s_1}^t a^0(x,\sigma)\,d\sigma\right] U_0(x,s_1)\,ds_1\dots ds_k \tag{3.12}$$

in which the domain of integration Δ is defined by $0\leqq s_1\leqq\dots\leqq s_k\leqq t$. Using (3.3), we obtain for $k\geqq 1$ the formula

$$U_k^0(x,t) = (-1)^k \int\cdot\int_\Delta M_k(x,s_1,\dots,s_k) \times \exp\left[-\int_{s_1}^t a^0(x,\sigma)\,d\sigma\right] U_0(x,s_1)\,ds_1\dots ds_k \tag{3.13}$$

and find that for $k\geqq 0$ the weak limit W_k^0 of $b^\varepsilon U_k^\varepsilon$ is given by

$$W_k^0(x,t) = (-1)^k \int\cdot\int_\Delta M_{k+1}(x,s_1,\dots,s_k,t) \times \exp\left[-\int_{s_1}^t a^0(x,\sigma)\,d\sigma\right] U_0(x,s_1)\,ds_1\dots ds_k. \tag{3.14}$$

Using (3.4)–(3.5), we can write formulas (3.13)–(3.14) as

$$U_k^0(x,t) = (-1)^k \int_0^t N_k(x,s,t)\exp\left[-\int_s^t a^0(x,\sigma)\,d\sigma\right] U_0(x,s)\,ds$$

for $k \geqq 1$ and

$$W_k^0(x, t) = (-1)^k \int_0^t P_{k+1}(x, s, t) \exp\left[-\int_s^t a^0(x, \sigma)\, d\sigma\right] U_0(x, s)\, ds$$

for $k \geqq 0$. The limit of equation (3.10) gives then

$$\frac{\partial U_k^0}{\partial t} + a^0 U_k^0 + W_{k-1}^0 = 0 \quad \text{in } \Omega\times(0, T) \text{ with } U_k^0(x, 0) = 0 \text{ in } \Omega$$

for $k \geqq 1$, so that by using (3.11) one deduces that

$$\frac{\partial}{\partial t}\left[\sum_{k=0}^{\infty} \gamma^k U_k^0\right] + a^0\left[\sum_{k=0}^{\infty} \gamma^k U_k^0\right] + \sum_{k=1}^{\infty} \gamma^{k+1} W_k^0 = f \quad \text{in } \Omega\times(0, T)$$

and our purpose is then to show a relation

$$\sum_{k=1}^{\infty} \gamma^{k+1} W_k^0(x, t) = -\int_0^t K(x, s, t; \gamma)\left[\sum_{k=0}^{\infty} \gamma^k U_k^0(x, s)\right] ds$$

with K being analytic in γ and having an expansion (3.7). The identification of the first terms in the expansion gives

$$K(x, s, t; \gamma) = \gamma^2 M_2(x, s, t) \exp\left[-\int_s^t a^0(x, \sigma)\, d\sigma\right] + O(\gamma^3)$$

showing (3.8). In order to identify the terms in γ^k for $k > 2$ we must verify the identity

$$(-1)^{k-1} \int_0^t P_k(x, s, t) \exp\left[-\int_s^t a^0(x, \sigma)\, d\sigma\right] U_0(x, s)\, ds$$

$$= -\int_0^t K_k(x, t, s)\, U_0(x, s)\, ds$$

$$-\sum_{j=2}^{k-1} \int_0^t K_j(x, s, t)\left\{(-1)^{k-j} \int_0^s N_{k-j}(x, \sigma, s) \exp\left[-\int_\sigma^s a^0(x, \tau)\, d\tau\right] U_0(x, \sigma)\, d\sigma\right\} ds$$

which holds for every choice of smooth functions U_0 if when $k \geqq 3$ the function K_k satisfies condition (3.9).

To show that we have indeed identified the kernel K and that we can take $\gamma = 1$; it only remains to verify the radius of convergence of the formal power series obtained. From (2.2) and (3.3) we deduce that the functions M_k satisfy the bounds

$$|M_k(x, s_1, \ldots, s_k)| \leqq (\beta - \alpha)^k$$

if $k \geqq 2$, and from (3.4) and (3.5) we then deduce the bounds

$$|N_k(x, s, t)| \leqq (\beta - \alpha)^k \frac{(t - s)^{k-1}}{(k - 1)!}$$

and

$$|P_k(x, s, t)| \leqq (\beta - \alpha)^k \frac{(t - s)^{k-2}}{(k - 2)!}$$

if $k \geqq 2$. Then (3.8) implies for K_2 the bound

$$|K_2(x, s, t)| \leqq (\beta - \alpha)^2 \exp [-\alpha(t-s)],$$

and when $k > 2$ we look for a bound of K_k of the form

$$|K_k(x, s, t)| \leqq C_k(\beta - \alpha)^k \frac{(t-s)^{k-2}}{(k-2)!} \exp [-\alpha(t-s)].$$

Putting this bound into (3.9) implies the inequality

$$C_k \leqq 1 + \sum_{j=2}^{k-1} C_j$$

if $k > 2$, which because $C_2 = 1$, gives

$$C_k \leqq 2^{k-2}$$

if $k \geqq 2$, showing that the power series (3.7) has then an infinite radius of convergence.

Remark 3.15. Notice that since a^ε does depend upon t, (3.8) does not necessarily imply that M_2 is nonnegative, and so one cannot expect that the kernel K be alway nonnegative.

Remark 3.16. When a^ε does not depend upon t, the preceding algorithm for computing the convolution kernel K differs from the one presented in [7] and seems more efficient for a practical purpose although it does not seem to imply easily the theoretical characterization (1.4)–(1.5).

Remark 3.17. Formula (3.8) giving the first term of the expansion of the kernel K can be considered as an example of small amplitude homogenization analogous to those described in [8, 9].

4. A nonlinear problem

We consider now a nonlinear equation:

$$\frac{\partial u^\varepsilon(x, t)}{\partial t} + a^\varepsilon(x, t)\{u^\varepsilon(x, t)\}^2 = f(x, t) \quad \text{in } \Omega \times (0, T) \text{ with } u^\varepsilon(x, 0) = 0 \text{ in } \Omega, \tag{4.1}$$

the sequence of coefficients a^ε in which satisfies the bounds (2.2)–(2.3), and f is measurable and satisfies the bound

$$0 \leqq f(x, t) \leqq F \quad \text{a.e. } (x, t) \in \Omega \times (0, T) \tag{4.2}$$

for some F. Let $u^\varepsilon = L^\varepsilon(f)$ be the solution which, because a^ε is nonnegative, is nonnegative and bounded on the interval $(0, T)$; the best bound is $(F/\alpha)^{\frac{1}{2}} \tanh((\alpha F)^{\frac{1}{2}} t)$ if $\alpha > 0$ and Ft if $\alpha = 0$.

For each f one can extract a subsequence such that u^ε converges weakly to a function u^0. In order to extract a subsequence such that for every f satisfying a condition (4.2) $L^\varepsilon(f)$ converges weakly to $u^0 = L^0(f)$, we notice that for a given F the mapping L^ε is uniformly continuous from $L^1(0, T)$ to $L^\infty(0, T)$ and even Lipschitz continuous with a bound depending only on F and β, and that there is a countable, dense set of functions f for the $L^1(0, T)$ topology.

Of course L^0 is nonanticipative because the value of each $u^\varepsilon(s)$ and thus the value of $u^0(s)$ depends only upon the values of f in the interval $(0, s)$. What we should wish now is to characterize a class of nonlinear equations with memory effects that would arise in a natural way to describe L^0.

What follows is only an attempt to grasp the complexity of what this nonlinear memory effect can be. Much more needs to be done in order to clarify this question.

By analogy with what was done above we will consider a problem

$$\frac{\partial U^\varepsilon}{\partial t} + [a^0 + \gamma b^\varepsilon]\{U^\varepsilon\}^2 = f \quad \text{in } \Omega \times (0, T) \text{ with } U^\varepsilon(\cdot, 0) = 0 \text{ in } \Omega \tag{4.3}$$

and do an expansion in the parameter γ. We look for an expansion

$$U^\varepsilon = U_0 + \sum_{k=1}^{\infty} \gamma^k U_k^\varepsilon \quad \text{in } \Omega \times (0, T) \tag{4.4}$$

where the first term U_0 is independent of ε and is a solution of

$$\frac{\partial U_0}{\partial t} + a^0\{U_0\}^2 = f \quad \text{in } \Omega \times (0, T) \text{ with } U(\cdot, 0) = f \text{ in } \Omega. \tag{4.5}$$

The following terms can be computed by induction, the terms U_1^ε and U_2^ε being solutions of

$$\frac{\partial U_1^\varepsilon}{\partial t} + 2a^0 U_0 U_1^\varepsilon + b^\varepsilon U_0^2 = 0 \quad \text{in } \Omega \times (0, T) \text{ with } U_1^\varepsilon(\cdot, 0) = 0 \text{ in } \Omega \tag{4.6}$$

and

$$\begin{gathered}\frac{\partial U_2^\varepsilon}{\partial t} + 2a^0 U_0 U_2^\varepsilon + a^0\{U_1^\varepsilon\}^2 + 2b^\varepsilon U_0 U_1^\varepsilon = 0 \\ \text{in } \Omega \times (0, T) \text{ with } U_2^\varepsilon(\cdot, 0) = 0 \text{ in } \Omega,\end{gathered} \tag{4.7}$$

and the general term U_k^ε satisfying

$$\begin{gathered}\frac{\partial U_k^\varepsilon}{\partial t} + 2a^0 U_0 U_k^\varepsilon + a^0 \sum_{j=1}^{k-1} U_j^\varepsilon U_{k-j}^\varepsilon + b^\varepsilon \left\{2U_0 U_{k-1}^\varepsilon + \sum_{j=1}^{k-2} U_j^\varepsilon U_{k-j-1}^\varepsilon\right\} = 0 \\ \text{in } \Omega \times (0, T) \text{ with } U_k^\varepsilon(\cdot, 0) = 0 \quad \text{in } \Omega\end{gathered} \tag{4.8}$$

if $k \geqq 3$.

Because b^ε converges weakly to 0, we deduce that U_1^ε converges weakly to 0, but we need its precise expression to compute other weak limits; from (4.6) we deduce that

$$U_1^\varepsilon(x, t) = - \int_0^t R(x, s, t)\, b^\varepsilon(x, s)\, \{U_0(x, s)\}^2\, ds \quad \text{in } \Omega \times (0, T) \tag{4.9}$$

where R is defined on $\Omega \times (0, T \times (0, T)$ by the formula

$$R(x, s, t) = \exp\left[- \int_s^t 2a^0(x, \sigma)\, U_0(x, \sigma)\, d\sigma\right] \quad \text{in } \Omega \times (0, T) \times (0, T). \tag{4.10}$$

The quantity $a^0\{U_1^\varepsilon\}^2 + 2b^\varepsilon U_0 U_1^\varepsilon$, which appears in the equation satisfied by U_2^ε, is then equal to

$$\begin{aligned} a^0(x, t) \int_0^t R(x, s_1, t)\, b^\varepsilon(x, s_1)\, \{U_0(x, s_1)\}^2\, ds_1 \int_0^t R(x, s_2, t)\, b^\varepsilon(x, s_2)\, \{U_0(x, s_2)\}^2\, ds_2 \\ - 2b^\varepsilon(x, t)\, U_0(x, t) \int_0^t R(x, s, t)\, b^\varepsilon(x, s)\, \{U_0(x, s)\}^2\, ds, \end{aligned} \tag{4.11}$$

and its weak limit is

$$\begin{aligned} a^0(x, t) \int_0^t \int_0^t M_2(x, s_1, s_2)\, R(x, s_1, t)\, R(x, s_2, t)\, \{U_0(x, s_1)\}^2\, \{U_0(x, s_2)\}^2\, ds_1\, ds_2 \\ - 2U_0(x, t) \int_0^t M_2(x, s, t)\, R(x, s, t)\, \{U_0(x, s)\}^2\, ds, \end{aligned} \tag{4.12}$$

in which the function M_2 is defined by (3.8) for the extracted subsequence.

Remark 4.13. The preceding computations show that if one has extracted a subsequence such that (3.3) holds, then U_2^ε converges weakly to U_2^0, which satisfies an equation involving M_2. From (4.8) one can easily verify by induction that U_k^ε converges weakly to U_k^0, which also satisfies an equation involving M_k. Unfortunately the algebraic structure of these terms is so intricate that some other idea seems necessary in order to push these computations to their end. If a^ε does not depend upon t the Young measure associated to a subsequence characterizes the limit equation.

Remark 4.14. One can summarize the preceding computations by writing a nonlinear delay equation satisfied at order 2 in γ. If we denote by U_* the truncated expansion

$$U_*(x, t) = U_0(x, t) + \gamma^2 U_2^0(x, t), \tag{4.15}$$

then U_* satisfies an equation

$$\frac{\partial U_*}{\partial t} + a^0\{U_*\}^2 + \gamma^2 a^0 \int_0^t \int_0^t M_*(\cdot, s_1, s_2)\, R_*(\cdot, s_1, t)\, R_*(\cdot, s_2, t)\, ds_1\, ds_2$$

$$- \gamma^2 \int_0^t 2M_*(\cdot, s, t)\, R_*(\cdot, s, t)\, ds = f + O(\gamma^3) \tag{4.16}$$

$$\text{in } \Omega \times (0, T) \text{ with } U_*(\cdot, 0) = 0 \text{ in } \Omega$$

in which the functions M_* and R_* depend upon U_* through the formulas

$$M_*(x, s, t) = M_2(x, s, t)\, U_*(x, s)\, U_*(x, t) \quad \text{in } \Omega \times (0, T) \times (0, T) \tag{4.17}$$

and

$$R_*(x, s, t) = \exp\left[- \int_s^t 2a^0(x, \sigma)\, U_*(x, \sigma)\, d\sigma\right] U_*(x, s) \quad \text{in } \Omega \times (0, T) \times (0, T). \tag{4.18}$$

Remark 4.19. One can imagine other ways to study (4.1) than by introducing a parameter γ like in (4.3). Pushing further the algebraic computations initiated above or others could shed some light on use of the diagrammatic techniques that theoretical physicists are so fond of. Indeed many difficult problems of physics require answering questions of homogenization for hyperbolic systems, and not many mathematical results are available, while some of the strange rules devised by theoretical physicists seem relatively efficient. The description of non-classical memory effects induced by homogenization may generate a complex hierarchy of nonlinear terms very similar to those that physicists attribute to creations and annihilations of particles or other strange diagrams. The preceding problem is but a simple and naive test for studying this idea.

Acknowledgements. This work was supported in part by NSF grant DMS-8803317.

References

1. COLEMAN, B. D., & V. J. MIZEL, "Norms and semigroups in the theory of fading memory," Arch. Rational Mech. Anal. **23** (1967), 87–123.
2. COLEMAN, B. D., & V. J. MIZEL, "On the general theory of fading memory," Arch. Rational Mech. Anal. **29** (1968), 18–31.
3. COLEMAN, B. D., & W. NOLL, "An approximation theorem for functionals with applications in continuum mechanics," Arch. Rational Mech. Anal. **6** (1960), 355–370.
4. COLEMAN, B. D., & W. NOLL, "Foundations of linear viscoelasticity," Reviews Mod. Phys. **33** (1961), 239–249; erratum: ibid. **36** (1964), 1103.
5. MASCARENHAS, L., "A linear homogenization problem with time dependent coefficient," Trans. A.M.S. **281**, 1 (1984), 179–195.
6. TARTAR, L., "Nonlinear constitutive relations and homogenization," *Contemporary developments in continuum mechanics and partial differential equations* (Proc. Internat.

Sympos., Inst. Mat., Univ. Fed. Rio de Janeiro, Rio de Janeiro, 1977), 472–484. North-Holland Math. Studies, 30, North-Holland, Amsterdam, 1978.
7. TARTAR, L., "Nonlocal effects induced by homogenization," *Partial Differential Equations and the Calculus of Variations, Essays in Honor of Ennio De Giorgi*, II, 925–938, Birkhäuser, Boston, 1989.
8. TARTAR, L., "*H*-measures, a new approach for studying homogenization, oscillations and concentration effects in partial differential equations," to appear in Proc. Roy. Soc. Edinburgh.
9. TARTAR, L., "*H*-measures and small amplitude homogenization," R. V. KOHN & G. MILTON, eds. Random Media and Composites, 89–99, SIAM, Philadelphia, 1989.

Department of Mathematics
Carnegie Mellon University
Pittsburgh

(Received August 30, 1989)

On Formation of Singularities in One-Dimensional Nonlinear Thermoelasticity

WILLIAM J. HRUSA & SALIM A. MESSAOUDI

Dedicated to Bernard D. Coleman on the occasion of his 60th birthday

1. Introduction

It is well known that smooth motions of nonlinear elastic bodies generally will break down in finite time due to the formation of shock waves. On the other hand, for thermoelastic materials, the conduction of heat provides dissipation that competes with the destabilizing effects of nonlinearity in the elastic response. The work of COLEMAN & GURTIN [2] on the growth and decay of acceleration waves provides a great deal of insight concerning the interplay between dissipation and nonlinearity in one-dimensional nonlinear thermoelastic bodies. Assuming that the elastic modulus, specific heat, and thermal conductivity are strictly positive, the stress-temperature modulus is nonzero, and that the elastic response is genuinely nonlinear they show that acceleration waves of small initial amplitude decay but waves of large initial amplitude can explode in finite time. In other words, thermal diffusion manages to restrain waves of small amplitudes but nonlinearity in the elastic response is dominant for waves of large amplitudes.

For smooth initial data that are close to equilibrium (in appropriate Sobolev norms) global existence and decay of classical solutions to the field equations of one-dimensional nonlinear thermoelasticity has been established by SLEMROD [11], ZHENG & SHEN [13, 14], JIANG [8], and HRUSA & TARABEK [7]. Further, assuming that the constitutive relations are of a special form, DAFERMOS & HSIAO [5] have shown that for certain smooth initial data with large gradients the solution of the Cauchy problem will develop singularities in finite time. However, the equations of motion studied in [5] are partially uncoupled and consequently do not satisfy the assumptions used to establish global existence when the data are close to equilibrium. In particular, for the equations of [5], if the initial temperature is spatially homogeneous and at equilibrium, then the temperature remains constant in time while the strain ε and the velocity v satisfy the system

$$\begin{aligned} \varepsilon_t(x, t) &= v_x(x, t), \\ v_t(x, t) &= p(\varepsilon(x, t))_x, \quad x \in \mathbb{R},\ t \geqq 0, \end{aligned} \tag{1.1}$$

of nonlinear elasticity. If the elastic response function p is nonlinear, then solutions of (1.1) generally will develop singularities in finite time even if the initial data are very smooth and small (*cf.* LAX [9] and MACCAMY & MIZEL [10]). Moreover, as DAFERMOS & HSIAO point out, their constitutive relations for the stress and internal energy are not consistent with the existence of a free energy.

In this paper we study a special class of nonlinear thermoelastic materials. Our constitutive relations are similar to those used by DAFERMOS & HSIAO [5]. Although we do not claim that our constitutive equations are appropriate to any real material, they do satisfy the assumptions used to establish global existence of smooth solutions when the data are close to the equilibrium. Moreover, they are fully compatible with the second law of thermodynamics; in particular there is a free energy. We consider the Cauchy problem in which the body occupies the entire real line and the initial values of the strain, velocity, and temperature are prescribed. As our main result we show that there are smooth initial data for which the solution will develop singularities in finite time. Our proof follows the argument of DAFERMOS & HSIAO [5] very closely. The primary difference is that our equations of evolution contain an additional term that prevents us from using the maximum principle to obtain an important *a priori* bound. This difficulty is overcome by exploiting some relations associated with the second law of thermodynamics.

2. Balance of Laws and Constitutive Equations

Consider a homogeneous one-dimensional body that occupies the interval B in a (fixed) reference configuration and has unit reference density. We denote by

ε the strain,

v the velocity,

σ the stress,

e the internal energy,

q the heat flux, and

T the absolute temperature.

Each of the above fields is assumed to be a smooth function of the reference position x and the time t; the strain and the temperature are required to satisfy

$$\varepsilon > -1, \quad T > 0. \tag{2.1}$$

In the absence of external heat supply and body force, the laws of balance of momentum and energy can be written as

$$\begin{aligned} \varepsilon_t &= v_x, \\ v_t &= \sigma_x, \\ e_t + q_x &= \sigma\varepsilon_t, \quad x \in B, \quad t \geqq 0. \end{aligned} \tag{2.2}$$

A thermoelastic material is described by constitutive relations of the form

$$\begin{aligned} \sigma(x,t) &= \hat{\sigma}(\varepsilon(x,t), T(x,t)), \\ e(x,t) &= \hat{e}(\varepsilon(x,t), T(x,t)), \\ q(x,t) &= \hat{q}(\varepsilon(x,t), T(x,t), T_x(x,t)). \end{aligned} \tag{2.3}$$

The second law of thermodynamics imposes restrictions[1] on $\hat{\sigma}$, $\hat{e}$, and $\hat{q}$. In particular, $\hat{\sigma}$ and $\hat{e}$ are required to satisfy the compatibility relation

$$\hat{e}_\varepsilon(\varepsilon, T) = \hat{\sigma}(\varepsilon, T) - T\hat{\sigma}_T(\varepsilon, T), \quad \forall\, \varepsilon > -1, \quad T > 0 \tag{2.4}$$

and $\hat{q}$ must obey the heat conduction inequality

$$g \cdot \hat{q}(\varepsilon, T, g) \leqq 0, \quad \forall\, \varepsilon > -1, \quad T > 0, \quad g \in \mathbb{R}. \tag{2.5}$$

We note that (2.4) is equivalent to the existence of a free energy,

$$\psi = \hat{\psi}(\varepsilon, T), \tag{2.6}$$

and a corresponding entropy,

$$\eta = \hat{\eta}(\varepsilon, T), \tag{2.7}$$

which satisfy

$$\psi = e - T\eta, \tag{2.8}$$

and

$$\hat{\sigma} = \hat{\psi}_\varepsilon, \quad \hat{\eta} = -\hat{\psi}_T. \tag{2.9}$$

An important consequence of (2.6), (2.7), (2.9) is the identity

$$\psi_t + \eta T_t - \sigma\varepsilon_t = 0, \tag{2.10}$$

which is equivalent to GIBBS's relation

$$e_t - T\eta_t - \sigma\varepsilon_t = 0, \tag{2.11}$$

by virtue of (2.8). In view of (2.11), the equation of balance, of energy $(2.2)_3$, can be rewritten as

$$T\eta_t + q_x = 0. \tag{2.12}$$

We refer to the recent book of DAY [6] for more information on one-dimensional nonlinear thermoelasticity.

---

[1] The idea of using the second law of thermodynamics to obtain restrictions on constitutive relations is due to COLEMAN & NOLL [4]. The restrictions (2.4), (2.5) follow from the results of [4]. COLEMAN & NOLL assume that the stress and internal energy are independent of the temperature gradient. A subsequent argument of COLEMAN & MIZEL [3] shows that compatibility with the second law requires that stress and internal energy (as well as the entropy and free energy) be independent of the temperature gradient.

Here we consider special constitutive equations of the form

$$\begin{aligned} \hat{\sigma}(\varepsilon, T) &= p(\varepsilon) + \beta(T - \overline{T}), \\ \hat{e}(\varepsilon, T) &= P(\varepsilon) + c(T - \overline{T}) - \beta \overline{T} \varepsilon, \\ \hat{q}(\varepsilon, T, T_x) &= -\varkappa T_x, \end{aligned} \tag{2.13}$$

where $p : (-1, \infty) \to \mathbb{R}$ is a given smooth function,

$$P(\varepsilon) := \int_0^\varepsilon p(\xi)\, d\xi, \quad \forall\, \varepsilon > -1, \tag{2.14}$$

while β, $\overline{T}$, c, $\varkappa$ are constants, and

$$\overline{T}, c, \varkappa > 0. \tag{2.15}$$

It is easy to verify that these constitutive equations satisfy (2.4) and (2.5) and that

$$\hat{\psi}(\varepsilon, T) = P(\varepsilon) + \beta\varepsilon(T - \overline{T}) + cT \log(\overline{T}/T) + c(T - \overline{T}) \tag{2.16}$$

is a free energy; moreover the corresponding entropy is given by

$$\hat{\eta}(\varepsilon, T) = -\beta\varepsilon - c \log(\overline{T}/T). \tag{2.17}$$

Remark 2.1. DAFERMOS & HSIAO [5] assume that the stress and the heat flux are given by $(2.13)_1$, $(2.13)_3$ and that

$$\hat{e}(\varepsilon, T) = P(\varepsilon) + c(T - \overline{T}).$$

As they point out, their constitutive relations comply with (2.4) only if $\overline{T} = 0$ (or $\beta = 0$).

3. Formation of Singularities

We assume that the body occupies the entire real line in its reference configuration (*i.e.* $B = \mathbb{R}$) and that the stress, internal energy, and heat flux are given by (2.3), where $\hat{\sigma}$, $\hat{e}$, and $\hat{q}$ have the special forms (2.13). We assume further that

$$p \in C^4(-1, \infty), \quad p(0) = 0, \quad p'(\xi) > 0 \quad \forall\, \xi > -1, \tag{3.1}$$

and that

$$\overline{T}, c, \varkappa > 0. \tag{3.2}$$

We seek a smooth solution of (2.2) when the initial values of the strain, velocity, and temperature are prescribed. Thus we consider the initial value problem

$$\begin{aligned} \varepsilon_t(x, t) &= v_x(x, t), \\ v_t(x, t) &= p'(\varepsilon(x, t))\, \varepsilon_x(x, t) + \beta\theta_x(x, t), \\ c\theta_t(x, t) &= \varkappa\theta_{xx}(x, t) + \beta(\theta + \overline{T})\, v_x(x, t), \quad x \in \mathbb{R},\ t \geqq 0, \end{aligned} \tag{3.3}$$

$$\varepsilon(x, 0) = \varepsilon_0(x), \quad v(x, 0) = v_0(x), \quad \theta(x, 0) = \theta_0(x), \quad x \in \mathbb{R},$$

where ε_0, v_0, and θ_0 are given functions and

$$\theta(x, t) := T(x, t) - \bar{T}. \tag{3.4}$$

The existence of a local solution can be established by a standard contraction-mapping argument and we omit the details. The relevant result is recorded below.

Proposition. *Assume that* (3.1) *and* (3.2) *hold,*

$$\varepsilon_0, v_0, \theta_0 \in H^3(\mathbb{R}), \tag{3.5}$$

and that

$$\varepsilon_0(x) > -1, \quad \theta_0(x) > -\bar{T} \quad \forall\, x \in \mathbb{R}. \tag{3.6}$$

Then the initial-value problem (3.3) *has a unique local solution* (ε, v, θ) *on the maximal interval* $[0, t_0)$, $t_0 > 0$, *with*

$$\begin{gathered} \varepsilon, v \in C([0, t_0); H^3(\mathbb{R})) \cap C^1([0, t_0); H^2(\mathbb{R})), \\ \theta \in C([0, t_0); H^3(\mathbb{R})) \cap C^1([0, t_0); H^1(\mathbb{R})), \end{gathered} \tag{3.7}$$

$$\varepsilon(x, t) > -1, \quad \theta(x, t) > -\bar{T} \quad \forall\, x \in \mathbb{R}, \quad t \in [0, t_0). \tag{3.8}$$

Moreover, if

$$\sup_{\substack{-\infty < x < \infty \\ 0 \leq t < t_0}} (\{|\varepsilon_x| + |\varepsilon_t| + |\theta_x| + |\theta|\}(x, t)) < \infty, \tag{3.9}$$

$$\inf_{\substack{-\infty < x < \infty \\ 0 \leq t < t_0}} \varepsilon(x, t) > -1, \qquad \inf_{\substack{-\infty < x < \infty \\ 0 \leq t < t_0}} \theta(x, t) > -\bar{T}, \tag{3.10}$$

then $t_0 = \infty$.

Remark 3.1. If (3.1), (3.2), (3.5), (3.6) hold and $\varepsilon_0, v_0, \theta_0 \in W^{2,1}(\mathbb{R})$, then the local solution (ε, v, θ) of (3.3) also satisfies

$$\varepsilon, v, \theta \in C([0, t_0); W^{2,1}(\mathbb{R})). \tag{3.11}$$

(See Theorem 2 of [12] and Section 2 of [13].)

Remark 3.2. Under suitable assumptions on $\hat{\sigma}$, $\hat{e}$, and $\hat{q}$, an analogous result on local existence holds for general constitutive equations of the form (2.3). In this case, a bound stronger than (3.9) is needed to continue the solution globally in time.

Remark 3.3. It follows from the results of [7], [13] that if $\beta \neq 0$, (3.6) holds, and if

$$\|\varepsilon_0\|_{H^3} + \|v_0\|_{H^3} + \|\theta_0\|_{H^3}$$

is sufficiently small, then the initial value problem (3.3) has a globally defined smooth solution. (See Remark 2 of [7].)

Theorem. *Assume that* (3.1) and (3.2), *hold and that* $p''(0) > 0$. *Let* $\gamma, L, J > 0$ *be given. Then there exist* $\delta, M > 0$ *(depending on* γ, L, J*) with the following* *property*: *For each* $\varepsilon_0, v_0, \theta_0 \in H^3(\mathbb{R}) \cap W^{2,1}(\mathbb{R})$ *satisfying* (3.6),

$$|\varepsilon_0(x)| \leqq \delta^{\frac{1}{2}}, \quad |v_0(x)| \leqq \delta^{\frac{1}{2}}, \quad |\theta_0(x)| \leqq \delta, \quad \forall\, x \in \mathbb{R},\ |\theta_0'(x)| \leq J \tag{3.12}$$

$$\int_{-\infty}^{\infty} (\varepsilon_0^2(x) + v_0^2(x) + \theta_0^2(x) + |\theta_0(x)|)\, dx \leqq \delta^2, \tag{3.13}$$

$$\min_{-\infty<x<\infty} \{v_0'(x) + p'(\varepsilon_0(x))^{\frac{1}{2}}\, \varepsilon_0'(x)\} + \min_{-\infty<x<\infty} \{v_0'(x) - p'(\varepsilon_0(x))^{\frac{1}{2}}\, \varepsilon_0'(x)\} \geqq -J, \tag{3.14}$$

and

$$\max_{-\infty<x<\infty} \{v_0'(x) + p'(\varepsilon_0(x))^{\frac{1}{2}}\, \varepsilon_0'(x)\} + \max_{-\infty<x<\infty} \{v_0'(x) - p'(\varepsilon_0(x))^{\frac{1}{2}}\, \varepsilon_0'(x)\} \geqq M, \tag{3.15}$$

the length t_0 *of the maximal interval of existence of a smooth solution* (ε, v, θ) *of* (3.3) *is less than* (*or equal to*) L; *moreover the local solution satisfies*

$$|\varepsilon(x,t)| \leqq \gamma, \quad |v(x,t)| \leqq \gamma, \quad |\theta(x,t)| \leqq \gamma, \quad \forall\, x \in \mathbb{R}, \quad t \in [0, t_0). \tag{3.16}$$

Remark 3.4. It is not difficult to show that for every $\delta, J, M > 0$ there exist $\varepsilon_0, v_0, \theta_0 \in H^3(\mathbb{R}) \cap W^{2,1}(\mathbb{R})$ *satisfying* (3.12) through (3.15).

Remark 3.5. An analogous theorem holds if the assumption $p''(0) > 0$ is replaced by $p''(0) < 0$.

4. Proof of the Theorem

The first part of the proof consists of establishing *a priori* estimates for the local solution of (3.3). After the estimates are obtained, we establish the formation of singularities in finite time by analyzing a differential inequality.

Let $t^*, \delta \in (0, 1]$ and $J > 0$ be given. Assume that (3.12), (3.13), (3.14) hold and that (3.3) has a solution (ε, v, θ) on $\mathbb{R} \times [0, t^*]$ satisfying

$$\begin{gathered} \varepsilon, v \in C([0,t^*]; H^3(\mathbb{R}) \cap W^{2,1}(\mathbb{R})) \cap C^1([0,t^*]; H^2(\mathbb{R})), \\ \theta \in C([0,t^*]; H^3(\mathbb{R}) \cap W^{2,1}(\mathbb{R})) \cap C^1([0,t^*]; H^1(\mathbb{R})) \end{gathered} \tag{4.1}$$

and

$$|\varepsilon(x,t)| \leqq \tfrac{1}{2}, \quad |\theta(x,t)| \leqq \tfrac{1}{2}\bar{T} \quad \forall\, x \in \mathbb{R},\ t \in [0, t^*]. \tag{4.2}$$

We shall derive several *a priori* bounds for (ε, v, θ) in terms of the constant δ appearing in (3.12), (3.13). Throughout this section we use Γ to denote a generic positive constant that depends only on β, c, $\varkappa$, $\bar{T}$, J, and properties of the function p.

Lemma 4.1. *There exists a constant* $\tilde{\Gamma}$ (*which depends only on* c, $\bar{T}$, $\varkappa$, *and the maximum and the minimum of* p' *on* $[-\frac{1}{2}, \frac{1}{2}]$) *such that*

$$\int_{-\infty}^{\infty} \{\varepsilon^2 + v^2 + \theta^2\}(x,t)\, dx + \int_0^t \int_{-\infty}^{\infty} \theta_x^2(x,s)\, dx\, ds \leqq \tilde{\Gamma}\, \delta^2 \quad \forall\, t \in [0, t^*]. \tag{4.3}$$

Proof. It follows from (2.12) that

$$\theta\eta_t = -\frac{\theta}{(\overline{T}+\theta)} q_x. \tag{4.4}$$

Using $(2.2)_1$, $(2.2)_2$, (2.10), and (4.4), we conclude that

$$\frac{\partial}{\partial t}(\psi + \theta\eta + \tfrac{1}{2}v^2) - \frac{\overline{T}}{(\overline{T}+\theta)^2} q\theta_x = \frac{\partial}{\partial x}\left(\sigma v - \frac{\theta}{(\overline{T}+\theta)} q\right). \tag{4.5}$$

We integrate (4.5) over $\mathbb{R}\times[0, t^*]$ and use (2.13), (2.16), and (2.17) to obtain

$$\begin{aligned}\int_{-\infty}^{\infty} &\left\{P(\varepsilon) + \tfrac{1}{2}v^2 - c\overline{T}\log\left(1+\frac{\theta}{\overline{T}}\right) + c\theta\right\}(x,t)\,dx \\ &+ \varkappa \int_0^t \int_{-\infty}^{\infty} \frac{\overline{T}}{(\overline{T}+\theta)^2}\theta_x^2(x,s)\,dx\,ds \\ &= \int_{-\infty}^{\infty} \left\{P(\varepsilon_0) + \tfrac{1}{2}v_0^2 - c\overline{T}\log\left(1+\frac{\theta_0}{\overline{T}}\right) + c\theta_0\right\}(x)\,dx. \end{aligned} \tag{4.6}$$

It is straightforward to verify that

$$\tfrac{1}{2}m_1\lambda^2 \leqq P(\lambda) \leqq \tfrac{1}{2}m_2\lambda^2 \quad \forall\, \lambda \in [-\tfrac{1}{2}, \tfrac{1}{2}], \tag{4.7}$$

$$\frac{2c}{9\overline{T}}\xi^2 \leqq -c\overline{T}\log\left(1+\frac{\xi}{\overline{T}}\right) + c\xi \leqq \frac{2c}{\overline{T}}\xi^2 \quad \forall\, \xi \in [-\tfrac{1}{2}\overline{T}, \tfrac{1}{2}\overline{T}], \tag{4.8}$$

where

$$m_1 := \min_{\lambda\in[-\frac{1}{2},\frac{1}{2}]} p'(\lambda), \qquad m_2 := \max_{\lambda\in[-\frac{1}{2},\frac{1}{2}]} p'(\lambda). \tag{4.9}$$

Consequently (4.6) yields the inequality (4.3). $\square$

Following DAFERMOS & HSIAO [5], we use the method of ALIKAKOS [1] to obtain a pointwise bound for θ in terms of δ.

Lemma 4.2. *The temperature θ obeys the estimate*

$$|\theta(x,t)| \leqq \Gamma\delta \quad \forall\, x\in\mathbb{R}, \quad t\in[0,t^*]. \tag{4.10}$$

Proof. The idea of the proof is to put

$$A_n := \max_{0\leqq t\leqq t^*} \int_{-\infty}^{\infty} \theta^{2^n}(x,t)\,dx, \quad n = 1,2,3,\ldots \tag{4.11}$$

and show that

$$A_n \leqq (\Gamma\delta)^{2^n}, \quad n = 1,2,3,\ldots. \tag{4.12}$$

The desired conclusion then follows by raising (4.12) to the 2^{-n} power and letting n tend to infinity.

To establish (4.12), we multiply $(3.3)_3$ by $2^n\theta^{2^n-1}$, and integrate with respect to x, using integration by parts, to obtain

$$\frac{d}{dt}\int_{-\infty}^{\infty}\theta^{2^n}(x,t)\,dx = -4(1-2^{-n})\frac{\varkappa}{c}\int_{-\infty}^{\infty}[\partial_x\theta^{2^{n-1}}(x,t)]^2\,dx$$

$$-2\cdot 2^n\frac{\beta}{c}\int_{-\infty}^{\infty}\{\partial_x\theta^{2^{n-1}}(x,t)\,\theta^{2^{n-1}}(x,t)\,v(x,t)\}\,dx$$

$$+2^n\overline{T}\frac{\beta}{c}\int_{-\infty}^{\infty}\theta^{2^n-1}(x,t)\,v_x(x,t)\,dx. \tag{4.13}$$

We note that (4.13) is similar to (2.9) of [5] except for the term $\int_{-\infty}^{\infty}\theta^{2^n-1}(x,t)\,v_x(x,t)\,dx$ which (if $\beta \neq 0$) can be handled as follows:

$$\left|\int_{-\infty}^{\infty}\theta^{2^n-1}(x,t)\,v_x(x,t)\,dx\right|$$

$$=(2^n-1)\left|\int_{-\infty}^{\infty}\theta_x(x,t)\,\theta^{2^n-2}(x,t)\,v(x,t)\,dx\right|$$

$$\leqq (2^n-1)\int_{|\theta|\leqq\delta}|\theta_x(x,t)\,\theta^{2^n-2}(x,t)\,v(x,t)|\,dx$$

$$+(2^n-1)\int_{|\theta|\geqq\delta}|\theta_x(x,t)\,\theta^{2^n-2}(x,t)\,v(x,t)|\,dx$$

$$\leqq (2^n-1)\,\delta^{2^n-2}\int_{-\infty}^{\infty}|\theta_x(x,t)|\,v(x,t)\,dx+(2^n-1)\,\delta^{-1}\int_{-\infty}^{\infty}|\theta_x(x,t)\,\theta^{2^n-1}(x,t)\,v(x,t)|\,dx$$

$$\leqq (2^n-1)\,\delta^{2^n-2}\left\{\tfrac{1}{2}\int_{-\infty}^{\infty}\theta_x(x,t)^2\,dx+\tfrac{1}{2}\int_{-\infty}^{\infty}v(x,t)^2\,dx\right\}$$

$$+(2^n-1)\,\delta^{-1}\,2^{-(n-1)}\int_{-\infty}^{\infty}|\partial_x\theta^{2^{n-1}}(x,t)\,\theta^{2^{n-1}}(x,t)\,v(x,t)|\,dx$$

$$\leqq \tfrac{1}{2}(2^n-1)\,\delta^{2^n-2}\int_{-\infty}^{\infty}\theta_x(x,t)^2\,dx+\tfrac{1}{2}\tilde{\Gamma}(2^n-1)\,\delta^{2^n}$$

$$+(2^n-1)\,\delta^{-1}\,2^{1-n}\left\{\frac{\varkappa}{\beta\overline{T}}\frac{\delta}{2\cdot 2^n}\int_{-\infty}^{\infty}(\partial_x\theta^{2^{n-1}}(x,t))^2\,dx\right.$$

$$\left.+\tfrac{1}{2}\frac{\beta\overline{T}}{\varkappa}\frac{2^n}{\delta}\max\theta^{2^n}(x,t)\int_{-\infty}^{\infty}v^2(x,t)\,dx\right\}$$

$$\leqq \tfrac{1}{2}(2^n-1)\,\delta^{2^n-2}\int_{-\infty}^{\infty}\theta_x(x,t)^2\,dx+\tfrac{1}{2}\tilde{\Gamma}(2^n-1)\,\delta^{2^n}$$

$$+\frac{\varkappa}{\beta\overline{T}}2^{-n}(1-2^{-n})\int_{-\infty}^{\infty}(\partial_x\theta^{2^{n-1}}(x,t))^2\,dx+\frac{\beta\overline{T}}{\varkappa}\tilde{\Gamma}(2^n-1)\max\theta^{2^n}(x,t), \tag{4.14}$$

where $\tilde{\Gamma}$ is the constant in (4.3).

Next we observe that

$$\left|\int_{-\infty}^{\infty} \partial_x \theta^{2^{n-1}}(x,t)\, \theta^{2^{n-1}}(x,t)\, v(x,t)\, dx\right|$$

$$\leqq \tfrac{1}{2}\frac{c}{\beta 2^n}\int_{-\infty}^{\infty} (\partial_x \theta^{2^{n-1}}(x,t))^2\, dx + \tfrac{1}{2}\frac{\beta 2^n}{c}\int_{-\infty}^{\infty} (v(x,t)\, \theta^{2^{n-1}}(x,t))^2\, dx$$

$$\leqq \tfrac{1}{2}\frac{c}{\beta 2^n}\int_{-\infty}^{\infty} (\partial_x \theta^{2^{n-1}}(x,t))^2\, dx + \tfrac{1}{2}\frac{\beta 2^n}{c}\max \theta^{2^n}(x,t)\int_{-\infty}^{\infty} v^2(x,t)\, dx. \tag{4.15}$$

By combining (4.13) through (4.15) we find that

$$\begin{aligned}\frac{d}{dt}\int_{-\infty}^{\infty} \theta^{2^n}(x,t)\, dx \leqq &-\frac{\varkappa}{2c}\int_{-\infty}^{\infty} (\partial_x \theta^{2^{n-1}}(x,t))^2\, dx \\ &+ 2\cdot 2^n\,\tfrac{1}{2}\frac{\beta 2^n}{c}\max \theta^{2^n}(x,t)\int_{-\infty}^{\infty} v^2(x,t)\, dx \\ &+ 2^n \overline{T}\frac{\beta}{c}\Big\{\tfrac{1}{2}(2^n-1)\,\delta^{2^n-2}\int_{-\infty}^{\infty} \theta_x(x,t)^2\, dx \\ &+ \tfrac{1}{2}\tilde{\Gamma}(2^n-1)\,\delta^{2^n} + \frac{\beta\overline{T}}{\varkappa}\tilde{\Gamma}(2^n-1)\max \theta^{2^n}(x,t)\Big\}.\end{aligned} \tag{4.16}$$

For every $v > 0$ we have

$$\begin{aligned}\max \theta^{2^n}(x,t) &\leqq 2\int_{-\infty}^{\infty} \theta^{2^{n-1}}(x,t)\,|\partial_x \theta^{2^{n-1}}(x,t)|\, dx \\ &\leqq v\int_{-\infty}^{\infty} (\partial_x \theta^{2^{n-1}}(x,t))^2\, dx + v^{-1}\int_{-\infty}^{\infty} \theta^{2^n}(x,t)\, dx \\ &\leqq v\int_{-\infty}^{\infty} (\partial_x \theta^{2^{n-1}}(x,t))^2\, dx + v^{-1}\max \theta^{2^{n-1}}(x,t)\int_{-\infty}^{\infty} \theta^{2^{n-1}}(x,t)\, dx \\ &\leqq v\int_{-\infty}^{\infty} (\partial_x \theta^{2^{n-1}}(x,t))^2\, dx + \tfrac{1}{2}\max \theta^{2^n}(x,t) \\ &\quad + \tfrac{1}{2}v^{-2}\left(\int_{-\infty}^{\infty} \theta^{2^{n-1}}(x,t)\, dx\right)^2,\end{aligned} \tag{4.17}$$

and consequently

$$\max \theta^{2^n}(x,t) \leqq 2v\int_{-\infty}^{\infty} (\partial_x \theta^{2^{n-1}}(x,t))^2\, dx + v^{-2}\left(\int_{-\infty}^{\infty} \theta^{2^{n-1}}(x,t)\, dx\right)^2. \tag{4.18}$$

By using (4.3), (4.16), and (4.18) (with v sufficiently small) we conclude that

$$\begin{aligned}\frac{d}{dt}\int_{-\infty}^{\infty} \theta^{2^n}(x,t)\, dx \leqq\; &2^{6n}\overline{\Gamma}\left(\int_{-\infty}^{\infty} \theta^{2^{n-1}}(x,t)\, dx\right)^2 + \tilde{\Gamma}\, 2^{2n-1}\,\delta^{2^n} \\ &+ 2^{2n-1}\,\delta^{2^n-2}\int_{-\infty}^{\infty} \theta_x(x,t)^2\, dx,\end{aligned} \tag{4.19}$$

where $\bar{\Gamma}$ is a constant that depends on β, c, $\varkappa$, and $\tilde{\Gamma}$. Integrating (4.19) with respect to time, recalling that $t^* \leqq 1$, and using (3.12), (3.13), and (4.3), we find that

$$A_n \leqq \delta^{2^n} + 2^{6n}\bar{\Gamma}(A_{n-1})^2 + \tilde{\Gamma}\, 2^{2n-1}\, \delta^{2^n} + \tilde{\Gamma}\, 2^{2n-1}\, \delta^{2^n}. \tag{4.20}$$

We choose Γ such that $\Gamma^{\frac{1}{2}} \geqq \max\{2^{14}\bar{\Gamma}, 16, \tilde{\Gamma}\}$. Using Lemma 4.1 and (4.20) we obtain

$$A_n \leqq 2^{-6n}\Gamma^{-\frac{1}{2}}(\Gamma\,\delta)^{2^n}, \quad n = 1, 2, 3, \ldots \tag{4.21}$$

by induction. In particular (4.12) holds and the proof is complete. □

We now proceed to obtain pointwise bounds for ε and v. For this purpose we introduce

$$r(x,t) := v(x,t) + \int_0^{\varepsilon(x,t)} p'(\xi)^{\frac{1}{2}}\, d\xi, \quad s(x,t) := v(x,t) - \int_0^{\varepsilon(x,t)} p'(\xi)^{\frac{1}{2}}\, d\xi \tag{4.22}$$

and the differential operators

$${}^{\backslash} := \frac{\partial}{\partial t} - p'(\varepsilon(x,t))^{\frac{1}{2}}\frac{\partial}{\partial x}, \quad {}^{/} := \frac{\partial}{\partial t} + p'(\varepsilon(x,t))^{\frac{1}{2}}\frac{\partial}{\partial x}. \tag{4.23}$$

A straightforward calculation yields

$$r^{\backslash} = \beta\theta_x, \quad s^{/} = \beta\theta_x. \tag{4.24}$$

In order to express θ_x in a convenient form, we define

$$\chi(x,t) := \int_{-\infty}^{x} \{P(\varepsilon(y,t)) + c\theta(y,t) + \tfrac{1}{2}v^2(y,t)\}\, dy. \tag{4.25}$$

We note that by virtue of $(3.3)_3$, θ admits the representation

$$\begin{aligned}\theta(x,t) = {} & \frac{c^{\frac{1}{2}}}{(4\pi\varkappa t)^{\frac{1}{2}}} \int_{-\infty}^{\infty} \exp\left\{-\frac{c}{4\varkappa}\frac{(x-y)^2}{t}\right\}\theta_0(y)\, dy \\ & + \frac{\beta}{(4\pi\varkappa c)^{\frac{1}{2}}} \int_0^t \int_{-\infty}^{\infty} (t-\tau)^{-\frac{1}{2}} \exp\left\{-\frac{c}{4\varkappa}\frac{(x-y)^2}{t-\tau}\right\} [\theta(y,\tau) + \bar{T}]\, v_y(y,\tau)\, dy\, d\tau.\end{aligned} \tag{4.26}$$

Integrating (4.26) with respect to x, using Lemma 4.1, and integrating by parts, we conclude that

$$\left|\int_{-\infty}^{x} \theta(y,t)\, dy\right| \leqq \Gamma\,\delta, \quad \forall\, x \in \mathbb{R}, \quad t \in [0, t^*] \tag{4.27}$$

and consequently

$$|\chi(x,t)| \leqq \Gamma\,\delta, \quad \forall\, x \in \mathbb{R}, \quad t \in [0, t^*]. \tag{4.28}$$

By virtue of $(2.2)_1$ and $(2.2)_2$, the equation of balance of energy, $(2.2)_3$, can be expressed in the form

$$(e + \tfrac{1}{2}v^2)_t + q_x = (\sigma v)_x. \tag{4.29}$$

We integrate (4.29) with respect to x over $(-\infty, x)$ and use (2.13), (4.25) to obtain

$$\begin{aligned}\varkappa\theta_x &= \chi_t(p(\varepsilon) + \beta\theta + \beta\bar{T})\, v, \\ &= \chi^{\backslash} + p'(\varepsilon)^{\frac{1}{2}}\,(P(\varepsilon) + c\theta + \tfrac{1}{2}v^2) - (p(\varepsilon) + \beta\theta + \beta\bar{T})\, v, \\ &= \chi^{/} - p'(\varepsilon)^{\frac{1}{2}}\,(P(\varepsilon) + c\theta + \tfrac{1}{2}v^2) - (p(\varepsilon) + \beta\theta + \beta\bar{T})\, v. \end{aligned} \tag{4.30}$$

Therefore, by setting

$$\Phi := r - \frac{\beta}{\varkappa}(\chi - \chi_0), \quad \Psi := s - \frac{\beta}{\varkappa}(\chi - \chi_0), \quad \chi_0(x) := \chi(x, 0), \tag{4.31}$$

we get

$$\Phi^{\backslash} = \frac{\beta}{\varkappa}p'(\varepsilon)^{\frac{1}{2}}\,(P(\varepsilon) + c\theta + \tfrac{1}{2}v^2) - (p(\varepsilon) + \beta\theta + \beta\bar{T})\, v + \frac{\beta}{\varkappa}p'(\varepsilon)^{\frac{1}{2}}\,\partial_x\chi_0, \tag{4.32}$$

$$\Psi^{/} = -\frac{\beta}{\varkappa}p'(\varepsilon)^{\frac{1}{2}}\,(P(\varepsilon) + c\theta + \tfrac{1}{2}v^2) - p((\varepsilon) + \beta\theta + \beta\bar{T})\, v + \frac{\beta}{\varkappa}p'(\varepsilon)^{\frac{1}{2}}\,\partial_x\chi_0. \tag{4.33}$$

The right-hand sides of (4.32), (4.33) can be expressed in terms of Φ and Ψ through the relations:

$$\varepsilon = E^{-1}(\Phi - \Psi), \quad v = \tfrac{1}{2}(\Phi + \Psi) + \frac{\beta}{\varkappa}(\chi - \chi_0), \tag{4.34}$$

where

$$E(z) := \int_0^z p'(\xi)^{\frac{1}{2}}\, d\xi \quad \forall\, z \in \mathbb{R}. \tag{4.35}$$

Lemma 4.3. *For every* $\lambda > 0$ *there is a* $\delta > 0$ *such that if* (3.12), (3.13) *hold then*

$$|\varepsilon(x, t)| < \lambda, \quad |v(x, t)| < \lambda, \quad \forall\, x \in \mathbb{R},\, t \in [0, t^*]. \tag{4.36}$$

Proof. By virtue of (4.34), (4.35), and (4.28), its suffices to show that given $\bar{\lambda} > 0$ there is a $\delta > 0$ such that if (3.12), (3.13) hold then

$$|\Phi(x, t)| < \bar{\lambda}, \quad |\Psi(x, t)| < \bar{\lambda}, \quad \forall\, x \in \mathbb{R}, \quad t \in [0, t^*]. \tag{4.37}$$

Following DAFERMOS & HSIAO [5], we define the nonnegative Lipschitz functions Φ^* and Ψ^* by

$$\Phi^*(t) := \max_x |\Phi(x, t)|, \quad \Psi^*(t) := \max_x |\Psi(x, t)| \quad \forall\, t \in [0, t^*]. \tag{4.38}$$

(The maxima in (4.38) exist because $\Phi(x, t)$ and $\Psi(x, t)$ tend to zeros as x tends to $\pm\infty$.)

We fix $t \in (0, t^*]$ and choose $\hat{x}$ and $\check{x}$ such that

$$\Phi^*(t) = |\Phi(\hat{x}, t)|, \quad \Psi^*(t) = |\Psi(\check{x}, t)|. \tag{4.39}$$

Then for every $h \in (0, t]$, we have

$$\begin{aligned} \Phi^*(t-h) &\geqq |\Phi(\hat{x} + hp'(\varepsilon(\hat{x}, t))^{\frac{1}{2}}, t-h)|, \\ \Psi^*(t-h) &\geqq |\Psi(\check{x} - hp'(\varepsilon(\hat{x}, t))^{\frac{1}{2}}, t-h)|. \end{aligned} \tag{4.40}$$

We subtract (4.40) from (4.39), divide the resulting inequalities by h, and let $h \downarrow 0$, to conclude that

$$D^-\Phi^*(t) \leqq |\Phi^{\backslash}(\hat{x}, t)|, \quad D^-\Psi^*(t) \leqq |\Psi^{/}(\check{x}, t)|. \tag{4.41}$$

It follows from (2.14), (4.34), (4.35), (4.10) and (4.28) that

$$\begin{aligned} &\left| \pm \frac{\beta}{\varkappa} p'(\varepsilon)^{\frac{1}{2}} \left(P(\varepsilon) + c\theta + \tfrac{1}{2} v^2\right) - (p(\varepsilon) + \beta\theta + \beta\bar{T})\, v + \frac{\beta}{\varkappa} p'(\varepsilon)^{\frac{1}{2}} \partial_x \chi_0 \right| \\ &\qquad \leqq \Gamma\{[|\Phi| + |\Psi|]^2 + [|\Phi| + |\Psi|] + \delta\}. \end{aligned} \tag{4.42}$$

By combining (4.41), (4.32), (4.38) and (4.42) we find that

$$\frac{d}{dt}[\Phi^*(t) + \Psi^*(t)] \leqq \Gamma\{[\Phi^*(t) + \Psi^*(t)]^2 + [\Phi^*(t) + \Psi^*(t)] + \delta\}, \tag{4.43}$$

for almost all $t \in [0, t^*]$. Moreover, by virtue of (4.38), (4.31), (4.28), (4.22), and (3.13), we have

$$\Phi^*(0) \leqq \Gamma\,\delta^{\frac{1}{2}}, \quad \Phi^*(0) \leqq \Gamma\,\delta^{\frac{1}{2}}. \tag{4.44}$$

Consequently if δ is sufficiently small, (4.43), (4.44) yield

$$\Phi^*(t) + \Psi^*(t) \leqq \bar{\lambda} \quad \forall\, t \in [0, t^*], \tag{4.45}$$

which implies (4.37). □

Our next task is to estimate the partial derivatives of ε, v, and θ. For this purpose we define

$$w := r_x, \quad \omega := s_x, \tag{4.46}$$

and note that

$$\varepsilon_x = \tfrac{1}{2} p'(\varepsilon)^{-\frac{1}{2}}(w - \omega), \quad v_x = \tfrac{1}{2}(w + \omega). \tag{4.47}$$

A simple computation yields

$$\varepsilon^{\backslash} = \omega, \quad \varepsilon^{/} = w. \tag{4.48}$$

Differentiation of (4.24) with respect to x gives

$$\partial_t r_x - p'(\varepsilon)^{\frac{1}{2}} r_{xx} - \tfrac{1}{2} p'(\varepsilon)^{-\frac{1}{2}} p''(\varepsilon)\, \varepsilon_x r_x = \beta\theta_{xx}, \tag{4.49}$$

$$\partial_t s_x + p'(\varepsilon)^{\frac{1}{2}} s_{xx} + \tfrac{1}{2} p'(\varepsilon)^{-\frac{1}{2}} p''(\varepsilon)\, \varepsilon_x s_x = \beta\theta_{xx}. \tag{4.50}$$

We substitute ε_x from (4.47) into (4.49), (4.50), use (4.48) to obtain

$$w^{\backslash} + \tfrac{1}{4} p'(\varepsilon)^{-1} p''(\varepsilon)\, \varepsilon^{\backslash} w - \tfrac{1}{4} p'(\varepsilon)^{-1} p''(\varepsilon)\, w^2 = \beta\theta_{xx}, \tag{4.51}$$

$$\omega^{/} + \tfrac{1}{4} p'(\varepsilon)^{-1} p''(\varepsilon)\, \varepsilon^{/} w - \tfrac{1}{4} p'(\varepsilon)^{-1} p''(\varepsilon)\, \omega^2 = \beta\theta_{xx}. \tag{4.52}$$

and multiply (4.51), (4.52) by $p'(\varepsilon)^{\frac{1}{4}}$ to arrive at

$$\left(p'(\varepsilon)^{\frac{1}{4}} w\right)^{\backslash} - \tfrac{1}{4} p'(\varepsilon)^{-\frac{5}{4}} p''(\varepsilon) \left(p'(\varepsilon)^{\frac{1}{4}} w\right)^2 = \beta p'(\varepsilon)^{\frac{1}{4}} \theta_{xx}, \tag{4.53}$$

$$\left(p'(\varepsilon)^{\frac{1}{4}} \omega\right)^{/} - \tfrac{1}{4} p'(\varepsilon)^{-\frac{5}{4}} p''(\varepsilon) \left(p'(\varepsilon)^{\frac{1}{4}} \omega\right)^2 = \beta p'(\varepsilon)^{\frac{1}{4}} \theta_{xx}. \tag{4.54}$$

We then define

$$f := p'(\varepsilon)^{\frac{1}{4}} \left(w - \frac{\beta c}{\varkappa}\theta\right), \qquad g := p'(\varepsilon)^{\frac{1}{4}} \left(\omega - \frac{\beta c}{\varkappa}\theta\right). \tag{4.55}$$

By using (4.47) through (4.54) and

$$\theta_{xx} = \frac{c}{\varkappa}\theta_t - \frac{\beta}{\varkappa}(\theta + \bar{T})\, v_x = \frac{c}{\varkappa}\theta^{\backslash} + \frac{c}{\varkappa} p'(\varepsilon)^{\frac{1}{4}}\theta_x - \frac{\beta}{2\varkappa}(\theta + \bar{T})(w + \omega)$$

$$= \frac{c}{\varkappa}\theta^{/} - \frac{c}{\varkappa} p'(\varepsilon)^{\frac{1}{4}}\theta_x - \frac{\beta}{2\varkappa}(\theta + \bar{T})(w + \omega) \tag{4.56}$$

we find that

$$f^{\backslash} = \tfrac{1}{4} p'(\varepsilon)^{-\frac{5}{4}} p''(\varepsilon) f^2 + \frac{\beta c}{4\varkappa} p'(\varepsilon)^{-1} p''(\varepsilon)\, \theta\, (2f - g) - \frac{\beta^2}{2\varkappa}\theta(f + g)$$

$$+ \frac{\beta c}{\varkappa} p'(\varepsilon)^{\frac{3}{4}}\theta_x - \frac{c\beta^3}{2\varkappa^2} p'(\varepsilon)^{\frac{1}{4}}\theta^2 - \frac{\beta^2}{2\varkappa}\bar{T} p'(\varepsilon)^{\frac{1}{4}}(f + g) - \frac{c\beta^3}{2\varkappa^2}\bar{T}\theta, \tag{4.57}$$

$$g^{/} = \tfrac{1}{4} p'(\varepsilon)^{-\frac{5}{4}} p''(\varepsilon) g^2 + \frac{\beta c}{4\varkappa} p'(\varepsilon)^{-1} p''(\varepsilon)\, \theta\, (2g - f) - \frac{\beta^2}{2\varkappa}\theta(f + g)$$

$$- \frac{\beta c}{\varkappa} p'(\varepsilon)^{\frac{3}{4}}\theta_x - \frac{\beta^3 c}{\varkappa^2} p'(\varepsilon)^{\frac{1}{4}}\theta^2 - \frac{\beta^2}{2\varkappa}\bar{T} p'(\varepsilon)^{\frac{1}{4}}(f + g) - \frac{c\beta^3}{2\varkappa^2}\bar{T}\theta. \tag{4.58}$$

In order to state our final estimate it is convenient to introduce

$$F(t) := \max_x |f(x, t)|, \qquad G(t) := \max_x |g(x, t)| \qquad \forall\, t \in [0, t^*]. \tag{4.59}$$

In the calculations that follow, the coefficient $\Gamma(\delta + \bar{T})$ could be replaced by Γ. The form of our coefficients was motivated by purposes of comparison with [5].

Lemma 4.4. *The temperature θ obeys the estimate*

$$|\theta_x(x, t)| \leqq \Gamma + \Gamma(\delta + \bar{T}) \int_0^t \frac{F(\tau) + G(\tau)}{(t - \tau)^{\frac{1}{2}}}\, d\tau \qquad \forall\, t \in [0, t^*]. \tag{4.60}$$

Proof. By using (4.26) and substituting for v_x in terms of f, g, and θ, differentiation of (4.26) with respect to x yields

$$\theta_x(x,t) = \frac{c^{\frac{1}{2}}}{(4\pi\varkappa t)^{\frac{1}{2}}} \int_{-\infty}^{\infty} \exp\left\{-\frac{c}{4\varkappa}\frac{(x-y)^2}{t}\right\} \theta_0'(y)\,dy$$
$$-\frac{\beta c^{\frac{1}{2}}}{(64\pi\varkappa^3)^{\frac{1}{2}}} \int_0^t \int_{-\infty}^{\infty} (x-y)(t-\tau)^{-\frac{3}{2}} \exp\left\{-\frac{c}{4\varkappa}\frac{(x-y)^2}{t-\tau}\right\}$$
$$\times (\theta(y,\tau)+\bar{T})\left(p'^{-\frac{1}{2}}(f+g)+2\frac{\beta c}{\varkappa}\theta\right) dy\,d\tau. \tag{4.61}$$

The assertion of the lemma follows easily from (4.61). □

We are now ready to establish the formation of a singularity in finite time. Let $L \in (0, 1]$ and $\varepsilon_0, v_0, \theta_0 \in H^3(\mathbb{R}) \cap W^{2,1}(\mathbb{R})$ satisfying (3.6) and (3.12) through (3.15) be given. Consider the solution (ε, v, θ) given by the proposition of Section 3. We shall show that if δ is sufficiently small and M is sufficiently large then $t_0 \leq L$, where t_0 is the length of the maximal interval. So as to get a contradiction we assume that $t_0 > L$. By virtue of Lemmas 4.2 and 4.3 we may choose δ small enough that for every $t^* \in (0, L]$ such that (4.2) holds, the sharper bound

$$|\varepsilon(x,t)| \leq \tfrac{1}{4}, \quad |\theta(x,t)| \leq \tfrac{1}{4}\bar{T}, \quad \forall\, x \in \mathbb{R},\ t \in [0, t^*] \tag{4.62}$$

also holds. Consequently, by continuity, if δ is sufficiently small then

$$|\varepsilon(x,t)| \leq \tfrac{1}{2}, \quad |\theta(x,t)| \leq \tfrac{1}{2}\bar{T}, \quad \forall\, x \in \mathbb{R},\ t \in [0, t^*]. \tag{4.63}$$

For the remainder of the proof, we assume that δ is small enough that (4.63) holds. We define, as in [5], the nonnegative Lipschitz functions

$$\begin{aligned} F^+(t) &= \max_x f(x,t), \quad G^+(t) = \max_x g(x,t), \\ F^-(t) &= \min_x f(x,t), \quad G^-(t) = \min_x g(x,t) \quad \forall\, t \in [0, L], \end{aligned} \tag{4.64}$$

and note that

$$F(t) \leq F^+(t) + F^-(t), \quad G(t) \leq G^+(t) + G^-(t) \quad \forall\, t \in [0, L]. \tag{4.65}$$

We fix $t \in [0, L)$ with $F^+(t) > 0$ and/or $G^+(t) > 0$ and choose $\hat{x}$ and $\check{x}$ such that

$$F^+(t) = f(\hat{x}, t) \quad \text{and/or} \quad G^+(t) = g(\check{x}, t). \tag{4.66}$$

For every $h \in (0, L - t]$ we then have

$$\begin{aligned} F^+(t+h) &\geq f(\hat{x} - hp'(\varepsilon(\hat{x}, t))^{\frac{1}{2}}, t+h), \\ G^+(t+h) &\geq g(\check{x} + hp'(\varepsilon(\check{x}, t))^{\frac{1}{2}}, t+h). \end{aligned} \tag{4.67}$$

Subtracting (4.66) from (4.67) and dividing through by h, as $h \downarrow 0$ we obtain

$$D^+F^+(t) \geqq f^{\backslash}(\hat{x}, t) \quad \text{and/or} \quad D^+G^+(t) \geqq g^{/}(\check{x}, t). \tag{4.68}$$

Next we fix $t \in [0, L)$ with $F^-(t) > 0$ and/or $G^-(t) > 0$ and choose $\hat{y}$ and $\check{y}$ such that

$$F^-(t) = -f(\hat{y}, t) \quad \text{and/or} \quad G^-(t) = -g(\check{y}, t). \tag{4.69}$$

For every $h \in (0, t]$ we then have

$$\begin{aligned} F^-(t+h) &\geqq -f(\hat{y} + hp'(\varepsilon(\hat{y}, t))^{\frac{1}{2}}, t-h), \\ G^-(t+h) &\geqq -g(\check{y} - hp'(\varepsilon(\hat{y}, t))^{\frac{1}{2}}, t-h). \end{aligned} \tag{4.70}$$

Subtracting (4.70) from (4.69) and dividing through by h as $h \downarrow 0$, we obtain

$$D^-F^-(t) \leqq -f^{\backslash}(\hat{y}, t) \quad \text{and/or} \quad D^-G^-(t) \geqq -g^{/}(\check{y}, t). \tag{4.71}$$

Since $p''(0) > 0$ we may choose constants α_1, α_2 such that

$$p''(\xi) \geqq \alpha_1 \quad \forall\, \xi \in [-\alpha_1, \alpha_2]. \tag{4.72}$$

By virtue of Lemma 4.3 we may choose δ small enough that

$$|\varepsilon(x, t)| \leqq \alpha_2 \quad \forall\, x \in \mathbb{R}, t \in [0, L]. \tag{4.73}$$

Thus there is a constant α such that

$$\tfrac{1}{4} p'(\varepsilon(x, t))^{-\frac{5}{4}} p''(\varepsilon(x, t)) \geqq 2\alpha > 0 \quad \forall\, x \in \mathbb{R}, t \in [0, L]. \tag{4.74}$$

Therefore, by combining the inequalities above, we get estimates of the form:

$$\begin{aligned} \frac{d}{dt}[F^+(t) + G^+(t)] \geqq{}& \alpha[F^+(t) + G^+(t)]^2 - \Gamma(\delta + \overline{T})\,[F^+(t) + G^+(t)] \\ & - \Gamma(\delta + \overline{T})\,[F^-(t) + G^-(t)] - \Gamma(\delta + \overline{T}) \int_0^t \frac{F^+(\tau) + G^+(\tau)}{(t-\tau)^{\frac{1}{2}}}\, d\tau \\ & - \Gamma(\delta + \overline{T}) \int_0^t \frac{F^-(\tau) + G^-(\tau)}{(t-\tau)^{\frac{1}{2}}}\, d\tau - \Gamma, \end{aligned} \tag{4.75}$$

$$\begin{aligned} \frac{d}{dt}[F^-(t) + G^-(t)] \leqq{}& \Gamma(\delta + \overline{T})\,[F^+(t) + G^+(t)] + \Gamma(\delta + \overline{T})\,[F^-(t) + G^-(t)] \\ & + \Gamma(\delta + \overline{T}) \int_0^t \frac{F^+(\tau) + G^+(\tau)}{(t-\tau)^{\frac{1}{2}}}\, d\tau \\ & + \Gamma(\delta + \overline{T}) \int_0^t \frac{F^-(\tau) + G^-(\tau)}{(t-\tau)^{\frac{1}{2}}}\, d\tau + \Gamma, \end{aligned} \tag{4.76}$$

for almost all $t \in [0, L]$.

In view of (4.38), (4.57), (4.58), (4.10), (3.14), and Lemma 4.3, (4.76) yields

$$[F^-(t) + G^-(t)] \leqq \Gamma + \Gamma(\delta + \overline{T}) \int_0^t K(t - \tau) [F^+(t) + G^+(t)] \, d\tau, \qquad (4.77)$$

where $K \in L^1(0, L)$. Finally, by combining (4.75) with (4.77) we obtain the inequality

$$\frac{d}{dt} [F^+(t) + G^+(t)] \geqq \alpha[F^+(t) + G^+(t)]^2 - \Gamma(\delta + \overline{T}) [F^+(t) + G^+(t)]$$

$$- \Gamma(\delta + \overline{T}) \int_0^t Z(t - \tau) [F^+(t) + G^+(t)] \, d\tau - \Gamma, \qquad (4.78)$$

where $Z \in L^1(0, L)$. By virtue of Lemmas 4.2 and 4.3 it follows that when M (in (3.15)) is sufficiently large then $[F^+ + G^+]$ blows up in a finite time $t^+ < L$, which contradicts the assumption that $t_0 > L$. This completes the proof. □

Acknowledgements. We are grateful to M. E. GURTIN and W. O. WILLIAMS for valuable discussions. This research was partially supported by the United States Air Force under Grant Nos. AFOSR-85-0307 and AFOSR-88-0265.

References

1. ALIKAKOS, N. D., An application of the invariance principle to reaction-diffusion equations, J. Diff. Equations **33** (1979), 201–225.
2. COLEMAN, B. D., & M. E. GURTIN, Waves in materials with memory III. Thermodynamic influences on the growth and decay of acceleration waves, Arch. Rational Mech. Anal. **19** (1965), 266–298.
3. COLEMAN, B. D., & V. J. MIZEL, Thermodynamics and departure from Fourier's law of heat conduction. Arch. Rational Mech. Anal. **13** (1963), 245–261.
4. COLEMAN, B. D., & W. NOLL, The thermodynamics of elastic materials with heat conduction and viscosity, Arch. Rational Mech. Anal. **13** (1963), 167–178.
5. DAFERMOS, C. M., & L. HSIAO, Development of singularities in solutions of the equations of nonlinear thermoelasticity, Q. Appl. Math. **44** (1986), 463–474.
6. DAY, W. A., *A Commentary on Thermodynamics*, Springer-Verlag, 1988.
7. HRUSA, W. J., & M. A. TARABEK, On smooth solutions of the Cauchy problem in one-dimensional nonlinear thermoelasticity, Q. Appl. Math. **47** (1989), 631–644.
8. JIANG, S., Global existence and asymptotic behavior of solutions in one-dimensional nonlinear thermoelasticity, Thesis, University of Bonn (1988).
9. LAX, P. D., Development of singularities in solutions of nonlinear hyperbolic partial differential equations, J. Math. Physics **5** (1964), 611–613.
10. MACCAMY, R. C., & V. J. MIZEL, Existence and Nonexistence in the large solutions of quasilinear wave equations, Arch. Rational Mech. Anal. **25** (1967), 299–320.
11. SLEMROD, M., Global existence, uniqueness, and asymptotic stability of classical solutions in one-dimensional thermoelasticity, Arch. Rational Mech. Anal. **76** (1981), 97–133.

12. ZHENG, S., & W. SHEN, L^p Decay estimates of solutions to the Cauchy problem of hyperbolic-parabolic coupled systems, Scientia Sinica (to appear).
13. ZHENG, S., & W. SHEN, Global solutions to the Cauchy problem of a class of hyperbolic-parabolic coupled systems, Scientia Sinica (to appear).
14. ZHENG, S., & W. SHEN, Global solutions to the Cauchy problem of a class of hyperbolic-parabolic coupled systems, in: S. T. XIAO & F. Q. PU (eds.), *International Workshop on Applied Differential Equations*, World Scientific Publishing, 1986, 335–338.

Department of Mathematics
Carnegie Mellon University
Pittsburgh, Pennsylvania

(Received July 3, 1989)

Published Works of Bernard D. Coleman

A. Scientific Papers

1. (with R. M. Fuoss) Quaternization Kinetics. I. Some Pyridine Derivatives in Tetramethylene Sulfone, *J. Am. Chem. Soc.* **77**, 5472–5476 (1955).
2. (with S. N. Timasheff, H. M. Dintzis, & J. G. Kirkwood) Studies of Molecular Interactions in Isoionic Protein Solutions by Light-Scattering, *Proc. National Acad. Sci.* **41**, 710–714 (1955).
3. Time Dependence of Mechanical Breakdown Phenomena, *J. Appl. Phys.* **27**, 862–866 (1956).
4. Application of the Theory of Absolute Reaction Rates to the Creep Failure of Polymeric Filaments, *J. Polymer Sci.* **20**, 447–455 (1956).
5. (with S. N. Timasheff, H. M. Dintzis, & J. G. Kirkwood) Light-Scattering Investigation of Charge Fluctuations in Isoionic Serum Albumin Solutions, *J. Am. Chem. Soc.* **79**, 782–791 (1957).
6. A Stochastic Process Model for Mechanical Breakdown, *Trans. Soc. Rheology* **1**, 153–168 (1957).
7. (with A. G. Knox) The Interpretation of Creep Failure in Textile Fibers as a Rate Process, *Textile Res. J.* **27**, 393–399 (1957).
8. Time Dependence of Mechanical Breakdown in Bundles of Fibers. I. Constant Total Load, *J. Appl. Phys.* **28**, 1058–1064 (1957).
9. On the Properties of Polymers with Random Stereo-Sequences, *J. Polymer Sci.* **31**, 155–164 (1958).
10. (with D. W. Marquardt) Time Dependence of Mechanical Breakdown in Bundles of Fibers. II. The Infinite Ideal Bundle Under Linearly Increasing Loads, *J. Appl. Phys.* **29**, 1065–1067 (1958).
11. Time Dependence of Mechanical Breakdown in Bundles of Fibers. III. The Power Law Breakdown Rule, *Trans. Soc. Rheology* **2**, 153–168 (1958).
12. (with D. W. Marquardt) Time Dependence of Mechanical Breakdown in Bundles of Fibers. IV. The Infinite Ideal Bundle Under Oscillating Loads, *J. Appl. Phys.* **29**, 1091–1099 (1958).
13. The Statistics and Time Dependence of Mechanical Breakdown in Fibers, *J. Appl. Phys.* **29**, 968–983 (1958).
14. On the Strength of Classical Fibers and Fiber Bundles, *J. Mech. Phys. Solids* **7**, 60–70 (1958).

15. (with A. G. KNOX & W. F. MCDEVIT) The Effect of Temperature on the Rate of Creep Failure for 66 Nylon, *Textile Res. J.* **28**, 393–399 (1958).
16. (with W. NOLL) Conditions for Equilibrium at Negative Absolute Temperatures, *Phys. Rev.* **115**, 262–265 (1959).
17. Time Dependence of Mechanical Breakdown in Bundles of Fibers. V. Fibers of Class A-2, *J. Appl. Phys.* **30**, 720–724 (1959).
18. (with W. NOLL) On Certain Steady Flows of General Fluids, *Arch. Rational Mech. Anal.* **3**, 289–303 (1959).
19. (with W. NOLL) On the Thermostatics of Continuous Media, *Arch. Rational Mech. Anal.* **4**, 97–128 (1959).
20. (with W. NOLL) Helical Flow of General Fluids, *J. Appl. Phys.* **30**, 1508–1512 (1959).
21. (with S. N. TIMASHEFF) On Light-Scattering Studies of Isoionic Proteins, *Arch. Biochem. & Biophys.* **87**, 63–69 (1960).
22. (with C. TRUESDELL) On the Reciprocal Relations of Onsager, *J. Chem. Phys.* **33**, 28–31 (1960).
23. (with R. M. FUOSS & M. WATANABE) Quaternization Kinetics of Poly-4-Vinylpyridine, *J. Polymer Sci.* **38**, 5–15 (1960).
24. (with W. NOLL) An Approximation Theorem for Functionals, with Applications in Continuum Mechanics, *Arch. Rational Mech. Anal.* **6**, 355–370 (1960).
25. (with W. NOLL) Recent Results in the Continuum Theory of Viscoelastic Fluids, *Annals New York Acad. Sciences* **89**, 672–714 (1961).
26. (with W. NOLL) Foundations of Linear Viscoelasticity, *Rev. Modern Phys.* **33**, 239–249 (1961).
27. (with W. NOLL) Normal Stresses in Second-Order Viscoelasticity, *Trans. Soc. Rheology* **5**, 41–46 (1961).
28. Mechanical and Thermodynamical Admissibility of Stress-Strain Functions, *Arch. Rational Mech. Anal.* **9**, 172–186 (1962).
29. Kinematical Concepts with Applications in the Mechanics and Thermodynamics of Incompressible Viscoelastic Fluids, *Arch. Rational Mech. Anal.* **9**, 273–300 (1962).
30. Substantially Stagnant Motions, *Trans. Soc. Rheology* **6**, 293–300 (1962).
31. (with W. NOLL) Steady Extension of Incompressible Simple Fluids, *Phys. Fluids* **5**, 840–843 (1962).
32. (with W. NOLL) The Thermodynamics of Elastic Materials with Heat Conduction and Viscosity, *Arch. Rational Mech. Anal.* **13**, 167–178 (1963).
33. (with V. J. MIZEL) Thermodynamics and Departures from Fourier's Law of Heat Conduction, *Arch. Rational Mech. Anal.* **13**, 245–261 (1963).
34. (with L. E. BRAGG) On Strain Energy Functions for Isotropic Elastic Materials, *J. Math. Phys.* **4**, 424–426 (1963).
35. (with L. E. BRAGG) A Thermodynamical Limitation on Compressibility, *J. Math. Phys.* **4**, 1074–1077 (1963).
36. (with T. G. FOX) General Theory of Stationary Random Sequences with Applications to the Tacticity of Polymers, *J. Polymer Sci.* **A1**, 3183–3196 (1963).
37. (with T. G. FOX) A Multistate Mechanism for Homogeneous Ionic Polymerization: I. The Diastereosequence Distribution, *J. Polymer Sci.* **38**, 1065–1067 (1963).
38. (with T. G. FOX) A Multistate Mechanism for Homogeneous Ionic Polymerization: II. The Molecular Weight Distribution, *J. Am. Chem. Soc.* **85**, 1241–1244 (1963).
39. (with T. G. FOX) On the Two-State Mechanism for Homogeneous Ionic Polymerization, *J. Polymer Sci.* **C4**, 345–360 (1963).
40. (with F. GORNICK and G. WEISS) Statistics of Irreversible Termination in Homogeneous Anionic Polymerization, *J. Polymer Sci.* **39**, 3233–3239 (1963).

41. (with W. NOLL) Material Symmetry and Thermostatic Inequalities in Finite Elastic Deformations, *Arch. Rational Mech. Anal.* **15**, 87–111 (1964).
42. Thermodynamics of Materials with Memory, *Arch. Rational Mech. Anal.* **17**, 1–46 (1964).
43. On Thermodynamics, Strain Impulses, and Viscoelasticity, *Arch. Rational Mech. Anal.* **17**, 230–254 (1964).
44. (with W. NOLL) Simple Fluids with Fading Memory, *Proceedings of the Symposium on Second-Order Effects in Elasticity, Plasticity, and Fluid Dynamics*, Haifa, April 1962; 530–552 (1964).
45. (with H. MARKOVITZ) Incompressible Second-Order Fluids, *Advances in Applied Mechanics* **8**, 69–101 (1964).
46. (with H. MARKOVITZ) Normal Stress Effects in Second-Order Fluids, *J. Appl. Phys.* **35**, 1–9 (1964).
47. (with H. MARKOVITZ) Nonsteady Helical Flows of Second-Order Fluids, *Phys. Fluids* **7**, 833–841 (1964).
48. (with V. J. MIZEL) Existence of Caloric Equations of State in Thermodynamics, *J. Chem. Phys.* **40**, 1116–1125 (1964).
49. (with R. J. DUFFIN & V. J. MIZEL) Instability, Uniqueness, and Nonexistence Theorems for the Equation $u_t = u_{xx} - u_{xtx}$ on a Strip, *Arch. Rational Mech. Anal.* **19**, 100–116 (1965).
50. (with M. E. GURTIN & I. HERRERA R.) Waves in Materials with Memory. I. The Velocity of One-Dimensional Shock and Acceleration Waves, *Arch. Rational Mech. Anal.* **19**, 1–19 (1965).
51. (with M. E. GURTIN) Waves in Materials with Memory. II. On the Growth and Decay of One-Dimensional Acceleration Waves, *Arch. Rational Mech. Anal.* **19**, 239–265 (1965).
52. (with M. E. GURTIN) Waves in Materials with Memory. III. Thermodynamic Influences on the Growth and Decay of Acceleration Waves, *Arch. Rational Mech. Anal.* **19**, 266–298 (1965).
53. (with M. E. GURTIN) Waves in Materials with Memory. IV. Thermodynamics and the Velocity of General Acceleration Waves, *Arch. Rational Mech. Anal.* **19**, 317–338 (1965).
54. Simple Liquid Crystals, *Arch. Rational Mech. Anal.* **20**, 41–58 (1965).
55. (with C. TRUESDELL) Homogeneous Motions of Incompressible Materials, *Z.A.M.M.* **45**, 547–551 (1965).
56. (with R. J. DUFFIN & V. J. MIZEL) Symposia on Applied Mathematics, Proceedings, Vol. XVII: Applications of Nonlinear Partial Differential Equations in Mathematical Physics, American Mathematical Society, p. 86 (1965) [Summary only].
57. (with T. G. FOX & M. REINMÖLLER) On Stereosequence Distributions from Anionic Polymerizations, *J. Polymer Sci.* Part B, **4**, 1029–1032 (1966).
58. (with J. M. GREENBERG & M. E. GURTIN) Waves in Materials with Memory. V. On the Amplitude of Acceleration Waves and Mild Discontinuities, *Arch. Rational Mech. Anal.* **22**, 333–354 (1966).
59. (with V. J. MIZEL) Norms and Semi-Groups in the Theory of Fading Memory, *Arch. Rational Mech. Anal.* **23**, 87–123 (1966).
60. (with M. E. GURTIN) Thermodynamics and One-Dimensional Shock Waves in Materials with Memory, *Proc. Roy. Soc. A* **292**, 562–574 (1966).
61. (with M. E. GURTIN) Thermodynamics and Wave Propagation, *Quarterly Applied Math.* **24**, 257–262 (1966).
62. (with M. E. GURTIN) Acceleration Waves in Nonlinear Materials, *Modern Developments in the Mechanics of Continua*, Academic Press (1966), pp. 165–174.

63. (with V. J. MIZEL) Breakdown of Laminar Shearing Flows for Second-Order Fluids in Channels of Critical Width, *Z.A.M.M.* **46**, 445–448 (1966).
64. (with J. M. GREENBERG) Thermodynamics and the Stability of Fluid Motion, *Arch. Rational Mech. Anal.* **25**, 321–341 (1967).
65. (with V. J. MIZEL) Existence of Entropy as a Consequence of Asymptotic Stability, *Arch. Rational Mech. Anal.* **25**, 243–270 (1967).
66. (with V. J. MIZEL) A General Theory of Dissipation in Materials with Memory, *Arch. Rational Mech. Anal.* **27**, 255–274 (1967).
67. (with M. E. GURTIN) Thermodynamics with Internal State Variables, *J. Chem. Phys.* **47**, 597–613 (1967).
68. (with M. E. GURTIN) On the Growth and Decay of Discontinuities in Fluids with Internal State Variables, *Phys. Fluids* **10**, 1454–1458 (1967).
69. (with M. E. GURTIN) Some Properties of Gases with Vibrational Relaxation when Viewed as Materials with Memory, *Meccanica* **2**, 125–140 (1967).
70. (with M. E. GURTIN) Equipresence and Constitutive Equations for Rigid Heat Conductors, *Z.A.M.P.* **18**, 199–208 (1967).
71. (with M. E. GURTIN) Thermodynamics and Wave Propagation in Non-Linear Materials with Memory, *Proceedings of the IUTAM Symposium on Irreversible Aspects of Continuum Mechanics*, Vienna, June 1966; 54–76 (1968).
72. (with M. E. GURTIN) On the Stability Against Shear Waves of Steady Flows of Non-Linear Viscoelastic Fluids, *J. Fluid Mech.* **33**, 165–181 (1968).
73. On the Use of Symmetry to Simplify the Constitutive Equations of Isotropic Materials with Memory, *Proc. Roy. Soc.* A **306**, 449–476 (1968).
74. (with V. J. MIZEL) On the General Theory of Fading Memory, *Arch. Rational Mech. Anal.* **29**, 18–31 (1968).
75. (with V. J. MIZEL) On Thermodynamic Conditions for the Stability of Evolving Systems, *Arch. Rational Mech. Anal.* **29**, 105–113 (1968).
76. (with V. J. MIZEL) On the Stability of Solutions of Functional-Differential Equations, *Arch. Rational Mech. Anal.* **30**, 173–196 (1968).
77. (with E. H. DILL) On the Stability of Certain Motions of Incompressible Materials with Memory, *Arch. Rational Mech. Anal.* **30**, 197–224 (1968).
78. On the Stability of Equilibrium States of General Fluids, *Arch. Rational Mech. Anal.* **36**, 1–32 (1970).
79. (with D. R. OWEN) On the Thermodynamics of Materials with Memory, *Arch. Rational Mech. Anal.* **36**, 245–269 (1970).
80. Some Recent Results in the Theory of Fading Memory, *Proceedings of the International Conference on Thermodynamics*, Cardiff, April 1970 (Vol. 22, pp. 321–333, Pure and Applied Chemistry).
81. (with E. H. DILL & R. A. TOUPIN) A Phenomenological Theory of Streaming Birefringence, *Arch. Rational Mech. Anal.* **39**, 358–399 (1970).
82. On the Dynamical Stability of Fluid Phases, *Proceedings of the IUTAM Symposium on Instability of Continuous Systems*, Herrenalb (Karlsruhe), September 1969; 272–283 (1971).
83. (with E. H. DILL) On Induced Birefringence in Viscoelastic Materials, NASA Contractor Report CR-1703 (1971).
84. (with E. H. DILL) Thermodynamic Restrictions on the Constitutive Equations of Electromagnetic Theory, *Z.A.M.P.* **22**, 691–702 (1971).
85. (with E. H. DILL) Theory of Induced Birefringence in Materials with Memory, *J. Mech. Phys. Solids* **19**, 215–243 (1971).
86. (with E. H. DILL) On the Thermodynamics of Electromagnetic Fields in Materials with Memory, *Arch. Rational Mech. Anal.* **41**, 132–162 (1971).
87. On Retardation Theorems, *Arch. Rational Mech. Anal.* **43**, 1–23 (1971).

88. A Mathematical Theory of Lateral Sensory Inhibition, *Arch. Rational Mech. Anal.* **43**, 79–100 (1971).
89. Thermodynamics of Discrete Mechanical Systems with Memory, *Advances in Chemical Physics* **24**, 95–154 (1973).
90. (with E. H. DILL) On Thermodynamics and the Stability of Motions of Materials with Memory, *Arch. Rational Mech. Anal.* **51**, 1–53 (1973).
91. Fading Memory and Functional-Differential Equations, *Proceedings of the Battelle Summer Institute on Non-Linear Problems in the Physical Sciences and Biology*, Seattle, July 1972; 78–86 (1973).
92. On the Energy Criterion for Stability, in *Nonlinear Elasticity*, Academic Press (1973), pp. 31–55; a symposium held at the Mathematics Research Center, Madison, April 1973.
93. (with G. H. RENNINGER) On the Integral Equations of the Linear Theory of Recurrent Lateral Interaction in Vision, *Mathematical Biosciences* **20**, 155–170 (1974).
94. The Energy Criterion for Stability in Continuum Thermodynamics, *Rendiconti del Seminario Matematico e Fisico di Milano* **43**, 85–96 (1973).
95. (with D. R. OWEN) A Mathematical Foundation for Thermodynamics, *Arch. Rational Mech. Anal.* **54**, 1–104 (1974).
96. (with G. H. RENNINGER) Theory of Delayed Lateral Inhibition in the Compound Eye of *Limulus*, *Proceedings of the National Academy of Sciences* **71**, 2887–2891 (1974).
97. (with G. H. RENNINGER) Consequences of Delayed Lateral Inhibition in the Retina of *Limulus*. I. Elementary Theory of Spatially Uniform Fields, *J. Theor. Biol.* **51**, 243–265 (1975).
98. (with G. H. RENNINGER) Consequences of Delayed Lateral Inhibition in the Retina of *Limulus*. II. Theory of Spatially Uniform Fields, Assuming the 4-Point Property, *J. Theor. Biol.* **51**, 267–291 (1975).
99. (with D. R. OWEN) On the Initial Value Problem for a Class of Functional-Differential Equations, *Arch. Rational Mech. Anal.* **55**, 275–299 (1974).
100. (with H. MARKOVITZ) Asymptotic Relations Between Shear Stresses and Normal Stresses in General Incompressible Fluids, *J. Polymer Sci., Polymer Phys. Ed.* **12**, 2195–2207 (1974).
101. (with G. H. RENNINGER) Periodic Solutions of a Nonlinear Functional Equation Describing Neural Interactions, *Istituto Lombardo Accad. Sci. Lett. Rendiconti* **A 109**, 91–111 (1975).
102. (with G. H. RENNINGER) Periodic Solutions of Certain Nonlinear Integral Equations with a Time-Lag, *SIAM Journal on Applied Mathematics* **31**, 111–120 (1976).
103. (with D. R. OWEN) On Thermodynamics and Elastic-Plastic Materials, *Arch. Rational Mech. Anal.* **59**, 25–51 (1975).
104. (with D. R. OWEN) On Thermodynamics and Intrinsically Equilibrated Materials, *Annali di Matematica pura ed applicata* (IV) **108**, 189–196 (1976).
105. On the Mechanics of Materials with Fading Memory, *Proceedings of the IUTAM/IMU Symposium on Applications of Methods of Functional Analysis to Problems of Mechanics*, Marseilles, September 1975; 290–294 (1976). Springer Lect. Notes Math. Vol. 503.
106. (with E. H. DILL) Photoviscoelasticity: Theory and Practice, *Proceedings of the Symposium on The Photoelastic Effect and Its Applications*, Ottingnies (Brussels), Sept. 1973; 456–505 (1975).
107. (with G. H. RENNINGER) Theory of the Response of the *Limulus* Retina to Periodic Excitation, *J. Math. Biol.* **3**, 103–120 (1976).
108. (with V. J. MIZEL) Generalization of the Perceptron Convergence Theorem, *Brain Theory Newsletter* **1**, 67–68 (1976).

109. (with D. R. OWEN) Thermodynamics of Elastic-Plastic Materials, *Rendiconti dell'Accademia Nazionale dei Lincei*, (VIII) **61**, 77–81 (1976).
110. Finitely Convergent Learning Programs for the Separation of Sets of Shaded Figures, *Biological Cybernetics* **25**, 57–60 (1976).
111. (with G. H. RENNINGER) Theory of Spatially Synchronized Oscillatory Responses in the Retina of *Limulus*, *Mathematical Biosciences* **38**, 123–140 (1978).
112. (with D. R. OWEN) On the Thermodynamics of Semi-systems with Restrictions on the Accessibility of States, *Arch. Rational Mech. Anal.* **66**, 173–181 (1977).
113. (with V. J. MIZEL) Methods of Constructing Non-Linear Functionals which Separate Sets in Hilbert Spaces, *Istituto Lombardo (Rend. Sci.)* **A 111**, 123–144 (1977).
114. On the Thermodynamics of Non-Classical Systems, *Contemporary Developments in Continuum Mechanics and Partial Differential Equations*, 135–142 (1978). Proceedings of a Symposium held in Rio de Janeiro, August, 1977.
115. (with R. J. DUFFIN & G. KNOWLES) A Class of Optimal Design Problems with Linear Inequality Constraints, *Istituto Lombardo (Rend. Sci.)* **A 113**, 64–83 (1979).
116. On the Growth of Populations with Narrow Spread in Reproductive Age: I. General Theory and Examples, *J. Math. Biology* **6**, 1–19 (1978).
117. (with C. V. COFFMAN) On the Growth of Populations with Narrow Spread in Reproductive Age: II. Conditions of Convexity, *J. Math. Biology* **6**, 285–303 (1978).
118. (with C. V. COFFMAN) On the Growth of Populations with Narrow Spread in Reproductive Age: III. Periodic Variations in the Environment, *J. Math. Biology* **7**, 281–301 (1979).
119. Nonautonomous Logistic Equations as Models of the Adjustment of Populations to Environmental Change, *Mathematical Biosciences* **45**, 159–173 (1979).
120. (with Y.-H. HSIEH & G. KNOWLES) On the Optimal Choice of *r* for a Population in a Periodic Environment, *Mathematical Biosciences* **46**, 71–85 (1979).
121. (with D. R. OWEN) On the Thermodynamics of Elastic-Plastic Materials with Temperature-Dependent Moduli and Yield Stresses, *Arch. Rational Mech. Anal.* **70**, 339–354 (1979).
122. On the Optimal Shape of a Flexible Cable, *Proceedings of the Symposium on Constructive Approaches to Mathematical Models*, Carnegie Mellon University, Pittsburgh, July 1978, 365–373 (1979).
123. (with Y.-H. HSIEH) Theory of the Dependence of Population Levels on Environmental History for Semelparous Species with Short Reproductive Seasons, *Proceedings of the National Academy of Sciences* **76**, 5407–5410 (1979).
124. (with L. J. ZAPAS) Theory of the Instability and Failure of Viscoelastic Materials in Tension, *J. Polymer Sci., Polymer Phys. Ed.* **17**, 2215–2224 (1979).
125. (with R. J. DUFFIN & G. KNOWLES) Theory of Monotonic Transformations Applied to Optimal Design Problems, *Arch. Rational Mech. Anal.* **72**, 381–393 (1979).
126. (with J. A. MAXWELL & G. H. RENNINGER) On Oscillatory Responses of the *Limulus* Retina, *Biological Cybernetics* **37**, 125–129 (1980).
127. On Random Placement and Species-Area Relations, *Mathematical Biosciences* **54**, 191–215 (1981).
128. On Optimal Intrinsic Growth Rates for Populations in Periodically Changing Environments, *J. Math. Biology* **12**, 343–353 (1981).
129. (with D. R. OWEN & J. SERRIN) The Second Law of Thermodynamics for Systems with Approximate Cycles, *Arch. Rational Mech. Anal.* **77**, 103–142 (1982).
130. (with M. FABRIZIO & D. R. OWEN) On the Thermodynamics of Second Sound in Dielectric Crystals, *Arch. Rational Mech. Anal.* **80**, 135–158 (1982).
131. (with M. FABRIZIO & D. R. OWEN) Il Secondo Suono nei Cristalli: Termodinamica ed Equazioni Costitutivi, *Rend. Sem. Mat. Univ. Padova* **58**, 208–227 (1982).

132. (with M. A. Mares, M. R. Willig, & Y.-H. Hsieh) Randomness, Area, and Species Richness, *Ecology* **63**, 1121–1133 (1982).
133. (with D. R. Owen) On the Nonequilibrium Behavior of Solids that Transport Heat by Second Sound, *Comp. & Math. with Appls.* **8**, 527–546 (1983).
134. Necking and Drawing in Polymeric Fibers Under Tension, *Arch. Rational Mech. Anal.* **83**, 115–137 (1983).
135. On the Cold Drawing of Polymers, *Comp. & Math. with Appls.* **11**, 35–65 (1985).
136. (with D. R. Owen) Recent Research on the Foundations of Thermodynamics, *Appendix G1* to the Second Edition of *Rational Thermodynamics* by C. Truesdell, Springer-Verlag (1984), pp. 479–493.
137. On Slow-Flow Approximations to Fluids with Fading Memory, in *Viscoelasticity and Rheology*, Academic Press (1985), pp. 125–156; a symposium held at the Mathematics Research Center, Madison, November 1984.
138. (with M. L. Hodgdon) On Shear Bands in Ductile Materials, *Arch. Rational Mech. Anal.* **90**, 219–247 (1985).
139. (with M. L. Hodgdon) A Constitutive Relation for Rate-Independent Hysteresis in Ferromagnetically Soft Materials, *Int. J. Engineering Science* **24**, 897–919 (1986).
140. (with M. Fabrizio & D. R. Owen) Thermodynamics and the Constitutive Equations for Second Sound in Crystals, in *New Perspectives in Thermodynamics* edited by J. Serrin, Springer-Verlag (1986) § 10, pp. 171–185. English version of 131. with added notes.
141. (with W. J. Hrusa & D. R. Owen) Stability of Equilibrium for a Nonlinear Hyperbolic System Describing Heat Propagation by Second Sound in Solids, *Arch. Rational Mech. Anal.* **94**, 267–289 (1986).
142. (with W. J. Hrusa, D. C. Newman, & D. R. Owen) Nonlinear Effects Associated with Second Sound in Solids, *Proceedings of the International Meeting on Macroscopic Theories of Superfluids*, Accademia Nazionale dei Lincei, Rome, May 1986, in press.
143. (with M. L. Hodgdon) On a Class of Constitutive Relations for Ferromagnetic Hysteresis, *Arch. Rational Mech. Anal.* **99**, 375–396 (1987).
144. (with Y.-H. Hsieh) On Species-Area Relations, Part I, *Soochow J. Math.* **12**, 11–22 (1986); Part II, *ibid.* **13**, 31–44 (1987).
145. (with D. C. Newman) Implications of a Nonlinearity in the Theory of Second Sound in Solids, *Phys. Rev.* B **37**, 1492–1498 (1988).
146. (with M. L. Hodgdon) A Theory of Shear Bands, in *Constitutive Models of Deformation* (Proceedings of the ARO Workshop, Blacksburg, March 1986) SIAM, Philadelphia, pp. 27–36 (1987).
147. (with D. C. Newman) Constitutive Relations for Elastic Materials Susceptible to Drawing, *Proceedings of the Symposium on Constitutive Modeling for Nontraditional Materials*, ASME Winter Annual Meeting, Boston, December 1987, AMD Volume **85**, 47–58 (1987).
148. (with D. C. Newman) On the Rheology of Cold Drawing: I. Elastic Materials, *J. Polym. Sci. B, Polymer Physics* **26**, 1801–1822 (1988).
149. (with M. L. Hodgdon) On the Localization of Strain in Shearing Motions of Ductile Materials, *Res Mechanica* **23**, 223–238 (1988).
150. (with M. L. Hodgdon) Implications of a Constitutive Model for Strain Localization, *Journal de physique* **9**, **C3**, 489–496 (1988).
151. Approssimazione di Fluidi con Memoria Evanescente in Moti Ritardati, *Rendiconti del Circolo Matematico di Palermo*, Serie II **36**, 182–193 (1987).
152. (with L. J. Zapas) A Phenomenological Theory of the Influence of Strain History on the Rate of Isothermal Stress Relaxation, *J. Rheology* **33**, 501–516 (1989).

153. (with D. C. NEWMAN) On Adiabatic Shear Bands in Rigid-Plastic Materials, *Acta Mechanica*, **78**, 263–279 (1989).
154. (with D. C. NEWMAN) On Waves in Slender Elastic Rods, *Arch. Rational Mech. Anal.*, **109**, 39–61 (1990).
155. On the Loss of Regularity of Shearing Flows of Viscoelastic Fluids, in *Nonlinear Evolution Equations that Change Type*, Springer-Verlag, in press.

B. Books

Wave Propagation in Dissipative Materials, by B. D. COLEMAN, M. E. GURTIN, I. HERRERA R., & C. TRUESDELL; (Springer-Verlag, 1965).

Viscometric Flows of Non-Newtonian Fluids, Theory and Experiment, by B. D. COLEMAN, H. MARKOVITZ, & W. NOLL; (Springer-Verlag, 1966).

C. Lecture and Course Notes

1. Osservazione sulla Termodinamica e sulla Teoria dell'Elasticità Finita. (Text of a lecture given at Bologna and Naples in the Spring of 1961; multiplied typescript.)
2. On Global and Local Forms of the Second Law of Thermodynamics, in *Proprietà di Media e Teoremi di Confronto in Fiscia Matematica.* (A course given at the Centro Internazionale Matematico Estivo, Bressanone, 1963.)
3. (with M. E. GURTIN) Thermodynamics and Wave Propagation in *Elastic and Viscoelastic Media, in Non-Linear Continuum Theories.* (A course given at the Centro Internazionale Matematico Estivo, Bressanone, 1965.)
4. Text of Lectures on Thermodynamics of Materials with Memory. Course No. 73, at the *International Centre for Mechanical Sciences*, Udine, June 1971. (Bound separately and published by Springer-Verlag, Vienna, 1972.)
5. Thermodynamics and Constitutive Relations, in *Thermodynamics and Constitutive Equations*, Springer Lecture Notes in Physics, Vol. 228, pp. 1–43. (These notes for a course of 10 Lectures at the 2nd 1982 Session of the Centro Internazionale Matematico Estivo, held in Noto, Italy, are preliminary draughts of two papers, one by B. D. COLEMAN & D. R. OWEN, and the other by B. D. COLEMAN, M. FABRIZIO, & D. R. OWEN.)
6. (with D. R. OWEN) Global and Local Versions on the Second Law of Thermodynamics, in *Categories in Continuum Physics*, Springer Lecture Notes in Mathematics, Vol. 1174 (1982) pp. 83–99 (preliminary draught).
7. A Phenomenological Theory of the Mechanics of Cold Drawing, in *Orienting Polymers*, Springer Lecture Notes in Mathematics, Vol. 1063 (1983) pp. 76–142 (preliminary draught).

Places of First Publication

of the articles reprinted above, all in
Archive for Rational Mechanics and Analysis